U0381539

NATURAL HISTORY
UNIVERSAL LIBRARY

西方博物学大系

主编：江晓原

PARADISI IN SOLE
PARADISUS TERRESTRIS

园艺大要

[英] 约翰·帕金森 著

华东师范大学出版社

图书在版编目（CIP）数据

园艺大要 = Paradisi in Sole Paradisus Terrestris：
英文 /（英）约翰·帕金森著. — 上海：华东师范
大学出版社, 2018
（寰宇文献）
ISBN 978-7-5675-7688-9

Ⅰ.①园… Ⅱ.①斐… Ⅲ.①园艺-英文
Ⅳ.①S6

中国版本图书馆CIP数据核字(2018)第094682号

园艺大要

Paradisi in Sole Paradisus Terrestris
（英）约翰·帕金森著

特约策划　黄曙辉　徐　辰
责任编辑　庞　坚
特约编辑　许　倩
装帧设计　刘怡霖

出版发行　华东师范大学出版社
社　　址　上海市中山北路3663号　邮编 200062
网　　址　www.ecnupress.com.cn
电　　话　021-60821666　行政传真　021-62572105
客服电话　021-62865537
门市（邮购）电话　021-62869887
地　　址　上海市中山北路3663号华东师范大学校内先锋路口
网　　店　http://hdsdcbs.tmall.com/

印 刷 者　虎彩印艺股份有限公司
开　　本　16开
印　　张　40.75
版　　次　2018年6月第1版
印　　次　2018年6月第1次
书　　号　ISBN 978-7-5675-7688-9
定　　价　698.00元（精装全一册）

出 版 人　王　焰

（如发现本版图书有印订质量问题，请寄回本社客服中心调换或电话021-62865537联系）

《西方博物学大系》总序

江晓原

　　《西方博物学大系》收录博物学著作超过一百种，时间跨度为 15 世纪至 1919 年，作者分布于 16 个国家，写作语种有英语、法语、拉丁语、德语、弗莱芒语等，涉及对象包括植物、昆虫、软体动物、两栖动物、爬行动物、哺乳动物、鸟类和人类等，西方博物学史上的经典著作大备于此编。

中西方"博物"传统及观念之异同

　　今天中文里的"博物学"一词，学者们认为对应的英语词汇是 Natural History，考其本义，在中国传统文化中并无现成对应词汇。在中国传统文化中原有"博物"一词，与"自然史"当然并不精确相同，甚至还有着相当大的区别，但是在"搜集自然界的物品"这种最原始的意义上，两者确实也大有相通之处，故以"博物学"对译 Natural History 一词，大体仍属可取，而且已被广泛接受。

　　已故科学史前辈刘祖慰教授尝言：古代中国人处理知识，如开中药铺，有数十上百小抽屉，将百药分门别类放入其中，即心安矣。刘教授言此，其辞若有憾焉——认为中国人不致力于寻求世界"所以然之理"，故不如西方之分析传统优越。然而古代中国人这种处理知识的风格，正与西方的博物学相通。

　　与此相对，西方的分析传统致力于探求各种现象和物体之间的相互关系，试图以此解释宇宙运行的原因。自古希腊开始，西方哲人即孜孜不倦建构各种几何模型，欲用以说明宇宙如何运行，其中最典型的代表，即为托勒密（Ptolemy）的宇宙体系。

　　比较两者，差别即在于：古代中国人主要关心外部世界"如何"运行，而以希腊为源头的西方知识传统（西方并非没有别的知识传统，只是未能光大而已）更关心世界"为何"如此运行。在线性发展无限进步的科学主义观念体系中，我们习惯于认为"为何"是在解决了"如何"之后的更高境界，故西方的分析传统比中国的传统更高明。

然而考之古代实际情形，如此简单的优劣结论未必能够成立。例如以天文学言之，古代东西方世界天文学的终极问题是共同的：给定任意地点和时刻，计算出太阳、月亮和五大行星（七政）的位置。古代中国人虽不致力于建立几何模型去解释七政"为何"如此运行，但他们用抽象的周期叠加（古代巴比伦也使用类似方法），同样能在足够高的精度上计算并预报任意给定地点和时刻的七政位置。而通过持续观察天象变化以统计、收集各种天象周期，同样可视之为富有博物学色彩的活动。

还有一点需要注意：虽然我们已经接受了用"博物学"来对译 Natural History，但中国的博物传统，确实和西方的博物学有一个重大差别——即中国的博物传统是可以容纳怪力乱神的，而西方的博物学基本上没有怪力乱神的位置。

古代中国人的博物传统不限于"多识于鸟兽草木之名"。体现此种传统的典型著作，首推晋代张华《博物志》一书。书名"博物"，其义尽显。此书从内容到分类，无不充分体现它作为中国博物传统的代表资格。

《博物志》中内容，大致可分为五类：一、山川地理知识；二、奇禽异兽描述；三、古代神话材料；四、历史人物传说；五、神仙方伎故事。这五大类，完全符合中国文化中的博物传统，深合中国古代博物传统之旨。第一类，其中涉及宇宙学说，甚至还有"地动"思想，故为科学史家所重视。第二类，其中甚至出现了中国古代长期流传的"守宫砂"传说的早期文献：相传守宫砂点在处女胳膊上，永不褪色，只有性交之后才会自动消失。第三类，古代神话传说，其中甚至包括可猜想为现代"连体人"的记载。第四类，各种著名历史人物，比如三位著名刺客的传说，此三名刺客及所刺对象，历史上皆实有其人。第五类，包括各种古代方术传说，比如中国古代房中养生学说，房中术史上的传说人物之一"青牛道士封君达"等等。前两类与西方的博物学较为接近，但每一类都会带怪力乱神色彩。

"所有的科学不是物理学就是集邮"

在许多人心目中，画画花草图案，做做昆虫标本，拍拍植物照片，这类博物学活动，和精密的数理科学，比如天文学、物理学等等，那是无法同日而语的。博物学显得那么的初级、简单，甚至幼稚。这种观念，实际上是将"数理程度"作为唯一的标尺，用来衡量一切知识。但凡能够使用数学工具来描述的，或能够进行物理实验的，那就是"硬"科学。使用的数学工具越高深越复杂，似乎就越"硬"；物理实验设备越庞大，花费的金钱越多，似乎就越"高端"、越"先进"……

这样的观念，当然带着浓厚的"物理学沙文主义"色彩，在很多情况下是不正确的。而实际上，即使我们暂且同意上述"物理学沙文主义"的观念，博物学的"科学地位"也仍然可以保住。作为一个学天体物理专业出身，因而经常徜徉在"物理学沙文主义"幻影之下的人，我很乐意指出这样一个事实：现代天文学家们的研究工作中，仍然有绘制星图，编制星表，以及为此进行的巡天观测等等活动，这些活动和博物学家"寻花问柳"，绘制植物或昆虫图谱，本质上是完全一致的。

这里我们不妨重温物理学家卢瑟福（Ernest Rutherford）的金句："所有的科学不是物理学就是集邮（All science is either physics or stamp collecting）。"卢瑟福的这个金句堪称"物理学沙文主义"的极致，连天文学也没被他放在眼里。不过，按照中国传统的"博物"理念，集邮毫无疑问应该是博物学的一部分——尽管古代并没有邮票。卢瑟福的金句也可以从另一个角度来解读：既然在卢瑟福眼里天文学和博物学都只是"集邮"，那岂不就可以将博物学和天文学相提并论了？

如果我们摆脱了科学主义的语境，则西方模式的优越性将进一步被消解。例如，按照霍金（Stephen Hawking）在《大设计》（The Grand Design）中的意见，他所认同的是一种"依赖模型的实在论（model-dependent realism）"，即"不存在与图像或理论无关的实在性概念（There is no picture- or theory-independent concept of reality）"。在这样的认识中，我们以前所坚信的外部世界的客观性，已经不复存在。既然几何模型只不过是对外部世界图像的人为建构，则古代中国人干脆放弃这种建构直奔应用（毕竟在实际应用中我们只需要知道七政"如何"运行），又有何不可？

传说中的"神农尝百草"故事，也可以在类似意义下得到新的解读："尝百草"当然是富有博物学色彩的活动，神农通过这一活动，得知哪些草能够治病，哪些不能，然而在这个传说中，神农显然没有致力于解释"为何"某些草能够治病而另一些则不能，更不会去建立"模型"以说明之。

"帝国科学"的原罪

今日学者有倡言"博物学复兴"者，用意可有多种，诸如缓解压力、亲近自然、保护环境、绿色生活、可持续发展、科学主义解毒剂等等，皆属美善。编印《西方博物学大系》也是意欲为"博物学复兴"添一助力。

然而，对于这些博物学著作，有一点似乎从未见学者指出过，而鄙意以为，当我们披阅把玩欣赏这些著作时，意识到这一点是必须的。

这百余种著作的时间跨度为 15 世纪至 1919 年，注意这个时间跨度，正是西方列强"帝国科学"大行其道的时代。遥想当年，帝国的科学家们乘上帝国的军舰——达尔文在皇家海军"小猎犬号"上就是这样的场景之一，前往那些已经成为帝国的殖民地或还未成为殖民地的"未开化"的遥远地方，通常都是踌躇满志、充满优越感的。

作为一个典型的例子，英国学者法拉在（Patricia Fara）《性、植物学与帝国：林奈与班克斯》（*Sex, Botany and Empire, The Story of Carl Linnaeus and Joseph Banks*）一书中讲述了英国植物学家班克斯（Joseph Banks）的故事。1768 年 8 月 15 日，班克斯告别未婚妻，登上了澳大利亚军舰"奋进号"。此次"奋进号"的远航是受英国海军部和皇家学会资助，目的是前往南太平洋的塔希提岛(Tahiti，法属海外自治领，另一个常见的译名是"大溪地")观测一次比较罕见的金星凌日。舰长库克（James Cook）是西方殖民史上最著名的舰长之一，多次远航探险，开拓海外殖民地。他还被认为是澳大利亚和夏威夷群岛的"发现"者，如今以他命名的群岛、海峡、山峰等不胜枚举。

当"奋进号"停靠塔希提岛时，班克斯一下就被当地美丽的土著女性迷昏了，他在她们的温柔乡里纵情狂欢，连库克舰长都看不下去了，"道德愤怒情绪偷偷溜进了他的日志当中，他发现自己根本不可能不去批评所见到的滥交行为"，而班克斯纵欲到了"连嫖妓都毫无激情"的地步——这是别人讽刺班克斯的说法，因为对于那时常年航行于茫茫大海上的男性来说，上岸嫖妓通常是一项能够唤起"激情"的活动。

而在"帝国科学"的宏大叙事中，科学家的私德是无关紧要的，人们关注的是科学家做出的科学发现。所以，尽管一面是班克斯在塔希提岛纵欲滥交，一面是他留在故乡的未婚妻正泪眼婆娑地"为远去的心上人绣织背心"，这样典型的"渣男"行径要是放在今天，非被互联网上的口水淹死不可，但是"班克斯很快从他们的分离之苦中走了出来，在外近三年，他活得倒十分滋润"。

法拉不无讥刺地指出了"帝国科学"的实质："班克斯接管了当地的女性和植物，而库克则保护了大英帝国在太平洋上的殖民地。"甚至对班克斯的植物学本身也调侃了一番："即使是植物学方面的科学术语也充满了性指涉。……这个体系主要依靠花朵之中雌雄生殖器官的数量来进行分类。"据说"要保护年轻妇女不受植物学教育的浸染，他们严令禁止各种各样的植物采集探险活动。"这简直就是将植物学看成一种"涉黄"的淫秽色情活动了。

在意识形态强烈影响着我们学术话语的时代，上面的故事通常是这样被描述的：库克舰长的"奋进号"军舰对殖民地和尚未成为殖民地的那些地方的所谓"访问"，其实是殖民者耀武扬威的侵略，搭载着达尔文的"小猎犬号"军舰也是同样行径；班克斯和当地女性的纵欲狂欢，当然是殖民者对土著妇女令人发指的蹂躏；即使是他采集当地植物标本的"科学考察"，也可以视为殖民者"窃取当地经济情报"的罪恶行为。

后来改革开放，上面那种意识形态话语被抛弃了，但似乎又走向了另一个极端，完全忘记或有意回避殖民者和帝国主义这个层面，只歌颂这些军舰上的科学家的伟大发现和成就，例如达尔文随着"小猎犬号"的航行，早已成为一曲祥和优美的科学颂歌。

其实达尔文也未能免俗，他在远航中也乐意与土著女性打打交道，当然他没有像班克斯那样滥情纵欲。在达尔文为"小猎犬号"远航写的《环球游记》中，我们读到："回程途中我们遇到一群黑人姑娘在聚会，……我们笑着看了很久，还给了她们一些钱，这着实令她们欣喜一番，拿着钱尖声大笑起来，很远还能听到那愉悦的笑声。"

有趣的是，在班克斯在塔希提岛纵欲六十多年后，达尔文随着"小猎犬号"也来到了塔希提岛，岛上的土著女性同样引起了达尔文的注意，在《环球游记》中他写道："我对这里妇女的外貌感到有些失望，然而她们却很爱美，把一朵白花或者红花戴在脑后的鬓髻上……"接着他以居高临下的笔调描述了当地女性的几种发饰。

用今天的眼光来看，这些在别的民族土地上采集植物动物标本、测量地质水文数据等等的"科学考察"行为，有没有合法性问题？有没有侵犯主权的问题？这些行为得到当地人的同意了吗？当地人知道这些行为的性质和意义吗？他们有知情权吗？……这些问题，在今天的国际交往中，确实都是存在的。

也许有人会为这些帝国科学家辩解说：那时当地土著尚在未开化或半开化状态中，他们哪有"国家主权"的意识啊？他们也没有制止帝国科学家的考察活动啊？但是，这样的辩解是无法成立的。

姑不论当地土著当时究竟有没有试图制止帝国科学家的"科学考察"行为，现在早已不得而知，只要殖民者没有记录下来，我们通常就无法知道。况且殖民者有军舰有枪炮，土著就是想制止也无能为力。正如法拉所描述的："在几个塔希提人被杀之后，一套行之有效的易货贸易体制建立了起来。"

即使土著因为无知而没有制止帝国科学家的"科学考察"行为，这事也很像一个成年人闯进别人的家，难道因为那家只有不懂事的小孩子，闯入者就可以随便打探那家的隐私、拿走那家的东西、甚至将那家的房屋土地据为己有吗？事实上，很多情况下殖民者就是这样干的。所以，所谓的"帝国科学"，其实是有着原罪的。

如果沿用上述比喻，现在的局面是，家家户户都不会只有不懂事的孩子了，所以任何外来者要想进行"科学探索"，他也得和这家主人达成共识，得到这家主人的允许才能够进行。即使这种共识的达成依赖于利益的交换，至少也不能单方面强加于人。

博物学在今日中国

博物学在今日中国之复兴，北京大学刘华杰教授提倡之功殊不可没。自刘教授大力提倡之后，各界人士纷纷跟进，仿佛昔日蔡锷在云南起兵反袁之"滇黔首义，薄海同钦，一檄遥传，景从恐后"光景，这当然是和博物学本身特点密切相关的。

无论在西方还是在中国，无论在过去还是在当下，为何博物学在它繁荣时尚的阶段，就会应者云集？深究起来，恐怕和博物学本身的特点有关。博物学没有复杂的理论结构，它的专业训练也相对容易，至少没有天文学、物理学那样的数理"门槛"，所以和一些数理学科相比，博物学可以有更多的自学成才者。这次编印的《西方博物学大系》，卷帙浩繁，蔚为大观，同样说明了这一点。

最后，还有一点明显的差别必须在此处强调指出：用刘华杰教授喜欢的术语来说，《西方博物学大系》所收入的百余种著作，绝大部分属于"一阶"性质的工作，即直接对博物学作出了贡献的著作。事实上，这也是它们被收入《西方博物学大系》的主要理由之一。而在中国国内目前已经相当热的博物学时尚潮流中，绝大部分已经出版的书籍，不是属于"二阶"性质（比如介绍西方的博物学成就），就是文学性的吟风咏月野草闲花。

要寻找中国当代学者在博物学方面的"一阶"著作，如果有之，以笔者之孤陋寡闻，唯有刘华杰教授的《檀岛花事——夏威夷植物日记》三卷，可以当之。这是刘教授在夏威夷群岛实地考察当地植物的成果，不仅属于直接对博物学作出贡献之作，而且至少在形式上将昔日"帝国科学"的逻辑反其道而用之，岂不快哉！

2018 年 6 月 5 日
于上海交通大学
科学史与科学文化研究院

约翰·帕金森

（1567-1650）

约翰·帕金森（John Parkinson）是英国著名药剂师、植物学家兼园艺家。他早年生活在约克郡，十四岁赴伦敦求学，终于成为国王詹姆士一世的御用药剂师。他还是一位拥有"御用首席植物学家"头衔的园艺大师，在伦敦打造了一处庞大的个人庭园，其中种植了很多委托种子商人从海外带回的珍奇植物，是当时英国和欧洲大陆植物学家、药剂师和园艺家探讨学习的必经之处。

十六、十七世纪，欧洲的精神世界仍以基督教为主导，教义中人与乐园的关系异常重要，从《创世记》第二章的内容便可见其端倪：

创造天地的来历，在耶和华神造天地的日子，乃是这样，

野地还没有草木，田间的菜蔬还没有长起来；因为耶和华神还没有降雨在地上，也没有人耕地，

但有雾气从地上腾，滋润遍地。

耶和华神用地上的尘土造人，将生气吹在他鼻孔里，他就成了有灵的活人，名叫亚当。

耶和华神在东方的伊甸立了一个园子，把所造的人安置在那里。

耶和华神使各样的树从地里长出来，可以悦人的眼目，其上的果子好作食物。园子当中又有生命树和分别善恶的树。

有河从伊甸流出来，滋润那园子，从那里分为四道：

第一道名叫比逊，就是环绕哈腓拉全地的。在那里有金子，

并且那地的金子是好的；在那里又有珍珠和红玛瑙。

第二道河名叫基训，就是环绕古实全地的。

第三道河名叫希底结，流在亚述的东边。第四道河就是伯拉河。

耶和华神将那人安置在伊甸园，使他修理，看守。

为此，当时的本草学著作也都强调园艺的超然性与重要性。1629 年，帕金森刊行了《园艺大要》，书名原文为 "Paradisi in Sole Paradisus Terrestris"，意思是"人间乐园"，从书名和扉页的铜版寓意画可以看出，本书的宗旨是教人们如何用自己的双手在人世间营造庭园、栽培植物，再现传说中伊甸园的美景。

本书分为三部分，分别为"花园""菜园"和"果树园"。内容涵盖庭园设计、堆土、播种、栽培和堆肥等方面，配有约 800 幅精美植物画。在《园艺

大要》问世之前，西方植物研究和著作仍以实用的蔬果栽培法和药用知识为主，尚属于本草学阶段，而本书更进一步，在实用性基础上探讨研究植物本身的性情与美感，是当代植物学研究的滥觞之作。自帕金森后，西方植物研究才逐渐褪去本草学的身影，踏入植物学的领域，正因如此，他也被称为"英国历史上最后的本草学家，兼第一位植物学家"。

TO
THE QVEENES
MOST EXCELLENT
MAIESTIE.

Madame,

Nowing your Maieſtie ſo much delighted with all the faire Flowers of a Garden, and furniſhed with them as farre beyond others, as you are eminent before
them; this my Worke of a Garden, long
before this intended to be publiſhed, and
but now only finiſhed, ſeemed as it were
deſtined, to bee firſt offered into your
Highneſſe hands, as of right challenging the proprietie of
Patronage from all others. Accept, I beſeech your Maieſtie,
this ſpeaking Garden, that may informe you in all the particulars of your ſtore, as well as wants, when you cannot ſee
any of them freſh vpon the ground : and it ſhall further en-
courage him to accompliſh the remainder; who, in praying
that your Highneſſe may enioy the heauenly Paradiſe after
the many yeares fruition of this earthly, ſubmitteth to be

Your Maieſties

in all

humble deuotion,

IOHN PARKINSON.

TO THE COVRTEOVS
READER.

Lthough the ancient Heathens did appropriate the first inuention of the knowledge of Herbes, and so consequently of Physicke, some vnto Chiron the Centaure, and others vnto Apollo or Æsculapius his sonne; yet wee that are Christians haue out of a better Schoole learned, that God, the Creator of Heauen and Earth, at the beginning when he created Adam, inspired him with the knowledge of all natural things (which successiuely descended to Noah afterwardes, and to his Posterity): for, as he was able to giue names to all the liuing Creatures, according to their seuerall natures; so no doubt but hee had also the knowledge, both what Herbes and Fruits were fit, eyther for Meate or Medicine, for Vse or for Delight. And that Adam might exercise this knowledge, God planted a Garden for him to liue in, (wherein euen in his innocency he was to labour and spend his time) which hee stored with the best and choysest Herbes and Fruits the earth could produce, that he might haue not onely for necessitie whereon to feede, but for pleasure also; the place or garden called Paradise importing as much, and more plainly the words set downe in Genesis the second, which are these; Out of the ground the Lord God made to grow euerie tree pleasant to the sight and good for meate; and in the 24. of Numbers, the Parable of Balaam, mentioning the Aloe trees that God planted; and in other places if there were neede to recite them. But my purpose is onely to shew you, that Paradise was a place (whether you will call it a Garden, or Orchard, or both, no doubt of some large extent) wherein Adam was first placed to abide; that God was the Planter thereof, hauing furnished it with trees and herbes, as well pleasant to the sight, as good for meate, and that hee being to dresse and keepe this place, must of necessity know all the things that grew therein, and to what vses they serued, or else his labour about them, and knowledge in them, had been in vaine. And although Adam lost the place for his transgression, yet he lost not the naturall knowledge, nor vse of them: but that, as God made the whole world, and all the Creatures therein for Man, so hee may vse all things as well of pleasure as of necessitie, to bee helpes vnto him to serue his God. Let men therefore, according to their first institution, so vse their seruice, that they also in them may remember their seruice to God, and not (like our Grand-mother Eue) set their affections so strongly on the pleasure in them, as to deserue the losse of them in this Paradise, yea and of Heauen also. For truly from all sorts of Herbes and Flowers we may draw matter at all times not only to magnifie the Creator that hath giuen them such diuersities of formes, sents and colours, that the most cunning

Worke-

Worke-man cannot imitate, and ſuch vertues and properties, that although wee know many, yet many more lye hidden and vnknowne, but many good inſtructions alſo to our ſelues: That as many herbes and flowers with their fragrant ſweete ſmels doe comfort, and as it were reuiue the ſpirits, and perfume a whole houſe; euen ſo ſuch men as liue vertuouſly, labouring to doe good, and profit the Church of God and the Common wealth by their paines or penne, doe as it were ſend forth a pleaſing ſauour of ſweet inſtructions, not only to that time wherein they liue, and are freſh, but being drye, withered and dead, ceaſe not in all after ages to doe as much or more. Many herbes and flowers that haue ſmall beautie or ſauour to commend them, haue much more good vſe and vertue: ſo many men of excellent rare parts and good qualities doe lye hid vnknown and not reſpected, vntill time and vſe of them doe ſet forth their properties. Againe, many flowers haue a glorious ſhew of beauty and brauery, yet ſtinking in ſmell, or elſe of no other vſe: ſo many doe make a glorious oſtentation, and flouriſh in the world, when as if they ſtinke not horribly before God, and all good men, yet ſurely they haue no other vertue then their outſide to commend them, or leaue behind them. Some alſo riſe vp and appear like a Lilly among Thornes, or as a goodly Flower among many Weedes or Graſſe, eyther by their honourable authoritie, or eminence of learning or riches, whereby they excell others, and thereby may doe good to many. The frailty alſo of Mans life is learned by the ſoone fading of them before their flowring, or in their pride, or ſoone after, being either cropt by the hand of the ſpectator, or by a ſudden blaſt withered and parched, or by the reuolution of time decaying of it owne nature: as alſo that the faireſt flowers or fruits firſt ripe, are ſooneſt and firſt gathered. The mutabilitie alſo of ſtates and perſons, by this, that as where many goodly flowers & fruits did grow this yeare and age, in another they are quite pulled or digged vp, and eyther weedes and graſſe grow in their place, or ſome building erected thereon, and their place is no more known. The Ciuill reſpects to be learned from them are many alſo: for the delight of the varieties both of formes, colours and properties of Herbes and Flowers, hath euer beene powerfull ouer dull, vnnurtured, ruſticke and ſauage people, led only by Natures inſtinct; how much more powerfull is it, or ſhould be in the mindes of generous perſons? for it may well bee ſaid, he is not humane, that is not allured with this object. The ſtudy, knowledge, and trauel in them, as they haue been entertained of great Kings, Princes and Potentates, without diſparagement to their Greatneſſe, or hinderance to their more ſerious and weighty Affaires: ſo no doubt vnto all that are capable thereof, it is not onely pleaſant, but profitable, by comforting the minde, ſpirits and ſenſes with an harmeleſſe delight, and by enabling the iudgement to conferre and apply helpe to many dangerous diſeaſes. It is alſo an Inſtructer in the verity of the genuine Plants of the Ancients, and a Correcter of the many errours whereunto the world by continuance hath bin diuerted, and almoſt therein fixed, by eradicating in time, and by degrees, the pertinacious wilfulneſſe of many, who becauſe they were brought vp in their errours, are moſt vnwilling to leaue them without conſideration of the good or euill, the right or wrong, they draw on therewith. And for my ſelfe I may well ſay, that had not mine owne paines and ſtudies by a naturall inclination beene more powerfull in mee then any others helpe (although ſome through an euill diſpoſition and ignorance haue ſo far traduced me as to ſay this was rather another mans worke then mine owne, but I leaue them to their folly) I had neuer done ſo much as I here publiſh; nor beene fit or prepared for a larger, as time may ſuddenly (by Gods permiſſion) bring to light, if the maleuolent diſpoſitions of degenerate ſpirits doe not hinder the accompliſhment.

But

But perſwading my ſelfe there is no ſhowre that produceth not ſome fruit, or no word but worketh ſome effect, eyther of good to perſwade, or of reproofe to cuince; I could not but declare my minde herein, let others iudge or ſay what they pleaſe. For I haue alwaies held it a thing vnfit, to conceale or bury that knowledge God hath giuen, and not to impart it, and further others therewith as much as is conuenient, yet without oſtentation, which I haue euer hated. Now further to informe the courteous Reader, both of the occaſion that led me on to this worke, and the other occurrences to it. Firſt, hauing peruſed many Herbals in Làtine, I obſerued that moſt of them haue eyther neglected or not knowne the many diuerſities of the flower Plants, and rare fruits are known to vs at this time, and (except Cluſius) haue made mention but of a very few. In Engliſh likewiſe we haue ſome extant, as Turner and Dodonæus tranſlated, who haue haue ſaid little of Flowers, Gerard who is laſt, hath no doubt giuen vs the knowledge of as many as he attained vnto in his time, but ſince his dates we haue had many more varieties, then he or they euer heard of, as may be perceiued by the ſtore I haue here produced. And none of them haue particularly ſeuered thoſe that are beautifull flower plants, fit to ſtore a garden of delight and pleaſure, from the wilde and vnfit: but haue enterlaced many, one among another, whereby many that haue deſired to haue faire flowers, haue not known either what to chooſe, or what to deſire. Diuers Bookes of Flowers alſo haue been ſet forth, ſome in our owne Country, and more in others, all which are as it were but handfuls ſnatched from the plentifull Treaſury of Nature, none of them being willing or able to open all ſorts, and declare them fully ; but the greateſt hinderance of all mens delight was, that none of them had giuen any deſcription of them, but the bare name only. To ſatisfie therefore their deſires that are louers of ſuch Delights, I took vpon me this labour and charge, and haue here ſelected and ſet forth a Garden of all the chiefeſt for choyce, and faireſt for ſhew, from among all the ſeuerall Tribes and Kindreds of Natures beauty, and haue ranked them as neere as I could, or as the worke would permit, in affinity one vnto another. Secondly, and for their ſakes that are ſtudious in Authors, I haue ſet down the names haue bin formerly giuen vnto them, with ſome of their errours, not intending to cumber this worke with all that might bee ſaid of them, becauſe the deciding of the many controuerſies, doubts, and queſtions that concerne them, pertaine more fitly to a generall Hiſtory : yet I haue beene in ſome places more copious and ample then at the firſt I had intended, the occaſion drawing on my deſire to informe others with what I thought was fit to be known, reſeruing what elſe might be ſaid to another time & worke; wherein (God willing) I will inlarge my ſelfe, the ſubiect matter requiring it at my hands, in what my ſmall ability can effect. Thirdly, I haue alſo to embelliſh this Worke ſet forth the figures of all ſuch plants and flowers as are materiall and different one from another : but not as ſome others haue done, that is, a number of the figures of one ſort of plant that haue nothing to diſtinguiſh them but the colour, for that I hold to be ſuperfluous and waſte. Fourthly, I haue alſo ſet down the Vertues and Properties of them in a briefe manner, rather deſiring to giue you the knowledge of a few certaine and true, then to relate, as others haue done, a needleſſ and falſe multiplicitie, that ſo there might as well profit as pleaſure be taken from them, and that nothing might be wanting to accompliſh it fully. And ſo much for this firſt part, my Garden of pleaſant and delightfull Flowers. My next Garden conſiſteth of Herbes and Rootes, fit to be eaten of the rich and poor as nouriſhment and food, as ſance or condiment, as ſallet or refreſhing, for pleaſure or profit; where I doe as well play the Gardiner, to ſhew you (in briefe, but not at large) the times

and

and manner of ſowing, ſetting, planting, replanting, and the like (although all theſe things, and many more then are true, are ſet down very largely in the ſeuerall bookes that others haue written of this ſubiect) as alſo to ſhew ſome of the Kitchen vſes (becauſe they are Kitchen herbes &c.) although I confeſſe but very ſparingly, not intending a treatiſe of cookery, but briefly to giue a touch thereof; and alſo the Phyſicall properties, to ſhew ſomewhat that others haue not ſet forth; yet not to play the Empericke, and giue you receipts of medicines for all diſeaſes, but only to ſhew in ſome ſort the qualities of Herbes, to quicken the minds of the ſtudious. And laſtly an Orchard of all ſorts of domeſticke or forraine, rare and good fruits, fit for this our Land and Countrey, which is at this time better ſtored and furniſhed then euer in any age before. I haue herein endeauoured, as in the other Gardens, to ſet forth the varieties of euery ſort in as briefe a manner as poſsibly could be, without ſuperfluous repetitions of deſcriptions, and onely with eſpeciall notes of difference in leaues, flowers and fruits. Some few properties alſo are ſet downe, rather the chiefeſt then the moſt, as the worke did require. And moreouer before euery of theſe parts I haue giuen Treatiſes of the ordering, preparing and keeping the ſeuerall Gardens and Orchard, with whatſoeuer I thought was conuenient to be known for euery of them.

Thus haue I ſhewed you both the occaſion and ſcope of this Worke, and herein haue ſpent my time, paines and charge, which if well accepted, I ſhall thinke well employed, and may the ſooner haſten the fourth Part, A Garden of Simples; which will be quiet no longer at home, then that it can bring his Maſter newes of faire weather for the iourney.

Thine in what he may,

IOHN PARKINSON.

Ioanni

Ioanni Parkinfono *Pharmacopoeo Londinenfi folertif-*
fimo Botanico confummatifsimo
T. D. M. S. P. D.

POema panegyricum Opus tuum indefeffi laboris, vtili-
tatis eximiæ poftulat, & meriti iure à me extorqueret
(mi Parkinfone) fi fauentibus Mufis, & fecundo Apol-
line in bicipiti fomniare Parnaffo, & repentè Poetæ mihi
prodire liceret. In fœtus tui bonis auibus in lucem editi,
& prolixiorem nepotum feriem promittentis laudes, alii
Deopleni Enthoufiaftæ carmine fuos pangant elenchos;
quos fub figmentis ampullata hyperbolicarum vocum mulcedine, vates
ferè auribus mentibufue infinuant. Veritas nudicatis amans, fuco natiuum
candorem obumbranti non illuftranti perpetuum indixit bellum : In fim-
plicitate, quam affertionum neruofa breuitas exprimit, exultat. Audi quid
de te fentiam, Tu mihi fis in pofterum Crateuas Brittannus; inter omnes,
quotquot mihi hic innotuerunt, peritifsimus, exercitatifsimus, oculatif-
fimus, & emunctifsimæ naris Botanicus : Cuius opera in fortunata hac
Infula rem herbariam tractari, emendari, augeri, & popularibus tuis ver-
naculo fermone ad amufsim tradi, non decentiæ modo, fed etiam necefsi-
tatis eft. Macte tua fedulitate (Vir optime) neque te laborum tam arduis
lucubrationibus datorum hactenus pœniteat, vel deinceps impendendo-
rum pigeat. Difficilia quæ pulchra. Leniet debitæ laudis dulcedo vigiliarum
acerbitatem, & Olympicum ftadium cito pede, à carceribus ad metas ala-
criter decurrentem nobile manet βραβεῖον. Sed memento Artem longam,
Vitam effe breuem. Μηδὲν ἀναβαλλόμενος. Vide quid ad antiquum illum, cuius
fi non animam, faltem genium induifti, Crateuam fcribat Hippocrates,
Τέχνης πάσης ἀλήθειαν ἀναβολὴ ἱστικῆς δὲ κỳ πάνυ· ἐν ᾗ ψυχῆς κίνδυνος ἢ ὑπέρβασις. Nobilifsimam
Medicinæ partem Botanicam effe reputa. Floræ nunc litafti & Pomonæ,
Apollini vt audio propediem HORTO MEDICO facturus. Amabò integræ
Veftæ facra conficito, eiufque variegatum multis fimplicium morbifugo-
rum myriadibus finum abfolutè pandito, quem fine velo nobis exhibeas.
Nulla dies abeat fine linea. Sic tandem fructus gloriæ referes vberrimos,
quos iuftè fudoribus partos, vt in cruda & viridi fenectute decerpas diu,
iifque longum fruaris opto. *Vale.* *Datum Londini Calendas Octobris anno*
falutis 1629.

Theodorus de Mayerne Eques aurat. in Aula
Regum Magnæ Britanniæ Iacobi & Caroli
P. & F. Archiatrorum Comes.

Ad eximium arte & vſu Pharmacopæum & Botanographum *I. Parkinſonum.*

Gu. Turne-
rus. M. D.

Io. Gerar-
dus Chirur-
gus.

*H*Erbarum vires, primus te (magne Britannæ)
 Edocuit medicas, inclytus arte ſophus.
Atque cluens herbis alter, Chironis alumnus,
 Deſcripſit plantas, neu cadat vlla ſalus.
 Fortunate ſenex, ſis tu nunc tertius Heros
Hortos qui reſeras, deliciaſque ſoli,
Et flores Veneris lætos, herbaſque virentes,
Arboreos fætus, pharmacum & arte potens.
Poſteritas iuſtos poſthac tibi ſolvet honores,
 Laudabitque tuæ dexteritatis opus.

Ottuellus Meuerell. D.M. & Collegiæ
Med. Lond. ſocius.

Amico ſuo *Ioanni Parkinſono.*

*E*Xtollunt alij quos (Parkinſone) labores
 Da mihi iam veniam comminuiſſe tuos.
Extremos poteris credi migraſſe per Indos:
 Cum liber haud aliud quam tuus hortus hic eſt:
Ipſe habitare Indos tecum facis, haud petis Indos
 I nunc, & tua me comminuiſſe refer.
Eſt liber Effigies, tuus hic qui pingitur hortus,
 Digna manu facies hæc, facieque manus!
Vidi ego ſplendentem varigatis vndique gemmis
 Una fuit Salomon, turba quid ergo fuit?
Vt vario ſplendent Pallacia regia ſumptu,
 Et Procerum turbis Atria tota nitent:
Tunc cum feſta dies veniam dedit eſſe ſuperbis
 Quoſque ficus texit, nunc tria rura tegunt:
Plena tuo pariter ſpectatur Curia in Horto,
 Hic Princeps, Dux hic, Sponſaque pulchra Ducis.
Quæque dies eſt feſta dies, nec parcius vnquam
 Luxuriant, lauta hæc; Quotidiana tamen.
Ecce velut Patriæ Paradiſi haud immemor Exul,
 Hunc naturali pingit amore ſibi.
Pingit & ad vivum ſub eodem nomine, & hic eſt
 Fronticuli ſudor quem cerebrique dedit:
Aſtat Adam medius Paradiſo noſter in iſto
 Et ſpecies nomen cuique dat ipſe ſuum.
Hos cape pro meritis, qui florem nomine donas
 Æternum florens tu tibi Nomen habe.

Guilielmus Atkins.

Ad Amicum *Ioannem Parkinſonum* Pharmacopæum, & Archibotanicum Londinenſem.

Frica quas profert Plantas, quas India mittit,
Quas tua dat tellus, has tuus hortus habet:
Atque harum Species, florendi tempora, vires,
Et varias formas iſte libellus habet:
Neſcio plus librum talem mirabor, an hortum
Totus ineſt horto mundus; at iſte libro.
Parkinſone tuus liber, & labor, & tua ſit laus,
Herbas dum nobis das; datur herba tibi.

Guilielmus Brodus Pharmacopæus
ac Philobotanicus Londinenſis.

Ad Amicum *Ioannem Parkinſonum* Pharmacopæum & Botanicum inſignem. Carmen.

Vam magno pandis Floræ penetralia nixu
Atque facis cœlo liberiore frui?
Omnibus vt placeas, ô quam propenſa voluntas,
Solicituſque labor noĉte dieque premit?
Quam magno cultum ſtudio conquirere in hortum
Herbarum quicquid mundus in orbe tenet,
Immenſus ſumptus, multoſque extenſus in annos
Te labor afficiunt? & data nulla quies.
Talia quærenti, ſurgit novus ardor habendi,
Nec tibi tot ſoli munera magna petis;
Deſcriptos vivâ profers ſub imagine flores,
Tum profers menſæ quicquid & hortus alit,
Laudatos nobis fruĉtus & promis honores,
Profers, quas celebrant nullibi ſcripta virum,
Herbarum ſpecies, quibus eſt quoque grata venuſtas;
Sic nos multiplici munere, Amice, beas.
Hoc cape pro meritis, florum dum gratia floret,
Suntque herbis vires; en tibi Nomen erit.
In ſerum ſemper tua gloria floreat ævum,
Gloria quæ in longum non peritura diem.

Thomas Iohnſon vtriuſque
Societatis conſors.

THE ORDERING OF THE
GARDEN OF PLEASVRE.

CHAP. I.

The situation of a Garden of pleasure, with the nature of soyles, and how to amend the defects that are in many sorts of situations and grounds.

HE seuerall situations of mens dwellings, are for the most part vnauoideable and vnremoueable; for most men cannot appoint forth such a manner of situation for their dwelling, as is most fit to auoide all the inconueniences of winde and weather, but must bee content with such as the place will afford them; yet all men doe well know, that some situations are more excellent than o-thers: according therfore to the seuerall situation of mens dwel-lings, so are the situations of their gardens also for the most part. And although diuers doe diuersly preferre their owne seuerall places which they haue chosen, or wherein they dwell; As some those places that are neare vnto a riuer or brooke to be best for the pleasantnesse of the water, the ease of transportation of them-selues, their friends and goods, as also for the fertility of the soyle, which is seldome bad neare vnto a riuers side; And others extoll the side or top of an hill, bee it small or great, for the prospects sake; And againe, some the plaine or champian ground, for the euen leuell thereof: euery one of which, as they haue their commodities accompa-nying them, so haue they also their discommodities belonging vnto them, according to the Latine Prouerbe, *Omne commodum fert suum incommodum.* Yet to shew you for eue-rie of these situations which is the fittest place to plant your garden in, and how to de-fend it from the iniuries of the cold windes and frosts that may annoy it, will, I hope, be well accepted. And first, for the water side, I suppose the North side of the water to be the best side for your garden, that it may haue the comfort of the South Sunne to lye vpon it and face it, and the dwelling house to bee aboue it, to defend the cold windes and frosts both from your herbes, and flowers, and early fruits. And so likewise I iudge for the hill side, that it may lye full open to the South Sunne, and the house aboue it, both for the comfort the ground shall receiue of the water and raine descending into it, and of defence from winter and colds. Now for the plaine leuell ground, the buil-dings of the house should be on the North side of the garden, that so they might bee a defence of much sufficiency to safeguard it from many iniurious cold nights and dayes, which else might spoyle the pride thereof in the bud. But because euery one cannot so appoint his dwelling, as I here appoint the fittest place for it to be, euery ones pleasure thereof shall be according to the site, cost, and endeauours they bestow, to cause it come nearest to this proportion, by such helpes of bricke or stone wals to defend it, or by the helpe of high growne and well spread trees, planted on the North side thereof, to keepe it the warmer. And euery of these three situations, hauing the fairest buildings of the house facing the garden in this manner before specified, besides the benefit of shelter it shall haue from them, the buildings and roomes abutting thereon, shall haue recipro-cally the beautifull prospect into it, and haue both sight and sent of whatsoeuer is ex-cellent, and worthy to giue content out from it, which is one of the greatest pleasures a garden can yeeld his Master. Now hauing shewed you the best place where this your

A

garden

garden fhould be, let me likewife aduife you where it fhould not be, at leaft that it is the worft place wherein it may be, if it be either on the Weft or Eaft fide of your houfe, or that it ftand in a moorifh ground, or other vnwholfome ayre (for many, both fruits, herbes, and flowers that are tender, participate with the ayre, taking in a manner their chiefeft thriuing from thence) or neare any common Lay-ftalles, or common Sewers, or elfe neare any great Brew-houfe, Dye-houfe, or any other place where there is much fmoake, whether it be of ftraw, wood, or efpecially of fea-coales, which of all other is the worft, as our Citie of London can giue proofe fufficient, wherein neither herbe nor tree will long profper, nor hath done euer fince the vfe of fea-coales beganne to bee frequent therein. And likewife that it is much the worfe, if it bee neare vnto any Barnes or Stackes of corne or hey, becaufe that from thence will continually with the winde bee brought into the garden the ftrawe and chaffe of the corne, the duft and feede of the hey to choake or pefter it. Next vnto the place or fituation, let mee fhew you the grounds or foyles for it, eyther naturall or artificiall. No man will deny, but the naturall blacke mould is not only the fatteft and richeft, but farre exceedeth any other either naturall or artificiall, as well in goodneffe as durability. And next thereunto, I hold the fandy loame (which is light and yet firme, but not loofe as fand, nor ftiffe like vnto clay) to be little inferiour for this our Garden of pleafure; for that it doth caufe all bulbous and tuberous rooted plants to thriue fufficiently therein, as likewife all other flower-plants, Rofes, Trees, &c. which if it fhall decay by much turning and working out the heart of it, may foone be helped with old ftable manure of horfes, being well turned in, when it is old and almoft conuerted to mould. Other grounds, as chalke, fand, grauell, or clay, are euery of them one more or leffe fertill or barren than other; and therefore doe require fuch helpes as is moft fit for them. And thofe grounds that are ouer dry, loofe, and duftie, the manure of ftall fedde beafts and cattell being buried or trenched into the earth, and when it is thorough rotten (which will require twice the time that the ftable foyle of horfes will) well turned and mixed with the earth, is the beft foyle to temper both the heate and drineffe of them. So contrariwife the ftable dung of horfes is the beft for cold grounds, to giue them heate and life. But of all other forts of grounds, the ftiffe clay is the very worft for this purpofe; for that although you fhould digge out the whole compaffe of your Garden, carry it away, and bring other good mould in the ftead thereof, and fill vp the place, yet the nature of that clay is fo predominant, that in a fmall time it will eate out the heart of the good mould, and conuert it to its owne nature, or very neare vnto it: fo that to bring it to any good, there muft bee continuall labour beftowed thereon, by bringing into it good ftore of chalke, lime, or fand, or elfe afhes eyther of wood or of fea-coales (which is the beft for this ground) well mixed and turned in with it. And as this ftiffe clay is the worft, fo what ground foeuer commeth neareft vnto the nature thereof, is neareft vnto it in badneffe, the fignes whereof are the ouermuch moyfture thereof in Winter, and the much cleauing and chapping thereof in Summer, when the heate of the yeare hath confumed the moyfture, which tyed and bound it faft together, as alfo the ftiffe and hard working therein: but if the nature of the clay bee not too ftiffe, but as it were tempered and mixed with fand or other earths, your old ftable foyle of horfes will helpe well the fmall rifting or chapping thereof, to be plentifully beftowed therin in a fit feafon. Some alfo do commend the cafting of ponds and ditches, to helpe to manure thefe ftiffe chapping grounds. Other grounds, that are ouermoift by fprings, that lye too neare the vpper face of the earth, befides that the beds thereof had need to be laid vp higher, and the allies, as trenches and furrowes, to lye lower, the ground it felfe had neede to haue fome good ftore of chalke-ftones beftowed thereon, fome certaine yeares, if it may be, before it be laid into a Garden, that the Winter frofts may breake the chalke fmall, and the Raine diffolue it into mould, that fo they may bee well mixed together; than which, there is not any better manure to foyle fuch a moift ground, to helpe to dry vp the moyfture, and to giue heate and life to the coldneffe thereof, which doth alwayes accompany thefe moift grounds, and alfo to caufe it abide longer in heart than any other. For the fandy and grauelly grounds, although I know the well mollified manure of beafts and cattell to be excellent good, yet I know alfo, that fome commend a white Marle, and fome a clay to be well fpread thereon, and after turned thereinto: and for the chalkie ground, *è conuerfo*, I commend fatte clay to helpe it. You muft vnderftand, that the leffe rich or more barren that your ground is, there needeth

deth the more care, labour, and coſt to bee beſtowed thereon, both to order it rightly, & ſo to preſerue it from time to time : for no artificiall or forc't ground can endure good any long time, but that within a few yeares it muſt be refreſhed more or leſſe, according as it doth require. Yet you ſhall likewiſe vnderſtand, that this Garden of pleaſure ſtored with theſe Out-landiſh flowers ; that is, bulbous and tuberous rooted plants, and other fine flowers, that I haue hereafter deſcribed, and aſſigned vnto it, needeth not ſo much or ſo often manuring with ſoyle, &c. as another Garden planted with the other ſorts of Engliſh flowers, or a Garden of ordinary Kitchin herbes doth. Your ground likewiſe for this Garden had neede to bee well cleanſed from all annoyances (that may hinder the well doing or proſpering of the flowers therein) as ſtones, weedes, rootes of trees, buſhes, &c. and all other things cumberſome or hurtfull ; and therefore the earth being not naturally fine enough of it ſelfe, is vſed to bee ſifted to make it the finer, and that either through a hurdle made of ſticks, or lathes, or through ſquare or round ſieues platted with fine and ſtrong thin ſtickes, or with wyers in the bottome. Or elſe the whole earth of the Garden being courſe, may be caſt in the ſame manner that men vſe to try or fine ſand from grauell, that is, againſt a wall ; whereby the courſer and more ſtony, falling downe from the fine, is to be taken away from the foote of the heape, the finer ſand and ground remaining ſtill aboue, and on the heape. Or elſe in the want of a wall to caſt it againſt, I haue ſeene earth fined by it ſelfe in this manner : Hauing made the floore or vpper part of a large plat of ground cleane from ſtones, &c. let there a reaſonable round heape of fine earth be ſet in the midſt thereof, or in ſtead thereof a large Garden flowerpot, or other great pot, the bottome turned vpwards, and then poure your courſe earth on the top or head thereof, one ſhouell full after another ſomewhat gently, and thereby all the courſe ſtuffe and ſtones will fall downe to the bottome round about the heape, which muſt continually be carefully taken away, and thus you may make your earth as fine as if it were caſt againſt a wall, the heape being growne great, ſeruing in ſtead thereof. Thoſe that will not prepare their grounds in ſome of theſe manners aforeſaid, ſhall ſoone finde to their loſſe the neglect thereof : for the traſh and ſtones ſhill ſo hinder the encreaſe of their roots, that they will be halfe loſt in the earth among the ſtones, which elſe might be ſaued to ſerue to plant whereſoeuer they pleaſe.

CHAP. II.
The frame or forme of a Garden of delight and pleaſure, with the ſeuerall varieties thereof.

ALthough many men muſt be content with any plat of ground, of what forme or quantity ſoeuer it bee, more or leſſe, for their Garden, becauſe a more large or conuenient cannot bee had to their habitation : Yet I perſwade my ſelfe, that Gentlemen of the better ſort and quality, will prouide ſuch a parcell of ground to bee laid out for their Garden, and in ſuch conuenient manner, as may be fit and anſwerable to the degree they hold. To preſcribe one forme for euery man to follow, were too great preſumption and folly : for euery man will pleaſe his owne fancie, according to the extent he deſigneth out for that purpoſe, be it orbicular or round, triangular or three ſquare, quadrangular or foure ſquare, or more long than broad. I will onely ſhew you here the ſeuerall formes that many men haue taken and delighted in, let euery man chuſe which him liketh beſt, or may moſt fitly agree to that proportion of ground hee hath ſet out for that purpoſe. The orbicular or round forme is held in it owne proper exiſtence to be the moſt abſolute forme, containing within it all other formes whatſoeuer ; but few I thinke will chuſe ſuch a proportion to be ioyned to their habitation, being not accepted any where I thinke, but for the generall Garden to the Vniuerſity at Padoa. The triangular or three ſquare is ſuch a forme alſo, as is ſeldome choſen by any that may make another choiſe, and as I thinke is onely had where another forme cannot be had, neceſſitie conſtraining them to be therewith content. The foure ſquare forme is the moſt vſually accepted with all, and doth beſt agree to any mans dwelling, being (as I ſaid before) behinde the houſe, all the backe windowes thereof opening into it. Yet if it bee longer than the breadth, or broader than the length, the proportion of walkes, ſquares, and knots may be ſoon brought to the ſquare forme, and be ſo caſt, as the beauty thereof may

bee no leffe than the foure fquare proportion , or any other better forme, if any be. To
forme it therfore with walks,croffe the middle both waies,and round about it alfo with
hedges, with fquares, knots and trayles, or any other worke within the foure fquare
parts, is according as euery mans conceit alloweth of it, and they will be at the charge :
For there may be therein walkes eyther open or clofe, eyther publike or priuate, a maze
or wildernesse, a rocke or mount, with a fountaine in the midft thereof to conuey water
to euery part of the Garden, eyther in pipes vnder the ground, or brought by hand,and
emptied into large Cifternes or great Turkie Iarres , placed in conuenient places, to
ferue as an eafe to water the neareft parts thereunto. Arbours alfo being both gracefull
and neceffary,may be appointed in fuch conuenient places,as the corners,or elfe where,
as may be moft fit, to ferue both for fhadow and reft after walking. And becaufe many
are defirous to fee the formes of trayles,knots, and other compartiments, and becaufe
the open knots are more proper for thefe Out-landifh flowers ; I haue here caufed fome
to be drawne, to fatisfie their defires, not intending to cumber this worke with ouer manie; in that it would be almoft endleffe, to expreffe fo many as might bee conceiued and
fet downe, for that euery man may inuent others farre differing from thefe,or any other
can be fet forth. Let euery man therefore, if hee like of thefe , take what may pleafe his
mind, or out of thefe or his own conceit, frame any other to his fancy, or caufe others to
be done as he liketh beft, obferuing this *decorum*,that according to his ground he do caft
out his knots,with conuenient roome for allies and walkes;for the fairer and larger your
allies and walkes be,the more grace your Garden fhall haue,the leffe harme the herbes
and flowers fhall receiue, by paffing by them that grow next vnto the allies fides, and the
better fhall your Weeders cleanfe both the beds and the allies.

Chap. III.

The many forts of herbes and other things, wherewith the beds and parts of knots are bordered
to fet out the forme of them, with their commodities and difcommodities.

IT is neceffary alfo, that I fhew you the feuerall materials,wherewith thefe knots and
trayles are fet forth and bordered ; which are of two forts : The one are liuing
herbes , and the other are dead materials ; as leade, boords, bones, tyles, &c. Of
herbes, there are many forts wherewith the knots and beds in a Garden are vfed to bee
fet, to fhew forth the forme of them, and to preferue them the longer in their forme, as
alfo to be as greene, and fweete herbes,while they grow,to be cut to perfume the houfe,
keeping them in fuch order and proportion,as may be moft conuenient for their feuerall
natures , and euery mans pleafure and fancy : Of all which , I intend to giue you the
knowledge here in this place ; and firft, to begin with that which hath beene moft anciently receiued, which is Thrift. This is an euerliuing greene herbe,which many take to
border their beds, and fet their knots and trayles,and therein much delight, becaufe it
will grow thicke and bufhie, and may be kept, being cut with a paire of Garden fheeres,
in fome good handfome manner and proportion for a time, and befides, in the Summer
time fend forth many fhort ftalkes of pleafant flowers, to decke vp an houfe among o-
ther fweete herbes : Yet thefe inconueniences doe accompany it ; it will not onely in a
fmall time ouergrow the knot or trayle in many places,by growing fo thicke and bufhie,
that it will put out the forme of a knot in many places : but alfo much thereof will dye
with the frofts and fnowes in Winter, and with the drought in Summer, whereby many
voide places will be feene in the knot, which doth much deforme it, and muft therefore
bee yearely refrefhed : the thickneffe alfo and bufhing thereof doth hide and fhelter
fnayles and other fmall noyfome wormes fo plentifully, that Gilloflowers, and other
fine herbes and flowers being planted therein, are much fpoyled by them,and cannot be
helped without much induftry, and very great and daily attendance to deftroy them.
Germander is another herbe, in former times alfo much vfed, and yet alfo in many places; and becaufe it will grow thicke,and may be kept alfo in fome forme and proportion
with cutting, and that the cuttings are much vfed as a ftrawing herbe for houfes, being
pretty and fweete, is alfo much affected by diuers : but this alfo will often dye and grow
out of forme, and befides that ; the ftalkes will grow too great, hard and ftubby, the
rootes doe fo farre fhoote vnder ground, that vpon a little continuance thereof, will

spread into many places within the knot, which if continually they be not plucked vp, they will spoile the whole knot it selfe; and therefore once in three or foure yeares at the most, it must be taken vp and new set, or else it will grow too roynish and cumbersome. Hyssope hath also been vsed to be set about a knot, and being sweet, will serue for strewings, as Germander: But this, although the rootes doe not runne or creep like it, yet the stalkes doe quickly grow great aboue ground, and dye often after the first yeares setting, whereby the grace of the knot will be much lost. Marierome, Sauorie, and Thyme, in the like manner being sweete herbes, are vsed to border vp beds and knots, and will be kept for a little while, with cutting, into some conformity; but all and euery of them serue most commonly but for one yeares vse, and will soone decay and perish: and therefore none of these, no more than any of the former, doe I commend for a good bordering herbe for this purpose. Lauander Cotton also being finely slipped and set, is of many, and those of the highest respect of late daies, accepted, both for the beauty and forme of the herbe, being of a whitish greene mealy colour, for his sent smelling somewhat strong, and being euerliuing and abiding greene all the Winter, will, by cutting, be kept in as euen proportion as any other herbe may be. This will likewise soone grow great and stubbed, notwithstanding the cutting, and besides will now and then perish in some places, especially if you doe not strike or put off the snow, before the Sunne lying vpon it dissolue it: The rarity & nouelty of this herbe, being for the most part but in the Gardens of great persons, doth cause it to be of the greater regard, it must therfore be renewed wholly euery second or third yeare at the most, because of the great growing therof. Slips of Iuniper or Yew are also receiued of some & planted, because they are alwayes green, and that the Iuniper especially hath not that ill sent that Boxe hath, which I will presently commend vnto you, yet both Iuniper and Yew will soone grow too great and stubbed, and force you to take vp your knot sooner, than if it were planted with Boxe. Which lastly, I chiefly and aboue all other herbes commend vnto you, and being a small, lowe, or dwarfe kinde, is called French or Dutch Boxe, and serueth very well to set out any knot, or border out any beds: for besides that it is euer greene, it being reasonable thicke set, will easily be cut and formed into any fashion one will, according to the nature thereof, which is to grow very slowly, and will not in a long time rise to be of any height, but shooting forth many small branches from the roote, will grow very thicke, and yet not require so great tending, nor so much perish as any of the former, and is onely receiued into the Gardens of those that are curious. This (as I before said) I commend and hold to bee the best and surest herbe to abide faire and greene in all the bitter stormes of the sharpest Winter, and all the great heates and droughts of Summer, and doth recompence the want of a good sweet sent with his fresh verdure, euen proportion, and long lasting continuance. Yet these inconueniences it hath, that besides the vnpleasing sent which many mislike, and yet is but small, the rootes of this Boxe do so much spread themselues into the ground of the knot, and doe draw from thence so much nourishment, that it robbeth all the herbes that grow neare it of their sap and substance, thereby making all the earth about it barren, or at least lesse fertile. Wherefore to shew you the remedy of this inconuenience of spreading, without either taking vp the Boxe of the border, or the herbes and flowers in the knot, is I thinke a secret knowne but vnto a few, which is this: You shall take a broad pointed Iron like vnto a Slife or Chessill, which thrust downe right into the ground a good depth all along the inside of the border of Boxe somewhat close thereunto, you may thereby cut away the spreading rootes thereof, which draw so much moisture from the other herbes on the inside, and by this meanes both preserue your herbes and flowers in the knot, and your Boxe also, for that the Boxe will be nourished sufficiently from the rest of the rootes it shooteth on all the other sides. And thus much for the liuing herbes, that serue to set or border vp any knot. Now for the dead materials, they are also, as I said before diuers: as first, Leade, which some that are curious doe border their knots withall, causing it to be cut of the breadth of foure fingers, bowing the lower edge a little outward, that it may lye vnder the vpper crust of the ground, and that it may stand the faster, and making the vpper edge either plain, or cut out like vnto the battlements of a Church: this fashion hath delighted some, who haue accounted it stately (at the least costly) and fit for their degree, and the rather, because it will be bowed and bended into any round square, angular, or other proportion as one listeth, and is not much to be misliked, in that the Leade

doth

doth not eafily breake or fpoile without much iniury, and keepeth vp a knot for a very long time in his due proportion : but in my opinion, the Leade is ouer-hot for Summer, and ouer-cold for Winter. Others doe take Oaken inch boords, and fawing them foure or fiue inches broad, do hold vp their knot therewith : but in that thefe boordes cannot bee drawne compaffe into any fmall fcantling, they muft ferue rather for long outright beds, or fuch knots as haue no rounds, halfe rounds, or compaffings in them. And befides, thefe boordes are not long lafting, becaufe they ftand continually in the weather, efpecially the ends where they are faftned together will fooneft rot and perifh, and fo the whole forme will be fpoyled. To preuent that fault, fome others haue chofen the fhanke bones of Sheep, which after they haue beene well cleanfed and boyled, to take out the fat from them, are ftucke into the ground the fmall end downewards, and the knockle head vpwards, and thus being fet fide to fide, or end to end clofe together, they fet out the whole knot therewith, which heads of bones although they looke not white the firft yeare, yet after they haue abiden fome frofts and heates will become white, and prettily grace out the ground : but this inconuenience is incident to them, that the Winter frofts will raife them out of the ground oftentimes, and if by chance the knockle head of any doe breake, or be ftrucke off with any ones foot, &c. going by, from your ftore, that lyeth by you of the fame fort, fet another in the place, hauing firft taken away the broken peece: although thefe will laft long in forme and order, yet becaufe they are but bones many miflike them, and indeed I know but few that vfe them. Tyles are alfo vfed by fome, which by reafon they may bee brought compaffe into any fafhion many are pleafed with them, who doe not take the whole Tyle at length, but halfe Tyles, and other broken peeces fet fomewhat deepe into the ground, that they may ftand faft, and thefe take vp but little roome, and keepe vp the edge of the beds and knots in a pretty comely manner, but they are often out of frame, in that many of them are broken and fpoiled, both with mens feete paffing by, the weather and weight of the earth beating them downe and breaking them, but efpecially the frofts in Winter doe fo cracke off their edges, both at the toppes and fides that ftand clofe one vnto another, that they muft bee continually tended and repaired, with frefh and found ones put in the place of them that are broken or decayed. And laftly (for it is the lateft inuention) round whitifh or blewifh pebble ftones, of fome reafonable proportion and bigneffe, neither too great nor too little, haue beene vfed by fome to be fet, or rather in a manner but laide vpon the ground to fafhion out the traile or knot, or all along by the large grauelly walke fides to fet out the walke, and maketh a pretty handfome fhew, and becaufe the ftones will not decay with the iniuries of any time or weather, and will be placed in their places againe, if any fhould be thruft out by any accident, as alfo that their fight is fo confpicuous vpon the ground, efpecially if they be not hid with the ftore of herbes growing in the knot, is accounted both for durability, beauty of the fight, handfomneffe in the worke, and eafe in the working and charge, to be of all other dead materials the chiefeft. And thus, Gentlemen, I haue fhewed you all the varieties that I know are vfed by any in our Countrey, that are worth the reciting (but as for the fafhion of Iawe-bones, vfed by fome in the Low-Countries, and other places beyond the Seas, being too groffe and bafe, I make no mention of them) among which euery one may take what pleafeth him beft, or may moft fitly be had, or may beft agree with the ground or knot. Moreouer, all thefe herbes that ferue for borderings, doe ferue as well to be fet vpon the ground of a leuelled knot ; that is, where the allies and foot-pathes are of the fame leuell with the knot, as they may ferue alfo for the raifed knot, that is, where the beds of the knot are raifed higher than the allies : but both Leade, Boordes, Bones, and Tyles, are only for the raifed ground, be it knot or beds. The pebble ftones againe are onely for the leuelled ground, becaufe they are fo fhallow, that as I faid before, they rather lye vpon the earth than are thruft any way into it. All this that I haue here fet downe, you muft vnderftand is proper for the knots alone of a Garden. But for to border the whole fquare or knot about, to ferue as a hedge thereunto, euery one taketh what liketh him beft ; as either Priuet alone, or fweete Bryer, and white Thorne enterlaced together, and Rofes of one, or two, or more forts placed here and there amongft them. Some alfo take Lauander, Rofemary, Sage, Southernwood, Lauander Cotton, or fome fuch other thing. Some againe plant Cornell Trees, and plafh them, or keepe them lowe, to

forme

forme them into an hedge. And some againe take a lowe prickly shrubbe, that abideth alwayes greene, described in the end of this Booke, called in Latine *Pyracantha*, which in time will make an euer greene hedge or border, and when it beareth fruit, which are red berries like vnto Hawthorne berries, make a glorious shew among the greene leaues in the Winter time, when no other shrubbes haue fruit or leaues.

Chap. IV.

The nature and names of diuers Out-landish flowers, that for their pride, beauty, and earlinesse, are to be planted in Gardens of pleasure for delight.

HAuing thus formed out a Garden, and diuided it into his fit and due proportion, with all the gracefull knots, arbours, walkes, &c. likewise what is fit to keepe it in the same comely order, is appointed vnto it, both for the borders of the squares, and for the knots and beds themselues; let vs now come and furnish the inward parts, and beds with those fine flowers that (being strangers vnto vs, and giuing the beauty and brauery of their colours so early before many of our owne bred flowers, the more to entice vs to their delight) are most beseeming it: and namely, with Daffodils, Fritillarias, Iacinthes, Saffron-flowers, Lillies, Flowerdeluces, Tulipas, Anemones, French Cowslips, or Beares eares; and a number of such other flowers, very beautifull, delightfull, and pleasant, hereafter described at full, whereof although many haue little sweete sent to commend them, yet their earlinesse and exceeding great beautie and varietie doth so farre counteruaile that defect (and yet I must tell you with all, that there is among the many sorts of them some, and that not a few, that doe excell in sweetnesse, being so strong and heady, that they rather offend by too much than by too little sent, and some againe are of so milde and moderate temper, that they scarce come short of your most delicate and dantiest flowers) that they are almost in all places with all persons, especially with the better sort of the Gentry of the Land, as greatly desired and accepted as any other the most choisest, and the rather, for that the most part of these Out-landish flowers, do shew forth their beauty and colours so early in the yeare, that they seeme to make a Garden of delight euen in the Winter time, and doe so giue their flowers one after another, that all their brauery is not fully spent, vntil that Gilliflowers, the pride of our English Gardens, do shew themselues: So that whosoeuer would haue of euery sort of these flowers, may haue for euery moneth seuerall colours and varieties, euen from Christmas vntill Midsommer, or after; and then, after some little respite, vntill Christmas againe, and that in some plenty, with great content and without forcing; so that euery man may haue them in euery place, if they will take any care of them. And because there bee many Gentlewomen and others, that would gladly haue some fine flowers to furnish their Gardens, but know not what the names of those things are that they desire, nor what are the times of their flowring, nor the skill and knowledge of their right ordering, planting, displanting, transplanting, and replanting; I haue here for their sakes set downe the nature, names, times, and manner of ordering in a briefe manner, referring the more ample declaration of them to the worke following. And first of their names and natures: Of Daffodils there are almost an hundred sorts, as they are seuerally described hereafter, euery one to be distinguished from other, both in their times, formes, and colours, some being eyther white, or yellow, or mixt, or else being small or great, single or double, and some hauing but one flower vpon a stalke, others many, whereof many are so exceeding sweete, that a very few are sufficient to perfume a whole chamber, and besides, many of them be so faire and double, eyther one vpon a stalke, or many vpon a stalke, that one or two stalkes of flowers are in stead of a whole nose-gay, or bundell of flowers tyed together. This I doe affirme vpon good knowledge and certaine experience, and not as a great many others doe, tell of the wonders of another world, which themselues neuer saw nor euer heard of, except some superficiall relation, which themselues haue augmented according to their owne fansie and conceit. Againe, let me here also by the way tell you, that many idle and ignorant Gardiners and others, who get names by stealth, as they doe many other things, doe call

some

some of these Daffodils Narcisses, when as all know that know any Latine, that Narcissus is the Latine name, and Daffodill the English of one and the same thing; and therefore alone without any other Epithite cannot properly distinguish seuerall things. I would willingly therefore that all would grow iudicious, and call euery thing by his proper English name in speaking English, or else by such Latine name as euery thing hath that hath not a proper English name, that thereby they may distinguish the seuerall varieties of things and not confound them, as also to take away all excuses of mistaking; as for example: The single English bastard Daffodill (which groweth wilde in many Woods, Groues, and Orchards in England.) The double English bastard Daffodill. The French single white Daffodill many vpon a stalke. The French double yellow Daffodill. The great, or the little, or the least Spanish yellow bastard Daffodill, or the great or little Spanish white Daffodill. The Turkie single white Daffodill, or, The Turkie single or double white Daffodill many vpon a stalke, &c. Of Fritillaria, or the checkerd Daffodill, there are halfe a score seuerall sorts, both white and red, both yellow and blacke, which are a wonderfull grace and ornament to a Garden in regard of the Checker like spots are in the flowers. Of Iacinthes there are aboue halfe an hundred sorts, as they are specified hereafter; some like vnto little bells or starres, others like vnto little bottles or pearles, both white and blew, sky-coloured and blush, and some starlike of many pretty various formes, and all to giue delight to them that will be curious to obserue them. Of Crocus or Saffron flowers, there are also twenty sorts; some of the Spring time, others flowring onely in the Autume or Fall, earlier or later than another, some whereof abide but a while, others indure aboue a moneth in their glorious beauty. The Colchicum or Medowe Saffron, which some call the sonne before the father, but not properly, is of many sorts also; some flowring in the Spring of the yeare, but the most in Autume, whereof some haue faire double flowers very delightfull to behold, and some party coloured both single and double so variable, that it would make any one admire the worke of the Creatour in the various spots and stripes of these flowers. Then haue wee of Lillies twenty seuerall sorts and colours, among whom I must reckon the Crowne Imperiall, that for his stately forme deserueth some speciall place in this Garden, as also the Martagons, both white and red, both blush and yellow, that require to be set by themselues apart, as it were in a small round or square of a knot, without many other, or tall flowers growing neare them. But to tell you of all the sorts of Tulipas (which are the pride of delight) they are so many, and as I may say, almost infinite, doth both passe my ability, and as I beleeue the skill of any other. They are of two especiall sorts, some flowring earlier, and others later than their fellowes, and that naturally in all grounds, wherein there is such a wonderfull variety and mixture of colours, that it is almost impossible for the wit of man to descipher them thoroughly, and to giue names that may be true & seuerall distinctions to euery flower, threescore seuerall sorts of colours simple and mixed of each kind I can reckon vp that I haue, and of especiall note, and yet I doubt not, but for euery one of them there are ten others differing from them, which may be seen at seuerall times, and in seuerall places: & besides this glory of variety in colors that these flowers haue, they carry so stately & delightfull a forme, & do abide so long in their brauery (enduring aboue three whole moneths from the first vnto the last) that there is no Lady or Gentlewoman of any worth that is not caught with this delight, or not delighted with these flowers. The Anemones likewise or Windeflowers are so full of variety and so dainty, so pleasant and so delightsome flowers, that the sight of them doth enforce an earnest longing desire in the minde of any one to be a possessour of some of them at the least: For without all doubt, this one kinde of flower, so variable in colours, so differing in forme (being almost as many sorts of them double as single) so plentifull in bearing flowers, and so durable in lasting, and also so easie both to preserue and to encrease, is of it selfe alone almost sufficient to furnish a garden with their flowers for almost halfe the yeare, as I shall shew you in a fit and conuenient place. The Beares eares or French Cowslips must not want their deserued commendations, seeing that their flowers, being many set together vpon a stalke, doe seeme euery one of them to bee a Nosegay alone of it selfe: and besides the many differing colours that are to be seene in them, as white, yellow, blush, purple, red, tawney, murrey, haire colour, &c. which encrease much delight in all sorts of the Gentry of the Land, they are not vnfurnished with a pretty sweete sent,

which

which doth adde an encreafe of pleafure in thofe that make them an ornament for
their wearing. Flowerdeluces alfo are of many forts, but diuided into two efpeciall
kindes; the one bearing a leafe like a flagge, whofe rootes are tuberous, thicke and
fhort (one kinde of them being the Orris rootes that are fold at the Apothecaries,
whereof fweete powders are made to lye among garments) the other hauing round
rootes like vnto Onions, and narrow long leaues fomewhat like graffe: Of both thefe
kindes there is much variety, efpecially in their colours. The greater Flagge kinde is
frequent enough and difperfed in this Land, and well doth ferue to decke vp both a
Garden and Houfe with natures beauties: But the chiefe of all is your Sable flower,
fo fit for a mourning habit, that I thinke in the whole compaffe of natures ftore,
there is not a more patheticall, or of greater correfpondency, nor yet among all the
flowers I know any one comming neare vnto the colour of it. The other kinde which
hath bulbous or Onion like rootes, diuerfifieth it felfe alfo into fo many fine colours,
being of a more neate fhape and fuccinct forme than the former, that it muft not bee
wanting to furnifh this Garden. The Hepatica or Noble Liuerwoort is another flower of account, whereof fome are white, others red, or blew, or purple, fomewhat refembling Violets, but that there are white threads in the middeft of their flowers,
which adde the more grace vnto them; and one kinde of them is fo double, that it refembleth a double thicke Dafie or Marigold, but being fmall and of an excellent blew
colour, is like vnto a Button: but that which commendeth the flower as much as the
beauty, is the earlineffe in flowring, for that it is one of the very firft flowers that open
themfelues after Chriftmas, euen in the midft of Winter. The Cyclamen or Sowebread is a flower of rare receipt, becaufe it is naturally hard to encreafe, and that the
flowers are like vnto red or blufh coloured Violets, flowring in the end of Summer or
beginning of Autumne: the leaues likewife hereof haue no fmall delight in their pleafant colour, being fpotted and circled white vpon greene, and that which moft preferreth it, is the Phyficall properties thereof for women, which I will declare when I
fhall fhew you the feuerall defcriptions of the varieties in his proper place. Many o-
ther forts of flowers there are fit to furnifh this Garden, as Leucoium or Bulbous Violet, both early and late flowring. Mufcari or Muske Grape flower. Starre flowers of diuers forts. Phalangium or Spiderwort, the chiefe of many is that fort whofe flowers
are like vnto a white Lilly. Winter Crowfoote or Wolfes bane. The Chriftmas flower like vnto a fingle white Rofe. Bell flowers of many kindes. Yellow Larkes fpurre,
the prettieft flower of a fcore in a Garden. Flower-gentle or Floramour. Flower of
the Sunne. The Maruaile of Peru or of the world. Double Marfh Marigold or double yellow Buttons, much differing and farre exceeding your double yellow Crowfoote, which fome call ~~Batchelours Buttons.~~ Double French Marigolds that fmell
well, and is a greater kinde than the ordinary, and farre furpaffeth it. The double red
Ranunculus or Crowfoote (farre excelling the moft glorious double Anemone) and is
like vnto our great yellow double Crowfoote. Thus hauing giuen you the knowledge of fome of the choifeft flowers for the beds of this Garden, let me alfo fhew you
what are fitteft for your borders, and for your arbours. The Iafmine white and yellow. The double Honyfockle. The Ladies Bower, both white, and red, and purple
fingle and double, are the fitteft of Outlandifh plants to fet by arbours and banqueting
houfes, that are open, both before and aboue to helpe to couer them, and to giue both
fight, fmell, and delight. The forts of Rofes are fitteft for ftandards in the hedges or
borders. The Cherry Bay or Laurocerafus. The Rofe Bay or Oleander. The white
and the blew Syringa or Pipe tree, are all gracefull and delightfull to fet at feuerall
diftances in the borders of knots; for fome of them giue beautifull and fweete flowers.
The Pyracantha or Prickly Corall tree doth remaine with greene leaues all the yeare,
and may be plafhed, or laid downe, or tyed to make a fine hedge to border the whole
knot, as is faid before. The Wilde Bay or Laurus Tinus, doth chiefly defire to be fhel-
tered vnder a wall, where it will beft thriue, and giue you his beautifull flowers in Winter for your delight, in recompence of his fenced dwelling. The Dwarfe Bay or Mefe-
reon, is moft commonly either placed in the midft of a knot, or at the corners thereof,
and fometimes all along a walke for the more grace. And thus to fit euery ones fancy,
I haue fhewed you the variety of natures ftore in fome part for you to difpofe of them
to your beft content.

CHAP. V.

The nature and names of thoſe that are called vſually Engliſh flowers.

THoſe flowers that haue beene vſually planted in former times in Gardens of this Kingdome (when as our forefathers knew few or none of thoſe that are re-cited before) haue by time and cuſtome attained the name of Engliſh flowers, although the moſt of them were neuer naturall of this our Land, but brought in from other Countries at one time or other, by thoſe that tooke pleaſure in them where they firſt ſaw them : and I doubt not, but many other ſorts than here are ſet downe, or now knowne to vs, haue beene brought, which either haue periſhed by their negligence or want of skill that brought them, or elſe becauſe they could not abide our cold Win-ters ; thoſe onely remaining with vs that haue endured of themſelues, and by their en-creaſing haue beene diſtributed ouer the whole Land. If I ſhould make any large diſ-courſe of them, being ſo well knowne to all, I doubt I ſhould make a long tale to ſmall purpoſe : I will therefore but briefly recite them, that you may haue them together in one place, with ſome little declaration of the nature and quality of them, and ſo paſſe to other matters. And firſt of Primroſes and Cowſlips, whereof there are many pret-tie varieties ; ſome better knowne in the Weſt parts of this Kingdome, others in the North, than in any other, vntill of late being obſerued by ſome curious louers of vari-eties, they haue been tranſplanted diuerſly, and ſo made more common : for although we haue had formerly in theſe parts about London greene Primroſes vſually, yet we neuer ſaw or heard of greene Cowſlips both ſingle and double but of late dayes, and ſo likewiſe for Primroſes to be both ſingle and double from one roote, and diuers vp-on one ſtalke of diuers faſhions, I am ſure is not vſuall : all which deſire rather to bee planted vnder ſome hedge, or fence, or in the ſhade, than in the Sunne. Single Roſe Campions, both white, red, and bluſh, and the double red Roſe Campion alſo is knowne ſufficiently, and will abide moderate Sunne as well as the ſhade. The flower of Briſtow or None-ſuch is likewiſe another kinde of Campion, whereof there is both white flowring plants and bluſh as well as Orange colour, all of them being ſingle flowers require a moderate Sunne and not the ſhadow : But the Orange colour None-ſuch with double flowers, as it is rare and not common, ſo for his brauery doth well deſerue a Maſter of account that will take care to keepe and preſerue it. Batchelours Buttons both white and red, are kindes of wilde Campions of a very double forme, and will reaſonably well like the Sunne but not the ſhade. Wall-flowers are common in euery Garden, as well the ordinary double as the ſingle, and the double kinde deſi-reth no more ſhade than the ſingle, but the greater kindes both double and ſingle muſt haue the Sunne. Stock-Gilloflowers likewiſe are almoſt as common as Wall-flowers, eſpecially the ſingle kindes in euery womans Garden, but the double kindes are much more rare, and poſſeſſed but of a few, and thoſe onely that will bee carefull to pre-ſerue them in Winter ; for beſides that the moſt of them are more tender, they yeeld no ſeede as the ſingle kindes doe to preſerue them, although one kinde from the ſow-ing of the ſeed yeeld double flowers : They will all require the comfort of the Sunne, eſpecially the double kindes, and to be defended from cold, yet ſo as in the Summer they doe not want water wherein they much ioy, and which is as it were their life. Queenes Gilloflowers (which ſome call Dames Violets, and ſome Winter Gilloflow-ers, are a kinde of Stock-Gilloflower) planted in Gardens to ſerue to fill vp the parts thereof for want of better things, hauing in mine opinion neither ſight nor ſent much to commend them. Violets are the Springs chiefe flowers for beauty, ſmell, and vſe, both ſingle and double, the more ſhadie and moiſt they ſtand the better. Snapdragon are flowers of much more delight, and in that they are more tender to keep, and will hardly endure the ſharpe Winters, vnleſſe they ſtand well defended, are ſcarce ſeene in many Gardens. Columbines ſingle and double, of many ſorts, faſhions, and co-lours, very variable both ſpeckled and party coloured, are flowers of that reſpect, as that no Garden would willingly bee without them, that could tell how to haue them, yet the rarer the flowers are, the more trouble to keepe ; the ordinary ſorts on the con-trary

trary part will not be loft, doe what one will. Larkes heeles, or fpurres, or toes, as in feuerall Countries they are called, exceed in the varietie of colours, both fingle and double, any of the former times; for vntill of late dayes none of the moft pleafant colours were feene or heard of: but now the fingle kindes are reafonable well difperft ouer the Land, yet the double kindes of all thofe pleafant colours (and fome other alfo as beautifull) which ftand like little double Rofes, are enioyed but of a few: all of them rife from feed, and muft be fowne euery yeare, the double as well as the fingle. Panfyes or Hartes eafes of diuers colours, and although without fent, yet not without fome refpect and delight. Double Poppies are flowers of a great and goodly proportion, adorning a Garden with their variable colours to the delight of the beholders, wherein there is fome fpeciall care to be taken, left they turne fingle; and that is, if you fee them grow vp too thicke, that you muft pull them vp, and not fuffer them to grow within leffe than halfe a yard diftance, or more one from another. Double Daifies are flowers not to be forgotten, although they be common enough in euery Garden, being both white and red, both blufh and fpeckled, or party coloured, befides that which is called Iacke an Apes on horfebacke, they require a moift and fhadowie place; for they are fcorched away, if they ftand in the Sunne in any dry place. Double Marigolds alfo are the moft common in all Gardens. And fo are the French Marigolds that haue a ftrong heady fent, both fingle and double, whofe glorious fhew for colour would caufe any to beleeue there were fome rare goodneffe or vertue in them. Thefe all are fometimes preferued in the Winter, if they bee well defended from the cold. But what fhall I fay to the Queene of delight and of flowers, Carnations and Gilloflowers, whofe brauery, variety, and fweete fmell ioyned together, tyeth euery ones affection with great earneftneffe, both to like and to haue them? Thofe that were knowne, and enioyed in former times with much acceptation, are now for the moft part leffe accounted of, except a very few: for now there are fo many other varieties of later inuention, that troubleth the other both in number, beauty, and worth: The names of them doe differ very variably, in that names are impofed and altered as euerie ones fancy will haue them, that carryed or fent them into the feuerall Countries from London, where their trueft name is to be had, in mine opinion. I will here but giue you the names of fome, and referre you to the worke enfuing for your further knowledge. The red and the gray Hulo. The old Carnation, differing from them both. The Gran Pere. The Camberfiue. The Sauadge. The Chriftall. The Prince. The white Carnation, or Delicate. The ground Carnation. The French Carnation. The Douer. The Oxford. The Briftow. The Weftminfter. The Daintie. The Granado, and many other Gilloflowers too tedious to recite in this place, becaufe I haue amply declared them ~~in the booke following~~. But there is another fort of great delight and varietie, called the Orange tawny Gilloflower, which for the moft part hath rifen from feed, and doth giue feed in a more plentifull manner than any of the former forts, and likewife by the fowing of the feed there hath been gained fo many varieties of that excellent worth and refpect, that it can hardly be expreffed or beleeued, and called by diuers names according to the marking of the flowers; as The Infanta. The Stript Tawny. The Speckled Tawny. The Flackt Tawny. The Grifeld Tawny, and many others, euery one to bee diftinguifhed from others: Some alfo haue their flowers more double and large than others, and fome from the fame feed haue fingle flowers like broad fingle Pinkes: the further relation of them, *viz.* their order to fowe, encreafe, and preferue them, you fhall haue in the fubfequent difcourfe in a place by it felfe. Pinkes likewife both fingle and double are of much variety, all of them very fweete, comming neare the Gilloflowers. Sweete Williams and Sweete Iohns, both fingle and double, both white, red, and fpotted, as they are kindes of wilde Pinkes, fo for their grace and beauty helpe to furnifh a Garden, yet defire not to ftand fo open to the Sunne as the former. Double and fingle Peonies are fit flowers to furnifh a Garden, and by reafon of their durability, giue out frefh pleafure euery yeare without any further trouble of fowing. And laftly, Hollihocks both fingle and double, of many and fundry colours, yeeld out their flowers like Rofes on their tall branches, like Trees, to fute you with flowers, when almoft you haue no other to grace out your Garden: the fingle and double doe both yeeld feed, and yet doe after their feeding abide many yeares. Thus haue I fhewed you moft of the Englifh, as well as (I did before) the Outlandifh landish flowers, that are fit to furnish the knots, trailes, beds, and borders of this Garden. Roses onely, as I said before, I reserue to circle or encompasse all the rest, because that for the most part they are planted in the outer borders of the quarters, and sometimes by themselues in the middle of long beds, the sorts or kindes whereof are many, as they are declared in their proper place : but the White Rose, the Red, and the Damaske, are the most ancient Standards in England, and therefore accounted naturall.

CHAP. VI.

The order and manner to plant and replant all the sorts of Out-landish flowers spoken of before, as well those with bulbous rootes, as others with stringie rootes.

WHereas it is the vsuall custome of most in this Land, to turne vp their Gardens, and to plant them againe in the Spring of the yeare, which is the best time that may bee chosen for all English flowers, yet it is not so for your Out-landish flowers. And herein indeede hath beene not onely the errour of a great many to hinder their rootes from bearing out their flowers as they should, but also to hinder many to take delight in them, because as they say they will not thriue and prosper with them, when as the whole fault is in the want of knowledge of the fit and conuenient time wherein they should bee planted. And because our English Gardiners are all or the most of them vtterly ignorant in the ordering of these Out-landish flowers, as not being trained vp to know them, I haue here taken vpon mee the forme of a new Gardiner, to giue instructions to those that will take pleasure in them, that they may be the better enabled with these helpes I shall shew them, both to know how they should be ordered, and to direct their Gardiners that are ignorant thereof, rightly to dispose them according to their naturall qualities. And I doe wish all Gentlemen and Gentlewomen, whom it may concerne for their owne good, to bee as carefull whom they trust with the planting and replanting of these fine flowers, as they would be with so many Iewels ; for the rootes of many of them being small, and of great value, may be soone conueyed away, and a cleanly tale faire told, that such a roote is rotten, or perished in the ground if none be seene where it should be, or else that the flower hath changed his colour, when it hath been taken away, or a counterfeit one hath beene put in the place thereof ; and thus many haue been deceiued of their daintiest flowers, without remedy or true knowledge of the defect. You shall therefore, if you will take the right course that is proper for these kindes of flowers, not set or plant them among your English flowers ; for that when the one may be remoued, the other may not be stirred : but plant those rootes that are bulbous, or round like Onions, eyther in knots or beds by themselues which is the best, or with but very few English or Out-landish flower plants that haue stringie rootes : For you must take this for a generall rule, that all those rootes that are like Lillies or Onions, are to bee planted in the moneths of Iuly or August, or vnto the middle or end of September at the furthest, if you will haue them to prosper as they should, and not in the Spring of the yeare, when other gardening is vsed. Yet I must likewise giue you to vnderstand, that if Tulipas, and Daffodils, and some other that are firme and hard rootes, and not limber or spongie, being taken vp out of the ground in their fit season, that is, in Iune, Iuly, and August, and likewise kept well and dry, may bee reserued out of the ground vntill Christmas or after, and then (if they could not be set sooner) being set, will thriue reasonable well, but not altogether so well as the former, being set long before : but if you shall remoue these bulbous rootes againe, either presently after their planting hauing shot their small fibres vnder the round rootes, and sprung likewise vpwards, or before they be in flower at the soonest (yet Tulipas, Daffodils, and many other bulbous, may be safely remoued being in flower, and transplanted into other places, so as they be not kept too long out of the ground) you shall much endanger them either vtterly to perish, or to be hindered from bearing out their flowers they then would haue

B   borne,

borne, and for two or three years after from bearing flowers againe. For the order of their planting there are diuers wayes, some whereof I will shew you in this place: Your knot or beds being prepared fitly, as before is declared, you may place and order your rootes therein thus, Eyther many rootes of one kind set together in a round or cluster, or longwise crosse a bed one by another, whereby the beauty of many flowers of one kinde being together, may make a faire shew well pleasing to many ; Or else you may plant one or two in a place dispersedly ouer the whole knot, or in a proportion or diameter one place answering another of the knot, as your store will suffer you, or your knot permit : Or you may also mingle these rootes in their planting many of diuers forts together, that they may giue the more glorious shew when they are in flower; and that you may so doe, you must first obserue the seuerall kindes of them, which doe flower at one and the same time, and then to place them in such order and so neare one vnto another, that their flowers appearing together of seuerall colours, will cause the more admiration in the beholders : as thus, The Vernall Crocus or Saffron flowers of the Spring, white, purple, yellow, and stript, with some Vernall Colchicum or Medow Saffron among them, some Deus Caninus or Dogges teeth, and some of the small early Leucoium or Bulbous Violet, all planted in some proportion as neare one vnto another as is fit for them, will giue such a grace to the Garden, that the place will seeme like a peece of tapestry of many glorious colours, to encrease euery ones delight : Or else many of one sort together, as the blew, white and blush Grape flowers in the same manner intermingled, doe make a maruellous delectable shew, especially because all of them rise almost to an equall height, which causeth the greater grace, as well neare hand as farre of. The like order may be kept with many other things, as the Hepatica, white, blew, purple, and red set or sowne together, will make many to beleeue that one roote doth beare all those colours : But aboue and beyond all others, the Tulipas may be so matched, one colour answering and setting of another, that the place where they stand may resemble a peece of curious needle-worke, or peece of painting : and I haue knowne in a Garden, the Master as much commended for this artificiall forme in placing the colours of Tulipas, as for the goodnesse of his flowers, or any other thing. The diuers forts and colours of Anemones or Winde-flowers may be so ordered likewise, which are very beautifull, to haue the seuerall varieties planted one neare vnto another, that their seuerall colours appearing in one place will be a very great grace in a Garden, or if they be dispersed among the other forts of flowers, they will make a glorious shew. Another order in planting you may obserue ; which is this, That those plants that grow low, as the Aconitum Hyemale or Winter-wolues bane, the Vernall Crocus or Saffron-flowers of diuers forts, the little early Leucoium or Bulbous Violet, and some such other as rise not vp high, as also some Anemones may be very well placed somewhat neare or about your Martagons, Lillies, or Crownes Imperiall, both because these little plants will flower earlier than they, and so will bee gone and past, before the other greater plants will rise vp to any height to hinder them ; which is a way may well be admitted in those Gardens that are small, to saue roome, and to place things to the most aduantage. Thus hauing shewed you diuers wayes and orders how to plant your rootes, that your flowers may giue the greater grace in the Garden, let mee shew you likewise how to set these kindes of rootes into the ground ; for many know not well eyther which end to set vpwards or downewards, nor yet to what depth they should be placed in the ground. Daffodils if they be great rootes, will require (as must bee obserued in all other great plants) to bee planted somewhat deeper then the smaller of the same kinde, as also that the tops or heads of the rootes be about two or three fingers breadth hid vnder ground. The Tulipas likewise if you set them deepe, they will be the safer from frosts if your ground be cold, which will also cause them to be a little later before they be in flower, yet vsually if the mould be good, they are to be set a good hand breadth deep within the ground, so that there may be three or foure inches of earth at the least aboue the head, which is the smaller end of the roote: for if they shall lye too neare the vpper face or crust of the earth, the colds & frosts will pierce and pinch them the sooner. After the same order and manner must Hyacinthes, whether great or small, and other such great rootes be planted. Your greater rootes, as Martagons, Lillies, and Crownes Imperiall, must be set much deeper then any other bulbous roote, because they are greater rootes then others, and by themselues also, as

is

is moſt vſuall either in ſome ſquare, round, triangle, or other ſmall part in the Garden, becauſe they ſpread and take vp a very great deale of ground. All of them likewiſe are to be ſet with the broad end of the roote downewards, and the ſmall end vpwards, that is, both Lillies, Daffodils, Hyacinthes, and Tulipas, and all other ſorts of round rootes, which ſhew one end to bee ſmaller than another. But the Colchicum or Me-dow-Saffron onely requireth an exception to this generall rule, in regard the roote thereof hath a ſmall eminence or part on the one ſide thereof, which muſt bee ſet or planted downeward, and not vpward; for you ſhall obſerue, if the roote lye a little moiſt out of the ground, that it will ſhoote fibres out at the ſmall long end thereof, al-though you may perceiue when you take it vp, that the fibres were at the other broad end or ſide of the roote. As for the Crowne Imperiall, which is a broad round roote and flat withall, hauing a hole in the middle, for the moſt part quite thorow, when it is taken vp in his due time out of the ground, you ſhall perceiue the ſcales or cloues of the rootes to bee a little open on the vpperſide, and cloſe and flat on the vnderſide, which will direct you which part to ſet vpward, as alſo that the hole is bigger aboue then it is below. The Perſian Lilly is almoſt like vnto the Crowne Imperiall, but that the roote thereof is not ſo flat, and that it hath a ſmaller head at the one part, whereby it may be diſcerned the plainer how to be ſet. The Fritillaria is a ſmall white root di-uided as it were into two parts, ſo that many haue doubted, as formerly in the Crowne Imperiall, what part to ſet vppermoſt; you ſhall therefore marke, that the two parts of the roote are ioyned together at the bottome, where it ſhooteth out fibres or ſmall ſtringie rootes, as all other ſorts of bulbous rootes doe, and withall you ſhall ſee, that betweene the two parts of the roote a ſmall head will appeare, which is the burgeon that will ſpring vp to beare leaues and flowers. In the rootes of Anemones there are ſmall round ſwelling heads, eaſie enough to be obſerued if you marke it, which muſt be ſet vpwards. All other ſorts of ſtringie rooted plants (and not bulbous or tuberous rooted) that loſe their greene leaues in Winter, will ſhew a head from whence the leaues and flowers will ſpring, and all others that keepe their greene leaues, are to bee planted in the ſame manner that other herbes and flower-plants are accuſtomed to bee. But yet for the better thriuing of the ſtringie rooted plants, when you will plant them, let me informe you of the beſt way of planting, and the moſt ſure to cauſe any plant to comprehend in the ground without failing, and is no common way with any Gardiner in this Kingdome, that euer I heard or knew, which is thus : Preſuming that the ſtringie rooted plant is freſh and not old gathered, and a plant that being remoued will grow againe, make a hole in the ground large enough where you meane to ſet this roote, and raiſe the earth within the hole a little higher in the middle then on the ſides, and ſet the roote thereon, ſpreading the ſtrings all abroad about the middle, that they may as it were couer the middle, and then put the earth gently round about it, preſſing it a little cloſe, and afterwards water it well, if it be in Summer, or in a dry time, or o-therwiſe moderately : thus ſhall euery ſeuerall ſtring of the roote haue earth enough to cauſe it to ſhoote forth, and thereby to encreaſe farre better than by the vſuall way, which is without any great care and reſpect to thruſt the rootes together into the ground. Diuers other flower plants are but annuall, to bee new ſowne euery yeare ; as the Maruaile of the world, the Indian Creſſes, or yellow Larkes heeles, the Flower of the Sunne, and diuers other : they therefore that will take pleaſure in them, that they may enioy their flowers the earlier in the yeare, and thereby haue ripe ſeede of them while warme weather laſteth, muſt nurſe vp their ſeedes in a bed of hot dung, as Me-lons and Cowcumbers are, but your bed muſt be prouided earlier for theſe ſeeds, than for Melons, &c. that they may haue the more comfort of the Summer, which are to be carefully tended after they are tranſplanted from the hot bed, and couered with ſtraw from colds, whereby you ſhall not faile to gaine ripe ſeed euery yeare, which other-wiſe if you ſhould miſſe of a very kindly & hot Summer, you ſhould neuer haue. Some of theſe ſeedes neede likewiſe to be tranſplanted from the bed of dung vnder a warme wall, as the Flower of the Sunne, and the Maruaile of the world, and ſome others, and that for a while after their tranſplanting, as alſo in the heate of Summer, you water them at the roote with water that hath ſtood a day or two in the Sunne, hauing firſt laid a round wiſpe of hay or ſuch other thing round about the roote, that ſo all helpes may further their giuing of ripe ſeede. One or two rules more I will giue you concerning

theſe

these dainty flowers, the first whereof is this, That you shall not bee carefull to water any of your bulbous or tuberous rooted plants at any time; for they all of them do better prosper in a dry ground than in a wet, onely all sorts of tuberous rooted Flowerdeluces vpon their remouall had neede of a little water, and some will doe so also to such Tulipas and other bulbous rootes as they transplant, when they are in flower, and this is I grant in some sort tolerable, if it bee not too much, and done onely to cause the stalke and flower to abide sometime the longer before they wither, but else in no other case to be permitted. The second rule is, That I would aduise you to water none of your dainty flowers or herbes, with any water that hath presently before been drawne out of a well or pumpe, but onely with such water that hath stood open in the Sunne in some cisterne, tubbe, or pot for a day at the least, if more the better : for that water which is presently drawne out of a well, &c. is so cold, that it presently chilleth & killeth any dainty plant be it younger or elder grown, wherof I haue had sufficient proofe: and therfore I giue you this caution by mine own experience. Thus haue I directed you from point to point, in all the particulars of preparing & planting that belong to this Garden, sauing only that yet I would further enforme you, of the time of the flowring of these Out-landish plants, according to the seuerall moneths in the yeare, that euery one may know what flowers euery moneth yeeldeth, and may chuse what them liketh best, in that they may see that there is no moneth, but glorieth in some peculiar sorts of rare flowers. I would likewise rather in this place shew you, the true and best manner & order to encrease and preserue all sorts of Gilloflowers & Carnations, then ioyne it with the Chapter of Gilloflowers in the worke following, because it would in that place take vp too much roome. And lastly, I must of necessity oppose three sundry errours, that haue possessed the mindes of many both in former and later times, which are, that any flower may be made to grow double by art, that was but single before by nature : And that one may by art cause any flower to grow of what colour they will : And that any plants may be forced to flower out of their due seasons, either earlier or later, by an art which some can vse. All which being declared, I then suppose enough is spoken for an introduction to this worke, referring many other things to the seuerall directions in the Chapters of the booke.

CHAP. VII.

The seuerall times of the flowring of these Out-landish flowers, according to the seuerall moneths of the yeare.

I Intend in this place onely to giue you briefly, the names of some of the chiefest of these Out-landish flowers, according to the seuerall moneths of the yeare wherein they flower, that euery one seeing what sorts of flowers euery moneth yeeldeth, may take of them which they like best. I begin with Ianuary, as the first moneth of the yeare, wherein if the frosts be not extreme, you shall haue these flowers of plants ; the Christmas flower or Helleborus niger verus, Winter wolues bane or Aconitum hyemale, Hepatica or Noble Liuer wort blew and red, and of shrubbes, the Laurus Tinus or Wilde Bay tree, and Mesereon or the dwarfe Bay : but because Ianuarie is oftentimes too deepe in frosts and snow, I therefore referre the Hepaticas vnto the moneth following, which is February, wherein the weather beginneth to be a little milder, and then they will flower much better, as also diuers sorts of Crocus or Saffron flower will appeare, the little early Summer foole or Leucoium bulbosum, and towards the latter end thereof the Vernall Colchicum, the Dogges tooth Violet or Deus Caminus, and some Anemones, both single and double, which in some places will flower all the Winter long. March will yeeld more varieties ; for besides that it holdeth some of the flowers of the former moneth, it will yeeld you both the double blew Hepatica, and the white and the blush single : then also you shall haue diuers other sorts of Crocus or Saffron flowers, Double yellow Daffodils, Orientall Iacinths and others, the Crowne Imperiall, diuers sorts of early Tulipas, some sorts of French Cowslips, both tawney, murry, yellow, and blush, the early Fritillaria or checkerd Daffodill,

dill, and some other forts of early Daffodils, and many forts of Anemones. In Aprill commeth on the pride of thefe ftrangers; for herein you may behold all the forts of Auricula Vrfi or Beares Eares, many forts of Anemones, both fingle and double, both the forts of Tulipas, the earlier vntill the middle of the moneth, and the later then beginning; which are of fo many different colours, that it is almoft impoffible to expreffe them, the white, red, blacke, and yellow Fritillarias, the Mufcari or Muske Grape flower, both afh colour and yellow. Diuers other forts of Iacinths and Daffo-dils, both fingle and double, the fmaller forts of Flowerdeluces, the Veluet Flower-deluce and double Honyfuckles, with diuers others. May likewife at the beginning feemeth as glorious as Aprill, although toward the end it doth decline, in regard the heate of the Sunne hath by this time drawne forth all the ftore of natures tendereft dainties, which are vfually fpent by the end of this moneth, and then thofe of ftronger conftitution come forward. Herein are to bee feene at the beginning the middle flow-ring Tulipas, and at the end the later fort: fome kindes of Daffodils, the Day Lillies, the great white Starre flower, the Flowerdeluce of Conftantinople or the mourning Sable flower, the other forts of Flowerdeluces. Single and double white Crowfoote, and fingle and double red Crowfoot, the glory of a Garden: the early red Martagon, the Perfian Lilly, the yellow Martagon, the Gladiolus or Corne flagge, both white, red, and blufh: the double yellow Rofe, and fome other forts of Rofes. In Iune doe flower the white and the blufh Martagon, the Martagon Imperiall, the mountaine Lil-lies, and the other forts of white and red Lillies, the bulbous Flowerdeluces of diuers forts, the red flowred Ladies bower, the fingle and double purple flowred Ladies bow-er, the white Syringa or Pipetree, for the blew Pipe tree flowreth earlier, the white and the yellow Iafmin. Iuly holdeth in flower fome of the Ladies bowers and Iaf-mines, and befides doth glory in the Female Balfame apple, the Indian Creffes or yel-low Larkes fpurres, the purple Flower-gentle and the Rofe Bay. In Auguft begin fome of the Autumne bulbous flowers to appeare, as the white and the purple Col-chicum or Medow Saffron, the purple mountaine Crocus or Saffron flower, the little Autumne Leucoium and Autumne Iacinth, the Italian Starrewort, called of fome the purple Marigold, the Meruaile of Peru or of the world, the Flower of the Sunne, the great blew Bell-flower, the great double French Marigold. September flourifheth with the Flower of the Sunne, the Meruaile of the world, the purple Marigold, and blew Bell-flower fpoken of before, and likewife the other forts of Medow Saffron, and the double kinde likewife, the filuer Crocus, the Autumne yellow Daffodill, Cycla-men alfo or Sowbread fhew their flowers in the end of this moneth. October alfo will fhew the flowers of Cyclamen, and fome of the Medow Saffrons. In Nouember, as alfo fometimes in the moneth before, the party coloured Medow Saffron may bee feene, that will longeft hold his flower, becaufe it is the lateft that fheweth it felfe, and the afh coloured mountaine Crocus. And euen December it felfe will not want the true blacke Hellebor or Chriftmas flower, and the glorious fhew of the Laurus Tinus or wilde Bay tree. Thus haue I fhewed you fome of the flowers for euery moneth, but I referre you to the more ample declarion of them and all the others, vnto the work following.

Chap. VIII.

The true manner and order to encreafe and preferue all forts of Gilloflowers, as well by flippes as feedes.

BEcaufe that Carnations and Gilloflowers bee the chiefeft flowers of account in all our Englifh Gardens, I haue thought good to entreate fomewhat amply of them, and that a part by it felfe, as I faid a little before, in regard there is fo much to be faid concerning them, and that if all the matters to be entreated of fhould haue beene inferted in the Chapter of Gilloflowers, it would haue made it too tedi-ous and large, and taken vp too much roome. The particular matters whereof I mean in this place to entreate are thefe: How to encreafe Gilloflowers by planting and by
fowing,

sowing, and how to preserue them being encreased, both in Summer from noysome and hurtfull vermine that destroy them, and in Winter from frosts, snowes, and windes, that spoile them. There are two wayes of planting, whereby to encrease these faire flowers; the one is by slipping, which is the old and ready vsuall way, best knowne in this Kingdome; the other is more sure, perfect, ready, and of later inuention, *videlicet*, by laying downe the branches. The way to encrease Gilloflowers by slipping, is so common with all that euer kept any of them, that I thinke most persons may thinke me idle, to spend time to set downe in writing that which is so well known vnto all : Yet giue me leaue to tell them that so might imagine, that (when they haue heard or read what I haue written thereof, if they did know fully as much before) what I here write, was not to informe them, but such as did not know the best, or so good a way as I teach them : For I am assured, the greatest number doe vse, and follow the most vsuall way, and that is not alwaies the best, especially when by good experience a better way is found, and may be learned; and therefore if some can doe a thing better than others, I thinke it is no shame to learne it of them. You shall not then (to take the surest course) take any long spindled branches, nor those branches that haue any young shootes from the ioynts on them, nor yet sliue or teare any slippe or branch from the roote; for all these waies are vsuall and common with most, which causeth so many good rootes to rot and perish, and also so many slippes to be lost, when as for the most part, not the one halfe, or with some, not a third part doth grow and thriue of those slippes they set. And although many that haue store of plants, doe not so much care what hauocke they make to gaine some, yet to saue both labour and plants, I doe wish them to obserue these orders : Take from those rootes from whence you intend to make your encrease, those shootes onely that are reasonable strong, but yet young, and not either too small and slender, or hauing any shootes from the ioynts vpon them; cut these slippes or shootes off from the stemme or roote with a knife, as conueniently as the shoote or branch will permit, that is, either close vnto the maine branch, if it be short, or leauing a ioynt or two behinde you, if it be long enough, at which it may shoote anew : When you haue cut off your slippes, you may either set them by and by, or else as the best Gardiners vse to doe, cast them into a tubbe or pot with water for a day or two, and then hauing prepared a place conuenient to set them in, which had neede to bee of the finest, richest, and best mould you can prouide, that they may thriue therein the better, cut off your slippe close at the ioynt, and hauing cut away the lowest leaues close to the stalke, and the vppermost euen at the top, with a little sticke make a little hole in the earth, and put your slippe therein so deep, as that the vpper leaues may be wholly aboue the ground, (some vse to cleaue the stalke in the middle, and put a little earth or clay within the cleft, but many good and skilfull Gardiners doe not vse it); put the earth a little close to the slippe with your finger and thumbe, and there let it rest, and in this manner doe with as many slippes as you haue, setting them somewhat close together, and not too farre in sunder, both to saue ground and cost thereon, in that a small compasse will serue for the first planting, and also the better to giue them shadow : For you must remember in any case, that these slippes new set, haue no sight of the Sunne, vntill they be well taken in the ground, and shot aboue ground, and also that they want not water, both vpon the new planting and after. When these slippes are well growne vp, they must be transplanted into such other places as you thinke meete; that is, either into the ground in beds, or otherwise, or into pots, which that you may the more safely doe, after you haue well watered the ground, for halfe a day before you intend to transplant them, you shall separate them seuerally, by putting down a broad pointed knife on each side of the slippe, so cutting it out, take euery one by it selfe, with the earth cleauing close vnto the root, which by reason of the moisture it had formerly, and that which you gaue presently before, will be sufficient with any care had, to cause it to hold fast vnto the roote, for the transplanting of it : for if the earth were dry, and that it should fall away from the roote in the transplanting, it would hazzard and endanger the roote very much, if it did thriue at all. You must remember also, that vpon the remouing of these slips, you shadow them from the heate of the Sunne for a while with some straw or other thing, vntill they haue taken hold in their new place. Thus although it bee a little more labour and care than the ordinary way is, yet it is surer, and will giue you plants that

will

will be fo ftrongly growne before Winter, that with the care hereafter fpecified, you
fhall haue them beare flowers the next yeare after, and yeeld you encreafe of flippes
alfo. To giue you any fet time, wherein thefe flippes will take roote, and begin to
fhoote aboue ground, is very hard to doe; for that euery flip, or yet euery kinde of Gilloflower is not alike apt to grow; nor is euery earth in like manner fit to produce and
bring forward the flippes that are fet therein: but if both the flippe be apt to grow, and
the earth of the beft, fit to produce, I thinke within a fortnight or three weekes, you
fhall fee them begin to put forth young leaues in the middle, or elfe it may be a moneth
and more before you fhall fee any fpringing. The beft time likewife when to plant, is
a fpeciall thing to be knowne, and of as great confequence as any thing elfe: For if you
flippe and fet in September, as many vfe to doe, or yet in Auguft, as fome may thinke
will doe well, yet (vnleffe they be the moft ordinary forts, which are likely to grow at
any time, and in any place) the moft of them, if not all, will either affuredly perifh, or
neuer profper well: for the more excellent and dainty the Gilloflower is, the more
tender for the moft part, and hard to nurfe vp will the flippes be. The beft time therefore is, that you cut off fuch flippes as are likely, and fuch as your rootes may fpare,
from the beginning of May vntill the middle of Iune at the furtheft, and order them
as I haue fhewed you before, that fo you may haue faire plants, plenty of flowers, and
encreafe fufficient for new fupply, without offence or loffe of your ftore. For the en-
riching likewife of your earth, wherein you fhall plant your flippes, that they may the
better thriue and profper, diuers haue vfed diuers forts of manure; as ftable foyle of
horfe, beafts or kine, of fheepe, and pigeons, all which are very good when they are
thoroughly turned to mould, to mixe with your other earth, or being fteeped in water,
may ferue to water the earth at times, and turned in with it. And fome haue likewife
proued Tanners earth, that is, their barke, which after they haue vfed, doth lye on
heapes and rot in their yards, or the like mould from wood-ftackes or yards; but efpecially, and beyond all other is commended the Willow earth, that is, that mould
which is found in the hollow of old Willow trees, to be the moft principall to mixe
with other good earth for this purpofe. And as I haue now giuen you directions for
the firft way to encreafe them by flipping, fo before I come to the other way, let mee
giue you a caueat or two for the preferuing of them, when they are beginning to runne
vtterly to decay and perifh: The one is, that whereas many are ouer greedy to haue
their plants to giue them flowers, and therefore let them runne all to flower, fo farre
fpending themfelues thereby, that after they haue done flowring, they grow fo weake,
hauing out fpent themfelues, that they cannot poffibly be preferued from the iniuries
of the fucceeding Winter; you fhall therefore keepe the kinde of any fort you are delighted withall, if you carefully looke that too many branches doe not runne vp and
fpindle for flowers, but rather either cut fome of them downe, before they are run vp
too high, within two or three ioynts of the rootes; or elfe plucke away the innermoft
leaues where it fpringeth forwards, which you fee in the middle of euery branch, before it be runne vp too high, which will caufe them to breake out the fafter into flips
and fuckers at the ioynts, to hinder their forward luxurie, and to preferue them
the longer: The other is, If you fhall perceiue any of your Gilloflower leaues to
change their naturall frefh verdure, and turne yellowifh, or begin to wither in anie
part or branch thereof, it is a fure figne that the roote is infeted with fome cancker
or rottenneffe, and will foone fhew it felfe in all the reft of the branches, whereby the
plant will quickly be loft: to preferue it therefore, you fhall betime, before it be runne
too farre, (for otherwife it is impoffible to faue it) either couer all or moft of the
branches with frefh earth, or elfe take the faireft flippes from it, as many as you can
poffibly, and caft them into a pot or tubbe with water, and let them there abide for
two or three daies at the leaft: the firft way hath recouered many, being taken in time.
Thus you fhall fee them recouer their former ftiffeneffe and colour, and then you may
plant them as you haue beene heretofore directed; and although many of them may
perifh, yet fhall you haue fome of them that will grow to continue the kinde againe.
The other or fecond way to encreafe Gilloflowers by planting, is, as I faid before, by
in-laying or laying downe the branches of them, and is a way of later inuention, and
as frequently vfed, not onely for the tawney or yellow Gilloflower, and all the varieties therof, but with the other kinds of Gilloflowers, whereof experience hath fhewed
that

that they will likewiſe take if they be ſo vſed ; the manner whereof is thus : You muſt
chooſe out the youngeſt, likelieſt, and loweſt branches that are neareſt the ground (for
the vpper branches will ſooner breake at the ioynt, than bend downe ſo low into the
earth, without ſome pot with earth raiſed vp vnto them) and cut it on the vnderſide
thereof vpwards at the ſecond ioynt next vnto the roote, to the middle of the branch,
and no more, and not quite thorough in any caſe, and then from that ſecond ioynt vnto the third, ſlit or cut the branch in the middle longwiſe, that ſo it may be the more
eaſily bended into the ground, the cut ioynt ſeeming like the end of a ſlippe, when
you haue bended downe the branch where it is cut into the ground (which muſt bee
done very gently for feare of breaking) with a little ſticke or two thruſt ſlopewiſe,
croſſe ouer it, keepe it downe within the earth, and raiſe vp ſufficient earth ouer it, that
there it may lye and take roote, which commonly will be effected within ſixe weekes
or two moneths in the Summer time, and then (or longer if you doubt the time too
ſhort for it to take ſufficient roote) you may take or cut it away, and tranſplant it where
you thinke good, yet ſo as in any caſe you ſhadow it from the heate of the Sunne, vntill it haue taken good hold in the ground. The other way to encreaſe Gilloflowers,
is by ſowing the ſeede : It is not vſuall with all ſorts of Gilloflowers to giue ſeede, but
ſuch of them as doe yeeld ſeede may be encreaſed thereby, in the ſame manner as is
here ſet downe. The Orange tawney Gilloflower and the varieties thereof is the moſt
vſuall kinde, (and it is a kinde by it ſelfe, how various ſoeuer the plants be that riſe
from the ſeede) that doth giue ſeede, and is ſowne, and from thence ariſeth ſo many
varieties of colours, both plaine and mixt, both ſingle and double, that one can hardly
ſet them downe in writing : yet ſuch as I haue obſerued and marked, you ſhall finde
expreſſed in the Chapter of Gilloflowers in the worke following. Firſt therefore
make choiſe of your ſeede that you intend to ſowe (if you doe not deſire to haue as
many more ſingle flowers as double) that it bee taken from double flowers, and not
from ſingle, and from the beſt colours, howſoeuer ſome may boaſt to haue had double
and ſtript flowers from the ſeede of a ſingle one ; which if it were ſo, yet one Swallow
(as we ſay) maketh no Summer, nor a thing comming by chance cannot bee reckoned
for a certaine and conſtant rule ; you may be aſſured they will not vſually doe ſo : but
the beſt, faireſt, and moſt double flowers come alwaies, or for the moſt part, from the
ſeede of thoſe flowers that were beſt, faireſt, and moſt double ; and I doe aduiſe you
to take the beſt and moſt double : for euen from them you ſhall haue ſingle ones e-
now, you neede not to ſowe any worſer ſort. And againe, ſee that your ſeede bee
new, of the laſt yeares gathering, and alſo that it was full ripe before it was gathered,
leſt you loſe your labour, or miſſe of your purpoſe, which is, to haue faire and double
flowers. Hauing now made choiſe of your ſeede, and prepared you a bedde to ſowe
them on, the earth whereof muſt be rich and good, and likewiſe ſifted to make it the
finer ; for the better it is, the better ſhall your profit and pleaſure bee : hereon, being
firſt made leuell, plaine, and ſmooth, ſowe your ſeede ſomewhat thinne, and not too
thicke in any caſe, and as euenly as you can, that they be not too many in one place,
and too few in another, which afterwards couer with fine ſifted earth ouer them about
one fingers thickneſſe ; let this be done in the middle of Aprill, if the time of the yeare
be temperate, and not too cold, or elſe ſtay vntill the end of the moneth : after they
are ſprung vp and growne to be ſomewhat bigge, let them bee drawne forth that are
too cloſe and neare one vnto another, and plant them in ſuch place where they ſhall
continue, ſo that they ſtand halfe a yard of ground diſtance aſunder, which after the
planting, let be ſhadowed for a time, as is before ſpecified ; and this may bee done in
the end of Iuly, or ſooner if there be cauſe. I haue not ſet downe in all this diſcourſe
of planting, tranſplanting, ſowing, ſetting, &c. any mention of watering thoſe ſlips
or plants, not doubting but that euery ones reaſon will induce them to thinke, that
they cannot proſper without watering: But let this Caueat be a ſufficient remembrance
vnto you, that you neuer water any of theſe Gilloflowers, nor yet indeede any other
fine herbe or plant with cold water, ſuch as you haue preſently before drawne out from
a pumpe or Well, &c. but with ſuch water as hath ſtood open in the aire in a ciſterne,
tubbe, or pot, for one whole day at the leaſt ; if it be two or three daies it will be neuer
the worſe, but rather the better, as I haue related before : yet take eſpeciall heede that
you doe not giue them too much to ouer-glut them at any time, but temperately to ir-
rorate

rorate, bedew or fprinkle them often. From the feedes of thefe Gilloflowers hath rifen both white, red, bluſh, ſtamell, tawny lighter and ſadder, marbled, ſpeckled, ſtriped, flaked, and that in diuers manners, both ſingle and double flowers, as you ſhall fee them ſet downe in a more ample manner in the Chapter of Gilloflowers. And thus much for their encreaſe by the two wayes of planting and ſowing : For as for a third way, by grafting one into or vpon another, I know none ſuch to be true, nor to be of any more worth than an old Wiues tale, both nature, reaſon, and experience, all conteſting againſt ſuch an idle fancy, let men make what oſtentation they pleaſe. It now reſteth, that we alſo ſhew you the manner how to preſerue them, as well in Summer from all noyſome and hurtfull things, as in the Winter and Spring from the ſharp and chilling colds, and the ſharpe and bitter killing windes in March. The hurtfull things in the Summer are eſpecially theſe, too much heate of the Sunne which ſcorcheth them, which you muſt be carefull to preuent, by placing boughes, boords, clothes or mats, &c. before them, if they bee in the ground ; or elſe if they bee in pots, to remoue into them into the ſhadow, to giue them refreſhing from the heate, and giue them water alſo for their life : too much water, or too little is another annoyance, which you muſt order as you ſee there is iuſt cauſe, by withholding or giuing them water gently out of a watering pot, and not caſt on by diſhfuls : Some alſo to water their Gilloflowers, vſe to ſet their pots into tubbes or pots halfe full of water, that ſo the water may ſoake in at the lower holes in each flower pot, to giue moiſture to the roots of the Gilloflowers onely, without caſting any water vpon the leaues, and aſſuredly it is an excellent way to moiſten the rootes ſo ſufficiently at one time, that it doth ſaue a great deale of paines many other times. Eatwickes are a moſt infeſtuous vermine, to ſpoyle the whole beauty of your flowers, and that in one night or day ; for theſe creatures delighting to creepe into any hollow or ſhadowie place, doe creepe into the long greene pods of the Gilloflowers, and doe eate away the white bottomes of their leaues, which are ſweete, whereby the leaues of the flowers being looſe, doe either fall away of themſelues before, or when they are gathered, or handled, or preſently wither within the pods before they are gathered, and blowne away with the winde. To auoide which inconuenience, many haue deuiſed many waies and inuentions to deſtroy them, as pots with double verges or brimmes, containing a hollow gutter betweene them, which being filled with water, will not ſuffer theſe ſmall vermine to paſſe ouer it to the Gilloflowers to ſpoile them. Others haue vſed old ſhooes, and ſuch like hollow things to bee ſet by them to take them in : but the beſt and moſt vſuall things now vſed, are eyther long hollow canes, or elſe beaſts hoofes, which being turned downe vpon ſtickes ends ſet into the ground, or into the pots of earth, will ſoone draw into them many Earwickes, lying hid therein from ſunne, winde, and raine, and by care and diligence may ſoone bee deſtroyed, if euery morning and euening one take the hoofes gently off from the ſtickes, and knocking them againſt the ground in a plain allie, ſhake out all the Earwicks that are crept into them, which quickly with ones foot may be trode to peeces. For ſodain blaſting with thunder and lightening, or fierce ſharpe windes, &c. I know no other remedy, vnleſſe you can couer them therefrom when you firſt foreſee the danger, but patiently to abide the loſſe, whatſoeuer ſome haue aduiſed, to lay litter about them to auoide blaſting ; for if any ſhall make tryall thereof, I am in doubt, he ſhall more endanger his rootes thereby, being the Summer time, when any ſuch feare of blaſting is, than any wiſe ſaue them from it, or doe them any good. For the Winter preſeruation of them, ſome haue aduiſed to couer them with Bee-hiues, or elſe with ſmall Willow ſtickes, prickt croſſewiſe into the ground ouer your flowers, and bowed archwiſe, and with litter laid thereon, to couer the Gilloflowers quite ouer, after they haue beene ſprinkled with ſope aſhes and lyme mixt together : and this way is commended by ſome that haue written thereof, to be ſuch an admirable defence vnto them in Winter, that neither Ants, nor Snailes, nor Earwickes ſhall touch them, becauſe of the ſope aſhes and lyme, and neyther froſts nor ſtormes ſhall hurt them, becauſe of the litter which ſo well will defend them ; and hereby alſo your Gilloflowers will bee ready to flower, not onely in the Spring very early, but euen all the Winter. But whoſoeuer ſhall follow theſe directions, may peraduenture finde them in ſome part true, as they are there ſet downe for the Winter time, and while they are kept cloſe and couered ; but let them bee aſſured,

that

that all such plants, or the most part of them, will certainely perish and dye before the Summer be at an end : for the sope ashes and lyme will burne vp and spoile any herbe, and againe, it is impossible for any plant that is kept so warme in Winter, to abide eyther the cold or the winde in the Spring following, or any heate of the Sun, but that both of them will scorch them, and carry them quite away. One great hurt vnto them, and to all other herbes that wee preserue in Winter, is to suffer the snow to lye vpon them any time after it is fallen, for that it doth so chill them, that the Sunne afterward, although in Winter, doth scorch them and burne them vp : looke therefore vnto your Gilloflowers in those times, and shake or strike off the snow gently off from them, not suffering it to abide on them any day or night if you can ; for assure your selfe, if it doth not abide on them, the better they will be. The frosts likewise is another great annoyance vnto them, to corrupt the rootes, and to cause them to swell, rot, and breake : to preuent which inconuenience, I would aduise you to take the straw or litter of your horse stable, and lay some thereof about euery roote of your Gilloflowers (especially those of the best account) close vnto them vpon the ground, but be as carefull as you can, that none thereof lye vpon the greene leaues, or as little as may be, and by this onely way haue they been better defended from the frosts that spoile them in Winter, then by any other that I haue seen or knowne. The windes in March, and Sunneshine dayes then, are one of the greatest inconueniences that happeneth vnto them : for they that haue had hundreds of plants, that haue kept faire and greene all the Winter vntill the beginning or middle of March, before the end thereof, haue had scarce one of many, that either hath not vtterly perished, or been so tainted, that quickly after haue not been lost ; which hath happened chiefly by the neglect of these cautions before specified, or in not defending them from the bitter sharpe windes and sunne in this moneth of March. You shall therefore for their better preseruation, besides the litter laid about the rootes, which I aduise you not to remoue as yet, shelter them somewhat from the windes, with eyther bottomlesse pots, pales, or such like things, to keep away the violent force both of windes and sun for that moneth, and for some time before & after it also. : yet so, that they be not couered close aboue, but open to receiue ayre & raine. Some also vse to wind withes of hey or straw about the rootes of their Gilloflowers, and fasten them with stickes thrust into the ground, which serue very well in the steadof the other. Thus haue I shewed you the whole preseruation of these worthy and dainty flowers, with the whole manner of ordering them for their encrease : if any one haue any other better way, I shall be as willing to learne it of them, as I haue beene to giue them or any others the knowledge of that I haue here set downe.

Chap. IX.

That there is not any art whereby any flower may be made to grow double, that was naturally single, nor of any other sent or colour than it first had by nature ; nor that the sowing or planting of herbes one deeper than other, will cause them to be in flower one after another, euery moneth in the yeare.

THe wonderfull desire that many haue to see faire, double, and sweete flowers, hath transported them beyond both reason and nature, feigning and boasting often of what they would haue, as if they had it. And I thinke, from this desire and boasting hath risen all the false tales and reports, of making flowers double as they list, and of giuing them colour and sent as they please, and to flower likewise at what time they will, I doubt not, but that some of these errours are ancient, and continued long by tradition, and others are of later inuention : and therefore the more to be condemned, that men of wit and iudgement in these dayes should expose themselues in their writings, to be rather laughed at, then beleeued for such idle tales. And although in the contradiction of them, I know I shall vndergoe many calumnies, yet notwithstanding, I will endeauour to set downe and declare so much, as I hope may by reason

perswade

perſwade many in the truth, although I cannot hope of all, ſome being ſo ſtrongly wedded to their owne will, and the errours they haue beene bred in, that no reaſon may alter them. Firſt therefore I ſay, that if there were any art to make ſome flowers to grow double, that naturally were ſingle, by the ſame art, all ſorts of flowers that are ſingle by nature, may be made to grow double: but the ſorts of flowers that are ſingle by nature, whereof ſome are double, were neuer made double by art; for many ſorts abide ſtill ſingle, whereof there was neuer ſeene double: and therefore there is no ſuch art in any mans knowledge to bring it to paſſe. If any man ſhall ſay, that becauſe there are many flowers double, whereof there are ſingle alſo of the ſame kinde, as for example, Violets, Marigolds, Daiſyes, Daffodils, Anemones, and many other, that therefore thoſe double flowers were ſo made by the art of man: *viz.* by the obſeruation of the change of the Moone, the conſtellations or coniunctions of Planets, or ſome other Starres or celeſtiall bodies. Although I doe confeſſe and acknowledge, that I thinke ſome conſtellations, and peraduenture changes of the Moone, &c. were appointed by the God of nature, as conducing and helping to the making of thoſe flowers double, that nature hath ſo produced; yet I doe deny, that any man hath or ſhall euer be able to proue, that it was done by any art of man, or that any man can tell the true cauſes and ſeaſons, what changes of the Moone, or conſtellations of the Planets, wrought together for the producing of thoſe double flowers, or can imitate nature, or rather the God of nature, to doe the like. If it ſhall bee demanded, From whence then came theſe double flowers that we haue, if they were not ſo made by art? I anſwer, that aſſuredly all ſuch flowers did firſt grow wilde, and were ſo found double, as they doe now grow in Gardens, but for how long before they were found they became double, no man can tell; we onely haue them as nature hath produced them, and ſo they remaine. Againe, if any ſhall ſay, that it is likely that theſe double flowers were forced ſo to be, by the often planting and tranſplanting of them, becauſe it is obſerued in moſt of them, that if they ſtand long in any one place, and not be often remoued, they will grow ſtill leſſe double, and in the end turne ſingle. I doe confeſſe, that *Facilior eſt deſcenſus quàm aſcenſus,* and that the vnfruitfulneſſe of the ground they are planted in, or the neglect or little care had of them, or the growing of them too thicke or too long, are oftentimes a cauſe of the diminiſhing of the flowers doubleneſſe; but withall you ſhall obſerue, that the ſame rootes that did beare double flowers (and not any other that neuer were double before) haue returned to their former doubleneſſe againe, by good ordering and looking vnto: ſingle flowers haue only beene made ſomewhat fairer or larger, by being planted in the richer and more fruitfull ground of the Garden, than they were found wilde by nature; but neuer made to grow double, as that which is naturally ſo found of it ſelfe. For I will ſhew you mine owne experience in the matter. I haue been as inquiſitiue as any man might be, with euery one I knew, that made any ſuch report, or that I thought could ſay any thing therein, but I neuer could finde any one, that could aſſuredly reſolue me, that he knew certainly any ſuch thing to be done: all that they could ſay was but report, for the obſeruation of the Moone, to remoue plants before the change, that is, as ſome ſay, the full of the Moone, others the new Moone, whereupon I haue made tryall at many times, and in many ſorts of plants, accordingly, and as I thought fit, by planting & tranſplanting them, but I could neuer ſee the effect deſired, but rather in many of them the loſſe of my plants. And were there indeed ſuch a certaine art, to make ſingle flowers to grow double, it would haue beene knowne certainly to ſome that would practiſe it, and there are ſo many ſingle flowers, whereof there were neuer any of the kinde ſeene double, that to produce ſuch of them to be double, would procure both credit and coyne enough to him that ſhould vſe it; but *Vltra poſſe non eſt eſſe:* and therefore let no man beleeue any ſuch reports, bee they neuer ſo ancient; for they are but meere tales and fables. Concerning colours and ſents, the many rules and directions extant in manie mens writings, to cauſe flowers to grow yellow, red, greene, or white, that neuer were ſo naturally, as alſo to be of the ſent of Cinamon, Muske, &c. would almoſt perſwade any, that the matters thus ſet downe by ſuch perſons, and with ſome ſhew of probability, were conſtant and aſſured proofes thereof: but when they come to the triall, they all vaniſh away like ſmoake. I will in a few words ſhew you the matters and manners of their proceedings to effect this purpoſe: Firſt (they ſay) if you ſhall ſteepe your

your feedes in the lees of red Wine, you fhall haue the flowers of thofe plants to be of a purple colour. If you will haue Lillies or Gilloflowers to be of a Scarlet red colour, you fhall put Vermillion or Cynaber betweene the rinde and the finall heads growing about the roote : if you will haue them blew, you fhall diffolue Azur or Byfe betweene the rinde and the heads : if yellow, Orpiment : if greene, Vardigreafe, and thus of any other colour. Others doe aduife to open the head of the roote, and poure into it any colour diffolued, fo that there be no fretting or corroding thing therein for feare of hurting the roote, and looke what colour you put in, iuft fuch or neare vnto it fhall the colour of the flower bee. Some againe doe aduife to water the plants you would haue changed, with fuch coloured liquor as you defire the flower to be of, and they fhall grow to be fo. Alfo to make Rofes to bee yellow, that you fhould graft a white Rofe (fome fay a Damaske) vpon a Broome ftalke, and the flower will be yellow, fuppofing becaufe the Broome flower is yellow, therefore the Rofe will be yellow. Some affirme the like, if a Rofe be grafted on a Barbery bufh, becaufe both the bloffome and the barke of the Barbery is yellow, &c. In the like manner for fents, they haue fet downe in their writings, that by putting Cloues, Muske, Cinamon, Benzoin, or any other fuch fweete thing, bruifed with Rofe water, betweene the barke and the body of trees, the fruit of them will fmell and tafte of the fame that is put vnto them ; and if they bee put vnto the toppe of the rootes, or elfe bound vnto the head of the roote, they will caufe the flowers to fmell of that fent the matter put vnto them is of : as alfo to fteep the feeds of Rofes, and other plants in the water of fuch like fweet things, and then to fowe them, and water them morning and euening with fuch like liquor, vntill they be growne vp ; befides a number of fuch like rules and directions fet downe in bookes, fo confidently, as if the matters were without all doubt or queftion : whenas without all doubt and queftion I will affure you, that they are all but meere idle tales & fancies, without all reafon or truth, or fhadow of reafon or truth : For fents and colours are both fuch qualities as follow the effence of plants, euen as formes are alfo; and one may as well make any plant to grow of what forme you will, as to make it of what fent or colour you will; and if any man can forme plants at his will and pleafure, he can doe as much as God himfelfe that created them. For the things they would adde vnto the plants to giue them colour, are all corporeall, or of a bodily fubftance, and whatfoeuer fhould giue any colour vnto a liuing and growing plant, muft be fpirituall : for no folide corporeall fubftance can ioyne it felfe with the life and effence of an herbe or tree, and the fpirituall part of the colour thereof is not the fame with the bodily fubftance, but is a meere vapour that rifeth from the fubftance, and feedeth the plant, whereby it groweth, fo that there is no ground or colour of reafon, that a fubftantiall colour fhould giue colour to a growing herbe or tree : but for fent (which is a meere vapour) you will fay there is more probability. Yet confider alfo, that what fweete fent foeuer you binde or put vnto the rootes of herbes or trees, muft be either buried, or as good as buried in the earth, or barke of the tree, whereby the fubftance will in a finall time corrupt and rot, and before it can ioyne it felfe with the life, fpirit, and effence of the plant, the fent alfo will perifh with the fubftance : For no heterogeneall things can bee mixed naturally together, as Iron and Clay; and no other thing but homogeneall, can be nourifhment or conuertible into the fubftance of man or beaft : And as the ftomach of man or beaft altereth both formes, fents, and colours of all digeftible things ; fo whatfoeuer fent or colour is wholfome, and not poyfonfull to nature, being receiued into the body of man or beaft, doth neither change the bloud or skinne into that colour or fent was receiued : no more doth any colour or fent to any plant; for the plants are onely nourifhed by the moifture they draw naturally vnto them, be it of wine or any other liquor is put vnto them, and not by any corporeall fubftance, or heterogeneall vapour or fent, becaufe the earth like vnto the ftomach doth foone alter them, before they are conuerted into the nature and fubftance of the plant. Now for the laft part I vndertooke to confute, that no man can by art make all flowers to fpring at what time of the yeare hee will ; although, as I haue here before fhewed, there are flowers for euery moneth of the yeare, yet I hope there is not any one, that hath any knowledge in flowers and gardening, but knoweth that the flowers that appeare and fhew themfelues in the feuerall moneths of the yeare, are not one and the fame, and fo made to flower by art ; but that they are feuerall forts of plants, which

will

will flower naturally and conſtantly in the ſame moneths one yeare, that they vſe to doe in another, or with but little alteration, if the yeares proue not alike kindly: As for example, thoſe plants that doe flower in Ianuary and February, will by no art or induſtry of man be cauſed to flower in Summer or in Autumne; and thoſe that flower in Aprill and May, will not flower in Ianuary or February; or thoſe in Iuly, Auguſt, &c. either in the Winter or Spring: but euery one knoweth their owne appointed naturall times, which they conſtantly obſerue and keepe, according to the temperature of the yeare, or the temper of the climate, being further North or South, to bring them on earlier or later, as it doth with all other fruits, flowers, and growing greene herbes, &c. except that by chance, ſome one or other extraordinarily may be hindered in their due ſeaſon of flowring, and ſo giue their flowers out of time, or elſe to giue their flowers twice in the yeare, by the ſuperaboundance of nouriſhment, or the mildneſſe of the ſeaſon, by moderate ſhowers of raine, &c. as it ſometimes alſo happeneth with fruits, which chance, as it is ſeldome, and not conſtant, ſo we then terme it but *Luſus naturæ:* or elſe by forcing them in hot ſtoues, which then will periſh, when they haue giuen their flowers or fruits. It is not then, as ſome haue written, the ſowing of the ſeedes of Lillies, or any other plants a foote deepe, or halfe a foote deepe, or two inches deepe, that will cauſe them to be in flower one after another, as they are ſowne euery moneth of the yeare; for it were too groſſe to thinke, that any man of reaſon and iudgement would ſo beleeue. Nor is it likewiſe in the power of any man, to make the ſame plants to abide a moneth, two, or three, or longer in their beauty of flowring, then naturally they vſe to doe; for I thinke that were no humane art, but a ſupernaturall worke. For nature ſtill bendeth and tendeth to perfection, that is, after flowring to giue fruit or ſeede; nor can it bee hindered in the courſe thereof without manifeſt danger of deſtruction, euen as it is in all other fruit-bearing creatures, which ſtay no longer, then their appointed time is naturall vnto them, without apparent damage. Some things I grant may be ſo ordered in the planting, that according to that order and time which is obſerued in their planting, they ſhall ſhew forth their faire flowers, and they are Anemones, which will in that manner, that I haue ſhewed in the worke following, flower in ſeuerall moneths of the yeare; which thing as it is incident to none or very few other plants, and is found out but of late, ſo likewiſe is it knowne but vnto a very few. Thus haue I ſhewed you the true ſolution of theſe doubts: And although they haue not beene amplified with ſuch Philoſophicall arguments and reaſons, as one of greater learning might haue done, yet are they truely and ſincerely ſet downe, that they may ſerue *tanquam galeatum,* againſt all the calumnies and obiections of wilfull and obdurate perſons, that will not be reformed. As firſt, that all double flowers were ſo found wilde, being the worke of nature alone, and not the art of any man, by planting or tranſplanting, at or before the new or full Moone, or any other obſeruation of time, that hath cauſed the flower to grow double, that naturally was ſingle: Secondly, that the rules and directions, to cauſe flowers to bee of contrary or different colours or ſents, from that they were or would be naturally, are meere fancies of men, without any ground of reaſon or truth. And thirdly, that there is no power or art in man, to cauſe flowers to ſhew their beauty diuers moneths before their naturall time, nor to abide in their beauty longer then the appointed naturall time for euery one of them.

C   THE

THE GARDEN
OF
PLEASANT FLOWERS.

CHAP. I.

Corona Imperialis. The Crowne Imperiall.

Ecaufe the Lilly is the more ftately flower among manie : and amongft the wonderfull varietie of Lillies,
knowne to vs in thefe daies, much more then in former
times, whereof fome are white, others blufh, fome purple, others red or yellow, fome fpotted , others without fpots , fome ftanding vpright , others hanging or
turning downewards, The Crowne Imperiall for his
ftately beautifulnefs, deferueth the firft place in this our
Garden of delight, to be here entreated of before all o-
ther Lillies : but becaufe it is fo well knowne to moft
perfons, being in a manner euery where common, I fhall
neede onely to giue you a relation of the chiefe parts
thereof, (as I intend in fuch other things) which are thefe: The roote is yellowifh on
the outfide, compofed of fewer, but much thicker fcales, then any other Lilly but the
Perfian, and doth grow fometimes to be as great as a pretty bigge childes head, but
fomewhat flat withall, from the fides whereof, and not from the bottome, it fhooteth
forth thicke long fibres, which perifh euery yeare, hauing a hole in the midft thereof,
at the end of the yeare, when the old ftalke is dry and withered, and out of the which
a new ftalke doth fpring againe (from a bud or head to be feen within the hollowneffe
on the one fide) the yeare following : the ftalke then filling vp the hollowneffe, rifeth
vp three or foure foote high, being great, round, and of a purplifh colour at the bottome, but greene aboue, befet from thence to the middle thereof with many long and
broad greene leaues, very like to the leaues of our ordinary white Lilly, but fomewhat
fhorter and narrower, confufedly without order, and from the middle is bare or naked without leaues, for a certaine fpace vpwards, and then beareth foure, fixe, or tenne
flowers, more or leffe, according to the age of the plant, and the fertility of the foyle
where it groweth : The buddes at the firft appearing are whitifh, ftanding vpright a-
mong a bufh or tuft of greene leaues, fmaller then thofe below, and ftanding aboue
the flowers, after a while they turne themfelues, and hang downewards euerio
one vpon his owne footeftalke, round about the great ftemme or ftalke, fometimes of
an euen depth, and other while one lower or higher than another, which flowers are
neare the forme of an ordinary Lilly, yet fomewhat leffer and clofer, confifting of
fixe leaues of an Orange colour, ftriped with purplifh lines and veines, which adde
a great grace to the flowers : At the bottome of the flower next vnto the ftalke, euery

C 2   leaie leafe thereof hath on the outfide a certaine bunch or eminence, of a darke purplifh colour, and on the infide there lyeth in thofe hollow bunched places, certaine cleare drops of water like vnto pearles, of a very fweete tafte almoft like fugar: in the midft of each flower is a long white ftile or pointell, forked or diuided at the end, and fixe white chiues tipt with yellowifh pendents, ftanding clofe about it: after the flowers are paft, appeare fixe fquare feede veffels ftanding vpright, winged as it were or welted on the edges, yet feeming but three fquare, becaufe each couple of thofe welted edges are ioyned clofer together, wherein are contained broad, flat, and thinne feedes, of a pale brownifh colour, like vnto other Lillies, but much greater and thicker alfo. The ftalke of this plant doth oftentimes grow flat, two, three, or foure fingers broad, and then beareth many more flowers, but for the moft part fmaller then when it beareth round ftalkes. And fometimes it happeneth the ftalke to be diuided at the top, carrying two or three tufts of greene leaues, without any flowers on them. And fometimes likewife, to beare two or three rowes or crownes of flowers one aboue another vpon one ftalke, which is feldome and fcarce feene, and befides, is but meere accidentall: the whole plant and euery part thereof, as well rootes, as leaues and flowers, doe fmell fomewhat ftrong as it were the fauour of a Foxe, fo that if any doe but come neare it, he cannot but fmell it, which yet is not vnwholfome.

I haue not obferued any variety in the colour of this flower, more then that it will be fairer in a cleare open ayre, and paler, or as it were blafted in a muddy or fmoakie ayre. And although fome haue boafted of one with white flowers, yet I could neuer heare that any fuch hath endured in one vniforme colour.

The Place.

This plant was firft brought from Conftantinople into thefe Chriftian Countries, and by the relation of fome that fent it, groweth naturally in Perfia.

The Time.

It flowreth moft commonly in the end of March, if the weather be milde, and fpringeth not out of the ground vntill the end of February, or beginning of March, fo quicke it is in the fpringing: the heads with feed are ripe in the end of May.

The Names.

It is of fome called *Lilium Perficum*, the Perfian Lilly: but becaufe wee haue another, which is more vfually called by that name, as fhall be fhewed in the next Chapter, I had rather with Alphonfus Pancius the Duke of Florence his Phyfitian, (who firft fent the figure thereof vnto M^r. Iohn de Brancion) call it *Corona Imperialis*, The Crowne Imperiall, then by any other name, as alfo for that this name is now more generally receiued. It hath been fent alfo by the name *Tufai*, and *Tufchai*, and *Turfani*, or *Turfanda*, being, as it is like, the Turkifh names.

The Vertues.

For any Phyficall Vertues that are in it, I know of none, nor haue heard that any hath been found out: notwithftanding the ftrong fent would perfwade it might be applyed to good purpofe.

CHAP. II.
Lilium Perficum. The Perfian Lilly.

THe roote of the Perfian Lilly is very like vnto the root of the Crowne Imperiall, and lofing his fibres in like maner euery yeare, hauing a hole therin likewife where the old ftalke grew, but whiter, rounder, and a little longer, fmaller, and not finking at all like it, from whence fpringeth vp a round whitifh greene ftalke, not
much

1 *Corona Imperialis.* The Crowne Imperiall   2 *Lilium Persicum.* The Persian Lilly.
3 *Martagon Imperiale.* The Martagon Imperiall.

much lower than the Crowne Imperiall, but much fmaller, befet from the bottome to the middle thereof, with many long and narrow leaues, of a whitifh or blewifh greene colour, almoft like to the leafe of a Tulipa : from the middle vpwards, to the toppe of the ftalke, ftand many flowers one aboue another round about it, with leaues at the foote of euery one of them, each whereof is pendulous or hanging downe the head, like vnto the Crowne Imperiall, and not turning vp any of the flowers againe, but fmaller than in any other kinde of Lilly, yea not fo bigge as the flower of a Fritillaria, confifting of fixe leaues a peece, of a dead or ouerworne purplifh colour, hauing in the midft a fmall long pointell, with certaine chiues tipt with yellow pendents : after the flowers are paft (which abide open a long time, and for the moft part flower by degrees, the loweft firft, and fo vpwards) if the weather be temperate, come fixe fquare heads or feede veffels, feeming to be but three fquare, by reafon of the wings, very like to the heads of the Crowne Imperiall, but fmaller and fhorter, wherein are contained fuch like flat feed, but fmaller alfo, and of a darker colour.

The Place.

This was, as it is thought, firft brought from Perfia vnto Conftantinople, and from thence, fent vnto vs by the meanes of diuers Turkie Merchants, and in efpeciall, by the procurement of M.r Nicholas Lete, a worthy Merchant, and a louer of all faire flowers.

The Time.

It fpringeth out of the ground very neare a moneth before the Crowne Imperiall, but doth not flower till it bee quite paft (that is to fay) not vntill the latter end of Aprill, or beginning of May : the feed (when it doth come to perfection, as it feldome doth) is not ripe vntill Iuly.

The Names.

It hath been fent by the name of *Pennachio Perfiano*, and wee thereupon doe moft vfually call it *Lilium Perficum*, The Perfian Lilly. Clufius faith it hath been fent into the Low-Countries vnder the name of Sufam giul, and he thereupon thinking it came from Sufis in Perfia, called it *Lilium Sufianum*, The Lilly of Sufis.

The Vertues.

Wee haue not yet heard, that this hath beene applyed for any Phyficall refpect.

CHAP. III.

Martagon Imperiale, fiue Lilium Montanum maius,
The Martagon Imperiall.

VNder this title of *Lilium Montanum*, or *Lilium Siluefire*, I do comprehend only thofe kindes of Lillies, which carry diuers circles of greene leaues fet together at certaine diftances, round about the ftalke, and not fparfedly as the two former, and as other kindes that follow, doe. And although there bee many of this fort, yet becaufe their chiefeft difference is in the colour of the flower, wee will containe them all in one Chapter, and begin with the moft ftately of them all, becaufe of the number of flowers it beareth vpon one ftalke. The Imperiall Lilly hath a fcaly roote, like vnto all the reft of the Lillies, but of a paler yellow colour, clofely compact or fet together, being fhort and fmall oftentimes, in comparifon of the greatneffe of the
ftemme

stemme growing from it. The ftalke is brownifh and round at the bottome, and fome-times flat from the middle vpwards, three foote high or more, befet at certaine diftan-ces with rondles or circles of many broad leaues, larger and broader for the moft part than any other of this kinde, and of a darke green colour : It hath two or three, and fometimes foure of thefe rondles or circles of leaues, and bare without any leafe be-tweene; but aboue toward the tops of the ftalkes, it hath here and there fome leaues vpon it, but fmaller than any of the other leaues : at the toppe of the ftalke come forth many flowers, fometime three or foure fcore, thicke thruft, or confufedly fet together, and not thinne or fparfedly one aboue another, as in the leffer of this kinde of Moun-taine Lilly. It hath been fometimes alfo obferued in this kinde, that it hath borne ma-nie flowers at three feuerall fpaces of the ftalke, one aboue another, which hath made a goodly fhew; each flower whereof is pendulous, hanging downe, and each leafe of the flower turning vp againe, being thicke or flefhy, of a fine delayed purple colour, fpotted with many blackifh or brownifh fpots, of a very pleafant fweet fent, which ma-keth it the more acceptable : in the middle of the flower hangeth downe a ftile or pointell, knobbed or buttoned at the end with fixe yellow chiues, tipt with loofe pen-dents of an Orient red or Vermillion colour, which will eafily fticke like duft vpon any thing that toucheth them : the heads or feede veffels are fmall and round, with fmall edges about them, wherein is contained flat browne feede like other Lillies, but leffer. The root is very apt to encreafe or fet of, as we call it, wherby the plant feldome commeth to fo great a head of flowers, but rifeth vp with many ftalkes, and then carry fewer flowers.

Of this kinde there is fometimes one found, that beareth flowers without any fpots: the leaues whereof and ftalke likewife are paler, but not elfe differing.   *Martagon Imperiale flore non pun-Etato.*

Martagon flore albo. The White Martagon.

We haue alfo fome other of this kind, the firft whereof hath his ftalke & leafe greener than the former, the ftalke is a little higher, but not bearing fo thicke a head of flowers, although much more plentifull than the leffer Mountaine Lilly, being altogether of a fine white colour, without any fpots, or but very few, and that but fometimes alfo : the pendents in the middle of this flower are not red, as the former, but yellow; the roote of this, and of the other two that follow, are of a pale yellow colour, the cloues or fcales of them being brittle, and not clofely compact, yet fo as if two, and fometimes three fcales or cloues grew one vpon the head or vpperpart of another; which diffe-rence is a fpeciall note to know thefe three kindes, from any other kinde of Mountaine Lilly, as in all old rootes that I haue feene, I haue obferued, as alfo in them that are rea-fonably well growne, but in the young rootes it is not yet fo manifeft.

Martagon flore albo maculato. The White fpotted Martagon.

The fecond is like vnto the firft in all things, faue in this, that the flowers hereof are not altogether fo white, and befides hath many reddifh fpots on the infide of the leaues of the flower, and the ftalke alfo is not fo greene but brownifh.

Martagon flore carneo. The blufh Martagon.

A third fort there is of this kinde, whofe flowers are wholly of a delayed flefh co-lour, with many fpots on the flowers, and this is the difference hereof from the former.

Lilium Montanum fiue filueftre minus. The leffer Mountaine Lilly.

The leffer Mountaine Lilly is fo like in root vnto the greater that is firft defcribed, that it is hard to diftinguifh them afunder; but when this is fprung vp out of the ground, which is a moneth after the firft : it alfo carrieth his leaues in rondles about the ftalke, although not altogether fo great nor fo many. The flowers are more thinly fet on the ftalkes one aboue another, with more diftance betweene each flower than the former, and are of a little deeper flefh colour or purple, fpotted in the fame manner. The buds

1 *Martagon flore albo.* The white Martagon.　2 *Martagon five Lilium Canadense maculatum.* The spotted Martagon, or Lilly of Canada.　3 *Martagon Pompomeum.* The Martagon Pompony, or early red Martagon.

or

or heads of flowers, in some of these before they be blowne, are hoary white, or hairie, whereas in others, there is no hoarinesse at all, but the buddes are smooth and purplish : in other things this differeth not from the former.

Of this sort also there is one that hath but few spots on the flowers, whose colour is *Lilium Montanum non maculatum.* somewhat paler than the other.

Martagon Canadense maculatum. The spotted Martagon of Canada.

Although this strange Lilly hath not his flowers hanging downe, and turning vp againe, as the former kinds set forth in this Chapter; yet because the green leaues stand at seuerall ioynts as they do, I must needs insert it here, not knowing where more fitly to place it. It hath a small scaly roote, with many small long fibres thereat, from whence riseth vp a reasonable great stalke, almost as high as any of the former, bearing at three or foure distances many long and narrow greene leaues ; but not so many or so broad as the former, with diuers ribbes in them : from among the vppermost rundle of leaues breake forth foure or fiue flowers together, euery one standing on a long slender foote stalke, being almost as large as a red Lilly, but a little bending downewards, and of a faire yellow colour, spotted on the inside with diuers blackish purple spots or strakes, hauing a middle pointell, and sixe chiues, with pendents on them.

The Place.

All these Lillies haue been found in the diuers Countries of Germany, as Austria, Hungaria, Pannonia, Stiria, &c. and are all made Denisons in our London Gardens, where they flourish as in their owne naturall places. The last was was brought into France from Canada by the French Colonie, and from thence vnto vs.

The Time.

They flower about the later end of Iune for the most part, yet the first springeth out of the ground a moneth at the least before the other, which are most vsually in flower before it, like vnto the Serotine Tulipas, all of them being early vp, and neuer the neere.

The Names.

The first is vsually called *Martagon Imperiale*, the Imperiall Martagon, and is *Lilium Montanum maius*, the greatest Mountaine Lilly; for so it deserueth the name, because of the number of flowers vpon a head or stalke. Some haue called it *Lilium Sarasenicum*, and some *Hemerocallis*, but neither of them doth so fitly agree vnto it.

The second is *Lilium Montanum maius flore albo*, and of some *Martagon Imperiale flore albo*, but most vsually *Martagon flore albo*, the white Martagon. The second sort of this second kinde, is called *Martagon flore albo maculato*, the spotted white Martagon. And the third, *Martagon flore carneo*, the blush Martagon.

The third kinde is called *Lilium Montanum*, the Mountaine Lilly, and some adde the title *minus*, the lesser, to know it more distinctly from the other. Some also *Lilium Siluestre*, as Clusius, and some others, and of Matthiolus *Martagon*. Of diuers women here in England, from the Dutch name, Lilly of Nazareth. The last hath his title *Americanum & Canadense*, and in English accordingly.

CHAP. IV.

Chap. IV.

1. Martagon Pomponeum siue Lilium rubrum præcox, vel Lilium Macedonicum.
The early red Martagon, or Martagon Pompony.

*A*S in the former Chapter we described vnto you such Lillies, whose flowers being pendulous, turne their leaues backe againe, and haue their greene leaues, set by spaces about the stalke: so in this wee will set downe those sorts, which carry their greene leaues more sparsedly, and all along the stalke, their flowers hanging downe, and turning vp againe as the former, and begin with that which is of greatest beauty, or at least of most rarity.

1. Martagon Pomponeum angusti folium præcox.

1. This rare Martagon hath a scaly root closely compact, with broader and thinner scales than others, in time growing very great, and of a more deepe yellow colour then the former, from whence doth spring vp a round greene stalke in some plants, and flat in others, two or three foote high, bearing a number of small, long, and narrow greene leaues, very like vnto the leaues of Pinkes, but greener, set very thicke together, and without order about the stalke, vp almost vnto the toppe, and lesser by degrees vpwards, where stand many flowers, according to the age of the plant, and thriuing in the place where it groweth; in those that are young, but a few, and more sparsedly, and in others that are old many more, and thicker set: for I haue reckoned threescore flowers and more, growing thicke together on one plant with mee, and an hundred flowers on another: these flowers are of a pale or yellowish red colour, and not so deep red as the red Martagon of Constantinople, hereafter set down, nor fully so large: yet of the same fashion, that is, euery flower hanging downe, and turning vp his leaues againe. It is not so plentifull in bearing of seede as the other Lillies, but when it doth, it differeth not but in being lesse.

2. Martagon angusti folium magis serotinum.

There is another, whose greene leaues are not so thicke set on the stalke, but else differeth not but in flowring a fortnight later.

3. Martagon Pomponeum latifolium præcox.

There is another also of this kind, so like vnto the former in root, stalk, flower, & maner of growing, that the difference is hardly discerned; but consisteth chiefly in these two points: First, that the leaues of this are a little broader and shorter then the former; and secondly, that it beareth his flowers a fortnight earlier than the first. In the colour or forme of the flower, there can no difference bee discerned, nor (as I said) in any other thing. All these Lillies doe spring very late out of the ground, euen as the yellow Martagons doe, but are sooner in flower then any others.

4. Martagon flore phæniceo.

A fourth kinde hereof hath of late been knowne to vs, whose leaues are broader and shorter then the last, and the flowers of a paler red, tending to yellow, of some called a golden red colour: but flowreth not so early as they.

2. Lilium rubrum Byzantinum, siue Martagon Constantinopolitanum.
The red Martagon of Constantinople.

1. The red Martagon of Constantinople is become so common euery where, and so well knowne to all louers of these delights, that I shall seeme vnto them to lose time, to bestow many lines vpon it; yet because it is so faire a flower, and was at the first so highly esteemed, it deserueth his place and commendations, howsoeuer encreasing the plenty hath not made it dainty. It riseth out of the ground early in the spring, before many other Lillies, from a great thicke yellow scaly root, bearing a round brownish stalke, beset with many faire greene leaues confusedly thereon, but not so broad as the common white Lilly, vpon the toppe whereof stand one, two, or three, or more flowers, vpon long footestalkes, which hang downe their heads, and turne vp their leaues againe, of an excellent red crimson colour, and sometimes paler, hauing a long pointell in the middle, compassed with sixe whitish chiues, tipt with loose yellow pendents, of a reasonable good sent, but somewhat faint. It likewise beareth seede in heads, like vnto the other, but greater.

Martagon Constantinopolitanum maculatum.
The red spotted Martagon of Constantinople.

We haue another of this kinde, that groweth somewhat greater and higher, with a larger flower, and of a deeper colour, spotted with diuers blacke spots, or strakes and lines, as is to be seene in the Mountaine Lillies, and in some other hereafter to be described; but is not so in the former of this kinde, which hath no shew of spots at all. The whole plant as it is rare, so it is of much more beauty then the former.

2. *Martagon Pannonicum, siue Exoticum flore spadiceo.*
The bright red Martagon of Hungarie.

Although this Martagon or Lilly bee of another Countrey, yet by reason of the neerenesse both in leafe and flower vnto the former, may more fitly be placed next vnto them, then in any other place. It hath his roote very like the other, but the leaues are somewhat larger, and more sparsedly set vpon the stalke, else not much vnlike: the flowers bend downe, and turne vp their leaues againe, but somewhat larger, and of a bright red, tending to an Orenge colour, that is, somewhat yellowish, and not crimson, like the other.

3. *Martagon Luteum punctatum.* The Yellow spotted Martagon.

1. This Yellow Martagon hath a great scaly or cloued roote, and yellow, like vnto all these sorts of turning Lillies, from whence springeth vp a round greene strong stalke, three foote high at the least, confusedly set with narrow long greene leaues, white on the edges vp to the very toppe thereof almost, hauing diuers flowers on the head, turning vp againe as the former doe, of a faint yellowish, or greenish yellow colour, with many blacke spots or strakes about the middle of the leafe of euery flower, and a forked pointell, with sixe chiues about it, tipt with reddish pendents, of a heauie strong smell, not very pleasant to many. It beareth seede very plentifully, in great heads, like vnto the other former Lillies, but a little paler.

2. *Martagon Luteum non maculatum.* The Yellow Martagon without spots.

The other yellow Martagon differeth in no other thing from the former, but onely that it hath no spots at all vpon any of the leaues of the flowers, agreeing with the former, in colour, forme, height, and all things else.

3. *Martagon Luteum serotinum.* The late flowring Yellow Martagon.

There is yet another yellow Martagon, that hath no other difference then the time of his flowring, which is not vntill Iuly; vnlesse in this, that the flower is of a deeper yellow colour.

The Place.

The knowledge of the first kindes of these early Martagons hath come from Italy, from whence they haue bin sent into the Low-Countries, and to vs, and, as it seemeth by the name, whereby they haue bin sent by some into these parts, his originall should be from the mountaines in Macedonia.

The second sort is sufficiently knowne by his name, being first brought from Constantinople, his naturall place being not farre from thence, as it is likely. But the next sort of this second kinde, doth plainly tell vs his place of birth to be the mountaines of Pannonia or Hungarie.

The third kindes grow on the Pyrenæan mountaines, where they haue been searched out, and found by diuers louers of plants, as also in the Kingdome of Naples.

The

The Time.

The first early Martagons flower in the end of May, or beginning of Iune, and that is a moneth at the least before those that come from Constantinople, which is the second kinde. The two first yellow Martagons flower somewhat more early, then the early red Martagons, and sometimes at the same time with them. But the third yellow Martagon, as is said, flowreth a moneth later or more, and is in flower when the red Martagon of Constantinople floweth. And although the early red and yellow Martagons, spring later then the other Martagons or Lillies, yet they are in flower before them.

The Names.

The first early red Lillies or Martagons haue beene sent vnto vs by seuerall names, as *Martagon Pomponeum*, and thereafter are called Martagon of Pompony, and also *Lilium* or *Martagon Macedonicum*, the Lilly or Martagon of Macedonia. They are also called by Clusius *Lilium rubrum præcox*, the one *angustiore folio*, the other *latiore folio*. And the last of this kinde hath the title *flore phænicco* added or giuen vnto it, that is, the Martagon or Lilly of Macedonia with gold red flowers.

The Martagons of Constantinople haue beene sent by the Turkish name *Zufihiare*, and is called *Martagon*, or *Lilium Byzantinum* by some, and *Hemerocallis Chalcedonica* by others; but by the name of the Martagon of Constantinople they are most commonly receiued with vs, with the distinction of *maculatum* to the one, to distinguish the sorts. The last kinde in this *classis*, hath his name in his title, as it hath been sent vnto vs.

The Yellow Martagons are distinguished in their seuerall titles, as much as is conuenient for them.

Chap. V.

Lilium Aureum & Lilium Rubrum. The Gold and Red Lillies.

THere are yet some other kindes of red Lillies to bee described, which differ from all the former, and remaine to be spoken of in this place. Some of them grow high, and some lowe, some haue small knots, which wee call bulbes, growing vpon the stalkes, at the ioynts of the leaues or flowers, and some haue none: all which shall be intreated of in their seuerall orders.

Lilium pumilum cruentum. The dwarfe red Lilly.

The dwarfe red Lilly hath a scaly roote, somewhat like vnto other Lillies, but white, and not yellow at all, and the cloues or scales thicker, shorter, and fewer in number, then in most of the former : the stalke hereof is not aboue a foote and a halfe high, round and greene, set confusedly with many faire and short greene leaues, on the toppe of which doe stand sometimes but a few flowers, and sometimes many, of a faire purplish red colour, and a little paler in the middle, euery flower standing vpright, and not hanging downe, as in the former, on the leaues whereof here and there are some blacke spots, lines or markes, and in the middle of the flower a long pointell, with some chiues about it, as is in the rest of these Lillies.

Lilium rubrum multiplici flore.   This kinde is sometimes found to yeeld double flowers, as if all the single flowers should grow into one, and so make it consist of many leaues, which notwithstanding

his

1 *Martagon rubrum sine luteum.* The red or the yellow Martagon. 2 *Lilium Bulbiferum.* The red bulbed Lilly. 3 *Lilium aureum.* The gold red Lilly. 4 *Lilium album.* The white Lilly.

his so continuing sundry yeares, vpon transplanting , will *redire ad ingenium* , that is, quickly come againe to his old byas or forme.

Lilium Aureum. The Gold red Lilly.

The second red Lilly without bulbes groweth much higher then the first , and almost as high as any other Lilly : the roote hereof is white and scaly, the leaues are somewhat longer, and of a darke or sad greene colour ; the flowers are many and large , standing vpright as all these sorts of red Lillies doe , of a paler red colour tending to an Orenge on the inside, with many blacke spots, and lines on them , as in the former, and more yellow on the outside : the seede vessels are like vnto the roundish heads of other Lillies, and so are the seedes in them likewise.

1. *Lilium minus bulbiferum.* The dwarfe bulbed Lilly.

The first of the Lillies that carrieth bulbes on the stalke , hath a white scaly roote like the former ; from whence riseth vp a small round stalke , not much higher then the first dwarfe Lilly, seeming to be edged, hauing many leaues thereon of a sad green colour set about it, close thrust together : the greene heads for flowers,will haue a kind of woollinesse on them, before the flowers begin to open , and betweene these heads of flowers, as also vnder them,and among the vppermost leaues,appeare small bulbes or heads, which being ripe if they be put into the ground,or if they fall of themselues, will shoote forth leaues, and beare flowers within two or three yeares like the mother plant , and so will the bulbes of the other hereafter described : the flowers of this Lilly are of a faire gold yellow colour, shadowed ouer with a shew of purple , but not so red as the first , or the next to bee described. This Lilly will shoote strings vnder ground, like as the last red Lilly will doe also, whereat will grow white bulbed roots, like the rootes of the mother plant, thereby quickly encreasing it selfe.

2. *Lilium Cruentum bulbiferum.* The Fierie red bulbed Lilly.

The second bulbed Lilly riseth vp with his stalke as high as any of these Lillies, carrying many long and narrow darke greene leaues about it , and at the toppe many faire red flowers, as large or larger then any of the former, and of a deeper red colour, with spots on them likewise , hauing greater bulbes growing about the toppe of the stalke and among the flowers, then any else.

Lilium Cruentum flore pleno. The Fierie red double Lilly.

The difference of this doth chiefly consist in the flower, which is composed of manie leaues, as if many flowers went to make one, spotted with black spots,and without any bulbes when it thus beareth,which is but accidentall , as the former double Lilly is said to be.

3. *Lilium maius bulbiferum.* The greater bulbed red Lilly.

The third red Lilly with bulbes, riseth vp almost as high as the last, and is the most common kinde we haue bearing bulbes. It hath many leaues about the stalke, but not of so sad a greene colour as the former : the flowers are of as pale a reddish yellow colour as any of the former,and comming neerest vnto the colour of the Gold red Lilly. This is more plentifull in bulbes, and in shooting strings , to encrease rootes vnder ground, then the others.

The Place.

These Lillies doe all grow in Gardens, but their naturall places of growing is the Mountaines and the Vallies neere them in Italy , as Matthiolus saith :

faith : and in many Countries of Germany , as Hungarie , Auſtria , Stiria, and Bohemia, as Cluſius and other doe report.

The Time.

They flower for the moſt part in Iune, yet the firſt of theſe is the earlieſt of all the reſt.

The Names.

All theſe Lillies are called *Lilia Rubra*, Red Lillies : Some call them *Lilium Aureum, Lilium Purpureum , Lilium Puniceum , & Lilium Cruentum.* Some alſo call them *Martagon Chimiſtarum.* Cluſius calleth theſe bulbed Lillies *Martagon Bulbiferum.* It is thought to be *Hyacinthus Poetarum* , but I referre the diſcuſſing thereof to a fitter time. Wee haue, to diſtinguiſh them moſt fitly (as I take it) giuen their proper names in their ſeuerall titles.

CHAP. VI.

Lilium Album. The White Lilly.

Now remaineth onely the White Lilly, of all the whole family or ſtocke of the Lillies, to bee ſpoken of, which is of two ſorts. The one is our common or vulgar White Lilly ; and the other, that which was brought from Conſtantinople.

Lilium Album vulgare. The ordinary White Lilly.

The ordinary White Lilly ſcarce needeth any deſcription , it is ſo well knowne, and ſo frequent in euery Garden ; but to ſay ſomewhat thereof, as I vſe to doe of euery thing, be it neuer ſo common and knowne ; it hath a cloued or ſcaly roote, yellower and bigger then any of the red Lillies : the ſtalke is of a blackiſh greene colour, and riſeth as high as moſt of the Lillies , hauing many faire, broad, and long greene leaues thereon, larger and longer beneath, and ſmaller vpon the ſtalke vpwards ; the flowers are many or few, according to the age of the plant , fertility of the ſoile, and time of ſtanding where it groweth : and ſtand vpon long greene footſtalkes, of a faire white colour, with a long pintell in the middle, and white chiues tipt with yellow pendents about it ; the ſmell is ſomewhat heady and ſtrong.

Lilium Album Byzantinum. The White Lilly of Conſtantinople.

The other White Lilly, differeth but little from the former White Lilly, either in roote, leafe, or flower, but only that this vſually groweth with more number of flowers, then euer we ſaw in our ordinary White Lilly : for I haue ſeene the ſtalke of this Lilly turne flat, of the breadth of an hand, bearing neere two hundred flowers vpon a head, yet moſt commonly it beareth not aboue a dozen, or twenty flowers , but ſmaller then the ordinary, as the greene leaues are likewiſe.

The Place.

The firſt groweth onely in Gardens , and hath not beene declared where it is found wilde, by any that I can heare of. The other hath beene ſent from Conſtantinople, among other rootes, and therefore is likely to grow in ſome parts neere thereunto.

The Time.

They flower in Iune or thereabouts , but ſhoote forth greene leaues in

Autumne,

Autumne, which abide greene all the Winter, the stalke springing vp betweene the lower leaues in the Spring.

The Names.

It is called *Lilium Album*, the White Lilly, by most Writers; but by Poets *Rosa Iunonis*, Iuno's Rose. The other hath his name in his title.

The Vertues.

This Lilly aboue all the rest, yea, and I thinke this onely, and none of the rest is vsed in medicines now adayes, although in former times Empericks vsed the red; and therefore I haue spoken nothing of them in the end of their Chapters, reseruing what is to be said in this. This hath a mollifying, digesting, and cleansing quality, helping to suppurate tumours, and to digest them, for which purpose the roote is much vsed. The water of the flowers distilled, is of excellent vertue for women in trauell of childe bearing, to procure an easie deliuery, as Matthiolus and Camerarius report. It is vsed also of diuers women outwardly, for their faces to cleanse the skin, and make it white and fresh. Diuers other properties there are in these Lillies, which my purpose is not to declare in this place. Nor is it the scope of this worke; this that hath been said is sufficient: for were it not, that I would giue you some taste of the qualities of plants (as I said in my Preface) as I goe along with them, a generall worke were fitter to declare them then this.

Chap. VII.

Fritillaria. The checkerd Daffodill.

Lthough diuers learned men do by the name giuen vnto this delightfull plant, thinke it doth in some things partake with a Tulipa or Daffodill, and haue therefore placed it betweene them; yet I, finding it most like vnto a little Lilly, both in roote, stalke, leafe, flower, and seede, haue (as you see here) placed it next vnto the Lillies, and before them. Hereof there are many sorts found out of late, as white, red, blacke, and yellow, besides the purple, which was first knowne; and of each of them there are also diuers sorts: and first of that which is most frequent, and then of the rest, euery one in his place and order.

1. *Fritillaria vulgaris.* The common checkerd Daffodill.

The ordinary checkerd Daffodill (as it is vsually called, but might more properly be called the small checkerd Lilly) hath a small round white roote, and somewhat flat, made as it were of two cloues, and diuided in a maner into two parts, yet ioyning together at the bottome or seate of the roote, which holdeth them both together: from betweene this cleft or diuision, the budde for the stalke &c. appeareth, which in time riseth vp a foote, or a foote and a halfe high, being round and of a brownish greene colour, especially neere vnto the ground, whereon there standeth dispersedly foure or fiue narrow long and greene leaues, being a little hollow : at the toppe of the stalke, betweene the vpper leaues (which are smaller then the lowest) the flower sheweth it selfe, hanging or turning downe the head, but not turning vp againe any of his leaues, as some of the Lillies before described doe; (sometimes this stalke beareth two flowers, and very seldome three) consisting of sixe leaues, of a reddish purple colour, spotted diuersly with great spots, appearing like vnto square checkers, of a deeper colour; the inside of the flower is of a brighter colour then the outside, which hath some greenesse at the bottome of euery leafe: within the flower there appeare

sixe

1 *Fritillaria vulgaris*. The common Fritillaria. 2 *Fritillaria flore atrorubente*. The darke red Fritillaria. 4 *Fritillaria alba*. The white Fritillaria. 7 *Fritillaria lutea puncta*. The yellow checkerd Fritillaria. 8 *Fritillaria lutea Italica*. The great yellow Italian Fritillaria. 10 *Fritillaria lutea Lusitanica*. The small yellow Fritillaria of Portugall. 11 *Fritillaria Pyrenea*. The blacke Fritillaria. 12 *Fritillaria umbellifera*, The Spanish blacke Fritillaria.

fixe chiues tipt with yellow pendents, and a three-forked ftile or pointell compaffing a greene head, which when the flower is paft, rifeth vpright againe, and becommeth the feede veffell, being fomewhat long and round, yet hauing a fmall fhew of edges, flat at the head, like the head of a Lilly, and without any crowne as the Tulipa hath, wherein is contained pale coloured flat feede, like vnto a Lilly, but fmaller.

Fritillaria vul-garis pallidior, præcox, & ferotina. There is fome variety to be feene in this flower; for in fome the colour is paler, and in others againe of a very high or deepe colour: fometimes alfo they haue eight leaues, and fometimes ten or twelue, as if two flowers were made one, which fome thereupon haue called a Double Fritillaria. Some of them likewife doe flower very early, euen with or before the early flowring Tulipas; and fome againe flower not vntill a moneth or more after the former.

2. *Fritillaria flore atrorubente.* The bloud red Fritillaria.

The roote of this Fritillaria is fomewhat rounder and clofer then the former, from whence the ftalke rifeth vp, being fhorter and lower then in any other of thefe kindes, hauing one or two leaues thereon, and at the top thereof two or three more fet clofer together, which are broader, fhorter, and whiter then any of them before, almoft like vnto the leaues of the yellow Fritillaria, from among which toppe leaues commeth forth the flower, fomewhat bending downe, or rather ftanding forth, being larger then any of the former, and almoft equall in bigneffe vnto the yellow Fritillaria, of a duskie gray colour all ouer on the outfide, and of a very darke red colour on the infide, diuerfly fpotted or ftraked: this very hardly encreafeth by the roote, and as feldome giueth ripe feede, but flowreth with the other firft forts, and before the blacke, and abideth leffe time in flower then any.

3. *Fritillaria maxima purpurea fiue rubra.* The great purple or red Fritillaria.

This great Fritillaria hath his roote equall to the bigneffe of the reft of his parts, from whence rifeth vp one, & oftentimes two ftalks, hauing one, two or three flowers a peece on them, as nature and the feafons are fitting: euery one of thefe flowers are larger and greater then any of the former defcribed, and pendulous as they are, of a fad red or purplifh colour, with many thwart lines on them, and fmall long markes, which hardly feeme checkerwife, nor are fo eminent or confpicuous as in the former: the ftalke is ftrong and high, whereon are fet diuers long whitifh greene leaues, larger and broader then thofe of the former.

4. *Fritillaria alba.* The white Fritillaria.

The white Fritillaria is fo like vnto the firft, that I fhall not neede to make another defcription of this: it fhall (I hope) be fufficient to fhew the chiefe differences, and fo proceed to the reft. The ftalke and leaues of this are wholly greene, whereby it may eafily be knowne from the former, which, as is faid, is brownifh at the bottome. The flower is white, without almoft any fhew of fpot or marke in it, yet in fome the markes are fomewhat more plainly to be feene, and in fome againe there is a fhew of a faint kinde of blufh colour to be feene in the flower, efpecially in the infide, the bottomes of the leaues of euery flower fometimes are greenifh, hauing alfo a fmall lift of greene, comming downe towards the middle of each leafe: the head or feede veffell, as alfo the feede and the roote, are fo like vnto the former, that the moft cunning cannot diftinguifh them.

5. *Fritillaria flore duplici albicante.* The double blufh Fritillaria.

This Fritillaria hath a round flattifh white roote, very like vnto the laft Fritillaria, bearing a ftalke with long greene leaues thereon, little differing from it, or the firft ordinary Fritillaria: the flower is faid to be conftant, compofed of many leaues, being ten at the leaft, and moft vfually twelue, of a pale whitifh purple colour, fpotted like vnto the paler ordinary Fritillaria that is early, fo that one would verily thinke it were

but

but an accidentall kinde thereof, whereas it is (as is said before) held to bee conſtant, continuing in this manner.

6. *Fritillaria flore luteo puro.*   The pure yellow Fritillaria.

The pure yellow Fritillaria hath a more round, and not ſo flat a whitiſh roote as the former kindes, and of a meane bigneſſe; from the middle riſeth vp a ſtalke a foote and a halfe high, and ſometimes higher, whereon are ſet without order diuers long and ſomewhat broad leaues of a whitiſh greene colour, like vnto the leaues of the blacke Fritillaria, but not aboue halfe ſo broad: the flower is ſomewhat ſmall and long, not much vnlike to the blacke for ſhape and faſhion, but that the leaues are ſmaller and rounder pointed, of a faint yellowiſh colour, without any ſhew of ſpots or checkers at all, eyther within or without the flower, hauing ſome chiues and yellow pendents in the middle, as is to be ſeene in all of them: the ſeede is like the firſt kinde.

7. *Fritillaria flore luteo vario ſiue punctato.*   The checkerd yellow Fritillaria.

This Fritillaria groweth not much lower then the former, and browniſh at the riſing vp, hauing his leaues whiter, broader, and ſhorter then it, and almoſt round pointed. The flower is greater, and larger ſpread then any other before, of a faire pale yellow colour, ſpotted in very good order, with fine ſmall checkers, which adde a wonderfull pleaſing beauty thereunto: it hath alſo ſome liſts of greene running downe the backe of euery leafe. It ſeldome giueth ſeede; the roote alſo is like the other, but not ſo flat.

8. *Fritillaria lutea maxima Italica.*   The great yellow Italian Fritillaria.

This kinde of Fritillaria riſeth vp with a round and browne greene ſtalke, whereon are ſet diuers leaues ſomewhat broad and ſhort, which compaſſe the ſtalke at the bottome of them, of a darke greene colour; at the toppe of the ſtalke, which bendeth a little downewards, doe moſt vſually ſtand three or foure leaues, betweene which commeth forth moſt vſually but one flower, which is longer then the laſt, hanging downe the head as all the others doe, conſiſting of ſixe leaues, of a darke yellowiſh purple colour, ſpotted with ſome ſmall red checkers. This kinde flowreth late, and not vntill all the reſt are paſt.

9. *Fritillaria Italorum polyanthos flore paruo.*   The ſmall Italian Fritillaria.

This ſmall Italian Fritillaria carrieth more ſtore of flowers on the ſtalke, but they are much ſmaller, and of a yellowiſh greene colour, ſpotted with long and ſmall darke red checkers or markes: the ſtalke hath diuers ſmall ſhort greene leaues thereon, vnto the very toppe.

10. *Fritillaria lutea Iuncifolia Luſitanica.*   The ſmall yellow Fritillaria of Portugall.

The leaues of this Fritillaria are ſo ſmall, narrow and long, that it hath cauſed them to take the name of ruſhes, as if you ſhould call it, The ruſh leafed Fritillaria, which ſtand on a long weake round ſtalke, ſet without order: the flower is ſmall and yellow, but thicker checkerd with red ſpots then any of the other yellow Fritillaria's; the ſtalk of the flower, at the head thereof, being alſo of a yellowiſh colour.

11. *Fritillaria Pyrenea ſiue Apenninea.*   The blacke Fritillaria.

The roote of this kinde doth often grow ſo great, that it ſeemeth like vnto the roote of a ſmall Crowne Imperiall; the ſtalke is ſtrong, round, and high, ſet without order, with broader and whiter greene leaues then any of the former, bearing one, two, or three flowers; ſometimes at the toppe, being not ſo large as thoſe of the ordinary purple Fritillaria, but ſmaller, longer, and rounder, ſometimes a little turning vp the brims or edges of the leaues againe, and are of a yellowiſh ſhining greene colour on

the

the infide, fometimes fpotted with red fpots almoft through the whole infide of the flower, vnto the very edge, which abideth of a pale yellow colour, and fometimes there are very few fpots to be feene, and thofe from the middle onely on the infide (for on the outfide there neuer appeareth any fpots at all in this kinde) and fometimes with no fhew of fpots at all, fometimes alfo of a more pale greene, and fometime of a more yellow colour: the outfide of the flowers doe likewife vary, for in fome the out-fide of the leaues are of a darke fullen yellow, &c. elfe more pale yellow, and in other of a darke purplifh yellow colour, which in fome is fo deepe, and fo much, that it ra-ther feemeth blacke then purple or yellow, and this efpecially about the bottome of the flower, next vnto the ftalke, but the edges are ftill of a yellowifh greene: the head of feede, and the feede likewife is like vnto the former, but bigger in all refpects.

12. *Fritillaria Hifpanica vmbellifera.* The Spanifh blacke Fritillaria.

This Fritillaria is no doubt of kindred to the laft recited, it is fo like, but greater in all parts thereof, as if growing in a more fruitfull foile, it were the ftronger and luftier to beare more ftore of flowers: the flowers grow foure or fiue from the head together, hanging downe round about the ftalke, like vnto a Crowne Imperiall, and are of a yellowifh greene colour on the infide, fpotted with a few red fpots, the outfide being blackifh as the former.

The Place.

The firft of thefe plants was firft brought to our knowledge from France, where it groweth plentifully about Orleance; the other forts grow in di-uers other Countries, as fome in Portugall, Spaine, Italy, &c. as their names doe import, and as in time they haue been obferued by thofe that were cu-rious fearchers of thefe rarities, haue been fent to vs.

The Time.

The early kindes doe flower in the beginning of Aprill or thereabouts, according to the mildeneffe or fharpeneffe of the precedent Winter. The other doe flower after the firft are paft, for a moneths fpace one after ano-ther, and the great yellow is very late, not flowring vntill about the middle or end of May.

The Names.

This hath receiued diuers names: fome calling it *Flos Meleagridis*, the Ginny Hen Flower, of the variety of the colours in the flower, agreeing with the feathers of that Bird. Some call it *Narciffus Caperonius*, of the name of the firft inuentor or finder thereof, called Noel Caperon, an Apothecary dwelling in Orleance, at the time he firft found it, and was fhortly after the finding thereof taken away in the Maffacre in France. It is now generally called *Fritillaria*, of the word *Fritillus*, which diuers doe take for the Cheffe borde or table whereon they play, whereunto, by reafon of the refemblance of the great fquares or fpots fo like it, they did prefently referre it. It is called by Lobel *Lilionarciffus purpureus variegatus, & teffulatus*, making it a kinde of Tulipa; but as I faid in the beginning of the Chapter, it doth moft neerely refemble a fmall pendulous Lilly, and might therefore rightly hold the name of *Lilium variegatum*, or in Englifh, the checkerd Lilly. But be-caufe the errour which firft referred it to a Daffodill, is growne ftrong by cuftome of continuance, I leaue to euery one their owne will, to call it in Englifh eyther Fritillaria, as it is called of moft, or the checkerd Daffodill, or the Ginnie Hen flower, or, as I doe, the checkerd Lilly. I fhall not neede in this place further to explaine the feuerall names of euery of them, hauing giuen you them in their titles.

The

The Vertues.

I haue not found or heard by any others of any property peculiar in this plant, to be applied either inwardly or outwardly for any difeafe : the chiefe or onely vfe thereof is, to be an ornament for the Gardens of the curious louers of thefe delights, and to be worne of them abroad, which for the gallant beauty of many of them, deferueth their courteous entertainment, among many other the like pleafures.

CHAP. VIII.

Tulipa. The Turkes Cap.

NExt vnto the Lillies, and before the Narciffi or Daffodils, the difcourfe of Tulipas deferueth his place, for that it partaketh of both their natures ; agreeing with the Lillies in leaues, flowers, and feede, and fomewhat with the Daffodils in rootes. There are not onely diuers kindes of Tulipas, but fundry diuerfities of colours in them, found out in thefe later dayes by many the fearchers of natures varieties, which haue not formerly been obferued : our age being more delighted in the fearch, curiofity, and rarities of thefe pleafant delights, then any age I thinke before. But indeede, this flower, aboue many other, deferueth his true commendations and acceptance with all louers of thefe beauties, both for the ftately afpect, and for the admirable varietie of colours, that daily doe arife in them, farre beyond all other plants that grow, in fo much, that I doubt, although I fhall in this Chapter fet downe the varieties of a great many, I fhall leaue more vnfpoken of, then I fhall defcribe ; for I may well fay, there is in this one plant no end of diuerfity to be expected, euery yeare yeelding a mixture and variety that hath not before been obferued, and all this arifing from the fowing of the feede. The chiefe diuifion of Tulipas, is into two forts : *Præcoces*, early flowring Tulipas, and *Serotinæ*, late flowring Tulipas. For that fort which is called *Mediæ* or *Dubiæ*, that is, which flower in the middle time betweene them both, and may be thought to be a kinde or fort by it felfe, as well as any of the other two : yet becaufe they doe neerer participate with the *Serotinæ* then with the *Præcoces*, not onely in the colour of the leafe, being of the fame greenneffe with the *Serotinæ*, and moft vfually alfo, for that it beareth his ftalke and flower, high and large like as the *Serotinæ* doe ; but efpecially, for that the feede of a *Media Tulipa* did neuer bring forth a *Præcox* flower (although I know Clufius, an induftrious, learned, and painfull fearcher and publifher of thefe rarities, faith otherwife) fo farre as euer I could, by mine owne care or knowledge, in fowing their feede apart, or the affurance of any others, the louers and fowers of Tulipa feede, obferue, learne, or know : and becaufe alfo that the feede of the *Serotinæ* bringeth forth *Medias*, and the feede of *Medias Serotinæ*, they may well bee comprehended vnder the generall title of *Serotinæ*: But becaufe they haue generally receiued the name of *Media*, or middle flowring Tulipas, to diftinguifh betweene them, and thofe that vfually doe flower after them ; I am content to fet them downe, and fpeake of them feuerally, as of three forts. Vnto the place and ranke likewife of the *Præcoces*, or early flowring Tulipas, there are fome other feuerall kinds of Tulipas to be added, which are notably differing, not only from the former *Præcox Tulipa*, but euery one of them, one from another, in fome fpeciall note or other : as the *Tulipa Bolonienfis flore rubro*, the red Bolonia Tulipa. *Tulipa Bolonienfis flore luteo*, the yellow Bolonia Tulipa. *Tulipa Perfica*, the Perfian Tulipa. *Tulipa Cretica*, the Candie Tulipa, and others : all which fhall bee defcribed and entreated of, euery one apart by it felfe, in the end of the ranke of the *Præcoces*, becaufe all of them flower much about their time. To begin then with the *Præcox*, or early flowring Tulipas, and after them with the *Medias* and *Serotines*, I fhall for the better method, diuide their flowers into foure primary or principall colours, that is to fay, White, Purple, Red, and Yellow, and vnder euery one of thefe colours, fet downe the feuerall varieties

ties of mix.ures we haue feene and ooferued in them, that fo they may be both the better defcribed by me, and the better conceiued by others, and euery one placed in their
proper ranke. Yet I fhall in this, as I intend to doe in diuers other plants that are variable, giue but one defcription in generall of the plant, and then fet downe the varietie of forme or colour afterwards briefly by themfelues.

Tulipa præcox. The early flowring Tulipa.

The early Tulipa (and fo all other Tulipas) fpringeth out of the ground with his
leaues folded one within another, the firft or loweft leafe rifeth vp firft, fharpe pointed, and folded round together, vntill it be an inch or two aboue the ground, which
then openeth it felfe, fhewing another leafe folded alfo in the bofome or belly of the
firft, which in time likewife opening it felfe, fheweth forth a third, and fometimes a
fourth and a fifth : the lower leaues are larger then the vpper, and are faire, thicke,
broad, long, and hollow like a gutter, and fometimes crumpled on the edges, which
will hold water that falleth thereon a long time, of a pale or whitifh greene colour,
(and the *Mediæ* and *Serotinæ* more greene) couered ouer as it were with a mealineffe or
hoarineffe, with an eye or fhew of rednefle towards the bottome of the leaues, and the
edges in this kinde being more notable white, which are two principall notes to know
a *Præcox Tulipa* from a *Media* or *Serotina* : the ftalke with the flower rifeth vp in the
middle, as it were through thefe leaues, which in time ftand one aboue another, compaffing it at certaine vnequall diftances, and is often obferued to bend it felfe crooked-
ly downe to the ground, as if it would thruft his head thereinto, but turning vp his
head (which will be the flower) againe, afterwards ftandeth vpright, fometimes but
three or foure fingers or inches high, but more often halfe a foote, and a foot high, but
the *Mediæ*, and *Serotinas* much higher, carrying (for the moft part) but one flower on
the toppe thereof, like vnto a Lilly for the forme, confifting of fixe leaues, greene at
the firft, and afterwards changing into diuers and fundry feuerall colours and varieties, the bottomes likewife of the leaues of thefe fometimes, but moft efpecially of the
Mediæ, being as variable as the flower, wh.ch are in fome yellow, or green, or blacke,
in others white, blew, purple, or tawnie ; and fometimes one colour circling another:
fome of them haue little or no fent at all, and fome haue a better then others. After it
hath been blowne open three or foure dayes or more, it will in the heate of the Sunne
fpread it felfe open, and lay it felfe almoft flat to the ftalke : in the middle of the flower
ftandeth a greene long head (which will be the feed veffell) compaffed about with fixe
chiues, which doe much vary, in being fometimes of one, and fometimes of another
colour, tipt with pendents diuerfly varied likewife : the head in the middle of the
flower groweth after the flower is fallen, to be long, round, and edged, as it were
three fquare, the edges meeting at the toppe, where it is fmalleft, and making as it
were a crowne (which is not feen in the head of any Lilly) and when it is ripe, diuideth
it felfe on the infide into fixe rowes, of flat, thinne, brownifh, griftly feede, very like
vnto the feede of the Lillies, but brighter, ftiffer, and more tranfparent : the roote being well growne is round, and fomewhat great, fmall and pointed at the toppe, and
broader, yet roundifh at the bottome, with a certaine eminence or feate on the one
fide, as the roote of the Colchicum hath ; but not fo long, or great, it hath alfo an
hollownefle on the one fide (if it haue borne a flower) where the ftalke grew, (for although in the time of the firft fpringing vp, vntill it fhew the budde for flower, the
ftalke with the leaues thereon rife vp out of the middle of the roote ; yet when the
ftalke is rifen vp, and fheweth the budde for flower, it commeth to one fide, making
an impreffion therein) couered ouer with a brownifh thin coate or skin, like an Onion,
hauing a little woollineffe at the bottome ; but white within, and firme, yet compofed
of many coates, one folding within another, as the roote of the Daffodils be, of a reafonable good tafte, neyther very fweete, nor yet vnpleafant. This defcription may
well ferue for the other Tulipas, being *Medias* or *Serotinas*, concerning their fpringing
and bearing, which haue not any other great variety therein worth the note, which is
not expreffed here ; the chiefe difference refting in the variety of the colours of the
flower, and their feuerall mixtures and markes, as I faid before : fauing onely, that the
flowers of fome are great and large, and of others fmaller, and the leaues of fome long

and

1 *Tulipa præcox alba fine rubra, &c, vnius coloris.* The early white or red Tulipa, &c. being of one colour.
2 *Tulipa præcox purpurea oris albis.* The early purple Tulipa with white edges, or the Prince. 3 *Tulipa præcox variegata.* The early stript Tulipa. 4 *Tulipa præcox rubra oris luteis.* The early red Tulipa with yellow edges, or the Duke.

and pointed, and of others broad and round, or bluntly pointed, as shall bee shewed in the end of the Chapter : I shall therefore onely expresse the colours, with the mixture or composure of them, and giue you withall the names of some of them, (for it is impossible I thinke to any man, to giue seuerall names to all varieties) as they are called by those that chiefly delight in them with vs.

Tulipa præcox Alba.	The early White Tulipa.
1 *Niuea tota interdum purpureis staminibus, vel saltem luteis, fundo puro haud luteo.*	1 The flower whereof is either pure snow white, with purple sometimes, or at least with yellow chiues, without any yellow bottome.
2 *Alba siue niuea fundo luteo.*	2 Or pure white with a yellow bottome.
3 *Albida.*	3 Or milk white that is not so pure white.
4 *Alba, venis cæruleis in dorso.*	4 White with blew veines on the outside.
5 *Alba purpureis oris.* ⎧*Harum flores vel*	5 White with purple edges. ⎧Some of the elea‐
6 *Alba carneis oris.* ⎨*constantes, vel*	6 White with blush edges. ⎨biding constant, & others sprea‐
7 *Alba sanguineis oris.* ⎩*dispergentes.*	7 White with red edges. ⎩ding or running.
8 *Alba oris magnu carneis, & venis intro respicientibus.*	8 White with great blush edges, and some strakes running from the edge inward.
9 *Alba extra, carnei vero coloris intus, oras habens carneas saturatiores.*	9 White without, and somewhat blush within, with edges of a deeper blush.
10 *Albida, oris rubris, vel oris purpureis.*	10 Whitish, or pale white with red or purple edges.
11 *Albida purpurascentibus maculis extra, intus vero carnei vinacissimi.*	11 Whitish without, with some purplish veins & spots, & of a liuely blush within.
12 *Alba, purpureis maculis aspersa extra, intus vero alba purpurantibus oris.*	12 White without, spotted with small purple spots, and white within with purple edges.
13 *Dux Alba, i. e. coccineis & albis variata flammis, à medio ad oras intercursantibus.*	13 A white Duke, that is, parted with white & crimson flames, from the middle of each leafe to the edge.
14 *Princessa, i.e. argentei coloris maculis purpurascentibus.*	14 The Princesse, that is, a siluer colour spotted with fine deepe blush spots.
15 *Regina pulcherrima, albis & sanguineis aspersa radijs & punctis.*	15 The Queen, that is, a fine white sprinkled with bloud red spots, and greater strakes.

Tulipa præcox purpurea.	The early purple Tulipa.
1 *Purpurea satura rubescens, vel violacea.*	1 A deep reddish purple, or more violet.
2 *Purpurea pallida, Columbina dicta.*	2 A pale purple, called a Doue colour.
3 *Persici coloris saturi.*	3 A deep Peach colour.
4 *Persici coloris pallidioris.*	4 A paler Peach colour.
5 *Pæonia floris coloris.*	5 A Peony flower colour.
6 *Rosea.*	6 A Rose colour.
7 *Chermesina peramæna.*	7 A Crimson very bright.
8 *Chermesina parum striata.*	8 A Crimson stript with a little white.

9 *Princeps, i.e. purpurea saturatior vel dilutior, oris albis magnis vel paruis, fundo luteo, vel albo orbe, quæ multum variatur, & colore, & oris, ita vt purpurea elegans oris magnis albis, dicta est, Princeps excellens, &*

10 *Princeps Columbina, purpurea dilutior.*

11 *Purpurea Chermesina, rubicandioris coloris, albidis vel albis oris.*

12 *Purpurea, vel obsoleta albidis oris Princeps Brancion.*

13 *Purpurea diluta, oris dilutioris purpurei coloris.*

14 *Purpurea in exterioribus, carnei vero ad medium intus, oris albis, fundo luteo.*

15 *Purpurea albo plumata extra, oris albis, purpurascens intus, fundo luteo, vel orbe albo.*

16 *Alia, minus elegans plumata, minoribusq́, oris albidis.*

9 A Prince or Bracklar, that is, a deepe or pale purple, with white edges, greater or smaller, and a yellow bottome, or circled with white, which varieth much, both in the purple & edges, so that a faire deep purple, with great white edges, is called, The best or chiefe Prince, and

10 A paler purple with white edges, called a Doue coloured Prince.

11 A Crimson Prince or Bracklar.

12 A Brancion Prince, or purple Brancion.

13 A purple with more pale purple edges.

14 Purple without, and blush halfe way within, with white edges, and a yellow bottome.

15 Purple feathered with white on the out side, with white edges, and pale purple within, the ground being a little yellow, or circled with white.

16 Another very neere vnto it, but not so fairely feathered, being more obscure, and the edges not so great or whitish.

Tulipa præcox rubra.

The early red Tulipa.

1 *Rubra vulgaris fundo luteo, & aliquando nigro.*

2 *Rubra satura oris luteis paruis, dicta Roan.*

3 *Baro, i.e. rubra magis intensa, oris luteis paruis.*

4 *Dux maior & minor, i.e. rubra magis aut minus elegans satura, oris luteis maximis vel minoribus, & fundo luteo magno. Alia alijs est magis amœna, in alijs etiam fundo nigro vel obscuro viridi.*

5 *Ducissa, i.e. Duci similis, at plus lutei quàm rubri, oris magnis luteis, & rubore magis aut minus intus in gyrum acto, fundo item luteo magno.*

6 *Testamentum Brancion, i.e. rubra sanguinea satura, aut minus rubra, oris pallidis, magnis vel paruis: alia alijs magis aut minus elegans diuersimodo.*

1 An ordinary red, with a yellow, & sometimes a blacke bottome.

2 A deep red, with a small edge of yellow, called a Roane.

3 A Baron, that is, a faire red with a small yellow edge.

4 A Duke, a greater and a lesser, that is, a more or lesse faire deep red, with greater or lesser yellow edges, and a great yellow bottome. Some of this sort are much more or lesse faire then others, some also haue a blacke or darke greene bottome.

5 A Dutchesse, that is like vnto the Duke, but more yellow then red, with greater yellow edges, and the red more or lesse circling the middle of the flower on the inside, with a large yellow bottome.

6 A Testament Brancion, or a Brancion Duke,

E

7 *Flambans, ex rubore & flauedine radiata,
vel striata fundo luteo.*

8 *Mali Aurantij coloris, ex rubore, & flauedine integrè, non separatim mixta, oris
luteis paruis, vel absq, oris.*

9 *Minij, siue Cinabaris coloris, i.e. ex purpurea, rubedine, & flauedine radiata, vnguibus luteis, & aliquando oris.*

10 *Rex Tuliparum, i. e. ex sanguineo & aureo
radiatim mixta, à flammea diuersa, fundo
luteo, orbe rubro.*

11 *Tunica Morionis, i.e. ex rubore & aureo
separatim diuisa.*

Duke, that is, a faire deepe red, or lesse
red, with a pale yellow or butter coloured edge, some larger others smaller: and some more pleasing then others,
in a very variable manner.

7 A Flambant, differing from the Dutchesse; for this hath no such great yellow edge, but streaks of yellow through
the leafe vnto the very edge.

8 An Orenge colour, that is, a reddish yellow, or a red and yellow equally mixed,
with small yellow edges, and sometimes without.

9 A Vermillion, that is, a purplish red,
streamed with yellow, the bottome
yellow, and sometimes the edges.

10 The Kings flower, that is, a crimson or
bloud red, streamed with a gold yellow, differing from the Flambant, the
bottome yellow, circled with red.

11 A Fooles coate, parted with red and
yellow guardes.

Tulipa præcox lutea.

1 *Lutea siue flaua.*

2 *Pallida lutea siue straminea.*

3 *Aurea, oris rubicundis.*

4 *Straminea, oris rubris.*

5 *Aurea, rubore perfusa extra.*

6 *Aurea, vel magis pallida, rubore in gyrum
acta simillima Ducissæ, nisi minus rube-
dinis habet.*

7 *Aurea, extremitatibus rubris, dici potest,
Morionis Pileus præcox.*

The early yellow Tulipa.

1 A faire gold yellow without mixture.

2 A strawe colour.

3 A faire yellow with reddish edges.

4 A strawe colour, with red edges.

5 A faire yellow, reddish on the out side
onely.

6 A gold or paler yellow, circled on the
inside a little with red, very like the
Dutchesse, but that it hath lesse red
therein.

7 A gold yellow with red toppes, and may
be called, The early Fooles Cap.

Tulipa de Caffa. The Tulipa of Caffa.

There is another fort or kinde of early Tulipa, differing from the former, whofe pale greene leaues being as broad and large as they, and fometimes crumpled or wa-ued at the edges, in fome haue the edges onely of the faid leaues for a good breadth, of a whitifh or whitifh yellow colour, and in others, the leaues are lifted or parted with whitifh yellow and greene : the ftalke rifeth not vp fo high as the former, and beareth a flower at the toppe like vnto the former, in fome of a reddifh yellow colour, with a ruffet coloured ground or bottome, and in others, of other feuerall colours : the feede and roote is fo like vnto others of this kinde, that they cannot be diftinguifhed.

There is (as I doe heare) of this kinde, both *Pracoces*, and *Serotina*, early flowring, and late flowring, whereof although wee haue not fo exact knowledge, as of the reft, yet I thought good to fpeake fo much, as I could hitherto vnderftand of them, and giue others leaue (if I doe not) hereafter to amplifie it.

Tulipa Bolonienfis, fiue Bombycina flore rubro major. The greater red Bolonia Tulipa.

There are likewife other kindes of early Tulipas to bee fpoken of, and firft of the red Bolonia Tulipa ; the roote whereof is plainly difcerned, to be differing from all o-thers : for that it is longer, and not hauing fo plaine an eminence at the bottome there-of, as the former and later Tulipas, but more efpecially becaufe the toppe is plenti-fully ftored with a yellowifh filke-like woollineffe : the outfide likewife or skinne is of a brighter or paler red, not fo eafie to be pilled away, and runneth vnder ground both downeright and fidewife (efpecially in the Countrey ground and ayre, where it will encreafe aboundantly, but not either in our London ayre, or foreʼt grounds) fomewhat like vnto the yellow Bolonia Tulipa next following. It fhooteth out of the ground with broad and long leaues, like the former ; but neither fo broad, nor of fo white or mealy a greene colour as the former, but more darke then the late flowring Tulipa, fo that this may bee eafily difcerned by his leafe from any other Tulipa aboue the ground, by one that is skilfull. It beareth likewife three or foure leaues vpon the ftalke, like the former, and a flower alfo at the toppe of the fame fafhion, but that the leaues hereof are alwayes long, and fomewhat narrow, hauing a large blacke bot-tome, made like vnto a cheuerne, the point whereof rifeth vp vnto the middle of the leafe, higher then any other Tulipa : the flower is of a pale red colour, nothing fo liuely as in the early or late red Tulipas, yet fweeter for the moft part then any of them, and neereft vnto the yellow Bolonia Tulipa, which is much about the fame fent.

Tulipa pumilio rubra, fiue Bergomenfis rubra media & minor. The dwarfe red Bergomo Tulipa, a bigger and a leffer.

There are two other forts hereof, and becaufe they were found about Bergomo, do carry that name, the one bigger or leffer then another, yet neither fo great as the for-mer, hauing very little other difference to bee obferued in them, then that they are fmaller in all parts of them.

Tulipa Bolonienfis flore luteo. The yellow Bolonia Tulipa.

The roote of this Tulipa may likewife bee knowne from the former red (or any o-ther Tulipa) in that it feldome commeth to bee fo bigge, and is not fo woolly at the toppe, and the skinne or outfide is fomewhat paler, harder, and fharper pointed : but the bottome is like the former red, and not fo eminent as the early or late Tulipas. This beareth much longer and narrower leaues then any (except the Perfian & dwarfe yellow Tulipas) and of a whitifh greene colour : it beareth fometimes but one flower on a ftalke, and fometimes two or three wholly yellow, but fmaller, & more open then the other kinds, and (as I faid) fmelleth fweete, the head for feede is fmaller then in o-thers, and hath not that crowne at the head thereof, yet the feed is like, but fmaller.

Tulipa

Tulipa Narbonensis, siue Monspeliensis vel pumilio. The French or dwarfe yellow Tulipa.

This Tulipa is very like vnto the yellow Bolonia Tulipa, both in roote, leafe, and flower, as also in the colour thereof, being yellow: the onely difference is, that it is in all things lesser and lower, and is not so apt to beare, nor so plentifull to encrease by the roote.

Tulipa Italica maior & minor. The Italian Tulipa the greater and the lesser.

Both these kindes of Tulipas doe so neere resemble the last kinde, that I might almost say they were the same, but that some difference which I saw in them, maketh mee set them apart; and consisteth in these things, the stalkes of neither of both these rise so high, as of the first yellow Bolonia Tulipa: the leaues of both sorts are writhed in and out at the edges, or made like a waue of the sea, lying neerer the ground, and the flower being yellow within, is brownish or reddish on the backe, in the middle of the three outer leaues the edges appearing yellow. Both these kindes doe differ one from the other in nothing, but in that one is bigger, and the other smaller then the other which I saw with Iohn Tradescante, my very good friend often remembred.

Tulipa Lusitanica, siue pumilio versicolor. The dwarfe stript Tulipa.

This dwarfe Tulipa is also of the same kindred with the three last described; for there is no other difference in this from them, then that the flower hath some red veins running in the leaues thereof.

There are two other sorts of dwarfe Tulipas with white flowers, whereof Lobel hath made mention in the Appendix to his *Aduersaria*; the one whereof is the same that Clusius setteth forth, vnder the title of *Pumilio altera*: but because I haue not seen either of them both, I speake no further of them.

Tulipa pumilio alba. The white dwarfe Tulipa.

But that white flower that Iohn Tradescante shewed me, and as hee saith, was deliuered him for a white Pumilio, had a stalke longer then they set out theirs to haue, and the flower also larger, but yet had narrower leaues then other sorts of white Tulipas haue.

Tulipa Bicolor. The small party coloured Tulipa.

Vnto these kindes, I may well adde this kinde of Tulipa also, which was sent out of Italy, whose leaues are small, long, and narrow, and of a darke greene colour, somewhat like vnto the leaues of an Hyacinth: the flower is small also, consisting of sixe leaues, as all other Tulipas doe, three whereof are wholly of a red colour, and the other three wholly of a yellow.

Tulipa Persica. The Persian Tulipa.

This rare Tulipa, wherewith we haue beene but lately acquainted, doth most fitly deserue to be described in this place, because it doth so neerely participate with the Bolonia and Italian Tulipas, in roote, leafe, and flower: the roote hereof is small, couered with a thicke hard blackish shell or skinne, with a yellowish woollinesse both at the toppe, and vnder the shell. It riseth out of the ground at the first, with one very long and small round leafe, which when it is three or foure inches high, doth open it selfe, and shew forth another small leafe (as long almost as the former) breaking out of the one side thereat, and after it a third, and sometimes a fourth, and a fift; but each shorter then other, which afterwards be of the breadth of the dwarfe yellow Tulipa, or somewhat broader, but much longer then any other, and abiding more hollow, and of the colour of the early Tulipas on the inside: the stalke riseth vp a foot and a halfe

high

1 *Tulipa Bombycina flore rubro.* The red Bolonia Tulipa. 2 *Tulipa Bolonienfis flore luteo.* The yellow Bolonia Tulipa. 3 *Tulipa pumilio intra se lutea.* Fine red or yellow dwarfe Tulipa. 4 *Folium Tulipæ de Caffa per totum striatum.* The leafe of the Tulipa of Caffa striped throughout the whole leafe. 5 *Folium Tulipæ de Caffa per oras striatum.* The leafe of the Tulipa of Caffa striped at the edges onely. 6 *Tulipa Persica.* The Persian Tulipa. 7 *Tulipa Cretica.* The Tulipa of Candie. 8 *Tulipa Armeniaca.* The Tulipa of Armenia.

E 3

high fometimes, bearing one flower thereon, compofed of fixe long and pointed leaues of the forme of other fmall Tulipas, and not fhewing much bigger then the yellow Italian Tulipa, and is wholly white, both infide and outfide of all the leaues, except the three outtermoft, which haue on the backe of them, from the middle toward the edges, a fhew of a brownifh blufh, or pale red colour, yet deeper in the midft, and the edges remaining wholly white : the bottomes of all thefe leaues are of a darke or dun tawnie colour, and the chiues and tippes of a darkifh purple or tawnie alfo. This doth beare feed but feldome in our Country, that euer I could vnderftand, but when it doth, it is fmall like vnto the Bolonia or dwarfe yellow Tulipas, being not fo plentifull alfo in parting, or fetting of by the root as they, and neuer groweth nor abideth fo great as it is brought vnto vs, and feldome likewife flowreth after the firft yeare : for the rootes for the moft part with euery one grow leffe and leffe, decaying euery yeare, and fo perifh for the moft part by reafon of the frofts and cold, and yet they haue been fet deepe to defend them, although of their owne nature they will runne downe deep into the ground.

Tulipa Byzantina duobus floribus Cluſij. The fmall Tulipa of Conftantinople.

The fmall Tulipa of Conftantinople, beareth for the moft part but two leaues on the ftalke, which are faire and broad, almoft like vnto the Candy Tulipa, next hereunto to be defcribed : the ftalke it felfe rifeth not aboue a foote high, bearing fometimes but one flower, but moft commonly two thereon, one below another, and are no bigger then the flowers of the yellow Bolonia Tulipa, but differing in colour ; for this is on the outfide of a purplifh colour, mixed with white and greene, and on the infide of a faire blufh colour, the bottome and chiues being yellow, and the tippes or pendents blackifh : the roote is very like the yellow Bolonia Tulipa.

Tulipa Cretica. The Tulipa of Candie.

This Tulipa is of later knowledge with vs then the Perfian, but doth more hardly thriue, in regard of our cold climate ; the defcription whereof, for fo much as wee haue knowledge, by the fight of the roote and leafe, and relation from others of the flower, (for I haue not yet heard that it hath very often flowred in our Country) is as followeth. It beareth faire broad leaues, refembling the leaues of a Lilly, of a greenifh colour, and not very whitifh : the ftalke beareth thereon one flower, larger and more open then many other, which is eyther wholly white, or of a deepe red colour, or elfe is variably mixed, white with a fine reddifh purple, the bottomes being yellow, with purplifh chiues tipt with blackifh pendents : the roote is fmall, and fomewhat like the dwarfe yellow Tulipa, but fomewhat bigger.

Tulipa Armeniaca. The Tulipa of Armenia.

This fmall Tulipa is much differing from all the former (except the fmall or dwarfe white Tulipas remembred by Lobel and Clufius, as is before fet downe) in that it beareth three or foure fmall, long, and fomewhat narrow greene leaues, altogether at one ioynt or place ; the ftalke being not high, and naked or without leaues from them to the toppe, where it beareth one fmall flower like vnto an ordinary red Tulipa, but fomewhat more yellow, tending to an Orenge colour with a blacke bottome : the roote is not much bigger then the ordinary yellow Bolonia Tulipa, before fet downe.
And thefe are the forts of this firft *Clafsu* of early Tulipas.

Tulipa media. The meaner or middle flowring Tulipa.

For any other, or further defcription of this kinde of Tulipa, it fhall not neede, hauing giuen it fufficiently in the former early Tulipa, the maine difference confifting firft in the time of flowring, which is about a moneth after the early Tulipas, yet fome more fome leffe : for euen in the *Praecoces*, or early ones, fome flower a little earlier, and later then others, and then in the colours of the flowers ; for wee haue obferued many

colours,

colo rs, and mixtures, or varieties of colours in the *Medias*, which we could neuer fee in the *Præcoces*, and so also some in the *Præcoces*, which are not in the *Medias*: yet there is farre greater varieties of mixture of colours in these *Medias*, then hath been obserued in all the *Præcoces*, (although Clusius saith otherwise) eyther by my selfe, or by any other that I haue conuersed with about this matter, and all this hath happened by the sowing of the seede, as I said before. I will therefore in this place not trouble you with any further circumstance, then to distinguish them, as I haue done in the former early Tulipas, into their foure primary colours, and vnder them, giue you their seuerall varieties and names, for so much as hath come to my knowledge; not doubting, but that many that haue trauelled in the sowing of the seed of Tulipas many yeares, may obserue each of them to haue some variety that others haue not: and therefore I thinke no one man can come to the knowledge of all particular distinctions.

Tulipa media alba.

1 *Niuea, fundo albo vel luteo.*
2 *Argentea, quasi alba cineracea fundo lutescente, purpureis staminibus.*
3 *Margaritina alba, carneo dilutissima.*
4 *Alba, fundo cæruleo vel nigro.*
5 *Albida.*
6 *Alba, oris rubris.*
7 *Alba, purpureis oris.* ⎫ *Hæc tria genera in aliquibus constanter tenent oras, in a'ys dispergunt.*
8 *Alba, oris coccineis.* ⎭
9 *Albida primum, deinde albidior, oris purpureis, & venis intrò respicientibus, dicta nobis Hackquenay.*
10 *Alba, sanguineo colore variata, fundo vel albissimo, vel alio.*
11 *Alba, radiatim disposita flammis, & maculis coccineis.*
12 *Alba, purpurea rubedine plumata, diuersarum specierum, quæ cum superiore, vel albo, vel luteo, vel paruo cæruleo constant fundo, quæ constanter tenent punctatos colores, & non dispergunt, sed post trium aut quatuor dierum spatium pulchriores apparent.*
13 *Panni argentei coloris, i.e. alba, plumata, punctata, striata, vel diuersimodè variata, rubedine dilutiore, vel saturatiore purpurea, interius vel exterius, vel vtrinq; diuersarum specierum.*
14 *Tunica morionis alba varia, i.e. ex albo & purpureo striata diuersimodè, fundo albo vel alio.*
15 *Holias alba vel albida, absq; fundo, vel fundo purpureocæruleo, vel cæruleo albo circundato, diuersè signata, vel variata intus ad medietatem foliorum, sursum in orbem vt plurimum, vel ad oras pertingens amplas & albas. Hæ species tantoperè multiplicantur, vt vix sint explicabiles.*

The white meane flowring Tulipa.

1 A snow white, with a white or yellow bottome.
2 A siluer colour, that is, a very pale or whitish ashe colour, with a yellowish bottome and purple chiues.
3 A Pearle colour, that is, white, with a wash or shew of blush.
4 A white, with a blew or black bottome.
5 A Creame colour.
6 A white, with red edges. ⎫ *These three forts doe hold their edges constant in some, but will spread in others.*
7 A white, with purple edges. ⎬
8 A white, with crimson edges. ⎭
9 A pale or whitish yellow, which after a few dayes groweth more white, with purplish red edges, and some streakes running inward from the edge, which we call an Hackney.
10 A white mixed with a bloud red very variably, and with a pure white, or other coloured bottome.
11 A white, streamed with crimson flames, and spots through the whole flower.
12 A white, speckled with a reddish purple, more or lesse, of diuers sorts, with white, yellow, or blew bottomes, all which doe hold their markes constant, and doe not spread their colours, but shew fairer after they haue stood blown three or foure dayes.
13 A cloth of siluer of diuers sorts, that is, a white spotted, striped, or otherwise marked with red or purple, in some paler, in some deeper, either on the inside, or on the outside, or on both.
14 A white Fooles coate of diuers sorts, that is, purple or pale crimson, and white, as it were empaled together, eyther with a white ground or other, whereof there is great variety.
15 A white Holias, that is, a faire white, or paler white, eyther without a bottome, or with a blewish purple bottome, or blew and white circling the bottome,

and

and from the middle vpwards, speckled and straked on the inside for the most part, with bloud red or purplish spots and lines vnto the very edges, which abide large and white. Of this kinde there are found very great varieties, not to be expressed.

Tanta est huius varietas, vel multitudine, vel striarum paucitate & distinctione, vel fundu variantibus, vt ad taedium esset perscribere.

Of this sort there is so much variety, some being larger or fairer marked then others, their bottomes also varying, that it is almost impossible to expresse them.

Talipa media purpurea.	The meane flowring purple Tulipa.
1 *Purpurea satura.*	1 A faire deep purple.
2 *Purpurea dilutior, diuersarum specierum, quarum Rosea vna, Carnea sit altera.*	2 A paler purple, of many sorts, whereof a Rose colour is one, a Blush another.
3 *Persici coloris, duarum aut trium specierum.*	3 A Peach colour of two or three sorts.
4 *Chermesina, obscura, aut pallida.*	4 A Crimson, deepe, or pale.
5 *Stamela, intensior aut remissior.*	5 A Stamell, darke or light.
6 *Xerampelina.*	6 A Murrey.
7 *Purpurea, striata.*	7 A purple, stript and spotted.
8 *Persici saturi, vel diluti coloris, vndulata, vel radiata.*	8 A Peach colour, higher or paler, waued or stript.
9 *Columbina, oris & radijs albis.*	9 A Doue colour, edged and straked with white.
10 *Purpurea rubra, oris albis, similis Praecoci, dicta Princeps.*	10 A faire red purple, with white edges, like vnto the early Tulipa, called a Prince
11 *Chermesina, vel Heluola, lineis albis in medio, & versus oras, fundo caeruleo, vel albo, itemq, albo orbe.*	11 A faire Crimson, or Claret wine colour, with white lines both in the middle, and towards the edges, most haue a blew bottome, yet some are white, or circled with white.
12 *Purpurea remissior, aut intensior, oris albis, paruis aut magnis, vt in Principe praecoci, fundo vel caeruleo orbe albo, vel albo orbe caeruleo amplo.*	12 A light or deepe purple, with white edges, greater or smaller, like the early Prince, the bottomes eyther blew circled with white, or white circled with a large blew.
13 *Holias Heluola, sanguineis guttis intus à medio sursum in orbem, fundo caeruleo.*	13 A purple Holias, the colour of a pale Claret wine, marked and spotted with bloud red spots, round about the middle of each leafe vpward on the inside onely, the bottome being blew.
14 *Tunica Morionis purpurea rubra satura, albido striata, quam in alba saturatior, fundo ex caeruleo & albo.*	14 A Crimson Fooles Coate, a darke crimson, and pale white empaled together, differing from the white Fooles Coate, the bottome blew and white.
15 *Purpurea rubra satura vel diluta, albo vel albedine, punctata vel striata diuersimode, dicta Cariophyllata.*	15 A deeper or paler reddish purple, spotted or striped with a paler or purer white, of diuers sorts, called the Gilloflower Tulipa.

Tulipa

1 *Tulipa rubra & lutea varia* The Fooles Coate red and yellow. 2 *Tulipa Holeas alba abſ̄; fundo.* The white Holeas without a bottome. 3 *Tulipa argentea, vel punctata, &c.* The cloth of ſilver, or other ſpotted Tulipa. 4 *Tulipa alba flammis coccineis.* The white Fooles Coate. 5 *Tulipa Holeas alba, &c. fundo purpureo, &c.* A white Holeas, &c. with a purple bottome, &c. 6 *Tulipa rubra & lutea flammea, &c.* A red and yellow flamed Tulipa, &c. 7 *Tulipa alba striata & punctata.* A white ſtriped and ſpotted Tulipa. 8 *Tulipa altera variata, &c.* Another variable Tulipa,

Tulipa media rubra.	The meane flowring red Tulipa.
1 Rubra communis, fundo luteo, vel nigro.	1 A faire red which is ordinary, with a yellow or blacke bottome.
2 Mali Aurantij coloris.	2 A deepe Orenge colour.
3 Cinabaris coloris.	3 A Vermillion.
4 Lateritij coloris.	4 A pale red, or Bricke colour.
5 Rubra, luteo aspersa.	5 A Gingeline colour.
6 Rubra, oris luteis.	6 A red with small yellow edges.
7 Testamentum Brancion rubra satura, oris pallidis, diuersarum specierum, rubore variantium, & orarum amplitudine.	7 A Testament Brancion of diuers sorts, differing both in the deepnesse of the red, and largenesse of the pale coloured edges.
8 Cinabaris radiata, magis aut minus serotina.	8 A Vermillion flamed, flowring later or earlier.
9 Rubra purpurascens obsoleta, exterioribus folijs, perfusa luteo intus, oris pallidis luteis.	9 A dead purplish red without, and of a yellowish red within, with pale yellow edges.
10 Rubra purpurascens elegans extra, & intus lutescens, oris pallidis luteis, fundo luteo vel viridi.	10 A bright Crimson red on the outside, more yellowish on the inside, with pale yellow edges, and a bottome yellow or greene.
11 Rubra flambans coccinea, crebris maculis luteis absq; fundo.	11 A red Flambant, spotted thicke with yellow spots without any bottome.
12 Flambans elegantior rubra, i.e. radijs luteis intercursantibus ruborem	12 A more excellent red Flambant, with flames of yellow running through the red.
13 Flambans remissior vtroq; colore.	13 A pale coloured Flambant.
14 Panni aurei coloris.	14 A cloth of gold colour.
15 Tunica Morionis verior, seu Palto du Sot. optima, tenijs amplis amœnis & crebris, ex rubro & flauo separatim diuisis & excurrentibus, flos constans.	15 A true Fooles Coate, the best is a faire red & a faire yellow, parted into guards euery one apart, varied through euery leafe to the very edge, yet in most abiding constant.
16 Tunica Morionis altera, tenijs minoribus & minus frequentibus, magis aut minus alia alijs inconstans.	16 Another Fooles Coate, not so fairely marked, nor so much, some of these are more or lesse constant in their marks, & some more variable then others.
17 Tunica Morionis pallida, i. e. tenijs vel strijs frequentioribus in vtroq; colore pallidis, flos est constans & elegans.	17 A pale Fooles Coate, that is, with pale red, and pale yellow guardes or stripes very faire and constant.
18 Pileus Morionis, radijs luteis, in medio foliorum latis, per ruborem excurrentibus, fundo luteo, apicibus luteis, & tribus exterioribus folijs luteis oris rubris, vel absq; oris.	18 A Fooles Cappe, that is, with lists or stripes of yellow running through the middle of euery leafe of the red, broader at the bottome then aboue, the bottome being yellow, the three outer leaues being yellow with red edges, or without.

1 *Tulipa tricolor.* A Tulipa of three colours.   2 *Tulipa Macedonica, siue de Cassauaria.* The Tulipa of Caffa purple, with pale white stripes. 3 *Tulipa Heluola charmesina versicolor.* A pure Claret wine colour variable. 4 *Tulipa Caryophyllata Wilmeri.* Mr. Wilmers Gilloflower Tulipa 5 *Tulipa vermelina flammis albis* A Crimson with white flames. 6 *Tulipa Goliah.* A kind of Zwitter called Goliah. 7 *Tulipa le Zwisse.* A Tulipa called the Zwitter. 8 *Tulipa alba flammis coccineis.* Another white Flambant or Fooles Coate. 9 *Tulipa Cinnabarina albo flammata* The Vermillion Painted 10 *Tulipa plumata rubra & lutea.* The feathered Tulipa red and yellow.

19 *Le Suisse, tenijs radiata magnis ex rubore & pallore.*

20 *Altera dicta Goliah à floris magnitudine, tenijs radiata simillima le Suisse, nisi rubor & albedo sint elegantiores.*

21 *Holias rubra, i.e. sanguinea argenteis radijs, & guttu in orbem dispositis, præsertim interiùs, fundo viridi saturo.*

22 *Holias coccinea, rubra coccinea, albo radiata in orbem, circa medium foliorum interiùs, fundo albo.*

23 *Alia huic similis, fundo albo & caruleo.*

19 A Swisse, paned with a faire red and pale white or strawe colour.

20 A Goliah, so called of the bignesse of the flower, most like to the Swisse in the marks and guardes, but that the red and white is more liuely.

21 A red Holias. A bloud red stript with siluer white veines and spots, with a darke green bottome.

22 A Crimson red Holias, that is, a faire purplish red, spotted with white circlewise about the middle of the inner leaues, and a white bottome.

23 Another like thereunto, with a blew and white bottome.

Tulipa media lutea.

1 *Lutea, siue Aurea vulgaris.*
2 *Straminea.*
3 *Sulphurea.*

4 *Mali Aurantij pallidi coloris.*
5 *Lutea dilutè purpurea striata, aurei panni pallidi instar,*
6 *Pallidè lutea fuscedine adumbrata.*
7 *Flaua, oris rubris magnis, aut parnis.*

8 *Straminea oris rubris magnis intensis, vel parnis remissis.*
9 *Obscura & fuliginosa lutea, instar Folij decidui, ideoq; Folium mortuum appellatur.*
10 *Flaua, rubore perfusa, etiamque striata per totum, dorso coccineo, oris pallidis.*

11 *Pallidè lutea, perfusa & magis aut minus rubore striata, fundo vel luteo, vel viridi.*
12 *Testamentum Clusij, i.e. lutea pallida fuligine obfusca, exteriùs & interiùs ad oras vsq; pallidas, per totum vero floris medium, maculis interiùs aspersa instar omnium aliarum Holias, dorso obscuriore, fundo viridi.*

The meane flowring yellow Tulipa.

1 A faire gold yellow.
2 A Strawe colour.
3 A Brimstone colour pale yellowish greene.
4 A pale Orenge colour.
5 A pale cloth of gold colour.
6 A Custard colour a pale yellow shadowed ouer with a browne.
7 A gold yellow with red edges, greater or smaller.
8 A Strawe colour with red edges, deeper or paler, greater or smaller.
9 A sullen or smoakie yellow, like a dead leafe that is fallen, and therefore called, *Fuëlle mort.*
10 A yellow shadowed with red, and striped also through all the leaues, the backside of them being of a red crimson, and the edges pale.
11 A pale yellow, shadowed and striped with red, in some more in some lesse, the bottomes being either yellow or green.
12 A *Testamentum Clusij*, that is, a shadowed pale yellow, both within & without, spotted round about the middle on the inside, as all other Holias are, the backe of the leaues being more obscure or shadowed with pale yellow edges, and a greene bottome.

13 *Flam-*

13 *Flambans lutea, diuersimodè intus magis aut minus striata, vel in alijs extra maculata rubore, fundo vt plurimum nigro, vel in alijs luteo.*

14 *Flambans pallidior & elegantior.*

15 *Holias lutea intensior vel remissior diuersimodè, in orbem radiata interius, rubris maculis ad supremas vsÿ oras, aliquoties crebrè, alias parcè, fundo viridi, vel tanetto obscuro.*

16 *Holias straminea rubore striata & punctata, instar alba Holias.*

17 *Tunica Morionis lutea, alijs dicta Flammea, in qua color flauus magis & conspicuus rubore, diuersimodè radiata.*
Huc reddenda esset viridium Tuliparum classis, quæ diuersarum etiam constat specierum. Vna viridis intensior, cuius flos semper ferè semiclausus manet staminibus fimbriatis. Altera remissior, instar Psittaci pennarum viridium, luteo variata oris albis. Tertia adhuc dilutiori viriditate oris purpureis. Quarta, cujus folia æqualiter purpura diluta, & viriditate diuisa sunt. Quinta, folijs longissimis stellamodo expansis, ex rubore & viriditate coacta.

13 A yellow Flambant of diuers forts, that is, the whole flower more or leſſe ſtreamed or ſpotted on the inſide, and in ſome on the outſide with red, the bottome in moſt being blacke, yet in ſome yellow.

14 A paler yellow Flambant more beautifull.

15 A yellow Holias, paler or deeper yellow very variable, ſpotted on the inſide round about the middle, with red ſometimes plentifully, or elſe ſparingly, with a green or dark tawny bottome.

16 A ſtrawe coloured Holias, ſpotted and ſtreamed with red, as is to bee ſeene in the white Holias.

17 A yellow Fooles coate, of ſome called a flame colour, wherein the yellow is more then the red, diuerſly ſtreamed.
Vnto theſe may be added the greene Tulipa, which is alſo of diuers ſorts. One hauing a great flower of a deepe green colour, ſeldome opening it ſelfe, but abiding alwaies as it were halfe ſhut vp and cloſed, the chiues being as it were feathered. Another of a paler or yellowiſh green, paned with yellow, and is called, The Parret, &c. with white edges. A third of a more yellowiſh green, with red or purpliſh edges. A fourth, hath the leaues of the flower equally almoſt parted, with greene and a light purple colour, which abiding a long time in flower, groweth in time to be fairer marked: for at the firſt it doth not ſhew it ſelfe ſo plainely diuided. Some call this a greene Swiſſer. A fifth hath the longeſt leaues ſtanding like a ſtarre, conſiſting of greene and purple.

Tulipa Serotina. The late flowring Tulipa.

The late flowring Tulipa hath had his deſcription expreſſed in the precedent diſcourſe, ſo that I ſhall not neede to make a repetition of what hath already beene ſet downe. The greateſt matter of knowledge in this kinde is this, That it hath no ſuch plentifull variety of colours or mixtures in his flowers, as are in the two former ſorts, but is confined within theſe limits here expreſſed, as farre as hath come to our knowledge.

Tulipa Serotina.

Rosea intensior, aut remissior.
Rubra vulgaris, aut saturatior, & quasi nigricans, fundo luteo vel nigro, vel nigro orbe, aureo incluso, dicta Oculus Solis.
Lutea communis.
Lutea oris rubris.
Lutea guttis sanguineis, fundo nigro vel varie.

The late flowring Tulipa.

A Roſe colour deeper or paler.
An ordinary red, or elſe a deeper red like blacke bloud, with a blacke or yellow bottome, or blacke circled with yellow, called the Suns eye.
An ordinary yellow.
A yellow with red edges.
A yellow with red ſpots and veines, the bottome blacke or diſcoloured.

F

Theſe

There yet remaine many obferuations, concerning thefe beautifull flowers, fit to be knowne, which could not, without too much prolixity, be comprehended within the body of the defcription of them ; but are referued to bee intreated of a part by them-felues.

All forts of Tulipas beare vfually but one ftalke , and that without any branches : but fometimes nature is fo plentifull in bearing, that it hath two or three ftalkes , and fometimes two, or more branches out of one ftalke (euery ftalke or branch bearing one flower at the toppe) but this is but feldome feene ; and when it doth happen once, it is hardly feene againe in the fame roote, but is a great figne , that the roote that doth thus, being an old roote, will the fame yeare part into diuers rootes, whereof euery one, being of a reafonable greatneffe, will beare both his ftalke and flower the next yeare, agreeing with the mother plant in colour, as all the of-fets of Tulipas doe for the moft part : for although the young of-fets of fome doe vary from the maine roote, euen while it groweth with them, yet being feparated, it will bee of the fame colour with the mother plant.

There groweth oftentimes in the *Media* , and fometimes alfo in the *Præcoces* , but more feldome, a fmall bulbe or roote , hard aboue the ground, at the bottome of the ftalke, and betweene it and the lower leafe, which when the ftalke is dry , and it ripe, being put into the ground, will bring forth in time a flower like vnto the mother plant, from whence it was taken.

The flowers alfo of Tulipas confift moft commonly of fixe leaues , but fometimes they are feene to haue eight or tenne, or more leaues ; but vfually , thofe rootes beare but their ordinary number of fixe leaues the next yeare : the head for feede then , is for the moft part foure fquare , which at all other times is but three fquare , or when the flower wanteth a leafe or two, as fometimes alfo it doth, it then is flat , hauing but two fides.

The forme of the flower is alfo very variable ; for the leaues of fome Tulipas are all fharpe pointed, or all blunt and round pointed , and many haue the three outer leaues fharpe pointed, and the three inner round or pointed, and fome contrariwife, the three outermoft round pointed, and the three inner fharpe pointed. Againe, fome haue all the leaues of the flowers long and narrow , and fome haue them broader and fhorter. Some *Præcoces* alfo haue their flowers very large and great, equall vnto eyther the *Media*, or *Serotina* , which moft commonly are the largeft , and others haue them as fmall as the Bolonia Tulipa.

The bottomes of the leaues of the flowers are alfo variably diuerfified , and fo are both the chiues or threeds that ftand vp about the head, and the tips or pendents that are hanging loofe on the toppes of them ; and by the difference of the bottomes or chiues, many flowers are diftinguifhed, which elfe are very like in colour , and alike alfo marked.

For the fmell alfo there is fome diuerfity ; for that the flowers of fome are very fweete, of others nothing at all, and fome betweene both, of a fmall fent, but not offen-fiue : and yet fome I haue obferued haue had a ftrong ill fent ; but how to fhew you to diftinguifh them , more then by your owne fenfe , I cannot : for the feedes of fweete fmelling Tulipas doe not follow their mother plant, no more then they doe in the colour.

And laftly, take this , which is not the leaft obferuation , worth the noting, that I haue obferued in many : When they haue beene of one entire colour for diuers yeares, yet in fome yeare they haue altered very much, as if it had not beene the fame , *viz.* from a purple or ftamell , it hath beene variably either parted , or mixed , or ftriped with white, eyther in part , or through the whole flower, and fo in a red or yellow flower, that it hath had eyther red or yellow edges , or yellow or red fpots , lines, veines, or flames, running through the red or yellow colour , and fometimes it hath happened, that three leaues haue been equally parted in the middle with red and yel-low, the other three abiding of one colour, and in fome the red had fome yellow in it, and the yellow fome red fpots in it alfo ; whereof I haue obferued , that all fuch flow-ers , not hauing their originall in that manner, (for fome that haue fuch or the like markes from the beginning, that is, from the firft and fecond yeares flowring, are con-ftant, and doe not change) but as I faid, were of one colour at the firft, doe fhew the

weakneffe

weakneſſe and decay of the roote, and that this extraordinary beauty in the flower, is but as the brightneſſe of a light, vpon the very extinguiſhing thereof, and doth plainly declare, that it can doe his Maſter no more ſeruice, and therefore with this iollity doth bid him good night. I know there is a common opinion among many (and very confidently maintained) that a Tulipa with a white flower, hath changed to beare a red or yellow, and ſo of the red or yellow, and other colours, that they are likewiſe inconſtant, as though no flowers were certaine : but I could neuer either ſee or heare for certaine any ſuch alteration, nor any other variation, but what is formerly expreſſed. Let not therefore any iudicious be carried away with any ſuch idle conceit, but rather ſuſpeſt ſome deceit in their Gardeners or others, by taking vp one, and putting in another in the place, or elſe their owne miſtaking.

Now for the ſowing, planting, tranſplanting, choiſe, and ordering of Tulipas, which is not the leaſt of regard, concerning this ſubieſt in hand, but (as I think) would be willingly entertained ; What I haue by my beſt endeauours learned, by mine owne paines in almoſt forty yeares trauell, or from others informations, I am willing here to ſet downe; not doubting, but that ſome may adde what hath not come to my knowledge.

Firſt, in the ſowing of ſeedes of Tulipas, I haue not obſerued (whatſoeuer others haue written) nor could of certainty learne of others, that there doth ariſe from the ſeedes of *Præcoces* any *Medias* or *Serotine* Tulipas, (or but very ſeldome) nor am certainly aſſured of any : but that the ſeedes of all *Præcoces* (ſo they be not doubtfull, or of the laſt flowring ſorts) will bring *Præcoces* : And I am out of doubt, that I neuer ſaw, nor could learne, that euer the ſeede of the *Medias* or *Serotines* haue giuen *Præcoces* ; but *Medias* or *Serotines*, according to their naturall kinde. But if there ſhould bee any degeneration, I rather incline to thinke, that it ſooner commeth to paſſe *(à meliore ad peius*, for *facilis eſt deſcenſus*, that is) that *Præoces* may giue *Medias*, then that *Medias* or *Serotines* ſhould giue *Præcoces*.

For the choiſe of your ſeede to ſowe. Firſt, for the *Præcoces*, Cluſius ſaith, that the *Præcox Tulipa*, that beareth a white flower, is the beſt to giue the greateſt variety of colours. Some among vs haue reported, that they haue found great variety riſe from the ſeede of the red *Præcox*, which I can more hardly beleeue: but Cluſius his experience hath the greater probability, but eſpecially if it haue ſome mixture of red or purple in it. The purple I haue found to be the beſt, next thereunto is the purple with white edges, and ſo likewiſe the red with yellow edges, each of them will bring moſt of their owne colours. Then the choiſe of the beſt *Medias*, is to take thoſe colours that are light, rather white then yellow, and purple then red; yea white, not yellow, purple, not red : but theſe againe to be ſpotted is the beſt, and the more the better; but withall, or aboue all in theſe, reſpeſt the ground or bottome of the flower, (which in the *Præcox Tulipa* cannot, becauſe you ſhall ſeldome ſee any other ground in them but yellow) for if the flower be white, or whitiſh, ſpotted, or edged, and ſtraked, and the bottome blew or purple (ſuch as is found in the Holias, and in the Cloth of ſiluer, this is beyond all other the moſt excellent, and out of queſtion the choiſeſt of an hundred, to haue the greateſt and moſt pleaſant variety and rarity. And ſo in degree, the meaner in beauty you ſowe, the leſſer ſhall your pleaſure in rarities be. Beſtowe not your time in ſowing red or yellow Tulipa ſeede, or the diuers mixtures of them ; for they will (as I haue found by experience) ſeldome be worth your paines. The *Serotina*, or late flowring Tulipa, becauſe it is ſeldome ſeene, with any eſpeciall beautifull variety, you may eaſily your ſelues gheſſe that it can bring forth (euen as I haue alſo learned) no raritie, and little or no diuerſity at all.

The time and manner to ſowe theſe ſeedes is next to be conſidered. You may not ſowe them in the ſpring of the yeare, if you hope to haue any good of them ; but in the Autumne, or preſently after they be thorough ripe and dry : yet if you ſowe them not vntill the end of Oſtober, they will come forward neuer the worſe, but rather the better ; for it is often ſeene, that ouer early ſowing cauſeth them to ſpring out of the ground ouer early, ſo that if a ſharpe ſpring chance to follow, it may goe neere to ſpoile all, or the moſt of your ſeede. Wee vſually ſowe the ſame yeares ſeede, yet if you chance to keepe of your owne, or haue from others ſuch ſeed, as is two years old, they will thriue and doe well enough, eſpecially if they were ripe and well gathered :

You

You must not sowe them too thicke, for so doing hath lost many a pecke of good seede, as I can tell; for if the seede lye one vpon another, that it hath not roome vpon the sprouting, to enter and take roote in the earth, it perisheth by and by. Some vse to tread downe the ground, where they meane to sowe their seede, and hauing sowne them thereon, doe couer them ouer the thicknesse of a mans thumbe with fine sifted earth, and they thinke they doe well, and haue good reason for it: for considering the nature of the young Tulipa rootes, is to runne downe deeper into the ground, euery yeare more then other, they thinke to hinder their quicke descent by the fastnesse of the ground, that so they may encrease the better. This way may please some, but I doe not vse it, nor can finde the reason sufficient; for they doe not consider, that the stiffenesse of the earth, doth cause the rootes of the young Tulipas to bee long before they grow great, in that a stiffe ground doth more hinder the well thriuing of the rootes, then a loose doth, and although the rootes doe runne downe deeper in a loose earth, yet they may easily by transplanting be holpen, and raised vp high enough. I haue also seene some Tulipas not once remoued from their sowing to their flowring; but if you will not lose them, you must take them vp while their leafe or stalke is fresh, and not withered: for if you doe not follow the stalke downe to the roote, be it neuer so deepe, you will leaue them behinde you. The ground also must be respected; for the finer, softer, and richer the mould is, wherein you sowe your seede, the greater shall be your encrease and varietie: Sift it therefore from all stones and rubbish, and let it be either fat naturall ground of it selfe, or being muckt, that it bee thoroughly rotten: but some I know, to mend their ground, doe make such a mixture of grounds, that they marre it in the making.

After the seede is thus sowne, the first yeares springing bringeth forth leaues, little bigger then the ordinary grasse leaues; the second yeare bigger, and so by degrees euery yeare bigger then other. The leaues of the *Præcoces* while they are young, may be discerned from the *Medias* by this note, which I haue obserued. The leaues of them doe wholly stand vp aboue the ground, shewing the small footstalkes, whereby euerie leafe doth stand, but the leaues of the *Medias* or *Serotines* doe neuer wholly appeare out of the ground, but the lower part which is broad, abideth vnder the vpper face of the earth. Those Tulipas now growing to bee three yeares old, (yet some at the second, if the ground and ayre be correspondent) are to bee taken vp out of the ground, wherein yee shall finde they haue runne deepe, and to be anew planted, after they haue been a little dryed and cleansed, eyther in the same, or another ground againe, placing them reasonable neare one vnto another, according to their greatnesse, which being planted and couered ouer with earth againe, of about an inch or two thicknesse, may be left vntaken vp againe for two yeare longer, if you will, or else remoued euery yeare after, as you please; and thus by transplanting them in their due season (which is still in the end of Iuly, or beginning of August, or thereabouts) you shall according to your seede and soyle, haue some come to bearing, in the fifth yeare after the flowring, (and some haue had them in the fourth, but that hath beene but few, and none of the best, or in a rich ground) some in the sixth and seuenth, and some peraduenture, not vntill the eighth or tenth yeare: but still remember, that as your rootes growe greater, that in re-planting you giue them the more roome to be distant one from another, or else the one will hinder, if not rot the other.

The seede of the *Præcoces*, doe not thriue and come forward so fast as the *Medias* or *Serotines*, nor doe giue any of-sets in their running downe as the *Medias* doe, which vsually leaue a small roote at the head of the other that is runne downe euery yeare; and besides, are more tender, and require more care and attendance then the *Medias*, and therefore they are the more respected.

This is a generall and certaine rule in all Tulipas, that all the while they beare but one leafe, they will not beare flower, whether they bee seedlings, or the of-sets of elder rootes, or the rootes themselues, that haue heretofore borne flowers; but when they shew a second leafe, breaking out of the first, it is a certaine signe, that it will then beare a flower, vnlesse some casualty hinder it, as frost or raine, to nip or spoile the bud, or other vntimely accident befall it.

To set or plant your best and bearing Tulipas somewhat deeper then other rootes, I hold it the best way; for if the ground bee either cold, or lye too open to the cold

Northerne

Northerne ayre, they will be the better defended therein, and not suffer the frosts or cold to pierce them so soone: for the deepe frosts and snowes doe pinch the *Præcoces* chiefly, if they bee too neare the vppermost crust of the earth; and therefore many, with good successe, couer ouer their ground before Winter, with either fresh or old rotten dung, and that will maruelloufly preserue them. The like courfe you may hold with seedlings, to caufe them to come on the forwarder, so it bee after the first yeares sowing, and not till then.

To remoue Tulipas after they haue shot forth their fibres or small strings, which grow vnder the great round rootes, (that is, from September vntill they bee in flower) is very dangerous; for by remouing them when they haue taken fast hold in the ground, you doe both hinder them in the bearing out their flower, and besides, put them in hazzard to perish, at least to bee put backe from bearing for a while after, as oftentimes I haue proued by experience: But when they are now risen to flower, and so for any time after, you may safely take them vp if you will, and remoue them without danger, if you haue any good regard vnto them, vnlesse it be a young bearing roote, which you shall in so doing much hinder, becaufe it is yet tender, by reafon it now beareth his first flower. But all Tulipa roots when their stalke and leaues are dry, may most safely then be taken vp out of the ground, and be so kept (so that they lye in a dry, and not in a moist place) for sixe moneths, without any great harme: yea I haue knowne them that haue had them nine moneths out of the ground, and haue done reafonable well, but this you must vnderstand withall, that they haue not been young but elder rootes, and they haue been orderly taken vp and preferued. The dryer you keep a Tulipa roote the better, so as you let it not lye in the sunne or winde, which will pierce it and spoile it.

Thus Gentlewomen for your delights, (for these pleasures are the delights of leafure, which hath bred your loue & liking to them, and although you are herein predominant, yet cannot they be barred from your beloued, who I doubt not, wil share with you in the delight as much as is fit) haue I taken this paines, to set downe, and bring to your knowledge such rules of art, as my small skill hath enabled mee withall concerning this subiect, which of all other, seemed fittest in this manner to be enlarged, both for the varietie of matter, and excellency of beautie herein; and also that these rules set forth together in one place, might faue many repetitions in other places, so that for the planting and ordering of all other bulbous rootes, and the sowing the seedes of them, you may haue recourfe vnto these rules, *(tanquam ad normam & examen)* which may serue in generall for all other, little diuersitie of particulars needing exception.

The Place.

The greater Tulipas haue first beene fent vs from Constantinople, and other parts of Turkie, where it is said they grow naturally wilde in the Fields, Woods, and Mountaines; as Thracia, Macedonia, Pontus about the Euxine Sea, Cappadocia, Bithynia, and about Tripolis and Aleppo in Syria also: the lesser haue come from other feuerall places, as their names doe defcipher it out vnto vs; as Armenia, Persia, Candye, Portugall, Spaine, Italy, and France. They are all now made Denizens in our Gardens, where they yeeld vs more delight, and more encreafe for their proportion, by reafon of the culture, then they did vnto their owne naturals.

The Time.

Thefe doe flower some earlier, some later, for three whole moneths together at the least, therein adorning out a Garden most glorioufly, in that being but one kinde of flower, it is so full of variety, as no other (except the Daffodils, which yet are not comparable, in that they yeeld not that alluring pleafant variety) doe the like besides. Some of the *Præcoces* haue beene in flower with vs, (for I speake not of their owne naturall places, where the Winters are milder, and the Spring earlier then ours) in the moneth of Ianuary, when the Winter before hath beene milde, but many in February,

and

and all the *Precoces*, from the beginning to the end of March, if the yeare be kindly : at what time the *Medias* doe begin, and abide all Aprill, and part of May, when the *Serotines* flower and fade; but this, as I said, if the yeare be kindly, or else each kinde will be a moneth later. The seede is ripe in Iune and Iuly, according to their early or late flowring.

The Names.

There haue beene diuers opinions among our moderne Writers, by what name this plant was knowne to the ancient Authors. Some would haue it be *Cosmosandalos*, of the Ancient. Dodonæus referreth it to πυπὰν of Theophraftus, in his feuenth Booke and thirteenth Chapter : but thereof he is so briefe, that besides the bare name, wee cannot finde him to make any further relation of forme, or quality. And Bauhinus, vpon Matthiolus Commentaries of Dioscorides, and in his Pinax also, followeth his opinion. Camerarius in his Hortus Medicus is of opinion, it may be referred to the Helychryfuni of Crateua. Gefner, as I thinke, firft of all, and after him Lobel, Camerarius, Clufius and many others, referre it to the Satyrium of Dioscorides : and furely this opinion is the moft probable for many reasons. Firft, for that this plant doth grow very frequent in many places of Greece, and the leffer Afia, which were no doubt fufficiently knowne both to Theophraftus, and Dioscorides, and was accounted among bulbous rootes, although by fundry names. And fecondly, as Dioscorides fetteth forth his Satyrium, fo this moft commonly beareth three leaues vpon a ftalke (although fometimes with vs it hath foure or fiue) like vnto a Lilly, whereof fome are often feen to be both red, in the firft fpringing, and alfo vpon the decaying, efpecially in a dry time, and in a dry ground : the flower likewife of fome is white, and like a Lilly; the roote is round, and as white within as the white of an egge, couered with a browne coate, hauing a fweetifh, but not vnpleafant tafte, as any man without danger many try. This defcription doth fo liuely fet forth this plant, that I thinke wee fhall not needeto be any longer in doubt, where to finde Dioscorides his Satyrium Triphyllum, feeing wee haue fuch plenty growing with vs. And thirdly, there is no doubt, but that it hath the fame qualities, as you fhall hereafter heare further. And laftly, that plant likewife that beareth a red flower, may very well agree with his Erythronium; for the defcriptions in Dioscorides are both alike, as are their qualities, the greateft doubt may be in the feede, which yet may agree vnto Lin or Flaxe as fitly, or rather more then many other plants doe, in many of his comparifons, which yet wee receiue for currant. For the feede of Tulipas are flat, hard, and fhining as the feede of *Linum* or Flaxe, although of another colour, and bigger, as Dioscorides himfelfe fetteth it downe. But if there fhould be a miftaking in the writing of λιν for λεν, in the Greeke Text, as the flippe is both eafie and likely, it were then out of all queftion the fame : for the feede is very like vnto the feede of Lillies, as any man may eafily difcerne that know them, or will compare them. It is generally called by all the late Writers, *Tulipa*, which is deriued from the name *Tulpan*, whereby the Turkes of *Dalmatia* doe entitle their head Tyres, or Caps; and this flower being blowne, laide open, and inuerted, doth very well refemble them. We haue receiued the early kinde from Conftantinople, by the name of *Cafa lale*, and the other by the name of *Cauala lale*. Lobel and others doe call it *Lilio-narciffus*, becaufe it doth refemble a Lilly in the leafe, flower, and feede, and a Daffodill in the roote. We call it in Englifh the Turkes Cap, but moft vfually *Tulipa*, as moft other Chriftian Countries that delight therein doe. Dalefchampius calleth it Oulada.

The Vertues.

Dioscorides writeth, that his firft Satyrium is profitable for them that haue a convulfion in their necke, (which wee call a cricke in the necke) if it be drunke in harfh (which we call red)wine.

That the roots of Tulipas are nourifhing,there is no doubt,the pleafant, or at leaft the no vnpleafant tafte,may hereunto perfwade; for diuers haue had them fent by their friends from beyond Sea, and miftaking them to bee Onions, haue vfed them as Onions in their pottage or broth, and neuer found any caufe of miflike, or any fenfe of euill quality produced by them, but accounted them fweete Onions.

Further, I haue made tryall of them my felfe in this manner. I haue pre-ferued the rootes of thefe Tulipas in Sugar, as I haue done the rootes of E-ringus, Orchis, or any other fuch like, and haue found them to be almoft as pleafant as the Eringus rootes,being firme and found, fit to be prefented to the curious; but for force of Venereous quality,I cannot fay,either from my felfe, not hauing eaten many, or from any other, on whom I haue beftowed them: but furely, if there be any fpeciall propertie in the rootes of Orchis, or fome other tending to that purpofe, I thinke this may as well haue it as they. It fhould feeme, that Diofcorides doth attribute a great Venereous faculty to the feede, whereof I know not any hath made any efpeciall ex-periment with vs as yet.

CHAP. IX.

Narciffus. The Daffodill.

THere hath beene great confufion among many of our moderne Writers of plants, in not diftinguifhing the manifold varieties of Daffodils; for euery one almoft, without confideration of kinde or forme, or other fpeciall note, gi-ueth names fo diuerfly one from another, that if any one fhall receiue from feuerall places the Catalogues of their names (as I haue had many) as they fet them down,and compare the one Catalogue with the other,he fhall fcarce haue three names in a dozen to agree together,one calling that by one name,which another calleth by another,that very few can tell what they meane. And this their confufion, in not diftinguifhing the name of *Narciffus* from *Pfeudonarciffus*, is of all other in this kinde the greateft and groffeft errour. To auoide therefore that gulfe, whereof I complaine that fo manie haue bin endrenched; and to reduce the Daffodils into fuch a methodicall order,that euery one may know,to what *Clafsis* or forme any one doth appertaine, I will firft diuide them into two principall or primary kindes : that is, into *Narciffos*, true Daffo-dils, and *Pfeudonarciffos*, baftard Daffodils: which diftinction I hold to be moft necef-farie to be fet downe firft of all,that euery one may be named without confufion vnder his owne primary kind, and then to let the other parts of the fubdiuifion follow, as is proper to them, and fitteft to exprefle them. Now to caufe you to vnderftand the difference betweene a true Daffodill and a falfe, is this ; it confifteth onely in the flow-er, (when as in all other parts they cannot bee diftinguifhed) and chiefly in the middle cup or chalice; for that we doe in a manner onely account thofe to bee *Pfeudonarciffos*, baftard Daffodils,whofe middle cup is altogether as long,and fometime a little longer then the outter leaues that doe encompaffe it, fo that it feemeth rather like a trunke or long nofe, then a cup or chalice, fuch as almoft all the *Narcifsi*, or true Daffodils haue; I fay almoft, becaufe I know that fome of them haue their middle cup fo fmall, that we rather call it a crowne then a cup; and againe, fome of them haue them fo long, that they may feeme to be of the number of the *Pfeudonarcifsi*, or baftard Daffodils:but yet may eafily be knowne from them, in that, although the cup of fome of the true Daffodils be great,yet it is wider open at the brim or edge,and not fo long and narrow all alike as the baftard kindes are ; and this is the chiefe and onely way to know how to feuer thefe kindes, which rule holdeth certaine in all, except that kinde which is called *Narciffus Iuncifolius reflexo flore*, whofe cup is narrow, and as long as the leaues that turne vp againe.

Secondly,

Secondly, I will subdiuide each of these again apart by themselues, into foure sorts; and first the *Narciffos*, or true Daffodils into

Latifolios, broad leaded Daffodils.

Anguftifolios, narrow leafed Daffodils.

Iuncifolios, Rushe Daffodils, and

Marinos, Sea Daffodils.

These sorts againe doe comprehend vnder them some other diuisions, whereby they may the better be diftinguished, and yet still bee referred to one of thofe foure former forts: as

Monanthos, that is, Daffodils that beare but one flower, or two at the most vpon a ftalke, and

Polyanthos, those that beare many flowers together vpon a ftalke: as alfo

Simplici flore, thofe that beare fingle flowers, and

Multiplici flore, or *flore pleno*, that is, haue double flowers.

Vernales, thofe that flower in the Spring, and among them some that are earlier, and therefore called

Pracoces, early flowring Daffodils, and

Autumnales, thofe that flower in Autumne onely.

And laftly, with the *Pfeudonarciffos*, or baftard Daffodils, I will keepe the same order, to diftinguish them likewife into their foure feuerall forts; and as with the true Daffodils, fo with thefe falfe, defcribe vnder euery fort: first, thofe that beare fingle flowers, whether one or many vpon a ftalke; and then thofe that beare double flowers, one or many alfo. As for the diftinctions of *maior* and *minor*, greater and leffer, and of *maximus* and *minimus*, greateft and leaft, they doe not onely belong to thefe Daffodils; and therefore muft be vfed as occafion permitteth, but vnto all other fort of plants. To begin therefore, I thinke fitteft with that ftately Daffodill, which for his excellency carrieth the name of None fuch.

1. *Narciffus latifolius omnium maximus, amplo calice flauo, fiue Nompareille.*
The great None fuch Daffodill, or Incomparable Daffodill.

This *Narciffus Nompareille* hath three or foure long and broad leaues, of a grayish greene colour, among which rifeth vp a ftalke two foote high at the leaft, at the toppe whereof, out of a thinne skinnie huske, as all Daffodils haue, commeth forth one large fingle flower, and no more vfually, confifting of fixe very pale yellow large leaues, almoft round at the point, with a large cuppe in the middle, fomewhat yellower then the leaues, the bottome whereof next vnto the ftalke is narrow and round, rifing wider to the mouth, which is very large and open, and vneuenly cut in or indented about the edges. The cup doth very well refemble the chalice, that in former dayes with vs, and beyond the Seas is ftill vfed to hold the Sacramentall Wine, that is with a narrower bottome, and a wide mouth. After the flower is paft, fometimes there commeth (for it doth not often) a round greene head, and blacke round feede therein, like vnto other Daffodils, but greater. The roote is great, as other Daffodils that beare large flowers, and is couered ouer with a brownish coate or skinne. The flower hath little or no fent at all.

Flore geminato   This doth fometimes bring forth a flower with ten or twelue leaues, and a cup much larger, as if it would be two, euen as the flower feemeth.

2. *Narciffus omnium maximus flore & calice flauo.*
The great yellow Incomparable Daffodill.

This other kinde differeth neither in forme, nor bigneffe of leafe or flower from the former, but in the colour of the circling leaues of the flower, which are of the same yellow colour with the cup.

Flore geminato.   This doth fometimes degenerate and grow luxurious alfo, bringing forth two flowers vpon a ftalke, each diftinct from other, and fometimes two flowers thruft together, as if they were but one, although it be but feldome; for it is not a peculiar kinde that is conftant, yearly abiding in the fame forme.

3. *Narciffus*

3. Narciſſus maximus griſeus calice flauo. The gray Peerleſſe Daffodill.

This Peerleſſe Daffodill well deſerueth his place among theſe kindes, for that it doth much reſemble them, and peraduenture is but a difference raiſed from the ſeede of the former, it is ſo like in leafe and flower, but that the leaues ſeeme to be ſomewhat greater, and the ſixe outer leaues of the flower to be of a gliſtering whitiſh gray colour, and the cup yellow, as the former, but larger.

4. Narciſſus latifolius flauo flore amplo calice, ſiue Matteneſſe.
The leſſer yellow Nompareille, or the Lady Matteneſſes Daffodill.

The leaues of this Daffodill, are ſomewhat like vnto the leaues of the firſt kind, but not altogether ſo long or broad: the ſtalke likewiſe riſeth not vp fully ſo high, and beareth one flower like the former, but leſſer, and both the cuppe and the leaues are of one colour, that is, of a pale yellow, yet more yellow then in the former: the cup of this alſo is leſſer, and a little differing; for it is neither fully ſo ſmall in the bottome, nor ſo large at the edges, nor ſo crumpled at the brimmes, ſo that all theſe differences doe plainly ſhew it to be another kinde, quite from the former.

The Place.

The places of none of theſe are certainly knowne to vs where they grow naturally, but we haue them onely in our Gardens, and haue beene ſent, and procured from diuers places.

The Time.

They flower ſometimes in the end of March, but chiefly in Aprill.

The Names.

The firſt and ſecond haue been ſent vs by the name of *Narciſſe Nompare-ille,* as it is called in French; and in Latine, *Narciſſus omnium maximus amplo calice flauo,* and *Narciſſus Incomparabilis,* that is, the Incomparable Daffodill, or the greateſt Daffodill of all other, with a large yellow cuppe: but aſſu-redly, although this Daffodill doth exceed many other, both in length and bigneſſe, yet the great Spaniſh baſtard Daffodill, which ſhall be ſpoken of hereafter, is in my perſwaſion oftentimes a farre higher and larger flower; and therefore this name was giuen but relatiuely, we may call it in Engliſh, The great None ſuch Daffodill, or the Incomparable Daffodill, or the great Peerleſſe Daffodill, or the Nompareille Daffodill, which you will: for they all doe anſwer either the French or the Latine name; and becauſe this name *Nompareille* is growne currant by cuſtome, I know not well how to alter it. The third kinde may paſſe with the title giuen it, without con-troule. The laſt is very well knowne beyond the Seas, eſpecially in the Low Countries, and thoſe parts, by the Lady Matteneſſe Daffodill, becauſe Cluſius receiued it from her. We may call it in Engliſh, for the correſpon-dency with the former, The leſſer yellow Nompareille, or Peerleſſe Daffo-dill, or the Lady Matteneſſe Daffodill, which you will.

Narciſſus Indicus flore rubro, dictus Iacobæus.
The Indian Daffodill with a red flower.

This Indian Daffodill is ſo differing, both in forme, not hauing a cuppe, and in co-lour, being red, from the whole Family of the Daffodils (except the next that follow-eth, and the Autumne Daffodils) that ſome might iuſtly queſtion the fitneſſe of his place here. But becauſe as all the plants, whether bulbous or other, that come from the

the Indies, either East or West (although they differ very notably, from those that grow in these parts of the world) must in a generall suruey and muster be ranked euery one, as neere as the surueiours wit will direct him, vnder some other growing with vs, that is of neerest likenesse; Euen so vntill some other can direct his place more fitly, I shall require you to accept of him in this, with this description that followeth, which I must tell you also, is more by relation then knowledge, or sight of the plant it selfe. This Daffodill hath diuers broad leaues, somewhat like vnto the common or ordinary white Daffodill, of a grayish greene colour; from the sides whereof, as also from the middle of them, rise vp sometimes two stalkes together, but most vsually one after another (for very often it flowreth twice in a Summer) and often also but one stalke alone, which is of a faint reddish colour, about a foote high or more, at the toppe whereof, out of a deepe red skinne or huske, commeth forth one flower bending downewards, consisting of sixe long leaues without any cup in the middle, of an excellent red colour, tending to a crimson; three of these leaues that turne vpwards, are somewhat larger then those three that hang downewards, hauing sixe threads or chiues in the middle, tipt with yellow pendents, and a three forked stile longer then the rest, and turning vp the end thereof againe: the roote is round and bigge, of a brownish colour on the outside, and white within. This is set forth by Aldinus, Cardinall Farnesius his Physitian, that at Rome it rose vp with stalkes of flowers, before any leaues appeared.

The Place, Time, and Names.

This naturally groweth in the West Indies, from whence it was brought into Spaine, where it bore both in Iune and Iuly, and by the Indians in their tongue named AZCAL XOCHITL, and hath beene sent from Spaine, vnto diuers louers of plants, into seuerall parts of Christendome, but haue not thriued long in these transalpine colder Countries, so far as I can heare.

Narcissus Trapezunticus flore luteo praecocissimus.
The early Daffodill of Trebizond.

Because this Daffodill is so like in flower vnto the former, although differing in colour, I thought it the fittest place to ioyne it the next thereunto. This early Daffodill hath three or foure short very greene leaues, so like vnto the leaues of the Autumne Daffodill, that many may easily bee deceiued in mistaking one for another, the difference consisting chiefly in this, that the leaues of this are not so broad or so long, not rise vp in Autumne: in the midst of these leaues riseth vp a short green stalke, an handfull high, or not much higher vsually, (I speake of it as it hath often flowred with mee, whether the cause be the coldnesse of the time wherein it flowreth, or the nature of the plant, or of our climate, I am in some doubt; but I doe well remember, that the stalkes of some plants, that haue flowred later with me then the first, haue by the greater strength, and comfort of the Sunne, risen a good deale higher then the first) bearing at the top, out of a whitish thinne skinne striped with greene, one flower a little bending downewards, consisting of sixe leaues, laid open almost in the same manner with the former Indian Daffodill, whereof some doe a little turne vp their points againe, of a faire pale yellow colour, hauing sixe white chiues within it, tipt with yellow pendents, and a longer pointell: the roote is not very great, but blackish on the outside, so like vnto the Autumne Daffodill, but that it is yellow vnder the first or outermost coate, that one may easily mistake one for another.

The Place.

It was sent vs from Constantinople among other rootes, but as wee may ghesse by the name, it should come thither from Trapezunte or Trebizond.

The Time.

It flowreth sometimes in December, if the former part of the Winter haue

1 *Narcissus Nompareille.* The incomparable Daffodill. 2 *Narcissus Mattenese.* The lesser yellow Nompareille Daffodill. 3 *Narcissus Iacobæus flore rubro.* The red Indian Daffodill. 4 *Narcissus Trapezuntitus.* The early Daffodill of Trabesond. 5 *Narcissus Montanus albus apophysibus praeditus.* The white winged Daffodill. 6 *Narcissus Montanus, sive Nompareille totus albus.* The white Nompareille, or Peerlesse Daffodill. 7 *Narcissus albus oblongo calice.* The white Daffodill with a long cup.

haue been milde ; but most vsually about the end of Ianuary , or else in Februarie the beginning or the end.

The Names.

Wee doe vsually call it from the Turkish name, *Narcissus Trapezunticus,*
and some also call it *Narcissus vernus præcox* , as Clusius doth , in English,
The early Daffodill of Trebizond.

Narcissus Montanus albus apophysibus præditus.
The white Mountaine Daffodill with eares, or
The white winged Daffodill.

This Mountaine Daffodill riseth vp with three or foure broad leaues , somewhat
long , of a whitish greene colour , among which riseth vp a stalke a foote and a halfe
high, whereon standeth one large flower, and sometimes two, consisting of sixe white
leaues a peece, not very broad , and without any shew of yellownesse in them , three
whereof haue vsually each of them on the backe part, at the bottome vpon the one side
of them, and not on both , a little small white peece of a leafe like an eare , the other
three hauing none at all : the cup is almost as large , or not much lesse then the small
Nompareille, small at the bottome, and very large, open at the brimme, of a faire yellow colour , and sometimes the edges or brimmes of the cup will haue a deeper yellow colour about it , like as if it were discoloured with Saffron : the flower is verie
sweete , the roote is great and white , couered with a pale coate or skinne , not verie
blacke, and is not very apt to encrease, seldome giuing of-sets ; neither haue I euer gathered seede thereof, because it passeth away without bearing any with me.

Narcissus Montanus, siue Nompareille totus albus amplo calice.
The white Nompareille Daffodill.

This white Nompareille Daffodill, is in roote and leafe very like vnto the former
mountain or winged Daffodill, but that they are a little larger : the stalke from among
the leaues riseth vp not much higher then it, bearing at the top one large flower, composed of sixe long white leaues, each whereof is as it were folded halfe way together,
in the middle whereof standeth forth a large white cup, broader at the mouth or brims
then at the bottome , very like vnto the lesser Nompareille Daffodill before remembred, which hath caused it to be so entituled : the sent whereof is no lesse sweete then
the former.

The Place.

The naturall places of these Daffodils are not certainly knowne to vs ;
but by the names they carry , they should seeme to bee bred in the Mountaines.

The Time.

These flower not so early as many other kindes doe, but rather are to bee
accounted among the late flowring Daffodils ; for they shew not their
flowers vntill the beginning of May , or the latter end of Aprill , with the
soonest.

The Names.

The names set downe ouer the heads of either of them be such, whereby
they are knowne to vs : yet some doe call the first *Narcissus auriculatus,* that
is to say, The Daffodill with eares: and the other , *Narcissus Nompareille
totus albus,* that is to say, The white Nompareille, or Peerlesse Daffodill.

1. *Narcissus*

1. *Narciſſus albus oblongo calice luteo præcox minor.*
The ſmall early white Daffodill with a long cup.

The leaues of this early Daffodill are broad, very greene, and not whitiſh as others, three or foure ſtanding together, about a foote long or better, among which riſeth vp a greene ſtalke, not full ſo high as the leaues, bearing one flower at the toppe thereof of a reaſonable bigneſſe, but not ſo great as the later kindes that follow are, conſiſting of ſix whitiſh leaues, but not perfect white, hauing a ſhew of a Creame colour appearing in them; in the middle is a long round yellow cup, about halfe an inch long or better. The ſmell of this flower is reaſonable ſweete, the roote is of a reaſonable bigneſſe, yet leſſer then the rootes of the later kindes.

2. *Narciſſus pallidus oblongo calice flauo præcox.*
The early Strawe coloured Daffodill with a long cup.

The leaues of this Daffodill are as greene as the former, but much narrower; and the leaues of the flower are more enclining to yellow, but yet very pale, as if it were a light ſtrawe colour, and ſeeme to bée a little more narrow and pointed then the former: the cup of this, is as long and yellow as the precedent. The ſmell whereof is very like the former, yet neither of them being ſo ſweete as thoſe that follow.

3. *Narciſſus albus oblongo calice luteo ſerotinus maior.*
The great late flowring white Daffodill with a long cup.

This later flowring Daffodill hath his leaues ſomewhat narrow & long, of a grayiſh or whitiſh greene colour, among which the ſtalke riſeth vp a foote and a halfe high, bearing one flower at the toppe, made of ſix white leaues, hauing the cup in the middle thereof as long as the former, and of a deepe yellow: the edges of this cuppe are ſometimes plaine, and ſometimes a little crumpled; they are often alſo circled at the brimmes with a Saffron colour, and often alſo without it, the ſmell whereof is very pleaſant, and not heady: the roote hereof is reaſonable bigge, and couered ouer rather with a pale then blackiſh skinne. This flower doth ſometimes alter his forme into eight leaues, which being narrow and long, ſeeme like a white ſtarre, compaſſing a yellow trunke.

4. *Narciſſus totus pallidus oblongo calice ſerotinus minor.*
The late pale coloured Daffodill with a long cup.

There is another of this kinde, whoſe flower is wholly of a pale white, or yellowiſh colour, differing neither in leafe nor roote from the former.

5. *Narciſſus pallidus oblongo calice flauo ſerotinus.*
The Strawe coloured late flowring Daffodill with a long yellow cup.

The chiefe difference of this Daffodill from the former, conſiſteth in the colour of the top of the flower, which is of a more yellow colour, and a little larger then the former, and the brimmes or edges of the cup of a deeper yellow, or Saffron colour. The ſmell of this is no leſſe ſweete then in the former.

6. *Narciſſus albus oblongo calice flauo ſerotinus, duobus floribus in caule.*
The late white Daffodill with a long cup, and two flowers on a ſtalke.

This Daffodill is ſurely a kinde of it ſelfe, although it be ſo like the former, abiding conſtant in his forme and manner of flowring, vſually bearing without miſſing two flowers vpon a ſtalke, very like vnto the former great white kinde, that one cannot know any greater matter of difference betweene them, then that it beareth two flowers on a ſtalke: the cuppes whereof are ſeldome touched with any ſhew of Saffron colour on them at the brimmes or edges, as ſome of the former haue.

G The

The Place.

All thefe Daffodils doe grow on the Pyrenæan mouñtaines, and haue been fought out, and brought into thefe parts, by thofe curious or couetous fearchers of thefe delights, that haue made vs partakers of them.

The Time.

The former kindes flower earlier by a fortnight then the later, the one in the later end of March, and the other not vntill the middle of Aprill.

The Names.

Their names are giuen to euery one of them in their feuerall titles, as fitly as may beft agree with their natures; and therefore I fhall not neede to fpeake any further of them.

Narciſſus medioluteus vulgaris.
The common white Daffodill called Primrofe Peerleſſe.

This Daffodill is fo common in euery Countrey Garden almoft through England, that I doubt I fhall but fpend my time in vaine, to defcribe that which is fo well knowne, yet for their fakes that know it not, I will fet downe the defcription of it in this manner. It hath long limber and broad leaues, of a grayifh greene colour, among which rifeth vp a ftalke, bearing at the toppe out of a skinnie huske fometimes but one flower, but moft commonly two flowers, and feldome three or more, but larger for the moft part, then any that beare many flowers vpon a ftalke, of a pale whitifh Creame colour, tending fomewhat neare vnto the colour of a pale Primrofe (which hath caufed our Countrey Gentlewomen, I thinke, to entitle it Primrofe Peerleffe) with a fmall round flat Crowne, rather then a cup in the middle, of a pale yellow colour, with fome pale chiues ftanding therein, being of a fweete, but ftuffing fent: the roote is reafonable great, and encreafing more then a better plant.

Narciſſus mediocroceus ferotinus. The late flowring white Daffodill.

This Daffodill hath much fmaller leaues, and fhorter then the laft, the ftalke alfo rifeth not fo high by much, and beareth but one flower thereon, of a pure white colour, made of fix fmall leaues, and fomewhat narrow, ftanding feuerally one from another, and not fo clofe together as the former, but appearing like a ftarre: the cup is fmall and round, of a pale yellow colour, but faffrony about the brims, hauing fix fmall pale chiues in the middle, the fmell whereof is much fweeter then in the former.

The Place.

The firft is thought to grow naturally in England, but I could neuer heare of his naturall place. I am fure it is plentifull enough in all Country Gardens, fo that wee fcarce giue it place in our more curious parkes. The fecond liueth onely with them that delight in varieties.

The Time.

The firft Daffodill flowreth in the middle time, being neither of the earlieft, nor of the lateft; but about the middle, or end of Aprill. The other flowreth with the lateft in May.

The Names.

I fhall not neede to trouble you with further repetitions of names, they hauing been fet downe in their titles, which are proper to them.

1 Narcissus vulgaris medio luteus. The common white Daffodill, or Primrose Peerlesse. 2 Narcissus medio purpureus maximus. The great white purple ringed Daffodill. 3 Narcissus medio purpureus praecox. The early purple ringed Daffodill. 4 Narcissus medio purpureus stellatus. The starry purple ringed Daffodill. 5 Narcissus Persicus. The Persian Daffodill. 6 Narcissus Autumnalis minor. The lesser Winter Daffodill. 7 Narcissus Autumnalis maior. The greater Winter Daffodill.

1. *Narciſſus medio purpureus præcox.* The early purple ringed Daffodill.

This early Daffodill hath many long grayiſh greene leaues, ſomewhat narrower and ſtiffer then the former common white Daffodill, among which riſeth vp a long naked hollow ſtalke (as all other Daffodils haue) bearing at the toppe one flower, and ſeldome two, made of ſixe long white leaues, ſtanding cloſe together about the ſtalke; the cup is yellow, and ſo flat, that it might rather bee called a crowne: for it ſtandeth very cloſe to the middle, and very open at the brimmes, circled with a reddiſh or purple coloured ring, hauing certaine chiues in the middle of it alſo. The ſmell hereof is very ſweete, exceeding many other.

2. *Narciſſus medio purpureus ſerotinus.* The late purple ringed Daffodill.

The leaues of this Daffodill are alwayes broader then the former early one, and ſome are very neare twice as broad: the flower is very like the former, being large, and his leaues ſtanding cloſe one to the ſide of another; the ring likewiſe that compaſſeth the yellow coronet, is ſometimes of a paler reddiſh purple, and ſometimes as deepe a red as the former: ſo that it differeth not in any other materiall point, then that it flowreth not vntill the other is paſt and gone. The ſent of this is like the former, the roote hereof is greater, as well as the leafe and flower.

3. *Narciſſus medio purpureus maximus.*
The great white purple ringed Daffodill.

There is another kinde, whoſe flower (as well as leaues and rootes) is larger then any other of this kinde, which onely maketh it a diſtinct ſort from the other: it flowreth alſo with the later ſort of theſe purple ringed Daffodils.

4. *Narciſſus medio purpureus ſtellaris.* The ſtarry purple ringed Daffodill.

This Daffodill hath his leaues a little narrower and greener then the former ſorts, the flower alſo of this hath his ſixe white leaues not ſo broad, but narrower, and ſeeming longer then they, not cloſing together, but ſtanding apart one from another, making it ſeeme like a white ſtarre: it hath alſo a yellow coronet in the middle, circled about with purple, like the former. This doth ſmell nothing ſo ſweete as the firſt, but yet hath a good ſent.

The Place.

The firſt, third, and fourth of theſe Daffodils, haue alwayes beene ſent vs from Conſtantinople among other bulbous rootes, ſo that wee know no further of their naturall places.

The ſecond groweth in many places of Europe, both in Germany, France, and Italy, as Cluſius hath noted.

The Time.

The firſt flowreth very early in March, euen with the firſt Daffodils. The ſecond, third, and fourth, about a moneth after.

The Names.

The early and ſtarre Daffodils, haue been ſent vs by the Turkiſh name of *Deuebohini*, and *Serincade.* But their names, they haue receiued ſince, to bee endenizond with vs, are ſet downe in their ſeuerall titles.

Narciſſus Perſicus. The Perſian Daffodill.

This Perſian Daffodill differeth from all other kindes of Daffodils in his manner of
growing,

growing, for it neuer hath leaues and flowers at one time together, wherein it is like vnto a Colchicum, yet in roote and leafe it is a Daffodill. The roote is a little blackish on the outside, somewhat like the roote of the Autumne Daffodill, from whence riseth vp a naked foote stalke, bearing one pale yellow flower, breaking through a thinne skinne, which first enclosed it, composed of six leaues, the three outermost being a little larger then the rest, in the middle of the flower there are six small chiues, and a longer pointell. The whole flower is of an vnpleasant sent : After the flower is past, come vp the leaues, sometimes before Winter, but most visually after the deepe of Winter is past with vs, in the beginning of the yeare, which are broad, long, and of a pale greene colour, like the leaues of other Daffodils, but not greene as the Autumne Daffodill is, and besides they doe a little twine themselues, as some of the Pancratium, or bastard Sea Daffodils doe.

Narcissus Autumnalis maior. The greater Autumne or Winter Daffodill.

The greater Autumne Daffodill riseth vp with three or foure faire broad and short leaues at the first, but afterwards grow longer, of a very deepe or darke greene colour, in the middle of which riseth vp a short, stiffe, round footestalke, bearing one faire yellow flower on the head thereof (inclosed at the first in a thinne skinne, or huske) and consisteth of six leaues as the former, with certaine chiues in the middle, as all or most other Daffodils haue, which passeth away without shew of any seed, or head for seed, although vnder the head there is a little greene knot, which peraduenture would beare seede, if our sharpe Winters did not hinder it. The roote is great and round, couered ouer with a blackish skinne or coate.

Narcissus Autumnalis minor. The lesser Autumne or Winter Daffodill.

Clusius setteth downe, that the manner of the flowring of this lesser Daffodill, is more like vnto the Persian Daffodill, then vnto the former greater Autumne kind; but I doe finde that it doth in the same sort, as the greater kinde, rise vp with his leaues first, and the flowers a while after : the flower of this is lesser, and a little paler then the flower of the greater kinde, but consisting in like sort of six leaues, narrow and sharpe pointed; the greene leaues also are almost of as deepe a greene colour, as the greater kinde, but smaller and narrower, and a little hollow in the middle. The roote is also alike, but lesser, and couered with a blackish skinne as the former. This hath sometimes borne blacke round seede in three square heads.

The Place.

The Persian Daffodill hath beene sent sometimes, but very seldome, among other rootes from Constantinople, and it is probable by the name whereby it was sent, that it should naturally grow in Persia.

The other two haue likewise beene sent from Constantinople, and as it is thought, grow in Thracia, or thereabouts.

The Time.

They all doe flower much about one time, that is, about the end of September, and in October.

The Names.

The first hath been sent by the name of *Serincade Persiana*, and thereupon is called *Narcissus Persicus*, The Persian Daffodill.

The other two haue been thought by diuers to be Colchica, and so haue they called them, vpon no other ground, but that their flower is in forme and time somewhat like Colchicum, when as if they had marked them better, they might plainly discerne, that in all other things they did resemble Daffodils; but now the names of *Colchicum luteum maius, & minus,* is quite

lost,

loft, time hauing worne them out, and they are called by most Herbarifts now adayes, *Narciffus Autumnalis maior & minor*, The greater and the leffer Autumne Daffodill.

Thus farre haue I proceeded with thofe Daffodils, that hauing broad leaues, beare but one fingle flower, or two at the moft vpon a ftalke : And now to proceed with the reft, that haue broad leaues, and beare fingle flowers, but many vpon a ftalke.

Narciffus Africanus aureus maior. The great yellow Daffodill of Africa.

This braue and ftately Daffodill hath many very long and broad leaues, of a better greene colour, then many others that are grayifh, among which appeareth a ftalke, not rifing to the height of the leaues, bearing at the toppe out of a skinnie hofe many faire, goodly, and large flowers, to the number of ten or twelue, if the roote bee well growne, and ftand in a warme place, euery one being larger then any of the French, Spanifh, or Turkie Daffodils, that beare many fingle flowers vpon a ftalke, and commeth neere vnto the bigneffe of the Englifh Daffodill, called Primrofe Peerleffe, before defcribed, or that French kinde hereafter defcribed, that beareth the largeft flowers, many vpon a ftalke (which fome would make to bee a kinde of that Englifh Daffodill, but bearing more flowers) and of a faire fhining yellow colour, hauing large, round, and open cups or boules, yellower then the outer leaues ; and is of fo exceeding fweete a fent, that it doth rather offend the fenfes by the aboundance thereof: the roote is great, and couered with a blackifh browne coate or skinne.

Narciffus Africanus aureus minor. The leffer Barbary Daffodill.

This leffer kinde is very neere the fame with the former, but that it lacketh fomewhat of his ftatelineffe of height, largeneffe of flower and cup (being of a paler yellow) and beauty of colour, for it beareth neither of thefe equall vnto the former, but is in them all inferiour. And thus by this priuatiue, you may vnderftand his pofitiue, and that fhall be fufficient at this time.

Narciffus Byzantinus totus luteus. The yellow Turkie Daffodill.

Whereas the laft defcribed, came fhort of the beauty of the former, fo this lacketh of that beauty is in the laft; for this, although it haue very long leaues, and a high ftalke, yet the flowers are neither fo many, as not being aboue foure or fiue, nor fo large, being not much greater then the ordinary French Daffodill hereafter defcribed, nor the colour fo faire, but much paler, and the cup alfo fmaller; and herein confifteth the chiefeft differences betweene this, and both the other, but that the fent of this is alfo weaker.

The Place.

The firft and the fecond grow in Barbary, about Argier, and Fez, as by the relation of them, that haue brought them into thefe parts, wee haue been enformed.

The laft hath been often brought from Conftantinople among other varieties of Daffodils, but from whence they receiued them, I could neuer learne.

The Time.

Thefe Daffodils do flower very early, euen with the firft fort of Daffodils, I meane after they haue been accuftomed vnto our climate : for oftentimes vpon their firft bringing ouer, they flower in Ianuary or February, efpecially if they be preferued from the frofts, and kept in any warme place; for they are very tender, and will foone perifh, being left abroad.

The Names.

The firft is called by diuers in French, *Narciffe d'Algiers*, and in many places

places of the Low Countries, *Narciſſen van Heck*, or *Narciſſus Heckius* ; by diuers others *Narciſſus Africanus aureus maior*, we may call it in Engliſh, The great African Daffodill, or the great Barbary Daffodill, or the great yellow Daffodill of Argiers, which you pleaſe.

The ſecond hath no other variation of name, then a diminutiue of the former, as is ſet downe in the title.

The third is no doubt the ſame, that Cluſius ſetteth downe in the twelfth Chapter of his ſecond Booke of the Hiſtory of more rare plants, and maketh the fourth ſort, which came from Conſtantinople, and may alſo be the ſame, which he maketh his fifth, which (as he ſaith) he receiued from Doctour *Simor Touar* of Seuill in Spaine. Wee call it, from the place from whence we receiued it, *Narciſſus Byzantinus*, with the addition of *totus luteus*, to put a difference from other ſorts that come from thence alſo : in Engliſh, The yellow ſingle Daffodill of Turkie.

Narciſſus Sulphureus maior. The greater Lemon coloured Daffodill.

The greater of theſe Daffodils, beareth three or foure greene and very long leaues, a foote and a halfe long at the leaſt, among which riſeth vp a round, yet creſted ſtalke, not ſo high as the leaues, bearing fiue or ſixe ſingle flowers thereon, euery one of them being greater then the ordinary French or Italian Daffodils, with many flowers vpon a ſtalke ; of a faint, but yet pleaſant yellow colour at the firſt, which after they haue been in flower a fortnight or thereabouts, change into a deeper, or more ſullen yellow colour : the cup in the middle is likewiſe larger, then in thoſe formerly named, and of a deeper yellow colour then the outer leaues, hauing onely three chiues within it. The ſmell is very pleaſant.

Narciſſus Sulphureus minor. The leſſer Lemon coloured Daffodill.

This leſſer Daffodill hath broader and ſhorter leaues then the former, of the colour of other Daffodils, and not greene like the former : the ſtalke of this riſeth vp higher then the leaues, bearing foure or fiue flowers vpon ſhorter footeſtalkes, and no bigger then the French Daffodill, of a pale yellow, which moſt doe call a Brimſtone colour, the cup or rather crowne in the middle, is ſmall, and broad open, of a little deeper yellow, hauing many chiues within it, and is as it were ſprinkled ouer with a kinde of mealineſſe. The ſmell of this is not full ſo pleaſant as the former.

The Place.

Both theſe haue been gathered on the Pyrenæan Mountaines, and both likewiſe haue been ſent out of Italy.

The Time.

They both flower in the middle time of the Daffodils flowring, that is, in Aprill.

The Names.

They haue their Latine names expreſſed in their titles, and ſo are their Engliſh alſo, if you pleaſe ſo to let them paſſe ; or elſe according to the Latine, you may call them, The greater and the leſſer Brimſtone coloured Daffodils ; ſome haue called them *Narciſſus Italicus*, but the Italians themſelues haue ſent them by the name of *Narciſſo Solfarigno.*

Narciſſus totus albus polyanthos. The milke white Daffodill many vpon a ſtalke.

The leaues of this Daffodill are of a meane ſize, both for length and breadth, yet ſomewhat greener then in the ordinary ſorts, that haue ſome whiteneſſe in them : the flowers are many vpon the ſtalke, as ſmall for the moſt part, as any of theſe kindes that beare many together, being wholly of a milke, or rather ſnow white colour, both the cuppe, which is ſmall, and the outer leaues that compaſſe it ; after which come ſmall heads, wherein is contained round blacke ſeede, as all other Daffodils doe, although ſome greater, and others leſſer, according to the proportion of the plants : the roote is couered ouer with a blackiſh skinne or coate ; the ſmell is very ſweete.

There are two other ſorts more of this kinde, the differences whereof are, that the one hath his leaues ſomewhat broader, and the flowers greater then the former : And the other ſmaller leaues and flowers alſo, whoſe cups being ſmall, are neuer ſeene fully open, but as it were halfe cloſed at the brimmes.

Narciſſus latifolius totus albus, mediocri calice reflexus.
The milke white Daffodill with the great cup.

There is yet another ſort of theſe milke white Daffodils, whoſe leaues are as broad as any of the former, and whoſe cup in the middle of the flower, is ſomewhat larger then in any of the leſſer ſorts, and leſſer then in the greater kinde : but the leaues of the flowers doe a little turne themſelues vpwards, which maketh a chiefe difference.

The Place.

Theſe Daffodils grow in Spaine, from whence I receiued many that flouriſhed a while, but periſhed by ſome fierce cold Winters : they likewiſe grow in France, from whence many alſo haue been brought vnto vs. They haue likewiſe been ſent from Conſtantinople to vs, among other kindes of Daffodils.

The Time.

They that come from Conſtantinople, for the moſt part doe flower earlier then the other, euen after they are accuſtomed to our ayre. Some of them flower notwithſtanding in the end of March, the reſt in Aprill.

The Names.

They are vſually called *Narciſſus totus albus polyanthos*, adding thereunto the differences of *maior, medius*, and *minor*, that is, The milke white Daffodill, the greater, the middle, and the leſſer ; for ſo ſome doe diſtinguiſh them. The laſt, for diſtinction, hath his name in his title ſufficient to expreſſe him.

1. *Narciſſus Narbonenſis, ſiue medio luteus pręcox,*
The early French Daffodill.

The leaues of this Daffodill, ſpring vp out of the ground a moneth or two ſometimes before the other of this kinde, that follow ; being alſo ſhorter, and narrower : the ſtalke likewiſe is not very high, bearing diuers flowers at the top, breaking through a thinne skinne, as is vſuall with all the Daffodils, euery one whereof is ſmall, conſiſting of ſix white leaues, and a ſmall yellow cup in the middle, which is of a prettie ſmall ſent, nothing ſo ſtrong as many others : the roote is great and round, and ſeldome parteth into of-ſets, euen as all the other that follow, bearing many ſingle flowers, doe.

2. *Narciſſus Narbonenſis vulgaris.* The ordinary French Daffodill.

This Daffodill hath long and broad greene leaues, a little hollowiſh in the middle, and edged on both ſides ; the ſtalke is a foote and a halfe high, bearing at the toppe diuers flowers, ſomewhat larger then the former, conſiſting of ſix white leaues, ſomewhat round ; the cup is yellow in the middle, ſmall and round, like vnto an Acorne cuppe, or a little fuller in the middle : this is the forme of that ſort which was firſt
brought

1 *Narcissus Africanus aureus maior.* The great yellow Daffodill of Africa. 2 *Narcissus Africanus luteus minor.* The lesser yellow Daffodill of Africa. 3 *Narcissus Narbonensis medio luteus.* The French Daffodill. 4 *Narcissus Pisanus, vel totus albus.* The Italian Daffodill, or the all white Daffodill. 5 *Narcissus Muscart.* Muskart his Daffodill. 6 *Narcissus Anglicus polyanthos.* The great English Daffodill.

brought vnto vs : But since there is found out some, whose cup is shorter, others flatter, some of a paler, others of a deeper yellow colour, and some that haue their cuppe longer then the rest. The rootes of them all are couered with a blackish skin or coate.

3. *Narcissus Narbonensis maior amplo flore.* The French Daffodill with great flowers.

The leaues of this Daffodill are somewhat like vnto the last, but not so broad, yet full as long, and spring sooner out of the ground, yet not so early as the first of these kindes: the stalke hereof is flatter, and riseth higher, bearing foure or fiue flowers, much larger then any of this kinde ; for euery one of them doth equall the English Daffodill, before described, but whiter then it, and the yellow cup larger, and more open then in any of the rest. The roote of this is not so great, or round, as the former, but is more plentifull in of-sets, then any other of these French, or Italian kindes.

4. *Narcissus Pisanus.* The Italian Daffodill.

This Italian Daffodill hath his leaues as large, or larger then the second French Daffodill, and his stalke somewhat higher, bearing many white flowers, very like vnto the common French Daffodill, but somewhat larger also; and the yellow cup in the middle likewise is larger, and rounder, then is vsually seen in any of the French kinds, except the last with the greatest flowers.

5. *Narcissus mediocroceus polyanthos.* The French Daffodill with Saffron coloured cups.

This French Daffodill hath diuers leaues of a grayish greene colour, not so broad or long as the last recited Daffodill, but comming neerer vnto the second French kinde, the flowers likewise are white, and many vpon a stalke, like thereunto, but the yellow cup is somewhat large, and circled with a Saffron like brimme or edge, which maketh the chiefest difference.

6. *Narcissus mediocroceus alter, dictus Massart.* Mussart his Daffodill.

The affinity between this & the last, (for it is not the same to be expressed vnder one title) hath made me ioyne it next vnto it, yet becaufe it hath a notable difference, it deserueth a place by himselfe. The leaues are large and long, and the flowers, being white, are larger also then in any other, except the greatest, but the cup hereof is small and short, rather seeming a coronet then a cup, of a deepe Saffron colour all about the brimmes or edges.

7. *Narcissus Anglicus polyanthos.* The great English Daffodill.

This Daffodill hath his leaues not much broader or longer, then the French kinde with great flowers, before described, the stalke with flowers riseth not fully so high as it, bearing many flowers thereon, not altogether so white, yet whiter then the former English Daffodill, called Primrose Peerlesse, but nothing so large, and with short, broad, and almost round leaues, standing close one vnto another : the yellow cup in the middle is bowle fashion, being somewhat deeper then in any of the former kinds, but not much greater : the smell hereof is very sweete and pleasant.

8. *Narcissus Narbonensis, siue medio luteus serotinus maior.* The greater late flowring French Daffodill.

The roote as well as the leaues of this Daffodill, are greater, larger, broader, and longer then in any other of the former French, or Italian kindes ; the stalke is as high as any of them, bearing at the toppe fiue or sixe white flowers, standing open spread like a starre, and not close together, euery one whereof is large, and round pointed,

the

the cup is yellow, small and short, yet not lying flat to the flower, but a little standing out with some threads in the middle, as all the former Daffodils haue. This is not so sweete as the earlier kindes.

9. *Narcissus medioluteus alter serotinus calice breui.*
The lesser late flowring French Daffodill.

This Daffodill is of the same kinde with the last described, the onely difference is, that it is lesser, and the yellow cuppe in the middle of the flower, is somewhat shorter then the former, although the former be shorter then many others, otherwise it differeth not, no not in time; for it flowreth late as the former doth.

The Place.

These Daffodils haue been brought vs from diuers places: The first and second grow naturally in many places of Spaine, that are open to the Sea: they grow likewise about Mompelier, and those parts in France. They haue been likewise sent among many other sorts of Daffodils from Constantinople, so that I may thinke, they grow in some places neere thereunto.

The fourth groweth plentifully in Italy, about Pisa in Tuscane, from whence we haue had plants to furnish our Gardens.

The seuenth is accounted beyond Sea to be naturall of our Country, but I know not any with vs that haue it, but they haue had it from them.

The rest haue been brought at diuers times, but wee know no further of their naturall places.

The Time.

The first flowreth earlier then any of the rest by a moneth, euen in the beginning of March, or earlier, if the weather be milde. The other in Aprill, some a little before or after another. The late kinds flower not vntill May.

The Names.

There can be no more said of the names of any of them, then hath beene set out in their titles; for they distinguish euery sort as fitly as we can: onely some doe call the first two sorts, by the name of *Donax Narbonensis.*

After all these Daffodils, that hauing broad leaues beare single flowers, either one or many vpon a stalke, I shall now goe on to set forth those broad leafed Daffodils, that carry double flowers, either one or many vpon a stalke together, in the same order that we haue vsed before.

1. *Narcissus albus multiplex.* The double white Daffodill.

The leaues of this Daffodill are not very broad, but rather of a meane size, being of the same largenesse with the leaues of the purple ringed Daffodill, the stalke riseth vp to be a foote and a halfe high, bearing out of a thinne white skinne or hose, one flower and no more, consisting of many leaues, of a faire white colour, the flower is larger then any other double white Daffodill, hauing euery leafe, especially the outermost, as large almost as any leafe of the single Daffodill with the yellow cup, or purple ring. Sometimes it happeneth, that the flower is very little double, and almost single, but that is either in a bad ground, or for that it hath stood long in a place without remouing; for then it hath such a great encrease of rootes about it, that it draweth away into many parts, the nourishment that should be for a few: but if you doe transplant it, taking away the of-sets, and set his rootes single, it will then thriue, and beare his flower as goodly and double, as I haue before described it: and is very sweete.

2. *Narcissus mediopurpureus multiplex.* The double purple ringed Daffodill.

There is little difference in the leaues of this kinde, from the leaues of the single purple

ple ringed Daffodill ; for it is probable it is of the same kinde, but by natures gift (and not by any humane art) made more plentifull, which abideth constant , and hath not that dalliance, which oftentimes nature sheweth, to recreate the senses of men for the present, and appeareth not againe in the same forme : the chiefest difference is, that the flower (being but sometimes one on a stalke , and sometimes two) consisteth of six white outer leaues, as large as the leaues of the single kinde, hauing many small yellow peeces, edged with purple circles round about them. instead of a cup ; and in the mid-dle of these peeces, stand other six white leaues, lesser then the former , and a yellow cup edged with a purple circle likewise , parted into peeces , and they comprehend a few other white leaues, smaller then any of the other , hauing among them some bro-ken peeces of the cup, with a few chiues also in the middle of the flower. The flower is very sweete.

There is of this kinde another, whose flower hath not so plaine a distinction , of a triple rowe of leaues in it : but the whole flower is confusedly set together , the outer leaues being not so large, and the inner leaues larger then the former ; the broken yel-low cuppe, which is tipt with purple, running diuersly among the leaues ; so that it sheweth a fairer, and more double flower then the former, as it is indeed.

3 . *Narcissus medioluteus corona duplici.*
The Turkie Daffodill with a double crowne.

This Daffodill hath three or foure leaues , as large and long almost , as the great double Daffodill of Constantinople next following hath : the stalke likewise is very neere as great, but as high altogether , bearing at the toppe foure or fiue flowers , the leaues whereof are as large, as of the first or second kinde of French Daffodils, before described, but not altogether of so pure a white colour; and being six in number, stand like the former single French Daffodils, but that the yellow cup in the middle of this is thicke and double, or as it were crumpled together, not standing very high to be con-spicuous , but abiding lowe and short , so that it is not presently marked , vnlesse one looke vpon it precisely ; yet is exceeding sweete. The roote is like vnto the roote of the purple ringed Daffodill, or somewhat bigger.

4 . *Narcissus Chalcedonicus flore pleno albo polyanthos.*
The double white Daffodill of Constantinople.

This beautifull and goodly Daffodill (wherewith all Florists greatly desire to bee acquainted , as well for the beauty of his double flowers , as also for his superaboun-ding sweete smell, one stalke with flowers being instead of a nosegay) hath many very broad, and very long leaues , somewhat greener then gray , among which riseth vp a strong round stalke, being sometimes almost flat, and ribbed, bearing foure or fiue, or more white flowers at the toppe, euery one being very great , large, and double , the leaues being confusedly set together , hauing little peeces of a yellow cup running a-mong them , without any shew of that purple ring that is in the former, and fall a-way without bearing seed, euen as all , or most other double flowers doe : the smell is so exceeding sweet and strong , that it will soone offend the senses of any, that shall smell much vnto it : the roote is great and thicke , couered with a blackish coate.

5 . *Narcissus Chalcedonicus fimbriatus multiplex polyanthos.*
The great double purple ringed Daffodill of Constantinople.

This Daffodill differeth very little or nothing in leafe from the former, the onely difference is in the flowers, which although they bee double , and beare many vpon a stalke, like vnto them, yet this hath the peeces of the yellow cuppes tipt with purple, as if they were shred or scattered among the white leaues, whereas the other hath only the yellow, without any shew of purple tips vpon them : the smell of this is as strong as of the other.

1 *Narcissus albus multiplex.* The double white Daffodill. 2 *Narcissus mediolureus corona duplici.* The Turkie Daffodill with a double crowne. 3 *Narcissus mediopurpureus multiplex.* The double purple ringed Daffodill. 4 *Narcissus chalcedonicus flore pleno albo polyanthos.* The double white Daffodill of Constantinople.

6. *Narcissus Cyprius flore pleno luteo polyanthos.*
The double yellow Daffodill of Cyprus.

The leaues of this Daffodill are almost as broad and long as the former, the stalke is a foot high and more, bearing foure or fiue flowers on the top, euery one very double, and of a fine pale yellow colour, of a strong heady sent. The root of this is also like the former.

The Place.

The first of these Daffodils, was first brought into England by Mr. Iohn de Franqueuille the elder, who gathered it in his owne Countrey of Cambray, where it groweth wilde, from whose sonne, Mr. Iohn de Franqueuille, now liuing, we all haue had it. The rest haue come from Constantinople at seuerall times ; and the last is thought to come from Cyprus. Wee haue it credibly affirmed also, that it groweth in Barbary about Fez and Argiers. Some of the double white kindes grow in Candy, and about Aleppo also.

The Time.

The Turkie kindes doe for the most part all flower early, in the end of March, or beginning of Aprill at the furthest, and the first double, about the middle or end of Aprill.

The Names.

All these Daffodils, except the first, haue had diuers Turkish names set vpon the packets, wherein they haue been sent, but there is small regard of certainty to be expected from them ; for that the name *Serincade*, without any more addition, which is a single Daffodill, hath beene imposed vpon that parcell of rootes, that haue borne most of them double flowers of diuers sorts ; and the name *Serincade Catamer lale*, which signifieth a double flowred Daffodill, hath had many single white flowers, with yellow cups, and some whose flowers haue been wholly white, cuppe and all, and some purple ringed, and double also among them. Their names, whereby they are knowne and called with vs, are, as fitly as may be, imposed in their titles : And this I hope shall suffice, to haue spoken of these sorts of Daffodils.

Hauing finished the discourse of the former sort of broad leafed Daffodils, it is fit to proceede to the next, which are *Angustifolios Narcissos*, those Daffodils that haue narrow leaues, and first to set downe those that beare single flowers, whether one or many flowers vpon a stalke, and then those that beare double flowers in the same manner.

Narcissus Virgineus. The Virginia Daffodill.

This plant I thought fittest to place here in the beginning of this *Classis*, not finding where better to shroud it. It hath two or three long, and very narrow leaues, as greene as the leaues of the great *Leucoium bulbosum*, and shining withall, which grow sometimes reddish, especially at the edges : the stalke riseth vp a spanne high, bearing one flower and no more on the head thereof, standing vpright like a little Lilly or Tulipa, made of six leaues, wholly white, both within and without, except that at the bottome next to the stalke, and a little on the backside of the three outer leaues, it hath a small dash or shew of a reddish purple colour : it hath in the middle a few chiues, standing about a small head pointed ; which head groweth to bee small and long, containing small blackish flat seede : the roote is small, long, and round, a little blackish on the outside, and white on the inside.

The

The Place.

This bulbous plant was brought vs from Virginia, where they grow a-boundantly ; but they hardly thriue and abide in our Gardens to beare flowers.

The Time.

It flowreth in May, and seldome before.

The Names.

The Indians in Virginia do call it *Attamusco,* some among vs do call it *Lilionarcissus Virginianus,* of the likenesse of the flower to a Lilly, and the leaues and roote to a Daffodill. Wee for breuity doe call it *Narcissus Virgineus,* that is, The Daffodill of Virginia, or else you may call it according to the former Latine name, The Lilly Daffodill of Virginia, which you will ; for both names may serue well to expresse the plant.

Narcissus angustifolius albidus præcox oblongo calice.
The early white narrow leafed Daffodill with a long cup.

This Daffodill hath three or foure narrow, long, and very greene leaues, a foote long for the most part : the stalke riseth not vp so high as the leaues, whereon standeth one flower, not altogether so great as the late flowring Daffodill, with a long cuppe, described before among the broad leafed ones, which consisteth of six pale coloured leaues, not pure white, but hauing a wash of light yellow among the white : the cuppe in the middle is round and long, yet not so long as to bee accounted a bastard Daffodill, within which is a middle pointell, compassed with six chiues, hauing yellow mealy pendents.

The Place.

This Daffodill groweth with the other sorts of broad leafed ones, on the Pyrenæan Mountaines, from whence they haue beene brought vnto vs, to furnish our Gardens.

The Time.

It flowreth early, a moneth before the other sorts of the same fashion, that is, in the beginning of March, if the time be milde, which the other before spoken of doe not.

The Names.

It hath no other name that I know, then is expressed in the title.

2. *Narcissus mediocroceus tenuifolius.* The small Daffodill with a Saffron crown.

This small Daffodill hath foure or fiue narrow leaues about a spanne long, among which riseth vp a stalke some nine inches high, bearing at the toppe one small white flower, made of six leaues, with a small yellow cup in the middle, shadowed ouer at the brimmes with a Saffron colour : the roote is small, round, and little long withall, couered with a blackish skinne or coate.

3. *Narcissus minimus mediopurpureus.* The least purple ringed Daffodill.

This little Daffodill hath small narrow leaues, shorter by much then any of the purple ringed Daffodils, before described : the stalke and flower keepe an equall proportion to the rest of the plant, being in forme and colour of the flower, like vnto the

H 2   Starre

Starre Daffodill before recited, but vnlike in the greatnesse : this also is to bee obserued, that the purple colour that circleth the brimmes of the cuppe, is so small, that
sometimes it is not well perceiued.

4. *Narcissus minimus Iuncifolij flore*. The least Daffodill of all.

This least Daffodill hath two or three whitish greene leaues, narrower then the two
last recited Daffodils, and shorter by halfe, being not aboue two or three inches long,
the stalke likewise is not aboue three or foure inches high, bearing one single flower at
the toppe, somewhat bigger then the smalnesse of the plant should seeme to beare,
very like vnto the least Rush Daffodill, and of the same bignesse, or rather somewhat
bigger, being of a faint yellow colour, both leaues, and cup, or crowne, (if you please
so to call it) ; for the middle part is spread very much, euen to the middle of the leaues
almost, and lyeth flat open vpon the flower : the roote is small, euen the smallest of any
Daffodill, and couered with a blackish skinne or coate.

The Place.

The first of these Daffodils haue beene brought vs from the Pyrenæan
Mountaines, among a number of other rare plants, and the last by a French
man, called Francis le Veau, the honestest roote-gatherer that euer came o-
uer to vs. The second was sent to Mr. Iohn de Franqueuille, before remembred, who imparted it to mee, as hee hath done many other good things;
but his naturall place wee know not.

The Time.

They all flower about the latter end of Aprill.

The Names.

Being brought without names, wee haue giuen them their names according to their face and fashion, as they are set downe in their titles.

Narcissus Autumnalis minor albus. The little white Autumne Daffodill.

This little Autumne Daffodill riseth with his flowers first out of the ground, without
any leaues at all. It springeth vp with one or two stalkes about a finger long, euery one
bearing out of a small huske one small white flower, laid open abroad like vnto the
Starre white Daffodill, before spoken of : in the middle of the flower is a small yellow cup of a meane size, and after the flower is past, there commeth in the same place
a small head, containing small, round, blacke seede, like vnto the Autumne Hyacinth:
the leaues come vp after the seede is ripe and gone, being small and narrow, not much
bigger then the Autumne Hyacinth : the roote is small and blackish on the outside.

The Place.

This Daffodill groweth in Spaine, where Clusius saw it, and brought it
into these parts.

The Time.

It flowreth in the beginning of Autumne, and his seede is ripe in the end
of October in those hot Countries, but in ours it will scarce abide to shew a
flower.

The Names.

The Spaniards, as Clusius reporteth, call it *Tomada*, and he vpon the sight
thereof,

1 *Narcissus Virgineus.* The Virginian Daffodill. 2 *Narcissus minimus luncifolij flore.* The least Daffodill of all. 3 *Narcissus Autumnalis minor simus.* The little white Autumne Daffodill. 4 *Narcissus albus Autumnalis medio obsoletus.* The white Autumne Daffodill with a sullen crown. 5 *Narcissus luncifolius maximus amplocalice.* The great Iunquilia with the largest flower or cup. 6 *Narcissus totus albus flore plena Virginianus.* The double white Daffodill of Virginia.

thereof, *Narcissus Autumnalis minor albus*, and wee in Englifh thereafter,
The little white Autumne Daffodill.

Narcissus albus Autumnalis medio obsoletus.
The white Autumne Daffodill with a fullen crowne.

This Autumne Daffodill hath two or three leaues at the moft, and very narrow, fo
that fome doe reckon it among the Rufh Daffodils, being fomewhat broad at the bottome, and more pointed at the toppe, betweene thefe leaues commeth vp the ftalke,
bearing vfually two flowers and no more at the toppe, made of fixe white leaues a
peece, pointed and not round : the cup is fmall and round, like vnto the cup or crowne
of the leaft Rufh Daffodill, of a yellow colour at the bottome, but toward the edge of
a dunne or fullen colour.

Narcissus angustifolius luteus semper florens Caccini.
The yellow Italian Daffodill of Caccini.

This Daffodill beareth a number of fmall, long, narrow, and very greene leaues,
broader then the leaues of any Rufh Daffodill, among which rife vp diuers ftalkes,
bearing at the head two or three flowers a peece, each of them being fmall and yellow,
the cup or crowne is fmall alfo, of a deeper yellow then the flower. The Nobleman of
Florence, who firft fent this plant to Chriftian Porret at Leyden, after the death of Carolus Clufius, writeth that euery ftalke doth beare with him more ftore of flowers,
then are formerly fet downe, and that it neuer ceafeth to beare flowers, but that after
one or moe ftalkes haue been in flower together, and are paft, there fucceed other in
their places.

The Place.

The firft is naturall of Spain, the naturall place of the other is not known
to vs.

The Time.

The times of the flowring, are fet downe both in the title and in the defcriptions ; the one to be in Autumne, the other to be all the Summer long.

The Names.

The Latine names are impofed on them, as are fitteft for them, and the
laft by that honourable man that fent it, which is moft fit to continue, and
not to bee changed. But wee, to let it bee knowne by an Englifh name to
Englifh people, haue entituled it, The yellow Italian Daffodill of Caccini:
if any man can giue it a more proper name, I fhall bee therewith right well
content.

Narcissus angustifolius, sine Iuncifolius maximus amplo calice.
The great Iunquilia with the large flower or cup.

Although this Daffodill importeth by his name, not to be of this family, but of the
next, confidering it is fo like vnto them, but bigger; yet I haue thought good to place
it in the end of thefe narrow leafed Daffodils, as being indifferent, whether it
fhould bee referred to this or to that. For this carrieth diuers long greene leaues,
like vnto the other Rufh Daffodils, but thicker and broader, fo that it may without any great errour, bee reckoned among thefe narrow leafed Daffodils, bearing at
the toppe two or three very faire large flowers, with a large and more open cuppe,
then in any other of the Rufh Daffodils, both of them of a faire yellow colour, yet
the cuppe a little deeper then the flower, and a little crumpled about the edges, and
hath a pretty fharpe fent : the roote is greater and longer then the other Rufh Daffodill, and couered likewife with a blackifh coate.

The

The Place.

We haue this in Gardens onely, and haue not heard of his naturall place.

The Time.

It flowreth in Aprill.

The Names.

I leaue it indifferent, as I said, whether you will call it *Narcissus angustifo-lius*, or *Iuncifolius magno calice*, or *maximus*, because it is the greatest of all the rest of that kinde.

Narcissus totus albus flore pleno Virginianus.
The double white Daffodill of Virginia.

The roote of this Daffodill, is very like vnto the former single Virginia Daffodill, set forth in the first place of this ranke of narrow leafed Daffodils, but that it is a little bigger and rounder, being a little long withall, and blackish also on the outside, as that is : from whence riseth vp two leaues, somewhat broader then the former : but of a like greennesse : the stalke riseth vp betweene these two leaues, about a span high, or not much higher, bearing one faire double snow white flower, very like in the fashion vnto the pale yellow double Daffodill, or bastard Daffodill of Robinus, hereafter de-scribed : For it is in the like manner laid open flat, and composed of six rowes of leaues, euery rowe lying in order iust opposite, or one before another, whereof those six leaues that make the first or outermost course, are the greatest, and all the rest lying, as I said, one vpon or before another, are euery rowe smaller then others from the middle of this flower, thrusteth forth a small long pointed forke or horne, white as the flower is.

The Place.

The place is named to be Virginia, but in what part it is not known to vs.

The Time.

It flowreth in the end of Aprill.

The Names.

It may be that this doth grow among the former single kinde, and called by the same name Attamusco, for that the plant is not much differing, yet hereof I am not certaine : But we, from the forme and countenance of the plant, doe call it *Narcissus Virginianus*, The Virginian Daffodill, and be-cause it beareth a double flower, it hath the title of double added vnto it.

The third order of Daffodils, I said in the beginning, was of *Iuncifolios*, Rush Daffo-dils, which are now next to be entreated of, I shall herein keepe the same order I vsed in the former ; but because I finde none of this order, that beare but one flower vpon a stalke, I must begin with those that beare many.

1. *Narcissus Iuncifolius albus.* The white Iunquilia.

This white Rush Daffodill hath small long leaues, a little broader, and of a whi-ter greene colour then the ordinary yellow Rush Daffodils : the stalke riseth vp halfe a foote high or more, bearing two or three small white flowers vpon a stalke, yet somewhat bigger then the common yellow Rush Daffodill, hauing a small round cuppe in the middle, white also as the leaues are. The seede is small, blacke,

blacke, and round, as other seedes of Daffodils are : the roote is small and round, couered with a blackish coate.

Narcissus Iuncifolius albus magno calice. The white Iunquilia with a great cup.

There is of this kinde another sort, that hath the cup in the middle of the flower, a
little larger then the other, but in all other things alike.

2. *Narcissus Iuncifolius flore albo reflexo.*
The white turning Iunquilia, or Rush Daffodill.

This turning white Daffodill hath foure or fiue long greene leaues, yet shorter and
broader then the ordinary yellow Iunquilia, and fully as greene also, from among
which riseth vp a slender greene stalke, a foote high, bearing out of a thinne skinnie
huske, three or foure, or more snow white flowers, standing vpon long greene foot-
stalkes, euery flower hanging downe his head, and turning vp his six narrow and long
leaues, euen to the very foot-stalke againe : from the middle of the flower hangeth
downe a long round cuppe, as white as the leaues, within which are contained three
small white chiues, tipt with yellow, and a small long pointell, thrusting out beyond
the brimmes of the cup : after the flowers are past, there come vp in their places small
three square heads, wherein is contained very small, round, and blacke shining seede :
the roote is small, round, and a little long withall, couered with a blackish browne
coate or skin. The flower is quite without any good sent, or indeed rather none at all.

3. *Narcissus Iuncifolius flore luteo reflexo.*
The yellow turning Iunquilia, or Rush Daffodill.

The leaues of this Rush Daffodill are greater and longer then the former, and of a
paler greene colour : the stalke riseth somewhat higher, bearing two or three flowers
thereon wholly of a gold yellow colour, both the cuppe and the leaues that turne vp
againe.

4. *Narcissus Iuncifolius calice albo reflexu folijs luteis.*
The yellow turning Iunquilia with a white cup.

This Daffodill hath his long rush-like leaues standing vpright as the former, betweene which riseth vp a greene stalke, about a foote high or more, bearing two or
three flowers thereon, whose turning leaues are of a faire pale yellow, and the cuppe
pale white, and not so pure a white as the former.

5. *Narcissus Iuncifolius calice luteo reflexis folijs albidis.*
The white turning Iunquilia with a yellow cup.

As the last had the leaues of the flower that turne vp againe yellow, and the cuppe
whitish, so this hath contrariwise the turning leaues of a whitish yellow, and the long
cup yellower, else in his long green leaues, or any other thing, there is small difference.

6. *Narcissus Iuncifolius luteus magno calice.*
The Iunquilia, or Rush Daffodill with a great cup.

This Rush Daffodill hath bigger leaues, and longer then the ordinary yellow Rush
Daffodill, being a little flat on the one side, and round on the other, but of the same
greennesse with all the rest : the stalke riseth vp two foote high, bearing two, and
sometimes three flowers thereon, being of a faire yellow colour, with a large open
cup in the middle, of a little deeper yellow colour, like vnto the great Iunquilia with
the large flower, before set downe, whereof this is a kinde, no doubt, but that is larger
and greater then this, both in leafe, flower, cup, &c. and this onely somewhat lesse in
all parts then that.

1 *Narcissus juncifolius albus.* The white Iunquila. 2 *Narcissus juncifolius flore albo reflexo.* The white turning Iunquila. 3 *Narcissus juncifolius calice luteo reflexis foliis albis.* The yellow turning Iunquila. 4 *Narcissus juncifolius luteus magno calice.* The yellow Iunquila with a great cuppe. 5 *Narcissus juncifolius luteus maior vulgaris.* The ordinary yellow Iunquila. 6 *Narcissus juncifolius Autumnalis flore viridi.* The greene Autumne Iunquila. 7 *Narcissus angustifolius aureus multiplex.* The golden double narrow leaued Daffodill. 8 *Narcissus juncifolius flore pleno.* The double Iunquila.

7. *Narciſſus Iuncifolius luteus vulgaris maior.*
The ordinary Iunquilia, or Ruſh Daffodill.

This ordinary Ruſh Daffodill hath foure or fiue long greene round leaues, like vnto Ruſhes, whereof it tooke the name : among theſe leaues riſeth vp the ſtalke, round and greene, a foote and a halfe high very often, bearing at the toppe three or foure flowers all yellow, but much ſmaller then the laſt, and ſo is the cup alſo : the ſeede is ſmall and blacke, incloſed in ſmall cornered heads; the roote is blackiſh on the outſide. The ſmell of the flower is very ſweete in all theſe ſorts of Ruſh Daffodils.

8. *Narciſſus Iuncifolius luteus medius.* The ſmaller Iunquilia, or Ruſh Daffodill.

The leaues of this Daffodill are like vnto the former, but ſmaller and rounder, the ſtalke riſeth not vp ſo high, nor are the flowers ſo great, but the leaues of the flower are a little rounder, and not ſo pointed as in the former, in all things elſe alike, ſauing leſſer.

9. *Narciſſus Iuncifolius luteus minor.* The leaſt Iunquilia, or Ruſh Daffodill.

This leaſt Daffodill hath fiue or ſix ſmall greene leaues, a little broader, and not ſo long as the laſt, among which riſeth vp a ſtalke almoſt a foote high, bearing one or two ſmall flowers at the toppe, of a paler yellow colour then the former, with a yellow open cuppe, or crowne rather in the middle, bigger then in either of the laſt two : the roote is very ſmall and blacke, like vnto the laſt in roundneſſe and colour.

10. *Narciſſus Iuncifolius luteus albicantibus lineis diſtinctus.*
The yellow Iunquilia, or Ruſh Daffodill with white lines.

This Ruſh Daffodill hath round, greene, and long leaues, like vnto the ordinary Ruſh Daffodill, with a ſtalke bearing two or three yellow flowers, hauing leaues ſome-what round at the point or end, with a line or ſtrake of white in the middle of euerie one of them, the cup is ſhort, and crowne faſhion, a little crumpled about the brims: the ſeede, roote, or any thing elſe differeth not.

11. *Narciſſus Iuncifolius Autumnalis flore viridi.*
The Autumne Ruſh Daffodill with a greene flower.

This ſtrange Ruſh Daffodill (I call it ſtrange, not onely becauſe it differeth from all others of this kinde, but alſo becauſe there are but few in theſe parts that haue had it, and fewer that doe ſtill enioy it, in that it is periſhed withall that had it) hath but one onely leafe, very long, round, and greene, in all that euer I ſaw growing, which bea-reth no flower while that greene leafe is freſh, and to bee ſeene : but afterwards the ſtalke riſeth vp, being like vnto the former greene leafe, round, naked, and greene vp to the toppe, where two or three flowers breake forth out of a ſmall thin skinne, euery one conſiſting of ſix ſmall and narrow greene leaues, very ſharpe pointed at the end, and as it were ending in a ſmall pricke or thorne : in the middle whereof is a ſmall round cup, or rather crowne, of the ſame colour with the leaues and ſtalke, which flower ſmelleth very ſweete, ſomewhat like vnto the reſt of the Ruſh Daffodils : this ſheweth not his flower vntill October, and the froſts quickly following after their flowring, cauſe them ſoone to periſh.

12. *Narciſſus anguſtifolius aureus multiplex.*
The golden double narrow leafed Daffodill.

The leaues of this Daffodill are very narrow, and of a whitiſh greene colour, not aboue foure or fiue inches long, from among which riſeth vp a ſtalke about a foote high, bearing at the top one flower, conſiſting of ſome outer leaues, which are of a yellow

low colour, and of many other leaues in the middle being smaller, and set thicke and round together of a more yellow gold colour, but with some whiter leaues among them, the middle part a little pointing forth: the flower standeth long before it doth perfect his colour, and abideth long in flower before the colour decay: the roote is in fashion almost like the ordinary Iunquilia, or Rush Daffodill. I acknowledge this Daffodill hath not his proper place; but because the figure is set in this table, let it thus passe at this time.

13. *Narcissus lancifolius luteus flore pleno.* The double Iunquilia, or Rush Daffodill.

The double Rush Daffodill hath his long greene leaues round, like the leaues of the common or ordinary Rush Daffodill, and of the same bignesse, among which riseth vp a long slender greene stalke, bearing two or three, seldome more small flowers, yellow and double, that is, with diuers rowes of leaues, hauing the yellow cup such as is in the single flower, broken into small shreads or peeces, running among the leaues of the flower, which peeces in some flowers are not so easily seene, being smaller then in others, this beareth no button or head vnder the flower for seede, his roote is round and blackish, browne on the outside, so like vnto the common Rush Daffodill, that it is almost impossible to know the one from the other.

There is another of this kinde, whose flowers are smaller, and not so double, one, two, or three at the most vpon a stalke, and of lesse beauty by much. *Altero minori flore.*

The Place.

All these Rush Daffodils, doe for the most part grow in Spaine and France, and on the Pyrenæan Mountaines, which are betweene Spaine and France, which Mountains are the Nourseries of many of the finest flowers, that doe adorne the Gardens of these louers of natures pride, and gathered in part by industrious, learned, generous men, inhabiting neare thereunto, and in part by such as make a gaine of their labours, bestowed vpon these things. Onely that with the greene flower was gathered in Barbary, and imparted vnto vs from France.

The Time.

They flower in the Spring, that is, in March and Aprill, except such whose time is set downe to be in Autumne.

The Names.

Their names are specified in their titles, and therefore I shall not need to set downe any further repetitions.

To conclude therefore this discourse of true Daffodils, there remaineth to speake of the Sea Daffodils, which (as I said in the beginning) is but one, that is frequent, and doth abide with vs. But there bee some others found about the Cape of good Hope, and in the West Indies, and brought into these parts rather for ostentation, then continuance, where they haue flowred onely once (if peraduenture so often) so that being such strangers, of so remote Countries, and of so diuers natures, I shall but shew you some of them, rather cursorily then curiously; and but onely for your satisfaction, giue you knowledge of two or three of them, that there haue beene seene such in flower, and that they are scarce to bee seene againe, except they bee fetcht a new euery yeare that they be seene.

Narcissus Marinus, siue tertius Matthioli.
The great white Sea Daffodill, or Matthiolus his third Daffodill.

The roote of this Daffodill by long continuance, standing in one place without being remoued, groweth to be much greater and larger, then any other Daffodill whatsoeuer, foeuer, and as bigge as any mean: Squilla or Sea Onion roote, hauing many long, thicke, and white fibres, or long rootes, diuerfly branched, and fpread vnder the vpper part of the earth, befide fome others that grow downward, and perifh not euery yeare, as the fibres of all, or moft of the other Daffodils doe; and therefore this plant will not thriue, and beare flowers, if it be often tranfplanted, but rather defire to abide in one place without remouing, as I faid, and that not to be ouerfhadowed, or couered with other herbes ftanding too neare it, which then will flourifh, and beare abboundantly: from this roote, which is couered with many blackifh coates, arifeth fix or feuen, or more leaues, twice fo broad almoft, as any of the former Daffodils, but not fo long by halfe as many of them, being but fhort, in comparifon of the breadth, and of a white greene colour, from the middle of which leaues, as alfo from the fides fometimes, fpringeth vp one or two, or more ftalkes, roundifh and thicke, and fometimes a little flat and cornered, a foote high or fomewhat more, bearing at the toppe, out of a skinnie huske, eight, ten, twelue, or more very large flowers, confifting of fix white leaues a peece, fpread or laid open, with a white fhort cuppe or crowne in the middle, lying flat vpon the leaues, cut or diuided into fix corners (and not whole, as the cuppe or crowne of any other fingle Daffodill) from euery of which edges, or corners of this cup or crowne, ftandeth one white long thread, a little crooked or turning vp at the end, tipt with a yellow pendent, and fome other white threads tipt with yellow pendents, ftanding alfo in the middle: after the flower is paft, there come vp great three fquare heads, wherein the feede is contained, which is great, blacke, and round, like vnto the feede of other Daffodils, but greater: the flower hath a reafonable good fent, but not very ftrong.

The Place.

It was firft found by the Sea fide, in the Ifle of Sardinia, and on the high Mountaines alfo of the fame Ifle, where it hath borne by report, thirty fiue flowers vpon a ftalke: it groweth likewife about Illyricum, and in diuers other places.

The Time.

It fpringeth later out of the ground then any other Daffodill, that is to fay, not vntill the later end of March, or beginning of Aprill, and flowreth in the end of May, or the beginning of Iune: the feede is ripe in the end of Iuly, or beginning of Auguft.

The Names.

The firft that hath made mention of this Daffodill, was Matthiolus, who placed it in the third place among his Daffodils, and is moft vfually now adayes called, *Narciffus tertius Matthioli*, Matthiolus his third Daffodill, the rather, becaufe Clufius vpon a more mature deliberation, firft referred it thereunto, but called it at the firft, *Lilionarciffus Hemerocallidis facie*, and, as hee faith, Iacobus Plateau (who firft fent him the figure hereof, with the defcription) called it *Lilionarciffus Orientalis*, but Clufius vpon certaine information, that it grew in the places aforefaid, mifliked the name of *Orientalis*, and added *Hemerocallis*, which yet is not fit, for that his *Hemerocallis Valentina*, is a plaine Pancration or Sea baftard Daffodill, whofe middle cup is longer then the cup of any true Daffodill, which (as I faid in the beginning of this Chapter) is the chiefeft note of difference, betweene a true and a baftard Daffodill. I receiued the feede of this Daffodill among many other feedes of rare plants, from the liberality of Mr. Doctor Flud, one of the Phyfitians of the Colledge in London, who gathered them in the Vniuerfity Garden at Pifa in Italy, and brought them with him, returning home from his trauailes into thofe parts, by the name of *Martagon rarifsimum*, (and hauing fowne them, expected fourteene yeares, before I faw them beare a flower, which the firft yeare that it did flower, bore foure ftalkes of

flowers,

1 *Narcissus tertius Matthioli.* The great white Sea Daffodill. 2 *Narcissus Indicus Autumnalis.* The Indian Autumne Daffodill. 3 *Narcissus marinus Africanus.* The Sea Daffodill of Africa. 4 *Narcissus marinus exoticus.* The strange Sea Daffodill.

flowers, with euery one of them eight or ten flowers on them) which of all other names, doth least anfwer the forme or qualities of this plant. It may moft fitly be called *Narciffus marinus maximus*, in Englifh, The great Sea Daffodill, both becaufe it is a true Daffodill, and the greateft of all other, and alfo becaufe it hath not been found, but in Iflands, or elfe in other places neare the Sea. Lobelius entituleth it *Pancratium Indicum alterum vernum, fiue Narciffus Indicus alter facie Pancratij Monfpeliaci*, but all this is wide from the matter, as may eafily be known, by that that hath been faid before. It is generally (as I faid before) called of all *Narciffus tertius Matthioli*, Matthiolus his third Daffodill, which may either fo paffe with vs, or as I called it, The great Sea Daffodill, which you will, & fo Clufius doth laftly entitle it.

1. *Pancratium Indicum, aut Narciffus Indicus Autumnalis quorundam Lobelij.* The Indian Autumne Daffodill of Lobel.

This plant hath in my opinion, a farre nearer refemblance vnto an Hyacinthus, then vnto any Daffodill: But becaufe Lobel hath fo fet it forth, I will fo publifh it vnto you, leauing it to iudgement. The roote is, as he faith, a fpan long, and of the thickneffe of a mans arme, couered with many white fhells, whereof the outermoft are of a darke red or Chefnut colour: the flowers rife vp in September, and October, being eight or ten in number, euery one by it felfe vpon a fmall footftalke, made of fix leaues a peece, fomewhat long, narrow, and pointed, like vnto the flowers of the Englifh Colchicum, or Medowe Saffron, of a whitifh yellow dunne colour, with fix long threads in the middle: the greene leaues are long and broad, and broad pointed.

2. *Narciffus Marinus Africanus, fiue Exoticus Lobelij.* The Sea Daffodill of Africa.

The roote of this ftrange plant(which of fome likeneffe is called a Daffodill) is very great, made as it were of many fcaly cloues, from whence rifeth vp a fmall fhort ftalke, bearing hard aboue the ground two faire broad greene pointed leaues, more long then broad, fo compaffing the ftalke at the bottome, that it feemeth to run through them: the ftalke is fpotted with diuers difcoloured fpots, and is bare or naked from thefe two leaues vnto the toppe, where it beareth one faire double flower, like vnto a double Auemone, of a delayed reddifh colour, tending to a blufh, with many threads fet about the middle head.

3. *Narciffus Marinus Exoticus.* The ftrange Sea Daffodill

This ftrange Sea Daffodill, hath fiue or fix large and long leaues of a pale greene colour, from among which rifeth vp a ftrong and bigge ftalke, bearing at the toppe, out of a thinne hofe or skinne, many very large flowers, made of fix long and pointed leaues apeece, of a blewifh purple colour, with a large round open cup in the middle, of a fadder colour then the leaues: the roote is very great, yet like vnto other great Daffodils, the outer skins whereof are of a darke browne colour.

The Place.

The Indian Daffodils grew in the vpper part of Hifpaniola in the Weft Indies, and brought hither, where they all foone perifhed.

The other grew neare the Cape of good Hope, and was brought into the parts of Holland and thereabouts, from whence we had it, & perifhed alfo.

The laft is vnknowne where it was gathered.

The Time.

The firft flowred in Autumne, as it is faid.

The other in the firft Summer of their bringing.

And fo did the laft, but the fame rootes will not flower with vs againe.

The

The Names.

So much hath been said of their names in their titles, as hath come to our knowledge; and therefore let that suffice.

Thus hauing gone through the whole Family of the true Daffodils, (for so much as hath come to our knowledge) and set them downe euery one by his name, and in his order; it is fit that we speake of their bastard brethren, and shew you them also, in the same order held with the former, as neare as the plenty of variety herein, which is not the like with the former, will giue leaue, that when you know them both by face and name, you may the better know to place or distinguish of others, that haue not passed vnder this rod.

Pseudonarcissus aureus Hispanicus maximus.
The great yellow Spanish bastard Daffodill.

The roote of this kinde of Daffodill is reasonable great, and blackish on the out-side, desiring to be deepe in the ground; and therefore will runne downe, where it will then encrease into many of-sets, from whence rise vp many thicke, long, and stiffe leaues, of a grayish greene colour, among which riseth vp a round strong stalke, some-times three foote high or better, bearing at the toppe one onely faire great yellow flower, standing forth right, and not pendulous, consisting of six short and somewhat broad leaues, with a very great, large, and long trunke, of an equall largenesse, but open at the mouth, and turning vp the brimmes a little, which are somewhat crum-pled: after the flower is past, there commeth in the place a three square head, contai-ning round blacke seede, like vnto other Daffodils.

Pseudonarcissus Pyrenaus Hispanico & Anglico similis.
The Mountaine bastard Daffodill of diuers kindes.

There is much variety in this kinde of bastard Daffodill: For one sort hath verie broad and whitish greene leaues, somewhat short in comparison of others, that are of that breadth: the flower is wholly yellow, but a little paler then the former Spanish kinde, hauing the leaues of his flower long, and somewhat narrow, standing like wings about the middle trunke, which is as long as the leaues, and smaller then in many other of this kinde, but a little yellower then the wings. Another sort hath narrower green leaues then this last, and longer, the flower is all yellow, but the trunke is larger, wider, and more open at the mouth then the former, and almost as large as the former Spa-nish, but not so high as the last. A third hath the wings of the flower of a Strawe co-lour, but the trunke is long and narrow, of a faire yellow. A fourth hath such like flowers, but that it is shorter, both the wings and the trunke: Some likewise haue the wings of the flower longer, then the long trunke, and some shorter. Some also are all yellow, and some haue their wings onely a little more pale or white, like the English kinde: Some againe haue their trunkes long and narrow, others haue them larger and wider open, and crumpled at the brimmes; so that it is needlesse, to spend a great deale of time and labour vpon such smally respected flowers, but that in the beholding of them, we may therein admire the worke of the Creatour, who can frame such diuersity in one thing: But this is beside the text, yet not impertinent.

Pseudonarcissus pallidus præcox. The early Strawe coloured bastard Daffodill.

The leaues of this Daffodill are of a meane size, betweene the broadest and the nar-rower kindes, of a grayish greene colour, and not very long: the stalke riseth vp a foot high or more, whereon standeth one large great flower, equalling the greatest Spanish bastard Daffodill, before described, in the largenesse of his trunke, and hauing the brimmes turned vp a little, which maketh it seeme the larger: the wings or outer leaues are in a maner as short, as they are in the greatest Spanish kinde, (and not long flagging down, like vnto the Mountain kinds) and stand straight outright: all the whole flower is

of

of one euen colour, that is, of a fine pale yellow, somewhat like vnto the colour of a Lemon peele or rinde, but somewhat whiter, which vsually we call a Strawe colour: the greatnesse of the flower, the earlinesse of the flowring, and the difference of colour from all the rest of this kinde, hath made me entreate of it apart by it selfe, as being no lesse worthy.

Pseudonarcissus Hispanicus flore albo maior.
The great white Spanish bastard Daffodill.

This bastard Daffodill hath diuers leaues rising vp together, long and broad, somewhat like vnto the first Spanish kinde, but a little broader, and of a whiter greene colour, yet not so white, as in the lesser Spanish white kindes, here after described: among these leaues riseth vp a round strong stalke, about two foote high, bearing one white flower at the toppe, bending downe the head, as all these white kindes doe, but is not of so pure a white, as the lesser kindes that follow, yet whiter then the greatest white Spanish kinde, next of all to be described: the whole flower, as well trunke as wings, is much larger then the lesser white kindes, and almost equalling the first Spanish yellow, but a little longer and narrower, a little crumpled and turning vp at the brimmes: the head and seede are like the first; the roote is greater and thicker then the first Spanish, and doth not encrease so much, nor is couered with a blacke, but rather with a whitish coate.

Pseudonarcissus Hispanicus maximus albidus.
The greatest Spanish white bastard Daffodill.

This kinde of bastard Daffodill is very like the last mentioned Daffodill, both in leaues and flowers, but larger in both: the flower of this is not full so white, but hath some shew of palenesse therein, and more vpon the first opening of the flower then afterwards, and is as great altogether, as the great Spanish yellow, at the least with a longer, and somewhat narrower trunke: the seede is like vnto the former, and so is the roote also, but greater, being white on the outside, and not blacke.

Pseudonarcissus Hispanicus flore albo medius & minor.
The two lesser white Spanish bastard Daffodils.

There are two other of these kindes of white Spanish Daffodils, one greater or lesser then the other, but neither of them so great as the former. The leaues of both are of a whitish greene colour, one a little broader then the other: the flowers of both are pure white, and bending downe the heads, that they almost touch the stalke againe, the greater flower hath the longer and narrower trunke; and the lesser flower, the shorter and wider open, yet both a little crumpled at the edges or brimmes: the rootes of both are like one vnto another, but differ in the greatnesse. From the seede of these haue sprung much variety, few or none keeping either colour or height with the mother plants.

Pseudonarcissus Anglicus vulgaris. Our common English wilde bastard Daffodill.

This bastard Daffodill is so common in all England, both in Copses, Woods, and Orchards, that I might well forbeare the description thereof, and especially, in that growing wilde, it is of little respect in our Garden: but yet, left I bee challenged of ignorance in common plants, and in regard of some variety therein worth the marking, I will set downe his description and variety as briefly as I may: It hath three or foure grayish greene leaues, long and somewhat narrow, among which riseth vp the stalke, about a span high or little higher, bearing at the toppe, out of a skinnie huske, as all other Daffodils haue, one flower (although sometimes I haue seene two together) somewhat large, hauing the six leaues that stand like wings, of a pale yellow colour, and the long trunke in the middle of a faire yellow, with the edges or brimmes a little crumpled or vneuen: after the flower is past, it beareth a round head, seeming three square, containing round blacke seede; the roote is somewhat blackish on the outside.

1 *Pseudonarcissus Hispanicus maximus aureus.* The great yellow Spanish bastard Daffodill. 2 *Pseudonarcissus Pyrenaus versiformis.* The Mountaine bastard Daffodill of divers kindes. 3 *Pseudonarcissus Hispanicus maior albus.* The greater white Spanish bastard Daffodill. 4 *Pseudonarcissus Hispanicus minor albus.* The little Spanish white bastard Daffodill. 5 *Pseudonarcissus tubo sexangulari.* The six cornered bastard Daffodill. 6 *Pseudonarcissus maximus aureus, sive Roseus Tradescanti.* Iohn Tradescants great Rose Daffodill. 7 *Pseudonarcissus aureus Anglicus maximus.* Master Wilmers great double Daffodill. 8 *Pseudonarcissus Hispanicus aureus flore pleno.* The double Spanish Daffodill, or Parkinsons double Daffodill. 9 *Pseudonarcissus Gallicus maior flore pleno.* The greater double French Daffodill. 10 *Pseudonarcissus Anglicus flore pleno,* The double English Daffodill, or Gerrards double Daffodill.

I 3

But there is another of this kinde like vnto the former, whose further description you haue here before; the wings of which flower are much more white then the former, and in a manner of a milke white colour, the trunke remaining almost as yellow as the former, and not differing in any thing else.

Pseudonarcissus tubo sexangulari. The six cornered bastard Daffodill.

This kinde of Daffodill hath two or three long, and somewhat broader leaues then the last, between which commeth forth a stalke, bearing one flower somewhat large, hauing the six outer leaues of a pale yellow colour, and the long trunke plaited or cornered all along vnto the very edge into six parts, of a little deeper yellow then the wings.

The Place.

The first great Spanish kinde was brought out of Spaine. The rest from the Pyrenæan Mountaines, onely the last sauing one is plentifull in our owne Countrey, but the white sort of that kinde came with the rest from the same Mountaines.

The Time.

The pale or third kinde, and the English bee the most early, all the rest flower in Aprill, and the greatest yellow somewhat earlier, then the other greater or lesser white.

The Names.

Their seuerall names are expressed in their titles sufficient to distinguish them, and therefore there needeth no more to be said of them.

1. *Pseudonarcissus aureus maximus flore pleno, siue Roseus Tradescanti.* The greatest double yellow bastard Daffodill, or Iohn Tradescant his great Rose Daffodill.

This Prince of Daffodils (belongeth primarily to Iohn Tradescant, as the first founder thereof, that we know, and may well bee entituled the Glory of Daffodils) hath a great round roote, like vnto other Daffodils, couered with a brownish outer skinne or peeling, from whence riseth vp foure or fiue somewhat large and broad leaues, of a grayish greene colour, yet not fully so long and large as the next following Daffodill: from the middle whereof riseth vp a stalke almost as high and great as it, bearing at the toppe (out of a skinnie huske) one faire large great flower (the budde, before it breake open, being shorter and thicker in the middle, and ending in a longer and sharper point then any of the other Daffodils) very much spread open, consisting of smaller and shorter leaues then the next, but more in number, and thicker and rounder set together, making it seeme as great and double as any Prouince Rose, and intermixt with diuers yellow and pale leaues, as it were in rowes one vnder another. It abideth long in flower, and spreadeth, by standing long, to be the broadest in compasse of any of the Daffodils, but falleth away at the last without giuing any seede, as all double Daffodils doe.

2. *Pseudonarcissus aureus Anglicus maximus.* Mr. Wilmers great double Daffodill.

The other great double Daffodill doth so neare resemble our ordinary English double kinde, that I doe not finde therein any greater difference, then the largenesse both of leaues and flowers, &c. and the statelinesse of growth. It beareth three or foure large, long, and broad leaues, somewhat longer and broader then the former, and of a whitish greene colour: the stalke riseth to bee two foote high, growing (in a fruitfull and fat soyle) strong, and somewhat round, bearing at the toppe, out of a thin skinne, one great and faire double flower, each leafe whereof is twice as large and

broad

broad as the former, diuersly intermixt with a rowe of paler, and a rowe of deeper yellow leaues, wholly disperfed throughout the flower, the pale colour as well as the deeper yellow, in this as in the other small Englifh kinde, growing deeper by ftanding: fometimes the leaues hereof are fcattered, and fpread wholly, making it fhew a faire, broad, open flower: and fometimes the outer leaues ftand feparate from the middle trunke, which is whole and vnbroken, and very thicke of leaues: and fometimes the middle trunke will bee halfe broken, neither expreffing a full open double flower, nor a clofe double trunke, as it is likewife feene in the fmall Englifh kinde, as fhall bee declared in his place: this beareth no feede; the roote hereof is thicke and great, and encreafeth as well as any other Daffodill.

3. *Pfeudonarciffus aureus Hifpanicus flore pleno.*
The great double yellow Spanifh baftard Daffodill, or Parkinfons Daffodill.

This double Spanifh Daffodill hath diuers leaues rifing from the roote, ftiffer, narrower, and not of fo whitifh a greene colour as the former, but more fullen or grayifh, plainely refembling the leaues of the fingle great kinde, from whence this hath rifen: the ftalke hereof likewife rifeth almoft as high as it, and neare the height of the laft recited double, bearing one double flower at the toppe, alwayes fpread open, and neuer forming a double trunke like the former, yet not fo faire and large as it, the outermoft leaues whereof being of a greenifh colour at the firft, and afterward more yellow, doe a little turne themfelues backe againe to the ftalke, the other leaues are fome of a pale yellow, and others of a more gold yellow colour, thofe that ftand in the middle are fmaller, and fome of them fhew as if they were hollow trunked, fo that they feeme to be greenifh, whitifh, yellow, and gold yellow, all mixed one among another: the root is great, round, and whitifh on the infide, couered with darke coloured skinnes or peelings. I thinke none euer had this kinde before my felfe, nor did I my felfe euer fee it before the yeare 1618. for it is of mine own raifing and flowring firft in my Garden.

4. *Pfeudonarciffus Gallicus maior flore pleno.*
The greater double French baftard Daffodill.

This greater double Daffodill, hath his whitifh greene leaues longer and broader then the fmaller French kinde, hereafter following, to bee defcribed, and broader, longer, and more limber then the double Englifh kinde: the ftalke rifeth vp not much higher, then the fmaller French kinde, but a little bigger, bearing at the top one great double flower, which when it is fully and perfectly blowne open (which is but feldome; for that it is very tender, the leaues being much thinner, and thereby continually fubiect, vpon any little diftemperature of the time, to cleaue fo faft one vnto another, that the flower cannot blow open faire) is a faire and a goodly flower, larger by halfe then the fmaller kinde, and fuller of leaues, of the fame pale whitifh yellow, or Lemon colour, with the leffer, or rather a little whiter, and not fet in the fame order of rowes as it is, but more confufedly together, and turning backe the ends of the outermoft leaues to the ftalke againe, and hauing the bottome of the flower on the backfide fomewhat greene, neither of which is found in the leffer kinde: the roote is very like vnto the leffer kinde, but a little bigger and longer.

5. *Pfeudonarciffus Anglicus flore pleno.*
The double Englifh baftard Daffodill, or Gerrards double Daffodill.

The leaues of this double Daffodill are very like vnto the fingle kinde, being of a whitifh greene colour, and fomewhat broad, a little fhorter and narrower, yet ftiffer then the former French kinde: the ftalke rifeth vp about a foote high, bearing at the toppe one very double flower, the outermoft leaues being of the fame pale colour, that is to bee feene in the wings of the fingle kinde; thofe that ftand next them, are fome as deepe a yellow as the trunke of the fingle, and others of the fame pale colour, with fome greene ftripes on the backe of diuers of the leaues: thus is the whole flower variably intermixt with pale and deepe yellow, and fome greene ftripes among them,

when

when it is fully open, and the leaues diſperſed and broken. For ſometimes the flower ſheweth a cloſe and round yellow trunke in the middle, ſeparate from the pale outer wings, which trunke is very double, ſhewing ſome pale leaues within it, diſperſed among the yellow: And ſometimes the trunke is more open, or in part broken, ſhewing forth the ſame colours intermixt within it: the flower paſſeth away without giuing any ſeede, as all other bulbous rootes doe that beare double flowers: the roote is ſmall, very like vnto the French double kindes, eſpecially the leſſer, that it is verie hard to know the one from the other.

The Place.

The firſt and greateſt kinde, we had firſt from Iohn Tradeſcante (as I ſaid before) whether raiſed from ſeed, or gained from beyond Sea, I know not.

The ſecond we firſt had from Vincent Sion, borne in Flanders, dwelling on the Banke ſide, in his liues time, but now dead; an induſtrious and worthy louer of faire flowers, who cheriſhed it in his Garden for many yeares, without bearing of any flowers vntill the yeare 1620. that hauing flowred with him, (and hee not knowing of whom hee receiued it, nor hauing euer ſeene the like flower before) he ſheweth it to Mr. Iohn de Franqueuille, of whom he ſuppoſed he had receiued it, (for from beyond Sea he neuer receiued any) who finding it to bee a kinde neuer ſeene or knowne to vs before, cauſed him to reſpect it the more, as it is well worthy. And Mr. George Wilmer of Stratford Bowe Eſquire, in his liues time hauing likewiſe receiued it of him (as my ſelfe did alſo) would needes appropriate it to himſelfe, as if he were the firſt founder thereof, and call it by his owne name Wilmers double Daffodill, which ſince hath ſo continued.

The third is of mine owne foſtering or raiſing, as I ſaid before; for aſſuredly, it is riſen from the ſeede of the great Spaniſh ſingle kinde, which I ſowed in mine owne Garden, and cheriſhed it, vntill it gaue ſuch a flower as is deſcribed.

The fourth is not certainly knowne where his originall ſhould be: Some thinke it to be of France, and others of Germany.

The laſt is aſſuredly firſt naturall of our owne Countrey, for Mr. Gerrard firſt diſcouered it to the world, finding it in a poore womans Garden in the Weſt parts of England, where it grew before the woman came to dwell there, and, as I haue heard ſince, is naturall of the Iſle of Wight.

The Time.

They doe all flower much about one time, that is, from the middle or end of March, as the yeare is forward, vnto the middle of Aprill.

The Names.

Vpon the three firſt I haue impoſed the names in Latine, as they are expreſſed in their titles: and for the Engliſh names, if you pleaſe, you may let them paſſe likewiſe as they are expreſſed there alſo, that thereby euery one may be truely diſtinguiſhed, and not confounded. The fourth, beſides the name in the title, is called of ſome *Narciſſus Germanicus*, which whether it be of Germany, or no, I know not; but that the name ſhould import ſo much. The laſt doth vſually carry Mr. Gerrards name, and called Gerrards double Daffodill.

1. *Pſeudonarciſſus anguſtifolius flore flaueſcente tubo quaſi abſciſſo.*
   The narrow leafed baſtard Daffodill with the clipt trunke.

This kinde of Daffodill hath long and narrow grayiſh greene leaues, bearing one ſingle flower at the toppe of his ſtalke, like vnto the former ſingle baſtard kindes, before

fore specified, hauing his outer leaues of a pale yellow colour, and his trunke of a deeper yellow : the chiefe differences in this from the former, is in the leaues, being narrow, and then in the trunke of the flower, which is not crumpled or turned vp, as most of the other are; and that the brimmes or edges of the flower is as if it had beene clipt off, or cut euen.

2. *Pseudonarcissus Hispanicus medius & minor luteus.*
The two lesser Spanish yellow bastard Daffodils.

These two lesser kindes of Spanish Daffodils, doe but differ in greatnesse the one from the other, and not in any thing else ; so that in declaring the one, you may vnderstand the other to bee a little greater. The lesser then hath three or foure narrow short whitish greene leaues, from among which commeth forth a short stalke, not aboue an hand breadth, or halfe a foote high, bearing one single flower, not fully standing outright, but a little bending downe, consisting of six small leaues, standing as wings about a small, but long trunke, a little crumpled at the brimmes : the whole flower, as well leaues as trunke, are of one deepe yellow colour, like vnto the great Spanish kinde : the roote is but small, and couered with a darkish coate. The other is in all parts greater, and (as I said) differeth not else.

3. *Pseudonarcissus Hispanicus luteus minimus.*
The least Spanish yellow bastard Daffodill.

The leaues of this small kinde are smaller and shorter then the former, seldome exceeding the length of three inches, and very narrow withall, but of the same grayish greene colour with the former : euery flower standeth vpon a small and short footestalke, scarce rising aboue the ground, so that his nose, for the most part, doth lye or touch the ground, and is made after the same fashion, and of the same colour with the former, but much smaller, as his roote is so likewise.

4. *Pseudonarcissus Gallicus minor flore pleno.*
The lesser French double bastard Daffodill.

The rootes of this lesser French kinde (if I may lawfully call it, or the greater kinde before specified, a bastard Daffodill ; for I somewhat doubt thereof, in that the flower of either is not made after the fashion of any of the other bastard Daffodils, but doth more nearely resemble the forme of the double white Daffodill, expressed before among the true Daffodils) are like vnto the double English kinde, as also to the former double greater French kinde, and the leaues are of the same whitish greene colour also, but narrower and not longer : the stalke riseth a little higher then the English, and not fully so high as the greater French, bearing one faire double flower thereon, of a pale yellow or Lemon colour, consisting of six rowes of leaues, euery rowe growing smaller then other vnto the middle, and so set and placed, that euery leafe of the flower doth stand directly almost in all, one vpon or before another vnto the middle, where the leaues are smallest, the outermost being the greatest, which maketh the flower seeme the more beautifull : this and the greater kinde hath no trunke, or shew of any other thing in the middle, as all or most of the other former double bastard Daffodils haue, but are flowers wholly composed of leaues, standing double euen to the middle.

The Place.

The first is vndoubtedly a naturall of the Pyrenæan Mountaines.
The Spanish kindes grew in Spaine, and
The French double kinde about Orleance in France, where it is said to grow plentifully.

The Time.

The first flowreth at the end of March.

The Spanish kindes are the most early, flowring betimes in March. The French double doth flower presently after.

The Names.

More cannot bee said or added, concerning the names of any of these Daffodils, then hath been set downe in their titles : onely the French kinde is most vsually called Robinus his Daffodill.

*Pseudonarcissus Iuncifolius albus.*The white bastard Rush Daffodill, or Iunquilia.

This bastard Rush Daffodill hath two or three long and very greene leaues, very like vnto the small yellow Rush Daffodill, formerly described, but not altogether so round, among which riseth vp a short stalke, seldome halfe a foote high, bearing at the toppe, out of a small skinnie huske, one small white flower, sometime declining to a pale colour, hauing six small and short leaues, standing about the middle of the trunke, which is long, and much wider open at the mouth, then at the bottome : the small outer leaues or wings are a little tending to greene, and the trunke (as I said) is either white, or whitish, hauing the brimmes a little vneuen : the seede is small, blacke, and round, like vnto other Rush Daffodils, but smaller.

Pseudonarcissus Iuncifolius luteus maior.
The greater yellow Iunquilia, or bastard Daffodill.

The leaues of this greater kinde are longer, greater, and a little broader then the former ; the stalke also is higher, and the flower larger, more open at the mouth and crumpled, then the white, but wholly of a yellow colour : the seede and the roots are bigger, according to the proportion of the plant.

*Pseudonarcissus Iuncifolius luteus minor.*The lesser yellow bastard Iunquilia.

This is so like vnto the last in all things, that I shall not neede to trouble you with repetitions of the same things formerly spoken ; the chiefest difference is the smalnesse of the plant in all parts.

*Pseudonarcissus Iuncifolius luteus serotinus.*The late yellow bastard Iunquilia.

There is likewise a third kinde, as great as the greater yellow, and in all his parts expressing and equalling it, but is accounted the fairer, and flowreth somewhat later.

The Place.

The Pyrenæan Hils haue afforded vs all these varieties, and wee preserue them carefully ; for they are all tender.

The Time.

All these flower in Aprill, except the last, which is a moneth later.

The Names.

The French and Lowe-Countrey men call them *Trompettes*, that is, Trumpets, from the forme of the trunke ; wee sometimes call them also by that name, but more vsually bastard Iunquilia's.

Pseudonarcissus marinus albus, Pancratium vulgo.
The white Sea bastard Daffodill.

The Sea bastard Daffodill (to conclude this Chapter, and the discourse of Daffo-
dills)

1 *Pseudonarcissus rubo quasi abscissе*.The bastard Daffodill with the clipt trunke.　2 *Pseudonarcissus Hispanicus minor*.The lesser Spanish bastard Daffodill.
3 *Pseudonarcissus Hispanicus minimus*.The least Spanish bastard Daffodill.　4 *Pseudonarcissus Gallicus maior flore pleno*.The lesser double French bastard Daffodill.　5 *Panoratum flore albo*.The white sea bastard Daffodill.　6 *Pseudonarcissus iuncifolius luteus maior*.The greater yellow bastard Iunquilia.　7 *Pseudonarcissus iuncifolius luteus minor*.The lesser yellow bastard Iunquilia.　8 *Pseudonarcissus iuncifolius luteus serotinus*.The late yellow bastard Iunquilia.　9 *Leucoium bulbosum praecox maius*.The great early bulbous Violet.　† *Leucoium bulbosum praecox minus*.The lesser early bulbous Violet.
10 *Leucoium bulbosum autumnale*.The small Autumne bulbous Violet.　11 *Leucoium bulbosum maius serotinum*.The great late flowring bulbous Violet.

dils) hath diuers broad whitish greene leaues , but not very long, among which riseth vp a stiffe round stalke, at the top whereof breaketh out of a great round skinny huske, fiue or six flowers, euery one made somewhat of the fashion of the great bastard Rush Daffodill, but greater, and wholly white ; the six leaues, being larger and longer then in the Rush kinde, and extending beyond the trunke , are tipt with greene at the point of each leafe , and downe the middle likewise on the backside. The trunke is longer, larger, and wider open at the mouth, cut in or indented at the brims or edges, and small at the bottome, with diuers white threeds in the middle, and is very sweet: vnder the flower is a round greene head, which groweth very great, hauing within it, when it is ripe, flat and blacke seede : the roote is great and white.

Flore luteo, & flore rubro. It is reported, that there are found other sorts; some that beare yellow flowers, and others that beare red: but we haue seene none such, and therefore I can say no more of them.

The Place.

This kinde groweth neare the Sea side, both in Spaine, Italy, and France, within the Straights, and for the most part, vpon all the Leuant shoare and Islands also, but will seldome either flower, or abide with vs in these colder Countries, as I haue both seene by those that I receiued from a friend, and heard by others.

The Time.

It flowreth in the end of Summer, that is, in August and September.

The Name.

Diuers doe call it *Pancratium*, as the learned of Mompelier, and others, with the addition of *flore Lilij*, after they had left their old errour, in taking it to be *Scylla*, and vsing it for *Scylla*, in the *Trochisces* that go into Andromachus Treakle. The learned of Valentia in Spaine, as Clusius saith, doe call it *Hemerocallis*, thinking it to be a Lilly; and Clusius doth thereupon call it, *Hemerocallis Valentina* : but in my opinion, all these are deceiued in this plant ; for it is neither a Lilly , to haue the name of *Hemerocallis* giuen vnto it, nor *Scylla*, nor *Pancratium*, as many doe yet call it : for certainly this is a kinde of Daffodill, the forme both of roote, leafe, and flower, doth assure me that haue seene it, and not *Pancratium*, which (as Dioscorides testifieth) is a kinde of *Scylla*, and in his time called *Scylla*, with a red roote, and a leafe like a Lilly , but longer, and was vsed both with the same preparation and quantity , and for the same diseases that *Scylla* was vsed , but that his force was weaker: all which doth plainly shew the errours that many learned men haue been conuersant in, and that all may see how necessary the knowledge of Herbarisme is to the practice of Physicke ; And left the roote of this Sea bastard Daffodill bee vsed in the stead of an wholsome remedy , which (as Clusius maketh mention) was deadly to him that did but cut his meate with that knife, which had immediately before cut this roote, and done in malice by him, that knew the force thereof, to kill his fellow, it working the more forceably by the euill attracting quality of the iron.

The Vertues of Daffodils in generall.

Howsoeuer Dioscorides and others, doe giue vnto some of them speciall properties, both for inward and outward diseases, yet know I not any in these dayes with vs, that apply any of them as a remedy for any griefe, whatsoeuer Gerrard or others haue written.

CHAP.

CHAP. X.

Leucoium bulbosum. The bulbous Violet.

Auing thus set downe the whole family, both of the true and baftard Daffodils, I fhould next fet in hand with the Hyacinths ; but becaufe *Leucoium bulbofum,* The bulbous Violet is a plant that doth challenge a place next vnto the
Daffodils, as moft nearly partaking with them, and a little with the Hyacinthes, I muft
of neceffity interpofe them, and fhew their defcriptions and differences, whereof fome
are early, of the firft Spring, others later, and fome of the Autumne.

Leucoium bulbofum præcox maius. The greater early bulbous Violet.

This bulbous Violet hath three or foure very greene, broad, flat, and fhort leaues,
among which rifeth vp a naked greene ftalke, bearing out of a fmall skinny hofe (as the
former Daffodils doe) one white flower, hanging downe his head by a very fmall
foot-ftalke, made of fix leaues, of an equall length, euery one whereof is tipt at the
end with a fmall greenifh yellow fpot : after the flower is paft, the head or feed-veffell
groweth to be reafonable great, fomewhat long and round, wherein is contained hard
round feede, which being dry, is cleare, and of a whitifh yellow colour : the roote is
fomewhat like a Daffodill roote, and couered with a blackifh outfide or skinne.

Leucoium bulbofum præcox minus. The leffer early bulbous Violet.

This leffer kinde rifeth vp with two narrow grayifh greene leaues, betweene which
commeth forth the ftalke, fiue or fix inches high, bearing one fmall pendulous flower,
confifting of three white leaues, which are fmall and pointed, ftanding on the outfide, and hauing three other fhorter leaues, which feeme like a cup in the middle, being
each of them round at the ends, and cut in the middle, making the forme of an heart,
with a greene tippe or fpot at the broad end or edge : the feede is whitifh, inclofed in
long and round heads, like the former, but leffer : the roote is like a fmall Daffodill,
with a blackifh gray coate, and quickly diuideth into many of-fets.
There is another of this kinde, that came among other bulbous rootes from Con- *Minus Byzan-*
ftantinople, and differeth in nothing from it, but that it is a little greater, both in root, *tinum.*
leafe, and flower.

The Place.

The two firft are found in many places of Germany, and Hungary. The
third, as I faid, was brought from Conftantinople.

The Time.

The two leffer forts doe moft commonly flower in February, if the weather be any thing milde, or at the furtheft in the beginning of March, but
the firft is feldome in flower, before the other be well neare paft, or altogether.

The Names.

Lobel and Dodonæus call the leffer kinde *Leucoium triphyllum,* and *Leu-
conarciffolirion triphyllum,* of the three leaues in the flower. Some doe call
it *Viola bulbofa alba.* The firft or greater kinde is called by Lobel, *Leuconar-
ciffolirion paucioribus floribus ;* and by Dodonæus, *Leucoium bulbofum hexa-
phyllum.* We doe moft vfually call them, *Leucoium bulbofum præcox maius, &
minus,* The greater, or the leffer early bulbous Violet. In Dutch, *Somer
Sottekens,* and not *Druifkens,* which are Grape-flowers, as fome haue
thought.

K   *Leucoium*

1. *Leucoium bulboſum Vernum minimum.*
The ſmall bulbous Violet of the Spring.

This ſmall *Leucoium* ſendeth forth his ſmall and long greene leaues, like haires in Autumne, and before Winter, which abide greene vntill Aprill, and then wither away quite, and about May there ariſeth vp a naked ſlender ſtalke, at the toppe whereof breake forth two ſmall white flowers, made of ſix leaues a peece, hanging downe their heads, the three inner leaues being a little larger then the three outward, a little reddiſh neare the ſtalke, and very ſweet : the root is ſmall and round, and couered with a darke coate.

2. *Leucoium bulboſum Autumnale.* The ſmall Autumne bulbous Violet.

As the former ſmall *Leucoium* ſprang vp with his leaues without flowers in Autumne, ſo this contrariwiſe, riſeth vp with his ſlender browniſh ſtalke of flowers in Autumne, before any greene leaues appeare, whereon ſtand two or three very ſmall ſnow white pendulous flowers, conſiſting of ſix leaues a peece, and a little reddiſh at the bottome of the flower next vnto the ſtalke, ſo like vnto the former, that one would take them to be both one : after which, there grow ſmall browne heads, containing ſmall, blacke, round ſeed ; after the flower is paſt, and the ſeede is ripening, and ſometimes after the heads are ripe, the leaues begin to ſpring vp, which when they are full growne, are long, greene, and as ſmall, or ſmaller then the leaues of the Autumne Hyacinth, which abide all the Winter, and Spring following, and wither away in the beginning of Summer : the roote is ſmall, long, and white.

3. *Leucoium maius bulboſum ſerotinum.*
The great late flowring bulbous Violet.

The late bulbous Violet hath three or foure broad flat greene leaues, very like vnto the firſt, but longer, among which riſeth vp a flattiſh ſtalke, being thicker in the middle then at both edges, on the toppe whereof ſtand three or foure flowers, hanging downe their heads, conſiſting of ſix leaues a peece, all of an equall length and bigneſſe, wholly white, except that each leafe hath a greene tippe at the end of them : the ſeede hereof is blacke and round ; the roote is reaſonable great and white.

The Place.

The two former ſmall ones were firſt found in Spaine, and Portugall, and ſent to me by Guillaume Boel ; but the firſt was ſo tender, that ſcarce one of a ſcore ſprang with me, or would abide. The greateſt haue beene found wilde in Germany and Auſtria.

The Time.

The ſmall ones haue their times expreſſed in their titles and deſcriptions; the laſt flowreth not vntill May.

The Names.

Theſe names that are ſet downe in their titles, doe paſſe with all Herbariſts in theſe daies.

The Vertues.

Wee haue not knowne theſe plants vſed Phyſically, either inwardly or outwardly, to any purpoſes in theſe dayes.

CHAP.

CHAP. XI.

Hyacinthus. The Hyacinth or Iacinth.

THe Iacinths are next to be entreated of, whereof there are many more kindes found out in thefe later times, then formerly were knowne, which for order and method fake, I will digeft vnder feuerall forts, as neare as I can, that a-uoiding confufion, by enterlacing one among another, I may the better put euery fort vnder his owne kinde.

Hyacinthus Indicus maior tuberofa radice.
The greater Indian knobbed Iacinth.

I haue thought fitteft to begin with this Iacinth, both becaufe it is the greateft and higheft, and alfo becaufe the flowers hereof are in fome likeneffe neare vnto a Daffo-dill, although his roote be tuberous, and not bulbous as all the reft are. This Indian Iacinth hath a thicke knobbed roote (yet formed into feuerall heads, fomewhat like vnto bulbous rootes) with many thicke fibres at the bottome of them ; from the di-uers heads of this roote arife diuers ftrong and very tall ftalkes, befet with diuers faire, long, and broad leaues, ioyned at the bottome clofe vnto the ftalke, where they are greateft, and grow fmaller to the very end, and thofe that grow higher to the toppe, being fmaller and fmaller, which being broken, there appeare many threeds like wooll in them : the toppes of the ftalkes are garniſhed with many faire large white flowers, each whereof is compofed of fix leaues, lying fpread open, as the flowers of the white Daffodill, with fome fhort threeds in the middle, and of a very fweete fcent, or rather ftrong and headie.

Hyacinthus Indicus minor tuberofa radice.
The fmaller Indian knobbed Iacinth.

The roote of this Iacinth is knobbed, like the roote of Arum or Wake Robin, from whence doe fpring many leaues, lying vpon the ground, and compaffing one another at the bottome, being long and narrow, and hollow guttered to the end, which is fmall and pointed, no leffe woolly, or full of threeds then the former : from the middle of thefe leaues rifeth vp the ftalke, being very long and flender, three or foure foot long, fo that without it be propped vp, it will bend downe, and lye vpon the ground, where-on are fet at certaine diftances many fhort leaues, being broad at the bottome, where they doe almoft compaffe the ftalke, and are fmaller toward the end where it is fharpe pointed: at the top of the ftalke ftand many flowers, with a fmall peece of a green leafe at the bottome of euery foot-ftalke, which feeme to bee like fo many white Orientall Iacinths, being compofed of fix leaues, which are much thicker then the former, with fix chiues or threeds in the middle, tipt with pale yellow pendents.

The Place.

They both grow naturally in the Weft Indies, from whence being firft brought into Spaine, haue from thence been difperfed vnto diuers louers of plants.

The Time.

They flower not in thefe cold Countries vntill the middle of Auguft, or not at all, if they bee not carefully preferued from the iniury of our cold Winters ; and then if the precedent Summer be hot, it may be flower a mo-neth fooner.

The Names.

Clufius calleth the leffer (for I thinke hee neuer faw the firft) *Hyacinthus*

K 2   *Indicus*

Indicus tuberosa radice, that is in English, The Indian Iacinth with a tuberous roote : Some would call these *Hyacinthus Eriophorus Indicus*, that is, The Indian woolly Iacinth, because they haue much wooll in them when they are broken; yet some doe doubt that they are not two plants seuerall, as of greater and lesser, but that the greatnesse is caused by the fertility of the soyle wherein it grew.

1. *Hyacinthus Botroides maior Moschatus, siue Muscari flore flauo.*
The great yellow Muske Grape-flower, or yellow Muscari.

This Muske Iacinth or Grape-flower, hath fiue or six leaues spread vpon the ground in two or three heads, which at the first budding or shooting forth out of the ground, are of a reddish purple colour, and after become long, thicke, hollow, or guttered on the vpperside, of a whitish greene colour, and round and darke coloured vnderneath : in the middle of these heads of leaues, rise vp one or two hollow weake brownish stalkes, sometimes lying on the ground with the weight of the flowers, (but especially of the seede) yet for the most part standing vpright, when they are laden towards the toppe, with many bottle-like flowers, which at their first appearing, and vntill the flowers begin to blow open, are of a browne red colour, and when they are blowne, of a faire yellow colour, flowring first below, and so vpwards by degrees, euery one of these flowers is made like vnto a little pitcher or bottle, being bigge in the belly, and small at the mouth, which is round, and a little turned vp, very sweete in smell, like vnto Muske, whereof it tooke the name *Muscari*; after the flowers are past, there come three square thicke heads, puffed vp as if it were bladders, made of a spongie substance, wherein are here and there placed blacke round seed : the roote is long, round, and very thicke, and white on the outside, with a little woollinesse on them, being broken, and full of a slimie iuice, whereunto are annexed thicke, fat, and long fibres, which perish not as most of the other Iacinths; and therefore desireth not to bee often remoued, as the other sorts may.

2. *Hyacinthus Botroides maior Moschatus, seu Muscari flore cineritio.*
The Ashcoloured Muske Grape-flower, or Muscari.

This Muscari differeth not in rootes, or forme of leaues or flowers from the former, the chiefe differences are these: the leaues hereof do not appeare so red at the first budding out of the ground, nor are so darke when they are fully growne; the stalke also most vsually hath more store of flowers thereon, the colour whereof at the first budding is a little duskie, and when they are full blowne, are of a bleake, yet bright ash-colour, with a little shew of purple in them, and by long standing change a little more gray; being as sweete, or as some thinke, more sweete then the former : the roote (as I said) is like the former, yet yeeldeth more encrease, and will better endure our cold clymate, although it doth more seldome giue ripe seede.

3. *Hyacinthus Botroides maior Moschatus, siue Muscari flore rubro.*
The red Muske Grape-flower.

This kinde (if there be any such, for I am in some doubt thereof) doth chiefly differ in the colour of the flower from the first, in that this should beare flowers when they are blowne, of a red colour tending to yellownesse.

4. *Hyacinthus Botroides maior Moschatus, siue Muscari flore albo.*
The white Muske Grape-flower.

This also is said to haue (if there bee such an one) his leaues like vnto the second kinde, but of a little whiter greene, and the flowers pale, tending to a white : the roots of these two last are said vsually not to grow to be so great as of the former two.

The Place.

The rootes of the two first sorts, haue been often sent from Constantinople,

1 *Hyacinthus Indicus maior tuberosa radice.* The greater Indian knobbed Iacinth. 2 *Hyacinthus Indicus minor tuberosa radice.* The lesser Indian knobbed Iacinth. 3 *Muscari flore flauo.* The yellow Muscari. 4 *Muscari flore einereitio.* The asheoloured Muscari. 5 *Hyacinthus Botroides caeruleus amoenus.* The skie coloured Grape-flower. 6 *Hyacinthus Botroides flore albo.* The white Grape flower. † *Hyacinthus Botroides ramosus.* The branched Grape-flower.

ple, among many other forts of rootes, and it may be come thither from beyond the Bofphorus in Afia; we haue them in our Gardens.

The other two forts are fprung (it is probable, if they be *in rerum natura*) from the feede of the two former; for we could neuer get fuch from Conftantinople, as if the Turkes had neuer knowledge of any fuch.

The Time.

They flower in March or Aprill, as the yeare is temperate, but the firft is fooneft vp out of the ground.

The Names.

The two former haue beene fent from Turkie by the name of *Mufchoromi* and *Dipcadi.* Matthiolus calleth it *Bulbus vomitorius,* faying that no root doth more prouoke vomit then it. Cafpar Bauhinus doth moft properly call it *Hyacinthus Mofchatus.* It is moft generally called *Mufcari,* by all Herbarifts and Florifts, yet becaufe it doth fo neerely refemble the Grape-flower, I haue named it *Hyacinthus Botroides maior Mufchatus,* to put a difference from the leffer Grape-flowers that follow; in Englifh, The great Muske Grape-flower, or Mufcari.

Hyacinthus Botroides minor cæruleus obfcurus. The darke blew Grape-flower.

This Grape-flower hath many fmall, fat, and weake leaues lying vpon the ground, which are fomewhat brownifh at their firft comming vp, and of a fad greene afterwards, hollow on the vpperfide, and round vnderneath, among which rife vp round, fmooth, weake ftalkes, bearing at the toppe many fmall heauie bottle-like flowers, in fhape like the former Mufcari, but very thicke thruft together, fmaller, and of a very darke or blackifh blew colour, of a very ftrong fmell, like vnto Starch when it is new made, and hot: the root is round, and blackifh without, being compaffed with a number of fmall rootes, or of-fets round about it, fo that it will quickly choke a ground, if it be fuffered long in it. For which caufe, moft men doe caft it into fome by-corner, if they meane to preferue it, or caft it out of the Garden quite.

Alter maior.   There is another of this kinde that is greater, both in leafe and flower, and differeth not in colour or any thing elfe.

Hyacinthus Botroides cæruleus amœnus. The skie coloured Grape-flower.

This Iacinth fpringeth vp with fewer leaues then the firft, and not reddifh, but green at his firft appearing; the leaues, when they are full growne, are long and hollow, like the former, but greener, fhorter, and broader, ftanding vpright, and not lying along vpon the ground as they doe: the flowers grow at the toppe of the ftalke, more fparfedly fet thereon, and not fo thicke together, but like a thinne bunch of grapes, and bottle-like as the former, of a perfect blew or skie-colour, euery flower hauing fome white fpots about the brimmes of them: this hath a very fweet fmell, nothing like the former: this roote is whiter, and doth not fo much encreafe as the former, yet plentifull enough.

Hyacinthus Botroides ramofus. The branched Grape-flower.

Of this kinde, there is another found to grow with many branches of flowers, breaking out from the fides of the greater ftalkes or branches: the leaues as all the reft of the plant is greater then the former.

Hyacinthus Botroides flore albo. The white Grape-flower.

The white Grape-flower hath his greene leaues a little whiter, then the blew or skie

skie coloured Grape-flower, his flowers are very pure white, alike sparsedly set on the stalkes, but a little lower and smaller then it, in all other things there is no difference.

Hyacinthus Botroides flore albo rubente. The blush Grape-flower.

The roote of this Grape-flower groweth greater, then either the skie coloured, or white Grape-flower, and seldome hath any small rootes or of-sets, as the other haue: his leaues also are larger, and somewhat broader; the flowers are of a pale, or bleake blush colour out of a white, and are a little larger, and grow a little higher and fuller of flowers then the white.

The Place.

They naturally grow in many places both of Germany and Hungary; in Spaine likewise, and on Mount Baldus in Italy, and Narbone in France, about the borders of the fields: we haue them in our Gardens for delight.

The Time.

These flower from the beginning of March, or sooner sometimes, vntill the beginning of May.

The Names.

They are most commonly called *Botroides*, but more truely *Botryodes*, of *Botrus* the Greeke word, which signifieth a bunch or cluster of grapes: Lobelius calleth the white one, *Dipcadi flore albo*, transferring the name *Dipcadi*, whereby the *Muscari* is called to this Iacinth, as if they were both one. Their seuerall names, whereby they are knowne and called, are set downe in their titles. The Dutchmen call them *Driuekens*, as I said before. Some English Gentlewomen call the white Grape-flower Pearles of Spaine.

1. *Hyacinthus Comosus albus.* The white haired Iacinth.

This Iacinth doth more neerly resemble the Grape-flowers, then the faire haired Iacinths that follow, whereof it beareth the name, in that it hath no haire or threeds at the toppe of the stalke or sides, as they: and therefore I haue placed it next vnto them, and the other to follow it, as being of another kinde. The root hereof is blackish, a little long and round, from whence rise vp three or foure leaues, being smooth and whitish, long, narrow, and hollow, like a trough or gutter on the vpperside: among which the stalke riseth vp a foote high or more, bearing at the toppe diuers small flowers, somewhat like the former, but not so thicke set together, being a little longer, and larger, and wider at the mouth, and as it were diuided into six edges, of a darke whitish colour, with some blacker spots about the brimmes on the inside: the heads or seedeuessels are three square, and somewhat larger, then the heads of any of the former lesser Grape-flowers, wherein is contained round blacke seede.

2. *Hyacinthus Comosus Byzantinus.* The Turkie faire haired Iacinth.

This other Iacinth which came from Constantinople, is somewhat like the former, but that it is bigger, both in roote, and leafe, and flower, and bearing greater store of flowers on the head of the stalke: the lower flowers, although they haue short stalkes at their first flowring, yet afterwards the stalkes grow longer, and those that are lower, stand out further then those that are highest, whose foot-stalkes are short, and almost close to the stemme, and of a more perfect purple then any below, which are of a duskie greenish purple colour: the whole stalke of flowers seem like a Pyramis, broad belowe, and small aboue, or as other compare it, to a water sprinkle; yet neither of both these Iacinths haue any threeds at the tops of the stalkes, as the other following haue.

3. *Hyacinthus Comoſus maior purpureus.*
The great purple faire haired Iacinth.

This faire haired Iacinth hath his leaues ſofter, longer, broader, and leſſe hollow then the former, lying for the moſt part vpon the ground: the ſtalke riſeth vp in the midſt of the leaues, being ſtronger, higher, and bearing a greater and longer head of flowers alſo then they: the flowers of this ſtand not vpon ſuch long foote-ſtalkes, but are ſhorter below, and cloſe almoſt to the ſtalke aboue, hauing many bright purpliſh blew threeds, growing higheſt aboue the flowers, as it were in a buſh together, euery one of theſe threeds hauing a little head at the end of them, ſomewhat like vnto one of the flowers, but much ſmaller: the reſt of the flowers below this buſh, are of a ſadder or deader purple, and not ſo bright a colour, and the loweſt worſt of all, rather encli-ning to a greene, like vnto the laſt Turkie kinde: the whole ſtalke with the flowers vp-on it, doth ſomewhat reſemble a long Purſe taſſell, and thereupon diuers Gentlewo-men haue ſo named it: the heads and ſeede are like vnto the former, but greater: the roote is great and white, with ſome redneſſe on the outſide.

4. *Hyacinthus Comoſus ramoſus purpureus.*
The faire haired branched Iacinth.

The leaues of this Iacinth are broader, ſhorter, and greener then of the laſt, not lying ſo weakly on the ground, but ſtanding ſomewhat more vpright: the ſtalke riſeth vp as high as the former, but branched out on euery ſide into many tufts of threeds, with knappes, as it were heads of flowers, at the ends of them, like vnto the head of threeds at the toppe of the former Iacinth, but of a little darker, and not ſo faire a blewiſh purple colour: this Iacinth doth ſomewhat reſemble the next Curld haire Ia-cinth, but that the branches are not ſo fairely compoſed altogether of curled threeds, nor of ſo excellent a faire purple or Doue colour, but more duskie by much: the roote is greater and ſhorter then of the next, and endreaſeth faſter.

5. *Hyacinthus Pennatus, ſiue Comoſus ramoſus elegantior.*
The faire Curld-haire Iacinth.

This admirable Iacinth riſeth vp with three or foure leaues, ſomewhat like vnto the leaues of the Muske Grape-flower, but leſſer; betweene which riſeth vp the ſtalke a-bout a foote high, or ſomewhat more, bearing at the toppe a buſh or tuft of flowers, which at the firſt appearing, is like vnto a Cone or Pineapple, and afterwards opening it ſelfe, ſpreadeth into many branches, yet ſtill retaining the forme of a Pyramis, be-ing broad ſpread below, and narrow vp aboue: each of theſe branches is againe diui-ded into many tufts of threeds or ſtrings, twiſted or curled at the ends, and of an ex-cellent purple or Doue colour, both ſtalkes and haires. This abideth a great while in his beauty, but afterwards all theſe flowers (if you will ſo call them) do fall away with-out any ſeede at all, ſpending it ſelfe as it ſhould ſeeme in the aboundance of the flow-ers: the roote is not ſo great as the laſt, but white on the outſide.

The Place.

The two firſt haue been ſent diuers times from Conſtantinople, the third is found wilde in many places of Europe, and as well in Germany, as in Italy. The two laſt are onely with vs in Gardens, and their naturall pla-ces are not knowne vnto vs.

The Time.

The three former kindes doe flower in Aprill, the two laſt in May.

The Names.

The firſt and ſecond haue no other names then are expreſſed in their titles.

1 *Hyacinthus Comosus albus.* The white haired Iacinth. 2 *Hyacinthus Comosus Byzantinus.* The Turkie faire haired Iacinth. 3 *Hyacinthus Comosus maior purpureus.* The purple faire haired Iacinth, or Purse taffels. 4 *Hyacinthus Comosus ramosus, siue Calamistratus.* The faire haired branched Iacinth. 5 *Hyacinthus Pennatus, siue Comosus elegantior.* The faire curld haire Iacinth.

tles. The third is called of some onely *Hyacinthus maior*, and of others *Hyacinthus comosus maior:* We call it in English, The purple faire haired Iacinth, becaufe of his tuft of purple threeds, like haires at the toppe, and (as I faid) of diuers Gentlewomen, purple taffels. The fourth is called by some as it is in the title, *Hyacinthus comosus ramosus*, and of others *Hyacinthus Calamistratus*. And the laft or fifth is diuerfly called by diuers, Fabius Columna in his *Phytobasanos* the fecond part, calleth it *Hyacinthus Sannesius*, becaufe hee firft faw it in that Cardinals Garden at Rome. Robin of Paris fent to vs the former of the two laft, by the name of *Hyacinthus Pennatus*, and *Hyacinthus Calamistratus*, when as others fent the laft by the name *Pennatus*, and the o- ther by the name of *Calamistratus*; but I thinke the name *Cincinnatus* is more fit and proper for it, in that the curled threeds which feeme like haires, are better expreffed by the word *Cincinnus*, then *Calamistrum*, this fignifying but the bodkin or inftrument wherewith they vfe to frifle or curle the haire, and that the bufh of haire it felfe being curled. Some alfo haue giuen to both thefe laft the names of *Hyacinthus Comosus Parnassi*, the one fairer then the other. Of all thefe names you may vfe which you pleafe; but for the laft kinde, the name *Cincinnatus*, as I faid, is the more proper, but *Pennatus* is the more common, and *Calamistratus* for the former of the two laft.

1. *Hyacinthus Orientalis Brumalis, sive præcox flore albo.* The white Winter Orientall Iacinth.

This early Iacinth rifeth vp with his greene leaues (which are in all refpects like to the ordinary Orientall Iacinths, but fomewhat narrower) before Winter, and fome- times it is in flower alfo before Winter, and is in forme and colour a plaine white Ori- entall Iacinth, but fomewhat leffer, differing onely in no other thing, then the time of his flowring, which is alwayes certaine to be long before the other forts.

2. *Hyacinthus Orientalis Brumalis, sive præcox flore purpureo.* The purple Winter Orientall Iacinth.

The difference of colour in this flower caufeth it to bee diftinguifhed, for elfe it is of the kindred of the Orientall Iacinths, and is, as the former, more early then the reft that follow: Vnderftand then, that this is the fame with the former, but hauing fine blewifh purple flowers.

3. *Hyacinthus Orientalis maior præcox, dictus Zumbul Indi.* The greateft Orientall Iacinth, or Zumbul Indi.

The roote of this Orientall Iacinth, is vfually greater then any other of his kinde, and moft commonly white on the outfide, from whence rife vp one or two great round ftalkes, fpotted from within the ground, with the lower part of the leaues alfo vpward to the middle of the ftalkes, or rather higher, like vnto the ftalkes of Dragons, but darker; being fet among a number of broad, long, and fomewhat hollow greene leaues, almoft as large as the leaues of the white Lilly: at the toppe of the ftalkes ftand more ftore of flowers, then in any other of this kinde, euery flower being as great as the greateft fort of Orientall Iacinths, ending in fix leaues, which turne at the points, of a faire blewifh purple colour, and all ftanding many times on one fide of the ftalkes, and many times on both fides.

4. *Hyacinthus Orientalis vulgaris diuersorum colorum.* The ordinary Orientall Iacinth.

The common Orientall Iacinth (I call it common, becaufe it is now fo plenti- full in all Gardens, that it is almoft not efteemed) hath many greene leaues, long, fomewhat broad and hollow, among which rifeth vp a long greene round ftalke, befet from the middle thereof almoft, with diuers flowers, ftanding on both fides

of the ſtalkes, one aboue another vnto the toppe, each whereof next vnto the foote-
ſtalke is long, hollow, round, and cloſe, ending in ſix ſmall leaues laid open,
and a little turning at the points, of a very ſweete ſmell : the colours of theſe
flowers are diuers, for ſome are pure white, without any ſhew of other colour in
them : another is almoſt white, but hauing a ſhew of blewneſſe, eſpecially at the brims
and bottomes of the flowers. Others againe are of a very faint bluſh, tending towards
a white : Some are of as deepe a purple as a Violet ; others of a purple tending to redneſſe, and ſome of a paler purple. Some againe are of a faire blew, others more watchet, and ſome ſo pale a blew, as if it were more white then blew : after the flowers
are paſt, there riſe vp great three ſquare heads, bearing round blacke ſeede, great and
ſhining : the roote is great, and white on the outſide, and oftentimes purpliſh alſo, flat
at the bottome, and ſmall at the head.

There is a kinde of theſe Iacinths, whoſe flowers are of a deepe purpliſh Violet co- *Flore purpureo
violaceo lineis*
lour, hauing whitiſh lines downe the backe of euery leafe of the flower, which turne *albicantis in
dorſo.*
themſelues a little backwards at the points.

There is another, whoſe flowers ſtand all opening one way, and not on all ſides, but *Floribus antror-
ſum reſpicienti-
bus.*
are herein like the great Zumbul Indi, before ſet out.

There is againe another kinde which flowreth later then all the reſt, and the flow- *Serotinus ere-
ctis floribus
diuerſorum co-
lorum.*
ers are ſmaller, ſtanding more vpright, which are either white or blew, or mixt with
white and purple.

5. *Hyacinthus Orientalis folioſo caule.* The buſhy ſtalked Orientall Iacinth.

This ſtrange Iacinth hath his rootes, leaues, and flowers, like vnto the former Orientall Iacinths : the onely difference in this is, that his ſtalke is not bare or naked, but
hath very narrow long leaues, growing diſperſedly, and without order, with the flowers thereon, which are blew, and hauing for the moſt part one leafe, and ſometimes
two at the foote, or ſetting on of euery flower, yet ſometimes it happeneth, ſome flowers to be without any leafe at the bottome, as nature, that is very variable in this plant,
liſteth to play : the heads and ſeede are blacke and round, like the other alſo.

6. *Hyacinthus Orientalis flore duplici.* The bleake Orientall Iacinth once double.

This double Iacinth hath diuers long leaues, like vnto the other Orientall Iacinths,
almoſt ſtanding vpright, among which riſeth vp a ſtalke, browniſh at the firſt, but
growing greene afterwards, bearing many flowers at the toppe, made like the flowers
of the former Iacinths, and ending in ſix leaues, greene at the firſt, and of a blewiſh
white when they are open, yet retaining ſome ſhew of greenneſſe in them, the brims
of the leaues being white ; from the middle of each flower ſtandeth forth another
ſmall flower, conſiſting of three leaues, of the ſame colour with the other flower, but
with a greene line on the backe of each of theſe inner leaues : in the middle of this little flower, there ſtand ſome threeds tipt with blacke : the ſmell of this flower is not ſo
ſweete as of the forme ; the heads, ſeede, and rootes are like the former.

7. *Hyacinthus Orientalis flore pleno cæruleo, vel purpuro violaceo.*
The faire double blew, or purple Orientall Iacinth.

The leaues of theſe Iacinths are ſmaller, then the leaues of moſt of the other former ſorts ; the ſtalkes are ſhorter, and ſmaller, bearing but three or foure flowers on
the heads of them for the moſt part, which are not compoſed like the laſt, but are more
faire, full, and double of leaues, where they ſhew out their full beauties, and of a faire
blew colour in ſome, and purple in others, ſmelling pretty ſweete ; but theſe doe ſeldome beare out their flowers faire ; and beſides, haue diuers other flowers that will be
either ſingle, or very little double vpon the ſame ſtalke.

8. *Hyacinthus Orientalis candidiſsimus flore pleno.*
The pure white double Orientall Iacinth.

This double white Iacinth hath his leaues like vnto the ſingle white Orientall Iacinth ;

cinth; his stalke is likewise long, slender, and greene, bearing at the toppe two or three flowers at the most, very double and full of leaues, of a pure white colour, without any other mixture therein, hanging downe their heads a little, and are reasonable sweete. I haue this but by relation, not by sight, and therefore I can giue no further assurance as yet.

The Place.

All these Orientall Iacinths, except the last, haue beene brought out of Turkie, and from Constantinople : but where their true originall place is, is not as yet vnderstood.

The Time.

The two first (as is said) flower the earliest, sometimes before Christmas, but more vsually after, and abide a great while in flower, in great beauty, especially if the weather be milde, when as few or no other flowers at that time are able to match them. The other greatest kinde floweth also earlier then the rest that follow, for the most part. The ordinary kindes flower some in March, and some in Aprill, and some sooner also ; and so doe the double ones likewise. The bushy stalked Iacinth flowreth much about the same time.

The Names.

The former two sorts are called *Hyacinthus Orientalis Brumalis*, and *Hyacinthus Orientalis præcox flore albo*, or *cæruleo*. The third is called of many *Zumbul Indicum*, or *Zumbul Indi*, and corruptly *Simboline* ; of others, and that more properly, *Hyacinthus Orientalis maior præcox*. The Turkes doe call all Iacinths *Zumbul*, and by adding the name of *Indi*, or *Arabi*, do shew from what place they are receiued. In English, The greatest Orientall Iacinth; yet some doe call it after the Turkish name *Zumbul Indi*, or *Simboline*, as is said before. The rest haue their names set downe in their titles, which are most fit for them.

Hyacinthus Hispanicus minor Orientalis facie.
The little Summer Orientall Iacinth.

This little Iacinth hath foure or fiue long narrow greene leaues, lying vpon the ground, among which riseth vp a slender smooth stalke, about a spanne high or more, *Flore cærulea.* bearing at the toppe many slender bleake blew flowers, with some white stripes and edges to be seene in most of them, fashioned very like vnto the flowers of the Orientall Iacinth, but much smaller : the flower hath no sent at all ; the seede is like the seede of the English Iacinth, or Harefbels : the roote is small and white.
Flore albo. There is another of this kinde, differing in nothing but in the colour of the flower, which is pure white.
Flore rubente. There is also another, whose flowers are of a fine delayed red colour, with some deeper coloured veines, running along the three outer leaues of the flower, differing in no other thing from the former.

The Place.

These plants haue been gathered on the Pyrenæan Mountaines, which are next vnto Spaine, from whence, as is often said, many rare plants haue likewise been gathered.

The Time.

They flower very late, euen after all or most of the Iacinths, in May for the most part.

The

1 *Hyacinthus Orientalis brumalis.* The Winter Orientall Iacinth. 2 *Zumbul Indi.* The greatest Orientall Iacinth. 3 *Hyacinthus Orientalis vulgaris.* The ordinary Orientall Iacinth. 4 *Hyacinthus Orientalis folioso
caule.* The bushy stalked Orientall Iacinth. 5 *Hyacinthus Orientalis flore duplici.* The Orientall Iacinth once
double. 6 *Hyacinthus Orientalis flore pleno cæruleo.* The faire double blew Orientall Iacinth.

The Names.

They are called eyther *Hyacinthus Hispanicus minor Orientalis facie*, as it is in the title, or *Hyacinthus Orientalis facie*, that is to say, The leſſer Spaniſh Iacinth, like vnto the Orientall : yet some haue called them, *Hyacinthus Orientalis ſerotinus minor*, The leſſer late Orientall Iacinth, that thereby they may be knowne from the reſt.

Hyacinthus Hispanicus obſoletus. The Spaniſh dunne coloured Iacinth.

This Spaniſh Iacinth ſpringeth very late out of the ground, bearing foure or fiue ſhort, hollow, aud ſoft whitiſh greene leaues, with a white line in the middle of euery one of them, among which riſe vp one or more ſtalkes, bearing diuers flowers at the toppes of them, all looking one way, or ſtanding on the one ſide, hanging downe their heads, conſiſting of ſix leaues, three whereof being the outermoſt, lay open their leaues, and turne back the ends a little again : the other three which are innermoſt, do as it were cloſe together in the middle of the flower, without laying themſelues open at all, being a little whitiſh at the edges : the whole flower is of a purpliſh yellow colour, with ſome white and green as it were mixed among it, of no ſent at all : it beareth blacke and flat ſeede in three ſquare, great, and bunched out heads : the roote is reaſonable great, and white on the outſide, with many ſtrong white fibres at it, which periſh not yearely, as the fibres of many other Iacinths doe, and as it ſpringeth late, ſo it holdeth his greene leaues almoſt vntill Winter.

Mauritanicus. There hath been another hereof brought from about Fez and Marocco in Barbary, which in all reſpects was greater, but elſe differed little.

Maximus Æthiopicus. There was another alſo brought from the Cape of good Hope, whoſe leaues were ſtronger and greener then the former, the ſtalke alſo thicker, bearing diuers flowers, confuſedly ſtanding vpon longer foote-ſtalkes, yet made after the ſame faſhion, but that the three inner leaues were whitiſh, and dented about the edges, otherwiſe the flowers were yellow and greeniſh on the inſide.

The Place.

Theſe plants grow in Spaine, Barbary, and Ethiopia, according as their names and deſcriptions doe declare.

The Time.

The firſt flowreth not vntill Iune ; for, as I ſaid, it is very late before it ſpringeth vp out of the ground, and holdeth his leaues as is ſaid, vntill September, in the meane time the ſeede thereof ripeneth.

The Names.

They haue their names according to the place of their growing ; for one is called *Hyacinthus Hispanicus obſoletioris coloris.* The other is called alſo *Hyacinthus Mauritanicus.* And the laſt, *Hyacinthus Æthiopicus obſoletus.* In Engliſh, The Spaniſh, Barbary, or Ethiopian Iacinth, of a dunne or duskie colour.

Hyacinthus Anglicus Belgicus, vel Hispanicus.
Engliſh Hareſ-bels, or Spaniſh Iacinth.

Our Engliſh Iacinth or Hareſ-bels is ſo common euery where, that it ſcarce needeth any deſcription. It beareth diuers long and narrow greene leaues, not ſtanding vpright, nor yet fully lying vpon the ground, among which ſpringeth vp the ſtalke, bearing at the toppe many long and hollow flowers, hanging downe their heads all
forward

forwards for the moſt part, parted at the brimmes into ſix parts, turning vp their points a little againe, of a ſweetiſh, but heady ſent, ſomewhat like vnto the Grape-flower: the heads for ſeede are long and ſquare, wherein is much blacke ſeede: the colour of the flowers are in ſome of a deeper blew, tending to a purple; in others of a paler blew, or of a bleake blew, tending to an aſh colour: Some are pure white, and ſome are party coloured, blew and white; and ſome are of a fine delayed purpliſh red or bluſh colour, which ſome call a peach colour. The rootes of all ſorts agree, and are alike, being white and very ſlimie; ſome whereof will be great and round, others long and ſlender, and thoſe that lye neare the toppe of the earth bare, will be greene.

Hyacinthus Hiſpanicus maior flore campanula inſtar. The greater Spaniſh bell-flowred Iacinth.

This Spaniſh bell-flowred Iacinth, is very like the former Engliſh or Spaniſh Ia-cinth, but greater in all parts, as well of leaues as flowers, many growing together at the toppe of the ſtalke, with many ſhort greene leaues among them, hanging downe their heads, with larger, greater, and wider open mouths, like vnto bels, of a darke blew colour, and no good ſent.

The Place.

The firſt groweth in many places of England, the Lowe-Countries, as we call them, and Spaine, but the laſt chiefly in Spaine.

The Time.

They flower in Aprill for the moſt part, and ſometimes in May.

The Names.

Becauſe the firſt is more frequent in England, then in Spain, or the Lowe-Countries, it is called with vs *Hyacinthus Anglicus*, The Engliſh Iacinth; but it is alſo called as well *Belgicus*, as *Hiſpanicus*: yet Dodonæus calleth it *Hyacinthus non ſcriptus*, becauſe it was not written of by any Authour be-fore himſelfe. It is generally knowne in England by the name of Hare-bels. The other Spaniſh Iacinth beareth his name in his title.

Hyacinthus Eriophorus. The Woolly Iacinth.

This Woolly Iacinth hath many broad, long, and faire greene leaues, very like vn-to ſome of the Iacinths, but ſtiffer, or ſtanding more vpright, which being broken, doe yeeld many threeds, as if a little fine cotton wooll were drawne out: among theſe leaues riſeth vp a long greene round ſtalke, a foote and a halfe high or more, whereon is ſet a great long buſh of flowers, which blowing open by degrees, firſt below, and ſo vpwards, are very long in flowring: the toppe of the ſtalke, with the flowers, and their little footſtalkes, are all blew, euery flower ſtanding outright with his ſtalke, and ſpreading like a ſtarre, diuided into ſix leaues, hauing many ſmall blew threeds, ſtan-ding about the middle head, which neuer gaue ripe ſeede, as farre as I can heare of: the root is white, ſomewhat like the root of a Muſcari, but as full of wooll or threeds, or rather more, then the leaues, or any other part of it.

The Place.

This hath been ſent diuers times out of Turkie into England, where it continued a long time as well in my Garden as in others, but ſome hard froſty Winters cauſed it to periſh with me, and diuers others, yet I haue had it againe from a friend, and doth abide freſh and greene euery yeare in my Garden.

The

The Time.

This flowred in the Garden of Mr. Richard Barnesley at Lambeth, onely once in the moneth of May, in the yeare 1606. after hee had there preserued it a long time : but neither he, nor any else in England that I know, but those that saw it at that time, euer saw it beare flower, either before or since.

The Names.

It is called by diuers *Bulbus Eriophorus,* or *Laniferus,* that is, Woolly Bulbous ; but because it is a Iacinth, both in roote, leafe, and flower, and not a *Narcissus,* or Daffodill, it is called *Hyacinthus Eriophorus,* or *Laniferus,* The Woolly Iacinth. It is very likely, that Theophrastus in his seuenth Book & thirteenth Chapter, did meane this plant, where hee declareth, that garments were made of the woolly substance of a bulbous roote, that was taken from between the core or heart of the roote (which, as hee saith, was vsed to be eaten) and the outermost shels or peelings ; yet Clusius seemeth to fasten this woolly bulbous of Theophrastus, vpon the next Iacinth of Spaine.

Hyacinthus Stellatus Baticus maior, vulgò Peruanus.
The great Spanish Starry Iacinth, or of Peru.

This Iacinth (the greatest of those, whose flowers are spread like a starre, except the two first Indians) hath fiue or six, or more, very broad, and long greene leaues, spread vpon the ground, round about the roote, which being broken are woolly, or full of threeds, like the former : in the middle of these leaues riseth vp a round short stalke, in comparison of the greatnesse of the plant (for the stalke of the Orientall Iacinth is sometimes twice so high, whose roote is not so great) bearing at the toppe a great head or bush of flowers, fashioned in the beginning, before they bee blowne or separated, very like to a Cone or Pineapple, and begin to flower belowe, and so vpwards by degrees, euery flower standing vpon a long blackish blew foote-stalke, which when they are blowne open, are of a perfect blew colour, tending to a Violet, and made of six small leaues, laid open like a starre ; the threeds likewise are blewish, tipt with yellow pendents, standing about the middle head, which is of a deeper blew, not hauing any good sent to be perceiued in it, but commendable only for the beauty of the flowers : after the flowers are past, there come three square heads, containing round blacke seede : the roote is great, and somewhat yellowish on the outside, with a knobbe or bunch at the lower end of the roote, (which is called the seate of the roote) like vnto the Muscari, Scylla, and many other bulbous rootes, at which hang diuers white, thicke, and long fibres, whereby it is fastened in the ground, which perish not euery yeare, but abide continually, and therefore doth not desire much remouing.

Hyacinthus Stellatus Baticus, siue Peruanus flore albo.
The great white Spanish starry Iacinth.

This other Spanish Iacinth is in most parts like vnto the former, but that his leaues are not so large, nor so deep a greene : the stalks of flowers likewise hath not so thicke a head, or bush on it, but fewer and thinner set : the flowers themselues also are whitish, yet hauing a small dash of blush in them : the threeds are whitish, tipt with yellow pendents : the seede and rootes are like vnto the former, and herein consisteth the difference betweene this and the other sorts.

Hyacinthus Stellatus Baticus, siue Peruanus flore carneo.
The great blush coloured Spanish Starry Iacinth.

This likewise differeth little from the two former, but onely in the colour of the
flowers;

1 *Hyacinthus Orientalis faie.* The little Summer Orientall Iacinth. 2 *Hyacinthus Mauritanicus.* The Barbary Iacinth. 3 *Hyacinthus Bætieus Hispaniteus* The Spanish duskie Iacinth. 4 *Hyacinthus Hispanicus flore campanula* The greater Spanish bel-flowred Iacinth. 5 *Hyacinthus Anglicus.* The English Iacinth or Harebels. 6 *Hyacinthus Eriophorus.* The Woolly Iacinth. 7 *Hyacinthus Stellaris Bæticus maior, sine Peruanus.* The great Spanish Starry Iacinth, or of Peru.

flowers ; for this being found growing among both the other , hath h is head of flow ers as great and large as the first, but the buds of his flowers, before they are open, are of a deepe blush colour, which being open, are more delayed , and of a pleasant pale purple, or blush colour, standing vpon purplish stalkes : the heads in the middle are whitish, and so are the threeds compassing it, tipt with yellow.

The Place.

These doe naturally grow in Spaine, in the Medowes a little off from the Sea, as well in the Island Gades, vsually called Cales, as likewise in other parts along the Sea side, as one goeth from thence to Porto Santa Maria, which when they be in flower, growing so thicke together , seeme to couer the ground, like vnto a tapistry of diuers colours , as I haue beene credibly enformed by Guillaume Boel , a Freeze-lander borne, often before and hereafter remembred, who being in search of rare plants in Spaine, in the yeare of our Lord 1607. after that most violent frosty Winter, which peri- shed both the rootes of this, and many other fine plants with vs , sent mee ouer some of these rootes for my Garden, and affirmed this for a truth, which is here formerly set downe, and that himselfe gathered those he sent mee, and many others in the places named , with his owne hands ; but hee saith, that both that with the white, and with the blush flowers, are farre more rare then the other.

The Time.

They flower in May, the seede is ripe in Iuly.

The Names.

This hath beene formerly named *Eriophorus Peruanus* , and *Hyacinthus Stellatus Peruanus,* The Starry Iacinth of Peru, being thought to haue grown in Peru, a Prouince of the West Indies ; but he that gaue that name first vn- to it, eyther knew not his naturall place, or willingly imposed that name, to conceale it, or to make it the better esteemed. It is most generally recei- ued by the name *Hyacinthus Peruanus,* from the first imposer thereof , that is, the Iacinth of Peru : but I had rather giue the name agreeing most fitly vnto it, and call it as it is indeede *Hyacinthus Stellatus Baticus,* The Spanish Starry Iacinth ; and because it is the greatest that I know hath come from thence, I call it, The great Starry Iacinth of Spaine, or Spanish Iacinth.

Hyacinthus Stellatus vulgaris, siue Bifolius Fuchsij.
The common blew Starry Iacinth.

This Starry Iacinth (being longest knowne, and therefore most common) riseth out of the ground, vsually but with two browne leaues, yet sometimes with three, inclo- sing within them the stalke of flowers , the buds appearing of a darke whitish colour, as soone as the leaues open themselues, which leaues being growne, are long, and hol- low , of a whitish greene on the vpper side, and browne on the vnder side, and halfe round, the browne stalke rising vp higher , beareth fiue or sixe small starre-like flowers thereon, consisting of six leaues, of a faire deepe blew, tending to a purple. The seede is yellowish, and round, contained in round pointed heads , which by rea- son of their heauinesse, and the weaknesse of the stalke, lye vpon the ground, and often perish with wet and frosts, &c. The roote is somewhat long, and couered with a yel- lowish coate.

Hyacinthus stellatus flore albo. The white Starry Iacinth.

The white Starry Iacinth hath his leaues like the former, but greene and fresh, not browne , and a little narrower also : the buddes for flowers at the first appeare a little blush, which when they are blowne, are white, but yet retaine in them a small shew of that blush colour.
                                                                        We

We haue another, whose flowers are pure white, and smaller then the other, the *Flore minore* leaues whereof are of a pale fresh greene, and somewhat narrower.

Hyacinthus Stellatus flore rubente. The blush coloured Starry Iacinth.

The difference in this from the former, is onely in the flowers, which are of a faire blush colour, much more eminent then in the others, in all things else alike.

Hyacinthus Stellatus Martius, siue pracox caruleus.
The early blew Starry Iacinth.

This Iacinth hath his leaues a little broader, of a fresher greene, and not browne at all, as the first blew Iacinth of Fuchsius last remembred: the buds of the flowers, while they are enclosed within the leaues, and after, when the stalke is gowne vp, doe remaine more blew then the buds of the former: the flowers, when they are blowne open, are like the former, but somewhat larger, and of a more liuely blew colour: the roote also is a little whiter on the outside. This doth more seldome beare seede then the former.

Hyacinthus Stellatus pracox flore albo. The white early Starry Iacinth.

There is also one other of this kinde, that beareth pure white flowers, the green leafe thereof being a little narrower then the former, and no other difference.

Hyacinthus Stellatus pracox flore suaue rubente.
The early blush coloured Starry Iacinth.

This blush coloured Iacinth is very rare, but very pleasant, his flowers being as large as the first of this last kinde, and somewhat larger then the blush of the other kinde: the leaues and rootes differ not from the last recited Iacinth.

The Place.

All these Iacinths haue beene found in the Woods and Mountaines of Germany, Bohemia, and Austria, as Fuchsius and Gesner doe report, and in Naples, as Imperatus and others doe testifie. Wee cherish them all with great care in our Gardens, but especially the white and the blush of both kindes, for that they are more tender, and often perish for want of due regard.

The Time.

The common kindes, which are first expressed, flower about the middle of February, if the weather bee milde, and the other kindes sometimes a fortnight after, that is, in March, but ordinarily much about the same time with the former.

The Names.

The first is called in Latine *Hyacinthus Stellatus vulgaris*, and *Hyacinthus Stellatus bifolius*, and *Hyacinthus Stellaris Fuchsij*, and of some *Hyacinthus Stellatus Germanicus*; wee might very well call the other kinde, *Hyacinthus Stellatus vulgaris alter*, but diuers call it *Pracox*, and some *Martius*, as it is in the title. In English they may bee seuerally called: the first, The common; and the other, The early Starry Iacinth (notwithstanding the first flowreth before the other) for distinction sake.

The *Hyacinthus* seemeth to be called *Vacinium* of Virgil in his Eclogues; for hee alwayes reckoneth it among the flowers that were vsed to decke Garlands, and neuer among fruits, as some would haue it. But in that hee calleth it *Vacinium nigrum*, in seuerall places, that doth very fitly answer the

common

common receiued cuſtome of thoſe times, that called all deepe blew colours, ſuch as are purples, and the like, blacke ; for the Violet it ſelfe is likewiſe called blacke in the ſame place, where he calleth the *Vacinium* blacke ;
ſo that it ſeemeth thereby, that he reckoned them to be both of one colour,
and we know the colour of the Violet is not blacke, as we doe diſtinguiſh of
blacke in theſe dayes. But the colour of this Starry Iacinth, being both of
ſo deepe a purple ſometimes, ſo neare vnto a Violet colour, and alſo more
frequent, then any other Iacinth with them, in thoſe places where Virgil
liued, perſwadeth me to thinke, that Virgil vnderſtood this Starry Iacinth
by *Vacinium* : Let others iudge otherwiſe, if they can ſhew greater probabilitie.

1. *Hyacinthus Stellatus Byzantinus nigra radice.*
The Starry Iacinth of Turkie with the blacke roote.

This Starry Iacinth of Conſtantinople hath three or foure freſh greene, thinne, and
long leaues, of the bigneſſe of the Engliſh Iacinth, but not ſo long, betweene which ri
ſeth vp a ſlender lowe ſtalke, bearing fiue or ſix ſmall flowers, diſperſedly ſet thereon,
ſpreading open like a ſtarre, of a pale or bleake blew colour : the leaues of the flowers
are ſomewhat long, and ſtand as it were ſomewhat looſly, one off from another, and
not ſo compactly together, as the flowers of other kindes : it ſeldome beareth ripe
ſeede with vs, becauſe the heads are ſo heauie, that lying vpon the ground, they rotte
with the wet, or are bitten with the froſts, or both, ſo that they ſeldome come to good:
the roote is ſmall in ſome, and reaſonable bigge in others, round, and long, white
within, but couered with deepe reddiſh or purpliſh peelings, next vnto it, and darker
and blacker purple on the outſide, with ſome long and thicke white fibres, like fingers
hanging at the bottome of them, as is to be ſeene in many other Iacinths : the roote it
ſelfe for the moſt part doth runne downewards, ſomewhat deep into the ground.

2. *Hyacinthus Stellatus Byzantinus maior.*
The greater Starry Iacinth of Conſtantinople.

This Iacinth may rightly be referred to the former Iacinth of Conſtantinople, and
called the greater, it is ſo like thereunto, that any one that knoweth that, will ſoone
ſay, that this is another of that ſort, but greater as it is in all his parts, bearing larger
leaues by much, and more ſtore, lying vpon the ground round about the roote : it
beareth many lowe ſtalkes of flowers, as bleake, and ſtanding as looſly as the former :
onely the roote of this, is not black on the outſide, as the other, but three times bigger.

3. *Hyacinthus Stellatus Byzantinus alter, ſiue flore boraginis.*
The other Starry Iacinth of Conſtantinople.

This other Iacinth hath for the moſt part onely foure leaues, broader and greener
then the firſt, but not ſo large or long as the ſecond : the ſtalke hath fiue or ſix flowers
vpon it, bigger and rounder ſet, like other ſtarry Iacinths, of a more perfect or deeper
blew then either of the former, hauing a whitiſh greene head or vmbone in the middle, beſet with ſix blew chiues or threeds, tipt with blacke, ſo cloſly compaſſing the
vmbone, that the threeds ſeeme ſo many prickes ſtucke into a clubbe or head ; ſome
therefore haue likened it to the flower of Borage, and ſo haue called it : after the flowers are paſt, come vp round white heads, wherein is contained round and white ſeede :
the roote is of a darke whitiſh colour on the outſide, and ſometimes a little reddiſh
withall.

The Place.

The firſt and the laſt haue beene brought from Conſtantinople ; the
firſt among many other rootes, and the laſt by the Lord Zouch, as Lobel
witneſſeth. The ſecond hath been ſent vs out of the Lowe-Countries, but
from whence they had it, we do not certainly know. They growe with vs in
our Gardens ſufficiently.   The

The Time.

These flower in Aprill, but the first is the earliest of the rest, and is in flower presently after the early Starry Iacinth, before described.

The Names.

The former haue their names in their titles, and are not knowne vnto vs by any other names that I know; but as I said before, the last is called by some, *Hyacinthus Boraginis flore.* The first was sent out of Turkie, by the name of *Susam giul,* by which name likewise diuers other things haue beene sent, so barren and barbarous is the Turkish tongue.

Hyacinthus Stellatus Æstiuus maior. The greater Summer Starry Iacinth.

This late Iacinth hath diuers narrow greene leaues, lying vpon the ground, somewhat like the leaues of the English Iacinth, but stiffer and stronger; among which riseth vp a round stiffe stalke, bearing many flowers at the toppe thereof, and at euery foote-stalke of the flowers a small short leafe, of a purplish colour: the flowers are starre-like, of a fine delayed purplish colour, tending to a pale blew or ash colour, striped on the backe of euery leafe, and hauing a pointed vmbone in the middle, with some whitish purple threeds about it, tipt with blew: the seede is blacke, round, and shining, like vnto the seede of the English Iacinth, but not so bigge: the roote is round and white, hauing some long thicke rootes vnder it, besides the fibres, as is vsuall in many other Iacinths.

Hyacinthus Stellatus Æstiuus minor. The lesser Summer Starry Iacinth.

This lesser Iacinth hath diuers very long, narrow, and shining greene leaues, spread vpon the ground round about the roote, among which riseth vp a very short round stalke, not aboue two inches high, carrying six or seuen small flowers thereon, on each side of the stalke, like both in forme and colour vnto the greater before described, but lesser by farre: the seede is blacke, contained in three square heads: the roote is small and white, couered with a browne coate, and hauing some such thicke rootes among the fibres, as are among the other.

The Place.

Both these Iacinths grow naturally in Portugall, and from thence haue been brought, by such as seeke out for rare plants, to make a gaine and profit by them.

The Time.

They both flower in May, and not before: and their seed is ripe in Iuly.

The Names.

Some doe call these *Hyacinthus Lusitanicus,* The Portugall Iacinth. Clusius, who first set out the descriptions of them, called them as is expressed in their titles; and therefore we haue after the Latine name giuen their English, according as is set downe. Or if you please, you may call them, The greater and the lesser Portugall Iacinth.

Hyacinthus Stellaris flore cinereo. The ash coloured Starry Iacinth.

This ash coloured Iacinth, hath his leaues very like vnto the leaues of the English Iacinth, and spreading vpon the ground in the same manner, among which rise vp one or two stalkes, set at the toppe with a number of small starre-like flowers, bushing bigger

ger below then aboue, of a very pale or white blew, tending to an aſh colour, and very ſweete in ſmell : the ſeede is blacke and round, like vnto the ſeede of the Engliſh Iacinth, and ſo is the roote, being great, round, and white; ſo like, I ſay, that it is hard to know the one from the other.

The Place.

The certaine originall place of growing thereof, is not knowne to vs.

The Time.

It flowreth in Aprill.

The Names.

Some doe call this *Hyacinthus Someri*, Somers Iacinth, becauſe as Lobel ſaith, he brought it firſt into the Lowe-Countries, eyther from Conſtantinople, or out of Italy.

Hyacinthus Stellatus Lilifolio & radice cæruleo.
The blew Lilly leafed Starre Iacinth.

This Iacinth hath ſix or ſeuen broad greene leaues, ſomewhat like vnto Lilly leaues, but ſhorter (whereof it tooke his name as well as from the roote) ſpread vpon the ground, and lying cloſe and round : before the ſtalke riſeth out from the middle of theſe leaues, there doth appeare a deepe hollow place, like a hole, to bee ſeene a good while, which at length is filled vp with the ſtalke, riſing thence vnto a foote or more high, bearing many ſtarre-like flowers at the toppe, of a perfect blew colour, neare vnto a Violet, and ſometimes of paler or bleake blew colour, hauing as it were a ſmall cuppe in the middle, diuided into ſix peeces, without any threeds therein : the ſeede is blacke and round, but not ſhining : the roote is ſomewhat long, bigge belowe, and ſmall aboue, like vnto the ſmall roote of a Lilly, and compoſed of yellow ſcales, as a Lilly, but the ſcales are greater, and fewer in number.

Hyacinthus Stellatus Lilifolius albus. The white Lilly leafed Starre Iacinth.

The likeneſſe of this Iacinth with the former, cauſeth me to be briefe, and not to repeate the ſame things againe, that haue already been expreſſed : You may therefore vnderſtand, that except in the colour of the flower, which in this is white, there is no difference betweene them.

Flore carneo. I heare of one that ſhould beare bluſh coloured flowers, but I haue not yet ſeene any ſuch.

The Place.

Theſe Iacinths haue been gathered on the Pyrenæan Hils, in that part of France that is called Aquitaine, and in ſome other places.

The Time.

Theſe flower in Aprill, and ſometimes later.

The Names.

Becauſe the roote is ſo like vnto a Lilly, as the leafe is alſo, it hath moſt properly beene called *Hyacinthus Stellatus Lilifolio & radice*, or for breuity *Lilifolius*, that is, The Starry Lilly leafed Iacinth. It is called *Sarabug* by the Inhabitants where it groweth, as Cluſius maketh the report from Venerius, who further ſaith, that by experience they haue found the cattell to ſwell and dye, that haue eaten of the leaues thereof.

Hyacinthus

1 *Hyacinthus stellatus præcox cæruleus.* The early blew starry Iacinth. 2 *Hyacinthus stellatus præcox albus,* The early white starry Iacinth. 3 *Hyacinthus stella-tus Byzantinus nigra radice.* The other starry Iacinth of Constantinople. 5 *Hyacinthus æstivus maior.* The greater Summer starry Iacinth. 6 *Hyacinthus stellatus floridus maior.* The ash colour of starry Iacinth. 7 *Hy-acinthus stellatus Liliifolius.* The Lilly leafed starry Iacinth. 8 *Hyacinthus Autumnalis.* The Autumne Iacinth. 9 *Scilla alba sive Hyacinthus marinus,* the Sea Onion or Squill.

Hyacinthus Autumnalis maior. The greater Autumne Iacinth.

The greater Autumne Iacinth hath fiue or six very long and narrow greene leaues, lying vpon the ground; the stalkes are set at the toppe with many starre-like flowers, of a pale blewish purple colour, with some pale coloured threeds, tipt with blew, standing about the head in the middle, which in time growing ripe, containeth therein small blacke seede, and roundish: the roote is great and white on the outside.

Hyacinthus Autumnalis minor. The lesser Autumne Iacinth.

This lesser Iacinth hath such like long and small leaues, but narrower then the former: the stalke is not full so high, but beareth as many flowers on it as the other, which are of a pale or bleake purple colour, very like vnto it also: the roote and seed are like the former, but smaller. These both for the most part, beare their flowers and seede before the greene leaues rise vp much aboue the ground.

Flore albo,   There is a kinde hereof found that beareth white flowers, not differing in any other thing from the smaller purple kinde last mentioned.

The Place.

The first and last are onely kept in Gardens, and not knowne to vs where their naturall place of growing wilde may be.
The second groweth wilde in many places of England. I gathered diuers rootes for my Garden, from the foote of a high banke by the Thames side, at the hither end of Chelsey, before you come at the Kings Barge-house.

The Time.

The greatest flowreth in the end of Iuly, and in August.
The other in August and September, you shall seldome see this plant with flowers and greene leaues at one time together.

The Names.

They haue their names giuen them, as they are expressed in their titles, by all former Writers, except Daleschampius, or hee that set forth that great worke printed at Lyons; for hee contendeth with many words, that these plants can bee no Iacinths, because their flowers appeare before their leaues in Autumne, contrary to the true Iacinth, as he saith: and therefore he would faine haue it referred to *Theophrastus bulbus in libro primo cap.* 12. and calleth it his *Tiphyum* mentioned in that place, as also *Bulbus asinum Daleschampij.* Howsoeuer these things may carry some probability in them, yet the likenesse both of rootes, and flowers especially, hath caused very learned Writers to entitle them as is set downe, and therefore I may not but let them passe in the like manner.

The Vertues.

Both the rootes and the leaues of the Iacinths are somewhat cold and drying, but the seede much more. It stayeth the loosnesse of the belly. It is likewise said to hinder young persons from growing ripe too soone, the roote being drunke in wine. It helpeth them also whose vrine is stopt, and is auaileable for the yellow Iaundise; but as you heare some are deadly to cattell, I therefore wish all to bee well aduised which of these they will vse in any inward physicke.

Scilla alba. The Sea Onion or Squill.

As I ended the discourse of both the true and the bastard Daffodils, with the Sea
kindes

kindes of both forts ; fo I thinke it not amiffe, to finifh this of the Iacinths with the defcription of a Sea Iacinth, which (as you fee) I take to be the *Scilla*, or Sea Onion, all his parts fo nearely refembling a Iacinth, that I know not where to ranke him better then in this place, or rather not any where but here. You fhall haue the defcription thereof, and then let the iudicious paffe their fentence, as they thinke meeteft.

The Squill or Sea Onion (as many doe call it)hath diuers thicke leaues,broad,long, greene, and hollowifh in the middle, and with an eminent or fwelling ribbe all along the backe of the leafe, (I relate it as I haue feene it, hauing fhot forth his leaues in the fhip by the way, as the Mariners that brought diuers rootes from out of the Straights, did fell them to mee and others for our vfe) lying vpon the ground, fomewhat like vnto the leaues of a Lilly : thefe fpring vp after the flowers are paft, and the feed ripe, they abiding all the Winter, and the next Spring, vntill the heate of the Summer hath fpent and confumed them, and then about the end of Auguft,or beginning of September, the ftalke with flowers arifeth out of the ground a foote and a halfe high, bearing many ftarre-like flowers on the toppe, in a long fpike one aboue another, flowring by degrees,the loweft firft, and fo vpwards, whereby it is long in flowring , very like, as well in forme as bigneffe, to the flowers of the great Starre of Bethlehem (thefe flowers I haue likewife feene fhooting out of fome of the rootes, that haue been brought in the like manner:) after the flowers are paft, there come vp in their places thicke and three fquare heads,wherin is contained fuch like flat,black,and round feed,as the Spanifh duskie Iacinth before defcribed did beare, but greater : the root is great & white, couered with many peelings or couerings, as is plainly enough feen to any that know them, and that fometimes wee haue had rootes, that haue beene as bigge as a pretty childes head, and fometimes two growing together, each whereof was no leffe then is faid of the other.

Scilla rubra fiue Pancratium verum. The red Sea Onion.

The roote of this Squill, is greater oftentimes then of the former, the outer coates or peelings being reddifh, bearing greater, longer,ftiffer,and more hollow leaues, in a manner vpright : this bringeth fuch a like ftalke and flowers, as the former doth , as Fabianus Ilges,Apothecary to the Duke of Briga, did fignifie by the figure thereof drawne and fent to Clufius.

The Place.

They grow alwayes neare the Sea, and neuer farre off from it, but often on the very baich of the Sea, where it wafheth ouer them all along the coafts of Spaine, Portugal, and Italy, and within the Straights in many places : it will not abide in any Garden farre from the Sea, no not in Italy, as it is related.

The Time.

The time wherein they flower, is expreffed to be in Auguft and September : the feede to be ripe in October and Nouember, and the greene leaues to fpring vp in Nouember and December.

The Names.

Thefe are certainly the true kindes of *Scilla* that fhould bee vfed in medicines, although (as Clufius reporteth) the Spaniards forbade him to tafte of the red Squill, as of a moft ftrong and prefent poifon. Pliny hath made more forts then can be found out yet to this day with vs : that *Scilla* that is called *Epimenidia*, becaufe it might be eaten, is thought to be the great *Ornithogalum*, or Starre of Bethlehem. *Pancratium* is, I know , and as I faid before, referred to that kinde of baftard Sea Daffodill , which is fet forth before in the end of the hiftory of the baftard Daffodils ; and diuers alfo would make the *Narciffus tertius Matthioli*, which I call the true Sea Daffodill,to be a *Pancratium*,but feeing Diofcorides (and no other is againft him) maketh

M

maketh *Pancratium* to be a kinde of Squill with reddish rootes, I dare not vphold their opinion against such manifest truth.

The Vertues.

The Squill or Sea Onion is wholly vsed physically with vs, becauſe wee can receiue no pleaſure from the ſight of the flowers. Pliny writeth, that Pithagoras wrote a volume or booke of the properties thereof, for the ſingular effects it wrought; which booke is loſt, yet the diuers vertues it hath is recorded by others, to be effectuall for the ſpleene, lungs, ſtomach, liuer, head and heart ; and for dropſies, old coughs, Iaundiſe, and the wormes ; that it cleareth the ſight, helpeth the tooth-ache, cleanſeth the head of ſcurfe, and running ſores ; and is an eſpeciall Antidote againſt poiſon : and therefore is vſed as a principall ingredient into the *Theriaca Andromachi,* which we vſually call Venice Treakle. The Apothecaries prepare hereof, both Wine, Vinegar, and Oxymel or Syrupe, which is ſingular to extenuate and expectorate tough flegme, which is the cauſe of much diſquiet in the body, and an hinderer of concoction, or diſgeſtion in the ſtomach, beſides diuers other wayes, wherein the ſcales of the rootes, being dryed, are vſed. And Galen hath ſufficiently explained the qualities and properties thereof, in his eight Booke of Simples.

Chap. XII.

Ornithogalum. Starre of Bethlehem.

After the Family of the Iacinths, muſt needes follow the kindes of Starre-flowers, or Starres of Bethlehem, as they are called, for that they doe ſo nearely reſemble them, that diuers haue named ſome of them Iacinths, and referred them to that kindred : all of them, both in roote, leafe, and flower, come nearer vnto the Iacinths, then vnto any other plant. They ſhall therefore bee next deſcribed, euery one in their order, the greateſt firſt, and the reſt following.

Ornithogalum Arabicum. The great Starre-flower of Arabia.

This Arabian Starre-flower hath many broad, and long greene leaues, very like vnto the leaues of the Orientall Iacinth, but lying for the moſt part vpon the ground, among which riſeth vp a round greene ſtalke, almoſt two foote high, bearing at the toppe diuers large flowers, ſtanding vpon long foote-ſtalkes, and at the bottome of euery one of them a ſmall ſhort pointed greene leafe : theſe flowers are made of ſix pure white leaues à peece, laid open as large as an ordinary Daffodill, but of the forme of a Starre Iacinth, or Starre of Bethlehem, which cloſe as they doe euery night, and open themſelues in the day time, eſpecially in the Sunne, the ſmell whereof is pretty ſweete, but weake : in the middle of the flower is a blackiſh head, compoſed with ſix white threeds, tipt with yellow pendents : the ſeede hath not beene obſerued with vs: the roote is great and white, with a flat bottome, very impatient of our cold Winters, ſo that it ſeldome proſpereth or abideth with vs ; for although ſometimes it doe abide a Winter in the ground, yet it often lyeth without ſpringing blade, or any thing elſe a whole yeare, and then periſheth : or if it doe ſpring, yet many doe not beare, and moſt after their firſt bearing doe decay and periſh. But if any be deſirous, to know how to preſerue the roote of this plant, or of many other bulbous rootes that are tender, ſuch as the great double white Daffodill of Conſtantinople, and other fine Daffodils, that come from hot Countries; let them keepe this rule : Let either the roote be planted in a large pot, or tubbe of earth, and houſed all the Winter, that ſo it may bee defended from the froſts ; Or elſe (which is the eaſier way) keepe the roote out of the ground euery yeare, from September, after the leaues and ſtalkes are paſt, vntill February, in some dry, but not hot or windy place, and then plant it in the ground vnder a South wall, or such like defended place, which will spring, and no doubt prosper well there, in regard the greatest and deepest frosts are past after February, so that seldome any great frosts come after, to pierce so deepe as the roote is to be set, or thereby to doe any great harme to it in such a place.

The Place.

This hath been often sent out of Turkie, and likewise out of Italy ; I had likewise two rootes sent mee out of Spaine by Guillaume Boel before re-membred, which (as hee said) hee gathered there, but they prospered not with me, for want of the knowledge of the former rule. It may be likely that Arabia is the place, from whence they of Constantinople receiue it.

The Time.

It flowreth in May, if it be of the first yeares bringing ; or in Iune, if it haue been ordered after the manner before set downe.

The Names.

It hath been sent out of Italy by the name of *Lilium Alexandrinum*, The Lilly of Alexandria, but it hath no affinity with any Lilly. Others call it *Hyacinthus Arabicus;* and the Italians, *Iacintho del pater nostro :* but it is no Ia-cinth neither, although the flowers be like some of them. Some also would referre it to a *Narcissus* or Daffodill, and it doth as little agree with it, as with a Lilly, although his flowers in largenesse and whitenesse resemble a Daffodill. Clusius hath most fitly referred it to the stocke or kindred of *Ornithogala*, or Starres of Bethlehem, as wee call them in English, and from the Turkish name, *Zumbul Arab*, entituled it *Ornithogalum Arabicum*, although *Zumbul*, as I haue before declared, is with them, a Iacinth, wee may call it in English, The Arabian Starre-flower, or Starre of Bethlehem, or the great Starre-flower of Arabia.

1. *Ornithogalum maximum album.*
The greatest white Starre-flower, or Starre of Bethlehem.

This great Starre-flower hath many faire, broad, long, and very fresh green leaues, rising vp very early, and are greater, longer, and greener then the leaues of any Ori-entall Iacint, which doe abide greene, from the beginning or middle of Ianuary, or before sometimes, vntill the end of May, at which time they begin to fade, and the stalke with the head of flowers beginneth to rise, so that it will haue either few or no leaues at all, when the flowers are blowne : the stalke is strong, round, and firme, rising two foote high or more, bearing at the toppe a great bush of flowers, seeming at the first to be a great greene eare of corne, for it is made spike-fashion, which when the flowers are blowne, doth rise to be very high, slender or small at the head aboue, and broad spread and bushing below, so that it is long in flowring ; for they flower below first, and so vpwards by degrees : these flowers are snow white, without any line on the backside, and is therein like vnto the former, as also in whitenesse, but nothing so large, with a white vmbone or head in the middle, beset with many white threeds, tipt with yellow : the seede is blacke and round, contained in three square heads : the roote is great, thicke, and short, and somewhat yellowish on the outside, with a flat bottome, both like the former, and the next that followeth.

2. *Ornithogalum maius spicatum album.*
The great white spiked Starre-flower.

This spiked Starre-flower in his growing, is somewhat like vnto the last described,

but

but springeth not vp so early, nor hath his leaues so greene, or large, but hath broad, long, whitish greene hollow leaues, pointed at the end, among which riseth vp the stalke, which is strong and high, as the former, hauing a great bush of flowers at the toppe, standing spike-fashion, somewhat like the former, flowring in the same maner by degrees, first below, and so vpwards ; but it is not so thicke set with flowers, nor so farre spread at the bottome as it, the flowers also are not so white, and each of the leaues of them haue a greene line downe the backe, leauing the edges on both sides white : after the flowers are past, the heads for seede grow three square, like the other, bearing such like blacke seede therein : the roote hereof is vsually bigger then the last, and whiter on the outside.

3. *Ornithogalum Pannonicum.* The Hungarian Starre-flower.

This Hungarian Starre-flower shooteth out diuers narrow, long, whitish greene leaues, spread vpon the ground before Winter, which are very like vnto the leaues of Gilloflowers, and so abide aboue ground, hauing a stalke rising in the middle of them the next Spring, about halfe a foote high or thereabouts, bearing many white flowers at the toppe, with greene lines downe the backe of them, very like vnto the ordinary Starres of Bethlehem : the roote is greater, thicker, and longer then the ordinary Starres, and for the most part, two ioyned together, somewhat grayish on the out side.

4. *Ornithogalum vulgare.* The Starre of Bethlehem.

The ordinary Starre of Bethlehem is so common, and well knowne in all countries and places, that it is almost needlesse to describe it, hauing many greene leaues with white lines therein, and a few white flowers set about the toppe of the stalke, with greenish lines downe the backe : the roote is whitish, and encreaseth aboundantly.

5. *Asphodelus bulbosus Galeni, siue Ornithogalum maius flore subnirescente.* The bulbous Asphodill, or greene Starre-flower.

Diuers haue referred this plant vnto the Asphodils, because (as I thinke) the flowers hereof are straked on the backe, and the leaues long and narrow, like vnto the Asphodils ; but the roote of this being bulbous, I rather (as some others doe) ioyne it with the *Ornithogala*, for they also haue strakes on the backe of the flowers. It hath many whitish greene leaues, long and narrow, spread vpon the ground, which spring vp in the beginning of the yeare, and abide vntill May, and then they withering, the stalke springeth vp almost as high as the first, hauing many pale yellowish greene flowers, but smaller, and growing more sparsedly about the stalke vpon short foot-stalkes, but in a reasonable long head spike-fashion : the seede is like vnto the second kinde, but smaller : the roote is somewhat yellowish, like the first great white kinde.

The Place.

The first is onely nursed in Gardens, his originall being not well knowne, yet some attribute it vnto *Pannonia* or Hungary. The second hath been found neare vnto Barcinone, and Toledo in Spaine. The third was found in Hungary by Clusius. Our ordinary euery where in the fields of Italy and France, and (as it is said) in England also. And the last groweth likewise by the corne fields in the vpper Hungary.

The Time.

They flower in Aprill and May, and sometimes in Iune.

The Names.

The first is called by Clusius *Ornithogalum maximum album*, because it is

greater

1 *Ornithogalum Arabicum.* The great starre-flower of Arabia. 2 *Ornithogalum maximum album* The greatest white starre flower. 3 *Ornithogalum maius spicatum album.* The great white spiked starre flower. 4 *Ornithogalum Pannonicum album.* The Hungarian starre flower. 5 *Asphodelus bulbosus, Galeni, sive Ornithogalum maius, subvirescente flore.* The bulbed Asphodill, or greene starre flower. 6 *Ornithogalum Hispanicum minus.* The little starre-flower of Spaine. 7 *Ornithogalum luteum.* The yellow starre-flower of Bethlehem. 8 *Ornithogalum Neapolitanum.* The starre-flower of Naples.

greater then the next, which hee tooke formerly for the greateſt : but it might more fitly, in my iudgement, bee called *Aſphodelus bulboſus albus* (if there be any *Aſphodelus bulboſus* at all) becauſe this doth ſo nearly reſemble that, both in the early ſpringing, and the decay of the greene leaues, when the ſtalkes of flowers doe riſe vp. Diuers alſo doe call it *Ornithogalum Pannonicum maximum album.*

The ſecond hath his name in his title, as moſt authors doe ſet it downe, yet in the great Herball referred to Dalechampius, it is called *Ornithogalum magnum Myconi.*

The third hath his name from the place of his birth, and the other from his popularity, yet Dodonæus calleth it *Bulbus Leucanthemos.*

The laſt is called by diuers *Aſphodelo-hyacinthinus*, and *Hyacintho-aſpho-delus Galeni.* Dodonæus calleth it *Aſphodelus fœmina*, and *Aſphodelus bul-boſus.* But Lobel, and Gerrard from him, and Dodonæus, doe make this to haue white flowers, whereas all that I haue ſeene, both in mine owne, and in others Gardens, bore greeniſh flowers, as Cluſius ſetteth it truely downe. Lobel ſeemeth in the deſcription of this, to confound the *Ornithogalum* of Mompelier with it, and calleth it *Aſphodelus hyacinthinus forte Galeni*, and ſaith that ſome would call it *Pancratium Monſpelienſe*, and *Aſphodelus Galeni.* But as I haue ſhewed, the *Ornithogalum ſpicatum* and this, doe plainly differ the one from the other, and are not both to be called by one name, nor to be reckoned one, but two diſtinct plants.

Ornithogalum Æthiopicum. The Starre-flower of Æthiopia.

The leaues of this plant are a foote long, and at the leaſt an inch broad, which being broken, are no leſſe woolly then the woolly Iacinth : the ſtalke is a cubit high, ſtrong and greene ; from the middle whereof vnto the toppe, ſtand large ſnow white flowers, vpon long, greene, thicke foot-ſtalkes, and yellowiſh at the bottome of the flower, in the middle whereof ſtand ſix white threeds, tipt with yellow chiues, compaſſing the head, which is three ſquare, and long containing the ſeede : the roote is thicke and round, ſomewhat like the *Aſphodelus Galeni.*

The Place.

This plant was gathered by ſome Hollanders, on the Weſt ſide of the Cape of good Hope.

The Time.

It flowred about the end of Auguſt with thoſe that had it.

The Names.

Becauſe it came from that part of the continent beyond the line, which is reckoned a part of Æthiopia, it is thereupon ſo called as it is ſet downe.

Ornithogalum Neopolitanum. The Starre-flower of Naples.

This beautifull plant riſeth out of the ground very early, with foure or fiue hollow pointed leaues, ſtanding round together, of a whitiſh greene colour, with a white line downe the middle of euery leafe on the inſide, ſomewhat narrow, but long, (Fabius Columna ſaith, three foot long in Italy, but it is not ſo with vs) in the middle of theſe leaues riſeth vp the ſtalke, a foote and a halfe high, bearing diuers flowers at the toppe, euery one ſtanding in a little cuppe or huſke, which is diuided into three or foure parts, hanging downe very long about the heads for ſeede : after the flower is paſt, theſe flowers doe all hang downe their heads, and open one way, although their little foot-ſtalkes come forth on all ſides of the greater ſtalke, being large, and compoſed of ſix long leaues, of a pure white on the inſide, and of a blewiſh or whitiſh greene colour

on

on the outside, leauing the edges of euery leafe white on both sides: in the middle of these flowers stand other small flowers, each of them also made of six small white leaues a peece, which meeting together, seeme to make the shew of a cuppe, within which are contained six white threeds, tipt with yellow, and a long white pointell in the middle of them, being without any sent at all: after the flowers are past, come vp great round heads, which are too heauie for the stalke to beare; and therefore lye downe vpon the leaues or ground, hauing certaine lines or stripes on the outside, wherein is contained round, blacke, rough seede: the roote is great and white, and somewhat flat at the bottome, as diuers of these kindes are, and doe multiply as plentifully into small bulbes as the common or any other.

The Place.

This Starre-flower groweth in the Medowes in diuers places of Naples, as Fabius Columna, and Ferrantes Imperatus doe testifie, from whence they haue been sent. And Matthiolus, who setteth out the figure thereof among his Daffodils, had (it should seeme) seene it grow with him.

The Time.

It flowreth in May, although it begin to spring out of the ground oftentimes in Nouember, but most vsually in Ianuary: the seede is ripe in Iuly.

The Names.

Matthiolus reckoneth this (as is said) among the Daffodils, for no other respect, as I conceiue, then that he accounted the middle flower to bee the cuppe or trunke of a Daffodill, which it doth somewhat resemble, and setteth it forth in the fourth place, whereupon many doe call it *Narcissus quartus Matthioli*, The fourth Daffodill of Matthiolus. Fabius Columna calleth it *Hyacinthus aruorum Ornithogalis flore*. Clusius (to whom Imperatus sent it, in stead of the Arabian which hee desired) calleth it of the place from whence he receiued it, *Ornithogalum Neopolitanum*, and we thereafter call it in English, The Starre-flower of Naples.

Ornithogalum Hispanicum minus. The little Starre-flower of Spaine.

Clusius hath set forth this plant among his *Ornithogala* or Starre-flowers, and although it doth in my minde come nearer to a *Hyacinthus*, then to *Ornithogalum*, yet pardon it, and let it passe as he doth. From a little round whitish roote, springeth vp in the beginning of the yeare, fiue or six small long green leaues, without any white line in the middle of them, among which rise vp one or two small stalkes, an hand length high or better, bearing seuen or eight, or more flowers, growing as it were in a tuft or vmbell, with small long leaues at the foote of euery stalke, the lower flowers being equall in length with the vppermost, of a pale whitish blew or ash colour, with a strake or line downe the backe of euery leafe of them, with some white threeds standing about a blewish head in the middle: these flowers passe away quickly, and giue no seed, so that it is not knowne what seede it beareth.

The Place.

This groweth in Spaine, and from thence hath been brought to vs.

The Time.

It flowreth in May.

The Names.

It hath no other name then is set down in the title, being but lately found out.

L. Orni-

*1.Ornithogalum album vnifolium.*The white ftarre-flower with one blade.

This little ftarre-flower I bring into this place, asthe fitteft in my opinion where to place it, vntill my minde change to alter it. It hath a very fmall round white roote, from whence fpringeth vp one very long and round greene leafe, like vnto a ruſh, but that for about two or three inches aboue the ground, it is a little flat, and from thence fpringeth forth a fmall ftalke not aboue three or foure inches high, bearing at the top thereof three or foure fmall white flowers, confifting of fix leaues a peece, within which are fix white chiues, tipt with yellow pendents, ftanding about a fmall three fquare head, that hath a white pointell fticking as it were in the middeft thereof : the flower is pretty and fweete, but not heady.

Ornithogalum luteum. The yellow Starre of Bethlehem.

This yellow Starre-flower rifeth vp at the firft,with one long,round,greeniſh leafe, which openeth it felfe fomewhat aboue the ground, and giueth out another fmall leafe, leſſer and ſhorter then the firft, and afterward the ftalke rifeth from thence alfo, being foure or fiue inches high, bearing at the toppe three or foure fmall green leaues, and among them foure or fiue fmall yellow ftarre-like flowers, with a greeniſh line or ftreake downe the backe of euery leafe, and fome fmall reddiſh yellow threeds in the middle : it feldome giueth feede : the roote is round, whitiſh, and fomewhat cleare, very apt to periſh, if it bee any little while kept dry out of the ground, as I haue twice tryed to my loſſe.

The Place.

The firft grew in Portugall, and Clufius firft of all others defciphers it.
The other is found in many places both of Germany and Hungary, in the moifter grounds.

The Time.

The firft flowreth in May : the other in Aprill, and fometimes in March.

The Names.

Carolus Clufius calleth the firft *Bulbus vnifolius*, or *Bolbine*, but referreth it not to the ftocke or kindred of any plant ; but (as you fee) I haue ranked it with the fmall forts of *Ornithogalum*, and giue it the name accordingly.
The other is referred for likeneſſe of forme,and not for colour,vnto the *Ornithogala*, or Starres of Bethlehem. It is called by Tragus and Fuchfius *Bulbus filueftris*, becaufe of the obuioufneſſe. Cordus taketh it to be *Sifyrinchium*. Lacuna calleth it *Bulbus efculentus*. Lobel and others in thefe dayes generally, *Ornithogalum luteum,* and wee thereafter in Engliſh, The yellow Starre-flower,or Starre of Bethlehem.

The Vertues.

The firft kinde being but lately found out, is not knowne to be vfed. The rootes of the common or vulgar, are (as Matthiolus faith) much eaten by poore people in Italy, either rawe or roafted, being fweeter in tafte then any Chefnut, and feruing as well for a neceffary food as for delight. It is doubtfull whether any of the reft may be fo vfed ; for I know not any in our Land hath made any experience.

There are many other forts of Starre-flowers, which are fitter for a generall then this Hiftory ; and therefore I referre them thereunto.

CHAP. XIII.

Moly. Wilde Garlicke.

VNto the former Starre-flowers, muſt needes bee ioyned another tribe or kind-red, which carry their ſtraked flowers Starre-faſhion, not ſpikewiſe, but in a tuft or vmbell thicke thruſt or ſet together. And although diuers of them ſmell not as the former, but moſt of their firſt Grandfathers houſe, yet all doe not ſo; for ſome of them are of an excellent ſent. Of the whole Family, there are a great many which I muſt leaue, I will onely ſelect out a few for this our Garden, whoſe flowers for their beauty of ſtatelineſſe, forme, or colour, are fit to bee entertained, and take place therein, euery one according to his worth, and are accepted of with the louers of theſe delights.

1. *Moly Homericum, vel potius Theophraſti.* The greateſt Moly of Homer.

Homers Moly (for ſo it is moſt vſually called with vs) riſeth vp moſt commonly with two, and ſometimes with three great, thicke, long, and hollow guttured leaues, of a whitiſh greene colour, very neare the colour of the Tulipa leafe, hauing ſome-times at the end of ſome of the leaues, and ſometimes apart by it ſelfe, a whitiſh round ſmall button, like vnto a ſmall bulbe, the like whereof alſo, but greater, doth grow be-tweene the bottome of the leaues and the ſtalke neare the ground, which being plan-ted when it is ripe, will grow into a roote of the ſame kinde : among theſe leaues riſeth vp a round, ſtrong, and tall ſtalke, a yard high or better, bare or naked vnto the toppe, where it beareth a great tuft or vmbell of pale purpliſh flowers; all of them almoſt ſtanding vpon equall foot-ſtalkes, or not one much higher then another, conſiſting of fiue leaues a peece, ſtriped downe the backe with a ſmall pale line, hauing a round head or vmbone with ſome threeds about it in the midſt : Theſe flowers doe abide a great while blowne before they vade, which ſmell not very ſtrong, like any Onion or Garlicke, but of a faint ſmell : and after they are paſt come the ſeede, which is blacke, wrapped in white cloſe huskes : the roote groweth very great, ſometimes bigger then any mans cloſed fiſt, ſmelling ſtrong like Garlicke, whitiſh on the outſide, and greene at the toppe, if it be but a while bare from the earth about it.

2. *Moly Indicum ſiue Caucaſon.* The Indian Moly.

The Indian Moly hath ſuch like thicke large leaues, as the Homers Moly hath, but ſhorter and broader, in the middle whereof riſeth vp a ſhort weake ſtalke, almoſt flat, not hauing any flowers vpon it, but a head or cluſter of greeniſh ſcaly bulbes, inclo-ſed at the firſt in a large thinne skinne, which being open, euery bulbe ſheweth it ſelfe, ſtanding cloſe one vnto another vpon his foot-ſtalke, of the bigneſſe of an Acorne, which being planted, will grow to bee a plant of his owne kinde : the roote is white and great, couered with a darke coate or skinne, which encreaſeth but little vnder ground; but beſides that head, it beareth ſmall bulbes aboue the ground, at the bot-tome of the leaues next vnto the ſtalke, like vnto the former.

The Place.

Both theſe doe grow in diuers places of Spaine, Italy, and Greece; for the laſt hath been ſent out of Turkie among other rootes. Ferrantes Impe-ratus a learned Apothecary of Naples, ſent it to diuers of his friends in theſe parts, and hath deſcribed it in his naturall hiſtory among other plants, printed in the Italian tongue. It grew alſo with Iohn Tradeſcante at Canterbury, who ſent me the head of bulbes to ſee, and afterwards a roote, to plant it in my Garden.

The

The Time.

The first flowreth in the end of May, and abideth vnto the midst of Iuly, and sometimes longer. The other beareth his head of bulbes in Iune and Iuly.

The Names.

We haue receiued them by their names expressed in their titles, yet the last hath also been sent by the name of *Ornithogalum Italicum*, but as all may easily see, it is not of that kindred.

1. *Moly montanum Pannonicum bulbiferum primum.*
The first bulbed Moly of Hungary.

This first Hungarian Moly hath three or foure broad and long greene leaues, folded together at the first, which after open themselues, and are carried vp with the stalke, standing thereon one aboue another, which is a foote high; at the toppe whereof doe grow a few sad reddish bulbes, and betweene them long footstalkes, bearing flowers of a pale purplish colour; after which followeth blacke seede, inclosed in roundish heads: the roote is not great, but white on the outside, very like vnto the roote of Serpents Moly, hereafter described, encreasing much vnder ground, & smelling strong.

2. *Moly montanum Pannonicum bulbiferum secundum.*
The second bulbed Moly of Hungary.

The second Moly hath narrower greene leaues then the former: the stalke is about the same height, and beareth at the toppe a great cluster of small greene bulbes, which after turne of a darker colour; from among which come forth long foot-stalks, whereon stand purplish flowers: the roote is couered with a blackish purple coate or skinne.

3. *Moly Serpentinum.* Serpents Moly.

This Moly must also be ioyned vnto the bulbous Molyes, as of kindred with them, yet of greater beauty and delight, because the bulbes on the heads of the small stalkes are redder, and more pleasant to behold: the stalke is lower, and his graßie winding leaues, which turne themselues (whereof it tooke the name) are smaller, and of a whiter greene colour: it beareth among the bulbes purplish flowers also, but more beautifull, the sent whereof is nothing so strong: the roote is small, round, and whitish, encreasing into a number of small rootes, no bigger then pease round about the greater roote.

4. *Moly caule & folijs triangularibus.* The three cornered Moly.

This three square Moly hath foure or fiue long, and somewhat broad pale greene leaues, flat on the vpper side, and with a ridge downe the backe of the leafe, which maketh it seeme three square: the stalke which riseth vp a foote and a halfe high or better, is three square or three cornered also, bearing at the toppe out of a skinnie huske diuers white flowers, somewhat large and long, almost bell-fashion, with stripes of greene downe the middle of euery leafe, and a few chiues tipt with yellow in the middle about the head, wherein when it is ripe, is inclosed small blacke seede: the roote is white on the outside, and very like the yellow Moly; both roote, leafe, and flower hath a smacke, but not very strong of Garlicke.

5. *Moly Narcissinis folijs.* Daffodill leafed Moly.

This Moly hath many long, narrow, and flat greene leaues, very like vnto the leaues of a Daffodill, from whence it tooke his name (or rather of the early greater *Leucoium bulbosum,*

1 *Moly Homericum vel potius Theophrasti.* The greatest Moly of Homer. 2 *Moly Indicum siue Caucasum.* The Indian Moly. 3 *Moly Pannonicum montanum.* The yellow Moly of Hungary. 4 *Moly Serpentinum.* Serpents Moly. 5 *Moly purpureum Neapolitanum.* The purplish Moly of Naples. 6 *Moly caule & folÿs triangularibus.* The three cornered Moly. 7 *Moly latifolium flore luteo.* The yellow Moly. 8 *Moly Dioscoridum Hispanicum.* The Spanish Moly of Dioscorides. 9 *Moly Zibettinum vel Moschatinum.* The sweete smelling Moly of Mompelier. 10 *Moly serotinum Coniferum.* The late Pineapple Moly

bulbosum, or bulbed Violet before described, ioyned next vnto the Daffodils, becaufe it is fo like them) among which rifeth vp two or three ftalkes fometimes, each of a foot and a halfe high, bearing at the toppe, inclofed in a skinny hofe, as all the Molyes haue, a number of fmall purplifh flowers, which doe not long abide, but quickly fade: the feede is blacke as others are; the roote is fometimes knobbed, and more often bulbed, hauing in the knobs fome markes of the old ftalkes to be feene in them, and fmelleth fomewhat like Garlicke, whereby it may be knowne.

6. *Moly montanum latifolium luteo flore.* The yellow Moly.

The yellow Moly hath but one long and broad leafe when it doth not beare flower, but when it will beare flower, it hath two long and broad leaues, yet one alwaies longer and broader then the other, which are both of the fame colour, and neare the bignefle of a reafonable Tulipa leafe : betweene thefe leaues groweth a flender ftalke, bearing at the toppe a tuft or vmbell of yellow flowers out of a skinnie hofe, which parteth three wayes, made of fix leaues a peece, laid open like a Starre, with a greenifh backe or outfide, and with fome yellow threeds in the middle : the feede is blacke, like vnto others : the roote is whitifh, two for the moft part ioyned together, which encreafeth quickly, and fmelleth very ftrong of Garlicke, as both flowers and leaues doe alfo.

7. *Moly Pyrenaum purpureum.* The purple mountaine Moly.

This purple Moly hath two or three leaues, fomewhat like the former yellow Moly, but not fo broad, nor fo white : the ftalke hath not fo many flowers thereon, but more fparingly, and of an vnpleafant purple colour : the roote is whitifh, fmelling fomewhat ftrongly of Garlicke, but quickly perifheth with the extremity of our cold Winters, which it will not abide vnleffe it be defended.

8. *Moly montanum latifolium purpureum Hifpanicum.* The purple Spanifh Moly.

This Moly hath two broad and very long greene leaues, like vnto the yellow Moly, in this, that they doe compaffe one another at the bottome of them, between which rifeth vp a ftrong round ftalke, two foote high or more, bearing at the toppe, out of a thinne huske, a number of faire large flowers vpon long foot-ftalkes, confifting of fix leaues a peece, fpread open like a Starre, of a fine delayed purple or blufh colour, with diuers threeds of the fame colour, tipt with yellow, ftanding about the middle head : betweene the ftalke and the bottome of the leaues it hath fome fmall bulbes growing, which being planted, will foone fpring and encreafe : the roote alfo being fmall and round, with many fibres thereat, hath many fmall bulbes fhooting from them; but neither roote, leafe, nor flower, hath any ill fent of Garlicke at all.

9. *Moly purpureum Neapolitanum.* The purple Moly of Naples.

The Neapolitane Moly hath three or foure fmall long greene leaues fet vpon the ftalke after it is rifen vp, which beareth a round head of very fine purple flowers, made of fix leaues a peece, but fo clofing together at the edge, that they feeme like vnto fmall cuppes, neuer laying themfelues open, as the other doe; this hath fome fent of his originall, but the roote more then any part elfe, which is white and round, quickly encreafing as moft of the Molyes doe.

10. *Moly pyxidatum argenteum Hifpanicum.* The Spanifh filuer cupped Moly.

This Spanifh Moly hath two or three very long rufh like leaues, which rife vp with the ftalke, or rather vanifh away when the ftalke is rifen vp to bee three foote high or more, bearing a great head of flowers, ftanding clofe at the firft, but afterwards fpreading much one from another, euery flower vpon a long foote-ftalke, being of a white filuer colour, with ftripes or lines on euery fide, and fafhioned fmall and hollow, like a cuppe or boxe : the feede I could neuer obferue, becaufe it flowreth fo late, that the Winter hindereth it from bearing feede with vs : the roote is fmall and round, white, and in a manner tranfparent, at leaft fo fhining, as if it were fo, and encreafeth nothing fo much, as many of the other forts : this hath no ill fent at all, but rather a pretty fmell, not to bee mifliked.

11. *Moly ferotinum Coniferum.* The late Pineapple Moly.

This late Moly that was fent me with the laft defcribed, and others alfo from Spain, rifeth vp with one long greene leafe, hollow and round vnto the end, towards this end on the one fide, breaketh out a head of flowers, enclofed in a thinne skinne, which after it hath fo ftood a good while, (the leafe in the meane time rifing higher, and growing harder, becommeth the ftalke) breaketh, and fheweth a great bufh or head of buds for flowers, thicke thruft together, fafhioned very like vnto the forme of a Pineapple (from whence I gaue it the name) of the bignefs of a Walnut: after this head hath ftood in this manner a moneth or thereabouts, the flowers fhew themfelues to bee of a fine delayed or whitifh purple colour, with diuers ftripes in euery of them, of the fame cup-fafhion with the former, but not opening fo plainly, fo that they cannot bee difcerned to bee open, without good heede and obferuation. It flowreth fo late in Autumne, that the early frofts doe quickly fpoile the beauty of it, and foone caufe it to rotte : the roote is fmall and round, and fhining like the laft, very tender alfo, as not able to abide our fharpe Winters, which hath caufed it vtterly to perifh with me.

12. *Moly Diofcorideum.* Diofcorides his Moly.

The roote of this fmall Moly is tranfparent within, but couered with a thicke yellowifh skinne, of the bigneffe of an Hafell Nut, or fomewhat bigger, which fendeth forth three or foure narrow graffie leaues, long and hollow, and a little bending downwards, of a whitifh greene colour, among which rifeth vp a flender weake ftalke, a foot and a halfe high, bearing at the toppe, out of a thinne skinne, a tuft of milke white flowers, very like vnto thofe of Ramfons, which ftand a pretty while in their beauty, and then paffe away for the moft part without giuing any feede : this hath little or no fent of Garlicke.
We haue another of this fort that is leffer, and the flowers rounder pointed.

13. *Moly Diofcorideum Hifpanicum.* The Spanifh Moly of Diofcorides.

This Moly came vnto me among other Molyes from Spaine, and is in all things like vnto the laft defcribed, but fairer, larger, and of much more beauty, as hauing his white flowers twice as great as the former ; but (as it feemeth) very impatient of our Winters, which it could not at any hand endure, but quickly perifhed, as fome others that came with it alfo.

14. *Moly Mofchatinum vel Zibettinum Monfpelienfe.*
The fweete fmelling Moly of Mompelier.

This fweete Moly, which I haue kept for the laft, to clofe vp your fenfes, is the fmalleft, and the fineft of all the reft, hauing foure or fiue fmall greene leaues, almoft as fine as haires, or like the leaues of the Feather-graffe : the ftalke is about a foote high, bearing fiue or fix or more fmall white flowers, laid open like Starres, made of fix leaues a peece, of an excellent fweete fent, refembling Muske or Ciuet ; for diuers haue diuerfly cenfured of it. It flowreth late in the yeare, fo that if the precedent Summer bee either ouer moift, or the Autumne ouer early cold, this will not haue that fweete fent, that it will haue in a hot drie time, and befides muft be carefully refpected : for it will hardly abide the extremity of our fharpe Winters.

The

The Place.

The places of thefe Molyes, are for the moft part expreffed in their ti-tles, or in their defcriptions.

The Ti..e.

The time is fet downe, for the moft part to bee in Iune and Iuly, the reft later.

The Names.

To make further relation of names then are expreffed in their tiles, were needleffe; let thefe therefore fuffice.

The Vertues.

All thefe forts of Molyes are fmall kindes of wilde Garlicke, and are to be vfed for the fame purpofes that the great Garden Garlicke is, although much weaker in their effects. For any other efpeciall property is in any of thefe, more than to furnifh a Garden of variety, I haue not heard at all.

And thus much may fuffice of thefe kindes for our Garden, referuing manie others that might be fpoken of, to a generall worke, or to my Garden of Simples, which as God fhall enable me, and time giue leaue, may fhew it felfe to the world, to abide the iudicious and criticke cenfures of all.

Chap. XIIII.

Afphodelus. The Afphodill.

THere remaine fome other flowers, like vnto the laft defcribed, to be fpecified, which although they haue no bulbous rootes, yet I thinke them fitteft to bee here mentioned, that fo I may ioyne thofe of neereft fimilitude together, vn-till I haue finifhed the reft that are to follow.

1. *Afphodelus maior albus ramofus.* The great white branched Afphodill.

The great white Afphodill hath many long, and narrow, hollow three fquare leaues, fharpe pointed, lying vpon the ground round about the roote: the ftalke is fmooth, round, and naked without leaues, which rifeth from the midft of them, di-uided at the toppe into diuers branches, if the plant bee of any long continuance, or elfe but into two or three fmall branches, from the fides of the maine great one, whereon doe ftand many large flowers Starre-fafhion, made of fix leaues a peece, whitifh on the infide, and ftraked with a purplifh line downe the backfide of euery leafe, hauing in the middle of the flowers fome fmall yellow threeds: the feede is blacke, and three fquare, greater then the feede of Bucke wheate, contained in roundifh heads, which open into three parts: the roote is compofed of many tube-rous long clogges, thickeft in the middle, and fmaller at both ends, faftened together at the head, of a darke grayifh colour on the outfide, and yellow within.

2. *Afphodelus albus non ramofus.* The white vnbranched Afpodill.

The vnbranched Afphodill is like vnto the former, both in leaues and flowers, but that the flowers of this are whiter, and without any line or ftrake on the backe fide, and

and the ſtalkes are without branches : the rootes likewiſe are ſmaller, and fewer, but made after the ſame faſhion.

3. *Aſphodelus maior flore carneo.* The bluſh coloured Aſphodill.

This Aſphodill is like to the laſt in forme of leaues and branches, and differeth in this, that his leaues are marked with ſome ſpots, and the flowers are of a bluſh or fleſh colour, in all other things alike.

4. *Aſphodelus minimus albus.* The leaſt white Aſphodill.

This leaſt Aſphodill hath foure or fiue very narrow long leaues, yet ſeeming three ſquare like the greateſt, bearing a ſmall ſtalke, of about a foote high among them, without any branches, and at the toppe a few white flowers, ſtraked both within and without, with a purpliſh line in the middle of euery leafe. The rootes are ſuch like tuberous clogges as are in the former, but much leſſer.

5. *Aſphodelus albus minor ſiue Fiſtuloſus.* The little hollow white Aſpnodill.

This little white Aſphodill hath a number of leaues growing thicke together, thicker and greener then thoſe of the ſmall yellow Aſphodill, or Kings Speare next following, among which riſeth vp diuers round ſtalkes, bearing flowers from the middle to the toppe, Starre-faſhion, with ſmall greene leaues among them, which are white on the inſide, and ſtriped on the backe with purple lines, like vnto the firſt deſcribed : the ſeede, and heads containing them, are three ſquare, like the ſeede of the little yellow Aſphodill : the rootes of this kinde are not glandulous, as the former, but ſtringie, long and white : the whole plant is very impatient of our cold Winters, and quickly periſheth, if it be not carefully preſerued, both from the cold, and much wet in the Winter, by houſing it ; and then it will abide many yeares : for it is not an annuall plant, as many haue thought.

6. *Aſphodelus luteus minor, ſiue Haſtula regia.* The ſmall yellow Aſphodill, or Kings ſpeare.

This ſmall yellow Aſphodill, which is vſually called the Kings ſpeare, hath many long narrow edged leaues, which make them ſeeme three ſquare, of a blewiſh or whitiſh greene colour : the ſtalke riſeth vp three foote high oftentimes, beſet with ſmall long leaues vp vnto the very flowers, which grow thicke together ſpike-faſhion one aboue another, for a great length, and wholly yellow, laid open like a Starre, ſomewhat greater then the laſt white Aſphodill, and ſmaller then the firſt, which when they are paſt yeeld round heads, containing blacke cornered ſeede, almoſt three ſquare : the rootes are many long yellow ſtrings, which ſpreading in the ground, doe much encreaſe.

The Place.

All theſe Aſphodils doe grow naturally in Spaine and France, and from thence were firſt brought vnto vs, to furniſh our Gardens.

The Time.

All the glandulous rooted Aſphodils doe flower ſome in May, and ſome in Iune ; but the two laſt doe flower, the yellow or laſt of them in Iuly, and the former white one in Auguſt and September, and vntill the cold and winter hinder it.

The Names.

Their ſeuerall names are giuen them in their titles, as much as is fit for

this

this difcourfe. For to fhew you that the Greekes doe call the ftalke of the great Afphodill Aῤϑεἶκ῾, and the Latines *Albucum*, or what elfe belongeth to them, is fitter for another worke, vnto which I leaue them.

The baftard Afphodils fhould follow next in place, if this worke were fit for them; but becaufe I haue tyed my felfe to expreffe onely thofe flowers and plants, that for their beauty, or fent, or both, doe furnifh a Garden of Pleafure, and they haue none, I leaue them to a generall Hiftory of plants, or that Garden of Simples before fpoken of, and will defcribe the Lilly Afphodils, and the *Phalangia* or Spider-worts, which are remaining of thofe, that ioyne in name or fafhion, and are to be here inferted, before I paffe to the reft of the bulbous rootes.

1. *Liliafphodelus phæniceus.* The gold red Day Lilly.

Becaufe the rootes of this and the next, doe fo nearely agree with the two laft reci-ted Afphodils, I haue fet them in this place, although fome doe place them next after the Lillies, becaufe their flowers doe come neareft in forme vnto Lillies; but whether you will call them Afphodils with Lilly flowers, as I thinke it fitteft, or Lillies with Afphodill rootes, or Lillies without bulbous rootes, as others doe, I will not contend. The red Day Lilly hath diuers broad and long frefh greene leaues, folded at the firft as it were double, which after open, and remaine a little hollow in the middle; among which rifeth vp a naked ftalke three foot high, bearing at the toppe many flow-ers, one not much diftant from another, and flowring one after another, not hauing lightly aboue one flower blown open in a day, & that but for a day, not lafting longer, but clofing at night, and not opening againe; whereupon it had his Englifh name, The Lilly for a day: thefe flowers are almoft as large as the flowers of the white Lilly, and made after the fame fafhion, but of a faire gold red, or Orange tawny colour. I could neuer obferue any feede to follow thefe flowers; for they feeme the next day after they haue flowred, (except the time be faire and dry) to bee fo rotten, as if they had lyen in wet to rotte them, whereby I thinke no feede can follow: the rootes are many thicke and long yellow knobbed ftrings, like vnto the fmall yellow Afphodill rootes, but fomewhat greater, running vnder ground in like fort, and fhooting young heads round about.

2. *Liliafphodelus luteus.* The yellow Day Lilly.

I fhall not neede to make a repetition of the defcription of this Day Lilly, hauing giuen you one fo amply before, becaufe this doth agree thereunto fo nearely, as that it might feeme the fame; thefe differences onely it hath, the leaues are not fully fo large, nor the flower fo great or fpread open, and the colour thereof is of a faire yel-low wholly, and very fweet, which abideth blowne many daies before it fade, and hath giuen blacke round feede, growing in round heads, like the heads of the fmall yellow Afphodill, but not fo great.

Clufius hath fet downe, that it was reported, that there fhould be another Liliafpho-dill with a white flower, but we can heare of none fuch as yet; but I rather thinke, that they that gaue that report might be miftaken, in thinking the Sauoye Spider-wort to be a white Liliafphodill, which indeede is fo like, that one not well experienced, or not well regarding it, may foone take one for another.

The Place.

Their originall is many moift places in Germany.

The Time.

They flower in May and Iune.

The Names.

They are called by fome *Liliago*, and *Lilium non bulbofum*, and *Liliafpho-delus*,

1 *Asphodelus maior albus ramosus.* The great white branched Asphodill. 2 *Asphodelus minor albus seu fistulosus.* The little hollow white
Asphodill. 3 *Asphodelus minor luteus, siue Hastula regia.* The small yellow Asphodill, or Kings speare. 4 *Liliasphodelus luteus.* The
yellow Day Lilly. 5 *Liliasphodelus phæniceus.* The gold red Day Lilly.

delus. In English we call them both Day Lillies , but the name doth not so well agree with the last, as with the first, for the causes aboue specified.

The Vertues.

The rootes of Asphodill hath formerly beene had in great account , but now is vtterly neglected ; yet by reason of their sharpenesse they open and cleanse, and therefore some haue of late vsed them for the yellow Iaundise. The Day Lillies haue no physicall vse that I know, or haue heard.

Chap. XV.

Phalangium. Spider-wort.

THese plants doe so nearely resemble those that are last set forth , that I thinke none that knowes them, will doubt, but that they must follow next vnto them, being so like vnto them, and therefore of the fairest of this kinde first.

1. *Phalangium Allobrogicum.* The Sauoye Spider-wort.

The Sauoye Spider-wort springeth vp with foure or fiue greene leaues , long and narrow, yet broader at the bottome, narrower pointed at the end , and a little hollow in the middle ; among which riseth vp a round stiffe stalke , a foote and a halfe high, bearing at the toppe one aboue another, seuen or eight, or more flowers , euery one as large almost as the yellow Day Lilly last described , but much greater then in any other of the Spider-worts, of a pure white colour, with some threeds in the middle, tipt with yellow, and a small forked pointell : after the flowers are past, the heads or seede vessels grow almost three square, yet somewhat round, wherein is contained blackish seede : the rootes are many white, round, thicke, brittle strings, ioyned together at the head , but are nothing so long, as the rootes of the other *Phalangia* or Spider-worts.

2. *Phalangium maius Italicum album.* The great Italian Spider-wort.

This great Spider-wort hath diuers long and narrow leaues spread vpon the ground, and not rising vp as the former , and not so broad also as the former, but somewhat larger then those that follow : the stalke is bigger, but seldome riseth vp so high as the next, whereof this is a larger kinde, hauing a long vnbranched stalke of white flowers, laid open like starres as it hath, but somewhat greater : the rootes are long and white, like the next, but somewhat larger.

3. *Phalangium non ramosum vulgare.* Vnbranched Spider-wort.

The leaues of this Spider-wort doe seeme to bee little bigger or longer then the leaues of grasse, but of a more grayish green colour, rising immediately from the head or tuft of rootes ; among which rise vp one or two stalkes, sometimes two or three foote long, beset toward the toppe with many white Starre-like flowers , which after they are past turne into small round heads, containing blacke seede, like vnto the seed of the little yellow Asphodill, but lesser : the rootes are long white strings, running vnder ground.

4. *Phalangium ramosum.* Branched Spider-wort.

The branched Spider-wort hath his leaues somewhat broader then the former, and of a more yellowish greene colour : the stalke hereof is diuersly branched at the top, bearing many white flowers, like vnto the former, but smaller : the seedes and rootes are like the former in all things.

The

1 *Phalangium Allobrogicum.* The Sauoye Spider-wort. 2 *Phalangium non ramosum.* Vnbranched Spider-wort. 3 *Phalangium ramosum.* Branched Spider-wort. 4 *Phalangium
Ephemerum Virginianum.* Iohn Tradescante's Spider-wort.

The Place.

The firft groweth on the Hils neare vnto Sauoye, from whence diuers, allured with the beauty of the flower, haue brought it into thefe parts.

The fecond came vp in my Garden, from the feede receiued out of Italy. The others grow in Spaine, France, &c.

The Time.

The vnbranched Spider-wort moft commonly flowreth before all the other, and the branched a moneth after it : the other two about one time, that is, towards the end of May, and not much after the vnbranched kinde.

The Names.

The firft (as I faid before) hath beene taken to be a white Lilliafphodill, and called *Liliafphodelus flore albo*; but Clufius hath more properly entituled it a *Phalangium*, and from the place of his originall, gaue him his other denomination, and fo is called of moft, as is fet downe in the title.

The other haue no other names then are expreffed in their titles, but only that Cordus calleth them *Liliago*; and Dodonæus, *lib.4. hift.plant.* would make the branched kinde to bee *Moly alterum Pliny*, but without any good ground.

The Vertues.

The names *Phalangium* and *Phalangites* were impofed on thefe plants, becaufe they were found effectuall, to cure the poyfon of that kinde of Spider, called *Phalangium*, as alfo of Scorpions and other Serpents. Wee doe not know, that any Phyfitian hath vfed them to any fuch, or any other purpofe in our dayes.

5. Phalangium Ephemerum Virginianum Ioannis Tradefcant.
The foon fading Spider-wort of Virginia, or Tradefcant his Spider-wort.

This Spider-wort is of late knowledge, and for it the Chriftian world is indebted vnto that painfull induftrious fearcher, and louer of all natures varieties, Iohn Tradefcant (fometimes belonging to the right Honourable Lord Robert Earle of Salisbury, Lord Treafurer of England in his time, and then vnto the right Honourable the Lord Wotton at Canterbury in Kent, and laftly vnto the late Duke of Buckingham) who firft receiued it of a friend, that brought it out of Virginia, thinking it to bee the Silke Graffe that groweth there, and hath imparted hereof, as of many other things, both to me and others ; the defcription whereof is as followeth :

From a ftringie roote, creeping farre vnder ground, and rifing vp againe in many places, fpringeth vp diuers heads of long folded leaues, of a grayifh ouer-worne greene colour, two or three for the moft part together, and not aboue compaffing one another at the bottome, and abiding greene in many places all the Winter; otherwhere perifhing, and rifing anew in the Spring, which leaues rife vp with the great round ftalke, being fet thereon at the ioynts, vfually but one at a ioynt, broad at the bottome where they compaffe the ftalke, and fmaller and fmaller to the end : at the vpper ioynt, which is the toppe of the ftalke, there ftand two or three fuch like leaues, but fmaller, from among which breaketh out a dozen, fixteene, or twenty, or more round green heads, hanging downe their heads by little foot-ftalkes, which when the flower beginneth to blow open, groweth longer, and ftandeth vpright, hauing three fmall pale greene leaues for a huske, and three other leaues within them for the flower, which lay themfelues open flat, of a deepe blew purple colour, hauing an vmbone or fmall head in the middle, clofely fet about with fix reddifh, hairy, or feathered threeds, tipt with yellow pendents : this flower openeth it felfe in the day, & fhutteth vfually at
night,

night, and neuer openeth againe, but perisheth, and then hangeth downe his head a-
gaine; the greene huske of three leaues, closing it selfe againe into the forme of a head,
but greater, as it was before, the middle vmbone growing to bee the seede vessell,
wherein is contained small, blackish, long seede: Seldome shall any man see aboue
one, or two at the most of these flowers blowne open at one time vpon the stalke,
whereby it standeth in flowring a longtime, before all the heads haue giuen out their
flowers.

The Place.

This plant groweth in some parts of Virginia, and was deliuered to Iohn
Tradescant.

The Time.

It flowreth from the end of May vntill Iuly, if it haue had greene leaues
all the Winter, or otherwise, vntill the Winter checke his luxuriousnesse.

The Names.

Vnto this plant I confesse I first imposed the name, by considering duely
all the parts thereof, which vntill some can finde a more proper, I desire
may still continue, and to call it *Ephemerum Virginianum Tradescanti*, Iohn
Tradescante's Spider-wort of Virginia, or *Phalangium Ephemerum Virginianum*, The soone fading or Day Spider-wort of Virginia.

The Vertues.

There hath not beene any tryall made of the properties since wee had it,
nor doe we know whether the Indians haue any vse thereof.

Chap. XVI.

Colchicum. Medowe Saffron.

TO returne to the rest of the bulbous and tuberous rooted plants, that remaine
to bee entreated of, the *Colchica* or Medowe Saffrons are first to bee handled,
whereof these later dayes haue found out more varieties, then formerly were
knowne; some flowring in the Spring, but the most in Autumne, and some bearing
double, but the greatest part single flowers: whereof euery one in their order, and
first of our owne Country kindes.

1. *Colchicum Anglicum album.* The white English Medowe Saffron.

It is common to all the Medowe Saffrons, except that of the Spring, and one other,
to beare their flowers alone in Autumne or later, without any green leaues with them,
and afterwards in February, their greene leaues: So that I shall not neede to make manie descriptions, but to shew you the differences that consist in the leaues, and colours
of the flowers; and briefly to passe (after I haue giuen you a full description of the
first) from one vnto another, touching onely those things that are note worthy. The
white English Medowe Saffron then doth beare in Autumne three or foure flowers at
the most, standing seuerally vpon weake foote-stalkes, a fingers length or more aboue
the ground, made of six white leaues, somewhat long and narrow, and not so large as
most of the other kindes, with some threeds or chiues in the middle, like vnto the
Saffron flowers of the Spring, wherein there is no colour of Saffron, or vertue to that
effect: after the flowers are past and gone, the leaues doe not presently follow, but
the roote remaineth in the ground without shew of leafe aboue ground, most part of
the Winter, and then in February there spring vp three or foure large and long greene
leaues,

leaues, when they are fully growne vp, standing on the toppe of a round, weake, green, and short foote-stalke, somewhat like the leaues of white Lillies, but not so large, and in the middest of these leaues, after they haue been vp some time, appeare two or three loose skinny heads, standing in the middle of the leaues vpon short, thicke, greene stalkes, and being ripe, conteine in them round small brownish seede, that lye as it were loose therein, and when the head is dry, may bee heard to rattle being shaken: the roote is white within, but couered with a thicke blackish skinne or coate, hauing one side thereof at the bottome longer then the other, with an hollownesse also on the one side of that long eminence, where the flowers rise from the bottome, and shooting downe from thence a number of white fibres, whereby it is fastened in the ground: the greene leaues afterwards rising from the top or head of the roote.

2. Colchicum *Anglicum purpureum*. The purple English Medowe Saffron.

There is no difference at all in this Medowe Saffron from the former, but only in the colour of the flowers, which as they were wholly white in the former, so in this they are of a delayed purple colour, with a small shew of veines therein.

3. Colchicum *Pannonicum album*. The white Hungary Medowe Saffron.

The greatest difference in this *Colchicum* from the former English white one, is, that it is larger both in roote, leafe, and flower, and besides, hath more store of flowers together, and continuing longer in beauty, without fading so soone as the former, and are also somewhat of a fairer white colour.

4. Colchicum *Pannonicum purpureum*. The purple Hungary Medowe Saffron.

This purple Medowe Saffron is somewhat like vnto the white of this kinde, but that it beareth not so plentifully as the white, nor doth the roote grow so great; but the flowers are in a manner as large as they, and of the like pale delayed purple colour, or somewhat deeper, as is in the purple English, with some veines or markes vpon the flowers, making some shew of a checker on the out side, but not so conspicuous, as in the true checkerd kindes. Wee haue a kinde hereof is party coloured with white streakes and edges, which abide constant, and hath been raised from the seede of the former.

5. Colchicum *Byzantinum*. Medowe Saffron of Constantinople.

This Medowe Saffron of Constantinople hath his leaues so broad and large, that hardly could any that neuer saw it before, iudge it to be a *Colchicum*; for they are much larger then any Lilly leaues, and of a darke greene colour: the flowers are correspondent to the leaues, larger and more in number then in any of the former purple kindes, of the same colour with the last purple kinde, but of a little deeper purple on the inside, with diuers markes running through the flowers, like vnto it, or vnto checkers, but yet somewhat more apparantly: the roote is in the middle greater and rounder then the others, with a longer eminence, whereby it may easily bee knowne from all other sorts.

6. Colchicum *Lusitanicum Fritillaricum*.
The checkerd Medowe Saffron of Portugall.

The flowers of this Medowe Saffron are larger and longer then the flowers of either the English or Hungarian, and almost as large as the last before mentioned, and of the same colour, but a little deeper, the spots and markes whereof are somewhat more easie to be seene euen a farre off, like vnto the flower of a Fritillaria, from whence it tooke his significatiue name: the leaues of this Medowe Saffron doe rise vp sooner then in any other of the Autumne kindes; for they are alwayes vp before Winter, and are foure or fiue in number, short rather then long, broad belowe, and pointed at the end, canaled or hollow, and standing round aboue the ground, one encompassing another at the bottome, like the great Spanish Starre Iacinth, called the Iacinth of Peru, but

1 *Colchicum Pannonicum.* The Hungarian Medow Saffron. 2 *Colchicum Byzantinum.* Medowe Saffron of Constantinople 3 *Colchicum Lusitanicum Fritillaricum.* The checkerd Medowe Saffron of Portugall 4 *Colchicum Neapolitanum Fritillaricum.* The checkerd Medowe Saffron of Naples. 5 *Colchicum Fritillaricum Chiense.* The checkerd Medowe Saffron of Chio or Sio. 6 *Colchicum Hermodactylum.* Physicall Medowe Saffron.

but shorter, and of a pale or grayish greene colour, differing from the colour of all the other Medowe Saffrons : the roote is like the roote of the English or Hungarian without any difference, but that it groweth somewhat greater. It is one of the first Medowe Saffrons that flower in the Autumne.

7. *Colchicum Neapolitanum Fritillaricum.*
The checkerd Medowe Saffron of Naples.

This checkerd Medowe Saffron of Naples, is very like vnto the last recited checkerd Saffron of Portugall, but that the flower is somewhat larger, yet sometimes very little, or not at all : the greatest marke to distinguish them is , that the flowers of this are of a deeper colour, and so are the spots on the flowers likewise , which are so conspicuous, that they are discerned a great way off, more like vnto the flowers of a deepe Fritillaria, then the former, and make a goodlier and a more glorious shew : the leaues of this doe rise vp early after the flowers, and are somewhat longer, of a darker greene colour, yet bending to a grayish colour as the other, not lying so neatly or round, but stand vp one by another, being as it were folded together : neither of both these last named checkerd Medowe Saffrons haue giuen any seede in this Countrey, that euer I could learne or heare of, but are encreased by the roote, which in this is like the former, but a little bigger.

8. *Colchicum Fritillaricum Chiense.*
The checkerd Medowe Saffron of Chio or Sio.

This most beautifull Saffron flower riseth vp with his flowers in the Autumne, as the others before specified doe, although not of so large a size, yet farre more pleasant and delightfull in the thicke, deepe blew, or purple coloured beautifull spots therein, which make it excell all others whatsoeuer : the leaues rise vp in the Spring, being smaller then the former , for the most part three in number, and of a paler or fresher greene colour, lying close vpon the ground, broad at the bottome , a little pointed at the end, and twining or folding themselues in and out at the edges , as if they were indented. I haue not seene any seede it hath borne : the roote is like vnto the others of this kinde, but small and long, and not so great : it flowreth later for the most part then any of the other, euen not vntill Nouember, and is very hard to be preserued with vs, in that for the most part the roote waxeth lesse and lesse euery yeare, our cold Country being so contrary vnto his naturall , that it will scarce shew his flower ; yet when it flowreth any thing early , that it may haue any comfort of a warme Sunne , it is the glorie of all these kindes.

9. *Colchicum versicolor.* The party coloured Medowe Saffron.

The flowers of this Medowe Saffron most vsually doe not appeare, vntill most of the other Autumne sorts are past, except the last, which are very lowe, scarce rising with their stalkes three fingers breadth aboue the ground , but oftentimes halfe hid within the ground : the leaues whereof are smaller, shorter, and rounder, then in any of the other before specified, some being altogether white , and others wholly of a very pale purple, or flesh colour ; and some againe parted , the one halfe of a leafe white, and the other halfe of the same purple, and sometimes striped purple and white, in diuers leaues of one and the same flower : and againe, some will be the most part of the leafe white, and the bottome purple, thus varying as nature list, that many times from one roote may be seene to arise all these varieties before mentioned : these flowers doe stand long before they fade and passe away ; for I haue obserued in my Garden some that haue kept their flower faire vntill the beginning of Ianuary, vntill the extremitie of the Winter frosts and snowes haue made them hide their heads : the leaues therefore accordingly doe rise vp after all other, and are of a brownish or darke greene colour at their first springing vp, which after grow to be of a deepe greene colour : the roote is like the former English or Hungarian kindes , but thicker and greater for the most part, and shorter also.

10. *Colchicum*

10. *Colchicum variegatum alterum.* Another party coloured Medowe Saffron.

There is another, whofe party coloured flowers rife a little higher, diuerfly ftriped and marked, with a deeper purple colour, and a pale or whitifh blufh throughout all the leaues of the flower.

11. *Colchicum montanum Hifpanicum minus.* The little Spanifh Medowe Saffron.

The flowers of this little Medowe Saffron are narrower and fmaller then any of the former, and of a deeper reddifh purple colour then either the Englifh or Hungarian kindes : the greene leaues alfo are fmaller then any other, lying on the ground, of a deepe or fad greene colour, rifing vp within a while after the flowers are paft, and doe abide greene all the Winter long : the roote is fmall and long, according to the reft of the plant, and like in forme to the others.

12. *Colchicum montanum minus verficolore flore.*
The fmall party coloured Medowe Saffron.

This little kinde differeth not from the Spanifh kinde laft fet forth, but in the varie-tie of the flower, which is as fmall as the former ; the three inner leaues being almoft all white, and the three outer leaues fome of them pale or blufh, and fome party co-loured, with a little greene on the backe of fome of them.

13. *Colchicum Hermodactilum.* Phyficall Medowe Saffron.

This Phyficall Medowe Saffron fpringeth vp with his leaues in Autumne, before his flowers appeare beyond the nature of all the former kindes, yet the flower doth, after they are vp, fhew it felfe in the middle of the greene leaues, confifting of fix white leaues, with diuers chiues in the middle, and paffeth away without giuing any feede that euer I could obferue : the greene leaues abide all the Winter and Spring follow-ing, decaying about May, and appeare not vntill September, when (as I faid) the flow-ers fhew themfelues prefently after the leaues are fprung vp.

14. *Colchicum atropurpureum.* The darke purple Medowe Saffron.

The greateft difference in this kinde confifteth in the flower, which at the firft ap-pearing is as pale a purple, as the flower of the former Hungarian kinde : but after it hath ftood in flower two or three dayes, it beginneth to change, and will after a while become to bee of a very deepe reddifh purple colour, as alfo the little foote-ftalke whereon it doth ftand : the flower is of the bigneffe of the Hungarian purple, and fo is the greene leafe : the feede and roote is like the Englifh purple kinde.

15. *Colchicum atropurpureum variegatum.*
The party coloured darke purple Medowe Saffron.

We haue of late gained another fort of this kinde, differing chiefly in the flower, which is diuerfly ftriped thorough euery leafe of the flower, with a paler purple co-lour, whereby the flower is of great beauty : this might feeme to bee a degeneration from the former, yet it hath abiden conftant with me diuers yeares, and giueth feede as plentifully as the former.

16. *Colchicum flore pleno.* Double flowred Medowe Saffron.

The double Medowe Saffron is in roote and leafe very like vnto the Englifh kinde : the flowers are of a fine pale or delayed purple colour, confifting of many leaues fet thicke together, which are fomewhat fmaller, as in the Englifh flower, being narrow and long, and as it were round at the points, which make a very double flower, hauing

O   fome some chiues with their yellow tips , disperssed as it were among the leaues in the middle : it flowreth in September , a little after the first shew of the earlier Medowe Saffrons are past.

17. *Colchicum variegatum pleno flore.*
The party coloured double Medowe Saffron.

We haue another of these double kinds (if it be not the very same with the former, varying in the flower as nature pleaseth oftentimes; for I haue this flower in my garden, as I here set it forth, euery yeare) whose flowers are diuersified in the partition of the colours, as is to be seene in the single party coloured Medowe Saffron before described, hauing some leaues white , and others pale purple, and some leaues halfe white and halfe purple, diuersly set or placed in the double flower , which doth consist of as many leaues as the former, yet sometime this party coloured flower doth not shew it selfe double like the former, but hath two flowers, one rising out of another , making each of them to be almost but single flowers, consisting of eight or ten leaues a peece: but this diuersity is not constant ; for the same roote that this yeare appeareth in that manner, the next yeare will returne to his former kinde of double flowers againe.

18. *Colchicum Vernum.* Medowe Saffron of the Spring.

This Medowe Saffron riseth vp very early in the yeare, that is , in the end of Ianuarie sometimes, or beginning, or at the furthest the middle of February, presently after the deepe Frosts and Snowes are past, with his flowers inclosed within three greene leaues, which opening themselues as soone almost as they are out of the ground, shew their buds for flowers within them very white oftentimes, before they open farre, and sometimes also purplish at their first appearing, which neuer shew aboue two at the most vpon one roote, and neuer rise aboue the leaues, nor the leaues much higher then they, while they last : the flower consisteth of six leaues, long and narrow, euery leafe being diuided, both at the bottome and toppe, each from other , and ioyned together onely in the middle, hauing also six chiues, tipt with yellow in the middle, euery chiue being ioyned to a leafe, of a pale red or deepe blush colour, when it hath stood a while blowne, and is a smaller flower then any Medowe Saffron, except the small Spanish kindes onely , but continueth in his beauty a good while , if the extremity of sharpe Frosts and Windes doe not spoile it : the leaues wherein these flowers are enclosed, at their first comming vp, are of a brownish greene colour, which so abide for a while, especially on the outside, but on the inside they are hollow, and of a whitish or grayish greene colour, which after the flowers are past, grow to bee of the length of a mans longest finger, and narrow withall : there riseth vp likewise in the middle of them the head or seede vessell, which is smaller and shorter, and harder then any of the former, wherein is contained small round browne seede : the roote is small, somewhat like vnto the rootes of the former , but shorter , and not hauing so long an eminence on the one side of the bottome.

19. *Colchicum Vernum atropurpureum.* Purple Medowe Saffron of the Spring.

The flower of this Medowe Saffron, is in the rising vp of his leaues and flowers together, and in all things else, like vnto the former, onely the flowers of this sort are at their first appearing of a deeper purple colour , and when they are blowne also are much deeper then the former, diuided in like manner, both at the bottome and toppe as the other , so that they seeme , like as if six loose leaues were ioyned in the middle part, to make one flower, and hath his small chiues tipt with yellow , cleauing in like manner to euery leafe.

The Place.

All these Medowe Saffrons, or the most part of them, haue their places expressed in their titles; for some grow in the fields and medowes of the champion grounds, others on the mountaines and hilly grounds. The English kindes grow in the West parts, as about Bathe, Bristow, Warmister,

1 *Colchicum montanum Hispanicum.* The little Spanish Medowe Saffron. 2 *Colchicum montanum minus versicolore flore.* The small party colour d Me owe
Saffron. 3 *Colchicum versicolor.* The party coloured Medowe Saffron. 4 *Colchicum variegatum alterum.* An other party colou ed Medowe Saffron. 5 *Colchicum atropurpureum.* The darke purple Medowe Saffron. 6 *Colchicum atropurpureum variegatum.* The variable darke purple Medowe Saffron. 7 *Colchicum vernum,* Medowe Saffron of the spring. 8 *Colchicum flore pleno,* Double Medowe Saffron.

ster, and other places also.   The double kindes are thought to come out of Germany.

The Time.

Their times likewise are declared in their seuerall descriptions : those that are earliest in Autumne, flower in August and September, the later in October, and the latest in the end of October, and in Nouember. The other are said to bee of the Spring, in regard they come after the deepe of Winter (which is most vsually in December and Ianuary) is past.

The Names.

The generall name to all these plants is *Colchicum*, whereunto some haue added *Ephemerum*, because it killeth within one dayes space ; and some *Strangulatorium*. Some haue called them also *Bulbus agrestis*, and *Filius ante Patrem*, The Sonne before the Father, because (as they thinke) it giueth seede before the flower : but that is without due consideration ; for the root of this (as of most other bulbous plants) after the stalke of leaues and seede are dry, and past, may be transplanted, and then it beginneth to spring and giue flowers before leaues, (and therein onely it is differing from other plants) but the leaues and seede follow successiuely after the flowers, before it may be remoued againe ; so that here is not seede before flowers, but contrarily flowers vpon the first planting or springing, and seede after, as in all other plants, though in a diuers manner.

The *Colchicum Hermodactilum* may seeme very likely to bee the *Colchicum Orientale* of Matthiolus, or the *Colchicum Alexandrinum* of Lobelius : And some thinke it to be the true *Hermodactilus*, and so call it, but it is not so. We doe generally call them all in English, Medowe Saffrons, or *Colchicum*, according to the Latine, giuing to euery one his other adiunct to know it by.

The Vertues.

None of these are vsed for any Physicall respect, being generally held to be deadly, or dangerous at the least. Only the true Hermodactile (if it be of this tribe, and not this which is here expressed) is of great vse, for paines in the ioynts, and of the hippes, as the *Sciatica*, and the like, to be taken inwardly. Costæus in his Booke of the nature of plants, saith, that the rootes of our common kindes are very bitter in the Spring of the yeare, and sweet in Autumne, which Camerarius contradicteth, saying, that he found them bitter in Autumne, which were (as he saith) giuen by some imposters to diuers, as an antidote against the Plague.

Chap. XVII.

Crocus. Saffron.

THere are diuers forts of Saffrons, whereof many doe flower in the Spring time, and some in Autumne, among whom there is but one onely kinde, that is called tame or of the Garden, which yeeldeth those blades or chiues that are vsed in meates and medicines, and many wayes profitable for other respects, none of the rest, which are all wilde kindes, giuing any blade equall vnto those of the tame kinde, or for any other vse, then in regard of their beautifull flowers of seuerall varieties, and as they haue been carefully sought out, and preserued by diuers, to furnish a Garden of dainty curiosity.   To entreate therefore of these, I must, to obserue an orderly declaration, diuide them into two primary families: the former shall be of those that yeeld their pleasant flowers in the Spring of the yeare, and the other that send out

their

their colours in the Autumne, among whom that *Rex pomarij* (as I may so call it) the tame or manured kinde, properly called of the Garden, is to be comprehended, for th it it giueth his pleasant flowers at that time among others. I shall againe distribute those of the Spring time into three chiefe colours, that is, into white, purple, and yellow, and vnder euery one of them, comprehend the seuerall varieties that doe belong vnto them ; which course I will also hold with those of the Autumne, that thus being rightly ranked, they may the more orderly be described.

1. *Crocus Vernus albus purus minor.*
The smaller pure white Saffron flower of the Spring.

This small Saffron flower springeth vp in the beginning of the yeare, with three or foure small greene leaues, somewhat broader, but much shorter then the true Saffron leaues, with a white line downe the middle of euery leafe : betweene these leaues, out of a white skinne, riseth vp one or two small flowers, made of six leaues a peece, as all the rest in generall are, of a pure white colour, without any mixture in it, which abide not in flower aboue a weeke, or rather lesse, so sodainly is the pleasure of this, and the purple lost : it flowreth not for the most part, vntill a moneth after the yellow Crocus appeareth in flower, and the ordinary stript Crocus is past : the seede is small, round, and reddish, yet not so red as the seede of the yellow, contained in three square heads, yet seldome beareth, but encreaseth by the roote plentifully enough, which is small, round, and flat at the bottome, somewhat white on the outside, but whiter within, shooting out small sprouts on euery side of the roote, which is the best note to know this kinde and the lesser purple, which are both alike, from all other rootes of Saffron flowers.

2. *Crocus albus maior multiflorus.* The great snow white Crocus.

This greater Saffron flower riseth vp vsually with three or foure greene leaues, larger then the former, with a white line in euery one of them : the flowers are greater, and more in number, rising together, but flowring one after another, of a pure snow white colour, and abiding but little longer in flower then the former.

3. *Crocus albus maior alter dictus Masiacus.*
The great white Saffron flower or Crocus of Mesia.

This great white Crocus of Mesia, riseth vp out of the ground, almost as early as the first sort of the yellow, with foure or fiue leaues, being very like vnto the leaues of the yellow Crocus, and as large, with white lines in them : the flowers also are as large as the flowers of the yellow, and many also rising one after another like vnto it, but not of so pure a white colour, as the former or last described, but rather tending to a Milky or Creame colour : the roote is not couered with any reddish, but rather pale skinnes or coates.

4. *Crocus albus Masiacus fundo violaceo.*
The great white Crocus of Mesia with a blew bottome.

There is another of this kinde, like vnto the former in all things, sauing that the bottomes of the flowers of this kinde, with some part of the stalke next the flower, are of a pale shining purple colour, and rising vp a pretty way into the flower; whereas another also of this kind, hath a little shew or marke of blew, and not purple, at the bottome of the flower onely, which maketh a difference.

5. *Crocus albus fundo purpureo.* The white Crocus with a purple bottome.

This Saffron flower is of the same kinde with the first, both in roote, leafe, and flower, in none of them differing from it, but in that the bottome of this flower, with that part of the short foote-stalke next vnto it, is of a violet or purple colour, and sometimes hauing here and there some purple small lines, or spots on the white leaues: it flowreth also with the first white, or somewhat later.

6. *Crocus*

6. *Crocus vernus albus striatus.* The white stript Crocus.

This stript Saffron flower is likewise neare the same first kind, or first white Crocus, hauing the like leaues and flowers, somewhat larger, but as soone fading almost as it : but herein this flower differeth, that it hath pale blewish lines and spots in all the leaues thereof, and more principally in the three outer leaues : the root is also white on the outside, like the first white, but greater, with young ones growing round about it.

7. *Crocus vernus albus polyanthos versicolor.* The greater party coloured white Crocus.

The greater party coloured Saffron flower, hath his greene leaues like vnto the second great white Crocus before mentioned, hauing more flowers then any of the former, except the first great white, the leaues whereof haue greater stripes then the last recited Crocus, but of a purple Violet colour, making each leafe seeme oftentimes to haue as much purple as white in them : the roote hereof is somewhat like the second white, but of a little more duskie colour on the outside, and not budding out on the sides at all, or very little.

8. *Crocus vernus albus versicolor.* The lesser party coloured white Crocus.

The leaues and flowers of this other party coloured Crocus, are for bignesse in a manner equall with the last, but hath not so many flowers rising together from the roote : the flower is finely marked with blew strakes on the white flower, but nothing so much as in the former : the roote also is like the last.

9. *Crocus Episcopalis.* The Bishops Crocus.

This party coloured or Bishops Saffron flower, is very like both in leaues and rootes vnto the Neapolitane blew Crocus, but somewhat greater : the flowers doe abide not so long time blowne, and hath all the leaues either wholly white, with blew stripes on both sides of them, or wholly of a fine delayed blew Violet colour, and the three innermost more blew and finely striped, both on the inside and outside of them, and sometimes it hath been seen to haue three leaues white, and three leaues of a pale blew.

10. *Crocus vernus striatus vulgaris.* The ordinary stript Crocus.

There is another sort of stript Saffron flower, which is most common and plentifull in most Gardens, which I must needes bring vnder the ranke of these white kinds, although it differre very notably, both in roote, leafe, and flower, from all of them : the leaues of this rise vp sooner then the yellow or white Crocus, lying spread vpon the ground for the most part, but narrower then any of the former : among these leaues spring vp diuers flowers, almost as large as the former great white Crocus, of a very bleake or pale purple colour, tending to white on the inside, and in many almost white, with some small whitish chiues tipt with yellow in the middle : the three outer leaues are of a yellowish white colour on the backe side of them, stript euery one of them with three broad stripes, of a darke murrey or purple colour, and a little sprinkled with some small purple lines, on both sides of those stripes ; but on the inside, of the same pale purple or white colour with the rest : the seede hereof is somewhat darker coloured then of the white, and is more liberall in bearing : the roote is differing from all the former, being rounder and bigger then any of them, except the kindes of Misia, yet somewhat flat withall, not hauing any shootes from the sides, but setting off into rootes plentifully, hauing a round circle compassing the bottome of the roote, which easily falleth away, when it is taken vp out of the ground, and couered with a browne coate, somewhat neare the colour of the yellow Crocus, but not altogether so bright : it flowreth vsually the first of all these sorts, or with the first of the early yellowes.

1. *Crocus vernus albus minor.* The small white Saffron flower of the spring. 2. *Crocus vernus Massicus albus.* The great white Crocus of Misia. 3. *Crocus vernus albus striatus.* The white stript Crocus. 4. *Crocus vernus albus polyanthos versicolor* The party coloured white Crocus. 5. *Crocus albus fundo purpureo.* The white Crocus with a purple bottome. 6. *Crocus vernus Neapolitanus.* The great blew Crocus of Naples. 7. *Crocus vernus purpureus maximus.* The great purple Crocus. 8. *Crocus vernus purpureus striatus.* The purple stript Crocus. 9. *Crocus vernus purpureus Capillorifolio.* The purple Crocus with small leaves. 10. *Crocus vernus flavus striatus.* The yellow stript Crocus. 11. *Crocus vernus luteus versicolor.* The cloth of gold Crocus.

11. *Crocus vernus striatus Turcicus.* The Turkie stript Crocus.

There is another of this kinde, whose flower is a little larger, and of a deeper purple colour, both on the inside and outside; the greene leafe also is bigger, and of a more whitish colour.

12. *Crocus vernus Capillarifolio albus.* The white Crocus with small leaues.

This white Crocus is in all things like vnto the purple of the same kinde, but that the flower of this is wholly white: the full description therefore hereof, you shall haue in that purple with small leaues, of this kinde hereafter set downe, whereunto I referre you.

13. *Crocus vernus purpureus minor.* The smaller purple Crocus.

The smaller purple Saffron flower of the Spring, hath his greene leaues so like vnto the first white flowred Saffron, that they can hardly be distinguished, onely they seem to bee a little narrower: the flower is also much about the same bignesse, or a little bigger, and seldome beareth aboue one flower from a roote, euen as the first doth, of a deepe purple Violet colour, the bottome of the flower, with the vpper part of the stalke next thereunto, being of a deeper or blacker purple; in the middle of the flower are some pale chiues tipt with yellow pendents, and a longer pointell, diuided or forked at the toppe: the roote of this is in all things so like vnto the first white, that it is impossible for the most cunning and conuersant in them, to know the one from the other. This beareth seede very sparingly, as the white doth, and is reddish like vnto it, but recompenseth that defect with a plentifull encrease by the roote: it likewise flowreth at the very same time with the white, and endureth as small a time.

14. *Crocus vernus purpureus maximus.* The greatest purple Crocus.

This great purple Crocus is of the same kinde with the next described, as well in roote as leafe, but greater; for the greene leaues hereof are the greatest and broadest of all other Crocus, with a large white line in the middle of euery one: it springeth vp much later then the former, and doth not shew his flower vntill the other bee past a good while: the flowers also are the largest of all these Crocus of the Spring time, and equalling, if not surpassing that purple kinde that flowreth in Autumne, hereafter set forth, of a very faire and deepe Violet colour, almost as deepe as the former: the seed vessels are large also and white, wherein is contained pale reddish seede, like vnto the next blew kinde, but somewhat greater: the roote is (as I said before) like vnto the next, that is, flat and round, with a duskie coloured outside, whose head for springing in it is as hardly discerned.

Alter Apiibus albidis. We haue one of this kinde, the toppes onely of whose purple flower are whitish, for the breadth of halfe the naile of a mans hand, which abideth constant euery yeare in that manner, and therefore is a difference fit to be remembred.

15. *Crocus vernus Neapolitanus siue caeruleus maior.*
The greater blew Crocus of Naples.

This great blew Crocus riseth vp with diuers greene leaues, broader then any of the former (except the last) with a white line running downe the inside of euery leafe, as in the former, among which riseth vp, out of diuers great long white skinnes, diuers large flowers, but not fully so great as the former, consisting of six leaues, of a palet blew or Violet colour then in the former, hauing in the middle of the flowers a few pale threeds, tipt with yellow, and a longer pointell of a gold yellow colour, forked or diuided at the toppe, smelling sweeter then in the former, and abiding a great while longer, being in flower vsually euen with the stript yellow Crocus, or before the former purple, and yeelding more plenty of seede: the roote hereof is not very great, but a little darke on the outside, being round and flat withall, that one can hardly know which is the vpperside thereof.

This

This kinde differeth very little from the former, either in roote, leafe, or flower, for the bignesse or colour, but that it seemeth to be a little bleaker or paler blew, because it flowreth a little earlier.

16. *Crocus vernus purpureus striatus.* The stript purple Crocus.

The leaues of this stript purple Saffron flower, are as large and broad as the last, or rather a little longer : the flowers also are as plentifull, and as large, of a fine delayed purple colour on the outside, with three broad strakes or lines downe the backe of the three outer leaues, and of a little deeper purple on the inside, as the other three leaues are also of a deeper purple colour, and are striped with the same deepe purple about the ground, or bottome of the leaues : this sometimes yeeldeth three square heads, containing in them brownish seede : the roote is like vnto the last, and flowreth much about the time of the former.

17. *Crocus vernus purpureus versicolor.* The siluer stript purple Crocus.

This stript Saffron flower, is in leaues and flowers somewhat like vnto the last stript purple, but a little smaller : the flowers are of a little deeper purple through the whole leaues, striped with white lines, both on the leaues, and towards the edges, which maketh a peculiar difference from all the rest : the roote of this is not so flat, though like it, and couered with a darke ash coloured skinne : it flowreth about the same time.

18. *Crocus purpureus flammeus maior.* The greater purple flame coloured Crocus.

The greene leaues of this Crocus or Saffron flower, are of a reasonable breadth and length, and of a pleasant fresh greennesse, with a faire broad white line downe the middle of them, but rising not out of the ground so early as the next described Crocus : the flowers are likewise of a meane bignesse, of a pale purple on the outside, somewhat whitish, especially the three outer leaues ; but on the inside of a deeper purple, and striped with great stripes like flames, hauing some chiues in the middle, and a longer one also feathered a little at the toppe : the roote is white on the outside, somewhat flat and round, but not so flat as the Neapolitane Crocus before described.

19. *Crocus purpureus flammeus minor.* The lesser purple flame coloured Crocus.

This Crocus hath almost as broad and long greene leaues as the former, and of the same verdure, which rise vp earlier then it, and is in flower likewise somewhat before it, being smaller for size by a little, but of as deepe a purple on the outside, as on the inside, flamed with faire broad stripes from the middle of the leaues, or somewhat lower vnto the edges : each of these giue seed that is of a pale reddish colour : the root is very like vnto the former, but a little lesser.

20. *Crocus vernus purpureus Capillarisolia.* The purple Crocus with small leaues.

This small kinde of Saffron flower riseth out of the ground, with two or three long and small green leaues, very like vnto the leaues of the fine Fether-Grasse hereafter described, standing vpright at the first, but afterwards lying vpon the ground ; among which come the flowers, sometimes three, but most vsually two vpon one stalke, if the roote be not young, which then will beare but one on a stalke, which is very short, so that the flowers scarce arise aboue the ground, yet laying themselues open in the day time, if it be faire, and the Sunne doe shine, otherwise they keepe close, and doe not open at all : and after one flower is past, which doth not last aboue three or foure dayes at the most, the others follow, which are of a bleake blewish purple in the middle of the flower, and of a deeper purple towards the ends or points of the leaues, but of a more sullen or darke purple on the outside of them, and yellowish at the bottome, with some yellow chiues in the middle : the seede is small and darker coloured then any of the former Crocus, contained also in smaller heads, standing one by another

vpon

vpon the fame fhort foote-ftalke, which then rifeth vp a little higher, fhewing the maner of the ftanding of the flowers, which in their flowring time could not fo eafily bee
difcerned : the roote is very fmall and round, hauing one fide at the bottome lower
then the other, very like the roote of a *Colchicum* or Medowe Saffron, and fomewhat
neare refembling alfo the hoofe of an horfe foote, couered with a very thicke skinne,
of a darke or blackifh browne colour : this flowreth the laft of all the former forts of
Saffron flowers, euen when they are all paft.

21. *Crocus vernus purpureus ftriatus Capillarifolio.*
The ftript purple Crocus with fmall leaues.

This fmall ftript purple Saffron flower hath fuch like leaues, as the laft defcribed
hath, betweene which rifeth the flower vpon as fhort a foote-ftalke, confifting of fix
leaues like the former, of a faire purple colour on the outfide of the three outer leaues,
with three lines or ftrakes downe euery leafe, of a deeper purple colour, and on the infide of a paler purple, as the other three leaues are alfo, with fome chiues tipt with yellow pendents, and a forked pointell in the middle : the roote of this is fomewhat
bigger then the former, and rounder, but couered with as thicke and as browne a
skinne : it flowreth about the fame time with the former.

22. *Crocus vernus luteus fine Mafiacus.* The yellow Crocus.

The yellow Crocus or Saffron flower, rifeth vp with three or foure leaues out of
the ground, being fomewhat neare the breadth of the great purple kindes, with a white
line in them, as in moft of the reft : the flowers ftand in the middle of thefe leaues, and
are very large, of a gold yellow colour, with fome chiues, and a forked point in the
middle : the feede hereof is of a brighter colour then in any of the other : the roote is
great and round, as great or greater then a Wall Nut fometimes, and couered with reddifh skinnes or coates, yeelding more ftore of flowers then moft of the former, and beginning to blowe with the firft forts, or prefently after, but outlaft many of them, and
are of a pleafant good fent.

Flore aureo.   Of this kinde we haue fome, whofe flowers are of a deeper gold yellow colour then
others, fo that they appeare reddifh withall.

Flore pallido.   And we haue alfo another fort, whofe flowers are very pale, betweene a white and a
yellow, not differing in any thing elfe.

Flore viridante luteo.   And another fmaller, whofe flower hath a fhew of greenneffe in the yellow, and
more greene at the bottome.

23. *Crocus vernus flauus ftriatus.* The yellow ftript Crocus.

This kinde of yellow ftript Crocus or Saffron flower, rifeth vp with more ftore of
narrower and greener leaues then the former, and after the leaues are fpread, there rife
vp many yellow flowers from among them, which are not of fo faire and bright a yellow colour, but more dead and fullen, hauing on the backfide of each of the three outtermoft leaues, three fmall ftripes, of an ouer-worne or dull purple colour, with fome
chiues and a pointell in the middle : the roote of this kinde, is very like the roote of
the former yellow, but fomewhat fmaller and fhorter, and couered with the like reddifh skinnes, but a little fadder : it flowreth not fo early as the former yellow, but abideth almoft as long as it.

24. *Crocus vernus luteus verficolor primus.* The beft cloth of gold Crocus.

The faireft cloth of gold Crocus or Saffron flower, rifeth vp very early, euen with
the firft, or the firft of all other Crocus, with three or foure very narrow and fhort
leaues, of a whiter colour then any of the former, which by and by after doe fhew
forth the flowers, rifing from among them out of the fame white skinne, which in-
cludeth the leaues, but are not fo plentifull as the former yellow, being but two or
three at the moft, of a faire gold yellow colour, yet fomewhat paler then the firft, hauing

uing on the backe of euery of the three outer leaues, three faire and great ſtripes, of a faire deepe purple colour, with ſome ſmall lines at the ſides or edges of thoſe purple ſtripes; on the inſide of theſe flowers, there is no ſigne or ſhew of any line or ſpot, but wholly of a faire gold yellow, with chiues and a fethertopt pointell in the middle: the ſeede hereof is like the former, but not ſo red: the roote of this kinde is eaſily knowne from the roote of any other Saffron flower, becauſe the outer peelings or ſhels being hard, are as it were netted on the outſide, hauing certaine ribbes, riſing vp higher then the reſt of the skinnes, diuided in the forme of a net-worke, of a darke browne colour, and is ſmaller and rounder then the former yellow, and not encreaſing ſo plentifully by the roote.

25. *Crocus vernus luteus verſicolor alter.*
The ſecond cloth of gold, or Duke Crocus.

There is no difference either in roote, leafe, or colour of flower, or time of flowring in this ſort from the laſt before mentioned; for the flower of this is of the ſame bigneſſe and colour, the only note of difference is in the marking of the three outer leaues, which haue not three ſtripes like the former, but are wholly of the ſame deepe purple colour on the backe of them, ſauing that the edges of them are yellow, which is the forme of a Duke Tulipa, and from thence it tooke the name of a Duke Crocus.

26. *Crocus vernus verſicolor pallidoluteus.* The pale cloth of gold Crocus.

We haue a third ſort of this kinde of cloth of gold Crocus, which hath leaues and flowers like the former, but differeth in this, that the colour of the flower is of a paler yellow by much, but ſtript in the ſame manner as the firſt, but with a fainter purple colour: the roote alſo is netted like them, to ſhew that this is but a variation of the ſame kinde.

27. *Crocus vernus verſicolor albidoluteus.* The cloth of ſiluer Crocus.

The chiefeſt note of difference in this Saffron flower is, that being as large a flower as any of the former of this kinde, it is of ſo pale a yellowiſh white, that it is more white then yellow, which ſome doe call a butter colour: the three outer leaues are ſtriped on the backe of them, with a paler purple blew ſhining colour, the bottome of the flower, and the vpper part of the ſtalke, being of the ſame purple blew colour: the roote of this is alſo netted as the other, to ſhew it is a variety of the ſame kinde.

And thus much for thoſe Saffron flowers that come in the Spring time; now to thoſe that flower in Autumne onely: and firſt of the true Saffron.

1. *Crocus verus ſatiuus Autumnalis.* The true Saffron.

The true Saffron that is vſed in meates and medicines, ſhooteth out his narrow long greene leaues firſt, and after a while the flowers in the middle of them appeare about the end of Auguſt, in September and October, according to the ſoile, and climate where they growe; theſe flowers are as large as any of the other former or later ſorts, compoſed of ſix leaues a peece, of a murrey or reddiſh purple colour, hauing a ſhew of blew in them: in the middle of theſe flowers there are ſome ſmall yellow chiues ſtanding vpright, which are as vnprofitable, as the chiues in any other of the wilde Saffrons, before or hereafter ſpecified; but beſides theſe, each flower hath two, three, or foure greater and longer chiues, hanging downe vpon or betweene the leaues, which are of a fierie red colour, and are the true blades of Saffron, which are vſed phyſically or otherwiſe, and no other: All theſe blades being pickt from the ſeuerall flowers, are laid and preſſed together into cakes, and afterwards dryed very warily on a Kill to preſerue them; as they are to be ſeene in the ſhops where they are ſold. I neuer heard that euer it gaue ſeede with any: the roote groweth often to be as great, or greater then a green Wall Nut, with the outer ſhell on it, couered with a grayiſh or aſh-coloured skin, which breaketh into long hairie threeds, otherwiſe then in any other roote of Crocus.

2. *Crocus*

2. *Crocus Byzantinus argenteus.* The ſiluer coloured Autumne Crocus.

This Saffron flower ſpringeth vp in October, and ſeldome before, with three or foure ſhort greene leaues at the firſt, but growing longer afterwards, and in the midſt of them, preſently after they haue appeared, one flower for the moſt part, and ſeldome two, conſiſting of ſix leaues, the three outermoſt whereof are ſomewhat larger then the other three within, and are of a pale bleake blew colour, almoſt white, which many call a ſiluer colour, the three innermoſt being of a purer white, with ſome yellow chiues in the middle, and a longer pointell ragged or fethered at the toppe: this very ſeldome beareth ſeede, but when the yeare falleth out to bee very milde; it is ſmall, round, and of a darke colour: the roote is pretty bigge, and rounder then any other Crocus, without any flat bottome, and couered with a darke ruſſet ſkinne.

3. *Crocus Pyrenaus purpureus.* The purple mountaine Crocus.

This purple Saffron flower of the Autumne, riſeth vp but with one flower vſually, yet ſometimes with two one after another, without any leaues at all, in September, or ſometimes in Auguſt, ſtanding vpon a longer foote-ſtalke then any kinde of Saffron flower, either of the Spring or Autumne, and is as large as the flower of the greateſt purple Saffron flower of the Spring, of a very deepe Violet purple colour, which decayeth after it hath ſtood blowne three or foure dayes, and becommeth more pale, hauing in the middle ſome yellow chiues, and a long fether topt pointell, branched, and riſing ſometimes aboue the edges of the flowers: about a moneth after the flowers are paſt, and ſometimes not vntill the firſt of the Spring, there riſeth vp three or foure long and broad greene leaues, with a white line in euery one of them, like vnto the firſt purple Vernall kindes, which abide vntill the end of May or Iune: the roote is ſmall and white on the outſide, ſo like vnto the roote of the leſſer Vernall purple or white Crocus, that it cannot be diſtinguiſhed, vntill about the end of Auguſt, when it doth begin to ſhoot, and then by the early ſhooting vp a long white ſprout for flower, it may be knowne. I neuer could obſerue it to giue any ſeede, the Winter (as I thinke) comming on it ſo quickly after the flowring, being the cauſe to hinder it.

4. *Crocus montanus Autumnalis.* The Autumne mountaine Crocus.

The mountaine Saffron flower ſpringeth vp later then any of the former, and doth not appeare vntill the middle or end of October, when all the flowers of the former are paſt, appearing firſt with three or foure ſhort greene leaues, like vnto the Byzantine Crocus, and afterwards the flowers betweene them, which are of a pale or bleake blew tending to a purple, the foote-ſtalkes of them being ſo ſhort, that they ſcarce appeare aboue ground at the firſt, but after two or three dayes they grow a little higher: the roote is very great and flat bottomed, couered with a grayiſh duskie coate or ſkinne, and encreaſeth very little or ſeldome.

The Place.

The ſeuerall places of theſe Saffron flowers, are in part ſet downe in their titles; the others haue beene found out, ſome in one Countrey, and ſome in another, as the ſmall purple and white, and ſtript white in Spaine: the yellow in Meſia about Belgrade, the great purple in Italy; and now by ſuch friends helpes as haue ſent them, they proſper as well in our Gardens, as in their naturall places. Yet I muſt giue you this to vnderſtand, that ſome of theſe formerly expreſſed, haue been raiſed vp vnto vs by the ſowing of their ſeede.

The Time.

Their ſeuerall times are likewiſe expreſſed in their deſcriptions; for ſome ſhew forth their pleaſant flowers in the Spring, wherein for the three firſt
monerhs,

1 *Crocus vernus luteus vulgaris.* The common yellow spring Crocus. 2 *Crocus vernus satiuus Autumnalis.* The true Saffron 3 *Crocus Byzantinus argenteus.* The siluer coloured Autumne Crocus. 4 *Crocus Pyreneus purpureus.* The purple mountaine Crocus. 5 *Crocus montanus Autumnalis,* The Autumne mountaine Crocus. 6 *Sisyrinchium maius,* The greater Spanish Nut.

P

moneths, our Gardens are furnished with the varietie of one sort or another: the rest in Autumne, that so they might procure the more delight, in yeelding their beauty both early and late, when scarce any other flowers are found to adorne them.

The Names.

I shall not neede to trouble you with an idle tale of the name of Crocus, which were to little purpose, nor to reiterate the former names imposed vpon them; let it suffice that the fittest names are giuen them, that may distinguish them one from another; onely this I must giue you to vnderstand, that the gold yellow *Crocus* or Saffron flower, is the true *Crocus Masiacus*, as I shewed before; and that neither the yellow stript, or cloth of gold (which wee so call after the Dutch name *Gaud Laken*) is the true *Masiacus*, as some suppose; and that the great white Saffron flower, by reason of his likenesse vnto the gold yellow, is called *Crocus albus Masiaci facie*, or *facie lutes*, that is, The white Saffron flower that is like the *Masiacus* or yellow.

The Vertues.

The true Saffron (for the others are of no vse) which wee call English Saffron, is of very great vse both for inward and outward diseases, and is very cordiall, vsed to expell any hurtfull or venemous vapours from the heart, both in the small Pockes, Measels, Plague, Iaundise, and many other diseases, as also to strengthen and comfort any cold or weake members.

Chap. XVIII.

Sisyrinchium. The Spanish Nut.

I Can doe no otherwise then make a peculiar Chapter of this plant, because it is neither a *Crocus*, although in the roote it come somewhat neare vnto that kinde that is netted; but in no other part agreeing with any the delineaments of a Saffron flower, and therefore could not be thrust into the Chapter amongst them: neither can I place it in the forefront of the Chapter of the *Iris bulbosa*, or bulbous Flowerdeluces, because it doth not belong to that Family: and although the flower thereof doth most resemble a Flowerdeluce, yet in that no other parts thereof doe fitly agree thereunto, I haue rather chosen to seate it by it selfe betweene them both, as partaking of both natures, and so may serue in stead of a bridge, to passe from the one to the other, that is, from the *Crocus* or Saffron flower, to the *Iris bulbosa* or bulbous rooted Flowerdeluce, which shall follow in the next Chapter by themselues.

The Spanish Nut hath two long and narrow, soft and smooth greene leaues, lying for the most part vpon the ground, and sometimes standing vp, yet bending downewards; betweene these leaues riseth vp a small stalke, halfe a foote high, hauing diuers smooth soft greene leaues vpon it, as if they were skinnes, through which the stalke passeth; at the toppe whereof stand diuers flowers, rising one after another, and not all flowring at once: for seldome shall you haue aboue one flower blowne at a time, each whereof doth so quickly passe and fade away, that one may well say, that it is but one dayes flower, or rather the flower of a few houres: the flower it selfe hath nine leaues, like vnto a Flowerdeluce, whereof the three that fall downe, haue in each of them a yellow spot: the other three, which in the Flowerdeluces are hollow and ridged, couering the other three that fall downe, in this stand vpright, and are parted at the ends: the three that stand vp in the middle are small and short: the whole flower is smaller then any Flowerdeluce, but of sundry colours; for some are of an excellent skie colour blew, others of a Violet purple, others of a darker purple colour, and some white, and many others mixed, either pale blew and deepe purple, or white and blew

mixed

mixed or striped together very variably, quickly fading as I said before : the seede is enclosed in small cods, so thinne and transparent, that one may easily see, and tell the seeds as they lye, which are of a brownish red colour : the roote is small, blackish and round, wrapped in a thicke skinne or huske, made like vnto a net, or somewhat like vnto the roote of the cloth of gold Crocus : when the plant is in flower, it is found to haue two rootes one aboue another, whereof the vppermost is firme and sound, and the vndermost loose and spongie, in like manner as is found in the rootes of diuers Orchides or Satyrions, Bee-flowers and the like, and without any good taste, or sweetnesse at all, although Clusius saith otherwise.

Sisyrinchium Mauritanicum. The Barbary Nut.

There is another of this kinde, not differing from the former in any other notable part, but in the flower, which in this is of a delayed purplish red colour, hauing in each of the three lower leaues a white spot, in stead of the yellow in the former, but are as soone fading as they.

The Place.

The former doe grow very plentifully in many parts both of Spaine and Portugall, where Guillaume Boel, a Dutch man heretofore remembred often in this Booke, found them ; of the sundry colours specified, whereas Clusius maketh mention but of one colour that he found.

The other was found in that part of Barbary, where Fez and Morocco do stand, and brought first into the Lowe-Countries : but they are both very tender, and will hardly abide the hard Winters of these colder regions.

The Time.

The first flowreth in May and Iune, the last not vntill August.

The Names.

The name *Sisyrinchium* is generally imposed vpon this plant, by all authors that haue written thereof, thinking it to bee the right *Sisyrinchium* of Theophrastus : but concerning the Spanish name *Nozelha*, which Clusius saith it is called by in Spaine, I haue beene credibly enformed by the aforenamed Boel, that this roote is not so called in those parts ; but that the small or common stript Crocus is called *Nozelha*, which is sweete in taste, and desired very greedily by the Shepheards and Children, and that the roote of this *Sisyrinchium* or Spanish Nut, is without any taste, and is not eaten. And againe, that there is not two kindes, although it grow greater, and with more flowers, in those places that are neare the Sea, where both the washing of the Sea water, and the moisture and ayre of the Sea, causeth the ground to bee more fertile. This I thought good, from the true relation of a friend, to giue the world to vnderstand, that truth might expell errour.

The Vertues.

These haue not been knowne to bee vsed to any Physicall purpose, but wholly neglected, vnlesse some may eate them, as Clusius reporteth.

CHAP. XIX.

Iris bulbosa. The bulbous Flowerdeluce.

THe Flowerdeluces that haue bulbous rootes are of two forts, the one greater then the other : the greater bearing larger and broader leaues and flowers, and the lesser narrower. But before I giue you the defcriptions of the vfuall greater kindes, I muft needes place one or two in the fore-front that haue no fellowes ; the one is called of Clufius, his broad leafed Flowerdeluce, and the other a Perfian, fomewhat like vnto it, which although they differ notably from the reft, yet they haue the neareft refemblance vnto thofe greater kindes, that come next after them.

Iris bulbofa prima latifolia Clufij.
Clufius his firft great bulbous Flowerdeluce.

This Flowerdeluce hath diuers long and broad leaues, not ftiffe, like all the other, but foft and greenifh on the vpperfide, and whitifh vnderneath ; among which rife vp fometimes feuerall fmall, fhort, flender ftalkes, and fometimes but one, not aboue halfe a foote high, bearing at the top one flower a peece, fomewhat like vnto a Flowerdeluce, confifting of nine leaues, whereofthofe three that ftand vpright, are fhorter and more clofed together, then in other forts of Flowerdeluces ; the other three that fall downe, turne vp their ends a little, and thofe three, that in other Flowerdeluces doe couer them at the bottome, ftand like the vpright leaues of other Flowerdeluces, but are parted into two ends, like vnto two fmall eares : the whole flower is of a faire blew, or pale fkie colour in moft, with a long ftripe in the middle of each of the three falling leaues, and in fome white, but more feldome : the roote is reafonable great, round and white, vnder the blackifh coates wherewith it is couered, hauing many long thicke white rootes in ftead of fibres, which make them feeme to be Afphodill rootes. The flower is very fweete.

Iris bulbofa Perfica. The Perfian bulbous Flowerdeluce.

This Perfian Flowerdeluce is fomewhat like vnto the former, both in roote and in leafe, but that the leaues are fhorter and narrower, and the flower being much about the fame fafhion, is of a pale blew ruffetifh colour, each of the three lower falling leaues are almoft wholly of a browne purple colour, with a yellow fpot in the middle of them : this as it is very rare, fo it feldome beareth flowers with vs.

The Place.

The firft groweth in many places of Spaine and Portugall, from whence I and others haue often had it for our Gardens, but by reafon of the tendernefle thereof, it doth hardly endure the fharpnefle of our cold Winters, vnlefle it be carefully preferued.

The other is faid to come from Perfia, and therefore it is fo entituled, and is as tender to be kept as the other.

The Time.

The firft flowreth moft vfually not vntill May with vs, yet many times fooner : but in Ianuary and February, as Clufius faith, in the naturall places thereof.

The other is as early oftentimes when it doth flower with vs.

The Names.

Becaufe Clufius by good iudgement referreth the firft to the greater
kindes

kindes of Flowerdeluces, and placeth it in the fore ranke, calling it *Iris bul-bofa latifolia prima*, that is, The firſt broad leafed Flowerdeluce, and all o-thers doe the like, I haue (as you ſee) in the like manner put it before all the other, and keepe the ſame name. The Spaniards, as he ſaith, called it *Lirio eſpadanal*, and they of Corduba, *Lirios azules*.

The other hath no other name then as it is in the title.

1. *Iris bulbofa maior ſiue Anglica cærulea.*
The blew Engliſh bulbous Flowerdeluce.

This bulbous Flowerdeluce riſeth vp early, euen in Ianuary oftentimes, with fiue or ſix long and (narrow, in compariſon of any great breadth, but in regard of the other kinde) broad whitiſh green leaues, creſted or ſtraked on the backſide, and halfe round, the inſide being hollow like a trough or gutter, white all along the inſide of the leafe, and blunt at the end ; among which riſeth vp a ſtiffe round ſtalke, a cubit or two foot high, at the toppe whereof, out of a skinnie huske, commeth forth one or two flowers, conſiſting of nine leaues a peece, three whereof that are turned downewards, are lar-ger and broader then the other, hauing in each of them a yellow ſpot, about the mid-dle of the leafe, other three are ſmall, hollow, ridged or arched, couering the lower part next the ſtalke of thoſe falling leaues, turning vp their ends, which are diuided into two parts, other three ſtand vpright, and are very ſmall at the bottome of them, and broader toward the toppe : the whole flower is of a faire blew colour ; after the flowers are paſt, come vp three ſquare heads, ſomewhat long, and lanke, or looſe, containing in them round yellowiſh ſeede, which when it is ripe, will rattle by the ſha-king of the winde in the dry huskes : the roote of this kinde is greater and longer then any of the ſmaller kindes with narrow leaues, couered with diuers browne skinnes, which ſeeme to be fraught with long threeds like haires, eſpecially at the ſmall or vp-per end of the roote, which thing you ſhall not finde in any of the ſmaller kindes.

2. *Iris bulbofa maior purpurea & purpuro violacea.*
The paler or deeper purple great bulbous Flowerdeluce.

Theſe purple Flowerdeluces differ not from the laſt deſcribed, either in roote or leafe : the chiefeſt difference conſiſteth in the flowers, which in theſe are ſomewhat larger then in the former, and in the one of a deepe blew or Violet purple colour, and in the other of a deepe purple colour, in all other things alike.

There is alſo another, in all other things like vnto the former, but only in the flower, *Flore cinerea* which is of a pale or bleake blew, which we call an aſh-colour.

3. *Iris bulbofa maior purpurea variegata ſiue ſtriata.*
The great purple ſtript bulbous Flowerdeluce.

There is another of the purple kinde, whoſe flower is purple, but with ſome veines or ſtripes of a deeper Violet colour, diuerſly running through the whole leaues of the flower.

And another of that bleake blew or aſh-colour, with lines and veines of purple in *Flore cinerea* the leaues of the flowers, ſome more or leſſe then other. *ſtriata purpu-rea.*

And againe another, whoſe flower is of a purple colour like vnto the ſecond, but *Fl re purpurea* that round about that yellow ſpot, in the middle of each of the three falling leaues (as *orbe cinerio* is vſuall in all the bulbous Flowerdeluces) there is a circle of a pale blew or aſh-colour, the reſt of the leafe remaining purple, as the other parts of the flower is.

4. *Iris bulbofa maior flore rubente.*
The great peach coloured bulbous Flowerdeluce.

There is another of theſe greater kindes, more rare then any of the former, not differing in roote, leafe, or flower, from the former, but onely that the flower in this is of a pale reddiſh purple colour, comming ſomewhat neare vnto the colour of a peach bloſſome.

5. Iris

5. Iris bulbosa maior siue latifolia alba.
The great white bulbous Flowerdeluce.

The great white bulbous Flowerdeluce, riseth not vp so early out of the ground as the blew or purple doth, but about a moneth or more after, whose leaues are somewhat larger, and broader then of the others: the stalke is thicker and shorter, bearing vsually two very large and great flowers, one flowring a little before the other, yet oftentimes both in flower together in the end, of a bleake blewish white colour, which wee call a siluer colour, while they are in the budde, and before they be blowne open, but then of a purer white, yet with an eye or shew of that siluer colour remaining in them, the three falling leaues being very large, and hauing that yellow spot in the middle of each of them: the seedes are likewise inclosed in heads, like vnto the blew or purple kindes, but larger, and are of a reddish yellow colour like them: the roote likewise is not differing, but greater.

6. Iris bulbosa maior alba variegata.
The great white stript bulbous Flowerdeluce.

This white stript Flowerdeluce, is in roote, leafe, and flower, and in manner of growing, like vnto the former white Flowerdeluce; the onely difference is in the marking of the flower, being diuers from it: for this hath in the white flower great veines, stripes, or markes, of a Violet blew colour, dispersed through the leaues of the flower very variably, which addeth a superexcellent beauty to the flower.

7. Iris bulbosa maior siue latifolia versicolor.
The great party coloured bulbous Flowerdeluce.

There is no difference in this from the former, but in the flower, which is of a whitish colour in the three falling leaues, hauing a circle of ash-colour about the yellow spot, the three rigged leaues being likewise whitish, but ridged and edged with that ash-colour, and the three vpright leaues of a pale blewish white colour, with some veines therein of a blewish purple.

Varietas.　There hath beene brought vnto vs diuers rootes of these kindes, with the dryed flowers remaining on them, wherein there hath beene seene more varieties, then I can well remember to expresse, which variety it is very probable, hath risen by the sowing of the seeds, as is truely obserued in the narrower leafed kinde of Flowerdeluce, in the Tulipa, and in some other plants.

Flore luteo.　Wee haue heard of one of this kinde of broad leafed Flowerdeluces, that should beare a yellow flower, in the like manner as is to be seene in the narrow leafed ones: but I haue not seene any such, and therefore I dare report no further of it, vntill time hath discouered the truth or falshood of the report.

The Place.

Lobelius is the first reporter, that the blew Flowerdeluce or first kinde of these broad leafed Flowerdeluces, groweth naturally in the West parts of England; but I am in some doubt of the truth of that report: for I rather thinke, that some in their trauels through Spaine, or other parts where it groweth, being delighted with the beauty of the flower, did gather the rootes, and bring them ouer with them, and dwelling in some of the West parts of England, planted them, and there encreasing so plentifully as they doe, they were imparted to many, thereby in time growing common in all Countrey folkes Gardens thereabouts. They grow also, and all the other, and many more varieties, about Tholouse, from whence Plantinianus Gassianus both sent and brought vs them, with many other bulbous rootes, and rare plants gathered thereabouts.

The

1 *Iris bulbosa latifolia prima Clusii.* Clusius his first great bulbous Flowerdeluce. 2 *Iris bulbosa maior cærulea sive Anglica.* The great blew or English Bulbous Flowerdeluce. 3 *Iris bulbosa maior purpurea variegata.* The great purple strip't bulbous Flowerdeluce. 4 *Iris bulbosa angustifolia maior alba.* The greater white narrow leafed bulbous Flowerdeluce. 5 *Iris bulbosa angustifolia versicolor.* The party coloured narrow leafed bulbous Flowerdeluce. 6 *Iris bulbosa angustifolia Africana.* The purple African bulbous narrow leafed Flowerdeluce.

The Time.

Thefe doe flower vfually in the end of May, or beginning of Iune, and their feede is ripe in the end of Iuly or Auguſt.

The Names.

Lobel calleth the firſt Engliſh blew Flowerdeluce, *Hyacinthus Poetarum flore Iridis, & propter Hyacinthinum colorem, id eſt violaceum dictus*: but I know not any great good ground for it, more then the very colour; for it is neither of the forme of a Lilly, neither hath it thofe mourning markes imprinted in it, which the Poet faineth to bee in his Hyacinth. It is moſt truely called an *Iris*, or Flowerdeluce (and there is great difference betweene a Lilly and a Flowerdeluce, for the formes of their flowers) becaufe it anfwereth thereunto very exactly, for the flower, and is therefore called vfually by moſt, either *Iris bulbofa Anglica*, or *Iris bulbofa maior ſiue latifolia*, for a difference betweene it, and the leffer with narrow leaues: In Engliſh, eyther The great Engliſh bulbous Flowerdeluce, or the great broad leafed bulbous Flowerdeluce, which you will, adding the other name, according to the colour.

And thus much for thefe broad leafed bulbous Flowerdeluces, fo much as hath come to our knowledge. Now to the feuerall varieties of the narrow leafed bulbous Flowerdeluces, fo much likewife as we haue been acquainted with.

Iris bulbofa minor ſiue anguſtifolia alba.
The fmaller white or narrow leafed bulbous Flowerdeluce.

This firſt Flowerdeluce, which beareth the fmaller flower of the two white ones, that are here to bee defcribed, fpringeth out of the ground alwaies before Winter, which after breaketh forth into foure or fiue fmall and narrow leaues, a foote long or more, of a whitiſh greene on the infide, which is hollow and chanalled, and of a blewiſh greene colour on the outfide, and round withall: the ſtalke of this kinde is longer and ſlenderer then the former, with fome ſhorter leaues vpon it, at the toppe whereof, out of ſhort skinny leaues, ſtand one or two flowers, fmaller, ſhorter, and rounder then the flowers of the former broad leafed Flowerdeluces, but made after the fame proportion with nine leaues, three falling downewards, with a yellow fpot in the middle, other three are made like a long arch, which couer the lower part next the ſtalke of thofe falling leaues, and turne vp at the ends of them, where they are diuided into two parts: the other three ſtand vpright, betweene each of the three falling leaues, being fomewhat long and narrow: the flower is wholly (fauing the yellow fpot) of a pure white colour, yet in fome hauing a ſhew of fome blew throughout, and in others towards the bottome of the three vpright leaues: after the flowers are paſt, there rife vp fo many long cods or feede veffels, as there were flowers, which are longer and fmaller then in the former, and a little bending like a Cornet, with three round fquares, and round pointed alfo, which diuiding it felfe when the feede is ripe into three parts, doe ſhew fix feuerall cells or places, wherein is contained fuch like round reddiſh yellow feedes, but fmaller then the former: the iroote is fmaller and ſhorter then the former, and without any haires or threeds, couered with browne thin skinnes, and more plentifull in giuing encreafe.

Iris bulbofa anguſtifolia alba flore maiore.
The greater white narrow leafed bulbous Flowerdeluce.

I ſhall not neede to make a feuerall defcription to euery one of thefe Flowerdeluces that follow, for that were but to make often repetition of one thing, which being once done, as it is, may well ferue to expreffe all the reſt, and but onely to adde the efpeciall

ciall differences, either in leafe or flower, for bigneſſe, colour, or forme, as is expedient to expreſſe and diſtinguiſh them ſeuerally. This greater white bulbous Flowerdeluce is like vnto the laſt deſcribed in all parts, ſauing that it is a little larger and higher, both in leafe, ſtalke, and flower, and much whiter then any of theſe mixed ſorts that follow, yet not ſo white as the former: the roote hereof is likewiſe a little bigger and rounder in the middle.

Albeſſent.
Milke white. There is another, whoſe falling leaues haue a little ſhew of yellowneſſe in them, and ſo are the middle ridges of the arched leaues, but the vpright leaues are more white, not differing in roote or leafe from the firſt white.

Argentea.
Siluer colour. And another, whoſe falls are of a yellowiſh white, like the laſt, the arched leaues are whiter, and the vpright leaues of a blewiſh white, which we call a ſiluer colour.

Albida.
Whitiſh. Another hath the fals yellowiſh, and ſometimes with a little edge of white about them, and ſometimes without; the vpright leaues are whitiſh, as the arched leaues are, yet the ridge yellower.

Albida labris
luteis.
White with
yellow fals. Another hath his fals yellow, and the vpright leaues white, all theſe flowers are about the ſame bigneſſe with the firſt.

Albida angu-
ſtior.
The narrow
white. But we haue another, whoſe flower is ſmaller, and almoſt as white as the ſecond, the lower leaues are ſmall, and doe as it were ſtand outright, not hauing almoſt any fal at all, ſo that the yellow ſpot ſeemeth to be the whole leafe, the arched leaues are not halfe ſo large as in the former, and the vpright leaues bowe themſelues in the middle, ſo that the tops doe as it were meete together.

And another of the ſame, whoſe falling leaues are a little more eminent and yellow, with a yellower ſpot.

Aurea ſiue lu-
tea Hiſpanica.
The Spaniſh
yellow. We haue another kinde that is called the Spaniſh yellow, which riſeth not vp ſo high, as ordinarily moſt of the reſt doe, and is wholly of a gold yellow colour.

Pallide lutea.
Straw colour. There is another, that vſually riſeth higher then the former yellow, and is wholly of a pale yellow, but deeper at the ſpot.

Albida lutea.
Pale Straw co-
lour. There is alſo another like vnto the pale yellow, but that the falling leaues are whiter then all the reſt of the flower.

Mauritanica
flaua ſerotina
minor.
The ſmall Bar-
bary yellow. There is a ſmaller or dwarfe kinde, brought from the backe parts of Barbary, neare the Sea, like vnto the yellow, but ſmaller and lower, and in ſtead of vpright leaues, hath ſmall ſhort leaues like haires: it flowreth very late, after all others haue almoſt giuen their ſeede.

Verſicolor Hiſ-
panica cærulea
labris albis.
The party co-
loured Spaniſh We haue another ſort is called the party coloured Spaniſh bulbous Flowerdeluce, whoſe falling leaues are white, the arched leaues of a whitiſh ſiluer colour, and the vpright leaues of a fine blewiſh purple.

Diuerſities.
The diuerſity Yet ſometimes this doth vary; for the falling leaues will haue either an
of variation of edge of blew, circling the white leaues, the arched leaues being a little
this flower. blewer, and the vpright leaues more purple.

Or the fals will be almoſt wholly blew, edged with a blewer colour, the arched leaues pale blew, and the vpright leaues of a purpliſh blew Violet colour.

Or the fals white, the arched leaues pale white, as the vpright leaues are. Or not of ſo faire a blewiſh purple, as the firſt ſort is.

Some of them alſo will haue larger flowers then others, and be more liberall in bearing flowers: for the firſt ſort, which is the moſt ordinary, ſeldome beareth aboue one flower on a ſtalke, yet ſometimes two. And of the others there are ſome that wil beare vſually two and three flowers, yet ſome again will beare but one. All theſe kindes ſmell ſweeter then many of the
Cærulea ſiue
purpurea minor other, although the moſt part be without ſent.
Luſitanica præ-
cox.
The ſmall There is another kinde, that is ſmaller in all the parts thereof then the
early purple former, the ſtalke is ſlender, and not ſo high, bearing at the toppe one or
Portugall. two ſmall flowers, all wholly of a faire blewiſh purple, with a yellow ſpot

in

in euery one of the three falling leaues, this vsually flowreth early, euen with the firft bulbous Flowerdeluces.

Purpurea ma-ior.
The greater purple,
We haue another purple, whofe flower is larger, and ftalke higher, and is of a very reddifh purple colour, a little aboue the ground, at the foote or bottome of the leaues and ftalke : this flowreth with the later fort of Flowerdeluces.

Purpurea ſerotina.
The late purple.
There is another, whofe flower is wholly purple, except the yellow fpot, and flowreth later then any of the other purples.

Purpura rubeſcens labris cæruleis.
A reddifh purple with blew fals.
There is yet another purple, whofe vpright leaues are of a reddifh purple, and the falling leaues of a blew colour.

Purpura rubeſcens labris albido cæru-leis.
A reddifh purple with whitifh blew fals.
And another of a reddifh purple, whofe falling leaues are of a whitifh blew colour, in nothing elfe differing from the laft.

Purpurea labris luteis.
Party coloured purple & yellow
Another hath his falling leaues of a faire gold yellow, without any ftripe, yet in fome there are veines running through the yellow leaues, and fome haue an edge of a fullen darke colour about them : the vpright leaues in euery of thefe, are of a Violet purple.

Purpurea labris ex albido cæruleo & luteo mixtis.
Party coloured purple with ftript yellow fals.
Another is altogether like this laft, but that the falling leaues are of a pale blew and yellow, trauerfing one the other, and the arched leaues of a pale purplifh colour.

Subpurpurea labris luteis.
Pale purple with yellow fals.
Another hath his vpright leaues of a paler purple, and the falling leaues yellow.

A paler pur, le-
And another little differing from it, but that the arched leaues are whitifh.

Subcærulea labris luteis.
Party coloured blew and yellow
Another whofe vpright leaues are of a pale blew, and the falling leaues yellow.

And another of the fame fort, but of a little paler blew.

Crinicoloris ela-gantior.
A faire haire co-lour.
We haue another fort, whofe vpright leaues are of a faire brownifh yellow colour, which fome call a *Fuille mort*, and others an haire colour ; the falling leaues yellow.

Altera obfoletior.
A dull haire co-lour.
And another of the fame colour, but fomewhat deader.

Iris bulbofa Africana ferpentaria caule.
The purple or murrey bulbous Barbary Flowerdeluce.

This Flowerdeluce as it is more ftrange (that is, but lately knowne and poffeffed by a few) fo it is both more defired, and of more beauty then others. It is in all refpects, of roote, leafe, and flower, for the forme like vnto the middle fort of thefe Flowerdeluces, onely the loweft part of the leaues and ftalke, for an inch or thereabouts, next vnto the ground, are of a reddifh colour, fpotted with many fpots, and the flower, being of a meane fize, is of a deepe purplifh red or murrey colour the whole flower throughout, except the yellow fpot in the middle of the three lower or falling leaues, as is in all others.

Purpura cæru-lea obfoleta la-bris fufcis.
The duskie party coloured purple,
And laftly, there is another fort, which is the greateft of all thefe narrow leafed Flowerdeluces, in all the parts of it ; for the roote is greater then any of the other, being thicke and fhort : the leaues are broader and longer, but of the fame colour: the ftalke is ftronger and higher then any of them, bearing two or three flowers, larger alfo then any of the reft, whofe falling leaues are of a duskie yellow, and fometimes with veines and borders about the brimmes, of another dunne colour, yet hauing that yellow fpot that is in all : the arched leaues are of a fullen pale purplifh yellow, and the vpright leaues of a dull or duskie blewifh purple colour : the heads or hornes for feede are likewife greater, and fo is the feede alfo a little.

The Place.

Thefe Flowerdeluces haue had their originall out of Spaine and Portugall, as it is thought, except thofe that haue rifen by the fowing, and thofe which are named of Africa.

The

The Time.

Thefe flower in Iune, and fometimes abide vnto Iuly, but vfually not fo early as the former broad leafed kindes, and are foone fpoiled with wet in their flowring.

The Names.

The feuerall names, both in Latine and Englifh, are fufficient for them as they are fet downe; for we know no better.

The Vertues.

There is not any thing extant or to be heard, that any of thefe kindes of Flowerdeluces hath been vfed to any Phyficall purpofes, and ferue onely to decke vp the Gardens of the curious.

And thus much for thefe forts of bulbous Flowerdeluces, and yet I doubt not, but that there are many differences, which haue rifen by the fowing of the feede, as many may obferue from their owne labours, for that euery yeare doth fhew forth fome variety that is not feene before. And now I will conuert my difcourfe a while likewife, to paffe through the feuerall rankes of the other kindes of tuberous rooted Flowerdeluces, called Flagges.

Chap. XX.

Iris latifolia tuberofa. The Flagge or Flowerdeluce.

There are two principall kindes of tuberous or knobby rooted Flowerdeluces, that is, the tall and the dwarfe, or the greater and the leffer; the former called *Iris maior* or *latifolia*, and the other *Iris minor*, or rather *Chamæiris*; and each of thefe haue their leffer or narrow leafed kindes to bee comprehended vnder them: Of all which in their order. And firft of that Flowerdeluce, which for his excellent beautie and raritie, deferueth the firft place.

Iris Chalcedonica fiue Sufiana maior. The great Turkie Flowerdeluce.

The great Turkie Flowerdeluce, hath diuers heads of long and broad frefh greene leaues, yet not fo broad as many other of thofe that follow, one folded within another at the bottome, as all other of thefe Flowerdeluces are: from the middle of fome one of thofe heads (for euery head of leaues beareth not a flower) rifeth vp a round ftiffe ftalke, two foote high, at the toppe whereof ftandeth one flower (for I neuer obferued it to beare two) the largeft almoft, but rareft of all the reft, confifting of nine leaues, like the others that follow, but of the colour almoft of a Snakes skinne, it is fo diuerfly fpotted; for the three lower falling leaues are very large, of a deepe or darke purple colour, almoft blacke, full of grayifh fpots, ftrakes, and lines through the whole leaues, with a blacke thrume or freeze in the middle of each of them: the three arched leaues that couer them, are of the fame darke purple colour, yet a little paler at the fides, the three vpper leaues are very large alfo, and of the fame colour with the lower leaues, but a little more liuely and frefh, being fpeckled and ftraked with whiter fpots and lines; which leaues being laid in water, will colour the water into a Violet colour, but if a little Allome be put therein, and then wrung or preffed, and the iuice of thefe leaues dryed in the fhadow, will giue a colour almoft as deepe as Indico, and may ferue for fhadowes in limming excellent well: the flower hath no fent that can be perceiued, but is onely commendable for the beauty and rarity thereof: it feldome beareth feedes in thefe cold Countries, but when it doth, it is contained in great heads,

being

being brownish and round, but not so floras in other sorts, the roots are more browne on the outside, and growing tuberous thicke, as all other that are kept in Gardens.

Iris Chalcedonica siue Susiana minor. The lesser Turkie Flowerdeluce.

There is another hereof little differing, but that the leafe is of a more yellowish greene colour, and the flower neither so large or faire, nor of so perspicuous markes and spots, nor the colour of that liuely (though darke) lustre.

The Place.

These haue been sent out of Turkie diuers times among other things, and it should seeme, that they haue had their originall from about Sufis, a chiefe Citie of Persia.

The Time.

They flower in May most vsually, before any of the other kindes.

The Names.

They haue been sent vnto vs, and vnto diuers other in other parts, from Constantinople vnder the name of *Alaia Susiana*, and thereupon it hath been called, both of them and vs, either *Iris Chalcedonica*, or *Susiana*, and for distinctio n *maior* or *minor*: In English, The Turkie Flowerdeluce, or the Ginnie Hen Flowerdeluce, the greater or the lesser.

Iris alba Florentina. The white Flowerdeluce.

The great white Flowerdeluce, hath many heads of very broad and flat long leaues, enclosing or folding one within another at the bottome, and after a little diuided one from another toward the top, thin edged, like a sword on both sides, and thicker in the middle: from the middle of some of these heads of leaues, riseth vp a round stiffe stalk, two or three foot high, bearing at the top one, two, or three large flowers, out of seuerall huskes or skins, consisting of nine leaues, as all the other do, of a faire white colour, hauing in the middle of each of the three falling leaues, a small long yellow frize or thrume, as is most vsuall in all the sorts of the following Flowerdeluces, both of the greater and smaller kindes: after the flowers are past, come the seed, inclosed in thicke short pods, full traught or stored with red roundish and flat seede, lying close one vpon another: the roote is tuberous or knobby, shooting out from euery side such like tuberous heads, lying for the most part vpon or aboue the ground, and fastened within the ground with long white strings or fibres, which hold them strongly, and encreaseth
flore pallido. fast. There is another like vnto this last in all things, sauing that the colour of the flower is of a more yellowish white, which we vsually call a Straw colour.

Iris alba maior Versicolor. The white party coloured Flowerdeluce.

This variable Flowerdeluce is like vnto the former, but that the leaues are not so large and broad, the flower hereof is as large almost, and as white as the former, but it hath a faire list or line of a blewish purple downe the backe of euery one of the three vpright leaues, and likewise round about the edges, both of the vpper and lower leaues, and also a little more purplish vpon the ridge of the arched leaues, that couer the falling leaues: the roote hereof is not so great as of the former white, but a little slenderer and browner.

Iris Dalmatica maior. The great Dalmatian Flowerdeluce.

This greater Flowerdeluce of Dalmatia, hath his leaues as large and broad as any of the Flowerdeluces whatsoeuer, his stalke and flower doe equall his other proportion,

tion, onely the colour of the flower is differing, being of a faire watchet or bleake blew colour wholly, with the yellow frize or thrum downe the middle of the lower or falling leaues, as before is said to be common to all these sorts of Flowerdeluces, in all other parts it little differeth, sauing onely this is obserued to haue a small shew of a purplish red about the bottome of the greene leaues.

Iris purpurea siue vulgaris. The common purple Flowerdeluce.

This Flowerdeluce, which is most common in Gardens, differeth nothing at all from those that are formerly described, either in roote, leafe, or flower for the forme of them, but onely that the leaues of this are not so large as the last, and the flower it selfe is of a deep purple or Violet colour, and sometimes a little declining to rednesse, especially in some places.

Sometimes this kinde of Flowerdeluce will haue flowers of a paler purple colour, *Purpurea palidior versicolor.* comming neare vnto a blew, and sometimes it will haue veines or stripes of a deeper blew, or purple, or ash colour, running through all the vpper and lower leaues.

There is another like vnto this, but more purple in the fals, and more pale in the *Carulea labris purpureis.* vpright leaues.

Iris Asiatica carulea. The blew Flowerdeluce of Asia.

This Flowerdeluce of Asia, is in largenesse of leaues like vnto the Dalmatian, but beareth more store of flowers on seuerall branches, which are of a deeper blew colour, and the arched leaues whitish on the side, and purplish on the ridges, but in other things like vnto it.

There is another neare vnto this, but that his leaues are a little narrower, and his *Purpurea.* flowers a little more purple, especially the vpper leaues.

Iris Damascena. The Flowerdeluce of Damasco.

This is likewise altogether like the Flowerdeluce of Asia, but that it hath some white veines in the vpright leaues.

Iris Lusitanica biflora. The Portugall Flowerdeluce.

This Portugall Flowerdeluce is very like the common purple Flowerdeluce, but that this is not so large in leaues, or flowers, and that it doth often flower twice in a yeare, that is, both in the Spring, and in the Autumne againe, and besides, the flowers haue a better or sweeter sent, but of the like purple or Violet colour as it is, and comming forth out of purplish skins or huskes.

Iris Camerarij siue purpurea versicolor maior.
The greater variable coloured purple Flowerdeluce.

The greater of the variable purple Flowerdeluces, hath very broad leaues, like vnto the leaues of the common purple Flowerdeluce, and so is the flower also, but differing in colour, for the three lower leaues are of a deepe purple colour tending to rednesse, the three arched leaues are of the colour with the vpper leaues, which are of a pale or bleake colour tending to yellownesse, shadowed ouer with a smoakie purplish colour, except the ridges of the arched leaues, which are of a more liuely purple colour.

Iris purpurea versicolor minor. The lesser variable purple Flowerdeluce.

This Flowerdeluce differeth not in any thing from the last, but onely that it hath narrower greene leaues, and smaller and narrower flowers, else if they be both conferred together, the colours will not seeme to varie the one from the other any whit at all.

There is another somewhat neare vnto these two last kindes, whose huskes from *Altera minus fuliginea.* whence

whence the flowers doe ſhoote forth, haue purple veines in them, and ſo haue the falling purpliſh leaues, and the three vpright leaues are not ſo ſmoakie, yet of a dun purple colour.

Iris cærulea verſicolor. The blew party coloured Flowerdeluce.

This party coloured Flowerdeluce hath his leaues of the ſame largeneſſe, with the leſſer variable purple Flowerdeluce laſt deſcribed, and his flowers diuerſly marked : for ſome haue the fals blew at the edges, and whitiſh at the bottome, the arched leaues of a yellowiſh white, and the vpright leaues of a whitiſh blew, with yellowiſh edges. Some againe are of a darker blew, with browniſh ſpots in them. And ſome are ſo pale a blew, that we may well call it an aſh-colour : And laſtly, there is another of this ſort, whoſe vpright leaues are of a faire pale blew, with yellowiſh edges, and the falling leaues parted into two colours, ſometimes equally in the halfe, each ſide ſutable to the other in colour : And ſometimes hauing the one leafe in that manner : And ſometimes but with a diuers coloured liſt in them ; in the other parts both of flower and leafe, like vnto the other.

Iris lutea variegata. The yellow variable Flowerdeluce.

This yellow variable Flowerdeluce loſeth his leaues in Winter, contrary to all the former Flowerdeluces, ſo that his roote remaineth vnder ground without any ſhew of leafe vpon it : but in the beginning of the Spring it ſhooteth out faire broad leaues, falling downwards at the points or ends, but ſhorter many times then any of the former, and ſo is the ſtalke likewiſe, not riſing much aboue a foote high, whereon are ſet two or three large flowers, whoſe falling leaues are of a reddiſh purple colour, the three that ſtand vpright of a ſmoakie yellow, the arched leaues hauing their ridges of a bleake colour tending to purple, the ſides being of the former ſmoakie yellow colour, with ſome purpliſh veines at the foote or bottome of all the leaues : the roote groweth ſomewhat more ſlender and long vnder ground, and of a darker colour then manie of the other.

Varietas.　　　Another ſort hath the vpright leaues of a reaſonable faire yellow, and ſtand more vpright, not bowing downe as moſt of the other, and the purple fals haue pale edges. Some haue their greene leaues party coloured, white and greene, more or leſſe, and ſo are the huskes of the flowers, the arched leaues yellow, as the vpright leaues are, with purpliſh veines at the bottome. And ſome haue both the arched and vpright leaues of ſo pale a yellow, that we may almoſt call it a ſtraw colour, but yellower at the bottome, with purple veines, and the falling leaues purple, with two purple ſpots in them.

. And theſe are the ſorts of the greater tuberous or Flagge Flowerdeluces that haue come to our knowledge : the next hereunto are the leſſer or narrow leafed kindes to be deſcribed ; and firſt of the greateſt of them.

1. *Iris anguſtifolia Tripolitana aurea.* The yellow Flowerdeluce of Tripoly.

This Flowerdeluce I place in the forefront of the narrow leafed Flowerdeluces, for the length of the leaues, compared with the breadth of them ; it may fitly bee called a narrow leafed Flowerdeluce, although they be an inch broad, which is broader then any of them that follow, or ſome of thoſe are ſet downe before, but as I ſaid, the length make them ſeem narrow, and therfore let it take vp his roome in this place, with the deſcription that followeth. It beareth leaues a yard long, or not much leſſe, and an inch broad, as is ſaid before, or more, of a ſad greene colour, but not ſhining : the ſtalke riſeth vp to be foure or fiue foote high, being ſtrong and round, but not very great, bearing at the toppe two or three long and narrow gold yellow flowers, of the faſhion of the bulbous Flowerdeluces, as the next to bee deſcribed is, without any mixture or variation therein : the heads for ſeede are three ſquare, containing within them many flat cornered ſeedes : the roote is long and blackiſh, like vnto the reſt that follow, but greater and fuller.

1 *Iris Chalcedonica sine Sufiana maior.* The great Turkie Flowerdeluce. 2 *Iris alba Florentine.* The white Flowerdeluce. 3 *Iris lati*
folia variegata. The variable Flowerdeluce. 4 *Chamæiris latifolia maior.* The greater dwarfe Flowerdeluce.

2. *Iris angustifolia maior cærulea.*
The greater blew Flowerdeluce with narrow leaues.

This kinde of Flowerdeluce hath his leaues very long and narrow, of a whitish greene colour, but neither so long or broad as the last, yet broader, thicker and stiffer then any of the rest with narrow leaues that follow : the stalke riseth sometimes no higher then the leaues, and sometimes a little higher, bearing diuers flowers at the top, successiuely flowring one after another, and are like vnto the flowers of the bulbous Flowerdeluces, but of a light blew colour, and sometimes deeper : after the flowers are past, rise vp six cornered heads, which open into three parts, wherein is contained browne seede, almost round : the roote is small, blackish and hard, spreading into many long heads, and more closely growing or matting together.

3. *Iris angustifolia purpurea marina.* The purple narrow leafed Sea Flowerdeluce.

This Sea Flowerdeluce hath many narrow hard leaues as long as the former, and of a darke greene colour, which doe smell a little strong : the stalke beareth two or three flowers like the former, but somewhat lesse, and of a darke purple or Violet colour : in seede and roote it is like the former.

4. *Iris angustifolia purpurea versicolor.*
The variable purple narrow leafed Flowerdeluce.

The leaues of this Flowerdeluce are very like the former Sea Flowerdeluce, and do a little stinke like them ; the flowers are differing, in that the vpper leaues are wholly purple or violet, and the lower leaues haue white veines, and purple running one among another : the seede and rootes differ not from the former purple Sea kinde.

5. *Iris angustifolia minor Pannonica siue versicolor Clusij.*
The small variable Hungarian Flowerdeluce of Clusius.

This Hungarian Flowerdeluce (first found out by Clusius, by him described, and of him tooke the name) riseth vp with diuers small tufts of leaues, very long, narrow, and greene, growing thicke together, especially if it abide any time in a place ; among which riseth vp many long round stalkes, higher then the leaues, bearing two or three, or foure small flowers, one aboue another, like the former, but smaller and of greater beauty : for the lower leaues are variably striped with white and purple, without any thrume or fringe at all ; the vpper leaues are of a blewish fine purple or Violet colour, & so are the arched leaues, yet hauing the edges a little paler : the heads for seede are smaller, and not so cornered as the other, containing seedes much like the former, but smaller : the roote is blacke and small, growing thicker and closer together then any other, and strongly fastened in the ground, with a number of hard stringie rootes the flowers are of a reasonable good sent.

6. *Iris angustifolia maior flore duplici.* The greater double blew Flowerdeluce.

This Flowerdeluce, differeth not either in roote or leafe from the first great blew Flowerdeluce of Clusius, but onely in that the leaues grow thicker together, and that the flowers of this kinde are as it were double with many leaues confusedly set together, without any distinct parts of a Flowerdeluce, and of a faire blew colour with many white veines and lines running in the leaues ; yet oftentimes the stalke of flowers hath but two or three small flowers distinctly set together, rising as it were out of one huske.

7. *Iris angustifolia minor alba Clusij.*
The small white Flowerdeluce of Hungary.

This likewise differeth little from the former Hungarian Flowerdeluce of Clusius,

but

1 *Iris angustifolia Tripolitana.* The yellow Flowerdeluce of Tripoli. 2 *Iris angustifolia maior cærulea.* The greater blew Flowerdeluce with narrow
leaues 3 *Iris angustifolia minor Pannonica siue variegata Clusij.* The small variable Hungarian Flowerdeluce of Clusius. 4 *Iris angustifolia maior flore aureolo.*
The greater double blew Flowerdeluce. 5 *Chamæiris angustifolia minor.* The lesser Grasse Flowerdeluce. 6 *Iris tuberosa.* The veluet Flowerdeluce.

but that the leafe is of a little paler greene colour, and the flower is of a faire whitish colour, with some purple at the bottome of the leaues.

Next after these narrow leafed Flowerdeluces, are the greater and smaller sorts of dwarfe kindes to follow; and lastly, the narrow or grasse leafed dwarfe kindes, which will finish this Chapter of Flowerdeluces.

1. *Chamæiris latifolia maior alba.* The greater white dwarfe Flowerdeluce.

This dwarfe Flowerdeluce hath his leaues as broad as some of the lesser kindes last mentioned, but not shorter; the stalke is very short, not aboue halfe a foote high or thereabouts, bearing most commonly but one flower, seldome two, which are in some of a pure white, in others paler, or somewhat yellowish through the whole flower, except the yellow frize or thrume in the middle of euery one of the falling leaues: after the flowers are past, come forth great heads, containing within them round pale seed: the roote is small, according to the proportion of the plant aboue ground, but made after the fashion of the greater kindes, with tuberous peeces spreading from the sides, and strong fibres or strings, whereby they are fastened in the ground.

2. *Chamæiris latifolia maior purpurea.* The greater purple dwarfe Flowerdeluce.

There is no difference either in roote, leafe, or forme of flower in this from the former dwarfe kinde, but onely in the colour of the flower, which in some is of a very deepe or blacke Violet purple, both the toppes and the fals: in others the Violet purple is more liuely, and in some the vpper leaues are blew, and the lower leaues purple, yet all of them haue that yellow frize or thrume in the middle of the falling leaues, that the other kindes haue.

Altera. There is another that beareth purple flowers, that might be reckoned, for the smalnesse and shortnesse of his stalke, to the next kinde, but that the flowers and leaues of this are as large as any of the former kindes of the smaller Flowerdeluces.

3. *Chamæiris latifolia minor alba.* The lesser white dwarfe Flowerdeluce.

There is also another sort of these Flowerdeluces, whose leaues and flowers are lesse, and wherein there is much variety. The leaues of this kinde, are all for the most part somewhat smaller, narrower, and shorter then the former: the stalke with the flower vpon it scarce riseth aboue the leaues, so that in most of them it may be rather called a foote-stalke, such as the Saffron flowers haue, and are therefore called of manie ἄχαυλοι without stalkes; the flowers are like vnto the first described of the dwarfe kindes, and of a whitish colour, with a few purplish lines at the bottome of the vpper leaues, and a list of greene in the falling leaues.

Straminea. Another hath the flowers of a pale yellow, called a Straw colour, with whitish stripes and veines in the fals, and purplish lines at the bottome of the vpper leaues.

4. *Chamæiris latifolia minor purpurea.* The lesser purple dwarfe Flowerdeluce.

The difference of this from the former, consisteth more in the colour then forme of the flower, which is of a deep Violet purple, sometimes paler, and sometimes so deepe, that it almost seemeth blacke: And sometimes the fals purplish, and the vpper leaues blew. Some of these haue a sweete sent, and some none.

Cærulea. There is another of a fine pale or delayed blew colour throughout the whole flower.

5. *Chamæiris latifolia minor suauerubens.* The lesser blush coloured dwarfe Flowerdeluce.

This Flowerdeluce hath the falling leaues of the flower of a reddish colour, and the thrumes blew: the vpper and arched leaues of a fine pale red or flesh colour, called a blush colour; in all other things it differeth not, and smelleth little or nothing at all.

6. *Chamæiris latifolia minor lutea versicolor.*
The lesser yellow variable dwarfe Flowerdeluce.

The falling leaues of this Flowerdeluce are yellowish, with purple lines from the middle downewards, sometimes of a deeper, and sometimes of a paler colour, and white thrumes in the middle, the vpper leaues are likewise of a yellowish colour, with purple lines in them : And sometimes the yellow colour is paler, and the lines both in the vpper and lower leaues of a dull or dead purple colour.

3. *Chamæiris latifolia minor cærulea versicolor.*
The lesser blew variable dwarfe Flowerdeluce.

The vpper leaues of this flower are of a blewish yellow colour, spotted with purple in the broad part, and at the bottome very narrow : the falling leaues are spread ouer with pale purplish lines, and a small shew of blew about the brimmes : the thrume is yellow at the bottome, and blewish aboue: the arched leaues are of a blewish white, being a little deeper on the ridge.
And sometimes the vpper leaues are of a paler blew rather whitish, with the yellow: both these haue no sent at all.

8. *Chamæiris marina purpurea.* The purple dwarfe Sea Flowerdeluce.

This small Flowerdeluce is like vnto the narrow leafed Sea Flowerdeluce before described, both in roote, leafe, and flower, hauing no other difference, but in the smalnesse and lownesse of the growing, being of the same purple colour with it.

9. *Chamæiris angustifolia maior.* The greater Grasse Flowerdeluce.

This Grasse Flowerdeluce hath many long and narrow darke greene leaues, not so stiffe as the former, but lither, and bending their ends downe againe, among which rise vp diuers stalkes, bearing at the toppe two or three sweete flowers, as small as any of them set downe before, of a reddish purple colour, with whitish yellow and purple strakes downe the middle of the falling leaues : the arched leaues are of a horse flesh colour all along the edges, and purple vpon the ridges and tips that turne vp againe : vnder these appeare three browne aglets, like vnto birds tongues : the three vpper leaues are small and narrow, of a perfect purple or Violet colour : the heads for seede haue sharper and harder cornered edges then the former: the seedes are somewhat grayish like the former, and so are the rootes, being small, blacke, and hard, growing thicke together, fastened in the ground with small blackish hard strings, which hardly shoote againe if the roote be remoued.

10. *Chamæiris angustifolia minor.* The lesser Grasse Flowerdeluce.

This Flowerdeluce is in leaues, flowers, and rootes so like the last described, that but onely it is smaller and lower, it is not to be distinguished from the other. And this may suffice for these sorts of Flowerdeluces, that furnish the Gardens of the curious louers of these varieties of nature, so farre forth as hath passed vnder our knowledge. There are some other that may be referred hereunto, but they belong to another history; and therefore I make no mention of them in this place.

The Place.

The places of most of these are set downe in their seuerall titles; for some are out of Turkie, others out of Hungaria, Dalmatia, Illyria, &c. as their names doe import. Those that grow by the Sea, are found in Spaine and France.

The

The Time.

Some of these do flower in Aprill, some in May, and some not vntill Iune.

The Names.

The names expressed are the fittest agreeing vnto them, and therefore it is needlesse againe to repeate them. Many of the rootes of the former or greater kindes, being dryed are sweete, yet some more then other, and some haue no sent at all: but aboue all the rest, that with the white flower, called of Florence, is accounted of all to be the sweetest root, fit to be vsed to make sweete powders, &c. calling it by the name of *Orris* rootes.

Iris tuberosa. The Veluet Flowerdeluce.

Vnto the Family of Flowerdeluces, I must needes ioyne this peculiar kinde, because of the neare resemblance of the flower, although it differ both in roote and leafe; lest therefore it should haue no place, let it take vp a roome here in the end of the Flowerdeluces, with this description following. It hath many small and foure square leaues, two foote long and aboue sometimes, of a grayish greene colour, stiffe at the first, but afterwards growing to their full length, they are weak and bend downe to the ground: out of the middle, as it were of one of these leaues, breaketh out the stalke, a foot high and better, with some leaues thereon, at the toppe whereof, out of a huske riseth one flower, (I neuer saw more on a stalke) consisting of nine leaues, whereof the three that fall downe are of a yellowish greene colour round about the edges, and in the middle of so deepe a purple, that it seemeth to be blacke, resembling blacke Veluet: the three arched leaues, that couer the lower leaues to the halfe, are of the same greenish colour that the edges and backside of the lower leaues are: the three vppermost leaues, if they may be called leaues, or rather short peeces like eares, are green also, but wherein a glimpse of purple may be seene in them: after the flower is past, there followeth a round knob or whitish seede vessell, hanging downe by a small foote-stalke, from betweene the huske, which is diuided as it were into two leaues, wherein is contained round white seede. The roote is bunched or knobbed out into long round rootes, like vnto fingers, two or three from one peece, one distant from another, and one longer then another, for the most part of a darkish gray colour, and reddish withall on the outside, and somewhat yellowish within.

The Place.

It hath beene sent out of Turkie oftentimes (as growing naturally thereabouts) and not knowne to grow naturally any where else.

The Time.

It flowreth in Aprill or May, sometimes earlier or later, as the Spring falleth out to be milde or sharpe.

The Names.

Matthiolus contendeth to make it the true *Hermodactylus*, rather from the shew of the rootes, which (as is said) are like vnto fingers, then from any other good reason: for the rootes hereof eyther dry or greene, do nothing resemble the true *Hermodactyli* that are vsed in Physicke, as any that knoweth them may easily perceiue, either in forme or vertue. It is more truely referred to the Flowerdeluces, and because of the tuberous rootes, called *Iris tuberosa*, although all the Flowerdeluces in this Chapter haue tuberous

rootes.

rootes, yet this much differing from them all. In Englifh it is vfually called, The Veluet Flowerdeluce, becaufe the three falling leaues feeme to be like fmooth blacke Veluet.

The Vertues.

Both the rootes and the flowers of the great Flowerdeluces, are of great vfe for the purging and cleanfing of many inward, as well as outward difea-fes, as all Authors in Phyficke doe record. Some haue vfed alfo the greene rootes to cleanfe the skinne, but they had neede to be carefull that vfe them, left they take more harme then good by the vfe of them. The dryed rootes called *Orris* (as is faid) is of much vfe to make fweete powders, or other things to perfume apparrell or linnen. The iuice or decoction of the green roots doth procure both neezing to be fnuft vp into the noftrils, and vomi-ting very ftrongly being taken inwardly.

Chap. XXI.

Gladiolus. Corne Flagge.

NExt vnto the Flagges or Flowerdeluces, come the *Gladioli* or Corne Flagges to bee entreated of, for fome refemblance of the leaues with them. There are hereof diuers forts, fome bigger and fome leffer, but the chiefeft difference is in the colour of the flowers, and one in the order of the flowers. Of them all in their feuerall orders.

Gladiolus Narbonenfis. The French Corne Flagge.

The French Corne Flagge rifeth vp with three or foure broad, long, and ftiffe greene leaues, one as it were out of the fide of another, being ioyned together at the bottome, fomewhat like vnto the leaues of Flowerdeluces, but ftiffer, more full of ribbes, and longer then many of them, and fharper pointed : the ftalke rifeth vp from among the leaues, bearing them on it as it rifeth, hauing at the toppe diuers huskes, out of which come the flowers one aboue another, all of them turning and opening themfelues one way, which are long and gaping, like vnto the flowers of Foxegloue, a little arched or bunching vp in the middle, of a faire reddifh purple colour, with two white fpots within the mouth thereof, one on each fide, made like vnto a Lozenge that is fquare and long pointed : after the flowers are paft, come vp round heads or feede veffels, wherein is contained reddifh flat feede, like vnto the feede of the Fritil-laria, but thicker and fuller : the roote is fomewhat great, round, flat, and hard, with a fhew as if it were netted, hauing another fhort fpongie one vnder it, which when it hath done bearing, and the ftalke dry, that the roote may be taken vp, fticketh clofe to the bottome, but may be eafily taken away, hauing vfually a number of fmall rootes encreafed about it, the leaft whereof will quickly grow, fo that if it be fuffered any long time in a Garden, it will rather choake and pefter it, then be an ornament vnto it.

Gladiolus Italicus binis floribus ordinibus. The Italian Corne Flagge.

The Italian Corne Flagge is like vnto the French in roote, leafe, and flower, with-out any other difference, then that the roote is fmaller and browner, the leafe and ftalke of a darker colour, and the flowers (being of a little darker colour like the for-mer, and fomewhat fmaller) ftand out on both fides of the ftalke.

Gladiolus Byzantinus. Corne Flagge of Conftantinople.

This Corne Flagge that came firft from Conftantinople, is in all things like vnto the French Corne Flagge laft defcribed, but that it is larger, both in rootes, leaues, and

flowers,

flowers, and likewise that the Flowers of this, which stand not on both sides, are of a deeper red colour, and flower later, after all the rest are past : the roote hereof being netted as plainly as any of the former, is as plentifull also to giue encrease, but is more tender and lesse able to abide our sharpe cold Winters.

Gladiolus flore rubente. Blush Corne Flagge.

This blush kinde is like vnto the French Corne Flagge in all respects , sauing onely that the flowers are of a pale red colour, tending to whitenesse, which wee vsually call a blush colour.

Gladiolus flore albo. White Corne Flagge.

This white Corne Flagge also differeth not from the last, but onely that the rootes are whiter on the outside, the leaues are greener, without any brownnesse or darknesse as in the former, and the flowers are snow white.

Gladiolus purpureus minor. The small purple Corne Flagge.

This also differeth not from any of the former, but onely in the smalnesse both of leafe, stalke, and flowers, which stand all on the one side , like vnto the French kinde, and of the same colour : the roote of this kinde is netted more then any other.

The Place.

They grow in France and Italy, the least in Spaine, and the Byzantine, as it is thought, about Constantinople, being (as is said) first sent from thence. Iohn Tradescante assured mee, that hee saw many acres of ground in Barbary spread ouer with them.

The Time.

They all flower in Iune and Iuly, and the Byzantine latest , as is said before.

The Names.

It hath diuers names ; for the Latines call it *Gladiolus*, of the forme of a sword, which the leafe doth resemble. The Romanes *Segetalis*, because it groweth in the Corne fields. Some call it *Victorialis rotunda*, to put a difference between it, and the *longa* , which is a kinde of Garlicke. Plinie saith, that *Gladiolus* is *Cypirus*, but to decide that controuersie, and many others, belongeth to another discourse, this being intended only for pleasure. Gerrard mistaketh the French kinde for the Italian.

The Vertues.

The roote being bruised, and applyed with Frankinsense (and often of it selfe without it) in the manner of a pultis or plaister, is held of diuers to be singular good to draw out splinters , thornes , and broken bones out of the flesh. Some take it to be effectuall to stirre vp Venerie, but I somewhat doubt thereof : For Galen in his eighth Booke of Simples, giueth vnto it a drawing, digesting, and drying faculty.

CHAP.

1 *Gladiolus Narbonensis.* The French Corne Flagge.   2 *Gladiolus Italicus.* The Italian Corne Flagge.   3 *Gladiolus Byzantinus,* Corne Flagge of Constantinople.   4 *Palma Christi mas.* The great male handed Satyrion.   5 *Orchis Hermaphroditica candida,* The white Butterflie Orchis   6 *Orchis Melittias siue apifera,* The Bee flower or Bee Orchis.   7 *Dens Caninus flore purpurante,* Dogges tooth Violet with a pale purplish flower.   8 *Dens Caninus flore albo,* Dogges tooth Violet with a white flower.

CHAP. XXII.

Orchis siue Satyrium. Bee flowers.

ALthough it is not my purpose in this place, to giue a generall history of all the sorts of Orchides, Satyrions, and the rest of that kinde ; yet because many of them are very pleasant to behold, and, if they be planted in a conuenient place, will abide some time in Gardens, so that there is much pleasure taken in them : I shall intrude some of them for curiosities sake, to make vp the prospect of natures beautifull variety, and only entreate of a few, leauing the rest to a more ample declaration.

1. *Satyrium Basilicum siue Palma Christi mas.*
The greater male handed Satyrion.

This handed Satyrion hath for the most part but three faire large greene leaues, neare vnto the ground, spotted with small blackish markes : from among which riseth vp a stalke, with some smaller leaues thereon, bearing at the toppe a bush or spike of flowers, thicke set together, euery one whereof is made like a body, with the belly broader belowe then aboue, where it hath small peeces adioyned vnto it : the flower is of a faire purple colour, spotted with deeper purple spots, and hauing small peeces like hornes hanging at the backes of the flowers, and a small leafe at the bottome of the foote-stalke of euery flower : the rootes are not round, like the other Orchides, but somewhat long and flat, like a hand, with small diuisions belowe, hanging downe like the fingers of a hand, cut short off by the knockles, two always growing together, with some small fibres or strings aboue the heads of these rootes, at the bottome of the stalke.

2. *Satyrium Basilicum siue Palma Christi femina.*
The female handed Satyrion.

This female Satyrion hath longer and narrower leaues then the former, and spotted with more and greater spots, compassing the stalke at the bottome like the other : this beareth likewise a bush of flowers, like vnto the other, but that each of these haue heads like hoods, whereas the former haue none : in some they are white with purple spots, and in others of a reddish purple, with deep or darke coloured spots: the roots are alike.

3. *Orchis Hermaphroditica candida.* The white Butterflie Orchis.

The rootes of this kinde take part with both the sorts of *Orchis* and *Satyrium*, being neither altogether round, nor fully handed, and thereupon it tooke the name, to signifie both kindes : the leaues are two in number, seldome more, being faire and broad, like vnto the leaues of Lillies, without any spot at all in them : at the toppe of the stalke stand many white flowers, not so thicke set as the first or second, euery one being fashioned like vnto a white Butterflie, with the wings spread abroad.

4. *Orchis Melitias siue apifera.* The Bee flower or Bee Orchis.

This is a small and lowe plant for the most part, with three or foure small narrow leaues at the bottome : the stalke is seldome aboue halfe a foote high, with foure or fiue flowers thereon one aboue another, hauing round bodies, and somewhat flat, of a kind of yellowish colour, with purple wings aboue them, so like vnto an honey Bee, that it might soone deceiue one that neuer had seene such a flower before : the roots are two together, round and white, hauing a certaine *muccilaginesse* or clamminesse within them, without any taste almost at all, as all or the most part of these kindes haue.

5. *Orchis Sphegodes.* Gnats Satyrion.

The leaues of this Orchis are somewhat larger then of the Bee flower, the stalke also
somewha:

somewhat higher : the flowers are fewer on the toppe, but somewhat larger then of the Bee flowers, made to the resemblance of a Gnat or great long Flie : the rootes are two round bulbes, as the other are.

6. *Orchis Myodes.* Flie Orchis.

The Flie Orchis is like vnto the last described, both in leafe and roote, the difference is in the flower, which is neither so long as the Gnat Satyrion, nor so great as the Bee Orchis, but the neather part of the Flie is blacke, with a list of ash-colour crossing the backe, with a shew of legges hanging at it : the naturall Flie seemeth so to bee in loue with it, that you shall seldome come in the heate of the day, but you shall finde one sitting close thereon.

The Place.

These grow in many places of England, some in the Woods, as the Butterflie, and the two former handed Satyrions : others on dry bankes and barren balkes in Kent, and many other places.

The Time.

They flower for the most part in the beginning or middle of May, or thereabouts.

The Names.

Their seuerall names are expressed in their titles, so much as may suffice for this discourse.

The Vertues.

All the kindes of Orchis are accounted to procure bodily lust, as well the flowers distilled, as the rootes prepared.

The rootes boyled in red Wine, and afterwards dryed, are held to bee a singular good remedie against the bloody Flixe.

CHAP. XXIII.

Dens Caninus. Dogs tooth Violet.

VNto the kindes of Orchides, may fitly be ioyned another plant, which by many is reckoned to be a *Satyrium*, both from the forme of roote and leafe, and from the efficacy or vertue correspondent thereunto. And although it cannot be the *Satyrium Erythronium* of Dioscorides, as some would entitle it, for that as I haue shewed before, his *Satyrium tryphillum* is the Tulipa without all doubt ; yet because it differeth very notably, and carrieth more beauty and respect in his flower then they, I shall entreate thereof in a Chapter by it selfe, and set it next vnto them.

Dens Caninus flore albo. Dogs tooth Violet with a white flower.

The white Dogs tooth hath for his roote a white bulbe, long and small, yet vsually greater then either of the other that follow, bigger belowe then aboue, with a small peece adioyning to the bottome of it, from whence rise vp in the beginning of the Spring, after the Winter frosts are past, two leaues for the most part (when it will flower, or else but one, and neuer three together that euer I saw) closed together when they first come vp out of the ground, which inclose the flower betweene them : the leaues when they are opened do lay themselues flat on the ground, or not much aboue it, one opposite vnto the other, with the stalke and the flower on it standing betweene them, which leaues are of a whitish greene colour, long and narrow, yet broader in the

R middle

middle then at both ends, growing leſſe by degrees each way, ſpotted and ſtriped all ouer the leaues with white lines and ſpots : the ſtalke riſeth vp halfe a foote high or more, bearing at the toppe one flower and no more, hanging downe the head, larger then any of the other of this kinde that follow, made or conſiſting of ſix white long and narrow leaues, turning themſelues vp againe, after it hath felt the comfort of the Sunne, that they doe almoſt touch the ſtalke againe, very like vnto the flowers of *Cyclamen* or Sowebread : it hath in the middle of the flower ſix white chiues, tipt with darke purple pendents, and a white three forked ſtile in the middle of them: the flower hath no ſent at all, but commendable onely for the beauty and forme thereof: after the flower is paſt, commeth in the place a round head ſeeming three ſquare, containing therein ſmall and yellowiſh ſeede.

Dens Caninus flore purpuraſcente. Dogs tooth with a pale purple flower.

This other Dogs tooth is like vnto the former, but leſſer in all parts, the leafe whereof is not ſo long, but broad and ſhort, ſpotted with darker lines and ſpots : the flower is like the other, but ſmaller, and of a delayed purple colour, very pale ſometimes, and ſometimes a little deeper, turning it ſelfe as the other, with a circle round about the vmbone or middle, the chiues hereof are not white, but declining to purple: the roote is white, and like vnto the former, but leſſer, as is ſaid before.

Dens Caninus flore rubro. Dogs tooth with a red flower.

This is in all things like vnto the laſt, both for forme and bigneſſe of flower and leafe : the chiefe difference conſiſteth in this, that the leaues hereof are of a yellowiſh mealy greene colour, ſpotted and ſtreaked with redder ſpots and ſtripes, and the flower of a deeper reddiſh purple colour, and the chiues alſo more purpliſh then the laſt, in all other things it is alike.

The Place.

The ſorts of *Dens Caninus* doe growe in diuers places; ſome in Italy on the Euganean Hils, others on the Apenine, and ſome about Gratz, the chiefe Citie of Stiria, and alſo about Bayonne, and in other places.

The Time.

They flower in March moſt vſually, and many times in Aprill, according to the ſeaſonableneſſe of the yeare.

The Names.

Cluſius did call it firſt *Dentali*, and Lobel, and from him ſome others *Satyrium*, and *Erytbronium*, but I haue ſaid enough hereof in the beginning of the Chapter. It is moſt commonly called *Dens Caninus*, and we in Engliſh, either Dogs tooth, or Dogs tooth Violet. Geſner called it *Hermodactylus*, and Matthiolus *Pſeudohermodactylus*.

The Vertues.

The roote hereof is held to bee of more efficacy for venereous effects, then any of the Orchides and Satyrions.

They of Stiria vſe the rootes for the falling ſickneſſe.

Wee haue had from Virginia a roote ſent vnto vs, that wee might well Iudge, by the forme and colour thereof being dry, to be either the roote of this, or of an Orchis, which the naturall people hold not onely to be ſingular to procure luſt, but hold it as a ſecret, loth to reueale it.

CHAP.

CHAP. XXIIII.

Cyclamen. Sowebread.

THe likenesse of the flowers, and the spotting of the leaues of the *Dens Caninus;* with these of the *Cyclamen* or Sowebread, maketh mee ioyne it next there-unto : as also that after the bulbous rooted plants I might begin with the tube-rous that remaine, and make this plant the beginning of them. Of this kinde there are diuers sorts, differing both in forme of leaues and time of flowring : for some doe flower in the Spring of the yeare, others afterwards in the beginning of Summer : but the most number in the end of Summer, or beginning of Autumne or Harue st, whereof some haue round leaues, others cornered like vnto Iuie, longer or shorter, greater or smaller. Of them all in order, and first of those that come in the Spring.

1. *Cyclamen Vernum flore purpureo.* Purple flowred Sowebread of the Spring.

This Sowebread hath a smaller roote then most of the others, yet round and blac-kish on the outside, as all or most of the rest are (I speake of them that I haue seene ; for Clusius and others doe report to haue had very great ones) from whence rise vp di-uers round, yet pointed leaues, and somewhat cornered withall, greene aboue, and spotted with white spots circlewise about the leafe, and reddish vnderneath, which at their first comming vp are folded together; among which come the flowers, of a red-dish purple colour and very sweete, euery one vpon a small, long, and slender reddish foote-stalke, which hanging downe their heads, turne vp their leaues againe : after the flowers are past, the head or seede vessell shrinketh downe, winding his footestalke, and coyling it selfe like a cable, which when it toucheth the ground, there abideth hid among the leaues, till it be growne great and ripe, wherein are contained a few small round seedes, which being presently sowne, will growe first into round rootes, and af-terwards from them shoote forth leaues.

2. *Cyclamen Vernum flore albo.* White flowred Sowebread of the Spring.

The white flowring Sowebread hath his leaues like the former, but not fully so much cornered, bearing small snow white flowers, as sweete as the other : and herein consisteth the chiefest difference, in all other things it is alike.

3. *Cyclamen Vernum Creticum flore albo.* White Candy Sowebread of the Spring.

This Sowebread is somewhat like the former white kinde, but that the leaues grow much larger and longer, with more corners at the edges, and more eminent spots on them : the flowers also somewhat longer and larger, and herein consisteth the whole difference.

4. *Cyclamen Æstivum.* Summer Sowebread.

Summer Sowebread hath round leaues like vnto the Romane Sowebread, but some-what cornered, yet with shorter corners then the Iuie leafed Sowebread, full of white spots on the vpperside of the leaues, and very purple vnderneath, sometimes they haue fewer spots, and little or no purple vnderneath : the flowers hereof are as small, as purple, and as sweete, as the purple Sowebread of the Spring time : the roote hereof is likewise small, blacke, and round.

5. *Cyclamen Romanum rotundifolium.* Romane Sowebread with round leaues.

The Romane Sowebread hath round leaues, somewhat like vnto the common Sowebread, but not fully so round pointed at the ends, a little cornered sometimes also, or as it were indented, with white spots round about the middle of the leaues,

and

and very conspicuous, which make it seeme the more beautifull : the flowers appeare in Autumne, and are shorter , and of a deeper purplish red colour then the Iuie Sowebread, rising vp before the leaues for the most part, or at least with them, and little or nothing sweete : the roote is round and blacke , vsually not so flat as it, but growing sometimes to bee greater then any other kinde of Sowebread. There is sometimes some variety to be seene , both in the leaues and flowers of this kinde ; for that sometime the leaues haue more corners , and either more or lesse spotted with white : the flowers likewise of some are larger or lesser, longer or rounder, paler or deeper coloured one then another. This happeneth most likely from the sowing of the seede, causing the like variety as is seene in the Iuie leafed Sowebread. It doth also many times happen from the diuersity of soyles and countries where they grow : the seed of this, as of all the rest, is small and round, contained in such like heads as the former , standing almost like the head of a Snake that is twined or folded within the body thereof. This and the other Autumnall kindes, presently after their sowing in Autumne, shoote forth leaues, and so abide all the Winter, according to their kinde.

Varietas.

6. *Cyclamen folio hederæ autumnale.* Iuie leafed Sowebread.

The Iuie leafed Sowebread groweth in the same manner that the former doth, that is, bringeth forth flowers with the leaues sometimes, or most commonly before them, whose flowers are greater then the common round leafed Sowebread, somewhat longer then the former Romane or Italian Sowebreads , and of a paler purple colour, almost blush, without that sweete sent as is in the first kinde of the Spring : the greene leaues hereof are more long then round, pointed at the ends , and hauing also one or two corners on each side, sometimes much spotted on the vpperside with white spots and marks,and sometimes but a little or not at all ; and so likewise sometimes more or lesse purple vnderneath : all the leaues and flowers doe stand vsually euery one seuerally by themselues , vpon their owne slender foote stalkes , as most of all the other kindes doe: but sometimes it happeneth, that both leaues and flowers are found growing from one and the same stalke, which I rather take to be accidentall, then naturall so to continue : the seede hereof is like the former kindes, which being sowne produceth variety, both in the forme of the leaues, and colour and smell of the flowers: some being paler or deeper, and some more or lesse sweete then others : the leaues also, some more or lesse cornered then others: the root groweth to be great, being round and flat, and of a blackish browne colour on the outside.

Varietas.

7. *Cyclamen autumnale hederæfolio flore albo.* Iuie leafed Sowebread with white flowers.

There is one of this kinde, whose leaues are rounder , and not so much cornered as the former , flowring in Autumne as the last doth , and whose flowers are wholly white, not hauing any other notable difference therein.

8. *Cyclamen autumnale angustifolium.* Long leafed Sowebread.

This kinde of Sowebread may easily be knowne from all the other kindes, becauſe his leafe is longer and narrower then others , faſhioned at the bottome thereof with points , ſomewhat like vnto *Arum* or Wake Robin leaues : the flowers are like the former ſorts for forme, but of a purple colour. There is alſo another of this kinde in all things like the former, but that the flowers are white.

9. *Cyclamen Antiochenum Autumnale flore purpureo duplici.* Double flowred Sowebread of Antioch.

This Sowebread of Antioch with double flowers, hath his leaues ſomewhat round, like vnto the leaues of the Summer Sowbread , but with leſſe notches or corners,& full of white ſpots on them: it beareth flowers on ſtalks, like vnto others,& likewiſe ſome ſtalks that haue two or three flowers on them,which are very large,with ten or twelue
leaues

1 *Cyclamen Vernum flore purpureo* Purple flowred Sowebread of the Spring. 2 *Cyclamen æstivum.* Summer Sowebread. 3 *Folium Cyclaminis Creticæ væ-uelæ flore candido.* A leafe of Candie Sowebread. 4 *Cyclamen Romanum Autumnale.* Romane Sowebread of the Autumne. 5 *Cyclamen hederæ folio Autumnale.* Iui-leafed Autumne Sowebread. 6 *Folium Cyclaminis Autumnalis flore albo.* A leafe of the Autumne Sowebread with a white flowr. 7 *Folium Cyclaminis angustifolij Autumnalis.* A leafe of the long leafed Sowebread. 8 *Cyclamen Antiochenum Autumnale, sive simplex purpureo duplici.* The double flowred Sowebread of Antioch. 9 *Cyclamen vulgare folio rotundo.* The common round leafed Sowebread.

R 3

leaues a peece, of a faire Peach colour, like vnto the flowers of purple Sowebread of the Spring, and deeper at the bottome.

There are of this kinde some, whose flowers appeare in the Spring, and are as large and double as the former, but of a pure white colour.

There are of these Sowebreads of Antioch, that haue but single flowers, some appearing in the Spring, and others in Autumne.

10. *Cyclamen vulgare folio rotundo.* The common Sowebread.

The common Sowebread (which is most vsed in the Apothecaries Shops) hath many leaues spread vpon the ground, rising from certaine small long heads, that are on the greater round rootes, as vsually most of the former sorts doe, being in the like manner folded together, and after spread themselues into round greene leaues, somewhat like vnto the leaues of *Asarum*, but not shining, without any white spots on the vpperside for the most part, or but very seldome, and reddish or purplish vnderneath, and very seldome greener: the flowers stand vpon small foot stalkes, and shew themselues open for the most part, before any leaues doe appeare, being smaller and shorter then those with Iuie leaues, and of a pale purple colour, yet sometimes deeper, hanging downe their heads, and turning vp their leaues againe, as all others doe, but more sweete then many other of the Autumne flowers: after the flowers are past, come the heads turning or winding themselues downe in like manner as the other do, hauing such like seede, but somewhat larger, and more vneuen, or not so round at the least: the roote is round, and not flat, of a browner colour, and not so blacke on the outside as many of the others.

The Place.

The Sowebreads of the Spring doe both grow on the Pyrenæan Mountaines in Italy, and in Candy, and about Mompelier in France; Antioch in Syria also hath yeelded some both of the Spring and Autumne. Those with round and Iuie leaues grow in diuers places both of France and Italy: and the common in Germany, and the Lowe-Countries. But that Autumne Sowebread with white flowers, is reported to grow in the Kingdome of Naples. I haue very curiously enquired of many, if euer they found them in any parts of England, neare or farther off from the places where they dwell: but they haue all affirmed, that they neuer found, or euer heard of any that haue found of any of them. This onely they haue assured, that there groweth none in the places, where some haue reported them to grow.

The Time.

Those of the Spring doe flower about the end of Aprill, or beginning of May. The other of the Summer, about the end of Iune or in Iuly. The rest some in August, and September, others in October.

The Names.

The Common Sowebread is called by most Writers in Latine, *Panis Porcinus*, and by that name it is knowne in the Apothecaries shops, as also by the name *Arthanita*, according to which name, they haue an ointment so called, which is to be made with the iuice hereof. It is also called by diuers other names, not pertinent for this discourse. The most vsuall name, whereby it is knowne to most Herbarists, is *Cyclamen* (which is the Greeke word) or as some call it *Cyclaminus*, adding thereunto their other seuerall titles. In English, Sowebread.

The Vertues.

The leaues and rootes are very effectuall for the spleene, as the Ointment before remembred plainly proueth, being vsed for the same purpose,

and

and that to good effect. It is vsed alfo for women in long and hard trauels, where there is danger, to accelerate the birth, either the roote or the leafe being applyed. But for any amorous effects, I hold it meere fabulous.

Chap. XXV.

Anemone. Windeflower and his kindes.

THe next tuberous rooted plants that are to follow (of right in my opinion) are the *Anemones* or Windeflowers, and although fome tuberous rooted plants, that is, the Afphodils, Spiderworts, and Flowerdeluces haue beene before inferted, it was, both becaufe they were in name or forme of flowers futable to them whom they were ioyned vnto, and alfo that they fhould not be feuered and entreated of in two feuerall places: the reft are now to follow, at the leaft fo many of them as be beautifull flowers, fit to furnifh a Florifts Garden, for natures delightfome varieties and excellencies. To diftinguifh the Family of *Anemones* I may, that is, into the wilde kindes, and into the tame or mannured, as they are called, and both of them nourfed vp in Gardens; and of them into thofe that haue broader leaues, and into thofe that haue thinner or more iagged leaues: and of each of them, into thofe that beare fingle flowers, and thofe that beare double flowers. But to defcribe the infinite (as I may fo fay) variety of the colours of the flowers, and to giue to each his true diftinction and denomination, *Hic labor, hoc opus eft,* it farre paffeth my ability I confeffe, and I thinke would grauell the beft experienced this day in Europe (and the like I faid concerning Tulipas, it being as contingent to this plant, as is before faid of the Tulipa, to be without end in yeelding varieties:) for who can fee all the varieties that haue fprung from the fowing of the feede in all places, feeing the variety of colours rifen from thence, is according to the variety of ayres & grounds wherein they are fowne, skill alfo helping nature in ordering them aright. For the feede of one and the fame plant fowne in diuers ayres and grounds, doe produce that variety of colours that is much differing one from another; who then can difplay all the mixtures of colours in them, to fet them downe in fo fmall a roome as this Book? Yet as I haue done (in the former part of this Treatife) my good will, to expreffe as many of each kinde haue come to my knowledge, fo if I endeauour the like in this, I hope the courteous wil accept it, and hold me excufed for the reft: otherwife, if I were or could be abfolute, I fhould take from my felf and others the hope of future augmentation, or addition of any new, which neuer will be wanting. To begin therefore with the wilde kinds (as they are fo accounted) I fhall firft entreate of the *Pulfatillas* or Pafque flowers, which are certainly kindes of wilde *Anemones,* both in leafe and flower, as may well be difcerned by them that are iudicious (although fome learned men haue not fo thought, as appeareth by their writings) the rootes of them making one fpeciall note of difference, from the other forts of wilde *Anemones.*

1. *Pulfatilla Anglica purpurea.* The purple Pafque flower.

The Pafque or Paffe flower which is of our owne Country, hath many leaues lying on the ground, fomewhat rough or hairie, hard in feeling, and finely cut into many fmall leaues, of a darke greene colour, almoft like the leaues of Carrets, but finer and fmaller, from among which rife vp naked ftalkes, rough or hairie alfo, fet about the middle thereof with fome fmall diuided leaues compaffing them, and rifing aboue thefe leaues about a fpanne, bearing euery one of them one pendulous flower, made of fix leaues, of a fine Violet purple colour, but fomewhat deepe withall, in the middle whereof ftand many yellow threeds, fet about a middle purple pointell: after the flower is paft, there commeth vp in the ftead thereof a bufhie head of long feedes, which are fmall and hoarie, hauing at the end of euery one a fmall haire, which is gray likewife: the roote is fmall and long, growing downewards into the ground, with a tuft of haire at the head thereof, and not lying or running vnder the vpper cruft thereof, as the other wilde *Anemones* doe.

2. Pulfa-

2. *Pulsatilla Danica.* The Passe flower of Denmarke.

There is another that was brought out of Denmarke, very like vnto the former, but that it is larger both in roote and leafe, and flower also, which is of a fairer purple colour, not so deepe, and besides, will better abide to bee mannured then our English kinde will, as my selfe haue often proued.

Vtriusque flore albo & flore duplici.

Of both these sorts it is said, that some plants haue bin found, that haue borne white flowers. And likewise one that bore double flowers, that is, with two rowes of leaues.

3. *Pulsatilla flore rubro.* The red Passe flower.

Lobel, as I take it, did first set forth this kinde, being brought him from Syria, the leaues whereof are finer cut, the flower smaller, and with longer leaues, and of a red colour.

4. *Pulsatilla flore luteo.* The yellow Passe flower.

The yellow Passe flower hath his leaues cut and diuided, very like vnto the leaues of the first kinde, but somewhat more hairie, greene on the vpperside, and hairie vnderneath: the stalke is round and hoary, the middle whereof is beset with some small leaues, as in the other, from among which riseth vp the stalke of the flower, consisting of six leaues of a very faire yellow colour on the inside, and of a hoary pale yellow on the outside; after which followeth such an head of hairie thrummes as in the former: the roote is of the bignesse of a mans finger.

5. *Pulsatilla flore albo.* The white Passe flower.

The white Passe flower (which Clusius maketh a kinde of *Anemone*, and yet as hee saith himselfe, doth more nearely resemble the *Pulsatilla*) hath, from amongst a tuft or head of haires, which grow at the toppe of a long blacke roote, many leaues standing vpon long stalkes, which are diuided as it were into three wings or parts, and each part finely cut and diuided, like vnto the Passe flower of Denmarke, but somewhat harder in handling, greenish on the vpperside, and somewhat gray vnderneath, and very hairie all ouer: among these leaues rise vp the stalkes, beset at the middle of them with three leaues, as finely cut and diuided as those belowe, from aboue which standeth the flower, being smaller, and not so pendulous as the former, but in the like manner consisting of six leaues, of a snow white colour on the inside, and a little browner on the outside, with many yellow thrums in the middle: after the flower is past, riseth vp such a like hoary head, composed as it were of many haires, each whereof hath a small seede fastened vnto it, like as the former Passe flowers haue.

The Place.

The first is found in many places of England, vpon dry bankes that lye open to the Sunne.

The second was first brought, as I take it, by Doctor Lobel from Denmarke, & is one of the two kinds, that Clusius saith are common in Germanie, this bearing a paler purple flower, and more early then the other, which is the same with our English, whose flower is so darke, that it almost seemeth blacke.

The red kinde, as Lobel saith, came from Syria.

The yellow Passe flower, which Clusius maketh his third wilde *Anemone*, was found very plentifully growing at the foote of St. Bernards Hill, neare vnto the Cantons of the Switzers.

The white one groweth on the Alpes neare Austria, in France likewise, and other places.

The

1 *Pulſatilla purpurea cum folio, ſemine, & radice.* The purple Paſque flower with leafe, feed, and root. 2 *Pulſatilla lutea flore.* The yellow Paſque flower. 3 *Pulſatilla rubra Syriaca Lobelij.* Red Paſque flower of Lobel. 4 *Pulſatilla rubra Swertij.* Swertz his red Paſque flower. 5 *Pulſatilla flore albo.* White Paſque flower. 6 *Anemone ſilueſtris alba Matthioli.* The wilde white broad leafed Windflower. 7 *Anemone ſilueſtris tenuifolia alba.* The wilde ſingle white Windflower. 8 *Anemone ſilueſtris tenuifolia lutea.* The yellow wilde thin leafed Windflower. 9 *Anemone ſilueſtris trifolia Dodonei.* The three leafed wilde Windflower. 10 *Anemone ſilueſtris flore pleno albo.* The double white wilde Windflower. 11 *Anemone ſilueſtris flore pleno purpureo.* The double purple wilde Windflower. * *Semen ſeparatim divulſum.* The feed ſeparated. † *Radix cum folio inferiore.* The roote with a lower leafe.

The Time.

All of them doe flower early in the yeare, that is, in the beginning of Aprill, about which time most commonly Easter doth fall.

The Names.

Their proper names are giuen to each in their seuerall titles, being all of them kindes of wilde *Anemones*, as I said in the beginning of the Chapter, and so for the most part all Authors doe acknowledge them. We call them in English, because they flower about Easter, Pasque Flower, which is the French name for Easter, or *Euphoniæ gratia,* Passe Flower, which may passe currant, without any further descant on the name, or else *Pulsatilla,* if you will, being growne old by custome.

The Vertues.

The sharpe biting and exulcerating quality of this plant, causeth it to be of little vse, notwithstanding Ioachimus Camerarius saith in his *Hortus Medicus,* that in Borussia, which is a place in Italy, as I take it, the distilled water hereof is vsed with good successe, to be giuen to them that are troubled with a Tertian Ague; for he saith that it is *medicamentum ἐνεργητικὸν,* that is, a medicine of force to helpe obstructions.

Anemone siluestris latifolia alba siue tertia Matthioli.
The white wilde broad leafed Windflower.

This Windflower hath diuers broad greene leaues, cut into diuisions, and dented about, very like vnto a broad leafed Crowfoote, among which riseth vp a stalke, hauing some such like cut leaues in the middle thereof, as growe below, but smaller; on the toppe whereof standeth one large white flower, consisting of fiue leaues for the most part, with some yellow threads in the middle, standing about such a greene head as is in the tame or garden *Anemones,* which growing greater after the flower is past, is composed of many small seedes, wrapped in white wooll, which as soone as they are ripe, raise themselues vp from the bottome of the head, and flye away with the winde, as the other tame or garden kindes doe: the roote is made of a number of long blacke strings, encreasing very much by running vnder ground, and shooting vp in diuers places.

Anemone siluestris tenuifolia lutea. The yellow wilde thin leafed Windflower.

The yellow wilde *Anemone* riseth vp with one or two small round naked stalkes, bearing about the middle of them, small, soft, and tender iagged leaues, deeply cut in and indented on the edges about, from aboue which doth grow the stalke, bearing small yellow flowers, standing vpon weake foote-stalkes, like vnto a small Crowfoot, with some threads in the middle: the roote is long and small, somewhat like vnto the roote of Pollipodie, creeping vnder the vpper crust of the earth: this kinde is lower, and springeth somewhat earlier then the other wilde kindes that follow.

Anemone siluestris tenuifolia alba simplex.
The single white thin leafed wilde Windflower.

This white wilde *Anemone* riseth vp with diuers leaues vpon seuerall long stalkes; which are somewhat like vnto the former, but that they are somewhat harder, and not so long, nor the diuisions of the leaues so finely snipt about the edges, but a little broader, and deeper cut in on euery side: the flowers hereof are larger and broader then the former, white on the inside, and a little purplish on the outside, especially at
the

the bottome of the flower next vnto the ftalke : the roote of this is very like vnto the laft.

There is another of this kinde, whofe flowers are purple, in all other things it is like *Purpurea.* vnto the white.

And likewife another, with a blufh or carnation coloured flower. *Coccinea fiue luteo rubens.*

There is one that is onely nurfed vp with vs in Gardens, that is fomewhat like vnto thefe former wilde *Anemones* in roote and leafe, but that the flower of this, being pure white within, and a little purplifh without, confifting of eight or nine fmall round *Peregrina alba.* pointed leaues, hath fometimes fome leaues vnder the flower, party coloured white and greene : the flower hath likewife a greene head, like a Strawberry, compaffed about with white threads, tipt with yellow pendents.

And another of the fame kinde with the laft, whofe flower confifting of eight or *Peregrina viridis.* nine leaues, is of a greenifh colour, except the foure outermoft leaues, which are a little purplifh, and diuided at the points into three parts ; the middle part is of a greenifh white colour, with a greene head in the middle as the other.

Anemone filueftris trifolia Dodonai. The three leafed wilde Windflower.

This wilde *Anemone* hath his rootes very like vnto the former kindes ; the leaues are alwaies three fet together at the toppe of flender ftalkes, being fmall and indented about, very like vnto a three leafed Graffe, but fmaller : the flower confifteth of eight fmall leaues, fomewhat like vnto a Crowfoote, but of a whitifh purple or blufh colour, with fome white threads, and a greene rough head in the middle.

Anemone filueftris flore pleno albo. The double white wilde Windflower.

This double kinde is very like vnto the fingle white kinde before defcribed, both in his long running rootes, and thin leaues, but fomewhat larger : the flowers hereof are very thicke and double, although they be fmall, and of a faint fweete fent, very white after it is full blowne for fiue or fix dayes, but afterwards it becommeth a little purplifh on the infide, but more on the outfide : this neuer giueth feede (although it haue a fmall head in the middle) like as many other double flowers doe.

Anemone filueftris flore pleno purpureo. The double purple wilde Windflower.

This double purple kinde hath fuch like iagged leaues as the laft defcribed hath, but more hoarie vnderneath : the flower is of a fine light purple toward the points of the leaues, the bottomes being of a deeper purple, but as thicke, and full of leaues as the former, with a greene head in the middle, like vnto the former : this kinde hath fmall greene leaues on the ftalkes vnder the flowers, cut and diuided like the lower leaues.

The Place.

The firft broad leafed *Anemone* groweth in diuers places of Auftria and Hungary. The yellow in diuers woods in Germany, but not in this Country that euer I could learne. The other fingle wilde kindes, fome of them are very frequent throughout the moft places of England, in Woods, Groues, and Orchards. The double kindes were found, as Clufius faith, in the Lowe-Countries, in a Wood neare Louaine.

The Time.

They flower from the end of March (that is the earlieft) and the beginning of Aprill, vntill May, and the double kindes begin within a while after the fingle kinds are paft.

The Names.

They are called *Ranunculi filuarum,* and *Ranunculi nemorum*, and as Clufius

sius would haue them, *Leimonia* of Theophraſtus; they are generally called of moſt Herbariſts *Anemones ſilueſtres*, Wilde *Anemones* or Windſflowers. The Italians call them *Gengeuo ſalnatico*, that is, Wilde Ginger, becauſe the rootes are, beſides the forme, being ſomewhat like ſmall Ginger, of a biting hot and ſharpe taſte.

Anemone Luſitanica ſiue hortenſis latifolia flore ſimplici luteo. The ſingle Garden yellow Windflower or Anemone.

This ſingle yellow Anemone or Windflower hath diuers broad round leaues, ſomewhat diuided and endented withall on the edges, browniſh at the firſt riſing vp out of the ground, and almoſt folded together, and after of a ſad greene on the vpperſide, and reddiſh vnderneath; among which riſe vp ſmall ſlender ſtalkes, beſet at the middle of them with two or three leaues, more cut and diuided then thoſe belowe, with ſmall yellow flowers at the toppe of them, conſiſting of ten or twelue leaues a peece, hauing a few yellow threads in the middle of them, ſtanding about a ſmall greene head, which in time growing ripe hath ſmall flat ſeede, incloſed within a ſoft wooll or downe, which is eaſily blowne away with the winde : the roote groweth downeward into the ground, diuerſly ſpread with branches here and there, of a browniſh yellow on the outſide, and whitiſh within, ſo brittle, that it can hardly bee touched without breaking.

Anemone latifolia flore luteo duplici. The double yellow Anemone or Windflower.

This double yellow Anemone hath ſuch broad round leaues as the ſingle kinde hath, but ſomewhat larger or ranker : the ſtalkes are beſet with larger leaues, more deeply cut in on the edges : the flowers are of a more pale yellow, with ſome purpliſh veines on the outſide, and a little round pointed; but they are all on the inſide of a faire yellow colour, conſiſting of two rowes of leaues, whereof the innermoſt is the narrower, with a ſmall greene head in the middle, compaſſed with yellow threads as in the former : the roote is like the roote of the ſingle; neither of theſe haue any good ſent, and this ſpringeth vp and flowreth later then the ſingle kinde.

Anemone latifolia purpurea ſtellata ſiue papaueracea. The purple Starre Anemone or Windflower.

The firſt leaues of this purple Anemone, which alwayes ſpring vp before Winter, (if the roote be not kept too long out of the ground,) are ſomewhat like the leaues of *Sanicle* or Selfe-heale, but the reſt that follow are more deeply cut in and iagged; among which riſe vp diuers round ſtalkes, beſet with iagged leaues as all other Anemones are, aboue which leaues, the ſtalkes riſing two or three inches high, beare one flower a peece, compoſed of twelue leaues or more, narrow and pointed, of a bleake purple or whitiſh aſh-colour, ſomewhat ſhining on the outſide, and of a fine purple colour tending to a murrey on the inſide, with many blackiſh blew threads or thrummes in the middle of the flower, ſet about a head, whereon groweth the ſeede, which is ſmall and blacke, incloſed in ſoft wooll or downe, which flieth away with the winde, carrying the ſeede with it, if it be not carefully gathered : the roote is blackiſh on the outſide, and white within, tuberous or knobby, with many fibres growing at it.

Anemone purpurea Stellata altera. Another purple Starre Anemone.

There is ſo great diuerſity in the colours of the flowers of theſe broad leafed kinds of Anemones or Windflowers, that they can very hardly be expreſſed, although in their leaues there is but little or no difference. I ſhall not neede therefore to make ſeuerall deſcriptions of euery one that ſhall be ſet downe, but it will be ſufficient, I thinke, to giue you the diſtinctions of the flowers : for as I ſaid, therein is the greateſt and chiefeſt difference. This other Starre Anemone differeth not from the former in leafe or flower, but onely that this is of a more pale ſullen colour on the outſide, and of a paler purple colour on the inſide.

　　　　　　　　　　　　　　　　　　　　　　　　　　　　　　　　There

Viola purpurea There is another, whose flower hath eight leaues, as many of them that follow haue (although diuers sorts haue but six leaues in a flower) and is of a Violet purple, and therefore is called, The Violet purple Anemone.

Varietas. Of all these three sorts last described, there be other that differ only in hauing white bottomes, some smaller and some larger.

Purpurea stri-ata. There is also another of the same Violet purple colour with the former, but a little paler, tending more to rednesse, whose flowers haue many white lines and stripes through the leaues, and is called, The purple stript Anemone.

Carnea viua-cissima simplex. There is another, whose greene leaues are somewhat larger, and so is the flower likewise, consisting of eight leaues, and sometimes of more, of the colour of Carnation silke, sometimes pale and sometimes deeper, with a whitish circle about the bottome of the leaues, which circle in some is larger, and more to be seene then in others, when the flower layeth it selfe open with the heate of the Sunne, hauing blewish threads in the middle. This may be called, the Carnation Anemone.

Persciuiolacea. We haue another, whose flower is betweene a Peach colour and a Violet, which is vsually called a Gredeline colour.

Cochenilla. And another of a fine reddish Violet or purple, which we call, The Cochenille Anemone.

Cardinalis. And another of a rich crimson red colour, and may be called, The Cardinall Anemone.

Sanguinea. Another of a deeper, but not so liuely a red, called, The bloud red Anemone.

Cramesina. Another of an ordinary crimson colour, called, The crimson Anemone.

Coccinea. Another of a Stamell colour, neere vnto a Scarlet.

Incarnata. Another of a fine delayed red or flesh colour, and may bee called, The Incarnadine Anemone.

Incarnata His-panica. Another whose flower is of a liuely flesh colour, shadowed with yellow, and may be called, The Spanish Incarnate Anemone.

Rubescens. Another of a faire whitish red, which we call, The Blush Anemone.

Moschutella. Another whose flower consisteth of eight leaues, of a darke whitish colour, stript all ouer with veines of a fine blush colour, the bottomes being white, this may be called, The Nutmegge Anemone.

Enfumata. Another whose flower is of a pale whitish colour, tending to a gray, such as the Monkes and Friers were wont to weare with vs, and is called, A Monkes gray.

Pauo maior simplici flore. There is another, whose leafe is somewhat broader then many or most of the Anemones, comming neere vnto the leafe of the great double Orenge coloured Anemone; the flower whereof is single, consisting of eight large or broad leaues, very neare vnto the same Orenge colour, that is in the double flower hereafter described, but somewhat deeper. This is vsually called in Latine, *Pauo maior simplici flore*, and we in English, The great single Orenge tawnie Anemone.

Pauo minor. There is likewise of this kinde another, whose flower is lesser, and called, The lesser Orenge tawnie Anemone.

Varietas mag-na ex seminio. There is besides these expressed, so great a variety of mixt colours in the flowers of this kinde of Anemone with broad leaues, arising euery yeare from the sowing of the seede of some of the choisest and fittest for that purpose, that it is wonderfull to obserue, not onely the variety of single colours, but the mixture of two or three colours in one flower, besides the diuersity of the bottomes of the flowers, some hauing white or yellowish bottomes, and some none, and yet both of the same colour: and likewise in the thrums or threads in the middle: But the greatest wonder of beauty is in variety of double flowers, that arise from among the other single ones, some hauing two or three rowes of leaues in the flowers, and some so thicke of leaues as a double Marigold, or double Crowfoote, and of the same seuerall colours that are in the single flowers, that it is almost impossible to expresse them seuerally, and (as is said before) some falling out to bee double in one yeare, which will proue single or lesse double in another,

other, yet very many abiding conftant double as at the firft ; and therefore let this briefe recitall be fufficient in ftead of a particular of all the colours.

Anemone Chalcedonica maxima verficolor.
The great double Windflower of Conftantinople.

This great Anemone of Conftantinople hath broader and greener leaues then any of the former kindes, and not fo much diuided or cut in at the edges, among which rife vp one or two ftalkes, (feldome more from one roote) hauing fome leaues about the middle of the ftalke, as other Anemones haue, and bearing at the toppes of the ftalkes one large flower a peece, very double, whofe outermoft leaues being broadeft, are greenifh at the firft, but afterwards red, hauing fometimes fome greene abiding ftill in the leaues, and the red ftriped through it : the other leaues which are within thefe are fmaller, and of a perfect red colour ; the innermoft being fmalleft, are of the fame red colour, but turned fomewhat inward, hauing no thrummes or threads in the middle, as the former haue, and bearing no feede : the roote is blackifh on the outfide, and white within, thicke and tuberous as the other kindes, but thicker fet and clofe together, not fhooting any long flender rootes as others doe. Some Gentlewomen call this Anemone, The Spanifh Marigold.

Anemone Chalcedonica altera fiue Pano maior flore duplici.
The great double Orenge tawney Anemone.

This other great Anemone of Conftantinople hath his large leaues fo like vnto the laft, that one can hardly diftinguifh them afunder ; the ftalke hath alfo fuch like leaues fet vpon it, bearing at the toppe a faire large flower, confifting of many leaues fet in two or three rowes at the moft, but not fo thicke or double as the laft, yet feeming to be but one thicke rowe of many fmall and long leaues, of an excellent red or crimfon colour, wherein fome yellow is mixed, which maketh that colour is called an Orengetawney ; the bottomes of the leaues are red, compaffed with a whitifh circle, the thrummie head in the middle being befet with many darke blackifh threads : the roote is like the former.

Anemone Superitica fiue Cyparifsia. The double Anemone of Cyprus.

This Anemone (which the Dutchmen call Superitz, and as I haue beene enformed, came from the Ifle of Cyprus) hath leaues very like the laft double Anemone, but not altogether fo large : the flower confifteth of fmaller leaues, of colour very neare vnto the laft double Orenge coloured Anemone, but more thicke of leaues, and as double as the firft, although not fo great a flower, without any head in the middle, or thrums about it as is in the laft, and differeth not in the roote from either of them both.

Somewhat like vnto this kinde, or as it were betweene this and the firft kinde of thefe great double Anemones, we haue diuers other forts, bearing flowers very thicke and double ; fome of them being white, or whitifh, or purple, deeper or paler, and fome of a reddifh colour tending to Scarlet or a Carnation colour, and fome alfo of a blufh or flefh colour, and diuers other colours, and all of them continue conftant in their colours.

Anemone Cacumeni Maringi fiue Perfica. The double Perfian Anemone.

This rare Anemone, which is faid to come out of Perfia to Conftantinople, and from thence to vs, is in leafe and roote very like vnto the former double Anemones before defcribed ; onely the flower hereof is rather like vnto the fecond great double Orenge coloured Anemone, vfually called *Pano maior flore pleno*, being compofed of three rowes of leaues, the outtermoft rowe confifting of ten or twelue larger leaues, and thofe more inward leffer and more in number, but all of them variably mixed with white, red, and yellow, hauing the bottomes of the leaues white : but inftead of a middle head with thrums about it, as the other hath, this hath a few narrow leaues, of a deepe yellow colour in the middle of the flower, ftanding vpright,

Hauing

Hauing thus farre proceeded in the two parts of the kindes of Anemones or Wind-flowers, it remaineth to entreate of the rest, which is those Anemones which haue thin cut leaues, whereof some haue reckoned vp thirty sorts with single flowers, which I confesse I haue not seene; but so many as haue come to my knowledge, I shall here set downe.

Anemone tenuifolia siue Geranifolia cærulea.
The Watchet Anemone or Storkes bill leafed Windflower.

This first Windflower with thin cut leaues, riseth not out of the ground vntil the great Winter frosts be past, that is, about the middle or end of February, and are somewhat brownish at their first appearing, but afterwards spread into wings of greene leaues, somewhat broader then the rest that follow, diuided into three parts, & each part into three leaues, euery one cut in about the edges, one standing against another vpon a long slender foote-stalke, and the end leafe by it selfe : among these riseth vp two or three greene stalkes, garnished with such like thin leaues as are at the bottome, from aboue which rise the flowers, but one vpon a stalke, consisting of fourteene or fifteene small pale blew or watchet leaues, lesser then any of the single kindes that follow, compassing many whitish threads, and a small greene head in the middle, somewhat like the head of the wilde Crowfoote, wherein is contained such like seede : the roote is blackish without, thrusting out into long tuberous peeces, somewhat like vnto some of the broad leafed Anemones.

Alba. Of this kinde there is another, whose leaues are not browne at their first rising, but greene, and the flowers are white, in other things not differing.

Anemone tenuifolia purpurea vulgaris.
The ordinary purple Anemone with thin leaues.

This purple Anemone which is most common, and therefore the lesse regarded, hath many winged leaues standing vpon feuerall stalkes, cut and diuided into diuers leaues, much like vnto the leaues of a Carrot ; among which rise vp stalkes with some leaues thereon (as is vsuall to the whole Family of Anemones, both wilde and tame, as is before said;) at the toppes whereof stand the flowers, made of six leaues most vsu-ally, but sometimes they will haue feuen or eight, being very large, and of a perfect purple Violet colour, very faire and liuely : the middle head hath many blackish thrums or threads about it, which I could neuer obserue in my Gardens to beare feed : the roote is smaller, and more spreading euery way into small long flat tuberous parts, then any other kindes of single or double Anemones.

carnea pallida. There is another very like in leafe and roote vnto the former, but the flower is nothing so large, and is whitish, tending to a blush colour, and of a deeper blush colour toward the bottome of the flower, with blackish blew thrums in the middle, and giueth no feede that I could euer obserue.

Carnea viuida There is likewisewise another like vnto the last in leafe and flower, but
unguibus albis. that the flower is larger then it, and is of a liuely blush colour, the leaues hauing white bottomes.

Alba venis pur- And another, whose flower is white, with purple coloured veines and
pureis. stripes through euery leafe, and is a lesser flower then the other.

Anemone tenuifolia coccinea simplex. The single Scarlet Anemone with thin leaues.

The leaues of this Scarlet Windflower are somewhat like vnto the former, but a lit-tle broader, and not so finely cut and diuided : the flower consisteth of six reasonable large leaues, of an excellent red colour, which we call a Scarlet ; the bottomes of the leaues are large and white, and the thrums or threads in the middle of a blackish pur-ple colour : the roote is tuberous, but consisting of thicker peeces, somewhat like vnto the rootes of the broad leafed Anemones, but somewhat browne, and not so blacke, and most like vnto the roote of the double Scarlet Anemone.

Coccinea absq; There is another of this kinde, whose flower is neare vnto the same co-
unguibus. lour, but this hath no white bottomes at all in his leaues.

We

*Flore bulose-
riceo.* We haue another which hath as large a flower as any single, and is of an Orient deepe red crimson Veluet colour.

Sanguinea. There is another of a deeper red colour, and is called, The bloud red single Anemone.

*Rubra fundo
luteo.* And another, whose flower is red with the bottomes yellow.

*Coccinea dilu-
tior.* Another of a perfect crimson colour, whereof some haue round pointed leaues, and others sharpe pointed, and some a little lighter or deeper then others.

*Alba stamini-
bus purpureis.* There is also one, whose flower is pure white with blewish purple thrums in the middle.

*Carnea Hispa-
nica.* And another, whose flower is very great, of a kinde of sullen blush colour, but yet pleasant, with blewish threads in the middle.

*Alba carneis
venis.* And another with blush veines in euery leafe of the white flower.

*Alba purpureis
vnguibus.* And another, the flower whereof is white, the bottomes of the leaues being purple.

Purpurascens. Another whose flower consisteth of many small narrow leaues, of a pale purple or blush colour on the outside, and somewhat deeper within.

*Facie florum
pomi simplex.* There is another like in leafe and roote vnto the first Scarlet Anemone, but the flower hereof consisteth of seuen large leaues without any bottomes, of a white colour, hauing edges, and some large stripes also of a carnation or flesh colour to bee seene in them, marked somewhat like an Apple blossome, and thereupon it is called in Latine, *Anemone tenuifolia simplex alba instar florum pomi*, or *facie florum pomi*, that is to say in English, The single thin leafed Anemone with Apple blossome flowers.

Multiplex. I haue heard that there is one of this kinde with double flowers.

1. *Anemone tenuifolia flore coccineo pleno vulgaris.*
The common double red or Scarlet Anemone.

The leaues of this double Anemone are very like vnto the leaues of the single Scarlet Anemone, but not so thin cut and diuided as that with the purple flower: the flower hereof when it first openeth it selfe, consisteth of six and sometimes of seuen or eight broad leaues, of a deepe red, or excellent Scarlet colour, the middle head being thick closed, and of a greenish colour, which after the flower hath stood blowne some time, doth gather colour, and openeth it selfe into many small leaues, very thicke, of a more pale red colour, and more Stamell like then the outer leaues : the root of this is thicke and tuberous, very like vnto the root of the single Scarlet Anemone.

2. *Anemone tenuifolia flore coccineo pleno variegata.*
The party coloured double Crimson Anemone.

We haue a kinde hereof, varying neither in roote, leafe, or forme of flower from the former, but in the colour, in that this will haue sometimes the outer broad leaues party coloured, with whitish or blush coloured great streakes in the red leaues both inside and outside ; as also diuers of the middle or inner leaues striped in the same manner : the roote hereof giueth fairer flowers in some yeares then in others, and sometimes giue flowers all red againe.

3. *Anemone tenuifolia flore coccineo saturo pleno.*
The double crimson Veluet Anemone.

Wee haue another also, whose flower is of a deepe Orenge tawny crimson colour, neare vnto the colour of the outer leaues, of the lesser French Marigold, and not differing from the former in any thing else.

4. *Anemone tenuifolia flore pleno suauerubente.* The greater double blush Anemone.

There is small difference to be discerned, either in the roote or leaues of this from the

the former double Scarlet Anemone, fauing that the leaues hereof are a little broader, and feeme to bee of a little frefher greene colour : the flower of this is as large almoft, and as double as the former, and the inner leaues likewife almoft as large as they, being of a whitifh or flefh colour at the firft opening of them, but afterwards become of a moft liuely blufh colour ; the bottomes of the leaues abiding of a deeper blufh , and with long ftanding, the tops of the leaues will turne almoft wholly white againe.

5. *Anemone tenuifolia flore albo pleno.* The double white Anemone.

This double white Anemone differeth little from the former blufh Anemone, but in that it is fmaller in all the parts thereof, and alfo that the flower hereof being wholly of a pure white colour, without any fhew of blufh therein, hath the middle thrummes much fmaller and fhorter then it, and not rifing vp fo high , but feeme as if they were chipped off euen at the toppes.

6. *Anemone tenuifolia flore pleno albicante.* The leffer double blufh Anemone.

This fmall double blufh Anemone differeth very little from the double white laft recited, but onely in the colour of the flower : for they are both much about the bigneffe one of another, the middle thrums likewife being as fmall and fhort, and as euen aboue, onely the flower at the firft opening is almoft white , but afterwards the outer leaues haue a more fhew of blufh in them, and the middle part a little deeper then they.

7. *Anemone tenuifolia flore pleno purpureo violaceo.* The double purple Anemone.

This double purple Anemone is alfo of the fame kindred with the firft double red or Scarlet Anemone for the form or doubleneffe of the flower, confifting but of fix or feuen leaues at the moft in this our Country, although in the hotter it hath ten or twelue, or more as large leaues for the outer border, and as large fmall leaues for the inner middle alfo, and almoft as double, but of a deepe purple tending toward a Violet colour, the outer leaues being not fo deepe as the inner : the roote and leafe commeth neare vnto the fingle purple Anemone before defcribed , but that the roote fpreadeth not fo fmall and fo much.

8. *Anemone tenuifolia flore pleno purpureo carnleo.* The double blew Anemone.

This Anemone differeth not in any thing from the former double purple, but onely that the flower is paler, and more tending to a blew colour.

9. *Anemone tenuifolia flore pleno rofeo.* The double Rofe coloured Anemone.

The double Rofe coloured Anemone differeth alfo in nothing from the former double purple, but onely in the flower, which is fomewhat fmaller, and not fo thicke and double, and that it is of a reddifh colour, neare vnto the colour of a pale red Rofe, or of a deepe coloured Damaske.

10. *Anemone tenuifolia flore pleno carneo viuacifsimo.* The double Carnation Anemone.

This Anemone, both in roote, leafe, and flower, commeth neareft vnto the former double white Anemone, for the largeneffe and doubleneffe of the flower, and in the fmalneffe of the middle thrums, and euenneffe at the toppes of them, being not fo large and great a flower as the double purple, either in the inner or outter leaues , but yet is very faire, thicke and double, and of a moft liuely Carnation filke colour, very deepe, both the outer leaues and middle thrums alfo fo bright, that it doth as it were amaze, and yet delight the minde of the beholder, but by long ftanding in the Sun, waxe a little paler, and fo paffe away as all the moft beautifull flowers doe.

11. *Anemone*

1 *Anemone tenuifolia simplex purpurea.* The single purple Anemone with thin cut leaues. 2 *Anemone tenuifolia simplex alba pura.* The single pure white Anemone. 3 *Anemone tenuifolia simplex chermesina.* The single bright Crimson Anemone. 4 *Anemone tenuifolia simplex sanguinea.* The single bloudred Anemone. 5 *Anemone tenuifolia simplex facie florum poni.* The single Apple bloome Anemone. 6 *Anemone tenuifolia simplex sanguinea flore.* The single white Anemone with thick bottome. 7 *Anemone tenuifolia simplex albo vnguibus carneis.* The single white Anemone with flesh bottome. 8 *Anemone tenuifolia flore pleno coccineo.* The double red or ordinary Scarlet Anemone. 9 *Anemone tenuifolia flore pleno ruberrima Amaranthina.* The double purple Veluet Anemone. 10 *Anemone tenuifolia flore pleno purpure violaceo.* The double blew with purple Anemone. 11 *Anemone tenuifolia flore pleno incarnadius coloris serici viuacissimi.* The double Carnation Anemone, or of a liuely Carnation silke colour.

11. *Anemone tenuifolia flore rubrofusco pleno coma Amarantina.*
The double purple Veluet Anemone.

This double Veluet Anemone is in all things like the laſt deſcribed Carnation A-
nemone, but ſomewhat larger, the difference conſiſteth in the colour of the flower,
which in this is of a deep or ſad crimſon red colour for the outer leaues, and of a deep
purple Veluet colour in the middle thrums, reſembling the colour of the leſſer *Amaranthus purpureus*, or Purple flower gentle hereafter deſcribed, whereof it tooke the
name, which middle thrums are as fine and ſmall, and as euen at the toppes as the
white or laſt Carnation Anemones.

12. *Anemone tenuifolia flore pleno tricolor.*
The double purple Veluet Anemone of three colours.

This double Anemone alſo is very like the laſt deſcribed Anemone, but that in the
middle of the purple thrums, there thruſteth forth a tuft of threads or leaues of a more
light crimſon colour.

And thus much for the kindes of Anemones or Windflowers, ſo farre forth as haue
hitherto come to our knowledge; yet I doubt not, but that more varieties haue beene
elſewhere collected, and will be alſo in our Countrey daily and yearly obſerued by
diuers, that raiſe them vp from ſowing the ſeede, wherein lyeth a pretty art, not yet familiarly knowne to our Nation, although it be very frequent in the Lowe-Countries,
where their induſtry hath bred and nouriſhed vp ſuch diuerſities and varieties, that
they haue valued ſome Anemones at ſuch high rates, as moſt would wonder at,
and none of our Nation would purchaſe, as I thinke. And I doubt not, if wee would
be as curious as they, but that both our ayre and ſoyle would produce as great variety,
as euer hath been ſeene in the Lowe-Countries; which to procure, if any of our Nation will take ſo much paines in ſowing the ſeedes of Anemones, as diuers haue done of
Tulipas: I will ſet them downe the beſt directions for that purpoſe that I haue learned,
or could by much ſearch and tryall attaine vnto; yet I muſt let them vnderſtand thus
much alſo, that there is not ſo great variety of double flowers raiſed from the ſeede of
the thin leafed Anemones, as from the broad leafed ones.

Firſt therefore (as I ſaid before) concerning Tulipas, there is ſome ſpeciall choice to
be made of ſuch flowers, whoſe ſeed is fitteſt to be taken. Of the *Latifolias*, the double
Orenge tawney ſeede being ſowne, yeeldeth pretty varieties, but the purples, and
reds, or crimſons, either *Latifolias* or *Tenuifolias*, yeeld ſmall variety, but ſuch as draw
neareſt to their originall, although ſome be a little deeper or lighter then others. But
the light colours be they which are the chiefe for choice, as white, aſh-colour, bluſh
or carnation, light orenge, ſimple or party coloured, ſingle or double, if they beare
ſeede, which muſt bee carefully gathered, and that not before it bee thorough ripe,
which you ſhall know by the head; for when the ſeede with the wollineſſe beginneth
to riſe a little of it ſelfe at the lower end, it muſt bee then quickly gathered, leſt the
winde carry it all away. After it is thus carefully gathered, it muſt be laid to dry for a
weeke or more, which then being gently rubbed with a little dry ſand or earth, will
cauſe the ſeede to be ſomewhat better ſeparated, although not thoroughly from the
woollineſſe or downe that compaſſeth it.

Within a moneth at the moſt after the ſeede is thus gathered and prepared, it muſt
be ſowne; for by that meanes you ſhall gaine a yeare in the growing, ouer that you
ſhould doe if you ſowed it in the next Spring.

If there remaine any woollineſſe in the ſeede, pull it in ſunder as well as you can,
and then ſowe your ſeede reaſonable thin, and not too thicke, vpon a plaine ſmooth
bed of fine earth, or rather in pots or tubbes, and after the ſowing, ſift or gently ſtraw
ouer them ſome fine good freſh mould, about one fingers thickneſſe at the moſt for the
firſt time: And about a moneth after their firſt ſpringing vp, ſift or ſtraw ouer them
in like manner another fingers thickneſſe of fine earth, and in the meane time if the
weather proue dry, you muſt water them gently and often, but not to ouerglut them
with moiſture; and thus doing, you ſhall haue them ſpring vp before Winter, and
grow

grow pretty strong, able to abide the sharpe Winter in their nonage, in vsing some little care to couer them loosely with some fearne, or furse, or beane hame, or straw, or any such, which yet must not lye close vpon them, nor too farre from them neither.

The next Spring after the sowing, if you will, but it is better if you stay vntill August, you may then remoue them, and set them in order by rowes, with sufficient distance one from another, where they may abide, vntill you see what manner of flower each plant will beare, which you may dispose of according to your minde.

Many of them being thus ordered (if your mould be fine, loose, and fresh, not stonie, clayish, or from a middin) will beare flowers the second yeare after the sowing, and most or all of them the third yeare, if the place where you sowe them, be not annoyed with the smoake of Brewers, Dyers, or Maultkils, which if it be, then will they neuer thriue well.

Thus much haue I thought good to set downe, to incite some of our owne Nation to be industrious; and to helpe them forward, haue giuen such rules of directions, that I doubt not, but they will vpon the tryall and view of the variety, proceede as well in the sowing of Anemones as of Tulipas.

I cannot (Gentlewomen) withhold one other secret from you, which is to informe you how you may so order Anemones, that after all others ordinarily are past, you may haue them in flower for two or three moneths longer then are to be seene with any other, that vseth not this course I direct you.

The ordinary time to plant Anemones, is most commonly in August, which will beare flower some peraduenture before Winter, but most vsually in February, March, and Aprill, few or none of them abiding vntill May; but if you will keepe some roots out of the ground vnplanted, vntill February, March, and Aprill, and plant some at one time, and some at another, you shall haue them beare flower according to their planting, those that shall be planted in February, will flower about the middle or end of May, and so the rest accordingly after that manner: And thus may you haue the pleasure of these plants out of their naturall seasons, which is not permitted to be enioyed in any other that I know, Nature being not so prone to bee furthered by art in other things as in this. Yet regard, that in keeping your Anemone rootes out of the ground for this purpose, you neither keep them too dry, nor yet too moist, for sprouting or rotting; and in planting them, that you set them not in too open a sunny place, but where they may be somewhat shadowed.

The Place.

I shall not need to spend much time in relating the seuerall places of these Anemones, but onely to declare that the most of them that haue not beene raised from seed, haue come from Constantinople to vs; yet the first broad leafed or yellow Anemone, was first found in Portugall, and from thence brought into these parts. And the first purple Starre Anemone in Germanie, yet was the same sent among others from Constantinople also. And the first thin cut leafed Anemone came first out of Italy, although many of that sort haue come likewise from Constantinople. And so haue the double red or Scarlet Anemones, and the great double blush, which I first had by the gift of M^r.Humfrey Packington of Worcestershire Esquire, at Haruington.

The Time.

The times of their flowring are sufficiently expressed in the descriptions, or in the rules for planting.

The Names.

The Turkish names whereby the great double broad leafed kindes haue beene sent vnto vs, were *Giul Catamer*, and *Giul Catamer lale*; And *Binizade*, *Binizante*, and *Galipoli lale* for the thinne cut leafed Anemones. All Authors haue called them *Anemones*, and are the true *Herba venti*.

We

Wee call them in English eyther Anemones, after the Greeke name, or Windflowers, after the Latine.

The Vertues.

There is little vse of these in Physicke in our dayes, eyther for inward or outward diseases; onely the leaues are vsed in the Ointment called *Marciatum*, which is composed of many other hot herbes, and is vsed in cold griefes, to warme and comfort the parts. The roote, by reason of the sharpenesse, is apt to drawe downe rheume, if it be tasted or chewed in the mouth.

CHAP. XXVI.

Aconitum. Wolfebane.

THere be diuers sorts of Wolfebanes which are not fit for this booke, but are reserued for a generall History or Garden of Simples, yet among them there are some, that notwithstanding their euill quality, may for the beauty of their flowers take vp a roome in this Garden, of whom I meane to entreate in this place: And first of the Winter Wolfesbane, which for the beauty, as well as the earlinesse of his flowers, being the first of all other, that shew themselues after Christmas, deserueth a prime place; and therefore for the likenesse of the rootes vnto the Anemones, I ioyne it next vnto them.

1. *Aconitum Hyemale.* The Winters Wolfesbane.

This little plant thrusteth vp diuers leaues out of the ground, in the deepe of Winter oftentimes, if there be any milde weather in Ianuary, but most commonly after the deepe frosts, bearing vp many times the snow vpon the heads of the leaues, which like vnto the Anemone, doe euery leafe rise from the roote vpon seuerall short foote-stalkes, not aboue foure fingers high, some hauing flowers in the middle of them, (which come vp first most vsually) and some none, which leaues stand as it were round, the stalke rising vp vnder the middle of the leafe, deeply cut in and gashed to the middle stalke almost, of a very faire deepe greene colour, in the middle whereof, close vnto the leafe, standeth a small yellow flower, made of six leaues, very like a Crowfoote, with yellow threads in the middle : after the flower is fallen, there rise vp diuers small hornes or cods set together, wherein are contained whitish yellow round seede. The roote is tuberous, so like both for shape and colour vnto the rootes of Anemones, that they will easily deceiue one not well experienced, but that it is browner and smoother without, and yellow within, if it be broken.

2. *Aconitum flore albido, siue Aconitum luteum Ponticum.* The whitish yellow Wolfesbane.

This Wolfesbane shooteth not out of the ground vntill the Spring be well begun, and then it sendeth forth great broad greene leaues, deeply cut in about the edges, not much vnlike the leaues of the great wilde Crowfoote, but much greater; from among which leaues riseth vp a strong stiffe stalke, three foote high, hauing here and there leaues set vpon it, like vnto the lowest, but smaller; the toppe of the stalke is diuided into three or foure branches, whereon are set diuers pale yellow flowers, which turne at the last to be almost white, in fashion like almost vnto the flowers of the Helmet flower, but much smaller, and not gaping so wide open : after the flowers are past come vp diuers short poddes, wherein is contained blacke seede : the roote is made of a number of darke browne strings, which spread and fasten themselues strongly in the ground.

3. *Napellus*

3. *Napellus verus flore cæruleo.* Blew Helmet flower or Monkes hood.

The Helmet flower hath diuers leaues of a fresh greene colour on the vpperside, and grayish vnderneath, much spread abroad and cut into many slits and notches, more then any of the Wolfebanes; the stalke riseth vp two or three foot high, beset to the top with the like leaues, but smaller: the toppe is sometimes diuided into two or three branches, but more vsually without, whereon stand many large flowers one aboue another, in forme very like vnto a Hood or open Helmet, being composed of fiue leaues, the vppermost of which and the greatest, is hollow, like vnto an Helmet or Headpeece, two other small leaues are at the sides of the Helmet, closing it like cheekes, and come somewhat vnder, and two other which are the smallest hang down like labels, or as if a close Helmet were opened, and some peeces hung by, of a perfect or faire blew colour, (but grow darker, hauing stood long) which causeth it be so nourished vp in Gardens, that their flowers, as was vsuall in former times, and yet is in many Countrey places, may be laid among greene herbes in windowes and roomes for the Summer time: but although their beauty may be entertained for the vses aforesaid, yet beware they come not neare your tongue or lippes, lest they tell you to your cost, they are not so good as they seeme to be: in the middest of the flower, when it is open and gapeth wide, are seene certaine small threads like beards, standing about a middle head, which when the flower is past, groweth into three or foure, or more small blackish pods, containing in them blacke seede: the rootes are brownish on the outside, and white within, somewhat bigge and round aboue, and small downewards, somewhat like vnto a small short Carrot roote, sometimes two being ioyned at the head together. But the name *Napellus* anciently giuen vnto it, doth shew they referred the forme of the roote vnto a small Turnep.

Authors. The wholsome Helmet flower, or counterpoison Monkes hood.

This wholsome plant I thought good to insert, not onely for the forme of the flower, but also for the excellent properties thereof, as you shall haue them related hereafter. The rootes hereof are small and tuberous, round and somewhat long, ending for the most part in a long fibre, and with some other small threads from the head downeward: from the head whereof riseth vp diuers greene leaues, euery one seuerally vpon a stalke, very much diuided, as finely almost as the leaues of Larkes heeles or spurres: among which riseth vp a hard round stalke, a foote high and better, with some such leaues thereon as grow belowe, at the toppe whereof stand many small yellowish flowers, formed very like vnto the former whitish Wolfesbane, bearing many blacke seedes in pods afterwards in the like manner.

Many more sorts of varieties of these kindes there are, but these onely, as the most specious, are noursed vp in Florists Gardens for pleasure; the other are kept by such as are Catholicke obseruers of all natures store.

The Place.

All these grow naturally on Mountaines, in many shadowie places of the Alpes, in Germany, and elsewhere.

The Time.

The first flowreth (as is said) in Ianuary, and February, and sometimes vntill March be well spent, and the seede is soone ripe after.
The other three flower not vntill Iune and Iuly.

The Names.

The first is vsually called *Aconitum hyemale Belgarum.* Lobelius calleth it
Bulbosum

Bulbosus vnifolius Batrachoides, Aconitum Elleboraceum, and *Ranunculus Monophyllos*, and some by other names. Most Herbarists call it *Aconitum hyemale*, and we in English thereafter, Winters Wolfesbane; and of some, Yellow Aconite.

The second is called by most Writers, *Aconitum luteum Ponticum*: Some also *Lupicida, Luparia*, and *Canicida*, of the effect in killing Wolues and Dogs: And some, because the flower is more white then yellow, doe call it *Aconitum flore albido*, we call it in English, The whitish yellow Aconite, or Wolfesbane, but some after the Latine name, The yellow Wolfesbane.

The third is called generally *Napellus*, and *Verus*, because it is the true *Napellus* of the ancient Writers, which they so termed from the forme of a Turnep, called *Napus* in Latine.

The fourth is called *Aconitum Salutiferum, Napellus Moysis, Antora* and *Anthora, quasi Antithora*, that is, the remedy against the poisonfull herbe *Thora*, in English according to the title, eyther wholsome Helmet flower, or counterpoison Monkes hood.

The Vertues.

Although the first three sorts of plants be very poisonfull and deadly, yet there may bee very good vse made of them for sore eyes (being carefully applyed, yet not to all sorts of sore eyes neither without discretion) if the distilled water be dropped therein.

The rootes of the counterpoison Monkes hood are effectuall not onely against the poison of the poisonfull Helmet flower, and all others of that kinde, but also against the poison of all venemous beasts, the plague or pestilence, and other infectious diseases, which raise spots, pockes, or markes in the outward skinne, by expelling the poison from within, and defending the heart as a most soueraigne Cordiall. It is vsed also with good successe against the wormes of the belly, and against the paines of the Wind collick.

Chap. XXVII.
Ranunculus. The Crowfoote.

NExt vnto the Aconites, of right are to follow the *Ranunculi*, or Crowfeete, for the neareneffe both of forme, of leaues, and nature of the plants, although leffe hurtfull, yet all of them for the most part being sharpe and exulcerating, and not without some danger, if any would be too bold with them. The whole Family of the *Ranunculi* is of a very large extent, and I am conftrained within the limits of a Garden of Pleasure; I must therefore select out onely such as are fit for this purpose, and set them here downe for your knowledge, leauing the rest for that other generall worke, which time may perfect and bring to light, if the couetous mindes of some that should be most affected towards it, doe not hinder it: or if the helpe of generous spirits would forward it.

1. Ranunculus montanus albus humilior. The lowe white mountaine Crowfoot.

This lowe Crowfoote hath three or foure broad and thicke leaues, almost round, yet a little cut in and notched about the edges, of a fine greene and shining colour on the vpperside, and not so green vnderneath, among which riseth a small short stalke, bearing one snow white flower on the toppe, made of fiue round pointed leaues, with diuers yellow threads in the middle, standing about a greene head, which in time groweth to be full of seede, in forme like vnto a small greene Strawberry: the roote is composed of many white strings.

Duplici flore.　　There is another of this lowe kinde, whose leaues are somewhat more deeply cut in on the edges, and the flower larger, and sometimes a little double, as it were with two rowes of leaues, in other things not differing from the former.

2. Ranunculus

2. *Ranunculus montanus albus maior vel elatior.* The great single white mountaine Crowfoote.

The leaues of this Crowfoote are large and greene, cut into three, and sometimes into fiue speciall diuisions, and each of them besides cut or notched about the edges, somewhat resembling the leaues of the Globe Crowfoote, but larger: the stalke is two foote and a halfe high, hauing three small leaues set at the ioynt of the stalke, where it brancheth out into flowers, which stand foure or fiue together vpon long foote-stalkes, made of fiue white leaues a peece, very sweete, and somewhat larger then the next white Crowfoote, with some yellow threads in the middle compassing a greene head, which bringeth seede like vnto other wilde Crowfeete: the roote hath many long thicke whitish strings, comming from a thicke head.

3. *Ranunculus montanus albus minor.* The lesser single white **Crowfoote.**

This Crowefoote hath faire large spread leaues, cut into fiue diuisions, and somewhat notched about the edges, greene on the vpperside, and paler vnderneath, hauing many veines running through the leaues: the stalke of this riseth not so high as the former, although this be reasonable tall, as being neare two foote high, spread into many branches, bearing such like white flowers, as in the former, but smaller: the seede of this is like the former, and so are the rootes likewise.

4. *Ranunculus albus flore pleno.* The double white Crowfoot.

The double white Crowfoote is of the same kinde with the last single white Crowfoote, hauing such like leaues in all respects: the onely difference is in the flowers, which in this are very thicke and double. Some doe make mention of two sorts of double white Crowfeete, one somewhat lower then another, and the lower likewise bearing more store of flowers, and more double then the higher: but I confesse, I haue neuer seene but one sort of double, which is the same here expressed, not growing very high, and reasonably well stored with flowers.

5. *Ranunculus praecox Rutaefolio siue Coriandrifolio.* The early Coriander leafed Crowfoote.

This Crowfoote hath three or foure very greene leaues, cut and diuided into many small peeces, like vnto the wing of leaues of Rue, or rather like the lower leaues of the Coriander(for they well resemble either of them)euery of them standing vpon a long purplish stalke, at the toppe whereof groweth the flower alone, being composed or made of twelue small white leaues, broad pointed, and a little endented at the ends, somewhat purplish on the outside, and white on the inside, sustained by diuers small greene leaues, which are in stead of a cup or huske: in the middle of the flower are many small white threads, tipt with yellow pendents, standing about a small greene head, which after groweth to bee full of seedes like a Strawberry, which knobs giue small blackish seede: the roote is white and fibrous.

6. *Ranunculus Thalictrifolio maior.* The great colombine leafed Crowfoot.

The lower leaues of this Crowfoote haue long stalkes, and are very like vnto the smaller leaues of Colombines, or the great Spanish *Thalictrum*, which hath his leaues very like vnto a Colombine, foure or fiue rising from the roote: the stalke riseth about a foote and a halfe high, somewhat reddish, beset here and there with the like leaues, at the toppe whereof stand diuers small white flowers, made of fiue leaues a peece, with some pale white threads in the middle: the seede is round and reddish, contained in small huskes or hornes: the roote is made of a bush or tuft of white strings.

T

7. *Ranunculus*

7. *Ranunculus Thalictrifolio minor Asphodeli radice.*
The small white Colombine leafed Crowfoote.

This small Crowfoote hath three or foure winged leaues spread vpon the ground, standing vpon long stalkes, and consisting of many small leaues set together, spreading from the middle ribbe, euery leafe somewhat resembling both in shape and colour the smallest and youngest leaues of Colombines : the flowers are white, standing at the toppe of the stalkes, made of fiue round leaues : the root hath three or foure thick, short, and round yellowish clogs hanging at the head, like vnto the Asphodill roote. The great Herball of Lyons, that goeth vnder the name of *Daleschampius,* saith, that Dr. Myconus found it in Spaine, and sent it vnder the name of Oenanthe; and therefore Ioannes Molineus who is thought to haue composed that booke, set it among the vmbelliferous plants, because the Oenanthes beare vmbels of flowers and seede, and haue tuberous or cloggy rootes; but with what iudgement, let others say, when they haue compared the vmbels of flowers and seede of the Oenanthes, with the flowers and seede of this plant, and whether I haue not more properly placed it among the *Ranunculi* or Crowfeete, and giuen it a denomination agreeable to his forme.

8. *Ranunculus Globosus.* The Globe Crowfoot.

This Crowfoote (which in the Northerne countries of England where it groweth plentifully, is called Locker goulous) hath many faire, broad, darke greene leaues next the ground, cut into fiue, sixe, or seuen diuisions, and iagged besides at the edges, among which riseth vp a stalke, whereon are set such like leaues as are belowe, but smaller, diuided toward the toppe into some branches, on the which stand seuerall large yellow flowers, alwayes folded inward, or as a close flower neuer blowing open, as other flowers doe, consisting of eleuen leaues for the most part, set or placed in three rowes, with many yellow threads in the middle, standing about a greene rough head, which in time groweth to be small knops, wherein are contained blacke seede : the roote is composed of many blackish strings.

9. *Ranunculus pratensis flore multiplici.* The double yellow field Crowfoot.

There is little or no difference in the leaues of this double Crowfoot, from those of the single kindes that growe in euery medowe, being large and diuided into foure or fiue parts, and indented about the edges, but they are somewhat smaller, and of a fresher greene : the flowers stand on many branches, much diuided or separated, being not very great, but very thicke and double : the roote runneth and creepeth vnder ground like as the single doth.

10. *Ranunculus Anglicus maximus multiplex.*
The Garden double yellow Crowfoot or Batchelours buttons.

This great double Crowfoote, which is common in euery Garden through England, hath many great blackish greene leaues, iagged and cut into three diuisions, each to the middle ribbe : the stalkes haue some smaller leaues on them, and those next vnder the branches long and narrow : the flowers are of a greenish yellow colour, very thicke and double of leaues, in the middle whereof riseth vp a small stalke, bearing another double flower, like to the other, but smaller : the roote is round, like vnto a small white Turnep, with diuers other fibres annexed vnto it.

11. *Ranunculus Gramineus.* Grasse leafed Crowfoot.

The leaues of this Crowfoote are long and narrow, somewhat like vnto Grasse, or rather like the leaues of single Gilloflowers or Pinckes, being small and sharpe pointed, a little hollow, and of a whitish greene colour : among these leaues rise vp diuers slender stalkes, bearing one small flower at the toppe of each, consisting of fiue yellow

leaues,

1 *Aconitum hyemale.* Winter Wolfesbane. 2 *Aconitum flore albido sive luteum Ponticum.* The whitish yellow Wolfesbane. 3 *Napellus verus.*
Blew Helmets or Monkes hood. 4 *Anthora.* The counterpoison Monkes hood. 5 *Ranunculus humilis albus simplex.* The single white low Crowfoot.
6 *Ranunculus humilis albus duplici flore.* The double lowe white Crowfoot. 7 *Ranunculus Coriandrifolius* The early Coriander leafed Crow foot.
8 *Ranunculus montanus elatior albus.* The great single white mountain Crowfoot. 9 *Ranunculus montanus albus flore pleno* The double white mountain
Crowfoot. 10 *Ranunculus Thalictrifolia minor.* The lesser Colombine leafed Crowfoot. 11 *Ranunculus globosus.* The globe Crowfoot.

leaues, with some threads in the middle : the roote is composed of many thicke, long, round white strings.

There is another of this kinde that beareth flowers with two rowes of leaues, as if it were double, differing in nothing else.

12. *Ranunculus Lusitanicus Autumnalis.* The Portugall Autumne Crowfoot.

This Autumne Crowfoote hath diuers broad round leaues lying on the ground, set vpon short foote-stalkes, of a faire greene colour aboue, and grayish vnderneath, snipt all about the edges, hauing many veines in them, and sometimes swelling as with blisters or bladders on them ; from among which rise vp two or three slender and hairy stalkes, bearing but one small yellow flower a peece, consisting of fiue and sometimes of six leaues, and sometimes of seuen or eight, hauing a few threads in the middle, set about a small greene head, like vnto many of the former Crowfeete, which bringeth small blacke seede : the roote is made of many thicke short white strings, which seeme to be grumous or kernelly rootes, but that they are somewhat smaller, and longer then any other of that kinde.

13. *Ranunculus Creticus latifolius.* The broad leafed Candy Crowfoot.

This Crowfoote of Candy, hath the greatest and broadest leaues of all the sorts of Crowfeete, being almost round, and without any great diuisions, but onely a few notches about the edges here and there, as large or larger sometimes then the palme of a mans hand ; among which riseth vp the stalke, not very high when it doth first flower, but afterwards, as the other flowers doe open themselues, the stalke groweth to be a foote and a halfe high, or thereabouts, hauing some leaues on it, deeply cut in or diuided, and bearing many faire yellow flowers, consisting of fiue leaues a peece, being somewhat whitish in the middle, when the flower hath stood blowne a little time : the roote is composed of a number of small kernelly knobs, or long graines, set thicke together. This flowreth very early, being vsually in flower before the end of March, and oftentimes about the middle thereof.

14. *Ranunculus Creticus albus.* The white Candy Crowfoote.

The leaues of this Crowfoote are very like vnto the leaues of the red Crowfoote of Tripoli or Asia, hereafter set downe, being somewhat broad and indented about the edges, some of the leaues being also cut in or gashed, thereby making it as it were three diuisions, of a pale greene colour, with many white spots in them : the stalke riseth vp a foote high, with some leaues on it, more diuided then the lower, and diuided at the toppe into two and sometimes into three branches, each of them bearing a faire snow white flower, somewhat large, included at the first in a brownish huske or cup of leaues, which afterwards stand vnder the flowers, consisting of fiue white large round pointed leaues, in the middle whereof is set many blackish purple thrums, compassing a small long greene head, composed of many scales or chaffie whitish huskes, when they are ripe, which are the seede, but vnprofitable in all that euer I could obserue : the rootes are many small graines or kernels, set together as in the former, and much about the same colour, that is, of a darke or duskie grayish colour, but much smaller.

Alba purpureis oris & venis.   There is another of this kinde, whose flowers haue purple edges, and sometimes some veines of the same purple in the leaues of the flowers, not differing in any other thing from the former.

Alba oris rubris.   And another, whose edges of the flowers are of a bright red colour.

15. *Ranunculus Creticus flore argenteo.* The Argentine, or cloth of siluer Crowfoot.

The greene leaues of this Crowfoote are as small and thinne, cut in or diuided on the edges, as the last two sorts ; the stalke riseth vp somewhat higher, and diuided into some branches, bearing at the toppe of euery of them one flower, somewhat smaller then the former, composed of six, seuen, and sometimes of eight small round pointed

leaues,

1 *Ranunculus gramineus flore simplici & duplici.* The single and the double grasse Crowfoot. 2 *Ranunculus Lusitanicus Auranticus.* The Portugall Autumne Crowfoot. 3 *Ranunculus Creticus latifolius.* The broad leafed Candy Crowfoot. 4 *Ranunculus Anglicus maximus multiplex.* The double English Crowfoot. 5 *Ranunculus pratensis flore multiplici.* The double yellow field Crowfoot. 6 *Ranunculus Creticus albus.* The white Candy Crowfoot. 7 *Ranunculus Asiaticus flore albo vel pallido varie.* The white or the straw coloured Crowfoot with red top. 8 *Ranunculus Tripolitanus flore rubro simplici.* The single red Crowfoot of Tripoli. 9 *Ranunculus Asiaticus flore rubro amplo.* The large single red Crowfoot of Asia. 10 *Ranunculus Asiaticus flore rubro pleno.* The double red Crowfoot of Asia. 11 *Caltha palustris flore pleno.* Double Marsh Marigold or Batchelours buttons.

T 3

leaues, of a whitish yellow blush colour on the inside wholly, except sometimes a little stript about the edges : but the outside of euery leafe is finely stript with crimson stripes, very thicke, somewhat like vnto a Gilloflower : in the middle riseth vp a small blacke head, compassed about with blackish blew threads or thrums, which head is as vnfruitfull for seede in our Countrey as the former. This flower hath no such greene leaues vnder it, or to enclose it before it be blowne open as the former : the rootes are in all things like the former.

16. *Ranunculus Asiaticus siue Tripolitanus flore rubro.*
The single red Crowfoote of Asia or Tripoli.

The lower leaues of this red Crowfoote are alwayes whole without diuisions, being onely somewhat deeply indented about the edges, but the other that rise after them are more cut in, sometimes into three, and sometimes into fiue diuisions, and notched also about the edges : the stalke riseth higher then any of the former, and hath on it two or three smaller leaues, more cut in and diuided then those belowe : at the toppe whereof standeth one large flower, made of fiue leaues, euery one being narrower at the bottome then at the toppe, and not standing close and round one to another, but with a certaine distance betweene, of a duskie yellowish red colour on the outside, and of a deepe red on the inside, the middle being set with many thrums of a darke purple colour : the head for seede is long, and scaly or chaffie, and idle in like manner as the rest : the roote is made of many graines or small kernels set together, and closing at the head, but spreading it selfe, if it like the ground, vnder the vpper crust of the earth into many rootes, encreasing from long strings, that runne from the middle of the small head of graines, as well as at the head it selfe.

17. *Ranunculus Asiaticus flore amplo rubro.* The large single red Crowfoot of Asia.

There hath come :o vs out of Turkie, together with the former, among many other rootes, vnder the same title, a differing sort of this Crowfoote, whose leaues weare broader, and much goaler, the flower also larger, and the leaues thereof broader, sometimes eight in a flower, standing round and close one to another, which maketh the fairer shew : in all other things it is like the former.

18. *Ranunculus Asiaticus flore rubro vario simplici.*
The red stript single Crowfoote of Asia.

This party coloured Crowfoote differeth not eyther in roote or leafe from the former, the chiefest difference is in the flower, which being red, somewhat like the former, hath yet some yellow stripes or veines through euery leafe, sometimes but little, and sometimes so much, that it seemeth to bee party coloured red and yellow : this sort is very tender ; for we haue twice had it, and yet perished with vs.

19. *Ranunculus Asiaticus flore luteo vario simplici.*
The yellow stript single Crowfoote of Asia.

There is little difference in the roote of this Crowfoote from the last described, but the leaues are much different, being very much diuided, and the flower is large, of a fine pale greenish yellow colour, consisting of six and seuen, and sometimes of eight or nine round leaues ; the toppes whereof haue reddish spots, and the edges sometimes also, with such purplish thrums in the middle that the other haue. None of these former Crowfeete with kernelly rootes, haue euer beene found to haue giuen so good seed in England, as that being sowne, any of them would spring vp ; for hereof tryall hath been often made, but all they haue lost their labour, that haue bestowed their paines therein, as farre as I know.

20. *Ranunculus Asiaticus flore rubro pleno.*
The double red Crowfoote of Asia.

The double red Crowfoote hath his rootes and leaues so like vnto the single red kinde, that none can perceiue any differ_nce, or know the one from the other, vntill the budde of the flower doe appeare, which after it is any thing forward, may be perceiued to be greater and fuller then the budde of the single kinde. This kinde beareth most vsually but one faire large double flower on the toppe of the stalke, composed of many leaues, set close together in three or foure rowes, of an excellent crimson colour, declining to Scarlet, the outter leaues being larger then the inner; and in stead of thrummes, hath many small leaues set together: it hath likewise six small narrow greene leaues on the backside of the flower, where the stalke is fastened to the flower.

There is of this double kinde another sort, whose flower is of the same colour with *Polifero flore.* the former, but out of the middle of the flower ariseth another double flower, but smaller.

The Place.

These plants grow naturally in diuers Countries; some in France, and Germany, and some in England, some in Spaine, Portugall, and Italy, and some haue been sent out of Turkie from Constantinople, and some from other parts, their titles for the most part descrying their Countries.

The Time.

Some of them flower early, as is set downe in their descriptions, or titles. The others in Aprill and May. The white Candy Crowfoote, and the other single and double sorts of Asia, about the same time, or somewhat later, and one in Autumne, as it is set downe.

The Names.

The names that are giuen seuerally to them may well serue this worke, that thereby they may bee distinguished one from another: For to set downe any further controuersie of names, how fitly or vnfitly they haue beene called, and how variably by diuers former Writers, is fitter for a generall History, vnto which I leaue what may be said, both concerning these and the rest: Onely this I would giue you to vnderstand, that the Turkie kindes haue been sent to vs vnder the names of *Terebolos* for the single, and *Terebolos Catamer lale* for the double, and yet oftentimes, those that haue been sent for double, haue proued single, so little fidelity is to bee found among them.

The Vertues.

All or most of these plants are very sharpe and exulcerating, yet the care and industry of diuers learned men haue found many good effects in many of them. For the rootes and leaues both of the wilde kindes, and of some of these of the Garden, stamped and applyed to the wrists, haue driuen away the fits in Feuers. The roote likewise of the double English kinde is applyed for pestilent sores, to helpe to breake them, by drawing the venome to the place. They helpe likewise to take away scarres and markes in diuers places of the body.

CHAP.

CHAP. XXVIII.

Caltha palustris flore pleno. Double Marsh Marigold.

AS an appendix to the Crowfeete, I must needes adde this plant, yet seuerally by it selfe, becauſe both it and his ſingle kinde are by moſt adioyned thereunto, for the neare reſemblance both in ſhape and ſharpeneſſe of quality. The ſingle kinde I leaue to the Ditch ſides, and moiſt grounds about them, as the fitteſt places for it, and onely bring the double kinde into my Garden, as fitteſt for his goodly proportion and beauty to be entertained, and haue place therein.

The double Marſh Marigold hath many broad and round greene leaues, a little endented about the edges, like vnto the ſingle kinde, but not altogether ſo large, eſpecially in a Garden where it ſtandeth not very moiſt : the ſtalkes are weake, round, hollow, and greene, diuided into three or foure branches at the toppe, with leaues at the ſeuerall ioynts, whereon ſtand very double flowers, of a gold yellow colour : the fiue outer leaues being larger then any of the reſt that are encompaſſed by them, which fall away after they haue ſtood blowne a great while (for it endureth in flower a moneth or more, eſpecially if it ſtand in a ſhadowie place) without bearing any ſeed : the rootes are compoſed of many thicke, long, and round whitiſh ſtrings, which runne downe deep into the ground, and there are faſtened very ſtrongly.

The Place.

This plant groweth naturally in diuers Marſhes, and moiſt grounds in Germany, yet in ſome more double then in others ; it hath long agoe beene cheriſhed in our Gardens.

The Time.

It flowreth in Aprill or May, as the yeare proueth earlier or later : all his leaues doe in a manner quite periſh in Winter, and ſpring anew in the end of February, or thereabouts.

The Names.

There is great controuerſie among the learned about the ſingle kinde, but thereof I ſhall not neede to ſpeake in this place ; if God permit I may in a fitter. This is called generally in Latine, *Caltha palustris multiplex*, or *flore pleno*. And wee in Engliſh (after the Latine, which take *Caltha* to be that which wee vſually call *Calendula*, a Marigold) The double Marſh Marigold.

The Vertues.

The roote hereof is ſharpe, comming neare vnto the quality of the Crowfeete, but for any ſpeciall property, I haue not heard or found any.

CHAP. XXIX.

Hepatica nobilis siue trifolia. Noble Liuerwort.

NExt vnto the Crowfeete are to follow the Hepaticas, becaufe of the likeneffe with them, feeming to be fmall Crowfeete in all their parts, but of another and more wholfome kinde. Their diuerfity among themfelues confifteth chiefly in the colour of the flowers, all of them being fingle, except one which is very thicke and double.

1. *Hepatica flore cæruleo fimplici maior.*
The great fingle blew Hepatica or noble Liuerwort.

The flowers of this Hepatica doe fpring vp, blow open, and fometimes fhed and fall away, before any leaues appeare or fpread open. The rootes are compofed of a bufh of blackifh ftrings, from the feuerall heads or buttons whereof, after the flowers are rifen and blowne, arife many frefh greene leaues, each feuerally ftanding vpon his foot-ftalke, folded together, and fomewhat browne and hairy at their firft comming, which after are broad, and diuided at the edges into three parts: the flowers likewife ftand euery one vpon his owne feuerall foote-ftalke, of the fame height with the leaues for the moft part, which is about foure or fiue fingers breadth high, made of fix leaues moft vfually, but fometimes it will haue feuen or eight, of a faire blew colour, with many white chiues or threads in the middle, ftanding about a middle greene head or vmbone, which after the flower is fallen groweth greater, and fheweth many fmall graines or feede fet clofe together (with three fmall greene leaues compaffing them vnderneath, as they did the flower at the bottome) very like the head of feed of manie Crowfeete.

2. *Hepatica minor flore pallido cæruleo.* The fmall blew Hepatica.

The leaues of this Hepatica are fmaller by the halfe then the former, and grow more aboundantly, or bufhing thicke together: the flowers (when it fheweth them, for I haue had the plant halfe a fcore yeares, and yet neuer faw it beare flower aboue once or twice) are of a pale or bleake blew colour, not fo large as the flowers of the former.

3. *Hepatica flore purpureo.* Purple Hepatica or noble Liuerwort.

This Hepatica is in all things like vnto the firft, but onely the flowers are of a deeper blew tending to a Violet purple: and therefore I fhall not neede to reiterate the former defcription.

4. *Hepatica flore albo minor.* The leffer white Hepatica.

The flowers of this Hepatica are wholly white, of the bigneffe of the red or purple, and the leaues fomewhat fmaller, and of a little whiter or paler greene colour, elfe in all other things agreeing with the former.

5. *Hepatica alba magno flore.* The great white Hepatica.

There is no other difference herein from the laft, but that the flower being as white, is as large as the next.

6. *Hepatica albida siue argentea.* Afh-coloured or Argentine Hepatica.

Both the leaues and the flowers of this Hepatica are larger then any of the former, except the laft: the flowers hereof at the firft opening feeme to bee a of blufh afh-colour, which doe fo abide three or foure dayes, decaying ftill vntill it turne almoft
white,

white, hauing yet still a shew of that blush ash-colour in them, till the very last.

7. *Hepatica alba straminibus rubris.* White Hepatica with red threads.

There is no difference between this Hepatica and the first white one, sauing that the threads in the middle of the flower, being white, as in the former, are tipt at the ends with a pale reddish colour, which adde a great beauty to the flowers.

8. *Hepatica flore rubro.* Red Hepatica or noble Liuerwort.

The leaues of this Hepatica are of a little browner red colour, both at their first comming vp, and afterwards, especially in the middle of the leafe more then any of the former: the flowers are in forme like vnto the rest, but of a bright blush, or pale red colour, very pleasant to behold, with white threads or chiues in the middle of them.

9. *Hepatica flore purpureo multiplici siue pleno.* The double purple Hepatica.

The double Hepatica is in all things like vnto the single purple kinde, sauing onely that the leaues are larger, and stand vpon longer foote-stalkes, and that the flowers are small buttons, but very thicke of leaues, and as double as a flower can be, like vnto the double white Crowfoote before described, but not so bigge, of a deepe blew or purple colour, without any threads or head in the middle, which fall away without giuing any seede.

10. *Hepatica flore caruleo pleno.* The double blew Hepatica.

In the colour of this flower, consisteth the chiefest difference from the last, except one may say it is a little lesse in the bignesse of the flower, but not in doublenesse of leaues.

The Place.

All these plants with single flowers grow naturally in the Woods, and shadowie Mountaines of Germany in many places, and some of them in Italy also. The double kinde likewise hath been sent from Alphonsus Pantius out of Italy, as Clusius reporteth, and was also found in the Woods, neare the Castle of Starnbeg in Austria, the Lady Heusenstains possession, as the same Clusius reporteth also.

The Time.

These plants doe flower very early, and are of the first flowers that shew themselues presently after the deepe frosts in Ianuary, so that next vnto the Winter Wolfesbane, these making their pride appeare in Winter, are the more welcome early guests. The double kinde flowreth not altogether so early, but sheweth his flower, and abideth when the others are past.

The Names.

They haue obtained diuers names; some calling them *Hepatica, Hepatica nobilis, Hepaticum trifolium, Trifolium nobile, Trifolium aureum,* and some *Trinitas,* and *Herba Trinitatis.* In English you may call them either Hepatica, after the Latine name, as most doe, or Noble Liuerwort, which you please.

The Vertues.

These are thought to coole and strengthen the liuer, the name importing as much; but I neuer saw any great vse of them by any the Physitians of our London Colledge, or effect by them that haue vsed them in Physicke in our Country.

Chap.

1 *Hepatica flore albo amplo simplici.* The large white Hepatica. 2 *Hepatica flore rubro simplici.* The red Hepatica. 3 *Hepatica flore purpureo pleno.* The double purple Hepatica. 4 *Geranium tuberosum.* Knobbed Cranes bill 5 *Geranium Batrachoides flore albo vel cæruleo.* The blew or white Crowfoote Cranes bill. 6 *Geranium Hæmatodes.* The red Rose Cranes bill. 7 *Geranium Romanum striatum.* The variable stript Cranes bill. 8 *Geranium Creticum.* Candy Cranes bill.

Chap. XXX.

Geranium. Storkes bill or Cranes bill.

AS was said before concerning the Crowfeet, of their large extent and restraint, the like may be said of the Storkes bils or Cranes bils ; for euen of these as of them, I must for this worke set forth the descriptions but of a few, and leaue the rest to a generall worke.

1. *Geranium tuberosum vel bulbosum.* Bulbous or knobbed Cranes bill.

The knobbed Cranes hath three or foure large leaues spread vpon the ground, of a grayish or rather dusty greene colour, euery one of them being as it were of a round forme, but diuided or cut into six or seuen long parts or diuisions, euen vnto the middle, which maketh it seeme to be so many leaues, each of the cuts or diuisions being deeply notched or indented on both sides ; among which riseth vp a stalke a foote high or better, bearing thereon diuers pale but bright purple flowers, made of fiue leaues a peece, after which come small heads with long pointed beakes, resembling the long bill of a Storke or Crane, or such like bird, which after it is ripe, parteth at the bottome where it is biggest, into foure or fiue seedes, euery one whereof hath a peece of the beake head fastened vnto it, and falleth away if it bee not gathered : the roote is tuberous and round, like vnto the roote of the *Cyclamen* or ordinary Sowbread almost, but smaller, and of a darke russet colour on the outside, and white within, which doth encrease vnder ground, by certaine strings running from the mother root into small round bulbes, like vnto the rootes of the earth Chesnut, and will presently shoote leaues, and quickly grow to beare flowers, but will not abide to be kept long dry out of the ground, without danger to be vtterly spoiled.

Geranium Batrachoides flore caruleo. The blew Crowfoote Cranes bill.

This Crowfoote Cranes bill hath many large leaues, cut into fiue or six parts or diuisions, euen to the bottome, and iagged besides on the edges, set vpon very long slender foote-stalkes, very like the leaues of the wilde Crowfoot ; from among which rise vp diuers stalkes with great ioynts, somewhat reddish, set with leaues like the former : the toppes of the stalkes are spread into many branches, whereon stand diuers flowers, made of fiue leaues a peece, as large as any of the wilde or field Crowfeete, round pointed, of a faire blew or watchet colour, which being past, there doe arise such heads or bils, as other of the Cranes bils haue : the roote is composed of many reddish strings, spreading in the ground, from a head made of diuers red heads, which lye oftentimes eminent aboue the ground.

Geranium Batrachoides flore albo. The white Crowfoote Cranes bill.

This Cranes bill is in leafe and flower altogether like the former, the onely difference betweene them consisteth in the colour of the flower, which in this is wholly white, and as large as the former : but the roote of this hath not such red heads as the other hath.

Geranium Batrachoides flore albo & caruleo vario. The party coloured Crowfoote Cranes bill.

The flowers of this Cranes bill are variably striped and spotted, and sometimes diuided, the one halfe of euery leafe being white, and the other halfe blew, sometimes with lesser or greater spots of blew in the white leafe, very variably, and more in some years then in others, that it is very hard to expresse all the varieties that may be obserued in the flowers, that blow at one time. In all other parts of the plant, it is so like vnto the former, that vntill it be in flower, the one cannot be knowne from the other.

Geranium

5. *Geranium Batrachoides alterum flore purpureo.*
Purple Crowfoote Cranes bill.

This purple Cranes bill hath many leaues rifing from the roote, fet vpon long foot-ftalkes, fomewhat like vnto the other, yet not fo broad, but more diuided or cut, that is, into feuen or more flits, euen to the middle, each whereof is likewife cut in on the edges more deeply then the former; the ftalkes are fomewhat knobbed at the ioynts, fet with leaues like vnto the lower, and bearing a great tuft of buds at the toppes of the branches, which breake out into faire large flowers, made of fiue purple leaues, which doe fomewhat refemble the flower of a Mallow, before it be too full blowne, each whereof hath a reddifh pointell in the middle, and many fmall threads compaffing it, this vmbell or tuft of buds doe flower by degrees, and not all at once, and euery flower abideth open little more then one day, and then fheddeth, fo that euery day yeeldeth frefh flowers, which becaufe they are fo many, are a long while before they are all paft or fpent: after the flowers are paft, there arife fmall beake heads or bils, like vnto the other Cranes bils, with fmall turning feede: the roote is compofed of a great tuft of ftrings, faftened to a knobby head.

6. *Geranium Romanum verficolor fiue ftriatum.* The variable ftript Cranes bill.

This beautifull Cranes bill hath many broad yellowifh greene leaues arifing from the roote, diuided into fiue or fix parts, but not vnto the middle as the firft kindes are: each of thefe leaues hath a blackifh fpot at the bottome corners of the diuifions, the whole leafe as well in forme as colour and fpots, is very like vnto the leafe of the *Geranium fufcum*, or fpotted Cranes bill, next following to be defcribed, but that the leaues of this are not fo large as the other: from among thefe leaues fpring vp fundry ftalkes a foote high and better, ioynted and knobbed here and there, bearing at the tops two or three fmall white flowers, confifting of fiue leaues a peece, fo thickly & variably ftriped with fine fmall reddifh veines, that no green leafe that is of that bigneffe can fhew fo many veines in it, nor fo thick running as euery leafe of this flower doth: in the middle of the flower ftandeth a fmall pointell, which when the flower is paft doth grow to be the feed veffell, wheron is fet diuers fmall feeds, like vnto the fmall feedes of other Cranes bils: the root is made of many fmall yellow threads or ftrings.

7. *Geranium fufcum fiue maculatum.* Swart tawny or fpotted Cranes bill.

The leaues of this Cranes bill are in all points like the laft defcribed, as well in the forme and diuifions as colour of the leaues, being of a yellowifh greene colour, but larger and ftronger by much: the ftalkes of this rife much higher, and are ioynted or knobbed with reddifh knees or ioynts, on the tops whereof ftand not many although large flowers, confifting of fiue leaues a peece, each whereof is round at the end, and a little fnipt round about, and doe bend or turne themfelues backe to the ftalkewards, making the middle to be higheft or moft eminent; the colour of the flower is of a darke or deepe blackifh purple, the bottome of euery leafe being whiter then the reft; it hath alfo a middle pointell ftanding out, which afterwards bring forth feede like vnto others of his kinde: the roote confifteth of diuers great ftrings, ioyned to a knobby head.

8. *Geranium Hæmatodes.* The red Rofe Cranes bill.

This Cranes bill hath diuers leaues fpread vpon the ground, very much cut in or diuided into many parts, and each of them againe flit or cut into two or three peeces, ftanding vpon flender long foote-ftalkes, of a faire greene colour all the Spring and Summer, but reddifh in Autumne: among thefe leaues fpring vp flender and weake ftalkes, befet at euery ioynt (which is fomewhat reddifh) with two leaues for the moft part, like vnto the lower: the flowers grow feuerally on the toppe of the ftalkes, and not many together in bunches or branches, as in all other of the Cranes bils, euery flower being as large as a fingle Rofe Campion flower, confifting of fiue large leaues, of a deeper red colour then in any other Cranes bill at the first opening, and will change more blewish afterwards : when the flower is past, there doth arise such like beakes as are in others of the same kinde, but small : the roote is hard, long, and thicke, with diuers branches spreading from it, of a reddish yellow colour on the outside, and whitish within, which abideth and perisheth not, but shooteth forth some new greene leaues, which abide all the Winter, although those that turne red doe fall away.

Geranium Creticum. Candy Cranes bill.

Candy Cranes bill beareth long and tender stalkes, whereon growe diuers broad and long leaues, cut in or ragged on the edges : the toppes of the stalkes are branched into many flowers, made of fiue leaues of a reasonable bignesse, and of a faire blew or watchet colour, with a purplish pointell in the middle, which being past, there follow beake heads like other Cranes bils, but greater, containing larger, greater, and sharper pointed seede, able to pierce the skinne, if one be not warie of it : the roote is white and long, with some fibres at it, and perisheth when it hath perfected his seede, and will spring of it owne sowing many times, if the Winter be not too sharpe, otherwise (being annuall) it must be sowne in the Spring of the yeare.

The Place.

Most of these Cranes bils are strangers vnto vs by nature, but endenizond in our English Gardens. It hath beene reported vnto mee by some of good credit, that the second or Crowfoot Cranes bill hath been found naturally growing in England, but yet I neuer saw it, although I haue seen many sorts of wilde kindes in many places. Matthiolus saith that the first groweth in Dalmatia and Illyria very plentifully. Camerarius, Clusius, and others, that most of the rest grow in Germany, Bohemia, Austria, &c. The last hath his place recorded in his title.

The Time.

All these Cranes bils doe for the most part flower in Aprill, and May, and vntill the middle of Iune. The variable or stript Cranes bill is vsually the latest of all the rest.

The Names.

The first is vsually called *Geranium tuberosum*, of some *Geranium bulbosum*, of the likenesse of the roote vnto a bulbe : It is without controuersie *Geranium primum* of Dioscorides. The second is called *Geranium Gratia Dei*, of others, *Geranium caeruleum*. The blew Cranes bill Lobel calleth it *Batrachoides*, because both leafe and flower are like vnto a Crowfoote, and the affinity with the Cranes bils in the seede causeth it rather to be referred to them then to the Crowfeete. The stript Cranes bill is called by some *Geranium Romanum*. The last sauing one is called *Geranium Haematodes*, or *Sanguineum*, of Lobel *Geranium Gruinale Haematodes supinum radice repente*. In English it may be called after the Greek and Latine, The bloudy Cranes bill, but I rather call it, The Rose Cranes bill, because the flowers are as large as single Roses, or as the Rose Campion. Some of them are called in many places of England Baffinets.

The Vertues.

All the kindes of Cranes bils are accounted great wound herbes, and effectuall to stay bleedings, yet some more then others. The Emperickes of Germanie, as Camerarius saith, extoll it wonderfully, for a singular remedie against the Stone, both in the reines and bladder.

CHAP.

CHAP. XXXI.

Sanicula guttata maior. Spotted Sanicle.

HAuing long debated with my selfe, where to place this & the other plants that follow in the two next Chapters, I haue thought it not amisse for this worke to set them downe here, both before the Beares eares, which are kindes of Sanicle, as the best Authors doe hold, and after the Cranes bils, both for some qualities somewhat resembling them, and for some affinity of the flowers with the former.

The spotted Sanicle hath many small round leaues, bluntly endented about the edges, somewhat like vnto the leaues of our white Saxifrage, of a full greene colour aboue, and whitish hairy, and somewhat reddish withall vnderneath: the stalkes are set here and there with the like leaues, rising a foote and a halfe high or more, very much diuided at the toppe into sundry small branches, bearing many very small white flowers, consisting of fiue small leaues, wherein are many small red spots to be seene, as small as pins points, of a pretty sweete sent, almost like Hawthorne flowers, in the middle whereof are many small threads compassing a head, which when it is ripe containeth small blacke seede: the roote is scaly, or couered with a chaffie matter, hauing many small white fibres vnderneath, whereby it is fastened in the ground.

There is another of this kinde, like both in roote, leafe, and flower to the former, *Minor non guttata.* the onely difference is, that this is lesser then the former, and hath no spots in the flower, as the other hath.

We haue also another smaller kinde then the last, both in leafe and flower, the leaues *Minus guttata.* whereof are smaller, but rounder, and more finely snipt or indented about the edges, like the teeth of a fine sawe: the stalke is little aboue a span high, hauing many small white flowers spotted as the first, but with fewer spots.

The Place.

These growe in the shadowie Woods of the Alpes, in diuers places, and with vs they more delight in the shade then the sunne.

The Time.

All these Sanicles doe flower in May, and continue flowring vntill Iune, and the seede soone ripeneth after: the rootes abide all the Winter, with some leaues on them, springing afresh in the beginning of the yeare.

The Names.

The former two are called by Clusius *Sanicula montana*, and by others *Sanicula guttata*: by Lobel *Geum Alpinum*. The third or last hath been sent vs vnder the name of *Sanicula montana altera minor*.

The Vertues.

The name imposed on these plants doe certainly assure vs of their vertues, from the first founders, that they are great healers, and from their taste, that they are great binders.

Chap. XXXII.

Cotyledon altera Matthioli. Spotted Nauelwort.

THis spotted Nauelwort, as many doe call it, hath many thicke small leaues, not so broad as long, of a whitish greene colour, lying on the ground in circles, after the manner of the heads of Houseleeke, and dented about the edges; from the middle whereof sometimes (for it doth not flower euery yeare in many places) ariseth vp a stalke, scarce a foote high, beset with such like leaues as are belowe, but somewhat longer : from the middle of the stalke vp to the top it brancheth forth diuersly, with a leafe at euery ioynt, bearing three or foure flowers on euery branch, consisting of fiue white leaues, spotted with small red spots, like vnto the spotted Sanicle, but with fewer and greater spots, hauing a yellowish circle or eye in the bottome of euery flower, and many whitish threads with yellowish tips in it : the seede is small and blacke, contained in small round heads : the roote is small, long, and threadie, shooting out such heads of leaues, which abide all the Winter, those that beare flower perishing.

Cotyledon altera minor. Small dented Nauelwort.

There is another like vnto that before described in most things, the differences be these : It hath shorter leaues then the former, and dented about the edges in the like manner : the flowers hereof are white, but greater, made of six leaues, and most vsually without any spots at all in them, some are seene to haue spots also : the heads or seede vessels are more cornered then the former.

Cotyledon altera flore rubro stellato. Small red flowred Nauelwort.

This hath also many heads of leaues, but more open, which are longer, greener, and sharper pointed then eyther of the former, somewhat reddish also, and not dented about the edges, but yet a little rough in handling : the stalke ariseth from among the leaues, being somewhat reddish, and the leaues thereon are reddish pointed, diuided at the toppe into many branches, with diuers flowers thereon, made of twelue small long leaues, standing like a starre, of a reddish purple colour, with many threads therein, set about the middle head, which is diuided at the toppe into many small ends, like pods or hornes, containing therein very small seede : the root is small like the former.

Sedum serratum flore rubente maculato. The Princes Feather.

This kinde of Sengreene is composed of heads of larger, broader, and thinner leaues then any of the former, of a sadder greene colour, somewhat vneuenly endented about the edges, and not so close set together, but spreading forth into seuerall heads like as the former sorts doe, although not so plentifully; from the middle of diuers of which heads rise vp brownish or reddish stalkes, set with smaller leaues thereon to the middle thereof, and then brancheth forth into seuerall sprigs, set with diuers small reddish flowers consisting of fiue leaues a peece, the innerside of which are of a pale red, somewhat whitish, spotted with many small bloud red spots, as small almost as pins points, with some small threads in the middle, standing about a small greene head, which turneth into the seede vessell, parted foure wayes at the head, wherein is contained small blackish seede : the rootes are small threads, which spread vnder the ground, and shoote vp seuerall heads round about it.

The Place.

All these growe in Germany, Hungarie, Austria, the Alpes, and other such like places, where they cleaue to the rocke it selfe, that hath but a crust of earth on it to nourish them. They will abide in Gardens reasonable well, if they be planted in shadowie places, and not in the sun.

The

1 *Sanicula guttata.* Spotted Sanicle. **2** *Cotyledon altera Matthioli.* Spotted Nauelwort. **3** *Cotyledon altera minor.* Small dented Nauelwort. **4** *Cotyledon altera flore rubro stellato.* Small red flowred Nauelwort. **5** *Sedum serratum flore rubente maculato.* The Princes Feather. **6** *Soldanella Alpina.* Blew Moonwort.

The Time.

They flower for the moſt part in the end of May, and ſometimes ſooner or later, as the yeare falleth out.

The Names.

The firſt is called by Matthiolus, *Cotyledon altera Dioſcoridis*, and *Vmbiliſ cus alter*, but it is not the true *Cotyledon altera* of Dioſcorides ; for *Sedum vulgare maius*, Our common Houſeleeke, by the conſent of the beſt moderne Writers, is the true *Cotyledon altera* of Dioſcorides, or *Vmbiliſcus Veneris alter*. I hold it rather to bee a kinde of ſmall Houſeleeke, as the other two likewiſe are. The ſecond is called by ſome *Aizoum* or *Sedum minus ſerratum*. The third hath his name in his title. Wee doe call them Nauelworts in Engliſh rather then Houſeleekes, *Euphonia gratia*. The laſt may be called dented Sengreene with reddiſh ſpotted flowers, but ſome of our Engliſh Gentlewomen haue called it, The Princes Feather, which although it be but a by-name, may well ſerue for this plant to diſtinguiſh it, and whereby to be knowne.

The Vertues.

They are all held to be cold and moiſt, like vnto other Houſeleekes.

CHAP. XXXIII.

Soldanella Alpina. Mountaine Soldanella or blew Moonewort.

THis beautifull plant hath many round and hard leaues, ſet vpon long footeſtalkes, a little vneuenly cut about the edges, greene on the vpperſide, and of a grayiſh greene vnderneath, and ſometime reddiſh like the leaues of Sowbread, which becauſe they doe ſomewhat reſemble the leaues of *Soldanella marina*, which is the Sea Bindweede, tooke the name thereof : the ſtalkes are ſlender, ſmall, round, and reddiſh, about a ſpan high, bearing foure or fiue flowers at the toppe, euery one hanging downe their heads, like vnto a Bell flower, conſiſting but of one leafe (as moſt of the Bindweeds doe) plated into fiue folds, each of them ending in a long point, which maketh the flower ſeem to haue fiue leaues, each whereof is deeply cut in on the edges, and hauing a round greene head in the middle, with a pricke or pointell at the end thereof : the flower is of a faire blew colour, ſometimes deeper or paler, or white, as nature liſteth without any ſmell at all : the middle head, after the flower is fallen, riſeth to be a long round pod, bearing that pricke it had at the end thereof, wherein is contained ſmall greeniſh ſeede : the roote hath many fibres ſhooting from a long round head or roote.

The Place.

This groweth on the Alpes, which are couered with ſnow the greateſt part of the yeare, and will hardly abide tranſplanting.

The Time.

In the naturall places it flowreth not vntill the Summer moneths, Iune, Iuly, and Auguſt, after the ſnow is melted from the Hils, but being brought into Gardens, it flowreth in the beginning of Aprill, or thereabouts.

The

The Names.

This plant, by reason of the likenesse of leaues with *Soldanella*, as was before said, is called by many *Soldanells*, but yet is no Bindweede ; and therefore I rather call it in English a Mountaine Soldanella, then as Gerrard doth, Mountaine Bindweede. It is likewise called by some, *Lunaria minor cærulea,* The lesser blew Lunary or Moonwort, and so I would rather haue it called.

The Vertues.

They that imposed the name of *Lunaria* vpon this plant, seeme to referre it to the wound or consolidating herbes, but because I haue no further relation or experience, I can say no more thereof vntill tryall hath taught it. Some also from the name *Soldanella,* which is giuen it, because of the likenesse of the leaues, haue vsed it to help the Dropsie, for which the Sea plant is thought to be effectuall.

CHAP. XXXIIII.

Auricula Vrsi. Beares eares.

THere are so many sundry and seuerall sorts of Beares eares, the variety consisting as well in the differing colours of the flowers, as the forme and colour of the leaues, that I shall not comprehend and set downe vnto you all the diuersities by many, that are risen vp to those that haue beene industrious in the sowing of the seedes of the seuerall sorts of them ; yet if you accept of these that I doe here offer vnto you, I shall giue you the knowledge of others, as time, occasion, and the view of them shall enable me. And because they are without all question kinds of Cowslips, I haue set them downe before them in the first place, as being of more beautie and greater respect, or at the least of more rarity vnto vs. To dispose them therefore into order, I shall ranke them vnder three principall colours, that is to say, Red or Purple, White, and Yellow, and shew you the varieties of each of them (for so many as are come to my knowledge) apart by themselues, and not promiscuously as many others haue done.

1. *Auricula Vrsi flore purpureo.* Purple Beares eare, or The Murrey Cowslip.

This purple Beares eare or Cowslip hath many greene leaues, somewhat long and smooth, narrow from the bottome of the leafe to the middle, and broad from thence to the end, being round pointed, and somewhat snipt or endented about the edges ; in the middle of these leaues, and sometimes at the sides also, doe spring round greene stalkes foure or fiue fingers high, bearing at the top many flowers, the buds whereof, before they are blowne, are of a very deepe purple colour, and being open, are of a bright, but deepe purple, vsually called a Murrey colour, consisting of fiue leaues a peece, cut in at the end as it were into two, with a whitish ring or circle at the bottome of each flower, standing in small greene cups, wherein after the flowers are fallen, are contained very small heads, not rising to the height of the cups, bearing a small pricke or pointell at the toppe of them, wherein is little blackish seede : the roote hath many whitish strings fastened to the maine long roote, which is very like vnto a Primrose or Cowslip roote, as it is in all other parts besides.

2. *Auricula Vrsi purpurea absq; orbe.* The murrey Cowslip without eyes.

There is another of this kinde, whose leafe is somewhat lesse, as the flower is also,

but

but of the same colour, and sometimes somewhat redder, tending to a Scarlet, without any circle at the bottome of the flower, in no other things differing from it.

3. *Auricula Vrsi minor flore tannetto.* Tawney Beares eares.

The leaues of this kinde haue a greater shew of mealinesse to be seene in them, and not much smaller then the former, yet snipt or endented about the ends like vnto them: the flowers are many, of the same fashion with the former, but smaller, each whereof is of as deepe a murrey or tawnie colour when it is blowne, as the buds of the former are before they are blowne, hauing a white circle at the bottome of the flower, and yellowish in the middle belowe the circle.

4. *Auricula Vrsi flore rubro saturo orbe luteo.* Deepe or bloud red Beares eares with eyes.

This kinde hath small and long greene leaues, nothing mealy, but snipt about the edges, from the middle of the leaues forwards to the ends: the flowers hereof are of a deepe red colour, tending to a bloud red, with a deepe yellow circle, or rather bottome in the middle.

Auricula Vrsi flore rubro saturo absque orbe.

There is another of this kinde, whose leaues are somewhat mealy, and smaller then any (that I haue seene) that haue mealy leaues: the flowers are of the same deepe red colour with the last described, yet hath no circle or bottome of any other colour at all.

5. *Auricula Vrsi flore purpuro caeruleo.* The Violet coloured Beares eare.

We haue another, whose leaues are somewhat mealy and large; the flowers whereof are of a paler purple then the first, somewhat tending to a blew.

6. *Auricula Vrsi flore obsoleto magno.* The Spaniards blush Beares eare.

This great Beares eare hath as large leaues as any other of this kindred whatsoeuer, and whitish or mealy withall, somewhat snipt about the edges, as many other of them are: the flowers stand at the toppe of a strong and tall stalke, larger then any of the other that I haue seene, being of a duskie blush colour, resembling the blush of a Spaniard, whose tawney skinne cannot declare so pure a blush as the English can; and therefore I haue called it the Spaniards blush.

7. *Auricula Vrsi flore rubello.* Scarlet or light red Beares eares.

The leaues of this kinde are very like the leaues of the first purple kinde, but that they are not so thicke, of a little paler greene colour, and little or nothing snipt about the edges: the flowers are of a bright, but pale reddish colour, not halfe so deepe as the two last with white circles in the bottomes of them, in other things this differeth not from others.

8. *Auricula Vrsi Roseo colore.* The Rose coloured Beares eare.

We haue another, whose leafe is a little mealy, almost as large as any of the former, whose flowers are of a light red colour, very neare the colour of an ordinary Damaske Rose, with a white eye at the bottome.

9. *Auricula Vrsi flore caeruleo folio Boraginis.* Blew Beares eares with Borage leaues.

This plant is referred to the kindred or family of the Beares eares, onely for the forme of the flower sake, which euen therein it doth not assimilate to the halfe; but because it hath passed others with that title, I am content to insert it here, to giue you

the

1. *Auricula Vrsi flore purpureo.* Purple Cowslips or Beares eares. 2. *Auricula Vrsi flore tannetto.* Tawney Beares eare. 3. *Auricula Vrsi flore & folio Boraginis.* Blew Beares eares with Borage leaues. 4. *Auricula Vrsi flore carneo* Blush Beares eare. 5. *Auricula Vrsi maxima lutea flore eleganti.* The greatest faire yellow Beares eares with eyes. 6. *Auricula Vrsi altera flore luteo.* The yellow Beares eare. 7. *Auricula Vrsi crinis coloris siue flore fusco.* The haire coloured Beares eare. 8. *Cortusa Matthioli.* Beares eare Sanicle.

the knowledge thereof, and rather to satisfie others then my selfe with the place thereof : the description whereof is as followeth : It hath diuers broad rough hairy leaues spread vpon the ground, somewhat like vnto the leaues of Borage for the roughnesse, but not for the largenesse; the leaues hereof being somewhat rent in some places at the edges : from among these leaues rise vp one, or two, or more brownish, round, and hairy stalkes, a span high or thereabouts, bearing at the toppes three or foure flowers a peece, consisting of fiue large pointed leaues, of a faire blew or light azur colour, with some small yellow threads in the middle, standing in small greene cups : the roote is long and brownish, hauing many small fibres annexed vnto it.

10. *Auricula Vrsi maior flore albo.* The great white Beares eare.

This white Beares eare hath many faire whitish greene leaues, somewhat paler then the leaues of any of the kindes of Beares eares, and a little snipt about the ends, as manie other are : among these leaues rise vp stalkes foure or fiue inches high, bearing at the toppe many flowers like vnto the small yellow Beares eare hereafter set downe, of a pale whitish colour, tending to yellow at the first opening of the flower, which after two or three dayes change into a faire white colour, and so continue all the while it flowreth : the roote is like the purple kinde, as all or most of the rest are, or very little differing.

11. *Auricula Vrsi minor flore albo.* The lesser white Beares eare.

The lesser Beares eare hath smaller leaues, of a little darker green colour : the stalke and flowers are likewise lesser then the former, and haue no shew of yellownesse at all, eyther in budde or flower, but is pure white, differing not in other things from the rest.

12. *Auricula Vrsi maxima lutea flore eleganti.* The greatest faire yellow Beares eare with eyes.

This yellow Beares eare hath many faire large thicke leaues, somewhat mealy or hoary vpon the greennesse, being larger then any other kinde, except the sixth, and the next yellow that followeth, smooth about the edges, and without any endenting at all: the stalke is great, round, and not higher then in other of the former, but bearing manie more flowers thereon then in any other kinde, to the number of thirty many times, standing so round and close together, that they seeme to be a Nosegay alone, of the same fashion with the former, but that the leaues are shorter and rounder, yet with a notch in the middle like the rest, of a faire yellow colour, neither very pale nor deepe, with a white eye or circle in the bottome, about the middle of euery flower, which giueth it the greater grace : the seede is of a blackish browne colour, like vnto others, but contained in greater round heads then any other, with a small pointell sticking in the middle : the roote is greater and thicker then any other, with long strings or fibres like vnto the other sorts, but greater.

13. *Auricula Vrsi maior lutea folio in cano.* The greater yellow Beares eare.

This greater yellow Beares eare hath his leaues larger, and more mealy or hoarie then the last, or any other of these kindes : the flowers are not so many, but longer, and not so thicke thrusting together as the first, but of a deeper yellow colour, without any eye or circle in the middle.

14. *Auricula Vrsi maior flore pallido.* The great Straw coloured Beares eare.

This hath almost as mealy leaues as the last, but nothing so large ; the flowers are of a faire strawe colour, with a white circle at the bottome of them, these three last haue no shew or shadow of any other colour in any part of the edge, as some others that follow haue.

15. *Auricula*

15. *Auricula Vrſi minor flore pallente.* The leſſer ſtraw coloured Beares eare.

We haue another, whoſe leafe is leſſe mealy, or rather pale green, and a little mealy withall; the flowers whereof are of a paler yellow colour then the laſt, and beareth almoſt as many vpon a ſtalke as the firſt great yellow.

16. *Auricula Vrſi minor lutea.* The leſſer yellow Beares eares.

The leaues of this Beares eare are nothing ſo large as either of the three former yellow kindes, but rather of the bigneſſe of the firſt white kinde, but yet a little larger, thicker, and longer then it, hauing vnder the greenneſſe a ſmall ſhew of mealineſſe, and ſomewhat ſnipt about the edges: the flowers are of a pale yellow colour, with a little white bottome in them: the ſeed and rootes are like vnto the other kindes.

17. *Auricula Vrſi flore flauo.* The deepe yellow or Cowſlip Beares eare.

This kinde hath ſomewhat larger leaues then the laſt, of a yellowiſh greene colour, without any mealineſſe on them, or endenting about the edges, but ſmooth and whole: the flowers are not larger but longer, and not laide open ſo fully as the former, but of as deepe a yellow colour as any Cowſlip almoſt, without any circle in the bottome: neither of theſe two laſt haue any ſhew of other colour then yellow in them, ſauing the white in the eye.

18. *Auricula Vrſi verſicolor prima ſiue flore rubeſcente.* The bluſh Beares eare.

The bluſh Beares eare hath his leaues as large, and as hoary or mealy as the third greater yellow, or ſtraw coloured Beares eare; among which riſeth vp a ſtalke about foure inches high, bearing from ſix to twelue, or more faire flowers, ſomewhat larger then the ſmaller yellow Beares eare before deſcribed, hauing the ground of the flower of a darke or dunne yellow colour, ſhadowed ouer a little with a ſhew of light purple, which therefore we call a bluſh colour, the edges of the flower being tipt with a little deeper ſhew of that purple colour, the bottome of the flower abiding wholly yellow, without any circle, and is of very great beauty, which hath cauſed me to place it in the forefront of the variable coloured Beares eares. And although ſome might thinke it ſhould be placed among the firſt ranke of Beares eares, becauſe it is of a bluſh colour, yet ſeeing it is aſſuredly gained from ſome of the yellow kindes by ſowing the ſeede, as many other ſorts are, as may be ſeene plainly in the ground of the flower, which is yellow, and but ſhadowed ouer with purple, yet more then any of the reſt that follow; I thinke I haue giuen it his right place: let others of skill & experience be iudges herein.

19. *Auricula Vrſi crinis coloris.* Haire coloured Beares eares.

The leaues of this kinde are more mealy like then the laſt bluſh kinde, but ſomewhat longer and larger, and ſnipt about the edges in the ſame manner, from the middle of the leafe forwards: the flower is vſually of a fine light browne yellow colour, which wee doe vſually call an Haire colour, and ſometimes browner, the edges of the flower haue a ſhew or ſhadow of a light purple or bluſh about them, but more on the outſide then on the inſide.

20. *Auricula Vrſi verſicolor lutea.* The yellow variable Beares eare.

This variable Beares eare hath his greene leaues ſomewhat like vnto the deepe yellow, or Cowſlip Beares eare before deſcribed, but ſomewhat of a freſher greene, more ſhining and ſmaller, and ſnipt about the edges towards the ends, as many of thoſe before are: the flowers are of a faire yellow colour, much laid open when it is full blowne, that it ſeemeth almoſt flat, daſht about the edges onely with purple, being more yellow in the bottome of the flower, then in any other part.

21. *Auricula*

21. *Auricula Vrsi versicolor lutescente viridi flore.* The variable green Beares eare.

This kinde of Beares eare hath greene leaues, very like vnto the last described, and snipt in the like manner about the edges, but in this it differeth, that his leaues do turne or fold themselues a little backwards: the flowers are of a yellowish greene colour, more closed then the former, hauing purplish edges, especially after they haue stood blowne some time, and haue little or none at the first opening: these haue no circles at all in them.

Many other varieties are to be found, with those that are curious conseruers of these delights of nature, either naturally growing on the mountaines in seuerall places, from whence they (being searched out by diuers) haue been taken and brought, or else raised from the seede of some of them, as it is more probable: for seuerall varieties haue beene obserued (and no doubt many of these before specified) to bee gotten by sowing of the seedes, euery yeare lightly shewing a diuersity, not obserued before, either in the leafe, diuers from that from whence it was taken, or in the flowers. I haue onely set downe those that haue come vnder mine owne view, and not any by relation, euen as I doe with all or most of the things contained in this worke.

The Place.

Many of these goodly plants growe naturally on mountaines, especially the Alpes, in diuers places; for some kindes that growe in some places, doe not in others, but farre distant one from the other. There hath likewise some beene found on the Pyrenæan mountaines, but that kinde with the blew flower and Borage leafe, hath beene gathered on the mountaines in Spaine, and on the Pyrenæans next vnto Spaine.

The Time.

They all flower in Aprill and May, and the seede is ripe in the end of Iune, or beginning of Iuly, and sometimes they will flower againe in the end of Summer, or in Autumne, if the yeare proue temperate, moist, and rainie.

The Names.

It is very probable, that none of these plants were euer knowne vnto the ancient Writers, becaufe we cannot be assured, that they may be truely referred vnto any plant that they name, vnlesse we beleeue Fabius Columna, that it should be *Alisma* of Dioscorides, for thereunto hee doth referre it. Diuers of the later Writers haue giuen vnto them diuers names, euery one according to his owne conceit. For Gesner calleth it *Lunaria arthritica*, and *Paralytica Alpina*. Matthiolus accounteth it to bee of the kindred of the Sanicles, and saith, that in his time it was called by diuers Herbarists, *Auricula Vrsi*, which name hath since bin receiued as most vsuall. We in English call them Beares eares, according to the Latine, or as they are called by diuers women, French Cowslips; they may be called Mountaine Cowslips, if you will, for to distinguish betweene them and other Cowslips, whereof these are seuerall kindes.

Sanicula Alpina siue Cortusa Matthioli. Beares eare Sanicle.

I cannot chuse but insert this delicate plant in the end of the Beares eares, for that it is of so neare affinity, although it differ much in the forme of the leaues, the description whereof is in this manner: The leaues that spring vp first are much crumpled, and as it were folded together, which afterwards open themselues into faire, broad, and roundish leaues, somewhat rough or hairy, not onely cut into fine diuisions, but somewhat notched also about the edges, of a darke greene colour on the vpperside, and

more

more whitish greene vnderneath ; amongst these leaues riseth vp one or two naked round stalkes, fiue or six inches high, bearing at the toppes diuers small flowers, somewhat sweete, like vnto the first purple Beares eare, hanging downe their heads, consisting of fiue small pointed leaues a peece, of a darke reddish purple colour, with a white circle or bottome in the middle, and some small threads therein : after the flowers are past, there come small round heads, somewhat longer then any of the Beares eares, standing vpright vpon their small foot-stalkes, wherein is contained small round and blackish seede : the roote consisteth of a thicke tuft of small whitish threads, rather then rootes, much enterlaced one among another : the leaues of this plant dye downe euery yeare, and spring vp a new in the beginning of the yeare, whereas all the Beares eares doe hold their leaues greene all the Winter, especially the middlemost, which stand like a close head, the outermost for the most part perishing after seed time.

The Place.

This groweth in many shadowie Woods both of Italy and Germany ; for both Clusius hath described it, finding it in the Woods of Austria and Stiria ; and Matthiolus setteth it downe, hauing receiued it from Anthonius Cortusus, who was President of the Garden at Padua, and found it in the woody mountaines of Vicenza, neare vnto Villestagna, whereon (as Matthiolus saith) there is found both with white flowers as well as with blew, but such with white flowers or blew we neuer could see or heare further of.

The Time.

It flowreth much about the time of the Beares eares, or rather a little later, and the seede is ripe with them.

The Names.

Clusius calleth it *Sanicula montana,* and *Sanicula Alpina,* and referreth it to the *Auricula Vrsi,* or Beares eare, which it doth most nearly resemble : but Matthiolus referreth it to the *Cariophyllata* or *Auens,* making it to be of that tribe or family, and calleth it *Cortusa* of him that first sent it him. Wee may call it eyther *Cortusa,* as for the most part all Herbarists doe, or Beares eare Sanicle as Gerrard doth.

The Vertues.

All the sorts of Beares eares are Cephalicall, that is, conducing helpe for the paines in the head, and for the giddinesse thereof, which may happen, eyther by the fight of steepe places subiect to danger, or otherwise. They are accounted also to be helping for the Palsey, and shaking of the ioynts ; and also as a Sanicle or wound-herbe. The leaues of the *Cortusa* taste a little hot, and if one of them bee laide whole, without bruising, on the cheeke of any tender skind woman, it will raise an orient red colour, as if some *fucus* had beene laide thereon, which will passe away without any manner of harme, or marke where it lay : This is Cortusus his obseruation. Camerarius in his *Hortus Medicus* saith, that an oyle is made thereof, that is admirable for to cure wounds.

X Chap.

Chap. XXXV.

Primula veris & Paralysis. Primroses and Cowslips.

WE haue so great variety of Primroses and Cowslips of our owne Country breeding, that strangers being much delighted with them, haue beene often furnished into diuers Countries, to their good content: And that I may set them downe in some methodicall manner, as I haue done other things, I will first set downe all the sorts of those we call Primroses, both single and double, and afterwards the Cowslips with their diuersities, in as ample manner as my knowledge can direct me. And yet I know, that the name of *Primula veris* or Primrose, is indifferently conferred vpon those that I distinguish for *Paralyses* or Cowslips. I doe therefore for your better vnderstanding of my distinction betweene Primroses and Cowslips, call those onely Primroses that carry but one flower vpon a stalke, be they single or double, except that of Master Hesket, and that with double flowers many vpon a stalke, set out in Gerards Herball, which is his onely, not found (as I thinke) *in rerum natura*, I am sure, such a one I could neuer heare of: And those Cowslips, that beare many flowers vpon a stalke together, constantly, be they single or double also. I might otherwise distinguish them also by the leafe; that all the Primroses beare their long and large broad yellowish greene leaues, without stalkes most vsually; and all the Cowslips haue small stalkes vnder the leaues, which are smaller, and of a darker greene, as vsually, but that this distinction is neither so certaine and generall, nor so well knowne.

1. *Primula veris flore albo.* The single white Primrose.

The Primrose that groweth vnder euery bush or hedge, in all or most of the Woods, Groues, and Orchards of this Kingdome, I may well leaue to his wilde habitation, being not so fit for a Garden, and so well knowne, that I meane not to giue you any further relation thereof: But we haue a kinde hereof which is somewhat smaller, and beareth milke white flowers, without any shew of yellownesse in them, and is more vsually brought into Gardens for the rarity, and differeth not from the wilde or ordinary kinde, either in roote or leafe, or any thing else, yet hauing those yellow spots, but smaller, and not so deepe, as are in the other wilde kinde.

2. *Primula veris flore viridi simplici.* The single greene Primrose.

The single greene Primrose hath his leaues very like vnto the greater double Primrose, but smaller, and of a sadder greene colour: the flowers stand seuerally vpon long foot-stalkes, as the first single kinde doth, but larger then they, and more laide open, of the same, or very neare the same yellowish greene colour that the huske is of, so that at the first opening, the huske and the flower seeme to make one double greene flower, which afterwards separating themselues, the single flower groweth aboue the huske, and spreadeth it selfe open much more then any other single Primrose doth, growing in the end to be of a paler greene colour.

3. *Primula veris flore viridante & albo simplici.*
The single greene and white Primrose.

The leaues of this differ in a manner nothing from the former, neither doth the flower but only in this, that out of the large yellowish green huskes, which contain the flowers of the former, there commeth forth out of the middle of each of them either a small peece of a whitish flower, or else a larger, sometimes making vp a whole flower, like an ordinary Primrose.

4. *Primula veris flore viridi duplici.* The double greene Primrose.

This double Primrose is in his leaues so like the former single greene kindes, that
the

1 *Primula veris flore albo.* The white Primrose. 2 *Primula veris flore viridi & albo simplici.* The green and white Primrose. 3 *Primula veris flore viridi duplici.* The double green Primrose. 4 *Primula veris Hesketi.* Master Heskets double Primrose. 5 *Primula veris pleno vulgaris.* The ordinary double Primrose. 6 *Paralysis veris flore viridante simplici.* The single green Cowslip. 7 *Paralysis flore geminato odorato* Double Cowslips or hose in hose. 8 *Paralysis inodora flore geminato.* Double Oxelips hose in hose. 9 *Paralysis flore & calice crispo.* Curld Cowslips or Gaskins. 10 *Paralysis minor angustifolia flore rubro.* Red Birds eyen. 11 *Paralysis hortensis flore pleno vulgaris.* Double Paigles. 12 *Paralysis fatua.* The foolish Cowslip, or Iacke an Apes on horse backe. 13 *Paralysis flore viridi roseo calamistrato.* The double greene feathered Cowslip.

the one cannot be knowne from the other vntill it come to flower, and then it beareth vpon euery stalke a double green flower, of a little deeper green colour then the flower of the former single kinde consisting but of two rowes of short leaues most visually, and both of an equall height aboue the huske, abiding a pretty time in flower, especially if it stand in any shadowed place, or where the Sun may come but a while vnto it.

5. *Primula veris Hesketi flore multiplici separatim diuiso.* Master Heskets double Primrose.

Master Heskets double Primrose is very like vnto the small double Primrose, both in leafe, roote, and heigth of growing, the stalke not rising much higher then it, but bearing flowers in a farre different manner; for this beareth not only single flowers vpon seuerall stalkes, but somtimes two or three single flowers vpon one stalk, and also at the same time a bigger stalke, and somewhat higher, hauing one greene huske at the toppe thereof, sometimes broken on the one side, and sometimes whole, in the middle whereof standeth sometimes diuers single flowers, thrust together, euery flower to be seene in his proper forme, and sometimes there appeare with some whole flowers others that are but parts of flowers, as if the flowers were broken in peeces, and thrust into one huske, the leaues of the flowers (being of a white or pale Primrose colour, but a little deeper) seldome rising aboue the height of the very huske it selfe; and sometimes as I haue obserued in this plant, it will haue vpon the same stalke, that beareth such flowers as I haue here described vnto you, a small flower or two, making the stalke seeme branched into many flowers, whereby you may perceiue, that it will vary into many formes, not abiding constant in any yeare, as all the other sorts doe.

6. *Primula hortensis flore pleno vulgaris.* The ordinary double Primrose.

The leaues of this Primrose are very large, and like vnto the single kind, but somewhat larger, becaufe it groweth in gardens: the flowers doe stand euery one seuerally vpon slender long footestalkes, as the single kinde doth, in greenish huskes of a pale yellow colour, like vnto the field Primrose, but very thicke and double, and of the same sweete sent with them.

7. *Primula veris flore duplici.* The small double Primrose.

This Primrose is both in leafe, roote, and flower, altogether like vnto the last double Primrose, but that it is smaller in all things; for the flower riseth not aboue two or three fingers high, and but twice double, that is, with two rowes of leaues, yet of the very same Primrose colour that the former is of.

8. *Paralysis vulgaris pratensis flore flauo simplici odorato.* The Common field Cowslip.

The common fielde Cowslip I might well forbeare to set downe, being so plentifull in the fields: but becaufe many take delight in it, and plant it in their gardens, I will giue you the description of it here. It hath diuers green leaues, very like vnto the wilde Primrose, but shorter, rounder, stiffer, rougher, more crumpled about the edges, and of a sadder greene colour, euery one standing vpon his stalke, which is an inch or two long: among the leaues rise vp diuers round stalkes, a foote or more high, bearing at the toppe many faire yellow single flowers, with spots of a deeper yellow, at the bottome of each leafe, smelling very sweete. The rootes are like to the other Primroses, hauing many fibres annexed to the great roote.

9. *Paralysis altera odorata flore pallido polyanthos.* The Primrose Cowslip.

The leaues of this Cowslip are larger then the ordinary fielde Cowslip, and of a darke yellowish greene colour: the flowers are many standing together, vpon the toppes of the stalkes, to the number of thirty sometimes vpon one stalke, as I haue counted them in mine owne Garden, and sometimes more, euery one hauing a longer

foote

foote ſtalke then the former, and of as pale a yellowiſh colour almoſt as the fielde
Primroſe, with yellow ſpots at the bottome of the leaues; as the ordinary hath, and of
as ſweet a ſent.

10. *Paralyſis flore viridante ſimplici.* The ſingle greene Cowſlip.

There is little difference in leafe or roote of this from the firſt Cowſlip, the chiefeſt
varietie in this kinde is this, that the leaues are ſomewhat greener, and the flowers being in all reſpects like in forme vnto the firſt kinde, but ſomewhat larger, are of
the ſame colour with the greene huskes, or rather a little yellower, and of a very
ſmall ſent; in all other things I finde no diuerſitie, but that it ſtandeth much longer in
flower before it fadeth, eſpecially if it ſtand out of the Sunne.

11. *Paralyſis flore & calice criſpo.* Curl'd Cowſlips or Gallegaskins.

There is another kinde, whoſe flowers are folded or crumpled at the edges, and the
huskes of the flowers bigger than any of the former, more ſwelling out in the middle,
as it were ribbes, and crumpled on the ſides of the huskes, which doe ſomewhat
reſemble mens hoſe that they did weare, and tooke the name of Gallegaskins from
thence.

12. *Paralyſis flore geminato odorato.*
Double Cowſlips one within another, or Hoſe in Hoſe.

The only difference of this kinde from the ordinary field Cowſlip is, that it beareth
one ſingle flower out of another, which is as a greene huske, of the like ſent that the
firſt hath, or ſomewhat weaker.

13. *Paralyſis flore flauo ſimplici inodoro abſque calicibus.* Single Oxe lippes.

This kinde of Cowſlip hath leaues much like the ordinary kinde, but ſomewhat
ſmaller : the flowers are yellow like the Cowſlip, but ſmaller, ſtanding many vpon
a ſtalke, but bare or naked, that is, without any huske to containe them, hauing but
little or no ſent at all; not differing in any thing elſe from the ordinary Cowſlip.

14. *Paralyſis flore geminato inodora.* Double Oxelips Hoſe in Hoſe.

As the former double Cowſlip had his flowers one within another, in the very like
manner hath this kinde of Cowſlip or Oxelippe, ſauing that this hath no huske to
containe them, no more then the former ſingle Oxelippe hath, ſtanding bare or naked,
of the very ſame bigneſſe each of them, and of the ſame deepe yellow colour with it,
hauing as ſmall a ſent as the former likewiſe.

Wee haue another of this kinde, whoſe leaues are ſomewhat larger, and ſo are the *Flore pallidiore*
flowers alſo, but of a paler yellow colour.

15. *Paralyſis inodora calicibus diſſectis.* Oxelips with iagged huskes.

This kinde differeth not from the firſt Oxelip in the ſmalneſſe of the greene leaues,
but in the flower, which ſtanding many together on a reaſonable high ſtalke, and being
very ſmall and yellow, ſcarce opening themſelues or layde abroade as it, hath a greene
huske vnder each flower, but diuided into ſixe ſeuerall ſmall long peeces.

16. *Paralyſis flore fatuo.* The Franticke, or Fooliſh Cowſlip :
Or Iacke an apes on horſe backe.

Wee haue in our gardens another kinde, not much differing in leaues from the former Cowſlip, and is called Fantaſticke, or Fooliſh, becauſe it beareth at the toppe of
the ſtalke a buſh or tuft of ſmall long greene leaues, with ſome yellow leaues, as it
were peeces of flowers broken, and ſtanding among the greene leaues. And ſometimes

some stalkes among those greene leaues at the toppe (which are a little larger then when it hath but broken peeces of flowers) doe carry whole flowers in huskes like the single kinde.

17. *Paralyſis minor flore rubro.* Red Birds eyes.

This little Cowſlippe(which will hardly endure in our gardens,for all the care and induſtrie we can vſe to keepe it) hath all the Winter long, and vntill the Spring begin to come on, his leaues ſo cloſed together, that it ſeemeth a ſmall white head of leaues, which afterwards opening it ſelfe, ſpreadeth round vpon the ground, and hath ſmall long and narrow leaues, ſnipt about the edges, of a pale greene colour on the vpper-ſide,& very white or mealy vnderneath, among theſe leaues riſe vp one or two ſtalks, ſmall & hoary,halfe a foot high,bearing at the top a buſh or tuft of much ſmaller flowers,ſtanding vpon ſhort foot ſtalkes, ſomewhat like vnto Cowſlips,but more like vnto the Beares eares,of a fine reddiſh purple colour, in ſome deeper,in others paler, with a yellowiſh circle in the bottomes of the flowers,like vnto many of the Beares eares, of a faint or ſmall ſent : the ſeede is ſmaller than in any of the former kindes, and ſo are the rootes likewiſe,being ſmall,white and threddy.

18. *Paralyſis minor flore albo.* White Birds eyes.

This kinde differeth very little or nothing from the former , ſauing that it ſeemeth a little larger both in leafe and flower, and that the flowers hereof are wholly white, without any great appearance of any circle in the bottome of them , vnleſſe it be well obſerued, or at leaſt being nothing ſo conſpicuous, as in the former.

Flore geminato. Theſe two kindes haue ſometimes , but very ſeldome, from among the middle of the flowers on the ſtalke,ſent out another ſmall ſtalke,bearing flowers theron likewiſe.

19. *Paralyſis hortenſis flore pleno.* Double Paigles or Cowſlips.

The double Paigle or Cowſlip hath ſmaller and darker greene leaues then the ſingle kinde hath, and longer ſtalkes alſo whereon the leaues doe ſtand : it beareth diuers flowers vpon a ſtalke , but not ſo many as the ſingle kinde , euery one whereof is of a deeper and fairer yellow colour then any of the former, ſtanding not much aboue the brimmes of the huskes that hold them, conſiſting of two or three rowes of leaues ſet round together, which maketh it ſhew very thicke and double, of a prettie ſmall ſent, but not heady.

20. *Paralyſis flore viridante pleno.* Double greene Cowſlips.

This double greene Cowſlip is ſo like vnto the ſingle greene kinde formerly ex-preſſed, that vntill they be neare flowring,they can hardly be diſtinguiſhed: but when it is in flower, it hath large double flowers , of the ſame yellowiſh greene colour with the ſingle, and more laid open then the former double Paigle.

21. *Paralyſis flore viridante ſiue calamiſtrato.*
The greene Roſe Cowſlip, or double greene feathered Cowſlip.

There is ſmall difference in the leaues of this double kinde from the laſt , but that they are not of ſo darke a greene : the chiefeſt difference conſiſteth in the flowers, which are many, ſtanding together at the toppes of the ſtalkes, but farre differing from all other of theſe kindes : for euery flower ſtanding vpon his owne ſtalke, is compoſed of many very ſmall and narrow leaues , without any huske to containe them , but ſpreading open like a little Roſe, of a pale yellowiſh greene colour , and without any ſent at all, abiding in flower, eſpecially if it ſtand in a ſhadowie place out of the ſunne, aboue two moneths, almoſt in as perfect beauty, as in the firſt weeke.

The Place.

All theſe kindes as they haue been found wilde, growing in diuers places

in

in England, so they haue been transplanted into Gardens, to be there nourished for the delight of their louers, where they all abide, and grow fairer
then in their naturall places, except the small Birds eyes, which will (as I
said) hardly abide any culture, but groweth plentifully in all the North
Countries, in their squally or wet grounds.

The Time.

These doe all flower in the Spring of the yeare, some earlier and some
later, and some in the midst of Winter, as they are defended from the colds
and frosts, and the mildnesse of the time will permit : yet the Cowslips doe
alwayes flower later then the Primroses, and both the single and double
greene Cowslips latest, as I said in their descriptions, and abide much after
all the rest.

The Names.

All these plants are called most vsually in Latine, *Primulæ veris, Primulæ
pratenses,* and *Primulæ siluarum,* because they shew by their flowring the new
Spring to bee comming on, they being as it were the first Embassadours
thereof. They haue also diuers other names, as *Herba Paralysis, Arthritica,
Herba Sancti Petri, Claues Sancti Petri, Verbasculum odoratum, Lunaria arthritica, Phlomis, Alisma siluarum,* and *Alismatis alterum genus,* as Fabius Columna calleth them. The Birds eyes are called of Lobel in Latine, *Paralytica Alpina, Sanicula angustifolia,* making a greater and a lesser. Others call
them *Sanicula angustifolia,* but generally they are called *Primula veris minor.*
I haue (as you see) placed them with the Cowslips, putting a difference betweene Primroses and Cowslips. And some haue distinguished them, by
calling the Cowslips, *Primula veris Elatior,* that is, the Taller Primrose, and
the other *Humilis,* Lowe or Dwarfe Primroses. In English they haue in like
manner diuers names, according to seuerall Countries, as Primroses, Cowslips, Oxelips, Palsieworts, and Petty Mulleins. The first kindes, which are
lower then the rest, are generally called by the name of Primroses (as I
thinke) throughout England. The other are diuersly named ; for in some
Countries they call them Paigles, or Palsieworts, or Petty Mulleins, which
are called Cowslips in others. Those are vsually called Oxelips, whose
flowers are naked, or bare without huskes to containe them, being not so
sweete as the Cowslip, yet haue they some little sent, although the Latine
name doth make them to haue none. The Franticke, Fantasticke, or Foolish
Cowslip, in some places is called by Country people, Iacke an Apes on
horse-backe, which is an vsuall name with them, giuen to many other
plants, as Daisies, Marigolds, &c. if they be strange or fantasticall, differing in the forme from the ordinary kinde of the single ones. The smallest
are vsually called through all the North Country, Birds eyen, because of
the small yellow circle in the bottomes of the flowers, resembling the eye
of a bird.

The Vertues.

Primroses and Cowslips are in a manner wholly vsed in Cephalicall diseases, either among other herbes or flowers, or of themselues alone, to case
paines in the head, and is accounted next vnto Betony, the best for that purpose. Experience likewise hath shewed, that they are profitable both for
the Palsie, and paines of the ioynts, euen as the Beares eares are, which
hath caused the names of *Arthritica, Paralysis,* and *Paralytica,* to bee giuen
them. The iuice of the flowers is commended to cleanse the spots or marks
of the face, whereof some Gentlewomen haue found good experience.

CHAP.

Pulmonaria. Lungwort, or Cowflips of Ierufalem.

ALthough thefe plants are generally more vfed as Pot-herbes for the Kitchen, then as flowers for delight, yet becaufe they are both called Cowflips, and are of like forme, but of much leffe beauty, I haue ioyned them next vnto them, in a diftinct Chapter by themfelues, and fo may paffe at this time.

1. *Pulmonaria maculofa.* Common fpotted Cowflips of Ierufalem.

The Cowflip of Ierufalem hath many rough, large, and round leaues, but pointed at the ends, ftanding vpon long foot-ftalkes, fpotted with many round white fpots on the vpperfides of the fad greene or browne leaues, and of a grayer greene vnderneath: among the leaues fpring vp diuers browne ftalkes, a foote high, bearing many flowers at the toppe, very neare refembling the flowers of Cowflips, being of a purple or red-difh colour while they are buds, and of a darke blewifh colour when they are blowne, ftanding in brownifh greene huskes, and fometimes it hath beene found with white flowers: when the flowers are paft, there come vp fmall round heads, containing blacke feed: the roote is compofed of many long and thicke blacke ftrings.

2. *Pulmonaria altera non maculofa.* Vnfpotted Cowflips of Ierufalem.

The leaues of this other kinde are not much vnlike the former, being rough as they are, but fmaller, of a fairer greene colour aboue, and of a whiter greene vnderneath, without any fpots at all vpon the leaues: the flowers alfo are like the former, and of the fame colour, but a little more branched vpon the ftalkes then the former: the rootes alfo are blacke like vnto them.

3. *Pulmonaria anguftifolia.* Narrow leafed Cowflips of Ierufalem.

The leaues hereof are fomewhat longer, but not fo broad, and fpotted with whitifh fpots alfo as the former: the ftalke hereof is fet with the like long hairy leaues, but fmaller, being a foote high or better, bearing at the toppe many flowers, ftanding in huskes like the firft, being fomewhat reddifh in the bud, and of a darke purplifh blew colour when they are blowne open: the feede is like the former, all of them doe well refemble Bugloffe and Comfrey in moft parts, except the roote, which is not like them, but ftringie, like vnto Cowflips, yet blacke.

The Place.

The Cowflips of Ierufalem grow naturally in the Woods of Germany, in diuers places, and the firft kinde in England alfo, found out by Iohn Goodier, a great fearcher and louer of plants, dwelling at Maple-durham in Hampfhire.

The Time.

They flower for the moft part very early, that is, in the beginning of Aprill.

The Names.

They are generally called in Latine, *Pulmonaria,* and *maculofa,* or *non ma-culofa,* is added for diftinctions fake. Of fome it is called *Symphitum macu-lofum,* that is, fpotted Comfrey. In Englifh it is diuerfly called; as fpotted Cowflips of Ierufalem, Sage of Ierufalem, Sage of Bethlehem, Lungwort, and

and spotted Comfrey , and it might bee as fitly called spotted Buglosse, whereunto it is as like as vnto Comfrey, as I said before.

The Vertues.

It is much commended of some , to bee singular good for vlcered lungs, that are full of rotten matter. As also for them that spit bloud, being boyled and drunke. It is of greatest vse for the pot , being generally held to be good, both for the lungs and the heart.

Chap. XXXVII.

1. *Buglossum & Borrago.* Buglosse and Borage.

ALthough Borage and Buglosse might as fitly haue been placed , I confesse, in the Kitchen Garden , in regard they are wholly in a manner spent for Physicall properties, or for the Pot , yet because anciently they haue been entertained into Gardens of pleasure, their flowers hauing been in some respect , in that they haue alwaies been enterposed among the flowers of womens needle-worke, I am more willing to giue them place here, then thrust them into obscurity, and take such of their tribe with them also as may fit for this place, either for beauty or rarity.

The Garden Buglosse and Borage are so well knowne vnto all, that I shall (I doubt) but spend time in waste to describe them ; yet not vsing to passe ouer any thing I name and appropriate to this Garden so sleightly , they are thus to bee knowne : Buglosse hath many long, narrow, hairy, or rough sad greene leaues, among which rise vp two or three very high stalks, branched at the top, whereon stand many blew flowers, consisting of fiue small round pointed leaues, with a small pointell in the middle, which are very smooth, shining, and of a reddish purple while they are buds, and not blowne open, which being fallen, there groweth in the greene huske, wherein the flower stood, three or foure roundish blacke seedes , hauing that thread or pointell standing still in the middle of them : the roote is blacke without, and whitish within, long, thicke, and full of slimie iuice (as the leaues are also) and perisheth not euery yeare, as the roote of Borage doth.

2. *Borrago.* Borage.

Borage hath broader, shorter, greener, and rougher leaues then Buglosse, the stalkes hereof are not so high, but branched into many parts , whereon stand larger flowers, and more pointed at the end then Buglosse , and of a paler blew colour for the most part (yet sometimes the flowers are reddish , and sometimes pure white) each of the flowers consisting of fiue leaues , standing in a round hairy whitish huske , diuided into fiue parts, and haue a small vmbone of fiue blackish threads in the middle, standing out pointed at the end, and broad at the bottome: the seed is like the other : the root is thicker and shorter then the roote of Buglosse , somewhat blackish without also , and whitish within, and perisheth after seede time , but riseth of it owne seede fallen, and springeth in the beginning of the yeare.

3. *Borrago semper virens.* Euerliuing Borage.

Euerliuing Borage hath many broad greene leaues, and somewhat rough, more resembling Comfrey then Borage , yet not so large as either ; the stalkes are not so high as Borage , and haue many small blew flowers on them , very like to the flowers of Buglosse for the forme, and Borage for the colour : the rootes are blacke, thicker then either of them , somewhat more spreading , and not perishing, hauing greene leaues all the Winter long, and thereupon tooke his name.

4. Anchusa.

4. *Anchusa.* Sea Buglosse or Alkanet.

The Sea Buglosse or Alkanet hath many long, rough, narrow, and darke greene leaues, spread vpon the ground (yet some that growe by the Sea side are rather hoarie and whitish) among these leaues riseth vp a stalke, spread at the toppe into many branches, whereon stand the flowers in tufts, like vnto the Garden Buglosse, or rather Comfrey, but lesser; in some plants of a reddish blew colour, and in others more red or purplish, and in others of a yellowish colour: after which come the seedes, very like vnto Buglosse, but somewhat longer and paler: the roote of most of them being transplanted, are somewhat blackish on the outside, vntill the later end of Summer, and then become more red: for those that grow wilde, will be then so red, that they will giue a very deepe red colour to those that handle them, which being dryed keepe that red colour, which is vsed to many purposes; the roote within being white, and hauing no red colour at all.

5. *Limonium Rauwolfij.* Marsh Buglosse.

This Limonium (which I referre here to the kindes of Buglosse, as presuming it is the fittest place whereto insert it)hath many long,narrow,and somewhat rough leaues lying vpon the ground, waued or cut in on both sides,like an Indenture,somewhat like the leaues of Ceterach or Miltwast, among which rise vp two or three stalkes,somewhat rough also, and with thin skinnes like wings,indented on both sides thereof also, like the leaues, hauing three small,long, rough, and three square leaues at euery ioynt where it brancheth forth; at the toppe whereof stand many flowers vpon their foote stalkes, in such a manner,as is not seene in any other plant, that I know : for although that some of the small winged foot stalkes are shorter, and some longer, standing as it were flatwise,or all on one side, and not round like an vmbell, yet are they euen at the toppe,and not one higher than another ; each of which small foote stalkes doe beare foure or fiue greenish heads or huskes, ioyned together, out of each of which doe arise other pale or bleake blew stiffe huskes, as if they were flowers,made as it were of parchment, which hold their colour after they are dry a long time; and out of these huskes likewise, doe come (at seuerall times one after another, and not all at one time or together) white flowers, consisting of fiue small round leaues, with some white threds in the middle : after these flowers are past, there come in their places small long seede, inclosed in many huskes, many of those heads being idle, not yeelding any good seede, but chaffe, especially in our Countrey, for the want of sufficient heate of the Sunne, as I take it : the roote is small, long,and blackish on the outside, and perisheth at the first approach of Winter.

The Place.

Borage and Buglosse grow onely in Gardens with vs, and so doth the *Semper virens*, his originall being vnknowne vnto vs. Alkanet or Sea Buglosse groweth neare the Sea, in many places of France, and Spaine, and some of the kindes also in England. But the Limonium or Marshe Buglosse groweth in Cales, and Malacca in Spaine, and is found also in Syria, as Rauwolfius relateth : and in other places also no doubt ; for it hath beene sent vs out of Italie, many yeares before eyther Guillaume Boel found it in Cales, or Clusius in Malacca.

The Time.

Borage and Buglosse doe flower in Iune, and Iuly, and sometimes sooner, and so doth the euer-liuing or neuer dying Borage,but not as Gerrard saith, flowring Winter and Summer, whereupon it should take his name, but leaueth flowring in Autumne, and abideth greene with his leaues all the Winter,

1 *Pulmonaria latifolia maculosa.* Cowſlips of Ieruſalem.　2 *Pulmonaria anguſtifolia.* Narrow leaſed Cowſlips of Ieruſalem.　3 *Borrago.* Borage.　4 *Borrago ſemper virens.* Euerliuing Borage.　5 *Anchuſa.* Sea Bugloſſe or Alkanet.　6 *Limonium Rauwolfij.* Marſh Bugloſſe.

ter, flowring the next Spring following. The other flower not vntill Iuly, and so continue, especially the Marshe Buglosse vntill September bee well spent, and then giueth seede, if early frosts ouertake it not; for it seldome commeth to be ripe.

The Names.

Our ordinary Borage by the consent of all the best moderne Writers, is the true *Buglossum* of Dioscorides, and that our Buglosse was vnknowne to the ancients. The *Borago semper virens*, Lobel calleth *Buglossum semper virens*, that is, Euer-liuing, or greene Buglosse : but it more resembleth Borage then Buglosse ; yet because Buglosse abideth greene, to auoyde that there should not be two *Buglossa semper virentia*, I had rather call it Borage then Buglosse. Anchusa hath diuers names, as Dioscorides setteth downe. And some doe call it *Fucus herba*, from the Greeke word, because the roote giuing so deepe a colour, was vsed to dye or paint the skinne. Others call it *Buglossum Hispanicum*, in English Alkanet, and of some Orchanet, after the French. Limonium was found by Leonhartus Rauwolfius, neere vnto Ioppa, which he setteth downe in the second Chapter of the third booke of his trauayles, and from him first knowne to these parts : I haue, as you see, referred it to the kindes of Buglosse, for that the flowers haue some resemblance vnto them, although I know that *Limonium genuinum* is referred to the Beetes. Let it therefore here finde a place of residence, vntill you or I can finde a fitter; and call it as you thinke best, eyther Limonium as Rauwolfius doth, or Marshe Buglosse as I doe, or if you can adde a more proper name, I shall not be offended.

The Vertues.

Borage and Buglosse are held to bee both temperate herbes, beeing vsed both in the pot and in drinkes that are cordiall, especially the flowers, which of Gentlewomen are candid for comfitts. The Alkanet is drying, and held to be good for wounds, and if a peece of the roote be put into a little oyle of Peter or Petroleum, it giueth as deepe a colour to the oyle, as the Hypericon doth or can to his oyle, and accounted to be singular good for a cut or greene wound.

The Limonium hath no vse that wee know, more then for a Garden; yet as Rauwolfius saith, the Syrians vse the leaues as sallats at the Table.

Chap. XXXVIII.

Lychnis. Campions.

THere bee diuers sorts of Campions, as well tame as wilde, and although some of them that I shall here entreate of, may peraduenture be found wilde in our owne Countrey, yet in regard of their beautifull flowers, they are to bee respected, and nourled vp with the rest, to furnish a garden of pleasure; as for the wilde kindes, I will leaue them for another discourse.

1. *Lychnis Coronaria rubra simplex.* The single red Rose Campion.

The single red Rose Campion hath diuers thicke, hoary, or woolly long greene leaues, abiding greene all the winter, and in the end of the spring or beginning of summer, shooteth forth two or three hard round woolly stalkes, with some ioynts thereon, and at euery ioynt two such like hoary greene leaues as those below, but smaller, diuersly branched at the toppe, hauing one flower vpon each seuerall long foot stalke,

consisting

confisting of fiue leaues, fomewhat broade and round pointed, of a perfect red crimfon colour, ftanding out of a hard long round huske, ridged or crefted in foure or fiue places; after the flowers are fallen there come vp round hard heads, wherein is contained fmall blackifh feed: the roote is fmall, long and wooddy, with many fibres annexed vnto it, and fhooteth forth anew oftentimes, yet perifheth often alfo.

2. *Lychnis Coronaria alba simplex.* The white Rofe Campion.

The white Rofe Campion is in all things like the red, but in the colour of the flower, which in this is of a pure white colour.

3. *Lichnis Coronaria albefcens fiue incarnata maculata & non maculata.* The blufh Rofe Campion fpotted and not fpotted.

Like vnto the former alfo are thefe other forts, hauing no other difference to diftinguifh them, but the flowers, which are of a pale or bleake whitifh blufh colour, efpecially about the brims, as if a very little red were mixed with a great deale of white, the middle of the flower being more white; the one being fpotted all ouer the flower, with fmall fpots and ftreakes, the other not hauing any fpot at all.

4. *Lychnis Coronaria rubra multiplex.* The double red Rofe Campion.

The double red Rofe Campion is in all refpects like vnto the fingle red kinde, but that this beareth double flowers, confifting of two or three rowes of leaues at the moft, which are not fo large as the fingle, and the whole plant is more tender, that is, more apt to perifh, then any of the fingle kindes.

5. *Lychnis Chalcedonica flore fimplici miniato.* Single Nonefuch, or Flower of Briftow, or Conftantinople.

This Campion of Conftantinople hath many broad and long greene leaues, among which rife vp fundry ftiffe round hairy ioynted ftalks three foot high, with two leaues euery ioynt: the flowers ftand at the toppes of them, very many together, in a large tuft or vmbell, confifting of fiue fmall long leaues, broade pointed, and notched-in in the middle, of a bright red orenge colour, which being paft, there come in their places fmall hard whitifh heads or feede veffels, containing blacke feede, like vnto the feede of fweet Williams, and hauing but a fmall fent; the roote is very ftringie, faftening it felfe very ftrongly in the ground, whereby it is much encreafed.

Of the fingle kinde there is alfo two or three other forts, differing chiefly in the co- *Flore albo.* lour of the flowers. The one is pure white. Another is of a blufh colour wholly, *Et carnea.* without variation. And a third is very variable; for at the firft it is of a pale red, and *Verficolor.* after a while groweth paler, vntill in the end it become almoft fully white; and all thefe diuerfities of the flowers are fometimes to bee feene on one ftalke at one and the fame time.

6. *Lychnis Chalcedonica flore miniato pleno.* Double Flower of Briftow, or Nonefuch.

This glorious flower being as rare as it is beautifull, is for rootes beeing ftringie, for leaues and ftalkes being hairy and high, and for the flowers growing in tufts, altogether like the firft fingle kinde: but herein confifteth the chiefeft difference, that this beareth a larger vmbell or tuft of flowers at the toppe of the ftalke, euery flower confifting of three or foure rowes of leaues, of a deeper orenge colour then it, which addeth the more grace vnto it, but paffeth away without bearing feede, as moft other double flowers doe, yet recompenceth that defect with encreafe from the roote.

7. *Lychnis plumaria filuefiris fimplex & multiplex.* The featherd wilde Campion fingle and double.

The leaues of this wilde Campion are fomewhat like the ordinary white wilde

Y   Campion

Campion, but not so large, or rather resembling the leaues of sweete Williams, but that they grow not so close, nor so many together: the stalkes haue smaller leaues at the ioynts then those belowe, and branched at the toppe, with many pale, but bright red flowers, iagged or cut in on the edges, like the feathered Pinke, whereof some haue taken it to be a kinde, and some for a kinde of wilde William, but yet is but a wilde Campion, as may be obserued, both by his huske that beareth the flowers, and by the grayish roundish seede, being not of the Family of Pinkes and Gillowers, but (as I said) of the Campions: the roote is full of strings or fibres.

Flore pleno. The double kinde is very like vnto the single kinde, but that it is lower and smaller, and the flowers very double.

8. *Lychnis siluestris flore pleno rubro.* Red Batchelours buttons.

The double wilde Campion (which of our Countrey Gentlewomen is called Batchelours buttons) is very like both in rootes, leaues, stalkes, and flowers vnto the ordinary wilde red Campion, but somewhat lesser, his flowers are not iagged, but smooth, and very thicke and double, so that most commonly it breaketh his short huske, wherein the flower standeth on the one side, seldome hauing a whole huske, and are of a reddish colour.

9. *Lychnis siluestris flore albo pleno.* White Batchelours buttons.

As the leaues of the former double Campion was like vnto the single kinde that had red flowers, so this hath his leaues like vnto the single white kinde, differing in no other thing from it, but in the doublenesse of the flowers, which by reason of the multiplicity of leaues in them thrusting forth together, breaketh his huskes wherein the flowers doe stand, as the other doth, and hath scarce one flower in many that is whole.

10. *Ocymoides arborea semper virens.* Strange Bassil Campion.

This Strange Campion (for thereunto it must bee referred) shooteth forth many round, whitish, wooddy, but brittle stalkes, whereon stand diuers leaues long, and somewhat thicke leaues, set by couples, narrow at the bottome, and broader toward the point, of a very faire greene and shining colour, so that there is more beauty in the greene leaues, which doe so alwaies abide, then in the flowers, which are of a pale red or blush colour, consisting of fiue small long broad pointed leaues, notched in the middle, which doe not lye close, but loosly as it were hanging ouer the huskes: after the flowers are past, there come heads that containe blackish seede: the roote is small, hard, white, and threadie.

11. *Muscipula Lobelij siue Ben rubrum Monspeliensium.* Lobels Catch Flie.

I must needes insert this small plant, to finish this part of the Campions, whereunto it belongeth, being a pretty toye to furnish and decke out a Garden. It springeth vp (if it haue beene once sowne and suffered to shed) in the later end of the yeare most commonly, or else in the Spring with fiue or six small leaues, very like vnto the leaues of Pinkes, and of the same grayish colour, but a little broader and shorter, and when it beginneth to shoote vp for flower, it beareth smaller leaues on the clammy or viscous stalkes (fit to hold any small thing that lighteth on it) being broad at the bottome compassing them, and standing two at a ioynt one against another: the toppes of the stalkes are diuersly branched into seuerall parts, euery branch hauing diuers small red flowers, not notched, but smooth, standing out of small, long, round, stript huskes, which after the flowers are past, containe small grayish seede: the roote is small, and perisheth after it hath giuen seede; but riseth (as is before said) of its owne seede, if it be suffered to shed.

The Place.

The Rose Campions, Flowers of Bristow, or None such, the Bassil Campion,

1 *Lychnis Coronaria simplex.* Single Rose Campion. 2 *Lychnis Coronaria rubra multiplex.* The double red Rose Campion. 3 *Lychnis Chalcedonica simplex.* Single None such, or flower of Bristow. 4 *Lychnis Chalcedonica flore pleno* Double None such, or flower of Bristow. 5 *Lychnis plumaria multiplex.* Pleasant in sight. 6 *Lychnis sylvestris flore pleno rubro.* Red Batchelours Buttons. 7 *Lychnis sylvestris flore pleno albo.* White Batchelours Buttons. 8 *Muscipula Lobelij.* Lobels Catch Flie.

pion, and the Catch Flie, haue been sent vs from beyond the Seas, and are onely nourfed vp in Gardens with vs; the other Campions that are double, haue been naturally so found double wilde (for no art or induftry of man, that euer I could be assured of to be true, be it by neuer so many repetitions of transplantations, and planeticall obseruations (as I haue said in the beginning of this worke) could bring any flower, single by nature, to become double, notwithstanding many affirmations to that purpose, but whatsoeuer hath been found wilde to be double, nature her selfe, and not art hath so produced it) and being brought into Gardens, are there encreafed by slipping, and parting the roote, because they giue no seede.

The Time.

All of them doe flower in the Summer, yet none before May.

The Names.

The first kindes are called *Lychnides satiua*, and *coronaria*, in Englifh generally Rofe Campions. The next is called *Lychnis Chalcedonica*, and *Byzantina*; in Englifh, of some Nonesuch, and of others Flower of Briftow, and after the Latine, Flower of Conftantinople, because it is thought the seede was first brought from thence; but from whence the double of this kinde came, we cannot tell. The names of the others of this kinde, both single and double, are set downe with their descriptions. The feathered Campions are called *Armoraria pratensis*, and *Flos Cuculi*, and of Clufius and others thought to be *Odontitis Plinij*. Some call them in Englifh Crowflowers, and Cuckowe-Flowers; and some call the double hereof, The faire Maide of France. The Baffil Campions were sent ouer among many other seedes out of Italy, by the name of *Ocimoides arborea semper virens*. *Arborea*, because the ftalke is more wooddy and durable then other Campions: And *semper virens*, because the leaues abide greene Winter and Summer. Clufius calleth it *Lychnis semper virens*, because it is certainly a Campion. The laft is diuerfly called of Authors; Lobel calleth it *Muscipula*: Others *Armoraria altera*: Dodonæus *Armerius flos quartus*. Clufius *Lychnis siluestris altera*, in his Spanifh obseruations, and *prima* in his Hiftory of plants, and saith, the learned of Salmantica in Spaine called it, *Ben rubrum*, as Lobel saith, they of Mompelier doe also: and by that name I receiued it firft out of Italy. It hath the name of Catch Flie, of *Muscipula* the Latine word, because the ftalkes in the hot Summer dayes haue a certaine viscous or clammy humour vpon them, whereby it eafily holdeth (as I said before) whatsoeuer small thing, as Flies, &c. lighteth vpon it.

The Vertues.

We know none in these dayes, that putteth any of these to any Phyficall vse, although some haue in former times.

Chap. XXXIX.

Keiri siue Leucoium luteum. Wall-flowers, or Wall Gilloflowers.

There are two forts of Wall-flowers, the one single, the other double, and of each of them there is likewife some differences, as shall be shewed in their descriptions.

1. *Keiri siue Leucoium luteum simplex vulgare.* Common single Wall-flowers.

The common single Wall-flower which groweth wilde abroad, and yet is brought into Gardens, hath sundry small, narrow, long, and darke greene leaues, set without order vpon small round whitish wooddy stalkes, which beare at the tops diuers single yellow flowers one aboue another, euery one hauing foure leaues a peece, and of a very sweete sent: after which come long pods, containing reddish seede: the roote is white, hard and thready.

2. *Keiri siue Leucoium luteum simplex maius.* The great single Wall-flower.

There is another sort of single Wall-flower, whose leaues as well as flowers are much larger then the former: the leaues being of a darker and shining greene colour, and the flowers of a very deepe gold yellow colour, and vsually broader then a twentie shilling peece of gold can couer: the spike or toppe of flowers also much longer, and abiding longer in flower, and much sweeter likewise in sent: the pods for seede are thicker and shorter, with a small point at the end: this is flower to encrease into branches, as also to be encreased by the branches, and more tender to be preserued; for the hard frosts doe cause it to perish, if it be not defended from them

3. *Keiri simplex flore albo.* White Wall-flower.

This Wall-flower hath his leaues as greene as the great kinde, but nothing so large: the flowers stand at the toppe, but not in so long a spike, and consisteth of foure leaues, of a very white colour, not much larger then the common kinde, and of a faint or weaker sent: the pods are nothing so great as the former great one: this is more easie to be propagated and encreased also, but yet will require some care in defending it from the colds of the Winter.

4. *Keiri siue Leucoium luteum vulgare flore pleno.* Common double Wall-flowers.

This ordinary double Wall-flower is in leaues and stalke very like vnto the first single kinde, but that the leaues hereof are not of so deepe a greene colour: the flowers stand at the top of the stalkes one aboue another, as it were a long spike, which flower by degrees, the lowest first, and so vpwards, by which it is a long time in flowring, and is very double, of a gold yellow colour, and very sweete.

5. *Keiri siue Leucoium luteum alterum flore pleno.* Pale double Wall-flowers.

Wee haue another sort of this kinde of double Wall-flower, whose double flowers stand not spike-fashion as the former, but more open spread, and doe all of them blowe open at one time almost, and not by degrees as the other doth, and is of a paler yellow colour, not differing in any thing else, except that the greene leaues hereof are of a little paler greene then it.

6. *Keiri siue Leucoium luteum maius flore pleno ferrugineo.*
Double red Wall-flowers.

We haue also another sort of double Wall-flower, whose leaues are as greene, and almost as large as the great single yellow kinde, or full as bigge as the leaues of the white Wall-flower: the flowers hereof are not much larger then the ordinary, but are of a darker yellow colour then the great single kinde, and of a more brownish or red colour on the vnderside of the leaues, and is as it were striped.

7. *Keiri siue Leucoium maximum luteum flore pleno.*
The greatest double yellow Wall-flower.

This great double Wall-flower is as yet a stranger in England, and therefore what I here write is more vpon relation (which yet I beleeue to be moſt true) then vpon ſight and ſpeculation. The leaues of this Wall flower are as greene and as large, if not larger then the great ſingle kinde : the flowers alſo are of the ſame deepe gold yellow colour with it, but much larger then any of the former double kindes, and of as ſweet a ſent as any, which addeth delight vnto beauty.

The Place.

The firſt ſingle kind is often found growing vpon old wals of Churches, and other houſes in many places of England, and alſo among rubbiſh and ſtones. The ſingle white and great yellow, as well as all the other double kindes, are nourſed vp in Gardens onely with vs.

The Time.

All the ſingle kindes doe flower many times in the end of Autumne, and if the Winter be milde all the Winter long, but eſpecially in the moneths of February, March, and Aprill, and vntill the heate of the Spring doe ſpend them : but the other double kindes doe not continue flowring in that manner the yeare throughout, although very early ſometimes, and very late alſo in ſome places.

The Names.

They are called by diuers names, as *Viola lutea*, *Leucoium luteum*, and *Keiri*, or *Cheiri*, by which name it is chiefly knowne in our Apothecaries ſhops, becauſe there is an oyle made thereof called *Cheirinum* : In Engliſh they are vſually called in theſe parts, Wall-flowers : Others doe call them Bee-flowers ; others Wall-Gilloflowers, Winter-Gilloflowers, and yellow Srocke-Gilloflowers ; but we haue a kinde of Stocke-Gilloflower that more fitly deſerueth that name, as ſhall be ſhewed in the Chapter following

The Vertues.

The ſweetneſſe of the flowers cauſeth them to be generally vſed in Noſegayes, and to decke vp houſes ; but phyſically they are vſed in diuers manners : As a Conſerue made of the flowers, is vſed for a remedy both for the Appoplexie and Palſie. The diſtilled water helpeth well in the like manner. The oyle made of the flowers is heating and reſoluing, good to eaſe paines of ſtrained and pained ſinewes.

Chap. XL.

Leucoium. Stocke-Gilloflower.

THere are very many ſorts of Stocke-Gilloflowers both ſingle and double, ſome of the fields and mountaines, others of the Sea marſhes and medowes ; and ſome nourſed vp in Gardens, and there preſerued by ſeede or ſlippe, as each kinde is apteſt to bee ordered. But becauſe ſome of theſe are fitter for a generall Hiſtory then for this our Garden of Pleaſure, both for that diuers haue no good ſent, others little or no beauty, and to be entreated of onely for the variety, I ſhall ſpare ſo many of them as are not fit for this worke, and onely ſet downe the reſt.

1. *Leucoium ſimplex ſativum diuerſorum colorum.*
Garden Stocke-Gilloflowers ſingle of diuers colours.

Theſe ſingle Stocke-Gilloflowers, although they differ in the colour of their flowers,

1 *Keiri sue Leucoium luteum vulgare.* Common Wall-flowers. 2 *Keiri sue Leucoium luteum maius simplex.* The great single
Wall flower. 3 *Keiri sue Leucoium luteum flore pleno vulgare.* Ordinary double Wall-flowers. 4 *Keiri maius flore pleno ferruginio.* The great double red Wall-flower. 5 *Leucoium satiuum simplex.* Single Stocke-Gilloflowers. 6 *Leucoium satiuum simplex flore striato.* Single striped Stocke-Gilloflowers.

ers, yet are in leafe and manner of growing, one so like vnto another, that vntill they come to flower, the one cannot be w.ll knowne that beareth red flowers, from another that beareth purple; and therfore one description of the plant shall serue, with a declaration of the sundry colours of the flowers. It riseth vp with round whitish woody stalkes, two, three, or foure foot high, whereon are set many long, and not very broad, soft, and whitish or grayish greene leaues, somewhat round pointed, and parted into diuers branches, at the toppes whereof grow many flowers, one aboue another, smelling very sweet, consisting of foure small, long, and round pointed leaues, standing in small long huskes, which turne into long and flat pods, sometimes halfe a foote long, wherein is contained flat, round, reddish seedes, with grayish ringes or circles about them, lying flat all along the middle rib of the pod on both sides: the roote is long, white, and woody, spreading diuers wayes. There is great variety in the colours of the flowers: for some are wholly of a pure white colour, others of a most excellent crimson red colour, others againe of a faire red colour, but not so bright or liuely as the other, some also of a purplish or violet colour, without any spot, marke, or line in them at all. There are againe of all these colours, mixed very variably, as white mixed with small or great spottes, strakes or lines of pure or bright red, or darke red, and white with purple spots and lines; and of eyther of them whose flowers are almost halfe white, and halfe red, or halfe white, and halfe purple. The red of both sorts, and the purple also, in the like manner spotted, striped, and marked with white, differing neyther in forme, nor substance, in any other point.

2. *Leucoium satinum albido luteum simplex.* The single pale yellow Stocke-Gilloflower.

There is very little difference in this kind from the former, for the manner of growing, or forme of leaues or flower. Only this hath greener leaues, and pale yellow almost white flowers, in all other things alike: this is of no great regard, but only for rarity, and diuersity from the rest.

3. *Leucoium Melancholicum.* The Melancholick Gentleman.

This wilde kinde of stocke gilloflower hath larger, longer and greener leaues then any of the former kindes, vneuenly gashed or sinuated on both edges lying on the ground, and a little rough or hairy withall: from among which rise vp the stalks, a yard high or more, and hairy likewise, bearing theron here and there some such like leaues as are below, but smaller, and at the top a great number of flowers, as large or larger then any of the former single kindes, made of 4. large leaues a peece also, standing in such like long huskes, but of a darke or sullen yellowish colour: after which come long roundish pods, wherein lye somewhat long but rounder and greater seede then any stocke gilloflower, and nearer both in pod and seede vnto the *Hesperis* or Dames Violet: this perisheth not vsually after seede bearing, although sometimes it doth.

4. *Leucoium marinum Syriacum.* Leuant stocke gilloflowers.

This kind of stocke gilloflower riseth vp at the first with diuers long and somewhat broad leaues, a little vneuenly dented or waued on the edges, which so continue the first yeare after the sowing: the stalke riseth vp the next yeare to bee two foot high or more, bearing all those leaues on it that it first had, which then do grow lesse sinuated or waued then before: at the top whereof stand many flowers, made of foure leaues a peece, of a delayed purple colour, but of a small sent which turne into very long and narrow flat pods, wherein are contained flat seed like the ordinary stocke gilloflowers, but much larger and of a darke or blackish browne colour: the root is white, and groweth deepe, spreading in the ground, but growing woody when it is in seede, and perisheth afterwards.

5. Leucoÿ alterum genus, flore tam multiplici quam simplici ex seminio oriundum.
Another sort of Stocke gilloflowers bearing as well double
as single flowers from seede.

This kinde of Stocke gilloflower differeth neyther in forme of leaues, stalkes, nor
flowers from the former, but that it oftentimes groweth much larger and taller; so that
whosoeuer shall see both these growing together, shall scarce discerne the difference,
onely it beareth flowers, eyther white, red or purple, wholly or entire, that is, of one
colour, without mixture of other colour in them (for so much as euer I haue obserued,
or could vnderstand by others) which are eyther single, like vnto the former, or very
thicke and double, like vnto the next that followeth; but larger, and growing with
more store of flowers on the long stalke. But this you must vnderstand withall, that
those plants that beare double flowers, doe beare no seede at all, and is very seldome
encreased by slipping or cutting, as the next kinde of double is: but the onely way to
haue double flowers any yeare, (for this kinde dyeth euery winter, for the most part,
after it hath borne flowers, and seldome is preserued) is to saue the seedes of those
plants of this kinde that beare single flowers, for from that seede will rise, some that
will beare single, and some double flowers, which cannot bee distinguished one from
another, I meane which will be single and which double, vntill you see them in flower,
or budde at the least. And this is the only way to preserue this kinde: but of the seed
of the former kinde was neuer known any double flowers to arise, and therefore you
must be carefull to marke this kinde from the former.

6. Leucoium flore pleno diuersorum colorum.
Double Stocke Gillowflowers of diuers colours.

This other kinde of Stock gilloflower that beareth onely double flowers, groweth
not so great, nor spreadeth his branches so farre, nor are his leaues so large, but is in
all things smaller, and lower, and yet is woody, or shrubby, like the former, bearing his flowers in the like manner, many vpon a long stalke, one aboue another, and
very double, but not so large as the former double, although it grow in fertile soyle,
which are eyther white, or red, or purple wholly, without any mixture, or else mixed
with spots and stripes, as the single flowers of the first kinde, but more variably, and
not in all places alike, neuer bearing seede, but must be encreased, only by the cutting
of the young sproutes or branches, taken in a fit season: this kinde perisheth not, as
the former double kinde doth, so as it bee defended in the winter from the extreame
frosts, but especially from the snow falling, or at the least remaining vpon it.

7. Leucoium satiuum luteum flore pleno.
The double yellow Stocke Gilloflower.

This double yellow Stockgilloflower is a stranger in England, as far as I can learne,
neyther haue I any further familiaritie with him, then by relation from Germany,
where it is affirmed to grow only in some of their gardens, that are curious louers of
these delights, bearing long leaues somewhat hoary or white, (and not greene like
vnto the Wallflower, whereunto else it might be thought to be referred) like vnto the
Stock gilloflowers, as the stalkes and branches also are, and bearing faire double
flowers, of a faire, but pale yellow colour. The whole plant is tender, as the double
Stock gilloflowers are, and must be carefully preserued in the winter from the coldes,
or rather more then the last double, lest it perish.

The Place.

The single kindes, especially some of them, grow in Italie, and some in
Greece, Candy, and the Isles adiacent, as may be gathered out of the verses
in Plutarches Booke *De Amore fraterno:*

Inter

Inter Echinopodas velut, asperam & inter Ononim,
Interdum crescunt mollia Leucoia.

Which sheweth, that the soft or gentle stocke gilloflowers doe sometimes grow among rough or prickely Furse and Cammocke. The other sorts are only to be found in gardens.

The Time.

They flower in a manner all the yeare throughout in some places, especially some of the single kindes, if they stand warme, and defended from the windes and cold : the double kindes flower sometimes in Aprill, and more plentifully in May, and Iune ; but the double of seed, flowreth vsually late, and keepeth flowring vnto the winter, that the frostes and colde mistes doe pull it downe.

The Names.

It is called *Leucoium, & Viola alba* : but the name Leucoium (which is in English the white Violet) is referred to diuers plants ; we call it in English generally, Stocke gilloflower, (or as others doe, Stocke gillouer) to put a difference betweene them, and the Gilloflowers and Carnations, which are quite of another kindred, as shall be shewne in place conuenient.

The Vertues.

These haue no great vse in Physick that I know : only some haue vsed the leaues of the single white flowred kinde with salt, to be laid to the wrests of them that haue agues, but with what good successe I cannot say, if it happen well I thinke in one (as many such things else will) it will fayle in a number.

CHAP. XLI.

1. *Hesperis, siue Viola Matronalis.* Dames Violets, or Queenes Gilloflowers.

THe ordinary Dames Violets, or Queene Gilloflowers, hath his leaues broader, greener, and sharper pointed, then the Stock gilloflowers, and a little endented about the edges : the stalkes grow two foot high, bearing many greene leaues vpon them, smaller then those at the bottome, and branched at the toppe, bearing many flowers, in fashion much like the flowers of stocke gilloflowers, consisting of foure leaues in like manner, but not so large, of a faint purplish colour in some, and in others white, and of a pretty sweet sent, especially towards night, but in the day time little or none at all : after the flowers are past, there doe come small long and round pods, wherein is contained, in two rowes, small and long blacke seede : the roote is wholly composed of stringes or fibres, which abide many yeares, and springeth fresh stalks euery yeare, the leaues abiding all the Winter.

2. *Hesperis Pannonica.* Dames Violets of Hungary.

The leaues of this Violet are very like the former, but smoother and thicker, and not at all indented, or cut in on the edges : the flowers are like the former, but of a sullen pale colour, turning themselues, and seldome lying plaine open, hauing many purple veines, and streakes running through the leaues of the flowers, of little or no sent in the day time, but of a very sweete sent in the euening and morning ; the seedes are alike also, but a little browner.

3 *Lysimachia*

1 *Leucoium Melancholicum.* Sullen Stocke-Gilloflowers. 2 *Leucoium sativum flore pleno.* Double Stocke-Gilloflowers. 3 *Leucoium sativum flore pleno vario.* Party coloured Stocke-Gilloflowers. 4 *Leucoium marinum Syriacum.* Leuant Stocke-Gilloflowers. 5 *Hesperis vulgaris.* Dames Violets or Winter Gilloflowers. 6 *Lysimachia lutea siliquosa Virginiana.* The tree Primrose of Virginia. 7 *Viola Lunaris siue Bolbonach.* The white Sattin flower.

3. *Lysimachia lutea siliquosa Virginiana.* The tree Primrose of Virginia.

Vnto what tribe or kindred I might referre this plant, I haue stood long in suspence, in regard I make no mention of any other *Lysimachia* in this work : lest therfore it should lose all place, let me ranke it here next vnto the Dames Violets, although I confesse it hath little affinity with them. The first yeare of the sowing the seede it abideth without any stalke or flowers lying vpon the ground, with diuers long and narrow pale greene leaues, spread oftentimes round almost like a Rose, the largest leaues being outermost, and very small in the middle : about May the next yeare the stalke riseth, which will be in Summer of the height of a man, and of a strong bigge size almost to a mans thumbe, round from the bottome to the middle, where it groweth crested vp to the toppe, into as many parts as there are branches of flowers, euery one hauing a small leafe at the foote thereof : the flowers stand in order one aboue another, round about the tops of the stalks, euery one vpon a short foot-stalke, consisting of foure pale yellow leaues, smelling somewhat like vnto a Primrose, as the colour is also (which hath caused the name) and standing in a greene huske, which parteth it selfe at the toppe into foure parts or leaues, and turne themselues downewards, lying close to the stalke : the flower hath some chiues in the middle, which being past, there come in their places long and cornered pods, sharpe pointed at the vpper end, and round belowe, opening at the toppe when it is ripe into fiue parts, wherein is contained small brownish seed : the roote is somewhat great at the head, and wooddy, and branched forth diuersly, which perisheth after it hath borne seede.

The Place.

The two first grow for the most part on Hils and in Woods, but with vs in Gardens onely.

The last, as may be well vnderstood by the title, came out of Virginia.

The Time.

They flower in May, Iune, and Iuly.

The Names.

The name of *Hesperis* is imposed by most Herbarists vpon the two first plants, although it is not certainly knowne to be the same that Theophrastus doth make mention of, in his sixth Booke and twenty fiue Chapter *de causis plantarum* : but because this hath the like effects to smell best in the euening, it is (as I said) imposed vpon it. It is also called *Viola Marina Matronalis, Hyemalis, Damascena* and *Muschatella* : In English, Dames Violets, Queens Gilloflowers, and Winter Gilloflowers.

The last hath his Latine name in the title as is best agreeing with it, and for the English, although it be too foolish I confesse, yet it may passe for this time till a fitter be giuen, vnlesse you please to follow the Latine, and call it Virginia Loose-strife.

The Vertues.

I neuer knew any among vs to vse these kindes of Violets in Physicke, although by reason of the sharpe biting taste, Dodonæus accounteth the ordinary sort to be a kinde of Rocket, and saith it prouoketh sweating, and vrine : and others affirme it to cut, digest, and cleanse tough flegme. The Virginian hath not beene vsed by any that I know, either inwardly or outwardly.

CHAP.

Chap. XLII.

Viola Lunaris ſiue Bolbonach. The Sattin flower.

VNto the kindes of Stocke-Gilloflowers I thinke fitteſt to adioyne theſe kindes of Sattin-flowers, whereof there are two ſorts, one frequent enough in all our Countrie, the other is not ſo common.

1. *Viola Lunaris vulgaris.* The common white Sattin flower.

The firſt of theſe Sattin flowers, which is the moſt common, hath his leaues broad belowe, and pointed at the end, ſnipt about the edges, and of a darke greene colour: the ſtalkes are round and hard, two foot high, or higher, diuided into many branches, ſet with the like leaues, but ſmaller: the tops of the branches are beſet with many purpliſh flowers, like vnto Dames Violets, or Stocke-Gilloflowers; but larger, being of little ſent: after the flowers are paſt, there come in their places round flat thin cods, of a darke colour on the outſide, but hauing a thinne middle skinne, that is white and cleare ſhining, like vnto very pure white Sattin it ſelfe, whereon lye flat and round browniſh ſeede, ſomewhat thicke and great: the rootes periſh when they haue giuen their ſeede, and are ſomewhat round, long, and thicke, reſembling the rootes of *Lilium non bulboſum*, or Day Lilly, which are eaten (as diuers other rootes are) for Sallets, both in our owne Country, and in many places beſide.

2. *Viola Lunaris altera ſeu peregrina.* Long liuing Sattin flower.

This ſecond kinde hath broader and longer leaues then the former, the ſtalkes alſo are greener and higher, branching into flowers, of a paler purple colour, almoſt white, conſiſting of foure leaues in like manner, and ſmelling pretty ſweete, bearing ſuch like pods, but longer and ſlenderer then they: the rootes are compoſed of many long ſtrings, which dye not as the former, but abide, and ſhoot out new ſtalkes euery yeare.

The Place.

The firſt is (as is ſaid) frequent enough in Gardens, and is found wilde in ſome places of our owne Country, as Maſter Gerard reporteth, whereof I neuer could be certainly aſſured, but I haue had it often ſent mee among other ſeedes from Italy, and other places. The other is not ſo common in Gardens, but found about Watford, as he ſaith alſo.

The Time.

They flower in Aprill or May, and ſometimes more early.

The Names.

It hath diuers names, as well in Engliſh as in Latine; for it is called moſt vſually *Bolbonach*, and *Viola Lunaris*: Of ſome *Viola latifolia*, and of others *Viola Peregrina*, and *Lunaria Græca*, *Lunaria maior*, and *Lunaria odorata*, and is thought to be *Thlaſpi Cratena*: In Engliſh, White Satten, or Satten flower: Of ſome it is called Honeſty, and Penny-flower.

The Vertues.

Some doe vſe to eate the young rootes hereof, before they runne vp to flower, as Rampions are eaten with Vinegar and Oyle; but wee know no Phyſicall vſe they haue.

Z Chap.

CHAP. XLIII.

Linum siluestre & Linaria. Wilde Flaxe and Tode Flaxe.

ALthough neither the manured Line or Flaxe is a plant fit for our Garden, nor many of the wilde sorts, yet there are some, whose pleasant and delightfull aspect doth entertaine the beholders eyes with good content, and these I will set downe here for varietie, and adioyne vnto them some of the *Linarias*, or Tode Flaxe, for the neare affinity with them.

1. *Linum siluestre flore albo.* Wilde Flaxe with a white flower.

This kinde of wilde Flaxe riseth vp with diuers slender branches, a foote high or better, full of leaues, standing without order, being broader and longer then the manured Flaxe: the tops of the branches haue diuers faire white flowers on them, composed of fiue large leaues a peece, with many purple lines or strikes in them: the seede vessell as well as the seede, is like vnto the heads and seede of the manured Flaxe: the rootes are white strings, and abide diuers yeares, springing fresh branches and leaues euery yeare, but not vntill the Spring of the yeare.

2. *Linum siluestre luteum.* Wilde Flaxe with a yellow flower.

This wilde Flaxe doth so well resemble a kinde of St. Iohns wort, that it will soone deceiue one that doth not aduisedly regard it: For it hath many reddish stalkes, and small leaues on them, broader then the former wilde Flaxe, but not so long, which are well stored with yellow flowers, as large as the former, made of fiue leaues a peece, which being past, there come small flattish heads, containing blackish seede, but not shining like the former: the rootes hereof dye not euery yeare, as many other of the wilde kindes doe, but abide and shoote out euery yeare.

3. *Linaria purpurea.* Purple Tode Flaxe.

This purple Tode Flaxe hath diuers thicke, small, long, and somewhat narrowish leaues, snipt about the edges, of a whitish greene colour, from among which rise vp diuers stalkes, replenished at the tops with many small flowers, standing together one aboue another spike-fashion, which are small and somewhat sweete, while they are fresh, fashioned somewhat like the common Tode flaxe that groweth wilde abroad almost euery where, but much smaller, with a gaping mouth, but without any crooked spurre behinde, like vnto them, sometimes of a sad purple neare vnto a Violet, and sometimes of a paler blew colour, hauing a yellow spot in the middle or gaping place: after the flowers are past, there come small, hard, round heads, wherein are contained small, flat, and grayish seede: the roote is small, and perisheth for the most part euery yeare, and will spring againe of it owne sowing, if it be suffered to shed it selfe, yet some hard Winters haue killed the seede it should seeme, in that sometimes it faileth to spring againe, and therefore had neede to be sowne anew in the Spring.

4. *Linaria purpurea odorata.* Sweete purple Tode Flaxe.

The lower leaues of this purple Tode Flaxe are nothing like any of the rest, but are long and broad, endented about the edges, somewhat resembling the leaues of the greater wilde white Daisie: the stalke is set at the bottome with such like leaues, but a little more diuided and cut in, and still smaller and smaller vpward, so that the vppermost leaues are very like the common Tode Flaxe, the toppe whereof is branched, hauing diuers small flowers growing along vpon them, in fashion and colour almost like the last described Tode Flaxe, but not altogether so deepe a purple: the heads and seedes are very like the former, but that the seede of this is reddish: the flowers

1 *Linum siluestre flore albo.* Wilde **Flaxe** with a white flower. 2 *Linaria purpurea siue cærulea.* Purple **Tode Flaxe.** 3 *Linaria purpurea odorata.* Sweete **purple Tode Flaxe.** 4 *Scoparia siue Beluidere Italorum* Broome **Tode Flaxe.** 5 *Antirrhinum maius.* The greater **Snapdragon.** 6 *Chamænerium flore delphinij.* The willowe flower.

in their naturall hot Countries haue a fine fent, but in thefe colder, little or none at all: the rootes are fmall and threadie, and perifh after they haue flowred and feeded.

5. *Linaria Valentina.* Tode Flaxe of Valentia.

This Spanifh Tode Flaxe hath three or fonre thicker and bigger ftalkes then the former, bearing fmall broad leaues, like vnto the fmall Centory, two or three together at a ioynt, round about the lower end of the ftalkes, but withont any order vpwards, at the toppes whereof ftand many flowers, in fafhion like vnto the common kinde, and almoft as large, of a faire yellow colour, but the gaping mouth is downie, and the fpurre behinde of a purplifh colour.

6. *Scoparia fiue Beluidere Italorum.* Broome Tode Flaxe.

Although this plant haue no beautifull flowers, yet becaufe the greene plant full of leaues is fo delightfull to behold, being in Italy and other places planted not onely in their Gardens, but fet likewife in pots to furnifh their Windowes, and euen with vs alfo hath growne to be fo dainty a greene bufh, that I haue thought it worthy to be a-mong the delights of my Garden; the defcription whereof is as followeth: This plea-fant Broome Flaxe rifeth vp moft vfually with one ftraight vpright fquare ftalke, three foote and a halfe high or better in our Gardens, branching it felfe out diuers waies, bearing thereon many long narrow leaues, like the Garden Line or Flaxe, very thicke fet together, like vnto a bufh, or rather like vnto a faire greene Cyprefſe tree, growing broad belowe, and fpire-fafhion vpwards, of a very faire greene colour : at the feue-rall ioynts of the branches, towards the tops, and among the leaues, there come forth fmall reddifh flowers, not eafily feene nor much regarded, being of no beauty, which turne into fmall round blackifh gray feede : the rootes are a number of blackifh ftrings fet together, and the whole plant perifheth euery yeare at the firft approach of any cold ayre, as if it neuer had beene fo faire a greene bufh.

The Place.

Thefe kindes of wilde Flaxe doe growe naturally in diuers places, fome in Germany, fome in Spaine, and fome in Italy. Thofe that delight in the beauty of natures variety, doe preferue them, to furnifh vp the number of pleafant afpecls.

The Time.

They all flower in the Summer moneths, and foone after perfect their feede.

The Names.

Their names are fufficiently expreffed in their titles, yet I muft giue you to vnderftand, that the laft is called of fome *Linaria magna*, and of others *Ofyris.*

The Vertues.

The wilde Flaxe hath no medecinable vertue appropriate vnto it that is knowne. The Tode Flaxe is accounted to be good, to caufe one to make water.

CHAP. XLIIII.

Antirrhinum. Snapdragon.

THere is some diuersity in the Snapdragons, some being of a larger, and others of a lesser stature and bignesse ; and of the larger, some of one, and some of another colour, but because the small kindes are of no beautie, I shall at this time onely entreate of the greater sorts.

1. *Antirrhinum album.* White Snapdragon.

The leaues of these Snapdragons (for I doe vnder one description comprehend the rest) are broader, longer, and greener then the leaues of the Garden Flaxe, or of the wilde Flaxe set confusedly vpon the tender greene branches, which are spread on all sides, from the very bottome, bearing at the toppes many flowers, somewhat resembling the former Tode Flaxe, but much larger, and without any heele or spurre, of a faire white colour, with a yellow spot in the mouth or gaping place : after the flowers are past, there come vp in their places hard round seede vessels, fashioned somewhat like vnto a Calues head, the snout being cut off, wherein is contained small blacke seede : the rootes are many white strings, which perish in most places after they haue giuen seede, notwithstanding any care or paines taken with them to preserue them a-liue, and yet they will abide in some places where they are defended in the Winter.

2. *Antirrhinum purpureum siue roseum.* Purple Snapdragon.

The purple Snapdragon is in stalkes, leaues, and flowers altogether like the former, and as large and great in euery part, or greater ; the only difference is, that this beareth pale Stammell or Rose coloured flowers, with a yellow spot in the mouth, and some-times of a paler colour, almost blush.

3. *Antirrhinum variegatum.* Variable Snapdragon.

This variable kinde is somewhat lesse, and tenderer then the last described, hauing also a reddish or blush coloured flower, lesser then the former, but much bigger then the middle kinde of Snapdragon (which is not set downe in this worke) the yellow spot in the mouth of it hath some white about it, and extending to both sides of the spot : the heads and seede are like the former : the rootes are smaller, but neuer will abide after they haue giuen flowers and seede.

4. *Antirrhinum luteum.* Yellow Snapdragon.

There is likewise another of these kindes, that beareth leaues as large as any of the former, & very faire yellow flowers, as large likewise as they, not differing in any thing else from the first ; let not any therefore imagine this to be a *Linaria* or Tode Flaxe : for all parts are answerable vnto the Snapdragons.

The Place.

All these are nourished with vs in our Gardens, although in Spaine and Italy they are found growing wilde.

The Time.

They flower for the most part the second yeare after the sowing, from A-prill vntill Iuly, and the seede is quickly ripe after.

Z 3

The

The Names.

The name *Antirrhinum* is vsually giuen to this plant, although it fully agreeth not eyther with the description of Dioscorides, or Theophrastus: It hath also diuers other names in Latine, as *Orontium*, *Canis cerebrum Os Leonis, Leo herba*, &c. In English Calues snout, from the forme of the seede vessels, and Snapdragon, or Lyons mouth, from the forme of the flowers.

The Vertues.

They are seldome or neuer vsed in Physicke by any in our dayes.

CHAP. XLV.

Chamænerium flore delphinij. The Willowe flower.

THis plant riseth vp with many strong, woddy, round, brownish great stalkes, three or foure foote high, beset here and there without order, with one broad and long whitish greene leafe at a ioynt, somewhat like vnto a *Lysimachia*, or Willow herbe, as also vnto a Peach leafe, but larger and longer : at the toppe of the branches stand many flowers one aboue another, of a pale reddish purple colour, consisting of fiue leaues, spread open with an heele or spurre behinde them, with many yellow threads in the middle, much larger then any flower of the Larkes spurres, and smelling somewhat sweete withall ; it beareth a shew of long pods with seede, but I could neuer obserue the seede : the rootes are like the rootes of *Lysimachia*, or the ordinary yellow Loose-strife, or Willowe herbe, but greater : running and spreading vnder ground, and shooting vp in many places, whereby it filleth a ground that it likes quickly : the stalkes dye downe euery yeare, and spring againe in many places farre asunder.

The Place.

Wee haue not knowne where this Willowe flower groweth naturally, but we haue it standing in an out corner of our Gardens, to fill vp the number of delightfull flowers.

The Time.

It flowreth not vntill May, and abideth a long while flowring.

The Names.

It may seeme to diuers, that this is that plant that Dodonæus called *Pseudolysimachium purpureum minus*, and Lobel seemeth by the name of *Delphinium buccinum* to aime at this plant, but withall calleth it *Chamænerium Gesneri*, and giueth the same figure that Dodonæus hath for his *Pseudolysimachium*: But that is one kinde of plant (which hath smaller and shorter stalkes, and very narrow long leaues, whose flowers stand vpon long slender cods, full of downe, with reddish seede, like vnto the *Lysimachia siliquosa siluestris*, and rootes that abide many yeares, but creepe not) and this is another, much greater, whose true figure is not extant in any Author that I know. It is vsually called *Chamænerium flore delphinij* ; but the name of *Delphinium buccinum* in my minde may not so conueniently be applyed vnto it. It is called in English, The Willowe flower, for the likenesse of the leaues, and the beauty and respect of the flowers.

The

The Vertues.

There is no vfe hereof in Phyficke that euer I could learne, but is onely cherifhed among other forts of flowers, that ferue to decke and fet forth a Garden of varieties.

Chap. XLVI.

Aquilegia. Colombines.

THere are many forts of Colombines, as well differing in forme as colour of the flowers, and of them both fingle and double carefully nourfed vp in our Gardens, for the delight both of their forme and colours.

1. *Aquilegia vulgaris flore fimplici.* Single Colombines.

Becaufe the whole difference of thefe Colombines ftandeth in the varieties of the forme, and colour of the flowers, and little in the leaues, I fhall not neede to make anie repetitions of the defcription of them, feeing one onely fhall fuffice for each peculiar kinde. The Colombine hath diuers large fpread leaues, ftanding on long ftalkes : euery one diuided in feuerall partitions, and roundly endented about the edges, in colour fomewhat like the leaues of Celondine, that is, of a darke blewifh greene colonr : the ftalkes rife vp fometimes two or three foote high, diuided vfually into many branches, bearing one long diuided leafe at the lower ioynt, aboue which the flowers growe, euery one ftanding on a long ftalke, confifting of fiue hollow leaues, crooked or horned at the ends, turning backward, the open flower fhewing almoft like vnto a Cinquefoile, but more hollow : after the flowers are paft, there arife fmall long cods, foure or fiue together, wherein are contained blacke fhining feede : the rootes are thicke and round, for a little fpace within the ground, and then diuided into branches, ending in many fmall fibres, abiding many yeares, and fhooting a frefh euery Spring from the round heads, that abide all the Winter. The variety of the colours of thefe flowers are very much, for fome are wholly white, fome of a blew or violet colour, others of a blufh or flefh colour, or deepe or pale red, or of a dead purple, or dead murrey colour, as nature lifteth to fhew it felfe.

2. *Aquilegia vulgaris flore pleno.* Double Colombines.

The double Colombines differ not in leafe or manner of growing from the fingle, fo that vntill they come to flower, they cannot bee difcerned one from another ; the onely difference is, it beareth very thicke and double flowers, that is, many horned or crooked hollow leaues fet together, and are not fo large as the leaues of the fingle flowers. The variety of colours in this double kinde is as plentifull, or rather more then in the fingle ; for of thefe there is party coloured, blew and white, and fpotted very variably, which are not in the fingle kinde, and alfo a very deepe red, very thicke and double, but a fmaller flower, and leffe plentifull in bearing then many of the other double forts. Thefe double kindes doe giue as good feede as the fingle kindes doe, which is not obferued in many other plants.

3. *Aquilegia inuerfis corniculis.* Double inuerted Colombines.

Thefe Colombines are not to be diftinguifhed eyther in roote, leaues, or feed from the former, the flowers onely make the difference, which are as double as the former, but that the heeles or hornes of thefe are turned inward, and ftand out in the middle of the flowers together : there is not that plentifull variety of colours in this kinde, as there is in the former : for I neuer faw aboue three or foure feuerall colours in this
kinde,

kinde, that is, white, purplish, reddish, and a dun or darke ouerworne purplish colour. These double flowers doe likewise turne into pods, bearing seede, continuing his kind, and not varying into the former.

4. *Aquilegia Rosea.* Rose Colombines.

¶. The leaues and other parts of this kinde of Colombine, differ little or nothing from the former, the diuersitie consisteth likewise in the flowers, which although they stand in the same manner seuerally vpon their small stalkes, somewhat more sparingly then the former doe, yet they haue no heeles or hornes, eyther inward or outward, or very seldome, but stand sometimes but with eight or tenne smooth small plaine leaues, set in order one by one in a compasse, in a double rowe, and sometimes with foure or fiue rowes of them, euery one directly before the other, like vnto a small thick double Rose layd open, or a spread Marigold : yet sometimes it happeneth, that some of these flowers will haue two or three of the first rowes of leaues without any heele, and the rest that are inward with each of them a peece of a small horne at them, as the former haue : the colours of these flowers are almost as variable, and as variably mixed as the former double kindes. This likewise giueth seede, preseruing his owne kinde for the most part.

5. *Aquilegia degener.* Degenerate Colombines.

This kinde of Colombine might seeme to some, to bee but a casuall degeneration, and no true naturall kinde, happening by some cause of transplanting, or otherwise by the art of man : but I haue not so found it, in that it keepeth, and holdeth his own proper forme, which is like vnto the double Rose Colombine, but that the outermost row of leaues are larger then any of the rest inwardes, and is of a greenish, or else of a purplish greene colour, and is not altogether so apt to giue good seed like the former.

The Place.

The single kindes haue beene often found in some of the wooddy mountaines of Germany, as Clusius saith, but the double kindes are chiefly cherished in gardens.

The Time.

They flower not vntill May, and abide not for the most part when Iune is past, and in the meane time perfecteth their seede.

The Names.

Costæus doth call this plant *Pothos* of Theophrastus, which Gaza translateth *Desiderium.* Dalechampius vpon Athenæus, calleth it *Diosanthos,* or *Iouis flos* of Theophrastus, who in his sixth Booke and seuenth Chapter reckoneth them both, that is, *Diosanthos* and *Pathos,* to be Summer flowers, but seuerally. Dodonæus *Leoherba,* and Gesner *Leontostomium.* Fabius Columna in his Phytobasanos, vnto whom Clusius giueth the greatest approbation, referreth it to the *Isopyrum* of Dioscorides. All later Writers doe generally call it, eyther *Aquileia, Aquilina,* or *Aquilegia* ; and we in English, generally (I thinke) through the whole Countrey, Colombines. Some doe call the *Aquilegia rosea,* *Aquilegia stellata.* The starre Colombine ; because the leaues of the flowers doe stand so directly one by another, besides the doublenesse, that they somewhat represent eyther a Rose or a Starre, and thereupon they giue it the name eyther of a Starre or Rose.

The Vertues.

Some in Spaine, as Camerarius saith, vse to eate a peece of the roote hereof

1 *Aquilegia simplex.* The single Colombine. 2 *Aquilegia flore multiplici.* The double Colombine. 3 *Aquilegia versicolor.* The party coloured Colombine. 4 *Aquilegia inuersis corniculis.* The double inuerted Colombine. 5 *Aquilegia Rosea siue Stellata.* The Rose or the Starre Colombine. 6 *Thalictrum Hispanicum album.* White Spanish tufts.

of faſting, many dayes together, to helpe them that are troubled with the ſtone in the kidneyes. Others vſe the decoction, of both herbe and roote in wine, with a little Ambargriſe, againſt thoſe kinds of ſwounings, which the Greekes call ἀϑυμία. The ſeede is vſed for the iaundiſe, and other obſtructions of the liuer. Cluſius writeth from the experience of Franciſcus Rapard, a chiefe Phyſician of Bruges in Flanders, that the ſeede beaten and drunke is effectuall to women in trauell of childe, to procure a ſpeedy deliuerie, and aduiſeth a ſecond draught thereof ſhould be taken if the firſt ſucceede not ſufficiently.

Chap. XLVII.

Thalictrum Hiſpanicum. Spaniſh Tufts, or Tufted Colombines.

From among the diuerſities of this plant, I haue ſelected out two ſorts for this my garden, as hauing more beautie then all the reſt; leauing the other to be entreated of, where all in generall may be included. I haue in this place inſerted them, for the likeneſſe of the leaues only, being in no other part correſpondent, and in a Chapter by themſelues, as it is moſt fit.

Thalictrum Hiſpanicum album. White Spaniſh tufted Colombines.

Theſe plants haue both one forme, in roote, leafe and flower, and therefore neede but one deſcription. The leaues are both for colour and forme ſo like vnto Colombines leaues (although leſſer and darker, yet more ſpread, and on larger ſtalkes) that they may eaſily deceiue one, that doth not marke them aduiſedly; for the leaues are much more diuided, and in ſmaller parts, and not ſo round at the ends : the ſtalkes are round, ſtrong, and three foote high at the leaſt, branching out into two or three parts, with leaues at the ſeuerall ioynts of them, at the toppes whereof ſtand many flowers, which are nothing but a number of threads, made like vnto a ſmall round tuft, breaking out of a white ſkinne, or leafe, which incloſeth them, and being vnblowne, ſhew like vnto little buttons : the colour of theſe threds or tufts in this are whitiſh with yellow tips on them, and ſomewhat purpliſh at the bottome, hauing a ſtrong but no good ſent, and abiding in their beautie (eſpecially if they grow in the ſhade, and not too hot in the ſun) a great while, and then fall away, like ſhort downe or threds : the ſeed veſſels are three ſquare, containing ſmall, long, and round ſeede; the rootes are many long yellow ſtringes, which endure and encreaſe much.

Thalictrum Montanum purpureum. Purple tufted Colombines.

This purple tufted Colombine differeth onely from the former, in that it is not ſo high nor ſo large, and that the colour of the flower or tuft is of a blewiſh purple colour with yellow tips, and is much more rare then the other.

The Place.

Theſe grow both in Spaine and Italie.

The Time.

They flower in the end of May, or in Iune, and ſometime later.

The Names.

Some doe call them *Thalietrum,* and ſome *Thalictrum.* Others *Ruta paiuſtris,* and *Ruta pratenſis,* and ſome *Rhabarbarum Monachorum,* or *Pſeudorhabarbarum, rbabarbarum,* by reason that the rootes being yellow, haue an opening qua-litie, and drying as Rubarbe. In English what other fit Names to giue thefe then I haue expreffed in the titles, I know not.

The Vertues.

The are a little hot and drying withall, good for old Vlcers, as Diofcori-des faith, to bring them to cicatrifing : in Italy they are vfed againft the Plague, and in Saxonye againft the Iaundife, as Camerarius faith.

Chap. XLVIII.

Radix caua. Hollow roote.

THe likeneffe of the leaues likewife of this plant with Colombines, hath caufed mee to infert it next the other, and although fome of this kinde bee of fmall refpect, being accounted but foolifh, yet let it fill vp a wafte corner, that fo no place be vnfurnifhed.

1. *Radix Caua maior flore albo.* The white Hollow roote.

The leaues of this hollow roote breake not out of the ground, vntill the end of March, or feldome before, and are both for proportion and colour fomewhat like vn-to the leaues of Colombines, diuided into fiue parts, indented about the edges, ftan-ding on fmall long footeftalkes of a whitifh greene colour, among which rife vp the ftalkes, without any leaues from the bottome to the middle, where the flowers fhoote forth one aboue another, with euery one a fmall fhort leafe at the foote thereof, which are long and hollow, with a fpurre behinde it, fomewhat like vnto the flowers of Larckes fpurres, but hauing their bellies fomwhat bigger, and the mouth not fo open, being all of a pure white colour : after the flowers are paft, arife fmall long and round cods, wherein are contained round blackifh feede : the roote is round and great, of a yellowifh browne colour on the outfide, and more yellow within, and hollow vnder-neath, fo that it feemeth but a fhell : yet being broken, euery part will grow : it abideth greene aboue ground but a fmall time.

2. *Radix Caua maior flore carneo.* Blufh colourd Hollow roote.

The blufh Hollow roote is in all things like vnto the former, but onely that the flowers hereof are of a delayed red or purple colour, which we call blufh : and fome-times of a very deepe red or purple colour ; but very rare to meete with.

3. *Radix Caua minor, feu Capnos fabacea radice.* Small hollow roote.

This fmall kinde hath his leaues of a blewifh greene colour, yet greener and fmal-ler then the former, growing more thicke together : the flowers are like in proportion vnto the former in all refpects, but leffer, hauing purplifh backes, and white bellyes : ftanding clofer and thicker together vpon the fhort ftalkes : the roote is folid or firme, round and a little long withall, two being vfually ioyned together, yellowifh both within and without : but I haue feene the dry roots that came from beyond Sea hither, that haue beene as fmall as hafell nuts, and fomewhat flat with the roundneffe, diffe-ring from thofe that growe with vs, whether the nature thereof is to alter by manuring, I know not.

The Place.

The greater kindes Clufius reporteth he found in many places of Hunga-rie,

rie, and the other parts neere thereunto : the lesser in the lower Germany, or Low Countries, as we call them.

The Time.

These are most truely to bee reckoned Vernall plants, for that they rise not out of the ground vntill the Spring bee come in, and are gone likewise before it be past, remaining vnder ground all the rest of the yeare, yet the lesser abideth longer aboue ground then the greater.

The Names.

Concerning the former of these, there is a controuersie among diuers, whether it should be *Thesium* of Theophrastus, or *Eriphium* of Galen, but here is no fit place to trauerse those opinions. Some would haue it to bee *Corydalis,* and some referre it to Plinie his *Capnos Cheledonia,* for the likenesse it hath both with Fumeterie and Celandine. It is generally called of all moderne Writers, *Radix Caua,* and we in English thereafter, Hollow roote. The lesser for the firmenesse of his round roote, is vsually called, *Capnos fabacea radice,* and the Dutch men thereafter, **Boonkens Hollwortell:** we of the likenesse with the former, doe call it the lesse Hollow roote.

The Vertues.

Some by the bitternesse doe coniecture (for little proofe hath beene had thereof, but in outward cases) that it clenseth, purgeth, and dryeth withall.

CHAP. XLIX.

Delphinium. Larkes heeles.

OF Larkes heeles there are two principall kindes, the wilde kinde, and the tame or garden ; the wilde kinde is of two sorts, one which is with vs noursed vp chiefly in gardens, and is the greatest ; the other which is smaller and lower, often found in our plowed landes, and elsewhere : of the former of these wilde sorts, there are double as well as single : and of the tame or more vpright, double also and single : and of each of diuers colours, as shall be set downe.

1. *Delphinium maius siue vulgare.* The ordinary Larkes heeles.

The common Larkes heele spreadeth with many branches much more ground then the other, rather leaning or bending downe to the ground, then standing vpright, whereon are set many small long greene leaues, finely cut, almost like Fennell leaues : the branches end in a long spike of hollow flowers, with a long spurre behinde them, very like vnto the flowers of the Hollow roote last described, and are of diuers seuerall colours, as of a blewish purple colour, or white, or ash colour or red, paler or deeper, as also party coloured of two colours in a flower : after the flowers are past, (which in this kinde abide longer then in the other) there come long round cods, containing very blacke seede : the root is hard after it groweth vp to seede, spreading both abroad and deepe, and perisheth euery yeare, vsually raising it selfe from it own sowing, as well as from the seede sowen in the spring time.

Varietas.

2. *Delphinium vulgare flore pleno.* Double common Larkes heeles.

Of this vulgar kinde there is some difference in the flower, although in nothing else : the flowers stand many vpon a stalke like the former, but euery one of them are as if
three

three or foure small flowers were ioyned together, with euery one his spurre behinde, the greatest flower being outermost, and as it were containing the rest, which are of a pale red, or deepe blush colour : Another of this kinde will beare his flowers with three or foure rowes of leaues in the middle, making a double flower with one spurre behinde onely : and of this kinde there is both with purple, blew, blush, and white flowers, and party coloured also; these doe all beare seed like the single, wherby it is encreased euery yeare.

3. *Delphinium aruense.* Wilde Larkes spurres.

This wilde Larkes spurre hath smaller and shorter leaues, smaller and lower branches, and more thinly or sparsedly growing vpon them, then any of the former : the flowers likewise are neyther so large as any of the former, nor so many growing together, the cods likewise haue smaller seede, and is harder to grow in gardens then any of the former, the most vsuall colour hereof is a pale reddish or blush colour, yet sometimes they are found both white and blew, and sometimes mixt of blew and blush, variably disposed, as nature can when she listeth ; but are much more rare.

4. *Delphinium elatius flore simplici diuersorum colorum.* Single vpright bearing Larkes heeles of many colours.

The difference betweene this and the last is, that the leaues of this are not fully so greene, nor so large; the stalkes grow vpright, to the height of a man, and sometimes higher, hauing some branches thereon, but fewer then the former, and standing likewise vpright, and not leaning downe as the former : the toppes of the stalkes are better stored with flowers then the other, being sometimes two foote long and aboue, of the same fashion, but not altogether so large, but of more diuers and seuerall colours, as white, pale, blush, redde deeper or paler, ashcoloured, purple or violet, and of an ouerworne blewish purple, or iron colour : for of all these we haue simple, without any mixture or spot : but we haue other sorts, among the simple colours, that rise from the same seede, and will haue flowers that wil be halfe white, and halfe blush or purple, or one leafe white, and another blush or purple, or else variably mixed and spotted : the seede and seede vessels are like the former but larger and harder.

5. *Delphinium elatius flore pleno diuersorum colorum.* Double vpright Larkes heeles of many colours.

These double Larkes heeles cannot bee knowne from the single of the same kinde, vntill they come towards flowring, for there appeare many flowers vpon the stalkes, in the same manner, and of as many colours almost as of the single, except the party coloured, which stand like little double Roses, layd or spread broade open, as the Rose Colombine without any heeles behinde them, very delightfull to behold, consisting of many small leaues growing together, and after they are fallen there come vp in their places three or foure small cods set together, wherein is contained here and there (for all are not full of seede, as the single kindes) blacke seede, like vnto all the rest, but smaller, which being sowen will bring plants that will beare both single and double flowers againe, and it often happeneth, that it variably altereth in colours from it owne sowing : for none of them hold constantly his owne colour, (so farre as euer I could obserue) but fall into others as nature pleaseth.

6. *Delphinium Hispanicum paruum.* Spanish wilde Larkes spurres.

This small Larkes spurre of Spaine, hath diuers long and broad leaues next the ground, cut-in on both sides, somewhat like vnto the leafe of a Scabious, or rather that kinde of Stœbe, which Lobel calleth *Crupina*, for it doth somewhat neerly resemble the same, but that this is smooth on the edges, and not indented besides the cuts, as the *Crupina* is, being of a whitish greene colour, and somewhat smooth and soft in handling : among the leaues riseth vp a whitish greene stalke, hauing many smaller

Aa   leaues

leaues vpon it that grow belowe, but not diuided, branching out into many small stalkes, bearing flowers like vnto the wilde Larkes heeles, but smaller, and of a bleake blewish colour, which being past, there come vp two or three small cods ioyned together, wherein is blacke seede, smaller and rounder then any of the former: the roote is small and thready, quickly perishing with the first cold that ouertaketh the plant.

The Place.

The greatest or first wilde kindes growe among corne in many countries beyond the Seas, and where corne hath beene sowne, and for his beauty brought and nourished in our Gardens: the lesser wilde kinde in some fields of our owne Country. The Spanish kinde likewise in the like places, which I had among many seedes that Guillaume Boel brought mee out of Spaine. The first double and single haue been common for many yeares in all countries of this Land, but the tall or vpright single kindes haue been entertained but of late yeares. The double kindes are more rare.

The Time.

These flower in the Summer onely, but the Spanish wilde kinde flowreth very late, so that oftentimes in our Country, the Winter taketh it before it can giue ripe seede: the double kindes, as well the vpright as the ordinary or wilde, are very choise and dainty many times, not yeelding good seede.

The Names.

They are called diuersly by diuers Writers, as *Consolida regalis*, *Calcaris flos*; *Flos regius*, *Buccinum Romanorum*, and of Matthiolus, *Cuminum siluestre alterum Dioscoridis*: but the most vsuall name with vs is *Delphinium*: but whether it be the true *Delphinium* of Dioscorides, or the Poets Hyacinth, or the flower of Aiax, another place is fitter to discusse then this. Wee call them in English Larkes heeles, Larkes spurres, Larkes toes or clawes, and Monkes hoods. The last or Spanish kinde came to mee vnder the name of *Delphinium latifolium trigonum*, so stiled eyther from the diuision of the leaues, or from the pods, which come vsually three together. Bauhinus vpon Matthiolus calleth it, *Consolida regalis peregrina paruo flore*.

The Vertues.

There is no vse of any of these in Physicke in these dayes that I know, but are wholly spent for their flowers sake.

Chap. L.

Balsamina foemina. The Female Balsam Apple.

I Haue set this plant in this place, for some likenesse of the flower, rather then for any other comparison, euen as I must also with the next that followeth. This plant riseth vp with a thicke round reddish stalke, with great and bunched ioynts, being tender and full of iuice, much like to the stalke of Purslane, but much greater, which brancheth it selfe forth from the very ground, into many stalkes, bearing thereon manie long greene leaues, snipt about the edges, very like vnto the Almond or Peach tree leaues; among which from the middle of the stalkes vpwards round about them, come forth vpon seuerall small short foot-stalkes many faire purplish flowers, of two or three colours in them, fashioned somewhat like the former Larkes heeles, or Monks hoods, but that they are larger open at the mouth, and the spurres behinde crooke or bend downewards: after the flowers are past, there come in their places round rough heads,

1 *Radix Cana maior flore albo.* The white flowred Hollow roots. 2 *Capnos fabacea radice.* The small Hollow roots. 3 *Delphinium flore simplici.*
Single Larkes spurs. 4 *Delphinium vulgare flore medio duplici.* Larkes spurs double in the middle. 5 *Delphinium vulgare flore pleno* Common
Larkes spurs double. 6 *Delphinium elatius flore pleno.* Double vpright Larkes spurs. 7 *Delphinium Hispanicum parvum.* Small Spanish Larkes spurs.
8 *Balsamina faemina.* The Female Balsam apple. 9 *Nasturtium Indicum,* Indian Crosses, or yellow Larkes spurs.

heads, pointed at the end, greene at the firſt, and a little yellower when they bee ripe, containing within them ſmall round blackiſh ſeede, which will ſoone skippe out of the heads, if they be but a little hardly preſſed betweene the fingers: the rootes ſpread themſelues vnder ground very much from the toppe, with a number of ſmall fibres annexed thereunto: this is a very tender plant, dying euery yeare, and muſt bee ſowne carefully in a pot of earth, and tended and watered in the heate of Summer, and all little enough to bring it to perfection.

The Place.

Wee haue alwaies had the ſeede of this plant ſent vs out of Italy, not knowing his originall place.

The Time.

It flowreth from the middle of Iuly, to the end of Auguſt: the ſeed doth ſeldome ripen with vs, eſpecially if the Summer be backward, ſo that wee are oftentimes to ſeeke for new and good ſeede from our friends againe.

The Names.

Some vſe to call it *Charantia fœmina*, *Balſamina fœmina*, *Balſamella*, and *Anguillara*, *Herba Sancta Katharina*. We haue no other Engliſh name to call it by, then the Female Balſame Apple, or *Balſamina*.

The Vertues.

Some by reaſon of the name, would attribute the property of Balme vnto this plant, but it is not ſufficiently knowne to haue any ſuch; yet I am well perſwaded, there may bee ſome extradinary quality in ſo beautifull a plant, which yet lyeth hid from vs.

Chap. LI.

Naſturtium Indicum. Indian Creſſes, or yellow Larkes heeles.

THe likeneſſe (as I ſaid before) of this flower likewiſe, hauing ſpurres or heeles maketh me ioyne it with the reſt, which is of ſo great beauty and ſweetneſſe withall, that my Garden of delight cannot bee vnfurniſhed of it. This faire plant ſpreadeth it ſelfe into very many long trayling branches, enterlaced one within another very confuſedly (yet doth it not winde it ſelfe with any claſpers about either pole or any other thing, but if you will haue it abide cloſe thereunto, you muſt tye it, or elſe it will lye vpon the ground) foure or fiue foot in length at the leaſt, wherby it taketh vp a great deale of ground: the leaues are ſmooth, greene, and as round as the Penniwort that groweth on the ground, without any cut or inciſure therein at all in any part, the ſtalkes whereof ſtand in the middle of each leafe, and ſtand at euery ioynt of the ſtalke, where they are a little reddiſh, and knobbed or bunched out: the flowers are of an excellent gold yellow colour, and grow all along theſe ſtalkes, almoſt at euery ioynt with the leaues, vpon pretty long foote-ſtalkes, which are compoſed of fiue leaues, not hollow or gaping, but ſtanding open each leafe apart by it ſelfe, two of them, that be larger and longer then the other, ſtand aboue, and the other two that are leſſer belowe, which are a little iagged or bearded on both ſides, and the fift loweſt: in the middle of each of the three lower leaues (yet ſometimes it is but in two of them) there is a little long ſpot or ſtreake, of an excellent crimſon colour, with a long heele or ſpurre behinde hanging downe: the whole flower hath a fine ſmall ſent, very pleaſing, which being placed in the middle of ſome Carnations or Gilloflowers

flowers (for they are in flower at the same time) make a delicate Tussimussie, as they call it, or Nosegay, both for sight and sent : After the flower is past, come the seede, which are rough or vneuen, round, greenish yellow heads, sometimes but one, and sometimes two or three standing together vpon one stalke, bare or naked of themselues, without any huske, containing a white pulpy kernell : the rootes are small, and spreading vnder ground, which perish with the first frosts, and must be sowne a new euery yeare ; yet there needeth no bed of horse-dung for the matter : the naturall ground will be sufficient, so as you defend it a little from those frosts, that may spoile it when it is newly sprung vp, or being yet tender.

The Place.

This goodly plant was first found in the West Indies, and from thence sent into Spaine vnto Monardus and others, from whence all other parts haue receiued it. It is now very familiar in most Gardens of any curiosity, where it yearly giueth ripe seed, except the yeare be very vnkindly.

The Time.

It flowreth sometimes in Iune, but vsually in Iuly (if it be well defended and in any good ground) and so continueth flowring, vntill the cold frosts and mistes in the middle or end of October, doe checke the luxurious nature thereof, and in the meane time the seede is ripe, which will quickly fall downe on the ground, where for the most part the best is gathered.

The Names.

Some doe reckon this plant among the *Clematides* or *Convolvuli*, the Clamberers or Bindweedes ; but (as I said) it hath no claspers, neither doth it winde it selfe : but by reason of the number of his branches, that run one within another, it may seeme to climbe vp by a pole or sticke, which yet doth but onely close it, as hauing something whereon to leane or rest his branches. Monardus and others call it *Flos sanguineus*, of the red spots in the flowers, as also *Mastnerzo de las Indias*, which is *Nasturtium Indicum*, by which name it is now generally knowne and called, and wee thereafter in English, Indian Cresses, yet it may bee called from the forme of the flowers onely, Yellow Larkes heeles.

The Vertues.

The Spaniards and others vse the leaues hereof in stead of ordinary Cresses, because the taste is somewhat sharpe agreeing thereunto, but other Physicall properties I haue heard of none attributed to it.

Chap. LII.

Viola. Violets.

THe Garden Violets (for the Wilde I leaue to their owne place) are so well knowne vnto all, that either keepe a Garden, or hath but once come into it, that I shall (I thinke) but lose labour and time to describe that which is so common. Yet because it is not onely a choise flower of delight, notwitsttstanding the popularity, and that I let not passe any thing without his particular description, I must also doe so by this. And hereunto I must adde that kinde of Violet, which, although it want that smell of the other, goeth beyond it in variety of dainty colours, called *Viola tricolor & flammea*, or Harts eases.

1. *Viola simplex Martia.* Single March Violets.

The single Garden Violet hath many round greene leaues, finely snipt or dented a-bout the edges, standing vpon seuerall small stalkes, set at diuers places of the many creeping branches, which as they runne, doe here and there take roote in the ground, bearing thereon many flowers seuerally at the ioynts of the leaues, which consist of fiue small leaues, with a short round tayle or spurre behinde, of a perfect blew purple colour, and of a very sweete sent, it bringeth forth round seede vessels, standing like-wise vpon their seuerall small stalkes, wherein is contained round white seede: but these heads rise not from where the flowers grew, as in all other plants that I know, but apart by themselues, and being sowne, will produce others like vnto it selfe, whereby there may be made a more speedy encrease to plant a Garden(as I haue done) or any other place, then by slipping, as is the vsuall manner: the rootes spread both deepe and wide, taking strong hold in the ground.

Flore albo.   Of this kinde there is another that beareth white flowers, not differing in smell or any thing else from the former.

Flore obsoleto.   And also another, that beareth flowers of a dead or sad reddish colour, in all other things alike, sauing that this hath not altogether so good a sent as the other.

2. *Viola Martia flore multiplici.* Double March Violets.

There is no difference betweene this Violet and the former, in any other thing then in the doublenesse of the flowers, which haue so many leaues set and thrust together, that they are like vnto hard buttons. There is of this double kinde both white and purple, as in the single; but the white sort is seldome so thicke and double as the pur-ple: but of the red colour to be double I neuer heard.

3. *Viola flammea siue tricolor.* Harts eases or Pansies.

The Harts ease hath his leaues longer, and more endented or cut in on the edges then the Violet hath, and somewhat round withall: the stalkes are vpright, yet weake, and ready to fall downe, and lye vpon the ground, set here and there with the like leaues, from whence come forth the flowers, of little or no sent at all, made like vnto a Violet, yet more open, and with larger leaues; but so variably mixed with blew or purple, white and yellow, that it is hard to set downe all the varieties: For some flow-ers will be more white, and but some spots of purple or blew in the two vpper leaues, and the lower leaues with some stripes of yellow in the middle: others will haue more purple in them then any other colour, both in the vpper and lower leaues, the side leaues blew, and the middle yellow, and others white and blew with yellow stripes, as nature listeth to distribute their colours: the seede is small, whitish, and round, con-tained in small round heads: the roote perisheth euery yeare, and raiseth it selfe vp plentifully by it owne sowing, if it be suffered.

4. *Viola tricolor flore duplici.* Double Harts ease.

We haue in our Gardens another sort, that beareth flowers with more leaues then the former, making it seeme to be twice double, and that onely in Autumne; for the first flowers are single that come in Summer: This is of that sort that beareth purple flowers: And it is to be obserued, that the seed of this kinde will not all bring double flowers, but only some, if the ground be fit and liking, so that if you haue once had of this double kinde, you shall seldome misse to haue double flowers againe euery yeare of it owne growing or sowing.

5. *Viola flammea lutea maxima.* The great yellow Pansie.

There is one other kinde of Harts ease, that decketh vp our Gardens not to be for-gotten, whose leaues and flowers are like the former, but more plentifull in stalkes and branches, and better abideth our Winters: the flowers are larger then any of the
former,

former, of a faire pale yellow colour, with some yellower ftripes now and then about the middle; for it is fometimes without any ftripes, and alfo of a little deeper yellow colour: this is to bee encreafed by flips, which will foone comprehend in a moift or moiftened ground, for that I neuer could obferue that it bore feede.

The Place.

Thefe plants were firft wilde, and by manuring brought to be both fairer in colour, and peraduenture of a better fent then when they grew wilde.

The Time.

The Violets flower in March, and fometimes earlier, and if the yeare be temperate and milde, in Autumne againe. The double Violets, as they are later before they flower then the fingle, fo they hold their flowers longer. The Harts eafe flowreth feldome vntill May; but then fome will abide to flower vntill the end of Autumne almoft, efpecially if the frofts be not early.

The Names.

The Violet is called *Viola nigra*, *purpurea*, and *Martia*: In Englifh, Violets, March Violets, and purple Violets. The Harts eafe is called *Viola flammea*, *Viola tricolor*, *Viola multicolor*, and of fome, *Iacea*, *Flos trinitatis*, and *Herba clauellata*: In Englifh, Harts eafe, and Panfies, of the French name *Penfees*. Some giue it foolifh names, as Loue in idleneffe, Cull mee to you, and Three faces in a hood. The great yellow Harts eafe is fo called, becaufe it is like in forme, and is the greateft of all other, although it haue not that diuerfity of colours in it that the other haue.

The Vertues.

The properties of Violets are fufficiently knowne to all, to coole and moiften: I fhall forbeare to recite the many vertues that may be fet downe, and onely let you know, that they haue in them an opening or purging quality, being taken either frefh and greene, or dryed, and made into powder, efpecially the flowers; the dryed leaues will doe the like, but in greater quantity. Coftæus in his booke of the nature of all plants faith, that the diftilled water of Harts eafe, is commended in the French difeafe, to be profitable, being taken for nine dayes or more, and fweating vpon it, which how true it is, I know not, and wifh fome better experience were made of it, before we put any great confidence in that affertion.

Chap. LIII.

Epimedium. Barrenwort.

This pretty plant rifeth vp out of the ground with vpright, hard, round, fmall ftalkes, a foote and a halfe high, or not two foote high at the higheft, diuided into three branches for the moft part, each branch whereof is againe diuided for the moft part into three other branches, and each of them beare three leaues (feldome either more or leffe) fet together, yet each vpon his owne foote-ftalke, each leafe being broad, round, and pointed at the end, fomewhat hard or dry in feeling, hayrie, or as it were prickly about the edges, but very tenderly, without harme, of a light greene colour on the vpperfide, and a little whiter vnderneath: from the middle of the ftemme or ftalke of leaues doth likewife come forth another long ftalke, not much higher then thofe with the leaues on them, diuided into other branches, each

whereof

whereof hath likewise three flowers, each vpon his owne footeftalke, confifting of eight fmall leaues a peece, yet feeming to be but of foure leaus fpread or layd open flat, for that the foure vppermoft, which are the fmaller and being yellow, doe lye fo clofe on the foure vndermoft, wch are a little broader and red, that they fhew as if they were yellow flowers with red edges, hauing yellow threds tipt with greene, ftanding in the middle of the flowers : the vnderfide of the lower leaues are of a pale yellowifh red, ftriped with white lines : after the flowers are paft, there come fmall long pods, wherein are contained flat reddifh feede : the rootes are fmall, reddifh and hard, fpreading, branching and enterlacing themfelues very much, and is fit to be placed on fome fhady fide of a garden : the whole plant is rather of a ftrong then any good fent, yet is cherifhed for the pleafant varietie of the flowers.

The Place.

Cæfalpinus faith it groweth on the mountaines of Liguria, that is nigh vnto Ligorne, in the Florentine Dominion. Camerarius faith, nigh vnto Vicenzo in Italie. Bauhinus on the Euganian hils, nigh vnto Padoa, and in Romania in fhadowie wet grounds.

The Time.

It flowreth from Iune vntill the end of Iuly, and to the middle of Auguft, if it ftand, as I faid it is fitteft, in a fhadowie place.

The Names.

It is of moft Writers accepted for the true *Epimedium* of Diofcorides, though he faith it is without flower or feede, being therein eyther miftaken, or mif-informed, as he was alfo in *Dictamnus* of Candy, and diuers other plants. From the triple triplicitie of the ftanding of the ftalkes and leaues, and quadriplicitie of the flowers, it might receiue another name in Englifh then is already impofed vpon it : but leſt I might be thought to be fingular or full of noueltie, let it paſſe with the name Barrenwort, as it is in the title.

The Vertues.

It is thought of diuers to agree in the propertie of caufing barrenneffe, as the ancients doe record of *Epimedium*.

CHAP. LIIII.

Papauer fatiuum. Garden Poppies.

O F Poppies there are a great many forts, both wilde and tame, but becaufe our Garden doth entertaine none, but thofe of beautie and refpect, I wil onely giue you here a few double ones, and leaue the reſt to a general furuey.

1.*Papauer multiplex album.* Double white Poppies.

The double white Poppy hath diuers broade, and long whitifh greene leaues, giuing milke (as all the reſt of the plant aboue ground doth, wherefoeuer it is broken) very much rent or torne in on the fides, and notched or indented befides, compaffing at the bottome of them a hard round brittle whitifh greene ſtalke, branched towards the toppe, bearing one faire large great flower on the head of euery branch, which before it breaketh out, is contained within a thin skinne, and being blowne open is very thick of leaues, and double, fomewhat iagged at the ends, and of a white colour ; in the

middle

1 *Viola Martia simplex.* Single March Violets. 2 *Viola Martia multiplex.* Double March Violets. 3 *Viola flammea sue tricolor.* Ordinary garden Pansies or Hartsease. 4 *Viola flammea lutea magna.* Great yellow Pansies. 5 *Viola tricolor duplex* Double Pansies or Hartseases. 6 *Epimedium.* Barrenwort. 7 *Papaver sativum flore pleno.* Double garden Poppies. 8 *Papaver sativum flore pleno laciniato.* Double feathered Poppies. 9 *Nigella Hispanica flore amplo.* Spanish Nigella or Fenell flower. 10 *Nigella multiplex caerulea* Double blew Nigella or Fenell flower. 11 *Nigella duplex flore albo.* Double white Nigella. 12 *Ptarmica flore pleno.* Double wilde Pelletory.

middle whereof ſtandeth a round head or bowle, with a ſtriped crowne on the heade of it, very like a ſtarre, compaſſed about with ſome threds, wherein when it is ripe, is contained ſmall, round, white ſeede, diſpoſed into ſeuerall cels : the roote is hard, wooddy, and long, periſhing euery yeare, and muſt bee new ſowne euery Spring, if they doe not ſpring of their own ſowing, which if it doe, the flowers are ſeldome ſo faire and double as they that are ſowne in the Spring : the whole plant is of a ſtrong heady ſmell.

2. *Papauer multiplex rubeſcens.* Double red or bluſh Poppies.

This other kind of double Poppy differeth not in any other thing from the former, but only in the colour of the flowers, which are of a bright red, tending to a bluſh colour, parted, paned or ſtriped in many places with white, and exceedingly more iagged then the former, almoſt like a feather at the ends, the bottomes of all the leaues being white : the ſeede hereof is white as the former, which is not ſo in any other Poppie, that beareth not a full white flower.

3. *Papauer multiplex nigrum ſiue purpureum.* Double purple or murry Poppies.

This kinde varyeth both in flowers and ſeede, although neyther in leaues or any other thing from the firſt : the flowers are thicke and double, and ſomewhat iagged at the ends, in ſome more, in ſome leſſe, eyther red or bluſh, or purpliſh red, more or leſſe, or of a ſad murrey or tawney, with browne, or blacke, or tawny bottomes : the ſeede is eyther of a grayiſh blew colour, or in others more blackiſh.

4. *Papauer Rhæas flore multiplici.* The double red field Poppie.

This double Poppie is like the wilde or fielde Poppie, which is well knowne to all to haue longer, narrower, and more iagged greene leaues then the former, the ſtalkes more hairy, and the flower of a deepe yellowiſh red colour, knowne to all. Now this differeth in nothing from it, but in the doubleneſſe of the flower, which is very thicke and double, but not ſo large as the former. This riſeth of ſeede in the like manner as they doe, and ſo to bee preſerued.

The Place.

From what place they haue beene firſt gathered naturally I cannot aſſure you, but we haue had them often and long time in our gardens, being ſent from Italie and other places. The double wilde kindes came from Conſtantinople, which whether it groweth neere vnto it or further off, we cannot tell as yet.

The Time.

They flower in the beginning or middle of Iune at the furtheſt, the ſeede is ripe within a ſmall while after.

The Names.

The generall knowne name to all, is *Papauer*, Poppie : the ſeuerall diſtinctions are according to their colours. Yet our Engliſh Gentlewomen in ſome places, call it by a by-name, Ione ſiluer pinne : *ſubauditur*, Faire without and fowle within.

The Vertues.

It is not vnknowne, I ſuppoſe to any, that Poppie procureth ſleepe, for which cauſe it is wholly and onely vſed, as I thinke : but the water of the
wilde

wilde Poppies, besides that it is of great vse in Pleuresies, and Rheumatick-
or thinne Distillations, is found by daily experience, to bee a soueraigne remedy against surfeits ; yet some doe attribute this propertie to the water
of the wilde Poppies.

Chap. LV.

Nigella. The Fenell flower, or Nigella.

A Mong the many sorts of Nigella, both wilde and tame, both single and double,
I will onely set downe three sorts, to be nourled vp in this garden, referring the
rest to a Physicke garden, or a generall Historie, which may comprehend all.

1. *Nigella Hispanica flore simplici.* The great Spanish Nigella.

Spanish Nigella riseth vp with diuers greene leaues, so finely cut, and into so many
parts, that they are finer then Fenell, and diuided somewhat like the leaues of Larkes
heeles, among which rise vp stalkes, with many such like leaues vpon them, branched
into three or foure parts, at the toppe of each whereof standeth one faire large flower,
like vnto other single Nigella's, consisting of fiue or six leaues sometimes, of a bleake
blew, or of a purplish blew colour, with a greene head in the middle, compassed about
with seuen or eight small blewish greene flowers, or peeces of flowers rather, made
like gaping hoodes, with euery of them a yellowish line thwart or crosse the middle
of them, with some threds also standing by them : after the flower is past the head
groweth greater, hauing sixe, seuen or eight hornes as it were at the toppe, greater and
longer, and standing closer together then any other Nigella, spreading very like a
starre, or the crowne of the Poppy head, but larger and longer, each whereof being
folded together, openeth a little when the head is ripe, which is greater aboue, and
smaller below, and not so round as the others are, containing within them small yellowish greene seede, or not so blacke as the other sorts : the rootes are small and yellow, perishing euery yeare as the others likewise doe.

2. *Nigella Damascena flore multiplici.* Double blew Nigella, or The Fenell flower.

The double Nigella is in leaues, stalkes and rootes, very like vnto the former Nigella, so that the one can very hardly bee discerned from the other before this rise vp
to flower, except it be that the leaues hereof are not fully so large as they : the flower
consisteth of three or foure rowes of leaues, layde one vpon another, of a pale blew
colour, with a greene round head compassed with diuers short threads in the middle,
and hauing fiue or sixe such small greene Fenell-like leaues vnder the flower, to beare
it vp (as it were) below, which adde a greater grace to the flowers, which at the first
sheweth sometimes white, but changeth quickely after : the horned heads hereof are
like vnto the heads of the other wilde kinde, which are somewhat rounder and greater, hauing within them blacke vneuen seedes, but without any sent.

3. *Nigella Catrina flore albo multiplici.* Double white Nigella.

This double white Nigella hath such like leaues as the last hath, but somewhat larger, of a yellower greene colour, and not so finely cut and iagged : the flowers are
somewhat lesse, and lesser double then the former, and in colour white, hauing no
greene leaues vnder the flower, as the former hath, the head whereof in the middle is
very like the head of the last double kinde, but not so great, wherein is contained black
seede for the most part, and sweete like the Romane Nigella, which only is sweet besides this : yet sometimes it is not so blacke, but rather a little more white or yellowish: the roote is yellow, and perisheth as the others euery yeare.

The

The Place.

All thefe, and the reft be found wilde in diuers Countreyes, as France, Spaine, Italie, &c. but wee onely cherifh them in our Gardens for our delight.

The Time.

They flower in the end of Iune, and in Iuly, or thereabouts.

The Names.

They are called *Melanthium, Gith,* and *Nigella,* and of fome *Flos Diuæ Catherinæ.* We may either call them *Nigella* according to the Latine name, or the Fenell flower, as fome doe, becaufe the double blew Nigella hath fmall Fenell-like leaues bearing vp the flower, as I fhewed before in the defcription.

The Vertues.

Thefe Nigella's are nothing fo hot in qualitie as the fingle Romane kind is, as may well be knowne by the fmell of the feede thereof; and therefore are not fit to be vfed in the fteed of it, as many ignorant perfons vfe to doe: for the fingle Romane feede is vfed to helpe paines, and cold diftillations in the head, and to dry vp the rheume. Pena faith, that the preffed oyle of the feede as well taken inwardly as vfed outwardly is an excellent remedy for the hardneffe and fwelling of the fpleene.

Chap. LVI.

Ptarmica filueftris flore pleno. Double wilde Pelletory.

THe double wilde Pelletorie hath ftraight and flender ftalkes, befet with long and narrow leaues, fnipt round about the edges, in all points like vnto the fingle wilde kinde, that groweth common with vs almoft euery where : on the toppes of the ftalkes ftand foure or fiue, or more white flowers, one aboue another, with a greene leafe at the bottome of the footeftalke of euery one of them, beeing fmall, thicke, and very double, with a little yellowifhneffe in the middle of euery flower, like both for forme and colour vnto the flower of the double Featherfew, but fmaller : the rootes are many long ftrings, running here and there in the ground : this hath no fmell at all, but is delightfome only for the double white flowers.

The Place.

It is only cherifhed in fome few Gardens, for it is very rare.

The Time.

It flowreth in the end of Iune or thereabouts.

The Names.

It is called of moft *Ptarmica,* or *Sternutamentoria,* of his qualitie to prouoke neefing ; and of fome *Pyrethrum,* of the hot biting tafte. We vfually call it Double wilde Pelletorie, and fome Sneefewort, but *Elleborus albus* is vfually fo called, and I would not two things fhould be called by one name, for the miftaking and mif-ufing of them.

The

The Vertues.

The properties hereof, no doubt, may well bee referred to the single kinde, beeing of the same qualitie, yet as I take it, a little more milde and temperate.

CHAP. LVII.

Parthenium flore pleno. Double Featherfew.

FEatherfew that beareth double flowers is so like vnto the single kinde, that the one cannot be discerned from the other, vntill it come to flower, bearing broad, pale or fresh greene leaues, much cut in on the sides : the stalkes haue such like leaues on them as grow below, from the toppes whereof come forth many double white flowers, like vnto the flowers of the former wilde Pelletory, but larger, and like also vnto the flowers of the double Camomill : the sent whereof is as strong as of the single.

The Place.

We haue this kinde only in Gardens, and as it is thought by others, is peculiar only to our owne Countrey.

The Time.

It flowreth in the end of May, and in Iune and Iuly.

The Names.

It is called diuersly by diuers : Some thinke it to be *Parthenium* of Dioscorides, but not of Galen; for his *Parthenium* is a sweet herbe, and is thought to bee *Amaracus*, that is Marierome : others call it *Matricaria*; and some *Amarella*. Gaza translateth it *Muraleum, Theoph. lib. 7. cap. 7.* It is generally in these parts of our Country called Double Feauerfew, or Featherfew.

The Vertues.

It is answerable to all the properties of the single kinde which is vsed for womens diseases, to procure their monthly courses chiefly. It is held to bee a speciall remedy to helpe those that haue taken *Opium* too liberally. In Italy some vse to eate the single kinde among other greene herbes, as Camerarius saith, but especially fryed with egges, and so it wholly loseth his strong and bitter taste.

CHAP. LVIII.

Chamæmelum. Camomill.

OVr ordinary Camomill is well knowne to all, to haue many smal trayling branches, set with very fine smal leaues, bushing and spreading thicke ouer the ground, taking roote still as it spreadeth : the toppes of the branches haue white flowers, with yellow thrummes in the middle, very like vnto the Featherfew, before described, but somewhat greater, not so hard, but more soft and gentle in handling, and the whole herbe to be of a very sweet sent.

Bb

1. *Chamæmelum nudum.* Naked Camomill.

We haue another sort of Camomill in some Gardens, but very rare, like vnto the former, but that it is whiter, finer, and smaller, and raiseth it selfe vp a little higher, and beareth naked flowers; that is, without that border of white leaues that is in the former, and consisteth onely of a yellow round thrummie head, smelling almost as sweete as the former.

2. *Chamæmelum flore pleno.* Double flowred Camomill.

The double Camomill groweth with his leaues vpon the ground, as the other single kinde doth, but of a little fresher greene colour, and larger withall : the stalkes with the flowers on them, doe raise themselues vp a little higher then the ordinary, and bearing one or two flowers vpon a stalk, which are composed of many white leaues set together in diuers rowes, which make a fine double flower, with a little yellow spot in the middle for the most part of euery one, and are much larger then any single kinde, smelling better, and more pleasing then the ordinary : this doth creepe vpon the ground as the other, but is more tender to be kept in the Winter. Yet if you saue the flowers hereof (and so will the double Featherfew also) when they haue stood long, and ready to fade, and keepe them dry vntill the Spring, and then breaking them or pulling them to peeces, sowe them, there will spring vp from them Camomill, and also Featherfew, that will againe beare double flowers.

The Place.

Our ordinary Camomill groweth wilde in many places of our Country, and as well neare London as in other places. The others are onely found in our Gardens, where they are cherished. Bauhinus saith, that the double flowred Camomill is found wilde about Orleance in France.

The Time.

The double kinde is vsually in flower in Iune, before the ordinary kinde, and most commonly past before it flowreth, which is not vntill Iuly or August. The naked Camomill flowreth betweene them both, or later.

The Names.

Camomill is called *Anthemis, Leucanthemis*, and *Leucanthemum*, of the whitenesse of the flowers; and *Chamæmalum* of the corrupted Italian name *Camomilla.* Some call the naked Camomill, *Chrysanthemum odoratum.* The double Camomill is called by some *Chamæmalum Romanum flore multiplici.*

The Vertues.

Camomill is put to diuers and sundry vses, both for pleasure and profit, both for inward and outward diseases, both for the sicke and the sound, in bathings to comfort and strengthen the sound, and to ease paines in the diseased, as also in many other formes applyed outwardly. The flowers boyled in Posset drinke prouoketh sweat, and helpeth to expell colds, aches, and other griefes. A Syrupe made of the iuice of the double Camomill, with the flowers and white wine, as Bauhinus saith, is vsed by some against the Iaundise and Dropsie, caused by the euill disposition of the splene.

CHAP.

1 *Parthenium flore pleno*. Double Featherfew. 2 *Chamæmelum nudum*. Naked Camomill. 3 *Chamæmelum flore pleno* Double Ca-
monill. 4 *Pyrethrum officinarum*. Pellitory of Spaine. 5 *Flos Adonis flore rubro & flore luteo*. Adonis flower both red & yellow.
6 *Helleborus niger ferulaceus siue Buphthalmum*. The great Oxe eye or the great yellow Anemone. 7 *Buphthalmum vulgare* The
common yellow Oxe eye.

CHAP. LIX.

Pyrethrum officinarum. Pelletory of Spaine.

I Must needes adioyne vnto the Camomils this fine and tender plant, for some neare resemblance it hath with them in face, though not in quality. It is a small and lowe plant, bearing many fine greene leaues vpon his slender branches, which leane or lye down vpon the ground, diuided into many parts, yet somewhat larger and broader then Camomill, the stalkes whereof are bigger, and more iuicie then it : the flowers that stand at the toppes of the stalkes are single, but much larger then any Camomill flower, hauing a pale or border of many leaues, white on the vpperside, and reddish vnderneath, set about the yellow middle thrumme; but not standing so close together ioyning at the bottome, as the Camomill flowers doe, but more seuered one from another : it beareth small whitish seede, which is hardly found and discerned from the chaffe : the roote is long, and growing downe right, of the bignesse of a mans finger or thumbe in our Countrey, but not halfe so great where it groweth naturally, with some fibres and branches from the sides thereof, of a very hot, sharpe, and biting taste, drawing much water into the mouth, after it hath been chewed a while : the plant with vs is very tender, and will hardly or not at all endure the hardnesse and extremities of our Winters, vnlesse it be very carefully preserued.

The Place.

It groweth in Spaine wilde in many places, and in other hot Countries, where it may feele no frosts to cause it perish.

The Time.

It flowreth so late with vs, that it is not vntill August, that oftentimes we cannot gather ripe seedes from it, before it perish.

The Names.

The name *Pyrethrum* (taken from πυρ, that is, *ignis*, fire) is giuen to this plant, because of the heate thereof, and that the roote is somewhat like in shew; but specially in property vnto the true *Pyrethrum* of Dioscorides, which is an vmbelliferous plant, whose rootes are greater, and more feruent a great deale, and haue a hayrie bush or toppe as *Meum*, and many other vmbelliferous plants haue. It is also called in Latine, *Salinaris*, of the effect in drawing much moisture into the mouth, to be spit out. We doe vsually call it Pelletory of Spaine.

The Vertues.

It is in a manner wholly spent to draw rheume from the teeth, by chewing it in the mouth, thereby to ease the tooth-ach, and likewise from the head, in the paines thereof.

CHAP.

CHAP. LX.

Flos Adonis flore rubro. Red Adonis flower.

ADonis flower may well be accounted a kinde of Camomill, although it hath some especiall differences, hauing many long branches of leaues lying vpon the ground, and some rifing vp with the stalke, so finely cut and iagged, that they much resemble the leaues of Mayweed, or of the former *Nigella* : at the top of the stalkes, which rise a foote high or better, stand small red flowers, consisting of six or eight round leaues, hauing a greene head in the middle, set about with many blackish threads, without any smell at all : after the flowers are past, there grow vp heads with many roundish white seedes at the toppes of them, set close together, very like vnto the heads of seede of the great Oxe eye, set downe in the next Chapter, but smaller : the rootes are small and thready, perishing euery yeare, but rifing of his owne seede againe, many times before Winter, which will abide vntill the next yeare.

Yellow Adonis flower is like vnto the red, but that the flower is somewhat larger, *Flore luteo:* and of a faire yellow colour.

The Place.

The first groweth wilde in the corn fields in many places of our own country, as well as in others, and is brought into Gardens for the beauties sake of the flower. The yellow is a stranger, but nourſed in our Gardens with other rarities.

The Time.

They flower in May or Iune, as the yeare falleth out to be early or late : the seed is soone ripe after, and will quickly fall away, if it be not gathered.

The Names.

Some haue taken the red kinde to be a kinde of Anemone, other to be *Eranthemum* of Dioscorides : the most vsuall name now with vs is *Flos Adonis,* and *Flos Adonidis* : In English, where it groweth wilde, they call it red Maythes, as they call the Mayweede, white Maythes ; and some of our English Gentlewomen call it Rofarubie : we vsually call it Adonis flower.

The Vertues.

It hath been certainly tryed by experience, that the seed of red Adonis flower drunke in wine, is good to ease the paines of the Collicke and Stone.

CHAP. LXI.

Buphthalmum. Oxe eye.

VNder the name *Buphthalmum,* or Oxe eye, are comprehended two or three seuerall plants, each differing from other, both in face and property, yet because they all beare one generall name, I thinke fittest to comprise them all in one Chapter, and first of that which in leafe & seed commeth nearest to the Adonis flower.

1. *Buphthalmum maius siue Helleborus niger ferulaceus.*
Great Oxe eye, or the yellow Anemone.

This great Oxe eye is a beautifull plant, hauing many branches of greene leaues

leaning

leaning or lying vpon the ground for the moſt part, yet ſome ſtanding vpright, which are as fine, but ſhorter then Fenell, ſome of them ending in a ſmall tuft of green leaues, and ſome hauing at the toppes of them one large flower apeece, ſomewhat reddiſh or browniſh on the outſide, while they are in bud, and a while after, and being open, ſhew themſelues to conſiſt of twelue or fourteene long leaues, of a faire ſhining yellow colour, ſet in order round about a greene head, with yellow thrums in the middle, laying themſelues open in the ſunne, or a faire day, but elſe remaining cloſe : after the flower is paſt, the head growing greater, ſheweth it ſelfe compact of many round whitiſh ſeede, very like vnto the head of ſeede of the Adonis flower laſt deſcribed, but much greater : the rootes are many long blackiſh fibres or ſtrings, ſet together at the head, very like vnto the rootes of the leſſer blacke Hellebor or Bearefoote, but ſomewhat harder, ſtiffer, or more brittle, and ſeeming without moiſture in them, which abide and encreaſe euery yeare.

2. *Buphthalmum minus, ſeu Anthemis flore luteo.* Small Oxe eye.

This plant might ſeeme to be referred to the Camomils, but that it is not ſweete, or to the Corne-Marigolds, but that the ſtalkes and leaues are not edible : it is therefore put vnder the Oxe eyes, and ſo we will deſcribe it ; hauing many weake branches lying vpon the ground, beſet with winged leaues, very finely cut and iagged, ſomewhat like vnto Mayweede, but a little larger : the flowers are like vnto the Corne Marigold, and larger then any Camomill, being wholly yellow, as well the pale or border of leaues, as the middle thrummes : the rootes are ſomewhat tough and long.

3. *Buphthalmum vulgare.* Common Oxe eye.

This Oxe eye riſeth vp with hard round ſtalkes, a foote and a halfe high, hauing many winged leaues vpon them, made of diuers long and ſomething broad leaues, ſnipt about the edges, ſet together ſomewhat like vnto Tanſie, but ſmaller, and not ſo much winged : the flowers ſtand at the toppes of the ſtalkes, of a full yellow colour, both the outer leaues and the middle thrum, and not altogether ſo large as the laſt : the rootes of this kinde periſh euery yeare, and require a new ſowing againe.

The Place.

The firſt groweth in diuers places of Auſtria, Bohemia, and thoſe parts, it hath beene likewiſe brought out of Spaine. The ſecond in Prouence, a country in France. The laſt in diuers places, as well of Auſtria as Morauia, and about Mentz and Norimberg, as Cluſius ſetteth downe. We haue them in our Gardens, but the firſt is of the greateſt reſpect and beauty.

The Time.

The firſt flowreth betimes, oftentimes in March, or at the furtheſt in Apill ; the ſeede is ripe in May, and muſt be quickly gathered, leſt it bee loſt. The other two flower not vntill Iune.

The Names.

The firſt is called *Buphthalmum* of Dodonæus, *Pſeudohelleborus* of Matthiolus, *Helleborus niger ferulaceus Theophraſti* by Lobel, of ſome others *Elleborus niger verus*, vſing it for the true blacke Ellebor, but it is much differing, as well in face as properties. Of others *Seſamoides minus.* Some haue thought it to be a yellow Anemone, that haue looked on it without further iudgement, and by that name is moſt vſually knowne to moſt of our Engliſh Gentlewomen that know it. But it may moſt fitly be called a *Buphthalmum,* as Dodonæus doth, and *Hiſpanicum* or *Auſtriacum,* for diſtinctions ſake. We doe moſt vſually call it *Helleborus niger ferulaceus,* as Lobel doth : Bauhinus calleth

calleth it *Helleborus niger tenuifolius Buphthalmi flore.* The second is called *Buphthalmum Narbonense* : In English, The French, or lesser Oxe eye, as the first is called, The great Oxe eye. The last, The common Oxe eye.

The Vertues.

The first hath been vsed in diuers places for the true blacke Ellebor, but now is sufficiently knowne to haue been an errour ; but what Physicall property it hath, other then Matthiolus hath expressed, to be vsed as Setterwort for cattell, when they rowell them, to put or draw the rootes hereof through the hole they make in the dewe lappe, or other places, for their coughes or other diseases, I know not, or haue heard or read of any. The others likewise haue little or no vse in Physicke now a dayes that I know.

CHAP. LXII.

Chrysanthemum. Corne Marigold.

ALthough the forts of Corne Marigolds, which are many, are fitter for another then this worke, and for a Catholicke Garden of Simples, then this of Pleasure and Delight for faire Flowers ; yet giue me leaue to bring in a couple : the one for a corner or by-place, the other for your choisest, or vnder a defenced wall, in regard of his statelinesse.

1. *Chrysanthemum Creticum.* Corne Marigold of Candy.

This faire Corne Marigold hath for the most part one vpright stalke, two foote high, whereon are set many winged leaues, at euery ioynt one, diuided and cut into diuers parts, and they againe parted into seuerall peeces or leaues : the flowers growe at the toppes of the stalkes, rising out of a scaly head, composed of ten or twelue large leaues, of a faire, but pale yellow colour, and more pale almost white at the bottome of the leaues, round about the yellow thrumme in the middle, being both larger and sweeter then any of the other Corne Marigolds : the seede is whitish and chaffie : the roote perisheth euery yeare.

2. *Chrysanthemum Peruuianum, siue Flos Solis.*
The golden flower of Peru, or the Flower of the Sunne.

This goodly and stately plant, wherewith euery one is now adayes familiar, being of many sorts, both higher and lower (with one stalke, without branches, or with many branches, with a blacke, or with a white seede, yet differing not in forme of leaues or flowers one from another, but in the greatnesse or smalnesse) riseth vp at the first like vnto a Pompion with two leaues, and after two, or foure more leaues are come forth, it riseth vp into a great stalke, bearing the leaues on it at seuerall distances on all sides thereof, one aboue another vnto the very toppe, being sometimes, and in some places, seuen, eight, or ten foote high, which leaues standing out from the stemme or stalke vpon their seuerall great ribbed foote-stalkes, are very large, broad belowe, and pointed at the end, round, hard, rough, of a sad greene colour, and bending downewards : at the toppe of the stalke standeth one great, large, and broad flower, bowing downe the head vnto the Sunne, and breaking forth from a great head, made of scaly greene leaues, like vnto a great single Marigold, hauing a border of manie long yellow leaues, set about a great round yellow thrumme, as it were in the middle, which are very like vnto short heads of flowers, vnder euery one whereof there is a seede, larger then any seede of the Thistles, yet somewhat like, and lesser, and rounder then any Gourd seede, set in so close and curious a manner, that when the seede is taken out, the head with the hollow places or cels thereof, seemeth very like vnto an hony combe ; which seede is in some plants very blacke, in the hotter countries, or very

white,

white, and great, or large, but with vs is neither fo large, blacke, or white; but fome-times blackifh or grayifh. Some fort rifeth not vp halfe the height that others doe, and fome againe beare but one ftemme or ftalke, with a flower at the toppe thereof; and others two or three, or more fmall branches, with euery one his flower at the end; and fome fo full of branches from the very ground almoft, that I haue accounted threefcore branches round about the middle ftalke of one plant, the loweft neare two yards long, others aboue them a yard and a halfe, or a yard long, with euery one his flower thereon; but all fmaller then thofe that beare but one or two flowers, and leffer alfo for the moft part then the flower on the middle ftalke it felfe. The whole plant, and euery part thereof aboue ground hath a ftrong refinous fent of Turpentine, and the heads and middle parts of the flowers doe oftentimes (and fometimes the ioynts of the ftalke where the leaues ftand) fweat out a moft fine thin & cleare Roffin or Turpentine, but in fmall quantity, and as it were in drops, in the heate and dry time of the year, fo like both in colour, fmell, and tafte vnto cleare Venice Turpentine, that it cannot be knowne from it: the roote is ftrongly faftened in the ground by fome greater roots branching out, and a number of fmall ftrings, which growe not deepe, but keepe vnder the vpper cruft of the earth, and defireth much moifture, yet dyeth euery yeare with the firft frofts, and muft be new fowne in the beginning of the Spring.

The Place.

Their places are fet downe in their titles, the one to come out of Candy, the other out of Peru, a Prouince in the Weft Indies.

The Time.

The firft flowreth in Iune, the other later, as not vntill Auguft, and fometimes fo late, that the early frofts taking it, neuer fuffer it to come to ripeneſs.

The Names.

The firft hath his name in his title. The fecond, befides the names fet downe, is called of fome *Planta maxima, Flos maximus, Sol Indianus*, but the moft vfuall with vs is, *Flos Solis*: In Englifh, The Sunne Flower, or Flower of the Sunne.

The Vertues.

There is no vfe of either in Phyficke with vs, but that fometimes the heads of the Sunne Flower are dreffed, and eaten as Hartichokes are, and are accounted of fome to be good meate, but they are too ftrong for my tafte.

Chap. LXIII.

Calendula. Marigolds.

SOme haue reckoned vp many forts of Marigolds, I had rather make but two, the fingle and the double; for doubtleffe, thofe that be moft double, rife from the beft feede, which are the middlemoft of the great double, and fome will be leffe double, whofe feede is greater then the reft, according to the ground where it groweth; as alfo thofe that be of a paler colour, doe come of the feed of the yellower fort.

1. *Calendula maxima.* The great Garden Marigold.

The Garden Marigold hath round greene ftalkes, branching out from the ground into many parts, whereon are fet long flat greene leaues, broader and rounder at the
point

1 *Chryfanthemum Creticum.* Corne Marigolds of Candy. 2 *Flos Solis.* The Flower of the Sunne. 3 *Calendula.* Marigolds. 4 *Aster Atticus five Italorum.* The purple Marigold. 5 *Pilofella maior.* Golden Moufe-eare. 6 *Scorfonera Hifpanica.* Spanifh Vipers grafle. 7 *Tragopogon.* Goates beard, or got to bed at noone.

point then any where elfe, and fmaller alfo at the fetting to of the ftalke, where it compaffeth it about: the flowers are fometimes very thicke and double (breaking out of a fcaly clammy greene head) compofed of many rowes of leaues, fet fo clofe together one within another, that no middle thrume can bee feene, and fometimes leffe double, hauing a fmall browne fpot of a thrume in the middle : and fometimes but of two or three rowes of leaues, with a large browne thrume in the middle; euery one whereof is fomewhat broader at the point, and nicked into two or three corners, of an excellent faire deepe gold yellow colour in fome, and paler in others, and of a pretty ftrong and refinous fweete fent : after the flowers are paft, there fucceede heads of crooked feede, turning inward, the outermoft biggeft, and the innermoft leaft : the roote is white, and fpreadeth in the ground, and in fome places will abide after the feeding, but for the moft part perifheth, and rifeth againe of his owne feede. Sometimes this Marigold doth degenerate, and beareth many fmall flowers vpon fhort ftalkes, compaffing the middle flower : but this happeneth but feldome, and therefore accounted but *lufus natura*, a play of nature, which fhe worketh in diuers other plants befides.

2. *Calendula fimplex.* The fingle Marigold.

There is no difference betweene this and the former, but that the flowers are fingle, confifting of one rowe of leaues, of the fame colour; eyther paler or deeper yellow, ftanding about a great browne thrumme in the middle : the feed likewife is alike, but for the moft part greater then in the double kindes.

The Place.

Our Gardens are the chiefe places for the double flowers to grow in ; for we know not of any other naturall place : but the fingle kinde hath beene found wilde in Spaine, from whence I receiued feede, gathered by Guillaume Boel, in his time a very curious, and cunning fearcher of fimples.

The Time.

They flower all the Summer long, and fometimes euen in winter, if it be milde, and chiefly at the beginning of thofe monethes, as it is thought.

The Names.

They are called *Caltha* of diuers, and taken to be that *Caltha*, wherof both Virgil and Columella haue written. Others doe call them *Calendula*, of the Kalendes, that is the firft day of the monethes, wherein they are thought chiefly to flower ; and thereupon the Italians call them, *Fiori di ogni mefe*, that is, The Flowers of euery moneth : We cal them in Englifh generally, eyther Golds, or Marigolds.

The Vertues.

The herbe and flowers are of great vfe with vs among other pot-herbes; and the flowers eyther greene or dryed, are often vfed in poffets, broths, and drinkes, as a comforter of the heart and fpirits, and to expel any malignant or peftilential quality, gathered neere thereunto. The Syrupe and Conferue made of the frefh flowers, are vfed for the fame purpofes to good effect.

CHAP. LXIIII.

Aster. Starre-wort.

Dioscorides and other of the ancient Writers, haue set forth but one kinde of Starre-wort, which they call *Aster Atticus*, of the place no doubt, where the greatest plentie was found, which was the Countrey of Athens : the later Writers haue found out many other plants which they referre to this kinde, calling them by the same name. It is not my purpose to entreate of them all, neyther doth this garden fitly agree with them : I shall therefore select out one or two from the rest, and giue you the knowledge of them, leauing the rest to their proper place.

1. *Aster Atticus flore luteo.* Yellow Starre-wort.

This Starre-wort riseth vp with two or three rough hairy stalkes, a foote and a halfe high, with long, rough or hairie, brownish, darke greene leaues on them, diuided into two or three branches : at the toppe of euery one wereof standeth a flat scaly head, compassed vnderneath with fiue or sixe long, broade, rough greene leaues, standing like a Starre, the flower it selfe standing in the middle thereof, made as a border of narrow, long, pale yellow leaues, set with a brownish yellow thrume : the roote dyeth euery yeare, hauing giuen his flower.

2. *Aster Atticus Italorum flore purpureo.* Purple Italian Starre-wort.

This Italian Starre-wort hath many wooddy, round brittle stalkes, rising from the roote, somewhat higher then the former, sometimes standing vpright, and other-whiles leaning downewards, whereon are set many somewhat hard, and rough long leaues, round pointed, without order vp to the toppe, where it is diuided into seuerall branches, whereon stand the flowers, made like vnto a single Marigold, with a border of blewish purple leaues, set about a browne middle thrume, the heads sustaining the flowers, are composed of diuers scaly greene leaues, as is to be seene in the Knap-weedes or Matfelons, which after the flowers are past yeelde a certaine downe, wherein lye small blacke and flat seedes, somewhat like vnto Lettice seede, which are carried away with the winde : the roote is composed of many white strings, which perisheth not as the former, but abideth, and springeth afresh euery yeare.

The Place.

The first is found in Spaine, as Clusius, and in France, as Lobel say. The other hath beene found in many places in Germany, and Austria : in Italie also, and other places ; we haue it plentifully in our Gardens.

The Time.

The first flowreth in Summer. And the other not vntill August or September.

The Names.

The first is called *Aster Atticus flore luteo, Bubonium, & Inguinalis,* and of many is taken to be the true *Aster Atticus* of Dioscorides : yet Matthiolus thinketh not so, for diuers good reasons, which hee setteth downe in the Chapter of *Aster Atticus,* as any man may vnderstand, if they will but reade the place, which is too long to bee inserted here. The other is thought by Matthiolus, to bee the truer *Aster Atticus,* (vnto whom I must also consent) and constantly also affirmed to be the *Amellus Virgilij,* as may be seene in the same place : but it is vsually called at this day, *Aster Italorum flore caruleo* or *purpureo,*

purpureo. Their English names are sufficiently expressed in their titles, yet some call the last, The purple Marigold, because it is so like vnto one in form.

The Vertues.

They are held, if they bee the right, to bee good for the biting of a mad dogge, the greene herbe being beaten with old hogs grease, and applyed, as also for swolne throats : It is likewise vsed for botches that happen in the groine, as the name doth import.

Chap. LXV.

Pilosella maior. Golden Mouse-eare.

SOme resemblance that the flowers of this plant hath with the former Golds, maketh me to insert it in this place, although I know it agreeth not in any other part, yet for the pleasant aspect thereof, it must bee in this my garden, whose description is as followeth : It hath many broade greene leaues spread vpon the ground, spotted with pale spots, yet more conspicuous at sometimes then at other ; somewhat hairy both on the vpper and vnderside, in the middle of these leaues rise vp one, two or more blackish hairy stalkes, two foote high at the least, bare or naked vp to the top, where it beareth an vmbell, or short tuft of flowers, set close together vpon short stalkes, of the forme or fashion of the Haukeweedes, or common Mouse-eare, but somewhat smaller, of a deep gold yellow, or orenge tawney colour, with some yellow threds in the middle, of little or no sent at all: after the flowers are past, the heads carry small, short, blacke seede, with a light downie matter on them, ready to bee carried away with the winde, as many other plants are, when they be ripe : the rootes spread vnder ground, and shoote vp in diuers other places, whereby it much encreaseth, especially if it be set in any moist or shadowie place.

The Place.

It groweth in the shadowie woods of France, by Lions, and Mompelier, as Lobell testifieth : we keepe it in our gardens, and rather in a shadowie then sunnie place.

The Time.

It flowreth in Somer, and sometimes againe in September.

The Names.

It is called by Lobell, *Pulmonaria Gallorum Hieratij facie* : and the Herbarists of France take it to be the true *Pulmonaria* of Tragus. Others call it *Hieratium flore aureo.* Pelleterius *Hieratium Indicum.* Some *Pilosella,* or *Auricula muris maior flore aureo.* And some *Chondrilla flore aureo.* Dalechamptus would haue it to bee *Corchorus,* but farre vnfitly. The fittest English name we can giue it, is Golden Mouse-eare, which may endure vntill a fitter bee imposed on it : for the name of Grim the Collier, whereby it is called of many, is both idle and foolish.

The Vertues.

The French according to the name vse it for the defects of the lunges, but with what good successe I know not.

CHAP. LXVI.

Scorfonera. Vipers graffe.

Lthough there be foure or fiue forts of *Scorfonera*, yet I fhall here defire you to be content with the knowledge only of a couple.

1. *Scorfonera Hifpanica maior.* The greater Spanifh Vipers graffe.

This Spanifh Vipers graffe hath diuers long, and fomewhat broad leaues, hard and crumpled on the edges, and fometimes vneuenly cut in or indented alfo, of a blewifh greene colour: among which rifeth vp one ftalke, and no more for the moft part, two foote high or thereabouts, hauing here and there fome narrower long leaues thereon then thofe below: the toppe of the ftalke brancheth it felfe forth into other parts, euery one bearing a long fcaly head, from out of the toppe whereof rifeth a faire large double flower, of a pale yellow colour, much like vnto the flower of yellow Goates beard, but a little leffer, which being paft, the feede fucceedeth, being long, whitifh and rough, inclofed with much downe, and among them many other long fmooth feedes, which are limber and idle, and are carryed away at the will of the winde: the roote is long, thicke and round, brittle and blacke, with a certaine roughneffe on the outfide: but very white within, yeelding a milkie liquor being broken, as euery other part of the plant doth befides, yet the roote more then any other part, and abideth many yeares without perifhing.

2. *Scorfonera Pannonica purpurea.* Purple flowred Vipers graffe.

This purple flowred Vipers graffe hath long and narrow leaues, of the fame blewifh greene colour with the former: the ftalke rifeth vp a foote and a halfe high, with a few fuch like leaues, but fhorter thereon, breaking at the toppe into two or three parts, bearing on each of them one flower, fafhioned like the former, and ftanding in the like fcaly knoppe or head, but of a blewifh purple colour, not fully fo large, of the fweeteft fent of any of this kinde, comming neereft vnto the fmell of a delicate perfume.

The Place.

The firft is of Spaine. The other of Hungarie and Auftrich: which now furnifh our gardens.

The Time.

They flower in the beginning of May: the feede is foone ripe after, and then perifhing downe to the roote for that yeare, fpringeth afrefh before Winter againe.

The Names.

They are called after the Spanifh name *Scorfonera*, which is in Latine *Viperaria*, of fome *Viperina*, and *Serpentina*: Wee call them in Englifh Vipers graffe, or *Scorfonera*.

The Vertues.

Manardus as I thinke firft wrote hereof, and faith that it hath been found to cure them that are bitten of a Viper, or other fuch like venemous Creature. The rootes hereof being preferued with fugar, as I haue done often, doe eate almoft as delicate as the Eringus roote, and no doubt is good to comfort and ftrengthen the heart, and vitall fpirits. Some that haue vfed the preferued roote haue found it effectuall to expelling winde out of the ftomacke, and to helpe fwounings and faintneffe of the heart.

Cc

CHAP.

Chap. LXVII.

Tragopogon. Goates beard.

I Muſt in this place ſet downe but two ſorts of Goates beards; the one blew or aſh-colour, the other red or purple, and leaue the other kindes : ſome to bee ſpoken of in the Kitchin Garden, and others in a Phyſicall Garden.

1. *Tragopogon flore cæruleo.* Blew Goates beard.

All the Goates beards haue long, narrow, and ſomewhat hollow whitiſh greene leaues, with a white line downe the middle of euery one on the vpperſide : the ſtalke riſeth vp greater and ſtronger then the Vipers graſſe, bearing at the toppe a great long head or huske, compoſed of nine or ten long narrow leaues, the ſharpe points or ends whereof riſe vp aboue the flower in the middle, which is thicke and double, ſome-what broad and large ſpread, of a blewiſh aſh-colour, with ſome whitiſh threads a-mong them, ſhutting or cloſing it ſelfe within the greene huske euery day, that it abi-deth blowing, vntill about noone, and opening not it ſelfe againe vntill the next mor-ning : the head or huske, after the flower is paſt, and the ſeede neare ripe, openeth it ſelfe; the long leaues thereof, which cloſed not before now, falling downe round a-bout the ſtalke, and ſhewing the ſeede, ſtanding at the firſt cloſe together, and the doune at the toppe of them : but after they haue ſtood a while, it ſpreadeth it ſelfe round, and is ready to be carried away with the winde, if it be not gathered : the ſeede it ſelfe is long, round, and rough, like the ſeede of the Vipers graſſe, but greater and blacker : the roote is long, and not very great, but periſheth as ſoone as it hath borne ſeede, and ſpringeth of the fallen ſeede, that yeare remaining greene all Winter, and flowring the next yeare following : the whole yeeldeth milke as the former, but ſome-what more bitter and binding.

2. *Tragopogon purpureum.* Purple Goates beard.

There is little difference in this kind from the former, but that it is a little larger, both in the leafe, and head that beareth the ſeed: the flowers alſo are a little larger, and ſpread more, of a darke reddiſh purple colour, with ſome yellow duſt as it were caſt vpon it, eſpecially about the ends : the roote periſheth in the like manner as the other.

The Place.

Both theſe haue been ſent vs from the parts beyond the Seas, I haue had them from Italy, where no doubt they grow naturally wilde, as the yellow doth with vs : they are kept in our Gardens for their pleaſant flowers.

The Time.

They flower in May and Iune : the ſeede is ripe in Iuly.

The Names.

Their generall name is after the Greeke word *Tragopogon,* which is in La-tine, *Barba hirci :* In Engliſh, Goates beard; the head of ſeede when it is rea-die to bee carried away with the winde, cauſing that name for the reſem-blance : and becauſe the flower doth euery day cloſe it ſelfe at noone (as I ſaid before) and openeth not againe vntill the next Sunne, ſome haue fitly called it, Goe to bed at noone.

The Vertues.

The rootes of theſe kindes are a little more bitter and more binding alſo
then

then the yellow kinde expressed in the Kitchin Garden; and therefore fitter for medicine then for meate, but yet is vsed as the yellow kinde is, which is more fit for meate then medicine. The distilled water is good to wash old sores and wounds.

Chap. LXVIII.

Flos Africanus. The French Marigold.

OF the French or African Marigolds there are three kindes as principall, and of each of them both with single and double flowers : of these, some diuersity is obserued in the colour of the flowers, as well as in the forme or largenesse, so that as you may here see, I haue expressed eight differences, and Fabius Columna nine or ten, in regard hee maketh a diuersity of the paler and deeper yellow colour : and although the lesser kinde, because of its euill sent, is held dangerous, yet for the beauty of the flower it findeth roome in Gardens.

1. *Flos Africanus maior siue maximus multiplex.*
The great double French Marigold.

This goodly double flower, which is the grace and glory of a Garden in the time of his beauty, riseth vp with a straight and hard round greene stalke, hauing some crests or edges all along the stalke, beset with long winged leaues, euery one whereof is like vnto the leafe of an Ash, being composed of many long and narrow leaues, snipt about the edges, standing by couples one against another, with an odde one at the end, of a darke or full greene colour : the stalke riseth to be three or foure foote high, and diuideth it selfe from the middle thereof into many branches, set with such like leaues to the toppes of them, euery one bearing one great double flower, of a gold yellow colour aboue, and paler vnderneath, yet some are of a pale yellow, and some betweene both, and all these rising from one and the same seede : the flower, before it be blowne open, hath all the leaues hollow; but when it is full blowne open, it spreadeth it selfe larger then any Prouince Rose, or equall vnto it at the least, if it be in good earth, and riseth out of a long greene huske, striped or furrowed, wherein after the flower is past, (which standeth in his full beauty a moneth, and oftentimes more, and being gathered, may be preserued in his full beauty for two moneths after, if it be set in water) standeth the seede, set thicke and close together vpright, which is blacke, somewhat flat and long : the roote is full of small strings, whereby it strongly comprehendeth in the ground : the flower of this, as well as the single, is of the very smell of new waxe, or of an honie combe, and not of that poisonfull sent of the smaller kindes.

2. *Flos Africanus maior simplex.* The great single French Marigold.

This single Marigold is in all things so like vnto the former, that it is hard to discerne it from the double, but by the flowers, onely the stalke will be browner then the double; and to my best obseruation, hath and doth euery yeare rise from the seede of the double flower : so that when they are in flower, you may see the difference (or not much before, when they are in bud) this single flower euer appearing with thrums in the middle, and the leaues, which are the border or pale standing about them, shewing hollow or fistulous, which after lay themselues flat and open (and the double flower appearing with all his leaues folded close together, without any thrum at all) and are of a deeper or paler colour, as in the double.

3. *Flos Africanus fistuloso flore simplex & multiplex.*
Single and double French Marigolds with hollow leafed flowers.

As the former two greatest sorts haue risen from the seede of one and the same (I

Cc 2

meane

meane the pod of double flowers) so doe these also, not differing from it in any thing, but that they are lower, and haue smaller greene leaues, and that the flower also being smaller, hath euery leafe abiding hollow, like vnto an hollow pipe, broad open at the mouth, and is of as deepe a yellow colour for the most part as the deepest of the former, yet sometimes pale also.

4. *Flos Africanus minor multiplex.* The lesser double French Marigold.

The lesser double French Marigold hath his leaues in all things like vnto the former, but somewhat lesser, which are set vpon round browne stalkes, not so stiffe or vpright, but bowing and bending diuers wayes, and sometimes leaning or lying vpon the ground : the stalkes are branched out diuersly, whereon are set very faire double flowers like the former, and in the like greene huskes, but smaller, and in some the outermost leaues will be larger then any of the rest, and of a deeper Orenge colour, almost crimson, the innermost being of a deepe gold yellow colour, tending to crimson : the whole flower is smaller, and of a stronger and more vnpleasant sauour, so that but for the beautifull colour, and doublenesse of the flower, pleasant to the eye, and not to any other sense, this kinde would finde roome in few Gardens : the rootes and seedes are like the former, but lesser.

5. *Flos Africanus minor simplex.* The small single French Marigold.

This single kinde doth follow after the last in all manner of proportion, both of stalkes, leaues, seedes, and rootes : the flowers onely of this are single, hauing fiue or six broad leaues, of a deepe yellow crimson colour, with deepe yellow thrummes in the middle, and of as strong a stinking sent, or more then the last.

The Place.

They growe naturally in Africa, and especially in the parts about Tunis, and where old Carthage stood, from whence long agoe they were brought into Europe, where they are onely kept in Gardens, being sowne for the most part euery yeare, vnlesse in some milde Winters. The last single and double kindes (as being more hardy) haue sometimes endured : but that kinde with hollow leafed flowers, as Fabius Columna setteth it downe, is accounted to come from Mexico in America.

The Time.

They flower not vntill the end of Summer, especially the greater kindes: but the lesser, if they abide all the Winter, doe flower more early.

The Names.

They haue been diuersly named by diuers men : Some calling them *Caryophyllus Indicus*, that is, Indian Gilloflowers, and *Tanacetum Peruvianum*, Tansie of Peru, as if it grew in Peru, a Prouince of America ; and *Flos Indicus*, as a flower of the Indies ; but it hath not beene knowne to haue beene brought from thence. Others would haue it to be *Othonna* of Plinie, and others ; some to be *Lycopersicum* of Galen. It is called, and that more truely, *Flos Tunetensis*, *Flos Africanus*, and *Caltha Africana*, that is, the flower of Tunis, the flower of Africa, the Marigold of Africa, and peraduenture *Pedua Pauorum*. We in English most vsually call them, French Marigolds, with their seuerall distinctions of greater or smaller, double or single. To that with hollow leafed flowers, Fabius Columna giueth the name of *Fistuloso flore*, and I so continue it.

The

1 *Flos Africanus maximus multiplex.* The greatest double French Marigold. 2 *Flos Africanus maior multiplex.* The greater double French Marigold
3 *Flos Africanus maximus simplex* The greatest single French Marigold. 4 *Flos Africanus multiplex fistulosus.* The double hollow French Marigold.
5 *Flos Africanus simplex fistulosus.* The single hollow French Marigold. 6 *Flos Africanus minor multiplex.* The smaller double French Marigold. 7 *Flos Africanus minor multiplex alter.* Another sort of the lesser double French Marigold. 8 *Flos Africanus minor simplex.* The lesser single French Marigold.

The Vertues.

We know no vſe they haue in Phyſicke, but are cheriſhed in Gardens for their beautifull flowers ſake.

Chap. LXIX.

Caryophyllus hortenſis. Carnations and Gilloflowers.

TO auoide confuſion, I muſt diuide Gilloflowers from Pinkes, and intreate of them in ſeuerall Chapters. Of thoſe that are called Carnations or Gilloflowers, as of the greater kinde, in this Chapter; and of Pinkes, as well double as ſingle, in the next. But the number of them is ſo great, that to giue ſeuerall deſcriptions to them all were endleſſe, at the leaſt needleſſe: I will therefore ſet downe onely the deſcriptions of three (for vnto theſe three may be referred all the other ſorts) for their faſhion and manner of growing, and giue you the ſeuerall names (as they are vſually called with vs) of the reſt, with their variety and mixture of colours in the flowers, wherein conſiſteth a chiefe difference. I account thoſe that are called Carnations to be the greateſt, both for leafe and flower, and Gilloflowers for the moſt part to bee leſſer in both; and therefore will giue you each deſcription apart, and the Orenge tawnie or yellow Gilloflower likewiſe by it ſelfe, as differing very notably from all the reſt.

1. *Caryophyllus maximus Harwicenſis ſiue Anglicus.* The great Harwich or old Engliſh Carnation.

I take this goodly great old Engliſh Carnation, as a preſident for the deſcription of all the reſt of the greateſt ſorts, which for his beauty and ſtatelineſſe is worthy of a prime place, hauing beene alwayes very hardly preſerued in the Winter; and therefore not ſo frequent as the other Carnations or Gilloflowers. It riſeth vp with a great thicke round ſtalke, diuided into ſeuerall branches, ſomewhat thickly ſet with ioynts, and at euery ioynt two long greene rather then whitiſh leaues, ſomewhat broader then Gilloflower leaues, turning or winding two or three times round (in ſome other ſorts of Carnations they are plaine, but bending the points downewards, and in ſome alſo of a darke reddiſh greene colour, and in others not ſo darke, but rather of a whitiſh greene colour:) the flowers ſtand at the toppes of the ſtalkes in long, great, and round greene huskes, which are diuided into fiue points, out of which riſe many long and broad pointed leaues, deeply iagged at the ends, ſet in order round and comely, making a gallant great double flower, of a deepe Carnation colour, almoſt red, ſpotted with many bluſh ſpots and ſtrakes, ſome greater and ſome leſſer, of an excellent ſoft ſweete ſent, neither too quicke as many others of theſe kinds are, nor yet too dull, and with two whitiſh crooked threads like hornes in the middle: this kinde neuer beareth many flowers, but as it is ſlow in growing, ſo in bearing, not to be often handled, which ſheweth a kinde of ſtatelineſſe, fit to preſerue the opinion of magnificence: the roote is branched into diuers great, long, wooddy rootes, with many ſmall fibres annexed vnto them.

2. *Caryophyllus hortenſis flore pleno rubro.* The red or Cloue Gilloflower.

The red Cloue Gilloflower, which I take as a preſident for the ſecond ſort, which are Gilloflowers, grow like vnto the Carnations, but not ſo thicke ſet with ioynts and leaues: the ſtalkes are more, the leaues are narrower and whiter for the moſt part, and in ſome doe as well a little turne: the flowers are ſmaller, yet very thicke and double in moſt, and the greene huskes wherein they ſtand are ſmaller likewiſe then the former: the ends of the leaues in this flower, as in all the reſt, are dented or iagged, yet in ſome more then in others; ſome alſo hauing two ſmall white threads, crooked at the ends like hornes, in the middle of the flower, when as diuers other haue none. Theſe kindes,

1 *Caryophyllus maximus rubro varius*. The great old Carnation or gray Hulo. 2 *Caryophyllus maior rubeo & albo varius*. The white Carnation. 3 *Caryophyllus albo rubens*. The Cambersine or the Poole flower. 4 *Caryophyllus Cantij fissus*. The faire made of Kent. 5 *Caryophyllus Subauditoie caroens*. The blush Sauadge. 6 *Caryophyllus Xerampelinus*. The Gredeline Carnation. 7 *Caryophyllus dictus Grimo's*. The Grimelo or Prince. 8 *Caryophyllus albus maior*. The great white Gilloflower. 9 *Elegans Heroina Bradshawi*. Master Bradshawes dainty Lady.

kindes, and especially this that hath a deepe red crimson coloured flower, doe endure the cold of our winters, and with lesse care is preserued : these sorts as well as the former doe very seldome giue any seede, as far as I could euer obserue or learne.

3. *Caryophyllus Silesiacus flore pleno miniato.*
The yellow or Orenge tawny Gilloflower.

This Gilloflower hath his stalkes next vnto the ground, thicker set, and with smaller or narrower leaues then the former for the most part : the flowers are like vnto the Cloue Gilloflowers, and about the same bignesse and doublenesse most vsually, yet in some much greater then in others; but of a pale yellowish Carnation colour, tending to an Orenge, with two small white threds, crooked at the ends in the middle, yet some haue none, of a weaker sent then the Cloue Gilloflower : this kinde is more apt to beare seede then any other, which is small, black, flat, and long, and being sowen, yeelde wonderfull varieties both of single and double flowers : some being of a lighter or deeper colour then the mother plants : some with stripes in most of the leaues : Others are striped or spotted, like a speckled Carnation or Gilloflower, in diuers sorts, both single and double : Some againe are wholly of the same colour, like the mother plant, and are eyther more or lesse double then it, or else are single with one row of leaues, like vnto a Pinck; and some of these likewise eyther wholly of a crimson red, deeper or lighter, or variably spotted, double or single as a Pinck, or blush eyther single or double, and but very seldome white : yet all of them in their greene leaues little or nothing varying or differing.

CARNATIONS.

Cariophylli maximi.

Caryophyllus maximus dictus Hulo rubro-varius.

THe gray *Hulo* hath as large leaues as the former old Carnation, and as deepely iagged on the edges : it hath a great high stalke, whereon stand the flowers, of a deepe red colour, striped and speckled very close together with a darkish white colour.

Caryophyllus maximus dictus Hulo ruber non varians.

The red *Hulo* is also a faire great flower, of a stamell colour, deeply iagged as the former, and groweth very comely without any spot at all in it, so that it seemeth to bee but a stamell Gilloflower, saue that it is much greater.

Caryophyllus maximus dictus Hulo caruleo purpureus.

The blew *Hulo* is a goodly faire flower, being of a faire purplish murrey colour, curiously marbled with white, but so smally to be discerned, that it seemeth only purple, it hath so much the Mastrie in it ; it resembleth the Brasill, but that it is much bigger.

Caryophyllus maximus dictus Grimelo siue Princeps.

The *Grimelo* or Prince is a faire flower also, as large as any Chrystall or larger, being of a faire crimson colour, equally for the most part striped with white, or rather more white then red, thorough euery leafe from the bottome, and standeth comely.

Caryophyllus maximus Incarnadinus albus.

The white Carnation or Delicate, is a goodly delightfull faire flower in his pride and perfection, that is, when it is both marbled and flaked, or striped and speckled with white vpon an incarnate crimson colour, beeing a very comely flower, but abideth not constant, changing oftentimes to haue no flakes or strakes of white, but marbled or speckled wholly.

Caryophyllus maximus Incarnadinus Gallicus.

The French Carnation is very like vnto the white Carnation, but that it hath more specks, and fewer stripes or flakes of white in the red, which hath the mastrie of the white.

Caryophyllus maximus Incarnadinus grandis.

The ground Carnation (if it be not the same with the graund or great old Carnation first set downe, as the alteration but of one letter giueth the coniecture) is a thicke flower, but spreadeth not his leaues abroade as others doe, hauing the middle standing higher then the outer leaues, and turning vp their brimmes or edges; it is a sad flower, with few stripes or spots in it : it is very subiect to breake the pod, that the flower seldome commeth faire and right ; the greene leaues are as great as the *Hulo* or Lombard red.

Caryophyllus maximus Chrystallinus.

The Chrystall or Chrystalline (for they are both one, howsoeuer some would make them differ) is a very delicate flower when it is well marked, but it is inconstant in the markes, being sometimes more striped with white and crimson red, and sometimes lesse or little or nothing at all, and changing also sometimes to be wholly red, or wholly blush.

Caryophyllus maximus flore rubro.

The red Chrystall, which is the red hereof changed, is the most orient flower of all other red Gilloflowers, because it is both the greatest, as comming from the Chrystall, as also that the red hereof is a most excellent crimson.

Caryophyllus maximus dictus Fragrans.

The Fragrant is a faire flower, and thought to come from the Chrystall, being as large, but of a blush red colour, spotted with small speckes, no bigger then pinnes points, but not so thicke as in the Pageant.

Caryophyllus maximus Sabaudicus varius.

The stript Sauadge is for forme and bignesse equall with the Chrystall or White Carnation, but as inconstant as eyther of them, changing into red or blush ; so that few branches with flowers containe their true mixtures, which are a whitish blush, fairely striped with a crimson red colour, thicke and short, with some spots also among.

Caryophyllus maximus Sabaudicus carneus.

The blush Sauadge is the same with the former, the same root of the stript Sauadge, as I said before, yeelding one side or part whose flowers will be eyther wholly blush, or hauing some small spots, or sometimes few or none in them.

Caryophyllus maximus Sabaudicus ruber.

The red Sauadge is as the blush, when the colour of the flower is wholly red without any stripes or spots in them, and so abideth long ; yet it is sometimes seene, that the same side, or part, or roote being separate from the first or mother plant, will giue striped and well marked flowers againe.

Caryophyllus maximus Oxoniensis.

The Oxeford Carnation is very like vnto the French Carnation, both for forme, largenesse and colour : but that this is of a sadder red colour, so finely marbled with white thereon, that the red hauing the maistry, sheweth a very sad flower, not hauing any flakes or stripes at all in it.

Caryophyllus maximus Regius, siue Bristoliensis maior.

The Kings Carnation or ordinary Bristow, is a reasonable great flower, deepely iagged, of a sad red, very smally striped and speckled with white : some of the leaues of the flower on the one side will turne vp their brimmes or edges : the greene leafe is very large.

Caryophyllus maximus Granatensis.

The greatest *Granado* is a very faire large flower, bigger then the Chrystall, and almost as bigge as the blew *Hulo* : it is almost equally diuided and stript with purple and white, but the purple is sadder then in the ordinary *Granado* Gilloflower, else it might bee said it were the same, but greater.   Diuers haue taken this flower to bee the *Gran Pere*, but you shall haue the difference shewed you in the next ensuing flower.

Caryophyllus maximus Gran Pere dictus.

The *Gran Pere* is a fair great flower, and comely for the forme, but of no great beautie for colour, because although it be stript red and white like the Queenes Gilloflower, yet the red is so sad that it taketh away all the delight to the flower.

Caryophyllus maximus Camberfine dictus.

The Camberfine is a great flower and a faire, beeing a redde flower, well marked or striped with white, somewhat like vnto a

Sauadge

Sauadge, say some, but that the red is not crimson as the Sauadge; others say the Daintie, but not so comely : the leaues of the flowers are many, and thrust together, without any due forme of spreading.

Caryophyllus maximus Longo-bardicus ruber.

The great Lombard red is a great sad red flower, so double and thick of leaues, that it most vsually breaketh the pod, and seldome sheweth one flower among twenty perfect : the blades or greene leaues are as large as the *Hulo.*

Caryophylli majores.

GILLOFLOWERS.

Caryophyllus maior Westmin-steriensis.

THe lustie Gallant or Westminster (some make them to be one flower, and others to bee two, one bigger then the other) at the first blowing open of the flower sheweth to be of a reasonable size and comelinesse, but after it hath stood blowen some time it sheweth smaller and thinner : it is of a bright red colour, much striped and speckled with white.

Caryophyllus maior Bristolien-sis purpureus.

The Bristow blew hath greene leaues, so large, that it would seeme to bring a greater flower then it doth, yet the flower is of a reasonable size, and very like vnto the ordinary *Granado* Gilloflower, striped and flaked in the same manner, but that the white of this is purer then that, and the purple is more light, and tending to a blew : this doth not abide constant, but changeth into purple or blush.

Caryophyllus maior Bristolien-sis carneus.

The Bristow blush is very like the last both in leafe and flower, the colour only sheweth the difference, which seldome varyeth to be spotted, or change colour.

Caryophyllus maior Dorobern-ensis ruber.

The red Douer is a reasonable great Gilloflower and constant, being of a faire red thicke poudered with white spots, and seemeth somewhat like vnto the ground Carnation.

Caryophyllus maior Doroberni-ensis dilatus siue albus.

The light or white Douer is for forme and all other things more comely then the former, the colour of the flower is blush, thicke spotted with very small spots, that it seemeth all gray, and is very delightfull.

Caryophyllus maior Cantii.

The faire maide of Kent, or Ruffling Robin is a very beautiful flower, and as large as the white Carnation almost : the flower is white, thicke poudered with purple, wherein the white hath the mastrie by much, which maketh it the more pleasant.

Caryophyllus maior Regineus.

The Queenes Gilloflower is a reasonable faire Gilloflower although very common, striped red and white, some great and some small with long stripes.

Caryophyllus maior elegans.

The Daintie is a comely fine flower, although it be not great, and for the smallnesse and thinnesse of the flower being red so finely marked, striped and speckled, that for the liuelinesse of the colours it is much desired, beeing inferiour to very few Gilloflowers.

Caryophyllus maior Brasiliensis.

The Brassill Gilloflower is but of a meane size, being of a sad purple colour, thicke poudered and speckled with white, the purple herein hath the mastrie, which maketh it shew the sadder, it is vnconstant, varying much and often to bee all purple : the greene leaues lye matting on the ground.

Caryophyllus maior Grana-tensis.

The *Granado* Gilloflower is purple and white, flaked and striped very much : this is also much subiect to change purple. There is a greater and a lesser of this kinde, besides the greatest that is formerly described.

The

Caryophyllus Turcicus.

The Turkie Gilloflower is but a small flower, but of great delight, by reason of the well marking of the flower, being most vsually equally striped with red and white.

Caryophyllus Cambrensis Poole.

The Poole flower, growing naturally vpon the rockes neare Cogshot Castle in the Isle of Wight, is a small flower, but very pleasant to the eye, by reason of the comely proportion thereof; it is of a bright pale red, thicke speckled, and very small with white, that it seemeth to bee but one colour, the leaues of the flower are but smally iagged about : it is constant.

Caryophyllus Pegma dilutior.

The light or pale Pageant is a flower of a middle size, very pleasant to behold, and is both constant and comely, and but that it is so common, would be of much more respect then it is : the flower is of a pale bright purple, thicke poudered, and very euenly with white, which hath the mastery, and maketh it the more gracefull.

Caryophyllus Pegma saturatior.

The sad Pageant is the same with the former in forme and bignesse, the difference in colour is, that the purple hath the mastery, which maketh it so sad, that it doth resemble the Brasill for colour, but is not so bigge by halfe.

Caryophyllus Heroina dictus elegans Magistri Bradshawy.

Master Bradshawe his dainty Lady may bee well reckoned among these sorts of Gilloflowers, and compare for neatenesse with most of them : the flower is very neate, though small, with a fine small iagge, and of a fine white colour on the vnderside of all the leaues, as also all the whole iagge for a pretty compasse, and the bottome or middle part of the flower on the vpperside also : but each leafe is of a fine bright pale red colour on the vpperside, from the edge to the middle, which mixture is of wonderfull great delight.

Caryophyllus albus optimus maior Londinensi & alius.

The best white Gilloflower groweth vpright, and very double, the blades growe vpright also, and crawle not on the ground.

The London white is greater and whiter then the other ordinary white, being wholly of one colour.

Caryophyllus maior rubens & minor.

The stamell Gilloflower is well knowne to all, not to differ from the ordinary red or cloue Gilloflower, but only in being of a brighter or light red colour : there is both a greater and a lesser of this kinde.

Caryophyllus purpureus maior & minor.

The purple Gilloflower a greater and a lesse : the stalke is so slender, and the leaues vpon them so many and thicke, that they lye and traile on the ground : the greatest is almost as bigge as a Chrystall, but not so double : the lesse hath a smaller flower.

Caryophyllus Persico violaceus.

The Gredeline Gilloflower is a very neate and handsome flower, of the bignesse of the Cloue red Gilloflower, of a fine pale reddish purple or peach colour, enclining to a blew or violet, which is that colour is vsually called a gredeline colour : it hath no affinity with eyther Purple, Granado, or Pageant.

Caryophyllus purpuro ceruleus.

The blew Gilloflower is neither very double nor great, yet round and handsome, with a deepe iagge at the edge, and is of an exceeding deepe purple colour, tending to a tawnie : this differeth from all other sorts, in that the leafe is as greene as grasse, and the stalkes many times red or purple : by the greene leaues it may be knowne in the Winter, as well as in the Summer.

Caryophyllus carneus.

The blush Gilloflower differeth not from the red or stamell, but only in the colour of the flower, which is blush.

Caryophyllus Silesiacus maximus Wittie.

Iohn Wittie his great tawny Gilloflower is for forme of growing, in leafe and flower altogether like vnto the ordinary tawny, the flower onely, because it is the fairest and greatest that any o-

the

ther hath nourced vp, maketh the difference, as also that it is of a faire deepe scarlet colour.

There are also diuers other Tawnies, either lighter or sadder, either lesse or more double, that they cannot be numbered, and all rising (as I said before) from sowing the seede of some of them : besides the diuersities of other colours both simple and mixed, euery yeare and place yeelding some variety was not seen with them before : I shall neede but onely to giue you the names of some of them we haue abiding with vs, I meane such as haue receiued names, and leaue the rest to euery ones particular denomination.

Of Blushes there are many sorts, as the deepe blush, the pale blush, the Infanta blush, a blush enclining to a red, a great blush, the fairest and most double of all the other blushes, and many others both single and double.

Of Reds likewise there are some varieties, but not so many as of the other colours; for they are most dead or deepe reds, and few of a bright red or stamell colour; and they are single like Pinkes, either striped or speckled, or more double striped and speckled variably, or else

There are neither purple nor white that rise from this seede that I haue obserued, except one white in one place.

Caryophyllus Silesiacus striatus. The striped Tawny are either greater or lesser, deeper or lighter flowers twenty sorts and aboue, and all striped with smaller or larger stripes, or equally diuided, of a deeper or lighter colour : and some also for the very shape or forme will bee more neate, close, and round; others more loose, vnequall, and sparsed.

Caryophyllus Silesiacus marmor-amulus. The marbled Tawny hath not so many varieties as the striped, but is of as great beauty and delight as it, or more : the flowers are greater or smaller, deeper or lighter coloured one then another, and the veines or markes more conspicuous, or more frequent in some then in others : but the most beautifull that euer I did see was with Master Ralph Truggie, which I must needes therefore call

Heroina Rodolphi florum Imperatoris. Master Tuggies Princesse, which is the greatest and fairest of all these sorts of variable tawnies, or seed flowers, being as large fully as the Prince or Chrystall, or something greater, standing comely and round, not loose or shaken, or breaking the pod as some other sorts will; the marking of the flower is in this manner : It is of a stamell colour, striped and marbled with white stripes and veines quite through euery leafe, which are as deeply iagged as the Hulo : sometimes it hath more red then white, and sometimes more white then red, and sometimes so equally marked, that you cannot discerne which hath the mastery; yet which of these hath the predominance, still the flower is very beautifull, and exceeding delightsome.

Caryophyllus Silesiacus assulosus. The Flaked Tawny is another diuersity of these variable or mixt coloured flowers, being of a pale reddish colour, flaked with white, not alwaies downeright, but often thwart the leaues, some more or lesse then others; the marking of them is much like vnto the Chrystall : these also as well as others will be greater or smaller, and of greater or lesse beauty then others.

Caryophyllus Silesiacus plumatus. The Feathered Tawny is more rare to meete with then many of the other; for most vsually it is a faire large flower and double, equalling the Lumbard red in his perfection : the colour hereof is vsually a scarlet, little deeper or paler, most curiously feathered and streamed with white through the whole leafe.

Caryophyllus Silesiacus phœniceus. The Speckled Tawny is of diuers sorts, some bigger, some lesse,

leffe, fome more, and fome leffe fpotted then others : Vfually it is a deepe fcarlet, fpeckled or fpotted with white, hauing alfo fome ftripes among the leaues.

Caryophyllus rofeus rotundus Magiftri Tuggie.

Mafter Tuggie his Rofe Gilloflower is of the kindred of thefe Tawnies, being raifed from the feede of fome of them, and onely poffeffed by him that is the moft induftrious preferuer of all natures beauties, being a different fort from all other, in that it hath round leaues, without any iagge at all on the edges, of a fine ftamell full colour, without any fpot or ftrake therin, very like vnto a fmall Rofe, or rather much like vnto the red Rofe Campion, both for forme, colour, and roundneffe, but larger for fize.

The Place.

All thefe are nourifhed with vs in Gardens, none of their naturall places being knowne, except one before recited, and the yellow which is *Silefia,* many of them being hardly preferued and encreafed.

The Time.

They flower not vntill the heate of the yeare, which is in Iuly (vnleffe it be an extraordinary occafion) and continue flowring, vntill the colds of the Autumne checke them, or vntill they haue wholly out fpent themfelues, and are vfually encreafed by the flips.

The Names.

Moft of our later Writers doe call them by one generall name, *Caryophyllus fativus,* and *flos Caryophylleus,* adding thereunto *maximus,* when wee meane Carnations, and *maior* when we would expreffe Gilloflowers, which name is taken from Cloues, in that the fent of the ordinary red Gilloflower efpecially doth refemble them. Diuers other feuerall names haue beene formerly giuen them, as *Vetonica,* or *Betonica altera,* or *Vetonica altilis,* and *coronaria. Herba Tunica, Viola Damafcena, Ocellus Damafcenus,* and *Barbaricus.* Of fome *Cantabrica Plinij.* Some thinke they were vnknowne to the Ancients, and fome would haue them to be *Iphium* of Theophraftus, wherof he maketh mention in his fixth and feuenth Chapters of his fixth booke, among Garland and Summer flowers; others to be his *Dios anthos,* or *Iouis flos,* mentioned in the former, and in other places. We call them in Englifh (as I faid before) the greateft kindes, Carnations, and the others Gilloflowers (*quafi* Iuly flowers) as they are feuerally expreffed.

The Vertues.

The red or Cloue Gilloflower is moft vfed in Phyficke in our Apothecaries fhops, none of the other being accepted of or vfed (and yet I doubt not, but all of them might ferue, and to good purpofe, although not to giue fo gallant a tincture to a Syrupe as the ordinary red will doe) and is accounted to be very Cordiall.

Chap. LXX.

Caryophylli filueftres. Pinkes.

There remaine diuers forts of wilde or fmall Gilloflowers (which wee vfually call Pinkes) to be entreated of, fome bearing fingle, and fome double flowers, fome fmooth, almoft without any deepe dents on the edges, and fome iagged, or as it were feathered. Some growing vpright like vnto Gilloflowers, others creeping

ping

1 *Caryophyllus siluestris simplex.* The vsuall single Pinke. 2 *Caryophyllus multiplex siluestris.* Double Pinkes. 3 *Caryophyllus siluestris plumarius.* Feathered or iagged Pinkes. 4 *Caryophyllus Stellatus.* Starre Pinkes. 5 *Caryophyllus retens.* Matted Pinkes. 6 *Caryophyllus mediterraneus.* The great Thistle or Sea Gilloflower. 7 *Caryophyllus marinus.* The ordinary Thistle or Sea Cushion.

ping or spreading vnder the toppe or crust of the ground, some of one colour, some of another, and many of diuers colours : As I haue formerly done with the Gilloflowers, so must I doe with these that are entertained in our Gardens, onely giue you the descriptions of some three or foure of them, according to their variety, and the names of the rest, with their distinctions.

1. Caryophyllus minor siluestris multiplex & simplex.
Double and single Pinkes.

The single and double Pinkes are for forme and manner of growing, in all parts like vnto the Gilloflowers before described, sauing onely that their leaues are smaller and shorter, in some more or lesse then in others, and so are the flowers also : the single kindes consisting of fiue leaues vsually (seldome six) round pointed, and a little snipt for the most part about the edges, with some threads in the middle, either crooked or straight : the double kindes being lesser, and lesse double then the Gilloflowers, hauing their leaues a little snipt or endented about the edges, and of diuers seuerall colours, as shall hereafter be set downe, and of as fragrant a sent, especially some of them as they : the rootes are long and spreading, somewhat hard and wooddy.

2. Caryophyllus plumarius. Feathered or iagged Pinkes.

The iagged Pinkes haue such like stalkes and leaues as the former haue, but somewhat shorter and smaller, or grasse-like, and of a whitish or grayish greene colour likewise : the flowers stand in the like manner at the toppes of the stalkes, in long, round, slender, greene huskes, consisting of fiue leaues, very much cut in on the edges, and iagged almost like a feather, of a light red, or bright purple colour, with two white threads standing in the middle, crooked like a horne at the end, and are of a very good sent. Some of these haue not those two crooked threads or hornes in the middle, but haue in their stead many small threads, not crooked at all : the seedes of them all are like vnto the seedes of Gilloflowers, or the other Pinkes, that is, small, blacke, long, and flat : the rootes are small and wooddy likewise.

3. Caryophyllus plumarius albus orbe rubro siue Stellatus. Starre Pinkes.

Of this kinde there is another sort, bearing flowers almost as deeply cut or iagged as the former, of a faire white colour, hauing a ring or circle of red about the bottome or lower part of the leaues, and are as sweete as the former : this being sowne of seede doth not giue the starre of so bright a red colour, but becommeth more dunne.

4. Caryophyllus plumarius Austriacus siue Superba Austriaca.
The feathered Pinke of Austria.

This kinde of Pinke hath his first or lower leaues, somewhat broader and greener then any of the former Pinkes, being both for breadth and greennesse more like vnto the Sweete Iohns, which shall bee described in the next Chapter : the leaues on the stalkes are smaller, standing by couples at euery ioynt, at the toppes whereof stand such like iagged flowers as the last described, and as large, but more deeply cut in or iagged round about, some of them of a purplish colour, but the most ordinary with vs are pure white, and of a most fragrant sent, comforting the spirits and senses a farre off: the seedes and rootes are like vnto the former. Some haue mistaken a kinde of wilde Campion, growing in our Woods, and by the paths sides in Hornsie Parke, and other places, to be this feathered Pinke : but the flowers declare the difference sufficiently.

5. Caryophyllus minor repens simplex & multiplex.
Single and double matted Pinkes.

The matted Pinke is the smallest, both for leafe and flower of all other Pinkes that are nourished in Gardens, hauing many short and small grassie greene leaues vpon the

stalkes,

stalkes, which as they grow and lye vpon the ground (and not ftanding fo vpright as the former) doe take roote againe, whereby it quickly fpreadeth, and couereth a great deale of ground in a little fpace : the flowers are fmall and round, a little fnipt about the edges, wherof fome are white, and fome red, and fome are white fpotted with red, and fome red fpotted with white, all of them being fingle flowers. But there is another of this kinde, not differing in leafe, but in flower : for that the firft flowers are but once double, or of two rowes of leaues, of a fine reddifh colour, fpotted with filuer fpots : but thofe that follow, are fo thicke and double, that they often-times doe breake the pod or huske ; being not altogether of fo deepe a red colour, but more pale.

6. *Caryophyllus Mediterraneus fiue Marinus maior.*
Great Sea Gilloflower or Great Thrift.

Vnto thefe kindes of Pinkes I muft needs adde, not only our ordinary Thrift (which is more frequent in gardens, to empale or border a knot, becaufe it abideth greene Winter and Summer, and that by cutting, it may grow thick, and be kept in what form one lift, rather then for any beautie of the flowers) but another greater kinde, which is of as great beautie and delight almoft as any of the former Pinkes, as well for that the leaues are like vnto Gilloflowers, being longer and larger then any Pinkes, and of a whitifh greene colour like vnto them, not growing long or by couples vpon the ftalkes as Pinkes and Gilloflowers doe, but tufting clofe vpon the ground, like vnto the common Thrift: as alfo that the ftalkes, rifing from among the leaues (being fome-times two foote high (as I haue obferued in my garden) are yet fo flender and weake, that they are fcarce able to beare the heads of flowers, naked or bare, both of leaues and ioynts, fauing only in one place, where at the ioynt each ftalke hath two fmall and very fhort leaues, not rifing vpwards as in all other Gilloflowers, Pinkes, and other herbes, but growing downewards) and doe beare each of them a tuft or vmbell of fmall purplifh, or blufh coloured flowers, at the toppes of them ftanding fomewhat like vnto fweete Williams, but more roundly together, each flower confifting of fiue fmall, round, ftiffe or hardifh leaues, as if they were made of paper, the bottome or middle being hollow, not blowing all at once as the ordinary Thrift, but for the moft part one after another, not fhewing vfually aboue foure or fiue flowers open at one time (fo farre as I could obferue in the plants that I kept) fo that it was long before the whole tuft of flowers were paft ; but yet the hoter and dryer the time was, the fooner it would be gone : the feede I haue not perfectly obferued, but as I remember, it was fomewhat like vnto the feede of Scabious ; I am fure nothing like vnto Gilloflowers or Pinkes : the roote is fomewhat great, long and hard, and not fo much fpreading in the ground as Gilloflowers or Pinkes.

Caryophyllus Marinus. Thrift, or Sea Cufhion.

Our common Thrift is well knowne vnto all, to haue many fhort and hard greene leaues, fmaller then many of the graffes, growing thicke together, and fpreading vpon the ground : the ftalkes are naked of leaues a fpanne high, bearing a fmall tuft of light purple, or blufh coloured flowers, ftanding round and clofe thrufting together.

Double Pinkes.	Single Pinkes.
THe double white Pinke is onely with more leaues in it then the fin-gle, which maketh the difference. The double red Pinke is in the fame manner double, differing from the fingle of the fame colour. The double purple Pinke differeth not	THe fingle white ordinary Pinke hath a fingle white flower of fiue leaues, finely iagged about the edges. The fingle red Pink is like the white, but that the leaues are not fo much iagged, and the flower is of a pale purplifh red colour.

from

from the single purple for colour, but onely in the doublenesse of the flower.

The *Granado* Pinke differeth not from the Gilloflower of the same name, but in the smalnesse both of leaues and flower.

The double Matted Pinke is before described.

The double blush Pinke is almost as great as the ordinary blush Gilloflower, and some haue taken it for one, but the greene leaues are almost as small as Pinks, and therefore I referre it to them.

The single purple Pinke is of a faire purple colour, like almost vnto the purple Gilloflower.

The great blush Pinke hath broader and larger leaues in the flower then any other Pinke, and of a faire blush colour.

The white Featherd Pinke hath the edges of the flower more finely and deeply cut in then the former.

The red or light purple featherd Pinke is like the former featherd Pinke, but only differeth in colour.

The Starre Pinke is a faire flower, finely iagged on the edges, with a faire red circle at the lower end of the leaues on the inside. The white featherd Pinke of Auftria is described before. The purple featherd Pinke of Auftria is so likewife. The single matted Pinke is before described. The speckled Pinke is a small flower hauing small spots of red here and there disperfed ouer the white flower.

Those single flowers being like vnto Pinkes that rife from the sowing of the orenge tawney, I bring not into this *classis*, hauing already spoken of them in the precedent Chapter.

The Place.

These are all like as the former, nourished in Gardens with vs, although many of them are found wilde in many places of Auftria, Hungarie, and Germany, on the mountaines, and in many other places, as Clufius recordeth. The ordinary Thrift groweth in the salt Marshes at Chattam by Rochefter, and in many other places in England : but the great kinde was gathered in Spaine, by Guillaume Boel that painefull searcher of simples, and the seede thereof imparted to me, from whence I had diuers plants, but one yeare after another they all perished.

The Time.

Many of these Pinkes both single and double, doe flower before any Gilloflower, and so continue vntill Auguft, and some, most of the Summer and Autumne.

The Names.

The feuerall titles that are giuen to these Pinkes, may fuffice for their particular names : and for their generall they haue beene expreffed in the former Chapter, beeing of the fame kindred, but that they are smaller, and more frequently found wilde. The two forts of Thrift are called *Caryophyllus Marinus.* The greater, *Maior & Mediterraneus ;* In English, The greater or Leuant Thrift, or Sea Gilloflower. The leffer *Minimus,* and is accounted of some to be a graffe, and therefore called *Gramen Marinum & Polyanthemum ;* In English, Thrift, Sea graffe, and our Ladies Cuffion, or Sea Cuffion.

The Vertues.

It is thought by diuers, that their vertues are answerable to the Gilloflowers, yet as they are of little vfe with vs, so I thinke of as small effect.

CHAP.

Chap. LXXI.

Armerius. Sweet Iohns, and sweet Williams.

THese kindes of flowers as they come neerest vnto Pinkes and Gilloflowers, though manifestly differing, so it is fittest to place them next vnto them in a peculiar Chapter.

1. *Armerius angustifolius rubens simplex.* Single red sweete Iohns.

The sweete Iohn hath his leaues broader, shorter and greener then any of the former Gilloflowers, but narrower then sweete Williams, set by couples, at the ioynts of the stalkes, which are shorter then most of the former, and not aboue a foote and a halfe high, at the tops whereof stand many small flowers, like vnto small Pinkes, but standing closer together, and in shorter huskes, made of fiue leaues, smaller then most of them, and more deeeply iagged then the Williams, of a red colour in the middle, and white at the edges, but of a small or soft sent, and not all flowring at once, but by degrees : the seede is blacke, somewhat like vnto the seede of Pinkes, the roote is disperfed diuersly, with many small fibres annexed vnto it.

2. *Armerius angustifolius albus simplex.* Single white sweet Iohns.

This white Iohn differeth not in any thing from the former, but onely that the leafe doth neuer change brownish, and that the flower is of a faire white colour, without any mixture.

3. *Armerius angustifolius duplex.* Double sweet Iohns.

There is of both those former kindes, some whose flowers are once double, that is, consisting of two or three rowes of leaues, and the edges not so deepely iagged ; not differing in any thing else.

4. *Armerius latifolius simplex flore rubro.* Single red sweet Williams.

The sweet Williams doe all of them spread into many very long trayling branches, with leaues lying on the ground, in the very like manner that the sweete Iohns doe : the chiefe differences betweene them are, that these haue broader, and darker greene leaues, somewhat brownish, especially towards the points, and that the flowers stand thicker and closer, and more in number together, in the head or tuft, hauing many small pointed leaues among them, but harmlesse, as all men know; the colour of the flower is of a deep red, without any mixture or spot at all.

5. *Armerius latifolius flore rubro multiplici.* Double red sweete Williams.

The double kinde differeth not from the single kinde of the same colour, but only in the doublenesse of the flowers, which are with two rowes of leaues in euery flower.

6. *Armerius latifolius variegatus siue versicolor.* Speckled sweete Williams, or London pride.

These spotted Williams are very like the first red Williams, in the forme or maner of growing, hauing leaues as broade, and browne sometimes as they, the flowers stand as thicke or thicker, clustring together, but of very variable colours : for some flowers will be of a fine delayed red, with few markes or spots vpon them, and others will

will bee full peckled or sprinkled with white or siluer spots, circlewise about the middle of the flowers, and some will haue many specks or spots vpon them dispersed: all these flowers are not blowne at one time, but some are flowring, when others are decaying, so that abiding long in their pride, they become of the more respect: The seede is blacke, as all the rest, and not to be distinguished one from another: the roots are some long, and some small and threddy, running vnder the vpper crust of the earth.

7. *Armerius latifolius flore rubro saturo holoserices.*
Sweet Williams of a deepe red or murrey colour.

The leaues of this kinde seeme to be a little larger, and the ioints a little redder then the former, but in the flower consisteth the chiefest difference, which is of a deepe red, or murrey purple colour, like vnto veluet of that colour, without any spots, but smooth, and as it were soft in handling, hauing an eye or circle in the middle, at the bottome of the leaues.

8. *Armerius latifolius simplex flore albo.*
Single white sweete Williams.

The white kinde differeth not in forme, but in colour from the former, the leaues are not browne at all, but of a fresh greene colour, and the flowers are wholly white, or else they are all one.

The Place.

These for the most part grow wilde in Italie, and other places: we haue them in our Gardens, where they are cherished for their beautifull varietie.

The Time.

They all generally doe flower before the Gilloflowers or Pinkes, or with the first of them: their seede is ripe in Iune and Iuly, and doe all well abide the extremitie of our coldest winters.

The Names.

They all generally are called *Armerius*, or *Armeria*, as some doe write, and distinguished as they are in their titles: Yet some haue called them *Vetonica agrestis*, and others *Herba Tunica, Scarlatea, & Caryophyllus siluestris:* Wee doe in English in most places, call the first or narrower leafed kindes, Sweet Iohns, and all the rest Sweete Williams; yet in some places they call the broader leafed kindes that are not spotted, Tolmeiners, and London tufts: but the speckled kinde is termed by our English Gentlewomen, for the most part, London pride.

The Vertues.

We haue not knowne any of these vsed in Physicke.

Chap. LXXII.

Bellis. Daisie.

Here be diuers sorts of Daisies, both great and small, both single and double, both wilde growing abroade in the fieldes, and elsewhere, and manured growing only in Gardens: of all which I intend not to entreate, but of those that are of most beautie and respect, and leaue the rest to their proper place.

1. Bellis

1 *Armerius angustifolius simplex.* Single sweete Iohns. 2 *Armerius angustifolius multiplex* Double sweet Iohns. 3 *Armerius latifolius simplex* Single sweete Williams. 4 *Armerius latifolius versicolor.* Spotted sweet Williams or pride of London. 5 *Armerius latifolius multiplex* Double sweet Williams. 6 *Bellis hortensis minor multiplex* Double Garden Daisies. 7 *Bellis minor hortensis flore vario.* Double red Daisies strict. 8 *Bellis minor hortensis prolifera* Double fruitfull Daisies or Iacke an Apes on horsebacke. 9 *Bellis cærulea sue Globularia.* Double blew Daisies or blew Globeflower. 10 *Bellis lutea montana sue Globularia lutea montana.* Double yellow Daisies or yellow Globeflower.

1. *Bellis maior flore albo pleno.* The great double white Daisie.

The great Daisie with the double white flower, is in all things so like vnto the great single kinde, that groweth by the high wayes, and in diuers medowes and fields, that there is no difference but in the flower, which is double. It hath many long, and somewhat broad leaues lying vpon the ground, deepely cut in on both sides, somewhat like vnto an oaken leafe ; but those that are on the stalkes are shorter, narrower, and not so deeply cut in, but onely notched on the edges : the flowers at the toppe are (as I said) white and double, consisting of diuers rowes of leaues, being greater in compasse then any of the double Daisies that follow, but nothing so double of leaues.

2. *Bellis minor flore rubro simplici.* Single red Daisies.

This single Daisie (like as all the rest of the small Daisies) hath many smooth, greene, round pointed leaues lying on the ground, a little snipt about the edges ; from among which rise many slender round foote-stalkes, rather then stalkes or stems, about an hand breadth high at the most, and oftentimes not halfe so high, bearing one flower a peece, consisting of many small leaues, as a pale or border set about a middle thrumme : the leaues of this kinde are almost wholly red, whereas in the wilde they are white or whitish, enclining to red on the edges, the middle being yellow in both sorts : the rootes are many small white threads or strings.

3. *Bellis minor hortensis flore pleno variorum colorum.* Double Garden Daisies of diuers colours.

The leaues of all the double Daisies are in forme like vnto the single ones, but that they are smaller, and little or nothing snipt or notched about the edges : the small stalkes likewise are smaller and lower, but bearing as double flowers as any that growe on the ground, being composed of many small leaues, thicke thrust together, of diuers colours ; for some are wholly of a pure white, others haue a little red, either dispersed vpon the white leaues, or on the edges, and sometimes on the backes of the leaues : some againe seeme to be of a whitish red, or more red then white, when as indeede they are white leaues dispersed among the red ; others of a deepe or darke red colour, and some are speckled or striped with white and red through the whole flower : and some the leaues will bee red on the vpperside, and white vnderneath ; and some also (but those are very rare) are of a greenish colour.

4. *Bellis minor hortensis prolifera.* Double double Daisies or childing Daisies.

There is no difference either in leafe or roote in this kinde from the former double Daisies : the chiefest variety consisteth in this, that it beareth many small double flowers, standing vpon very short stalkes round about the middle flower, which is vsually as great and double as any of the other double kindes, and is either wholly of a deepe red colour, or speckled white and red as in some of the former kindes, or else greenish, all the small flowers about it being of the same colour with the middlemost.

5. *Bellis cærulea siue Globularia.* Blew Daisies.

The likenesse and affinity that this plant hath with the former, both in the forme of leafe and flower, as also in the name, hath caused me to insert it, and another rare plant of the same kinde, in this place, although they be very rare to be met with in our English Gardens. This beareth many narrower, shorter, and blacker greene leaues then the former, lying round about vpon the ground ; among which rise vp slender, but stiffe and hard stalks, halfe a foot high or more, set here and there with small leaues, and at the top a small round head, composed of many small blew leaues, somewhat like vnto the head of a Scabious : It hath bin found likewise with a white head of flowers : the roote is hard and stringie : the whole plant is of a bitter taste.

6. *Globularia*

6. *Globularia lutea montana.* Yellow Daisies.

This mountaine yellow Daisie or Globe-flower hath many thicke, smooth, round pointed leaues, spread vpon the gronnd like the former; among which spring diuers small round rushie stalkes, a foote high, bearing about the middle of them two small leaues at the ioynts, and at the toppes round heads of flowers thrust thicke together, standing in purplish huskes, euery of which flowers do blow or spread into fiue leaues, starre-fashion, and of a faire yellow colour, smelling like vnto broome flowers, with many small threads in the middle compassing a flat pointell, horned or bended two wayes: after the flowers are past rise vp the seede vessels, which are round, swelling out in the middle, and diuided into foure parts at the toppes, containing within them round, flat, blacke seede, with a small cut or notch in them: the roote is a finger long, round and hard, with a thicke barke, and a woddy pith in the middle, of a sharpe drying taste and strong sent: the leaues are also sharpe, but bitter.

The Place.

The small Daisies are all planted, and found onely in Gardens, and will require to be replanted often, lest they degenerate into single flowers, or at least into lesse double. The blew Daisie is naturall of Mompelier in France, and on the mountaines in many places of Italy, as also the yellow kinde in the Kingdome of Naples.

The Time.

The Daisies flower betimes in the Spring, and last vntill May, but the last two flower not vntill August or September.

The Names.

They are vsually called in Latine *Bellides*, and in English Daisies. Some call them *Herba Margarita*, and *Primula veris*, as it is likely after the Italian names, of *Marguerite*, and *Fior di prima vera gentile*. The French call them *Pasquettes*, and *Marguerites*, and the Fruitfull sort, or those that beare small flowers about the middle one, *Margueritons*: our English women call them Iacke an Apes on horse-backe, as they doe Marigolds before recited, or childing Daisies: but the Physitians and Apothecaries doe in generall call them, especially the single or Field kindes, *Consolida minor*. The blew Daisie is called *Bellis cærulea*, and *Globularia*, of some *Scabiosa pumilum genus*. The Italians call it *Botanaria*, becaufe the heads are found like buttons. The yellow, *Globularia montana*, is onely described by Fabius Columna, in his last part of *Phytobasanos*, and by him referred vnto the former *Globularia*, although it differ in some notable points from it.

The Vertues.

The properties of Daisies are certainly to binde, and the roote especially being dryed, they are vsed in medicines to that purpose. They are also of speciall account among those herbes, that are vsed for wounds in the head.

CHAP.

CHAP. LXXIII.

Scabiosa. Scabious.

THe sorts of Scabious being many, yeeld not flowers of beauty or respect, fit to bee cherished in this our Garden of delight; and therefore I leaue them to the Fields and Woods, there to abide. I haue onely two or three strangers to bring to your acquaintance, which are worthy this place.

1. *Scabiosa flore albo.* White flowred Scabious.

This white Scabious hath many long leaues, very much iagged or gasht in on both sides, of a meane bignesse, being neither so large as many of the field, nor so small as any of the small kindes : the stalkes rise about a foote and a halfe high, or somewhat higher, at the tops whereof grow round heads, thicke set with flowers, like in all points vnto the field Scabious, but of a milke white colour.

2. *Scabiosa rubra Austriaca.* Red Scabious of Austria.

This red Scabious hath many leaues lying vpon the ground, very like vnto Deuils bit, but not so large, being shorter and snipt, not gashed about the edges, of a light greene colour; yet (there is another of a darker greene colour, whose flower is of a deeper red) the stalkes haue diuers such leaues on them, set by couples at the ioynts as grow belowe, and at the tops small heads of flowers, each consisting of fiue leaues, the biggest flowers standing round about in the outer compasse, as is vsuall almost in all the kinds of Scabious, of a fine light purple or red colour : after the flowers are past, come the seede, which is somewhat long and round, set with certaine haires at the head thereof, like vnto a Starre : the roote is composed of a number of slender strings, fastened at the head.

3. *Scabiosa rubra Indica.* Red flowred Indian Scabious.

This (reputed Indian) Scabious hath many large faire greene leaues lying on the ground, iagged or cut in on both sides to the middle ribbe, euery peece whereof is narrower then that at the end, which is the broadest : among these leaues rise vp sundry slender and weake stalkes, yet standing vpright for the most part, set with smaller and more iagged leaues at certaine distances, two or three at euery ioynt, branching forth at the toppe into other smaller branches, bearing euery one head of flowers, like in forme vnto other Scabiouses, but of an excellent deepe red crimson colour (and sometimes more pale or delayed) of no sent at all : after which doe come small roundish seede, like vnto the field Scabious : the roote is long and round, compassed with a great many small strings, and perisheth vsually as soone as it hath borne out his flowers and seede : otherwise if it doe not flower the first yeare of the sowing, if it be carefully defended from the extremity of Winter, it will flower the sooner the next yeare, as I my selfe haue often found by experience.

The Place.

The first is sometimes found wilde in our owne Countrey, but it is very geason, and hath been sent among other rare seedes from Italy.

The second was first found and written of by Clusius, in Pannonia and Austria, where it is very plentifull.

The third hath been sent both from Spaine & Italy, and is verily thought to grow naturally in both those parts.

The

1 *Scabiofa flore albo.* White flowred Scabious 2 *Scabiofa rubra Auſtriaca.* Red Scabious of Auſtria. 3 *Scabiſa rubra Indica.*
Red flowred Indian Scabious. 4 *Cyanus vulgaris minor.* Corn-flower of diuers colours. 5 *Cyanus Baticus.* Spaniſh Corn-flower.
6 *Cyanus floridus Turcicus.* The braue Sultans flower. 7 *Carthamus ſatiuus.* Spaniſh Saffron.

The Time.

The first and second flower earlier then the laft, for that it flowreth not vntill September or October, (vnleffe it be not apt to beare the firft yeare as I before faid) fo that many times (if none be more forward) it perifheth without bearing ripe feede, whereby we are oftentimes to feeke new feede from our friends in other parts.

The Names.

They haue all one generall name of Scabious, diftinguifhed eyther by their flower, or place of growing, as in their titles : yet the laft is called of diuers *Scabiofa exotica*, becaufe they thinke the name *Indica*, is not truely impofed vpon it.

The Vertues.

Whether thefe kindes haue any of the vertues of the other wilde kinds, I know none haue made any experience, and therefore I can fay no more of them.

Chap. LXXIIII.

Cyanus. Corne flower, or blew Bottles.

VNder the name of *Cyanus* are comprehended, not onely thofe plants which from the excellent blew colour of their flowers (furnifhing or rather peftering the Corne fieldes) haue peculiarly obtained that name, and which doth much vary alfo, in the colour of the flowers, as fhallbe fhewed; but fome other plants alfo for their neere refemblance, but with feuerall diftinctions. The *Cyanus maior, Ptarmica Auftriaca, Ptarmica Imperati,* and many others which may be adioyned vnto them, do more fitly belong to the Garden of Simples, whereunto I leaue them, and will here only entreate of thofe that may moft pleafe the delight of our Gentle Florifts, in that I labour and ftriue, to furnifh this our garden, with the chiefeft choyfe of natures beauties and delights.

1. *Cyanus vulgaris diuerforum colorum.* Corne flower of diuers colours.

All thefe forts of Corne flowers are for the moft part alike, both in leaues and flowers one vnto another for the forme: the difference betweene them confifteth in the varying colour of the flowers : For the leaues are long, and of a whitifh greene colour, deeply cut in on the edges in fome places, fomewhat like vnto the leaues of a Scabious : the ftalkes are two foote high or better, befet with fuch like leaues but fmaller, and little or nothing flit on the edges: the toppes are branched, bearing many fmal greene fcaly heads, out of which rife flowers, confifting of fiue or fixe, or more long and hollow leaues, fmall at the bottome, and opening wider and greater at the brims, notched or cut in on the edges, and ftanding round about many fmall threds in the middle : the colours of thefe flowers are diuers, and very variable ; for fome are wholly blew, or white, or blufh, or of a fad, or light purple, or of a light or dead red, or of an ouerworne purple colour, or elfe mixed of thefe colours, as fome, the edges white, and the reft blew or purple, or the edges blew or purple, and the reft of the flower white, or ftriped, fpotted, or halfed, the one part of one colour, and the other of another, the threds likewife in the middle varying in many of them ; for fome will haue the middle thrume of a deeper purple then the outer leaues, and fome haue white or blufh leaues, the middle thrume being reddifh, deeper or paler : After the flowers are paft, there come fmall, hard, white and fhining feede in thofe heads, wrapped

ped or set among a deale of fluckie matter, as is most vsuall, in all plants that beare scaly heads: the rootes are long and hard, perishing euery yeare when it hath giuen seede.

2. *Cyanus floridus Turcicus.* The Sultans flower.

As a kinde of these Corne flowers, I must needes adioyne another stranger, of much beautie, and but lately obtained from Constantinople, where, because (as it is said) the great Turke, as we call him, saw it abroade, liked it, and wore it himselfe; all his vassals haue had it in great regard, and hath been obtained from them, by some that haue sent it into these parts. The leaues whereof are greener, and not only gashed, but finely snipt on the edges: the stalkes are three foote high, garnished with the like leaues as are below, and branched as the former, bearing large scaly heads, and such like flowers but larger, hauing eight or nine of those hollow gaping leaues in euery flower, standing about the middle threds (if it be planted in good and fertile ground and be well watered, for it soone starueth and perisheth with drought) the circling leaues are of a fine delayed purple or blush colour, very beautifull to behold; the seede of this is smaller and blacker, and not enclosed in so much dounie substance, as the former (yet in our Countrey the seede is not so blacke, as it came vnto vs, but more gray) the roote perisheth likewise euery yeere.

3. *Cyanus Bæticus supinus.* The Spanish Corne-flower.

This Spanish kinde hath many square low bending or creeping stalkes, not standing so vpright as the former, but branching out more diuersly; so that one plant will take vp a great deale of ground: the leaues are broader then any of the rest, softer also, of a pale or whitish greene colour, and not much gashed on the edges: the flowers stand in bigger heads, with foure or fiue leaues vnder euery head, and are of a light pale purple or blush colour; after which come seede, but not so plentifully, yet wrapped in a great deale of fluckie matter, more then any: the roote groweth downe deepe into the ground, but perisheth euery yeare as they doe.

The Place.

The first or former kindes, grow many times in the Corne fields of our own Countrey, as well as of others, especially that sort with a blew flower: but the other sorts or colours are not so frequent, but are nourished in gardens, where they will vary wonderfully.

The second as is before set downe, groweth in Turkie: and the last in Spaine, found out and first sent to vs by that industrious searcher of simples, Guillaume Boel before remembred.

The Time.

The first doe flower in the end of Iune, and in Iuly, and somtimes sooner. The other two later, and not vntill August most commonly, and the seede is soone ripe after.

The Names.

The first is generally called *Cyanus,* and some following the Ditch name, call it *Flos frumenti.* The olde Writers gaue it the name of *Bapti secuba,* which is almost worne out. We doe call them in English, Blew Bottles, and in some places, Corne flowers, after the Ditch names. The second hath beene sent by the name of *Ambreboi,* which whether it be a Turkie or Arabian name, I know not. I haue called it from the place, from whence we had it, *Turcicus,* and for his beauty, *Floridus.* The Turkes themselues as I vnderstand, doe call it The Sultans flower, and I haue done so likewise, that it may bee distinguished from all the other kindes, or else you may call it,

The Turkey blush Corne flower, which you pleafe. The laft was fent by the name of *Iacea Batica*, but I had rather to referre it to the *Cyanus*, or Corne flowers, becaufe the flowers are like vnto the Corne flowers, and not vnto the Iaceas or Knapweedes.

The Vertues.

Thefe had no vfe in Phyficke in Galen and Diofcorides time, in that (as it is thought) they haue made no mention of them : We in thefe dayes doe chiefly vfe the firft kindes (as alfo the greater fort) as a cooling Cordiall, and commended by fome to be a remedy, not onely againft the plague and peftilentiall difeafes, but againft the poifon of Scorpions and Spiders.

CHAP. LXXV.

Iacea Marina Batica. Spanifh Sea Knapweede.

THere are a great many forts of Knapweedes, yet none of them all fit for this our Garden, but this only ftranger, which I haue beene bold to thruft in here, for that it hath fuch like gaping or open flowers, as the former Corne flowers haue, but notably differing, and therefore deferueth a peculiar Chapter, as partaking both with *Cyanus* and *Iacea*. It hath many long and narrow leaues vneuenly dented or waued on both edges (and not notched, gafhed or indented, as many other herbes are) being thicke, flefhie and brittle, a little hairy, and of an ouerworne darke greene colour, among which rife lowe weake ftalkes, with fuch like leaues as grow at the bottome, but fmaller, bearing but here and there a flower, of a bright reddifh purple colour, like in forme vnto the Corne flowers, but much larger, with many threds or thrumes in the middle of the fame colour, ftanding vp higher then any of the former: this flower rifeth out of a large fcaly head, all fet ouer with fmall fharpe (but harmeleffe) white prickles : the feedes are blackifh, like vnto the Knapweedes, and larger then any of the former Corne flowers : the roote is great and thicke, growing deepe into the ground, flefhie and full of a flimie or clammy iuice, and eafie to bee broken, blackifh on the outfide, and whitifh within, enduring many yeares, like as the other Knapweedes, or Matfelons doe, growing in time to be very thicke and great.

The Place.

It groweth naturally by the Sea fide in Spaine, from whence I receiued the feedes of Guillaume Boel, and did abide well in my garden a long time, but is now perifhed.

The Time.

It flowreth in the beginning of Iuly, or thereabouts, and continueth not long in flower : but the head abideth a great while, and is of fome beauty after the flower is paft; yet feldome giueth good feed with vs.

The Names.

It hath no other name then is fet down in the title, being altogether a Nouelift, and not now to be feene with any fauing my felfe.

The Vertues.

We haue not yet kuown any vfe hereof in Phyfick.

Chap. LXXVI.

Cnicus siue Carthamus satiuus. Bastard or Spanish Saffron.

THere are two or three sorts of *Cnicus* or bastard Saffrons which I passe ouer, as not fit for this Garden, and onely set downe this kinde, whose flowers are of a fairer and more liuely colour in our Country, then any hath come ouer from Spaine, where they manure it for the profit they make thereof, seruing for the dying of Silke especially, and transporting great quantities to diuers Countries. It hath large broad leaues, without any prickes at all vpon them in our Country, growing vpon the stalke, which is strong, hard, and round, with shorter leaues thereon vp to the toppe, where they are a little sharpe pointed, and prickly about the edges sometimes, which stalke riseth three or foure foote high, and brancheth it selfe toward the toppe, bearing at the end of euery branch one great open scaly head, out of which thrusteth out many gold yellow threads, of a most orient shining colour, which being gathered in a dry time, and kept dry, will abide in the same delicate colour that it bare when it was fresh, for a very long time after: when the flowers are past, the seede when it is come to maturity, which is very seldome with vs, is white and hard, somewhat long, round, and a little cornered: the roote is long, great, and wooddy, and perisheth quickly with the first frosts.

The Place.

It groweth in Spaine, and other hot Countries, but not wilde, for that it is accounted of the old Writers, Theophrastus and Dioscorides, to be a manured plant.

The Time.

It floweth with vs not vntill August, or September sometimes, so that it hardly giueth ripe seede (as I said) neither is it of that force to purge, which groweth in these colder Countries, as that which commeth from Spaine, and other places.

The Names.

The name *Cnicus* is deriued from the Greekes, and *Carthamus* from the Arabians, yet still *satiuus* is added vnto it, to shew it is no wilde, but a manured plant, and sowne euery where that wee know. Of some it is called *Crocus hortensis*, and *Sarasenicus*, from the Italians which so call it. We call it in English Bastard Saffron, Spanish Saffron, and Catalonia Saffron.

The Vertues.

The flowers are vsed in colouring meates, where it groweth beyond Sea, and also for the dying of Silkes: the kernels of the seede are onely vsed in Physicke with vs, and serueth well to purge melancholicke humours.

Chap. LXXVII.

Carduus. Thistles.

YOu may somewhat maruaile, to see mee curious to plant Thistles in my Garden, when as you might well say, they are rather plagues then pleasures, and more trouble to weede them out, then to cherish them vp, if I made therein no distinction or choise; but when you haue viewed them well which I bring in, I will then

then abide your cenfure, if they be not worthy of fome place, although it be but a corner of the Garden, where fomething muft needes be to fill vp roome. Some of them are fmooth, and without prickes at all, fome at the heads onely, and fome all ouer; but yet not without fome efpeciall note or marke worthy of refpect: Out of this difcourfe I haue the Artichoke, with all his kindes, and referue them for our Kitchin Garden, becaufe (as all know) they are for the pleafure of the tafte, and not of the fmell or fight.

1. *Acanthus fativus.* Garden Beares breech.

The leaues of this kinde of fmooth thiftle (as it is accounted) are almoft as large as the leaues of the Artichoke, but not fo fharp pointed, very deeply cut in and gafhed on both edges, of a fad green & fhining colour on the vpperfide, and of a yellowifh green vnderneath, with a great thicke rib in the middle, which fpread themfelues about the root, taking vp a great deale of ground. After this plant hath ftood long in one place, and well defended from the iniury of the cold, it fendeth forth from among the leaues one or more great and ftrong ftalkes, three or foure foote high, without any branch at all, bearing from the middle to the top many flowers one aboue another, fpike-fafhion round about the ftalke, with fmaller but not diuided greene leaues at euery flower, which is white, and fafhioned fomewhat like vnto a gaping mouth; after which come broad, flat, thicke, round, brownifh yellow feede (as I haue well obferued by them haue beene fent me out of Spaine, and which haue fprung vp, and doe grow with me; for in our Countrey I could neuer obferue any feede to haue growne ripe) the rootes are compofed of many great and thicke long ftrings, which fpread farre in and vnder the ground, fomewhat darkifh on the outfide, and whitifh within, full of a clammy moifture (whereby it fheweth to haue much life) and doe endure our Winters, if they be not too much expofed to the fharpe violence thereof, which then it will not endure, as I haue often found by experience.

2. *Acanthus filueftris.* Wilde or prickly Beares breech.

This prickly Thiftle hath diuers long greenifh leaues lying on the ground, much narrower then the former, but cut in on both fides, thicke fet with many white prickes and thornes on the edges: the ftalke rifeth not vp fo high, bearing diuers fuch like thornie leaues on them, with fuch a like head of flowers on it as the former hath: but the feede hereof (as it hath come to vs from Italy and other places, for I neuer faw it beare feed here in this Country) is blacke and round, of the bigneffe of a fmall peafe: the roote abideth reafonable well, if it be defended fomewhat from the extremity of our Winters, or elfe it will perifh.

3. *Eringium Pannonicum fiue Montanum.* Hungary Sea Holly.

The lower leaues of this Thiftle that lye on the ground, are fomewhat large, round, and broad, hard in handling, and a little fnipt about the edges, euery one ftanding vpon a long foote-ftalke: but thofe that growe vpon the ftalke, which is ftiffe, two or three foote high, haue no foote-ftalke, but encompaffe it, two being fet at euery ioynt, the toppe whereof is diuided into diuers branches, bearing fmall round rough heads, with fmaller and more prickly leaues vnder them, and more cut in on the fides then thofe belowe: out of thefe heads rife many blew flowers, the foote-ftalkes of the flowers, together with the toppes of the branches, are likewife blew and tranfparent, or fhining.

flore albo. We haue another of this kinde, the whole toppes of the ftalkes, with the heads and branches, are more white then blew: the feede contained in thefe heads are white, flat, and as it were chaffie: the roote is great and whitifh, fpreading farre into many branches, and fomewhat fweete in tafte, like the ordinary Sea Holly rootes.

4. *Carduus mollis.* The gentle Thiftle.

The leaues of this foft and gentle Thiftle that are next vnto the ground, are greene

1 *Acanthus sativus*, Garden Beares breech. 2 *Acanthus sylvestris*. Wilde Beares breech. 3 *Eringium Pannonicum* Mountaine Sea Holly. 4 *Carlina humilis*. The lowe Carline Thistle. 5 *Carduus sphaerocephalus maior*. The greater Globe-Thistle. 6 *Carduus sphaerocephalus minor*. The lesser Globe-Thistle. 7 *Carduus Eriocephalus*. The Friers crowne. 8 *Fraxinella*. Bastard Dittanie.

on the vpperside, and hoary vnderneath, broad at the bottome, somewhat long pointed, and vneuenly notched about the edges, with some soft hairie prickles, not hurting
the handler, euery one standing vpon a short foote-stalke; those that growe about the
middle stalke are like the former, but smaller and narrower, and those next the toppe
smallest, where it diuideth it selfe into small branches, bearing long and scaly heads,
out of which breake many reddish purple threads : the seede is whitish and hard, almost as great as the seede of the greater Centory : the roote is blackish, spreading vnder the ground, with many small fibres fastened vnto it, and abideth a great while.

5. *Carlina humilis.* The lowe Carline Thistle.

This lowe Thistle hath many iagged leaues, of a whitish greene colour, armed with
small sharp white prickles round about the edges, lying round about the root vpon the
ground, in the middle whereof riseth vp a large head, without any stalke vnder it, compassed about with many small and long prickly leaues, from among which the flower
sheweth it selfe, composed of many thin, long, whitish, hard shining leaues, standing
about the middle, which is flat and yellow, made of many thrums or threads like small
flowers, wherein lye small long seede, of a whitish or siluer colour : the roote is somewhat aromaticall, blackish on the outside, small and long, growing downewards into
the ground. There is another of this kinde that beareth a higher stalke, and a redder
flower, but there is a manifest difference betweene them.

6. *Carduus Sphærocephalus siue Globosus maior.* The greater Globe Thistle.

The greatest of these beautifull Thistles, hath at the first many large and long leaues
lying on the ground, very much cut in and diuided in many places, euen to the middle
ribbe, set with small sharpe (but not very strong) thornes or prickles at euery corner of
the edges, greene on the vpperside, and whitish vnderneath : from the middle of these
leaues riseth vp a round stiffe stalke, three foote and a halfe high, or more, set without
order with such like leaues, bearing at the toppe of euery branch a round hard great
head, consisting of a number of sharpe bearded huskes, compact or set close together,
of a blewish greene colour, out of euery one of which huskes start small whitish blew
flowers, with white threads in the middle of them, and rising aboue them, so that the
heads when they are in full flower, make a fine shew, much delighting the spectators :
after the flowers are past, the seede encreaseth in euery one, or the most part of the
bearded huskes, which doe still hold their round forme, vntill that being ripe it openeth it selfe, and the huskes easily fall away one from another, containing within
them a long whitish kernell : the roote is great and long, blackish on the outside, and
dyeth euery yeare when it hath borne seede.

7. *Carduus Globosus minor.* The lesser Globe Thistle.

The lesser kinde hath long narrow leaues, whiter then the former, but cut in and
gashed on the edges very much with some small prickes on them; the stalke is not
halfe so long, nor the heads halfe so great, but as round, and with as blew flowers as the
greater : this seldome giueth ripe seede, but recompenseth that fault, in that the roote
perisheth not as the former, but abideth many yeares.

8. *Carduus Eriocephalus siue Tomentosus.* The Friers Crowne.

This woolly Thistle hath many large and long leaues lying on the ground, cut in on
both sides into many diuisions, which are likewise somewhat vnequally cut in or diuided againe, hauing sharpe white prickles at euery corner of the diuisions, of a dead
or sad greene colour on the vpperside, and somewhat woolly withall, and grayish vnderneath : the stalke is strong and tall, foure or fiue foote high at the least, branching
out into diuers parts, euery where beset with such like leaues as growe below; at the
toppe of euery branch there breaketh out a great whitish round prickly head, flattish at
the toppe, so thicke set with wooll, that the prickles seeme but small spots or haires,

and

and doth so well resemble the bald crowne of a Frier, not onely before it be in flower, but especially after it hath done flowring, that thereupon it deseruedly receiued the name of the Friers Crowne Thistle : out of these heads riseth forth a purple thrumme, such as is to be seene in many other wilde Thistles, which when they are ripe, are full of a flockie or woolly substance, which breake at the toppe shedding it, and the seede which is blackish, flat, and smooth : the roote is great and thicke, enduring for some yeares, yet sometimes perishing, if it be too much exposed to the violence of the frosts in Winter.

The Place.

The first groweth naturally in Spaine, Italy, and France, and in many other hot Countries, and growe onely in Gardens in these colder climates, and there cherished for the beautifull aspect both of the greene plants, and of the stalkes when they are in flower. The Carline Thistle is found both in Germany and Italy in many places, and as it is reported, in some places of the West parts in England. The others are found some in France, some in Hungary, and on the Alpes, and the last in Spaine.

The Time.

They doe all flower in the Summer moneths, some a little earlier or later then others.

The Names.

The first is called *Acanthus satiuus* (because the other that is prickly, is called *siluestris* or *spinosus*) and *Branca vrsina* ; In English, Branck vrsine, and Beares breech. The third is called *Eringium montanum, Alpinum*, and *Pannonicum latifolium*: In English, Mountaine or Hungary Sea Holly. The fourth is called *Carduus mollis*, The gentle Thistle, because it hath no harmfull prickles, although it seeme at the first shew to be a Thistle. The fifth is called of diuers *Chamæleo albus*, and *Carlina*, as if they were both but one plant ; but Fabius Columna hath in my iudgement very learnedly descided that controuersie, making *Carlina* to be *Ixine* of Theophrastus, and *Chamæleo* another differing Thistle, which Gaza translateth *Vernilago*. We call it in English, The Carline Thistle. The other haue their names in their titles, as much as is conuenient for this discourse.

The Vertues.

The first hath alwaies been vsed Physically, as a mollifying herbe among others of the like slimie matter in Glisters, to open the body ; yet Lobel seemeth to make no difference in the vse of them both (that is, the prickly as well as the smooth.) The Carline Thistle is thought to bee good against poysons and infection. The rest are not vsed by any that I know.

Chap. LXXVIII.

Fraxinella. Bastard Dittany.

Hauing finished those pleasing Thistles, I come to other plants of more gentle handling, and first bring to your consideration this bastard Dittany, whereof there are found out two especiall kindes, the one with a reddish, the other with a whitish flower, and each of these hath his diuersity, as shall be presently declared.

1. *Fraxinella flore rubente.* Bastard Dittany with a reddish flower.

This goodly plant riseth vp with diuers round, hard, brownish stalkes, neare two foote high, the lower parts whereof are furnifhed with many winged leaues, fomewhat like vnto Liquerice, or a fmall young Afhe tree, confifting of feuen, nine, or eleuen leaues fet together, which are fomewhat large and long, hard and rough in handling, of a darkifh greene colour, and of an vnpleafant ftrong refinous fent: the vpper parts of the ftalkes are furnifhed with many flowers, growing fpike fafhion, at certaine diftances one aboue another, confifting of fiue long leaues a peece, whereof foure that ftand on the two fides, are fomewhat bending vpwards, and the fift hanging downe, but turning vp the end of the leafe a little againe, of a faint or pale red colour, ftriped through euery leafe with a deeper red colour, and hauing in the middle a taffell of fiue or fix long purplifh threds, that bowe downe with the lower leafe, and turne vp alfo the ends againe, with a little freefe or thrume at the ends of euery one: after the flowers are paft, arife hard, ftiffe, rough, clammy huskes, horned or pointed at the end, foure or fiue ftanding together, fomewhat like the feede veffels of the Wolfes-banes, or Colombines, but greater, thicker and harder, wherein is contained round fhining blacke feede, greater then any Colombine feede by much, and fmaller then Peony feede: the roote is white, large, and fpreading many wayes vnder ground, if it ftand long: the whole plant, as well roots as leaues and flowers, are of a ftrong fent, not fo pleafing for the fmell, as the flowers are beautifull to the fight.

2. *Fraxinella flore rubro.* Baftard Dittaine with a red flower.

This differeth not from the former eyther in roote, leafe or flower for the forme, but that the ftalkes and leaues are of a darker greene colour, and that the flowers are of a deeper red colour, (and growing in a little longer fpike) wherein the difference chiefly confifteth, which is fufficient to diftinguifh them.

3. *Fraxinella flore albo.* Baftard Dittanie with a white flower.

The white flowred *Fraxinella* hath his leaues and ftalkes of a frefher greene colour then any of the former; and the flowers are of a pure white colour, in forme differing nothing at all from the other.

4. *Fraxinella flore albo cæruleo.*
Baftard Dittanie with an afh coloured flower.

The colour of the flower of this *Fraxinella* onely putteth the difference betweene this, and the laft recited with a white flower: for this beareth a very pale, or whitifh blew flower, tending to an afh colour.

The Place.

All thefe kindes are found growing naturally, in many places both of Germany, and Italie: and that with the white flower, about Franckford, which being fent me, perifhed by the way by long and euill carriage.

The Time.

They flower in Iune and Iuly, and the feede is ripe in Auguft.

The Names.

The name *Fraxinella* is moft generally impofed on thofe plants, becaufe of the refemblance of them vnto young Afhes, in their winged leaues. Yet fome doe call them *Dictamus albus*, or *Dictamnus albus*, and *Diptamus albus*, as a difference from the *Dictamnus Creticus*, which is a farre differing plant. Some would haue it to be *Tragium* of Diofcorides, but befide other things wherein this differeth from *Tragium*, this yeeldeth no milkie iuice, as Diofcorides faith *Tragium* doth: We in Englifh doe eyther call it *Fraxinella*, or after the other corrupted name of *Dictamus*, Baftard Dittanie.

The

The Vertues.

It is held to be profitable against the stingings of Serpents, against contagious and pestilent diseases, to bring downe the feminine courses, for the paines of the belly and the stone, and in Epilepticall diseases, and other cold paines of the braines : the roote is the most effectuall for all these, yet the seede is sometimes vsed.

Chap. LXXIX.

Legumina. Pulse.

IF I should describe vnto you all the kindes of Pulse, I should vnfold a little world of varieties therein, more knowne and found out in these dayes, then at any time before, but that must bee a part of a greater worke, which will abide a longer time before it see the light. I shall only select those that are fit for this garden, and set them downe for your consideration. All sorts of Pulse may be reduced vnder two generall heads, that is, of Beanes and Pease, of each whereof there is both tame and wilde : Of Beanes, besides the tame or vsuall garden Beane, and the French or Kidney Beane, (whereof I meane to entreate in my Kitchen garden, as pertinent thereto) there is the Lupine or flat Beane, whereof I meane to entreate here, and the blacke Beane and others which must bee reserued for the Physicke Garden. And of the kindes of Pease some are fit for this Garden, (whereunto I will adioyne two or three other plants as neerest of affinitie, the flowers of some, and the fruit of others being delightfull to many, and therefore fit for this garden) some for the Kitchen, the rest for the Physicke garden. And first of Lupines or flat Beanes, accepted as delightfull to many, and therefore fit for this garden.

1. *Lupinus satiuus albus.* The white garden Lupine.

The garden Lupine riseth vp with a great round stalke, hollow and somewhat woolly, with diuers branches, whereon grow vpon long footestalkes many broade leaues, diuided into seuen or nine parts, or smaller leaues, equally standing round about, as it were in a circle, of a whitish greene colour on the vpperside, and more woolly vnderneath : the flowers stand many together at seuerall ioynts, both of the greater stalke, and the branches, like vnto beanes, and of a white colour in some places, and in others of a very bleake blew tending to white : after the flowers are past, there come in their places, long, broade, and flat rough cods, wherein are contained round and flat seede, yellowish on the inside, and couered with a tough white skin, and very bitter in taste : the rootes are not very great, but full of small fibres, whereby it fasteneth it selfe strongly in the ground, yet perisheth euery yeare, as all the rest of these kindes doe.

2. *Lupinus caruleus maximus.* The greater blew Lupine.

The Stemme or stalke of this Lupine is greater then the last before recited, as also the leaues more soft and woolly, and the flowers are of a most perfect blew colour, with some white spots in the middle : the long rough greenish cods are very great and large, wherein are contained hard, flat and round seede, not so white on the outside as the former, but somewhat yellower, greater also, and more rough or hard in handling.

3. *Lupinus caruleus minor.* The lesser blew Lupine.

This kinde of wilde Lupine differeth not in the forme of leafe or flower from the former, but only that it is much smaller, the leaues are greener, and haue fewer diuisions in them : the flower is of as deepe a blew colour as the last ; the cods likewise are small and long, containing small round seede, not so flat as the former, but more
discoloured

Minimus. difcoloured or fpotted on the outfide, then the greater kinde is. There is a leffer kind then this, not differing in any thing from this, but that it is leffer.

4. *Lupinus flore luteo.* The yellow Lupine.

The yellow Lupine groweth not vfually fo high, but with larger leaues then the fmall blew Lupine ; the flowers grow in two or three rundles or tufts, round about the ftalke and the branches at the ioynts, of a delicate fine yellow colour, like in fafhion vnto the other kindes, being larger then the laft, but nothing fo large as the greater kindes, and of a fine fmall fent : the feede is round, and not flat, but much about the forme and bigneffe of the fmall blew, or fomewhat bigger, of a whitifh colour on the outfide, fpotted with many fpots.

The Place.

The firft groweth in many places of Greece, and the Eafterne Countries beyond it, where it hath beene anciently cherifhed for their foode, being often watered to take away the bitterneffe. It groweth alfo in thefe Weftern parts, but ftill where it is planted. The great blew Lupine is thought to come from beyond the parts of Perfia, in Caramania. The leffer blew is found very plentifully wilde, in many places both of Spaine and Italy. The laft hath beene brought vs likewife out of Spaine, whereas it is thought it groweth naturally. They all grow now in the gardens of thofe, that are curious louers of thefe delights.

The Time.

They flower in Summer, and their feede is ripe quickly after.

The Names.

They are generally called *Lupini.* Plautus in his time faith, they were vfed in Comedies in ftead of money, when in any Scene thereof there was any fhew of payment, and therefore he calleth them *Aurum Comicum.* And Horace hath this Verfe,

Nec tamen ignorant, quid diſtent æra Lupinis,

to fhew that counterfeit money (fuch as counters are with vs, or as thefe Lupines were vfed in thofe times) was eafily knowne from true and currant coine. In Englifh wee vfually call them after the Latine name, Lupines ; and fome after the Dutch name, Figge-beanes, becaufe they are flat and round as a Figge that is preffed ; and fome Flat-beanes for the fame reafon. Some haue called the yellow Lupine, Spanifh Violets : but other foolifh names haue beene giuen it, as Virginia Rofes, and the like, by knauifh Gardiners and others, to deceiue men, and make them beleeue they were the finders out, or great preferuers of rarities, of no other purpofe, but to cheate men of their money : as you would therefore auoyde knaues and deceiuers, beware of thefe manner of people, whereof the skirts of our towne are too pitifully peftered.

The Vertues.

The firft or ordinary Lupine doth fcoure and cleanfe the skin from fpots, morphew, blew markes, and other difcolourings thereof, beeing vfed eyther in a decoction or pontker. Wee feldome vfe it in inward medicines, not that it is dangerous, but of neglect, for formerly it hath beene much vfed for the wormes, &c.

1 · *Lupinus maior.* The great Lupine. 2 *Lupinus luteus.* The yellow Lupine. 3 *Lathyrus latifolius seu Pisum perenne.* Pease euerlasting 4 *Pisum quadratum.* The crimson blossomd or square Pease. 5 *Medica cochleata vulgaris* Snailes or Barbary buttons. 6 *Medica spinosa.* Prickly Snailes 7 *Medica spinosa altera.* Another sort of prickly Snailes. 8 *Medica folliculata.* Broad buttons or Snailes. 9 *Medica lunata.* Halfe Moons 10 *Hedysarum clypeatum.* The red Sattin flower, or French Honysuckle. 11 *Scorpioides minus.* The lesser Caterpiller. 12 *Scorpioides maius.* The greater Caterpiller. 13 *Orobus Venetus.* Blew vpright Pease euerlasting.

1. *Lathyrus latifolius, siue Pisum perenne.* Peafe euerlafting.

This kinde of wilde Peafe that abideth long, and groweth euery yeare greater then other, fpringeth vp with many broade trayling branches, winged as it were on both the fides, diuerfly diuided into other fmaller branches, at the feuerall ioynts whereof ftand two hard, not broad, but fomewhat long greene leaues, and diuers twining clafpers, in fundry places with the leaues, from betweene the branches and the leaues, at the ioynts towards the toppes, come forth diuers purplifh peafe like bloffomes, ftanding on a long ftemme or ftalke, very beautifull to behold, and of a pretty fent or fmell : after which come fmall, long, thin, flat, hard skind cods, containing fmall round blackifh feede : the roote is great and thicke, growing downe deepe into the ground, of the thickneffe fometimes of a mans arme, blackifh on the outfide, and whitifh within, with fome branches and a few fibres annexed thereunto.

2. *Orobus Venetus.* Blew vpright euerlafting Peafe.

This pretty kinde of Peafe bloffome beareth diuers flender, but vpright greene branches fomewhat cornered, two foote high or thereabouts, hauing at feuerall diftances on both fides of them certaine winged leaues, fet together vpon long footeftalkes one againft another, confifting of fix or eight leaues, fomewhat broade and pointed, and without any odde one at the end : at the ioynts toward the toppes, betweene the leaues and the ftalkes, come forth many flowers fet together at the end of a pretty long footeftalke, of the fafhion of the former Peafe bloffome, but fomewhat fmaller, and of a purplifh violet colour : after which come flender and long pointed pods rounder then they, wherein is contained fmall round grayifh peafe : the roote is blacke, hard or woody, abiding after feede bearing as the former doth, and fhooting afrefh euery yeare.

3. *Lathyrus annuus filiquis orobi.* Partie coloured Cichelings.

This fmall Pulfe or wild Peafe, hath two or three long flender winged branches, with fmaller leaues theron then the former, and without any clafpers at all on them : the flowers ftand fingle, euery one by it felfe, or two at the moft together, the middle leaues whereof that clofe together are white, and the vpper leaues of a reddifh purple colour : after which come long round flattifh cods, bunched out in the feuerall places where the feedes lye, like vnto the cods of *Orobus* or the bitter Vetch, but greater : the roote is fmall and dyeth euery yeare.

4. *Pifum quadratum.* The crimfon bloffomd or fquare codded Peafe.

This pretty kinde of Pulfe might very well for the forme of the leaues, be referred to the kindes of *Lotus* or Trefoiles : but becaufe I haue none of that kindred to entreate of in this Worke, I haue thought fitteft to place it here before the Medica's, becaufe both pods and feedes are like alfo. It hath three or foure fmall weake ftalkes, diuided into many branches, hauing two ftalkes of leaues at euery ioynt, and three fmall foft leaues ftanding on a very fmall ftalke, comming from the ioynts : the flowers ftand for the moft part two together, of a perfect red or crimfon colour, like in forme almoft vnto a Peafe bloffome, after which come long thicke and round cods, with two skinnes or filmes, running all along the cod at the backe or vpperfide, and two other fuch like filmes, all along the belly or vnder fide, which make it feeme foure fquare, wherein there lye round difcoloured Peafe, fomewhat fmaller and harder then ordinary Peafe : the roote is fmall and perifheth euery yeare.

5. *Medica Cochleata vulgaris.* Snailes or Barbary buttons.

The plant that beareth thefe pretty toyes for Gentlewomen, is fomewhat like vnto a Threeleafed graffe or Trefoile, hauing many long trayling branches lying vpon the ground, whereon at diuers places are three fmall greene leaues, fet together at the end of a little footeftalke, each of them a little fnipt about the edges : at feuerall diftances, from the middle of thefe branches to the ends of them, come forth the flowers, two for the moft part ftanding together vpon a little footftalke, which are of a pale yellow colour, very fmall, and of the forme of a Peafe bloffome : after which come fmooth heads, which are turned or writhen round, almoft like a Snaile, hard and greene at the firft, fomewhat like a greene button (from the formes of both which came their names) but afterwards growing whiter, more foft and open, wherein lyeth yellowifh round and flat feede, fomewhat like vnto the Kidney beane : the roote is fmall and ftringie, dying downe euery yeare, and muft be new fowne in the fpring, if you defire to haue it.

6. *Medica fpinofa maior.* Prickly or thorny Snailes, or Buttons.

This kinde of *Medica* is in all things very like vnto the former, both in the long tray-ling branches, & three leaues alwaies growing together, but a little greater pale yellow flowers, and crooked or winding heads : but herein chiefly confifteth the difference, that this kinde hath his heads or buttons harder, a little greater, more clofed together, and fet with fhort and fomewhat hard prickles, all the head ouer, which being pulled open, haue thofe prickles ftanding on each fide of the filme or skinne, whereof the head confifteth, fomewhat like vnto a fifh bone, and in this kinde goeth all one way ; in which are contained fuch like feedes for the forme, as are in the former, but great and blacke, and fhining withall.

7. *Medica fpinofa altera.* Small thorney Buttons, or Snailes.

This other kinde is alfo like vnto the laft defcribed in all other things, except in the heads or buttons, which are a little fmaller, but fet with longer and fofter prickes vpon the filmes, and may eafily bee difcerned to goe both forwards and backewards, one enterlacing within another, wherein are contained fuch like flat and blacke fhining feede, made after the fafhion of a kidney, as are in the former, but fomewhat fmaller : the roote perifheth in like manner euery yeare.

8. *Medica lata.* Broade Buttons.

This kinde differeth not from the firft in leafe or flower, the fruite onely hereof is broade and flat, and not fo much twined as it.

9. *Medica Lunata.* Halfe Moones.

This is alfo a kinde of thefe Medicke fodders, hauing a trefoyle leafe and yellow flowers like the former forts, but both fomewhat larger, the chiefeft difference con-fifteth in the head or fruite, which is broade and flat, and not twined like the reft, but abideth halfe clofed, refembling a halfe Moone (and thereupon hath affumed both the Latine and Englifh name) wherein is contained flat feede, kidney fafhion like the for-mer.

10. *Hedyfarum clypeatum.* The red Sattin flower.

This red flowred Fitchling, hath many ftalkes of winged faire greene leaues, that is, of many fet on both fides a middle ribbe, whereof that at the end is the greateft of the reft : from the ioynts where the leaues ftand, come forth pretty long fmall ftalkes, bearing on them very many flowers, vp to the toppe one aboue another, of an excel-lent fhining red or crimfon colour, very like vnto Sattin of that colour, and fome-times of a white colour, (as Mafter William Coys, a Gentleman of good refpect in Effex, a great and ancient louer and cherifher of thefe delights, and of all other rare plants, in his life time affured me, he had growing in his garden at Stubbers by North Okenden) which are fomewhat large, and more clofed together, almoft flat and not open, as in moft of the other forts : after the flowers are paft, there come rough, flat, round huskes, fomewhat like vnto the old fafhioned round bucklers with-out pikes, three or foure ftanding one vpon or aboue another, wherein are contained

fmall

small brownish seede : the roote perisheth the same yeare it beareth seede, for often-times it flowreth not the first yeare it is sowne.

11. *Scorpioides maius & minus.*
Great and small Caterpillers.

Vnder one description I comprehend both these sorts of Scorpions grasse, or Cater-pillers, or Wormes, as they are called by many, whereof the greater hath been known but of late yeares ; and ioyne them to these pulses, not hauing a fitter place where to insert them. It is but a small low plant, with branches lying vpon the ground, and somewhat long, broad, and hard leaues theron, among which come forth small stalkes, bearing at the end for the most part, two small pale yellowish flowers, like vnto Tares or Vetches, but smaller, which turne into writhed or crooked tough cods; in the grea-ter sort they are much thicker, rounder and whiter, and lesser wound or turned toge-ther then in the smaller, which are slenderer, more winding, yet not closing like vnto the Snailes, and blacker more like vnto a Caterpiller then the other, wherein are contained brownish yellow seede, much like vnto a *Medica :* the rootes of both are small and fibrous, perishing euery yeare.

The Place.

These are found seuerally in diuers and seuerall places, but wee sow and plant them vsually to furnish our gardens.

The Time.

They doe all flower about the moneths of Iune and Iuly, and their seede is ripe soone after : but the second is earlier then the rest.

The Names.

The first is called *Clymenum* of Matthiolus, and *Lathyrus* of Lobel and o-thers : but *Lathyris* in Greeke is *Cataputia* in Latine, which is our Spurge, farre differing from this Pulse ; and therefore *Lathyrus* is more proper to distinguish them asunder, that two plants so farre vnlike should not bee called by one name : this is also called *Lathyrus latifolius,* because there is another called *angustifolius,* that differeth from it also : It is most vsually called with vs, *Pisum perenne,* and in English Pease blossome, or Pease euer-lasting. The second is called by Clusius, *Orobus venetus,* because it was sent him from Venice, with another of the same kinde that bore white flowers ; yet differeth but little or nothing from that kinde he found in Hungary, that I thinke the seuerall places of their growing only cause them to beare seue-rall names, and to be the same in deede. Although I yeeld vnto Clusius the Latine name which doth not sufficiently content mee ; yet I haue thought good to giue it a differing English name, according as it is in the title. The third, because I first receiued it among other seeds from Spaine, I haue giuen it the name, as it is entituled. The fourth is called of some *Sandalida Cretica, & Lotus siliquosus flore rubello, Lotus tetragonolobus, Pisum rubrum, & Pisum quadratum :* We vsually call it in English, Crimson Pease, or square Pease. The *Medica Cochleata* is called of Dodonæus *Trifolium Cochleatum,* but not iudged to be the true *Medica.* Wee call it in English, Medick fodder, Snailes Clauer, or as it is in the title, and so the rest of the Medica's accordingly. The *Hedysarum clypeatum* or *Securidaca* is called of Dodonæus *Onobrichis altera,* and we in English for the likenesse, The red Sattin flower, although some foolishly call it, the red or French Honysuckle. The last is called by Lobel, *Scorpioides bupleurifolio,* I haue called it *minus,* because the greatest sort which came to me out of Spaine was not knowne vnto him : in English they are generally called Caterpillers.

The

The Vertues.

The Medica's are generally thought to feede cattell fat much more then the Medow Trefoile, or Clauer graffe, and therefore I haue known diuers Gentlemen that haue plowed vp fome of their pafture grounds, and fowen them with the feedes of fome Medica's to make the experience. All the other forts are pleafures to delight the curious, and not any way profitable in Phyficke that I know.

CHAP. LXXX.

Pæonia. Peonie.

THere are two principall kindes of Peonie, that is to fay, the Male and the Female. Of the male kinde, I haue onely known one fort, but of the Female a great many; which are thus to be diftinguifhed. The Male his leafe is whole, without any particular diuifion, notch or dent on the edge, & his rootes long & round, diuided into many branches, fomewhat like to the rootes of Gentian or Elecampane, and not tuberous at all. The Female of all forts hath the leaues diuided or cut in on the edges, more or leffe, and hath alwaies tuberous rootes, that is, like clogs or Afphodill rootes, with many great thick round peeces hanging, or growing at the end of fmaller ftrings, and all ioyned to the toppe of the maine roote.

1. *Pæonia mas.* The Male Peonie.

The Male Peonie rifeth vp with many brownifh ftalkes, whereon doe grow winged leaues, that is, many faire greene, and fometimes reddifh leaues, one fet againft another vpon a ftalke, without any particular diuifion in the leafe at all : the flowers ftand at the toppes of the ftalkes, confifting of fiue or fix broade leaues, of a faire purplifh red colour, with many yellow threds in the middle, ftanding about the head, which after rifeth to be the feede veffels, diuided into two, three or foure rough crooked pods like hornes, which when they are ful ripe, open and turn themfelues down one edge to another backeward, fhewing within them diuers round black fhining feede, which are the true feede, being full and good, and hauing alfo many red or crimfon graines, which are lancke and idle, intermixed among the blacke, as if they were good feede, whereby it maketh a very pretty fhew: the roots are great, thick and long, fpreading in the ground, and running downe reafonable deepe.

2. *Pæonia fæmina vulgaris flore fimplici.*
The ordinary fingle Female Peonie.

This ordinary Female Peonie hath many ftalkes, with more ftore of leaues on them then the Male kinde hath, the leaues alfo are not fo large, but diuided or nicked diuerfly on the edges, fome with great and deepe, and others with fmaller cuts or diuifions, and of a darke or dead greene colour : the flowers are of a ftrong heady fent, moft vfually fmaller then the male, and of a more purple tending to a murrey colour, with yellow thrumes about the head in the middle, as the male kinde hath : the heads or hornes with feed are like alfo but fmaller, the feede alfo is blacke, but leffe fhining: the rootes confift, as I faid, of many thicke and fhort tuberous clogs, faftened at the ends of long ftrings, and all from the head of the roote, which is thicke and fhort, and tuberous alfo, of the fame or the like fent with the male.

3. *Pæonia fæmina vulgaris flore pleno rubro.*
The double red Peonie.

This double Peonie as well as the former fingle, is fo frequent in euerie Garden of note, through euery Countrey, that it is almoft labour in vaine to defcribe it : but yet becaufe I vfe not to paffe ouer any plant fo flightly , I will fet down the defcription briefly ,in regard it is fo common.It is very like vnto the former fingle female Peony, both in ftalkes and leaues, but that it groweth fomewhat higher, and the leaues are of a frefher greene colour : the flowers at the tops of the ftalkes are very large, thicke, and double (no flower that I know fo faire, great, and double ; but not abiding blowne aboue eight or ten daies) of a more reddifh purple colour then the former female kinde, and of a fweeter fent : after thefe flowers are paft,fometimes come good feed,which being fowne, bring forth fome fingle flowers , and fome double : the rootes are tuberous, like vnto the former female.

4. *Paonia fæmina flore carneo fimplici.* The fingle blufh Peony.

The fingle blufh Peony hath his ftalkes higher , and his leaues of a paler or whiter greene colour then the double blufh, and more white vnderneath(fo that it is very probable it is of another kinde, and not rifen from the feede of the double blufh , as fome might thinke) with many veines , that are fomewhat difcoloured from the colour of the leafe running through them : the flowers are very large and fingle , confifting of fiue leaues for the moft part, of a pale flefh or blufh colour, with an eye of yellow difperfed or mixed therewith, hauing many whitifh threads , tipt with yellow pendents ftanding about the middle head : the rootes are like the other female Peonies.

5. *Peonia famina flore pleno albicante.* The double blufh Peony.

The double blufh Peony hath not his ftalkes fo high as the double red , but fomewhat lower and ftiffer,bearing fuch like winged leaues,cut in or diuided here and there in the edges, as all thefe female kindes are, but not fo large as the laft : the flowers are fmaller,and leffe double by a good deale then the former double red, of a faint fhining crimfon colour at the firft opening , but decaying or waxing paler euery day : fo that after it hath ftood long (for this flower fheddeth not his leaues in a great while) it will change fomewhat whitifh ; and therefore diuers haue ignorantly called it, the double white Peony : the feedes , which fometimes it beareth , and rootes, are like vnto the former female kindes, but fomewhat longer, and of a brighter colour on the outfide.

6. *Peonia fæmina Byzantina.* The fingle red Peony of Conftantinople.

This red Peony of Conftantinople is very like in all things vnto the double red Peonie, but that the flowers hereof are fingle, and as large as the laft , and that is larger then either the fingle female, or the male kinde, confifting of eight leaues, of a deeper red colour then either the fingle or double Peonies, and not purplifh at all, but rather of the colour of an ordinary red Tulipa,ftanding clofe and round together : the roots of this kinde haue longer clogs, and not fo fhort as of the ordinary female kinde, and of a paler colour on the outfide.

The Place.

All thefe Peonies haue beene fent or brought from diuers parts beyond the Seas ; they are endenized in our Gardens , where wee cherifh them for the beauty and delight of their goodly flowers,as well as for their Phyficall vertues.

The Time.

They all flower in May , but fome (as I faid) abide a fmall time, and others many weekes.

The Names.

The name *Peonia* is of all the later Writers generally giuen to thefe plants, although they haue had diuers other names giuen by the elder Writers, as *Rofa fatuina, Idæus dactylus, Aglaophotis,* and others , whereof to fet
downe

1 *Pæonia mas cum semine.* The male Peony & the seed 2 *Pæonia fæmina byzantina.* The female red Peony of Constantinople. 3 *Pæonia fæmina flore pleno vulgaris.* The ordinary double Peony. 4 *Pæonia flore pleno albicante.* The double white Peony. 5 *Helleborus vernus atrorubente flore.* The early white Ellebor with a darke red flower. 6 *Helleborus niger verus.* The Christmas flower. 7 *Calceolus Mariæ.* Our Ladies Slipper.

downe the caufes, reafons, and errours, were to fpend more time then I intend for this worke. Wee call them in Englifh, Peonie, and diftinguifh
them according to their titles.

The Vertues.

The male Peony roote is farre aboue all the reft a moft fingular approued remedy for all Epilepticall difeafes, in Englifh, The falling ficknesse
(and more efpecially the greene roote then the dry) if the difeafe be not too
inueterate, to be boyled and drunke, as alfo to hang about the neckes of the
younger fort that are troubled herewith, as I haue found it sufficiently experimented on many by diuers. The feede likewife is of efpeciall vfe for women, for the rifing of the mother. The feede of the female kinde, as well
as the rootes, are moft vfually fold, and may in want of the other be (and
fo are generally) vfed.

Chap. LXXXI.

Helleborus niger. Beares foote.

THere are three forts of blacke Hellebor or Beares foote, one that is the true and
right kinde, whofe flowers haue the moft beautifull afpect, and the time of his
flowring moft rare, that is, in the deepe of Winter about Chriftmas, when no
other can bee feene vpon the ground : and two other that are wilde or baftard kindes,
brought into many Gardens for their Phyficall properties ; but I will only ioyne one
of them with the true kinde in this worke, and leaue the other for another.

1. *Helleborus niger verus.* The true blacke Hellebor,
or Chriftmas flower.

The true blacke Hellebor (or Beare foote as fome would call it, but that name doth
more fitly agree with the other two baftard kindes) hath many faire greene leaues rifing from the roote, each of them ftanding on a thicke round flefhly ftiffe green ftalke,
about an hand breadth high from the ground, diuided into feuen, eight, or nine parts
or leaues, and each of them nicked or dented, from the middle of the leafe to the point-
ward on both fides, abiding all the Winter, at which time the flowers rife vp on fuch
fhort thicke ftalkes as the leaues ftand on, euery one by it felfe, without any leafe
thereon for the moft part, or very feldome hauing one fmall fhort leafe not much vnder the flower, and very little higher then the leaues themfelues, confifting of fiue
broad white leaues, like vnto a great white fingle Rofe (which fometimes change to
be either leffe or more purple about the edges, as the weather or time of continuance
doth effect) with many pale yellow thrummes in the middle, ftanding about a greene
head, which after groweth to haue diuers cods fet together, pointed at the ends like
hornes, fomewhat like the feede veffels of the *Aconitum hyemale*, but greater & thicker,
wherein is contained long, round, and blackifh feede, like the feede of the baftard
kindes : the rootes are a number of brownifh ftrings running downe deepe into the
ground, and faftened to a thicke head, of the bigneffe of a finger at the toppe manie
times, and fmaller ftill downewards.

2. *Helleboraster minor.* The leffer baftard blacke Hellebor, or Beare foote.

The fmaller Beare foote is in moft things like vnto the former true blacke Hellebor;
for it beareth alfo many leaues vpon fhort ftalkes, diuided into many leaues alfo, but
each of them are long and narrow, of a blacker greene colour, fnipt or dented on both
edges, which feele fomewhat hard or fharpe like prickes, and perifh euery yeare, but
rife againe the next Spring : the flowers hereof ftand on higher ftalkes, with fome
leaues on them alfo, although but very few, and are of a pale greene colour, like in
forme

forme vnto the flowers of the former, but smaller, hauing also many greenish yellow threads or thrums in the middle, and such like heads or seede vessels, and blackish seed: the rootes are stringie and blackish like the former.

The Place.

The first groweth onely in the Gardens of those that are curious, and delight in all sorts of beautifull flowers in our Countrey, but wilde in many places of Germany, Italy, Greece, &c.

The other groweth wilde in many places of England, as well as the other greater sort, which is not here described; for besides diuers places within eight or ten miles from London, I haue seen it in the Woods of Northamptonshire, and in other places.

The Time.

The first of these plants doth flower in the end of December, and beginning of Ianuary most vsually, and the other a moneth or two after, and sometime more.

The Names.

The first is called _Helleborus_, or _Elleborus niger verus_, and is the same that both Theophrastus and Dioscorides haue written of, and which was called _Melampodion_, of Melampus the Goateheard, that purged and cured the mad or melancholicke daughters of Prætus with the rootes thereof. Dodonæus calleth it _Veratrum nigrum primum_, and the other _secundum_: Wee call it in English, The true blacke Hellebor, or the Christmas flower, because (as I said) it is most commonly in flower at or before Christmas. The second is a bastard or wilde kinde thereof, it so nearely resembleth the true, and is called of most of the later Writers, _Pseudoelleborus niger minor_, or _Helleboraster minor_, for a distinction betweene it and the greater, which is not here described: and is called in English, The smaller or lesser Beare foote, and most vsed in Physicke, because it is more plentifull, yet is more churlish and strong in operation then the true or former kinde.

The Vertues.

The rootes of both these kindes are safe medecines, being rightly prepared, to be vsed for all Melancholicke diseases, whatsoeuer others may feare or write, and may be without danger applied, so as care and skill, and not temerary rashnesse doe order and dispose of them.

The powder of the dryed leaues, especially of the bastard kinde, is a sure remedy to kill the wormes in children, moderately taken.

Chap. LXXXII.

Elleborus albus. White Ellebor or Neesewort.

THere are two sorts of great white Ellebors or Neeseworts, whereas there was but one kinde knowne to the Ancients; the other being found out of later dayes: And although neither of both these haue any beauty in their flowers, yet because their leaues, being faire and large, haue a goodly prospect, I haue inserted them in this place, that this Garden should not be vnfurnished of them, and you not vnacquainted with them.

1. Elleborus

1. *Elleborus albus vulgaris.* White Ellebor or Neesing roote.

The first great white Ellebor riseth at the first out of the ground, with a whitish greene great round head, which growing vp, openeth it selfe into many goodly faire large greene leaues, plaited or ribbed with eminent ribbes all along the leaues, compassing one another at the bottome, in the middle whereof riseth vp a stalke three foot high or better, with diuers such like leaues thereon, but smaller to the middle thereof; from whence to the toppe it is diuided into many branches, hauing many small yellowish, or whitish greene starre-like flowers all along vpon them, which after turne into small, long, three square whitish seede, standing naked, without any huske to containe them, although some haue written otherwise: the roote is thicke and reasonable great at the head, hauing a number of great white strings running downe deepe into the ground, whereby it is strongly fastened.

2. *Elleborus albus præcox siue atrorubente flore.*
The early white Ellebor with reddish flowers.

This other Ellebor is very like the former, but that it springeth vp a moneth at the least before it, and that the leaues are not fully so thicke or so much plaited, but as large or larger, and doe sooner perish and fall away from the plant: the stalke hereof is as high as the former, bearing such like starry flowers, but of a darke or blackish red colour: the seede is like the other: the roote hath no such head as the other (so farre as I haue obserued, both by mine own and others plants) but hath many long white strings fastened to the top, which is as it were a long bulbous scaly head, out of which spring the leaues.

The Place.

The first groweth in many places of Germany, as also in some parts of Russia, in that aboundance, by the relation of that worthy, curious, and diligent searcher and preseruer of all natures rarities and varieties, my very good friend, Iohn Tradescante, often heretofore remembred, that, as hee said, a good ship might be loaden with the rootes hereof, which hee saw in an Island there.

The other likewise groweth in the vpland wooddy grounds of Germanie, and other the parts thereabouts.

The Time.

The first springeth vp in the end or middle of March, and flowreth in Iune. The second springeth in February, but flowreth not vntill Iune.

The Names.

The first is called *Elleborus albus*, or *Helleborus albus*, the letter *H*, as all Schollers know, being but *aspirationis nota*: and *Veratrum album flore viridante*, of some *Sanguis Herculis*. The other is called *Elleborus albus præcox*, and *flore atrorubente*, or *atropurpurante*. We call the first in English, White Ellebor, Neesewort, or Neesing roote, because the powder of the roote is vsed to procure neesing; and I call it the greater, in regard of those in the next Chapter. The other hath his name according to the Latine title, most proper for it.

The Vertues.

The force of purging is farre greater in the roote of this Ellebor, then in the former; and therefore is not carelesly to bee vsed, without extreame danger; yet in contumacious and stubborne diseases it may bee vsed with
good

good caution and aduice. There is a Syrupe or Oxymel made hereof in the Apothecaries shops, which as it is dangerous for gentle and tender bodies, so it may be very effectuall in stronger constitutions. Pausanias *in Phocicis,* recordeth a notable stratagem that Solon vsed in besieging the Citie of Cirrheus, *viz.* That hauing cut off the riuer Plistus from running into the Citie, he caused a great many of these rootes to be put into a quantity thereof, which after they had steeped long enough therein, and was sufficiently infected therewith, he let passe into the Citie againe : whereof when they had greedily drunke, they grew so weake and feeble by the superpurgation thereof, that they were forced to leaue their wals vnmand, and nor guarded, whereby the Amphyctions their enemies became masters of their Citie. The like stratagems are set downe by diuers other Authors, performed by the helpe of other herbes.

Chap. LXXXIII.

Elleborine. Small or wilde white Ellebor.

THe likenesse of the leaues of these plants, rather then any other faculty with the former white Ellebor, hath caused them to be called *Elleborine,* as if they were smaller white Ellebors. And I for the same cause haue ioyned them next, whereof there are found many sorts : One which is the greater kinde, is of greatest beauty ; the other which are lesser differ not much one from another, more then in the colour of the flowers, whereof I will onely take three, being of the most beautie, and leaue the rest to another worke.

1. *Helleborine vel Elleborine maior, siue Calceolus Mariæ.*
Our Ladies Slipper.

This most beautifull plant of all these kindes, riseth vp with diuers stalkes, a foote and a halfe high at the most, bearing on each side of them broad greene leaues, somewhat like in forme vnto the leaues of the white Ellebor, but smaller and not so ribbed, compassing the stalke at the lower end ; at the tops of the stalkes come forth one, or two, or three flowers at the most, one aboue another, vpon small short foote-stalkes, with a small leafe at the foote of euery stalke : each of these flowers are of a long ouall forme, that is, more long then round, and hollow withall, especially at the vpper part, the lower being round and swelling like a belly : at the hollow part there are two small peeces like eares or flippets, that at the first doe couer the hollow part, and after stand apart one from another, all which are of a fine pale yellow colour, in all that I haue seene (yet it is said there are some found, that are more browne or tending to purple) there are likewise foure long, narrow, darke coloured leaues at the setting on of the flower vnto the stalke, wherein as it were the flower at the first standeth : the whole flower is of a pretty small sent : the seede is very small, very like vnto the seede of the *Orchides* or Satyrions, and contained in such like long pods, but bigger : the roots are composed of a number of strings enterlacing themselues one within another, lying within the vpper crust of the earth, & not spreading deep, of a darke brownish colour.

2. *Elleborine minor flore albo.*
The small or wilde white Ellebor with a white flower.

This smaller wilde white Ellebor riseth vp in the like manner vnto the former, and not much lower, bearing such like leaues, but smaller, and of a whiter greene colour, almost of the colour and fashion of the leaues of Lilly Conually ; the top of the stalke hath many more flowers, but lesser, growing together, spike-fashion, with small short leaues at the stalke of euery flower, which consisteth of fiue small white leaues, with a small close hood in the middle, without any sent at all : the seede and seede vessels are
like

like vnto the former, but smaller : the rootes are many small strings, dispersing them-selues in the ground.

3. *Elleborine minor flore purpurante.*
The small or wilde white Ellebor with blush flowers.

The leaues of this kinde are like vnto the last described, but somewhat narrower: the stalkes and flowers are alike, but smaller also, and of a pale purplish or blush co-lour, which causeth the difference.

The Place.

The first groweth in very many places of Germany, and in other Coun-tries also. It groweth likewise in Lancashire, neare vpon the border of Yorkeshire, in a wood or place called the Helkes, which is three miles from Ingleborough, the highest Hill in England, and not farre from Ingleton, as I am enformed by a courteous Gentlewoman, a great louer of these de-lights, called Mistris Thomasin Tunstall, who dwelleth at Bull-banke, neare Hornby Castle in those parts, and who hath often sent mee vp the rootes to London, which haue borne faire flowers in my Garden. The se-cond groweth in many places of England, and with the same Gentlewoman also before remembred, who sent me one plant of this kinde with the other. The last I haue not yet knowne to growe in England; but no doubt many things doe lye hid, and not obserued, which in time may bee discouered, if our Country Gentlemen and women, and others, in their seuerall places where they dwell, would be more carefull and diligent, and be aduertised either by themselues, or by others capable and fit to be imployed, as occa-sion and time might serue, to finde out such plants as growe in any the cir-cuits or limits of their habitations, or in their trauels, as their pleasures or affaires leade them. And because ignorance is the chiefe cause of neglect of many rare things, which happen to their view at sometimes, which are not to be seene againe peraduenture, or not in many yeares after, I would heartily aduise all men of meanes, to be stirred vp to bend their mindes, and spend a little more time and trauell in these delights of herbes and flowers, then they haue formerly done, which are not onely harmlesse, but pleasu-rable in their time, and profitable in their vse. And if any would be better enformed, and certified of such things they know not, I would be willing and ready to my best skill to aduertise them, that shall send any thing vp to me where I dwell in London. Thus farre I haue digressed from the matter in hand, and yet not without some good vse I hope, that others may make of it.

The Time.

The two first flower earlier then the last, and both the first about one time, that is, in the end of Aprill, or beginning of May. The last in the end of May, or in Iune.

The Names.

The first is called *Elleborine recentiorum maior*, and *Calceolus Mariae* : Of some thought to be *Cosmosandalos*, because it is *Sandali forma*. In English we call it our Ladies Slipper, after the Dutch name. The other two lesser kinds haue their names in their titles: I haue thought it fit to adde the title of small white Ellebors vnto these, for the forme sake, as is before said.

The Vertues.

There is no vse of these in Physicke in our dayes that I know.

CHAP.

Chap. LXXXIIII.

Lilium Conuallium. Lilly Conually.

The remembrance of the Conuall Lilly, spoken of in the precedent Chapter, hath caused me to insert these plants among the rest, although differing both in face and properties; but lest it should lose all place, let it keepe this. It is of two sorts, differing chiefly in the colour of the flowers, the one being white, and the other reddish, as shall be shewed in their descriptions following.

1. *Lilium Conuallium flore albo.* The white Lilly Conually.

The white Conuall or May Lilly, hath three or foure leaues rising together from the roote, one enclosed within another, each whereof when it is open is long and broad, of a grayish shining greene colour, somewhat resembling the leaues of the former wil ie Neesewort, at the side whereof, and sometime from the middle of them, riseth vp a small short naked foote-stalke, an hand breadth high or somewhat more, bearing at the toppe one aboue another many small white flowers, like little hollow bottles with open mouths, nicked or cut into fiue or six notches, turning all downewards one way, or on one side of the stalke, of a very strong sweete sent, and comfortable for the memory and senses, which turne into small red berries, like vnto Asparagus, wherein is contained hard white seede: the rootes runne vnder ground, creeping euery way, consisting of many small white strings.

2. *Lilium Conuallium flore rubente.* May Lillies with red flowers.

This other May Lilly differeth neither in roote, leafe, nor forme of flower from that before, but onely in the colour of the flower, which is of a fine pale red colour, being in my iudgement not altogether so sweet as the former.

The Place.

The first groweth aboundantly in many places of England. The other is a stranger, and groweth only in the Gardens of those that are curious louers of rarities.

The Time.

They both flower in May, and the berries are ripe in August.

The Names.

The Latines haue no other name for this plant but *Lilium Conuallium*, although some would haue it to be *Lilium vernum* of Theophrastus, and others *Oenanthe* of the same Author. Gesner thinketh it to be *Callionymus*. Lonicerus to be *Catalia*, and Fuchsius to be *Ephemerum non lethale*: but they are all for the most part mistaken. We call it in English Lilly Conually, May Lilly, and of some Liriconfancie.

The Vertues.

The flowers of the white kinde are often vsed with those things that help to strengthen the memory, and to procure ease to Apoplecticke persons. Camerarius setteth downe the manner of making an oyle of the flowers hereof, which he saith is very effectuall to ease the paines of the Goute, and such like diseases, to be vsed outwardly, which is thus: Hauing filled a glasse with the flowers, and being well stopped, set it for a moneths space in an Ants hill, and after being drayned cleare, set it by to vse.

Gg

Chap. LXXXV.

Gentiana. Gentian or Fell-wort.

THere are diuers forts of Gentians or Fell-wortes, fome greater, others leffer, and fome very fmall ; many of them haue very beautifull flowers, but becaufe fome are very fuddenly paft, before one would thinke they were blowne open, and others will abide no culture and manuring, I will onely fet forth vnto you two of the greater forts, and three of the leffer kindes, as fitteft, and more familiarly furnifhing our gardens, leauing the reft to their wilde habitations, and to bee comprehended in a generall Worke.

1. *Gentiana maior flore flauo.* The great Gentian.

The great Gentian rifeth vp at the firft, with a long, round and pointed head of leaues, clofing one another, which after opening themfelues, lye vpon the ground, and are faire, long and broad, fomewhat plaited or ribbed like vnto the leaues of white Ellebor or Neefeworte, but not fo fairely or eminently plaited, neyther fo ftiffe, but rather refembling the leaues of a great Plantane : from among which rifeth vp a ftiffe round ftalke, three foote high or better, full of ioynts, hauing two fuch leaues, but narrower and fmaller at euery ioynt, fo compaffing about the ftalke at the lower end of them, that they will almoft hold water that falleth into them : from the middle of the ftalke to the toppe, it is garnifhed with many coronets or rundles of flowers, with two fuch greene leaues likewife at euery ioynt, and wherein the flowers doe ftand, which are yellow, layd open like ftarres, and rifing out of fmall greenifh huskes, with fome threds in the middle of them, but of no fent at all, yet ftately to behold, both for the order, height and proportion of the plant : the feede is browne and flat, contained in round heads, fomewhat like vnto the feede of the *Fritillaria*, or checkerd Daffodill, but browner : the rootes are great, thicke and long, yellow, and exceeding bitter.

2. *Gentiana maior folio Afclepiadis.* Swallow-wort Gentian.

This kinde of Gentian hath many ftalkes rifing from the roote, neere two foote high, whereon grow many faire pale greene leaues, fet by couples, with three ribs in euery one of them, and doe fomewhat refemble the leaues of *Afclepias* or Swallow-wort, that is, broade at the bottome, and fharpe at the point : the flowers grow at the feuerall ioynts of the ftalkes, from the middle vpwards, two or three together, which are long and hollow, like vnto a bell flower, ending in fiue corners, or pointed leaues, and folded before they are open, as the flowers of the Bindeweedes are, of a faire blew colour, fometimes deeper, and fometimes paler : the heads or feede veffels haue two points or hornes at the toppes, and containe within them flat grayifh feed, like vnto the former, but leffe : the rootes hereof are nothing fo great as the former, but are yellow, fmall and long, of the bigneffe of a mans thumbe.

3. *Gentiana minor Cruciata.* Croffe-wort Gentian.

This fmall Gentian hath many branches lying vpon the ground, fcarce lifting themfelues vpright, and full of ioynts, whereat grow vfually foure leaues, one oppofite vnto another, in manner of a Croffe, from whence it tooke his name, in fhape very like vnto *Saponaria* or Sopewort, but fhorter, and of a darker greene colour: at the tops of the ftalkes ftand many flowers, thick thrufting together, and likewife at the next ioynt vnderneath, euery one of them ftanding in a darke blewifh greene huske, and confifting of fiue fmall leaues, the points or ends whereof only appeare aboue the huskes wherein they ftand, and are hardly to be feene, but that they are of a fine pale blew colour, and that many grow together : the feed is fmall and brown, hard, and fomewhat

like

1. *Lilium Convallium.* Lilliconstancy or Lilly Convally. 2 *Gentiana maior.* The great Gentian. 3 *Gentianella verna.* Small Gentian of the Spring. 4 *Gentiana cruciata.* Crossewort Gentian. 5 *Pneumonanthe seu Gentiana Autumnalis* Autumne Gentian. 6 *Saponaria flore duplici.* Double flowred Sopewort. 7 *Plantago Rosea.* Rose Platane.

like vnto the ſeed of the Marian Violets, or Couentry bels: the roots are ſmall and whitiſh, diſperſing themſelues diuerſly in the ground, of as bitter a taſte almoſt as the reſt.

4. *Gentianella Verna.* Small Gentian of the Spring.

The ſmall Gentian of the Spring hath diuers ſmall hard greene leaues, lying vpon
the ground, as it were in heads or tufts, ſomewhat broade below, and pointed at the
end, with fiue ribs or veines therein, as conſpicuous as in the former Gentians, among
which riſeth vp a ſmall ſhort ſtalke, with ſome ſmaller leaues thereon, at the toppe
whereof ſtandeth one faire, large, hollow flower, made bell faſhion, with wide open
brimmes, ending in fiue corners or diuiſions, of the moſt excellent deepe blew colour
that can be ſeene in any flower, with ſome white ſpots in the bottome on the inſide :
after the flower is paſt, there appeare long and round pods, wherein are contained
ſmall blackiſh ſeede : the rootes are ſmall, long, pale yellow ſtrings, which ſhoot forth
here and there diuers heads of leaues, and thereby encreaſe reaſonable well, if it finde
a fit place, and ground to grow, or elſe will not be nourſed vp, with all the care and diligence can be vſed: the whole plant is bitter, but not ſo ſtrong as the former.

5. *Gentiana Autumnalis ſiue Pneumonanthe.* Calathian Violet or Autumne Gentian.

This Gentian that flowreth in Autumne, hath in ſome places higher ſtalkes then in
others, with many leaues thereon, ſet by couples as in other Gentians, but long and
narrow, yet ſhewing the three ribbes or veines that are in each of them : the toppes of
the ſtalkes are furniſhed euery one with a flower or two, of an excellent blew purple
colour, ending in fiue corners, and ſtanding in long huskes : the rootes are ſomewhat
great at the top, and ſpreading into many ſmall yellow ſtrings, bitter as the reſt are.

6. *Saponaria flore duplici.* Double flowred Sopeworte.

Vnto theſe kindes of **Gentians**, I muſt needes adde theſe following plants, for that
the former is of ſome neere reſemblance in leafe with ſome of the former. And becauſe the ordinary Sopeworte or Bruiſeworte with ſingle flowers is often planted in
Gardens, and the flowers ſerue to decke both the garden and the houſe; I may vnder
the one deſcribe them both : for this with double flowers is farre more rare, and of
greater beautie. It hath many long and ſlender round ſtalkes, ſcarce able to ſuſtaine
themſelues, and ſtand vpright, being ful of ioynts and ribbed leaues at them, euery one
ſomewhat like a ſmall Gentian or Plantane leafe : at the toppes of the ſtalkes ſtand
many flowers, conſiſting of two or three rowes of leaues, of a whitiſh or pale purple
colour, and of a ſtrong ſweet ſent, ſomewhat like the ſmell of Iaſmin flowers, ſtanding
in long and thicke pale greene huskes, which fall away without giuing any ſeede, as
moſt other double flowers doe that encreaſe by the roote, which ſpreadeth within the
ground, and riſeth vp in ſundry diſtant places like the ſingle.

7 *Plantago Roſea.* Roſe Plantane.

This other plant is in all things like vnto the ordinary Plantane or Ribworte, that
groweth wilde abroade in many places, whoſe leaues are very large : but inſtead of
the long ſlender ſpike, or eare that the ordinary hath, this hath eyther a thicke long
ſpike of ſmall greene leaues vpon ſhort ſtalkes, or elſe a number of ſuch ſmall greene
leaues layd round-wiſe like vnto a Roſe, and ſometimes both theſe may be ſeene vpon
one and the ſame roote, at one and the ſame time, which abide a great while freſh vpon
the roote, and ſometimes alſo giueth ſeede, eſpecially from the more long and ſlender
ſpikes.

The Place.

Some of theſe Gentians grow on the toppes of hils, and ſome on the ſides
and foote of them in Germany and other Countreyes : ſome of them alſo
vpon barren heaths in thoſe places, as alſo in our owne Countrey, eſpecially

ly the Autumne Gentian, and as it is reported, the Vernall likewise. The single or ordinary Sopeworte is found wilde in many places with vs, but the double came to vs from beyond the Sea, and is scarce known or heard of in England. The Rose Plantaine hath beene long in England, but whether naturall thereof or no, I am not assured.

The Time.

They flower for the most part in Iune and Iuly, but the small Gentian of the Spring flowreth somewhat earlier, and that of the Autumne in August and September.

The Names.

Gentiana is the generall name giuen to the Gentians. We call them in English Gentian, Fellworte, Bitterwort, and Baldmoney. *Saponaria* taketh his name from the scouring qualitie it hath : Wee call it in English Sopewort, and in some places Bruisewort. Some haue thought it to bee *Struthium* of Dioscorides, or at least haue vsed it for the same causes, but therein they are greatly deceiued, as Matthiolus hath very well obserued thereon, and so is Dodonæus, that thought it to be *Alisma*. The Rose Plantaine is so called of the double spikes it catricth.

The Vertues.

The wonderfull wholsomnesse of Gentian cannot bee easily knowne to vs, by reason our daintie taltes refuse to take thereof, for the bitternesse sake : but otherwise it would vndoubtedly worke admirable cures, both for the liuer, stomacke and lunges. It is also a speciall counterpoison against any infection, as also against the violence of a mad dogges tooth : wilde Sopewort is vsed in many places, to scoure the countrey womens treen, and pewter vessels, and physically some make great boast to performe admirable cures in Hydropicall diseases, because it is diureticall, and in *Lue Veneria,* when other Mercuriall medicines haue failed. The Rose Plantaine no doubt hath the same qualities that the ordinary hath.

Chap. LXXXVI.

Campanula. Bell-flowers.

VNder the title of Bell-flowers are to bee comprehended in this Chapter, not only those that are ordinarily called *Campanula,* but *Viola Mariana,* and *Trachelium* also, whereof the one is called Couentry, the other Canterbury Bells.

1. *Campanula Persicifolio alba, vel cærulea.*
Peach-leafed Bell-flowers white or blew.

The Peach-leafed Bell-flower hath many tufts, or branches of leaues lying vpon the ground, which are long and narrow, somewhat like vnto the leafe of an Almond or Peach tree, being finely nicked about the edges, and of a sad greene colour, from among which rise vp diuers stalkes, two foote high or more, set with leaues to the middle, and from thence vpwards, with many flowers standing on seuerall small footestalkes, one aboue another, with a small leafe at the foote of euery one : the flowers stand in small greene huskes, being small and round at the bottome, but wider open at the brimme, and ending in fiue corners, with a three forked clapper in the middle, set about with some small threds tipt with yellow, which flowers in some plants are pure

white

white, and in others of a pale blew or watchet colour, hauing little or no sent at all : the seede is small, and contained in round flat heads, or seede vessels : the roote is very small, white and thteddy, creeping vnder the vpper crust of the ground, so that oftentimes the heat and drought of the Summer wil goe near to parch and wither it vtterly: it requireth therefore to be planted in some shadowie place.

2. *Campanula maior, siue Pyramidalis.*
The great or steeple Bell-flower.

This great Bell-flower hath diuers stalkes, three foote high or better, whereon grow diuers smooth, darke, greene leaues, broade at the bottome, and small at the point, somewhat vneuenly notched about the edges, and standing vpon longer footestalkes below then those aboue : the flowers are blew, and in some white, not so great or large as the former, but neare of the same fashion, growing thicker and more plentifully together, with smaller leaues among them, bushing thicke below, and rising smaller and thinner vp to the toppe, in fashion of a *Pyramis* or speere Steeple : the roote is thicke and whitish, yeelding more store of milke being broken (as the leaues and stalks also doe) then any other of the Bell-flowers, euery one whereof doe yeelde milke, some more and some lesse.

3. *Viola Mariana flore albido vel purpureo.*
Couentry Bels white or purple.

The leaues of Couentry Bels are of a pale or fresh greene colour, long, and narrow next vnto the bottome, and broader from the middle to the end, and somewhat round pointed, a little hairy all ouer, and snipt about the edges : the stalkes rise vp the yeare after the sowing, being somewhat hairy also, and branching forth from the roote, into diuers parts, whereon stand diuers leaues, smaller then the former, and of a darker greene colour : at the end of euery branch stand the flowers, in greene huskes, from whence come large, round, hollow Bels, swelling out in the middle, and rising somewhat aboue it, like the necke of a pot, and then ending in fiue corners, which are either of a faire or faint white, or of a pale blew purplish colour, and sometimes of a deeper purple or violet: after the flowers are past, there rise vp great square, or cornered seede vessels, wherein is contained in diuers diuisions, small, hard, shining, browne, flat seeds: the roote is white, and being young as in the first yeares sowing, is tender, and often eaten as other Rampions are ; but the next yeare, when it runneth vp to seede, it groweth hard, and perisheth : so that it is to be continued by euery other yeares sowing.

4. *Trachelium maius flore albo vel purpureo.*
Great Canterbury Bels white or purple.

The greater Canterbury Bels, or Throateworte, hath many large rough leaues, somewhat like vnto Nettle leaues, being broad and round at the bottome, and pointed at the end, notched or dented on the edges, and euery one standing on a long footstalk: among these leaues rise vp diuers square rough stalkes, diuided at the toppe into diuers branches, whereon grow the like leaues as grow below, but lesser; toward the ends of the branches stand the flowers, mixed with some longer leaues, euery one in his seuerall huske, which are hollow, long and round, like a bell or cup, wide open at the mouth, and cut at the brimme into fiue corners, or diuisions, somewhat lesser then the Couentry Bels, in some of a pure white, and others of a faire deepe purple violet colour, and sometimes paler : after the flowers are past, come smaller and rounder heades then in the former, containing flat seede, but blacker, and not so redde as the last : the roote is hard and white, disperfing it selfe into many branches vnder ground, not perishing euery yeare as the former (although it loseth all the leaues in winter) but abiding many yeares, and encreasing into diuers heades or knobs, from whence spring new leaues and branches.

5. *Trache-*

1 *Campanula persicifolia.* Peach leafed Bell-flower. 2 *Trachelium maius simplex.* Canterbury Bels. 4 *Trachelium flore duplici.* Double
Canterbury Bels. 3 *Viola Mariana.* Couentry Bels. 4 *Trachelium Giganteum.* Giants Throtewort. 5 *Trachelium minus.* The lesser
Throtewort. 6 *Trachelium Americanum siue Cardinalis planta.* The rich crimson Cardinals flower.

5. *Trachelium maius flore duplici albo & cæruleo.*
Canterbury Bels with double flowers both white and blew.

Of this kinde of Throateworte or Canterbury Bels, there is another fort, not differing in any thing from the former, but in the doubleneſſe of the flower : For there is of both the kindes, one that beareth double white flowers, and the other blew : Of each whereof I receiued plants from friends beyond the Sea, which grow well with me.

6. *Trachelium Giganteum flore purpurante.*
Pale purple Giants Throateworte.

This Bell-flower, although it hath a Gigantine name, yet did I neuer perceiue it in my Garden, to riſe vp higher then the former, the epithite beeing in my perſwaſion, only giuen for difference ſake : the leaues whereof are not ſo rough, but as large, and dented about the edges, ſomewhat larger pointed, and of a treſher greene colour : the ſtalkes beare ſuch like leaues on them, but more thinly or diſperſedly ſet, hauing a flower at the ſetting on of euery one of the leaues, from the middle vpwards, and are ſomewhat like the great Throateworte in forme, but of a pale or bleake reddiſh purple colour, turning the brims or corners a little backwards, with a forked clapper in the middle, ſufficient eminent and yellow : the ſeede hereof is white, and plentifull in the heads, which will abide all the winter vpon the ſtalkes, vntill all the ſeede being ſhed, the heads remaining ſeeme like torne rags, or like thin peeces of skin, eaten with wormes : the roote is great, thicke and white, abiding long without periſhing.

Flore albo. There is another which differeth not any thing but in the flower, which is white.

7. *Trachelium minus flore albo & purpureo.*
Small Throateworte or Canterbury Bells both white and purple.

The leſſer Throateworte hath ſmaller leaues, nothing ſo broade or hard as the former great kinde, but long, and little or nothing dented about the edges : the ſtalkes are ſquare and browniſh, if it beare purple flowers, and greene if it beare white flowers, which in forme are alike, and grow in a buſh or tuft, thicke ſet together, more then any of the former, and ſmaller alſo, being not much bigger then the flowers of the fielde, or garden Rampions : the roote is laſting, and ſhooteth afreſh euery yeare.

8. *Trachelium Americanum flore ruberrimo, ſiue Planta Cardinalis.*
The rich crimſon Cardinals flower.

This braue plant, from a white roote ſpreading diuers wayes vnder ground, ſendeth forth many greene leaues, ſpread round about the head thereof, each whereof is ſomewhat broade and long, and pointed at the end, finely alſo ſnipt about the edges : from the middle whereof ariſeth vp a round hollow ſtalke, two foote high at the leaſt, beſet with diuers ſuch leaues as grow below, but longer below then aboue, and branching out at the toppe aboundantly, euery branch bearing diuers greene leaues on them, and one at the foote of euery of them alſo, the toppes whereof doe end in a great large tuft of flowers, with a ſmall greene leafe at the foote of the ſtalke of euery flower, each footeſtalke being about an inch long, bearing a round greene huske, diuided into fiue long leaues or points turned downwards, and in the midſt of euery of them a moſt rich crimſon coloured flower, ending in fiue long narrow leaues, ſtanding all of them foreright, but three of them falling downe, with a long vmbone ſet as it were at the backe of them, bigger below, and ſmaller aboue, and at the toppe a ſmall head, being of a little paler colour then the flower, but of no ſent or ſmell at all, commendable only for the great buſh of ſo orient red crimſon flowers : after the flowers are paſt, the ſeede commeth in ſmall heads, cloſed within thoſe greene husks that held the flowers, which is very like vnto the ſeede veſſels of the *Viola Mariana,* or Couentry Bels, and is ſmall and browniſh.

The

The Place.

All thefe Bell-flowers do grow in our Gardens, where they are cherifhed for the beautie of their flowers. The Couentry Bels doe not grow wilde in any of the parts about Couentry, as I am credibly informed by a faithfull Apothecary dwelling there, called Mafter Brian Ball, but are nourfed in Gardens with them, as they are in other places. The laft groweth neere the riuer of Canada, where the French plantation in America is feated.

The Time.

They flower from May vntill the end of Iuly or Auguft, and in the mean time the feed is ripe : But the Peache-leafed Bell-flowers, for the moft part, flower earlier then the other.

The Names.

The firft is generally called *Campanula Perficifolia*, in Englifh Peach-leafed Bell-flower. The fecond is called *Campanula maior, Campanula lactefcens Pyramidalis*, and *Pyramidalis Lutetiana* of Lobel, in Englifh, Great or Steeple Bell-flower. The third is vfually called *Viola Mariana*, and of fome *Viola Marina*. Lobel putteth a doubt whether it be not *Medium* of Diofcorides, as Matthiolus and others doe thinke ; but in my opinion the thicknefle of the roote, as the text hath it, contradicteth all the reft. We call it generally in Englifh Couentry Bels. Some call it Marian, and fome Mercuries Violets. The fourth and fift are called *Trachelium* of *Ceruicaria*, of fome *Vuularia*, becaufe many haue vfed it to good purpofe, for the paines of the *Vuula*, or Throate : Yet there is another plant, called alfo by fome *Vuularia*, which is *Hippogloffum*, Horfe tongue, or Double tongue. The fixt hath his title to defcipher it out fufficiently, as is declared. The feuenth is called *Trachelium minus*, and *Ceruiaria minor*, of fome *Saponaria altera* ; in Englifh, Small Throateworte, or Small Canterbury Bels. The laft hath his name in the title, as it is called in France, from whence I receiued plants for my Garden with the Latine name : but I haue giuen it in Englifh.

The Vertues.

The Peach-Bels as well as the others may fafely bee vfed in gargles and lotions for the mouth, throate, or other parts, as occafion ferueth. The rootes of many of them, while they are young, are often eaten in fallets by diuers beyond the Seas.

Chap. LXXXVII.

Campana Cærulea fiue Convolvulus Cæruleus.
Blew Bell flowers, or blew Bindeweede.

There are two other kindes of Bell-flowers, much differing from the Tribe or Familie of the former, becaufe of their climbing or winding qualitie, which I muft needes place next them, for the likenefle of the flowers, although otherwife they might haue beene placed with the other clamberers that follow. Of thefe there is a greater, and a leffer, and of each likewife fome difference, as fhall be declared.

I. *Com-*

1. *Convolvulus caruleus maior rotundifolius.*
The greater blew Bindweede, or Bell-flower with round leaues.

This goodly plant riseth vp with many long and winding branches, whereby it climbeth and windeth vpon any poles, herbes, or trees, that stand neare it within a great compasse, alwaies winding it selfe contrary to the course of the Sunne : on these branches doe growe many faire great round leaues, and pointed at the end, like vnto a Violet leafe in shape, but much greater, of a sad greene colour : at the ioynts of the branches, where the leaues are set, come forth flowers on pretty long stalkes, two or three together at a place, which are long, and pointed almost like a finger, while they are buds, and not blowne open, and of a pale whitish blew colour, but being blowne open, are great and large bels, with broad open mouths or brims ending in fiue corners, and small at the bottome, standing in small greene huskes of fine leaues : these flowers are of a very deepe azure or blew colour, tending to a purple, very glorious to behold, opening for the most part in the euening, abiding so all the night and the next morning, vntill the Sunne begin to growe somewhat hot vpon them, and then doe close, neuer opening more : the plant carrieth so many flowers, if it stand in a warme place, that it will be replenished plentifully, vntill the cold ayres and euenings stay the luxury thereof : after the flowers are past, the stalkes whereon the flowers did stand, bend downwards, and beare within the huskes three or foure blacke seedes, of the bignesse of a Tare or thereabouts : the rootes are stringy, and perish euery yeare.

2. *Convolvulus trifolius siue hederaceus purpureus.*
The greater purple Bindeweede, or Bell-flower with cornered leaues.

The growing and forme of this Bindeweede or Bell-flower, is all one with the former, the chiefest differences consisting in the forme of the leafe, which in this is three cornered, like vnto an Iuie leafe with corners ; and in the flower, which is of a deeper blew, tending more to a deepe purple Violet, and somewhat more reddish in the fiue plaites of each flower, as also in the bottomes of the flowers.

3. *Convolvulus tennifolius Americanus.* The red Bell-flower of America.

Although this rare plant (because wee seldome haue it, and can as hardly keepe it) be scarce knowne in these cold Countries, yet I could not but make mention of it, to incite those that haue conueniencie to keepe it, to be furnished of it. It springeth vp at the first from the seede with two leaues, with two long forked ends, which abide a long time before they perish, betweene which riseth vp the stalke or stemme, branching forth diuers waies, being of a brownish colour, which windeth it selfe as the former great Bell-flower doth, whereon are set at seuerall ioynts diuers winged leaues, that is to say, many small narrow and long leaues set on both sides of the middle ribbe, and one at the end : from these ioynts arise long stalkes, at the ends whereof stand two or three small, long, hollow flowers, fashioned very like vnto the flowers of a Bindeweede, or the flowers of Tabacco, and ending in the like manner in fiue points, but not so much laid open, being of a bright red colour, plaited as the Bindeweedes or Bell-flowers before they be open, with some few threads in the middle, which turne into long pointed cods, wherein is contained long and blacke seede, tasting hot like Pepper : the roote is small and stringy, perishing euery yeare, and with vs will seldome come to flower, because our cold nights and frosts come so soone, before it cannot haue comfort enough of the Sun to ripen it.

4. *Convolvulus caruleus minor Hispanicus.*
The Spanish small blew Bindeweede.

This small Bindeweede hath small long leaues, somewhat broader then the next that followeth, and not so broad as the common small Bindeweede (that groweth
euery

euery where wilde on the bankes of fields abroad) let vpon the small trayling branches, which growe aboue two or three foote high : from the middle of thefe branches, and fo vnto the toppes of them, come forth the flowers at the ioynts with the
leaues, folded together at the firft into fiue plaites, which open into fo many corners,
of a most excellent faire skie coloured blew (fo pleafant to behold, that often it ama-
zeth the fpectator) with white bottomes, and yellowifh in the middle, which turne
into small round white heads, wherein are contained small blackifh cornered feede,
fomewhat like the former, but fmaller : the roote is small and threddy, perifhing as
the former euery yeare : this neuer windeth it felfe about any thing, but leaneth by
reafon of the weakneffe of the branches, and dyeth euery yeare after feede time, and
not to be fowne againe vntill the next Spring.

5. *Convolvulus purpureus Spicafolius*. Lauander leafed Bindeweede.

This small purple Bindeweede, where it naturally groweth, is rather a plague then
a pleafure, to whatfoeuer groweth with it in the fields ; yet the beauty of the flower
hath caufed it to be receiued into Gardens, bearing longer and fmaller leaues then the
laft, and fuch like small Bell-flowers, but of a fad purple colour : the roote is liuing,
as the common kinds are, and fpringeth againe where it hath been once fowne, without feare of perifhing.

The Place.

The firft two greater kindes haue beene fent vs out of Italy, but whether
they had them from the Eaft Indies, or from fome of the Eafterne Countries on this fide, wee know not : but they thriue reafonable well in our
Country, if the yeare be any thing kindly. The next came out of America,
as his name teftifieth. The leffer blew kinde groweth naturally in many
places both of Spaine and Portugall (from whence I firft receiued feedes
from Guillaume Boel, heretofore remembred.) The laft groweth wilde in
the fields, about Dunmowe in Effex, and in many other places of our
owne Countrey likewife.

The Time.

The three firft greater kindes flower not vntill the end of Auguft, or
thereabouts, and the feede ripeneth in September, if the colds and frofts
come not on too fpeedily. The leffer kindes flower in Iune and Iuly.

The Names.

The firft is called of fome *Campana Lazura*, as the Italians doe call it, or
Campana cærulea, of others *Convolvulus cæruleus maior, fiue Indicus*, and *Flos
noctis*. Of fome *Nil Auicenna*. The fecond is called *Convolvulus trifolius*,
or *hæderaceus*, for the diftinction of the leaues. In Englifh wee call them eyther Great blew Bell flowers, or more vfually, Great blew Bindeweedes.
That of America is diuerfly called by diuers. It is called *Quamoclit* of the
Indians, and by that name it was fent to Ioachinus Camerarius out of Italy,
where it is fo called ftill, as Fabius Columna fetteth it downe, and as my
felfe alfo can witneffe it, from thence being fo fent vnto mee : but Andræas
Cæfalpinus calleth it, *Iafminum folio Millefolij*, fuppofing it to be a Iafmine.
Camerarius faith, it may not vnfitly be called *Convolvulus tenuifolius*, accounting it a kinde of Bindeweede. Columna entituleth it *Convolvulus pennatus exoticus rarior*, and faith it cannot bee referred to any other kinde of
plant then to the Bindeweedes. Hee that publifhed the *Curæ pofteriores* of
Clufius, giueth it the name of *Iafminum Americanum*, which I would doe
alfo, if I thought it might belong to that Family ; but feeing the face and
forme of the plant better agreeing with the Bindeweedes or Bell-flowers,

I haue

I haue (as you fee) inferted it among them , and giuen it that name may bee
moft fit for it, efpecially becaufe it is but an annuall plant. The leffer kindes
haue their names fufficiently expreffed in their titles.

The Vertues.

We know of no vfe thefe haue in Phyficke with vs , although if the firft
be *Nil* of Auicen, both he and Serapio fay it purgeth ftrongly.

Chap. LXXXVIII.

Stramonium. Thorne-Apple.

VNto the Bell-flowers, I muft adioyne three other plants , in the three feuerall
Chapters following, for fome affinity of the flowers : and firft of the Thorne-
Apples, whereof there are two efpeciall kindes , that is, a greater and a leffer,
and of each fome diuerfity, as fhall be fet downe.

1. *Stramonium maius album.* The great white flowred Thorne-Apple.

The greater Thorne-Apple hath a great, ftrong, round greene ftalke, as high as any
man, if it be planted in good ground, and of the bigneffe of a mans wreft almoft at the
bottome, fpreading out at the toppe into many branches , whereon ftand many very
large and broad darke greene leaues, cut in very deeply on the edges, and hauing manie points or corners therein : the flowers come forth at the ioynts , betweene two
branches towards the toppe of them, being very large, long, and wide open , ending
in fiue points or corners, longer and larger then any other Bell-flowers whatfoeuer:
after the flowers are paft, come the fruit, which are thorny long heads, more prickly
and greene then the leffer kindes, which being ripe openeth it felfe into three or foure
parts , hauing a number of flat blackifh feede within them : the roote is aboundant in
fibres, whereby it ftrongly taketh hold in the ground, but perifheth with the firft
frofts ; yet the feede that is fhed when the fruit is ripe, commeth vp the next yeare.

2. *Stramonium maius purpureum.* The great purple flowred Thorne-Apple.

This purple Thorne-Apple is in largeneffe of leaues, thickneffe and height of ftalke,
greatneffe and forme of flowers and fruit , euery way equall and correfpondent vnto
the former, the chiefe differences be thefe : the ftalke is of a darke purple colour ; the
leaues are of a darker greene, fomewhat purplifh , and the flowers are of light purple
or pale Doue colour, enclining to white, and whiter at the bottome.

3. *Stramonium minus feu Nux Metel flore albo.* The fmaller Thorne-Apple with a white flower.

The fmaller Thorne-Apple rifeth vp with one round ftalke, of the bigneffe of a
mans finger, and neuer much aboue two foote high with vs, bearing a few large, broad,
fmooth leaues thereon, without any branches at all, which are vneuenly rent or torne
about the edges, with many ribs, and fmaller veines running through them, yet leffer
by much then the greater kinde : at the ioynts where the leaues ftand, come forth
long and large white flowers, with broad or wide open brims, folded together before
their opening, as the other former Bell-flowers or Bindeweedes, but hauing their fiue
corners more pointed or horned then either they, or the former Thorne-Apples : after the flowers are paft, fucceed fmall fruit, rounder and harder , fet with harder , but
blunt prickes then the former, wherein is contained brownifh yellow flat feede,
 	 ficking

1 *Convolvulus maior cœruleus.* The greater blew Bindweed or Bell-flower. 2 *Convolvulus trifolius seu hederaceus* The great purple Bindweed. 3 *Convolvulus cœruleus Hispanicus.* The Spanish small blew Bindweed. 4 *Stramonium marinum seu Pomum spinosum.* The great Thorne Apple. 5 *Datura seu Stramonium fore duplici.* The double flowred Thorne Apple. 7 *Stramonium fore geminato.* Double Thornae-Apple one out of another 8 *Tabacco latifolium,* Broad leafed Tabacco. 9 *Mirabilis Peruana.* The Meruaile of the world.

H h

sticking to the inward pulpe : the roote is not very great, but full of strings, and quickly perisheth with the first frosts.

4. *Stramonium minus flore geminato purpurante.*
The small double flowred purple Thorne-Apple.

In the flower of this plant, consisteth the chiefest difference from the former, which is as large as the last, pointed into more hornes or corners, and beareth two flowers, standing in one huske, one of them rising out from the middle of the other, like vnto those kindes of Cowslips and Oxelips, called double, or Hose in hose, before described, which are of a pale purplish colour on the outside, and almost white within : the fruit is round like the last, and beareth such like seede, so that vntill it bee in flower, their difference can hardly bee discerned : this is more tender then the last, although euen it is so tender, that it seldome beareth ripe seede with vs.

Flore duplici. Sometimes (for I think it is not another kind) the flower will haue as it were double rowes of leaues, close set together, and not consisting of two, rising so distinctly one aboue another.

The Place.

All these kindes haue been brought or sent vs out of Turkie and Egypt; but Garcias, and Christopherus Acosta, with others, affirme that they grow in the East Indies. The lesser kindes are very rare with vs, because they seldome come to maturity; and therefore we are still to seeke of new seede to sowe. The greater kindes are plentifull enough in our Gardens, and will well abide, and giue ripe fruit.

The Time.

The smaller kindes flower later then the greater; and therefore their fruit are the sooner spoiled with the cold ayres, dewes, and frosts, that come at the latter end of the yeare : but the greater kinds neuer misse lightly to ripen.

The Names.

Both the greater and smaller kindes are generally called *Stramonium, Stramonia, Pomum spinosum*, and *Datura.* Bauhinus vpon Matthiolus his Comentaries on Dioscorides, calleth it *Solanum fœtidum spinosum.* Some learned men haue referred it to *Nux Metel*, of the Arabian Authors. Wee call them generally in English, Thorne-Apples, and distinguish them by their titles of greater and lesser, single and double.

The Vertues.

The East Indian lasciuious women performe strange acts with the seed (of the smaller kinde, as I suppose, or it may be of either) giuing it their husbands to drinke. The whole plant, but especially the seed, is of a very cold and soporiferous quality, procuring sleep and distraction of senses. A few of the seeds steeped and giuen in drinke, will cause them that take it to seem starke drunke or dead drunke, which fit will within a few houres weare away, and they recouer their senses againe, as a drunken man raysed after sleep from his wine. It may therefore (in my opinion) be of safe and good vse to one, that is to haue a legge or an arme cut off, or to be cut for the stone, or some other such like cure to be performed, to take away the sense of paine for the time of doing it; otherwise I hold it not fit to be vsed without great caution. But the greene leaues of the greater kindes (as also of the lesser, but that with vs they are not so plentifull) are by tryed experience, found to be excellent good for any scalded or burned part, as also to take away any hot inflammations, being made vp into a salue or ointment with suet, waxe, and rossin, &c. or with *Axungia*, that is, Hogs larde.

Chap. LXXXIX.

Tabacco. Indian Henbane, or Tabacco.

There hath beene formerly but three kindes of Tabacco knowne vnto vs, two of them called Indian, and the third Englifh Tabacco. In thefe later yeares, we haue had in our gardens about London (before the fuppreffing of the planting) three or foure other forts at the leaft, and all of the Indian kinde, hauing fome efpeciall difference, eyther in leafe, or flower, or both : And in regard the flowers of fome of thefe carry a pretty fhew, I fhall only entreate of them, and not of the Englifh kind.

Tabacco latifolium. Broade leafed Tabacco.

The great Indian Tabacco hath many very large, long, thicke, fat and faire greene leaues, ftanding foreright for the moft part, and compaffing the ftalkes at the bottome of them, being fomewhat pointed at the end : the ftalke is greene and round, fixe or feuen foote high at fometimes, and in fome places, in others not paft three or foure foote high, diuided towards the toppe into many branches, with leaues at euery ioynt, and at the toppes of the branches many flowers, the bottomes hereof are long and hollow, and the toppes plaited or folded before they are open, but being open, are diuided fometimes into foure, or more vfually into fiue corners, fomewhat like vnto other of the Bell-flowers, but lying a little flatter open, of a light carnation colour. The feede is very fmall and browne, contained in round heads, that are clammy while they are greene, and pointed at the end : the roote is great, whitifh, and woody at the head, difperfing many long branches, and fmall fibres vnder the ground, whereby it is ftrongly faftened, but perifheth with our violent frofts in the winter, if it be left abroad in the garden, but if it be houfed, or fafely prouided for againft the froftes, the rootes will liue, and fpring afrefh the next yeare.

There is of this kinde another fort, whofe leaues are as large and long as the former, but thicker, and of a more dead greene colour, hanging downe to the ground-ward, and fcarce any ftanding forth-right, as the former, vnleffe they bee very young : the flowers of this kinde are almoft whole, without any great fhew of corners at the brims or edges, in all other things there is no difference.

There is another, whofe large and thicke flat leaues doe compaffe the ftalke at the bottome, and are as it were folded together one fide vnto another : the flowers are of a deeper blufh, or carnation colour, and with longer points and corners then in any of the former ; and in thefe two things confifteth the difference from the others, and is called Verines Tabacco.

Another hath his leaues not fo large and long as the firft, and thefe haue fhort footeftalkes, whereon they ftand, and doe not compaffe the ftalke as the other doe : the flower hereof is like the firft, but fmaller, and of a little paler colour.

Tabacco angustifolium. Narrow leafed Tabacco.

This kinde of Tabacco hath fomewhat lower, and fmaller ftalkes, then any of the former : the leaues hereof are fmaller and narrower, and not altogether fo thicke, but more pointed, and euery one ftanding vpon a footftalke, an inch and a halfe long at the leaft : the flowers hereof ftand thicker together, vpon the fmall branches, fomewhat larger, of a deeper blufh colour, and more eminent corners then in any the former : the feed and roots are alike, and perifh in like manner, vnleffe it be brought into a cellar, or other fuch couert, to defend it from the extremitie of the Winter.

The Place.

America or the Weft Indies is the place where all thefe kindes doe grow naturally, fome in one place, and fome in another, as in Peru, Trinidado,

Hifpani-

Hispaniola, and almost in euery Iland and Countrey of the continent thereof: with vs they are cherished in gardens, as well for the medicinable qualities, as for the beauty of the flowers.

The Time.

It flowreth in August, seldome before, and the seede is ripe quickly after. If it once sowe it selfe in a Garden, it will giue next year after young plants: but for the most part they will spring vp late, and therefore they that would haue them more early, haue sowen the seede vpon a bed of dung, and transplanted them afterwards.

The Names.

This plant hath gotten many names. The Indians call it in some places *Petum*, in others *Picielt*, and *Perebecenuc*, as Ouiedus and others doe relate. The Spaniards in the Indies first called it *Tabacco*, of an Iland where plenty of it grew. It hath in Christendome receiued diuers other names, as *Nicotiana*, of one Nicot a French man, who seeing it in Portugall, sent it to the French Queene, from whom it receiued the name of *Herba Regina*. Lobel calleth it *Sancta herba, & Sana sancta Indorum*. Some haue adiudged it to be an *Hioscyamus*, and therefore call it *Peruvianus*. The most vsuall name wherby we call it in English, is Tabacco.

The Vertues.

The herbe is, out of question, an excellent helpe and remedy for diuers diseases, if it were rightly ordered and applyed, but the continuall abuse thereof in so many, doth almost abolish all good vse in any. Notwithstanding if men would apply their wits to the finding out of the vertues, I make no doubt but many strange cures would bee performed by it, both inward and outward. For outward application, a Salue made hereof (as is before recited of the Thorne apple leaues) cureth vlcers, and wounds of hard curation: And for inward helpes, a Syrupe made of the iuice and sugar, or honey, procureth a gentle vomit (but the dryed leafe infused in wine much more) and is effectuall in astmaticall diseases, if it bee carefully giuen. And likewise cleanseth cankers and fistulaes admirably, as hath beene found by late experience. The ashes of Tabacco is often vsed, and with good successe, for cuts in the hands, or other places, and for other small greene wounds.

Chap. XC.

Mirabilia Peruviana. The Meruaile of Peru.

THis plant yeeldeth in our Gardens fiue or sixe seuerall varieties of beautifull flowers, as pure white, pure yellow, pure red, white and red spotted, and red and yellow spotted. But besides these, I haue had some other sorts, among which was one, of a pale purple or peach colour: all which, comming vnto mee out of Spaine with many other, seedes in an vnkindly yeare (an early winter following a cold summer) perished with mee; yet I plainely might discerne by their leaues, and manner of growing, to be diuers from them that we now haue and keepe. I shall need therefore (because the chiefest difference consisteth almost in the flowers) to giue only one description of the plant, and therein shew the varieties as is before declared.

Admirabilis. The Meruaile of the World.

The stalke of this meruellous plant is great and thick, bigger then any mans thumbe, bunched out or fwelling at euery ioynt, in fome the ftalkes will bee of a faire greene colour, and thofe will bring white, or white and red flowers : in others they will bee reddifh, and more at the ioynts, and thofe giue red flowers ; and in fome of a darker greene colour, which giue yellow flowers ; the ftalkes and ioynts of thofe that will giue red and yellow flowers fpotted, are fomewhat brownifh, but not fo red as thofe that giue wholly red flowers : vpon thefe ftalkes that fpread into many branches, doe grow at the ioynts vpon feuerall footeftalkes, faire greene leaues, broad at the ftalke, and pointed at the end : at the ioynts likewife toward the vpper part of the branches, at the foote of the leaues, come forth feuerall flowers vpon fhort footeftalkes, euery one being fmall, long and hollow from the bottome to the brimme, which is broade fpread open, and round; and confift but of one leafe without diuifion, like vnto a Bell flower, but not cornered at all : which flowers, as I faid, are of diuers colours, and diuerfly marked and fpotted, fome being wholly white, without any fpot in them for the moft part, through all the flowers of the plant ; fo likewife fome being yellow, and fome wholly red ; fome plants againe being mixed and fpotted, fo variably either white and red or purple, (except here and there fome may chance to be wholly white, or red or purple among the reft) or red and yellow through the whole plant, (except as before fome may chance in this kinde to be eyther wholly red, or wholly yellow) that you fhall hardly finde two or three flowers in a hundred, that will bee alike fpotted and marked, without fome diuerfitie, and fo likewife euery day, as long as they blow, which is vntill the winters, or rather autumnes cold blaftes do ftay their willing pronenefle to flower : And I haue often alfo obferued, that one fide of a plant will giue fairer varieties then another, which is moft commonly the Eafterne, as the more temperate and fhadowie fide. All thefe flowers doe open for the moft part, in the euening, or in the night time, and fo ftand blowne open, vntill the next mornings fun beginne to grow warme vpon them, which then clofe themfelues together, all the brims of the flowers fhrinking into the middle of the long necke, much like vnto the blew Bindeweede, which in a manner doth fo clofe vp at the funnes warme heate : or elfe if the day be temperate and milde, without any funne fhining vpon them, the flowers will not clofe vp for the moft part of that day, or vntill toward night : after the flowers are paft, come feuerall feedes, that is, but one at a place as the flowers ftood before, of the bigneffe (fometimes) of fmall peafe, but not fo round, ftanding within the greene huskes, wherein the flowers ftood before, being a little flat at the toppe, like a crowne or head, and round where it is faftened in the cup, of a blacke colour when it is ripe, but elfe greene all the while it groweth on the ftalke, and being ripe is foone fhaken downe with the wind, or any other light fhaking : the roote is long and round, greater at the head, and fmaller downwards to the end, like vnto a Reddifh, fpreading into two or three, or more branches, blackifh on the outfide and whitifh within. Thefe rootes I haue often preferued by art a winter, two or three (for they will perifh if they be left out in the garden, vnleffe it be vnder a houfe fide) becaufe many times, the yeare not falling out kindely, the plants giue not ripe feede, and fo we fhould be to feeke both of feede to fow, and of rootes to fet, if this or the like art to keep them, were not vfed ; which is in this manner : Within a while after the firft frofts haue taken the plants, that the leaues wither and fall, digge vp the rootes whole, and lay them in a dry place for three or foure dayes, that the fuperfluous moyfture on the outfide, may be fpent and dryed, which done, wrap them vp feuerally in two or three browne papers, and lay them by in a boxe, cheft or tub, in fome conuenient place of the houfe all the winter time, where no winde or moift ayre may come vnto them; and thus you fhall haue thefe rootes to fpring a frefh the next yeare, if you plant them in the beginning of March, as I haue fufficiently tryed. But fome haue tryed to put them vp into a barrell or firkin of fand, or afhes, which is alfo good if the fand and afhes be thorough dry, but if it bee any thing moift, or if they giue againe in the winter, as it is vfuall, they haue found the moifture of the rootes, or of the fand, or both, to putrefie the rootes, that they haue beene nothing worth, when they haue taken them forth. Take this note alfo for the fowing of your feede, that if you would haue variable flowers, and not all of one colour, you muft choofe out fuch flowers as be variable while they grow, that you may haue the feede of them : for if the flowers bee of one entire colour, you fhall haue for the moft part from thofe feedes, plants that will bring flowers all of that colour, whether it be white, red or yellow.

The

The Place.

These plants grow naturally in the West Indies, where there is a perpetuall summer, or at the least no cold frosty winters, from whence the seede hath been sent into these parts of Europe, and are dispersed into euery garden almost of note.

The Time.

These plants flower from the end of Iuly sometimes, or August, vntill the frosts, and cold ayres of the euenings in October, pull them down, and in the meane time the seed is ripe.

The Names.

Wee haue not receiued the seedes of this plant vnder any other name, then *Mirabilia Peruuiana,* or *Admirabilis planta.* In English wee call them, The meruaile of Peru, or the meruaile of the world : yet some Authors haue called it *Gelseminum,* or *Iasminum rubrum, & Indicum* : and Bauhinus *Solanum Mexicanum flore magno.*

The Vertues.

We haue not knowne any vse hereof in Physicke.

Chap. XCI.

Malua. Mallowes.

OF the kindred of Mallowes there are a great number, some of the gardens, others wilde, some with single flowers, others with double, some with whole leaues, others with cut or diuided : to entreate of them all is not my purpose, nor the scope of this worke, but onely of such whose flowers, hauing beautie and respect, are fit to furnish this garden, as ornaments thereunto. And first of those single kindes, whose flowers come neerest vnto the fashion of the former Bell-flowers, and after to the double ones, which for their brauery, are entertained euery where into euery Countrey womans garden.

1. *Malua Hispanica flore carneo amplo.*
The Spanish blush Mallow.

The Spanish Mallow is in forme and manner of growing, very like vnto our common fielde Mallow, hauing vpright stalkes two or three foote high, spread into diuers branches, and from the bottome to the toppe, beset with round leaues, like vnto our Mallowes, but somewhat smaller, rounder, and lesse diuided, yet larger below then aboue : the flowers are plentifully growing vpon the small branches, folding or writhing their leaues one about another before they bee blowne, and being open consist of fiue leaues, with a long forked clapper therein, of the same colour with the flower : the chiefest difference from the common consisteth in this, that the leaues of these flowers are longer, and more wide open at the brimmes (almost like a Bell-flower) and of a faire blush or light carnation colour, closing at night, and opening all the day : after the flowers are past, there come such like round heads, with small blacke seede, like vnto the common kinde, but somewhat smaller : the roote is small and long, and perisheth euery yeare.

2. *Alcea vulgaris flore carneo.* Veruaine Mallow with blush flowers.

There is a Mallow that hath long stalkes, and flowers like vnto the common wilde

Mal-

1 *Malua Hispanica flore carneo amplo.* The Spanish Mallow. 2 *Alcea Veneta.* The Venicee Mallow. 3 *Alcea Americana.* Thorney Mallow. 4 *Alcea Ægyptia.* The Mallow of Egypt. 5 *Altheat frutex.* The shrubbe Mallow. 6 *Malua horiensis simplex.* Single Hollihockes. 7 *Malua rosea multiplex.* Double Hollihockes.

Mallow, and of the fame deepe colour with it, so that you can hardly know it from the ordinary kinde, which is found growing wilde together with it, but onely by the leafe, which is as round and as large as the former, but cut into many fine diuisions, euen to the stalke that vpholdeth it, that it seemeth to consist onely of ragges, or peeces of leaues : Of this kinde I take a plante for this garden, growing in all respects like vnto it, but differing onely in the colour of the flowers, which are of the same blush or light carnation colour, or not much differing from the former Spanish kinde, with some veines therein of a deeper colour: the root hereof liueth, as the root of the common wilde kinde doth.

3. *Alcea peregrina siue vesicaria.* Venice Mallow, or Good night at noone.

The Venice Mallow hath long and weake stalkes, most vsually lying or leaning vpon the ground, hauing here and there vpon them long leaues and somewhat broad, cut in or gashed very deepely on both edges, that it seemeth as if they were diuers leaues set together, euery one standing on a long footestalke : at the ioynts of these stalkes, where the leaues are set, come forth seuerall flowers, standing vpon long footstalkes, which are somewhat larger then any of the former flowers, consisting of fiue leaues, small at the bottome, and wide at the brimmes, of a whitish colour tending to a blush, and sometimes all white, with spots at the bottomes of the leaues on the inside, of a very deepe purple or murrey colour, which addeth a great grace to the flower, and hauing also a long pestle or clapper in the middle, as yellow as gold : these flowers are so quickly faded and gone, that you shall hardly see any of them blowne open, vnlesse it bee betimes in the morning before the Sunne doe grow warme vpon them, for as soone as it feeleth the Sunnes warme heate, it closeth vp and neuer openeth againe, so that you shall very seldome see a flower blowne open in the day time, after nine a clocke in the morning : after these flowers are past, there rise vp in their places thinne, round, shining or transparent bladders, pointed at the toppe, and ribbed down all along, wherein are contained small, round, blackish seede : the roote is long and small, and perisheth euery yeare.

4. *Alcea fruticosa pentaphyllea.* Cinquefoile Mallow.

The stalkes of this Mallow are very long, hard or wooddy, more then of any of the other Mallowes : at the lower part whereof, and vp to the middle, stand diuers leaues vpon long footestalkes, parted or diuided into fiue parts or leaues, and dented about the edges ; but vpwards from the middle to the toppe, the leaues haue but three diuisions : among these leaues stand large wide open flowers, of the colour of the common Mallow : the seede is smaller then in any other Mallow, but the rootes are great and long, spreading in the ground like vnto the roots of Marsh Mallowes, springing vp afresh euery yeare from the roote.

5. *Sabdarisa seu Alcea Americana.* Thorney Mallowe.

This Thorney Mallowe hath greene leaues next vnto the ground, that are almost round, but pointed at the end, and dented very much about the edges; the other leaues that growe vpon the stalke are diuided into three parts, like vnto a trefoile, and some of them into fiue diuisions, all of them dented about the edges : the stalke is reddish, with some harmelesse prickles in sundry places thereon, and riseth vp three or foure foote high in a good ground, a fit place, and a kindly yeare, bearing plenty of flowers vpon the stalkes, one at the foote of euery leafe, the toppe it selfe ending in a long spike, as it were of buddes and leaues together : the flowers are of a very pale yellow, tending to a white colour, spotted in the bottome of each of the fiue leaues, with a deepe purple spot, broad at the lower part, and ending in a point about the middle of the leafe, which are quickly fading, and not abiding aboue one day, with a long pestle in the middle diuided at the toppe : after the flower is past, commeth vp a short prickly podde, set within a small greene huske or cup that bore the flower, wherein is contained

ned whitiſh, or rather browniſh yellow ſeede, flat and ſomewhat round, like vnto the ſeedes of Hollyhocke: the roote is ſtringie, and quickly periſheth; for it will hardly endure in our cold Country to giue flowers, much leſſe ſeede, vnleſſe (as I ſaid before) it happen in a kindly yeare, and be well planted and tended.

6. *Bamia ſeu Alcea Ægyptia.* The Mallow of Egypt.

This Mallow is alſo as tender to nourſe vp as the laſt, hauing the lower leaues broad like a Marſh Mallow, and of a freſh greene colour; but thoſe that growe vpon the ſtalke, and vp to the toppe, are diuided into fiue parts or points, but are not cut in to the middle ribbe, like the former Thorney Mallow, yet dented about the edges like vnto them: the flowers growe at the ſetting to of the leaues, like vnto a Mallow for forme, but of a whitiſh colour; after which come long fiue ſquare pointed pods, with hard ſhels, wherein are contained round blackiſh gray ſeede, as bigge as a Vetch or bigger: the roote periſheth quickly with vs, euen with the firſt froſts.

7. *Althæa frutex flore albo vel purpureo.* Shrubbe Mallow with a white or purple flower.

There are diuers ſorts of ſhrubbe Mallowes, whereof ſome that haue their ſtemmes or ſtalkes leſſe wooddy, dye downe to the ground euery yeare, and others that abide alwayes, are more wooddy: Of the former ſorts I intend not to ſpeake, referring them to a fitter place; and of the other, I will onely giue you the knowledge of one or two in this place, although I doe acknowledge their fitteſt place had been to be among the ſhrubbes; but becauſe they are Mallowes, I pray let them paſſe with the reſt of their kindred, and their deſcriptions in this manner: Theſe wooddy kindes of ſhrub Mallowes haue ſomewhat large, long, and diuided leaues, of a whitiſh greene colour, ſoft alſo, and as it were woolly in handling, ſet diſperſedly on the whitiſh hard or wooddy ſtalkes: their flowers are large, like vnto a ſingle Roſe or Hollyhocke, in the one being white with purple ſpots in the bottome; in the other either of a deepe red colour, or elſe of a paler purple, with a deeper bottome, and with veines running in euery leafe: they are ſomewhat tender, and would not be ſuffered to be vncouered in the Winter time, or yet abroad in the Garden, but kept in a large pot or tubbe, in the houſe or in a warme cellar, if you would haue them to thriue.

8. *Malua hortenſis roſea ſimplex & multiplex diuerſorum colorum.* Hollihockes ſingle and double of ſeuerall colours.

I ſhall not neede to make many deſcriptions of Hollihockes, in regard the greateſt difference conſiſteth in the flowers, which are in ſome ſingle, in ſome double, in ſome of one colour, and in others of other colours: for the loweſt leaues of Hollihockes are all round, and ſomewhat large, with many corners, but not cut in or diuided, ſoft in handling; but thoſe that growe vp higher are much more diuided into many corners: the ſtalkes ſometimes growe like a tree, at the leaſt higher then any man, with diuers ſuch diuided leaues on them, and flowers from the middle to the toppe, where they ſtand as it were a long ſpike of leaues and buds for flowers together: the flowers are of diuers colours, both ſingle and double, as pure white, and pale bluſh, almoſt like a white, and more bluſh, freſh and liuely, of a Roſe colour, Scarlet, and a deeper red like a crimſon, and of a darke red like blacke bloud; theſe are the moſt eſpeciall colours both of ſingle and double flowers that I haue ſeene: the ſingle flowers conſiſt of fiue broad and round leaues, ſtanding round like vnto ſingle Roſes, with a middle long ſtile, and ſome chiues aboue them: the double flowers are like vnto double Roſes, very thicke, ſo that no ſtile or vmbone is ſeene in the middle, and the outermoſt rowe of leaues in the flowers are largeſt, the innermoſt being ſmaller and thicke ſet together: after the flowers are paſt, there come vp as well in the double as ſingle, flat round heads, like flat cakes, round about the bottomes whereof growe flat whitiſh ſeede: the roote is long and great at the head, white and tough, like the roote of the common Mallowes, but greater, and will reaſonably well abide the Winter.

The

The Place.

The firſt groweth wilde in Spaine. The ſecond in our owne Countrey. The third is thought to growe in Italy and Venice; but Lobel denieth it, ſaying, that it is there onely in Gardens, and is more plentifull in theſe parts then with them. The fourth Cluſius ſaith he found in many places of Germany. The fifth is ſuppoſed to be firſt brought out of the Weſt Indies, but an Arabicke name being giuen it, maketh me ſomewhat doubtfull how to beleeue it. The ſixth groweth in Egypt, where it is of great vſe, as Proſper Alpinus hath ſet downe in his Booke of Egyptian plants. The ſeuenth groweth in ſome parts both of Spaine and France. The laſt is not found but in Gardens euery where.

The Time.

The firſt, ſecond, third, fourth, and laſt, doe flower from Iune vntill the end of Iuly and Auguſt. The reſt flower very late, many times not vntill September or October.

The Names.

The firſt and ſecond haue their names ſufficiently expreſſed in their titles. The third is diuerſly called, as *Malua boraria*, *Alcea veſicaria*, *Alcea Veneta*, *Alcea Peregrina*, and of Matthiolus, *Hypecoum*. The moſt vſuall Engliſh name is Venice Mallow. The fourth is called *Alcea fruticoſa pentaphyllea*, and *Cannabinifolio*, or *Pentaphyllifolio*: In Engliſh, Cinquefoile Mallow. The fifth hath been ſent vnder the name of *Sabdariſa*, and *Sabdariffa*, and (as I ſaid) is thought to be brought from America, and therefore it beareth the name of that Country. The ſixth is called in Egypt, *Bamia*, or *Bammia*, and by that name ſent with the addition *del Cayro* vnto it: In Engliſh, Egyptian Mallow, or Mallow of Egypt. The ſeuenth is called *Althea frutex*, and of ſome *Althea arborea*: In Engliſh, Shrubbe Mallow, becauſe his ſtemme is wooddie, and abideth as ſhrubbes and trees doe. The eight and laſt is called *Malua hortenſis*, *Malua Roſea*, and of ſome *Roſa vltramarina*: In Engliſh, of ſome Hockes, and vſually Hollihockes.

The Vertues.

All ſorts of Mallowes, by reaſon of their viſcous or ſlimie quality, doe helpe to make the body ſoluble, being vſed inwardly, and thereby helpe alſo to eaſe the paines of the ſtone and grauell, cauſing them to be the more eaſily voided: being outwardly applyed, they mollifie hard tumors, and helpe to eaſe paines in diuers parts of the body; yet thoſe that are of moſt vſe, are moſt common. The reſt are but taken vpon credit.

Chap. XCII.

Amaranthus. Flower-gentle.

WE haue foure or fiue ſorts of Flower-gentle to trimme vp this our Garden withall, which doe differ very notably one from another, as ſhall be declared in their ſeuerall deſcriptions; ſome of which are very tender, and muſt be carefully regarded, and all little enough to cauſe them beare ſeede with vs, or elſe wee ſhall bee to ſeeke euery yeare: others are hardy enough, and will hardly be loſt out of the Garden.

1. Amaranthus

1. *Amaranthus purpureus minor.* The small purple Flower-gentle.

This gallant purple Veluet flower, or Flower-gentle, hath a crested stalke two foote high or more, purplish at the bottome, but greene to the toppe, whereout groweth many small branches, the leaues on the stalkes and branches are somewhat broad at the bottome, and sharpe pointed, of a full greene colour, and often somewhat reddish withall, like in forme vnto the leaues of Blites (whereof this and the rest are accounted *species*, or sorts) or small Beetes : the flowers are long, spikie, soft, and gentle tufts of haires, many as it were growing together, broad at the bottome, and small vp at the toppe, pyramis or steeple-fashion, of so excellent a shining deepe purple colour, tending to a murrey, that in the most excellent coloured Veluet, cannot be seene a more orient colour, (and I thinke from this respect, the French call it *Passe velours*, that is to say, passing Veluet in colour) without any smell at all, which being bruised giueth the same excellent purple colour on paper, and being gathered in his full strength and beauty, will abide a great time (if it be kept out of the winde and sunne in a dry place) in the same grace and colour : among these tufts lye the seede scattered, which is small, very blacke, and shining : the rootes are a few thready strings, which quickly perish, as the whole plant doth, at the first approach of Winter weather.

2. *Amaranthus Coccineus.* Scarlet Flower-gentle.

The leaues of this Flower-gentle are longer, and somewhat narrower then the former ; the stalke groweth somewhat higher, bearing his long tufts at seuerall leaues, as also at the toppe of the stalkes, many being set together, but separate one from another, and each bowing or bending downe his head, like vnto a Feather, such as is worn in our Gallants and Gentlewomens heads, of an excellent bloudy Scarlet colour : the seede is blacke, like vnto the former : the roote perisheth quicklier, because it is more tender.

3. *Amaranthus tricolor.* Spotted or variable Flower-gentle.

The chiefest beauty of this plant consisteth in the leaues, and not in the flowers ; for they are small tufts growing all along the stalke, which is nothing so high as the former, especially with vs, and at the ioynts with the leaues : the leaues hereof are of the same fashion that the former are, and pointed also ; but euery leafe is to be seene parted into greene, red, and yellow, very orient and fresh (especially if it come to his full perfection, which is in hot and dry weather) diuided not all alike, but in some leaues, where the red or yellow is, there will be greene, and so varying, that it is very pleasant to behold : the seede hereof is blacke and shining, not to bee knowne from the former.

4. *Amaranthus Carnea spica.* Carnation Flower-gentle.

There is another more rare then all the rest, whose leaues are somewhat longer, and narrower then the first, and like vnto the second kinde : the spikes are short, many set together, like branches full of heads or eares of corne, euery one whereof hath some long haires sticking out from them, of a deep blush, tending to a carnation colour.

5. *Amaranthus purpureus maior panniculis sparsis.* Great Floramour, or purple Flower-gentle.

The great Floramour hath one thicke, tall, crested, browne red stalke, fiue or six foote high, from whence spring many great broad leaues, like vnto the former for the forme, but much larger & redder for the most part, especially the lowest, which brancheth forth into diuers parts, & from between these leaues, & the stalks or branches, as also at the tops of them, stand long, spikie, round, & somewhat flat tufts, of a more reddish purple colour then the first, and diuided also into seuerall parts, wherin when they

are

are full ripe, are to be ſeen an innumerable company of white ſeed, ſtanding out among the ſhort thrums, and do then eaſily fall away with a little touching; euery one of theſe white ſeed hath as it were an hole halfe bored through therin: the root is a great buſh of ſtrings, ſpreading in the ground, whereby it is ſtrongly faſtened, yet periſheth euery yeare, after it hath giuen his ſeede.

The Place.

All theſe plants growe in the Eaſterne Countries, as Perſia, Syria, Arabia, &c. except the greateſt, which hath been brought out of the Weſt Indies, where it is much vſed, eſpecially the ſeede: they are all, except it, nourſed vp with much care in our Gardens, and yet in a backward or cold yeare they will not thriue, for that they deſire much heate: but the greateſt doth alwayes giue ripe ſeede euery yeare.

The Time.

They beare their gallant tufts or ſpikes for the moſt part in Auguſt, and ſome not vntill September.

The Names.

The name *Amaranthus* is giuen to all theſe plants, taken from the Greeke word *ἀμάραντος*, *non marceſcens*, or *non ſeneſcens*, that is, neuer waxing old, and is often alſo impoſed on other plants, who haue the ſame property, that is, that their flowers being gathered in a fit ſeaſon, will retaine their natiue colour a long time, as ſhall be ſhewed in the Chapter following. Diuers do thinke the firſt to be *Phlox*, or *Flamma* of Theophraſtus. The third is called *Geloſia*, or *Celoſia* of Tragus. Spigelius in his *Iſagoges* ſaith, it is generally taken to be *Sephonia*, whereof Plinie maketh mention; and Lobel, to bee the Perſians *Theombroton* of Plinie. The Italians, from whom I had it (by the meanes of Mr. Doctor Iohn More, as I haue had many other rare ſimples) call it, *Blito di tre colori*, A three coloured Blite. The fifth, which is the greateſt, hath been ſent from the Weſt Indies by the name of *Quinua*, as Cluſius reporteth. The name Flower-gentle in Engliſh, and *Floramour*, which is the French, of *Flos amoris*, and *Paſſe velours*, as is before ſaid, or Veluet flower, according to the Italian, *Fior veluto*, are equally giuen to all theſe plants, with their ſeuerall diſtinctions, as they are expreſſed in their titles.

The Vertues.

Diuers ſuppoſe the flowers of theſe plants doe helpe to ſtay the fluxe of bloud in man or woman, becauſe that other things that are red or purple doe performe the ſame. But Galen diſproueth that opinion very notably, *in lib. 2. & 4. de ſimpl. medicament. facultatibus.*

CHAP. XCIII.

Helichryſum, ſiue Amaranthus luteus.
Golden Flower-gentle, Goldilockes, or Gold-flower.

THe propinquity of property (as I before ſaid) hath cauſed the affinity in name, and ſo in neighbourhood in theſe plants, wherein there are ſome diuerſity; and although they differ from them before in many notable points, yet they all agree with themſelues in the golden, or ſiluer heads or tufts they beare; and therefore I haue

1 *Amaranthus purpureus maior.* The great purple Flower-gentle. 2 *Amaranthus purpureus minor.* The lesser purple Flower-gentle. 3 *Amaranthus tricolor.* Variable Flower-gentle. 4 *Amaranthus coccineus.* Scarlet Flower gentle. 5 *Amaranthus carnea spica.* Carnation Flower-gentle. 6 *Flos Americanus Indicus Cassidinae.* 7 *Helichrysum Creticum.* Candy Goldilockes. 8 *Gnaphalium Americanum.* Liue-long or Life-euerlasting. 9 *Gnaphalium montanum.* Cats foote. 10 *Gnaphalium roseum.* The Cotton Rose.

haue comprifed them in one Chapter, and will begin with that which commeth nea-reft vnto the *Helichryfum* of Diofcorides, or *Aurelia* (as Gaza tranflateth it) of Theo-phraftus.

1. *Heliochryfum.* The Golden flower of life.

This firft Golden tuft rifeth vp with many hard, round, white ftalkes, a foote and a halfe high, whereon at certaine diftances ftand many fine cut leaues, or rather one leafe cut into many fmall fine parts, almoft as fmall as Fenell, but grayifh, like vnto the Cud-weedes or Cotton-weedes (whereof certainly thefe are fpeciall kindes) at the toppes of the ftalkes ftand many round flowers, of a pale gold colour, in an vmbell clofe to-gether, yet euery flower vpon his owne ftalke, and all of an euen height, which will keepe the colour, being gathered, and kept dry for a long time after, and are of a hot and quicke fent: the roote is fmall and wooddy, fpreading vnder the vpper cruft of the earth, and liueth long in his owne naturall place, but very hardly endureth the cold of our Winters, vnleffe they be milde, or it be well defended.

2. *Helichryfum Creticum.* Candy Goldilockes.

Candy Goldilockes hath two or three fmall flender white branches, fet here and there very fcatteringly, with fmall, long, and narrow hoary leaues, hauing yellow heads of flowers at the tops made into vmbels or tufts, not fo round and euen as the former, but longwife one aboue another, the heads being made as it were of fcales, loofly, and not fo clofely fet together, as in the next following, which when they are full ripe, doe paffe into doune, and are blowne away with the winde, hauing a fmall reddifh feede at the end; but will abide a long time, as the other in his beauty, being gathered in time, as the reft will doe.

3. *Helichryfum Orientale fiue Amaranthus luteus.* Golden Flower-gentle.

This moft beautifull plant is very like vnto the former Candy Goldilockes laft de-fcribed, but growing vp higher, with many more branches, and more hoary, white, and woolly, hauing alfo long and narrow white leaues, but fomewhat broader, and thicker fet on the branches: the tufts of flowers or vmbels likewife doe confift of longer and larger heads, more fcaly, and clofer compact together, of an excellent pale gold yellow colour, and fhining, with fome yellow threads or thrummes in the middle: the roote dyeth not euery yeare, but liueth long, efpecially in the South and Eaft Countries, where no colds or frofts are felt; but will require extraordinary care and keeping, and yet fcarce fufficient to preferue it in thefe cold Countries.

4. *Chryfocome fiue Stachas Citrina.* Golden tufts or Golden Caffidony.

This Golden flower is fomewhat like the former of thefe two laft defcribed, hauing hoary ftalkes and leaues, ftanding confufedly on them, being long, and narrower then any of the former: the tops of the ftalkes are diuided into many parts, each bearing a fmall long yellow head or flower at the toppe, with fome yellow thrummes in them, which heads being many, are diffufedly fet together, like a loofe or fparfed vmbell, keeping their colour long before they wither, and when they are ripe, haue thinne fmall reddifh feede, like Mariorome feede, but fmaller; the roote is fmall and blacke: the whole plant, as well leaues and flowers, as rootes, are of a ftrong fharpe fent, yet pleafant.

5. *Argyrocome fiue Gnaphalium Americanum.* Liue long or Life euerlafting.

This filuer tuft or Indian Cotton weede, hath many white heads of leafes at their firft fpringing out of the ground, couered with a hoary woollineffe like cotton, which rifing into hard, thicke round ftalkes, containe ftill the fame hoarineffe vpon them, as alfo vpon the long and narrow leaues which are fet thereon, efpecially on the vnder fide,

fide, for the vpper fides are of a darke fhining greene colour : the ftalkes are diuided at the toppe into many fmall branches, each whereof haue many fcaly tufted heads fet together, couered ouer with cotton before their opening, and then diffeuering one from another, abiding very white on the outfide, when they are fully growne, but with a fmall yellow thrume in the midde of euery flower, which in time turne into yellow doune, apt to be blowne away with euery winde : the roots are long and black on the outfide, creeping vnder ground very much.

6. Gnaphalium montanum flore albo & flore purpureo.
White and purple Cats foote.

This fmall Cudweede or Cottonweede, hath many fmall white woolly leaues growing from the roote, which is compofed of a few fmall blackifh threds, and lying vpon the ground fomewhat like vnto the leaues of a fmall Moufe-eare, but fmaller; from among which rifeth vp a fmall ftalke of halfe a foote high or thereabouts, befet here and there with fome few leaues, at the top whereof commeth forth a tuft of fmall flowers, fet clofe together, in fome of a pure white, in others of a purple or reddifh colour, in fome of a pale red or blufh, and in others of a white and purple mixt together, which for the beauty is much commended and defired, but will hardly abide to be kept in Gardens, fo vnwilling they are to leaue their naturall abiding.

7. Gnaphalium Rofeum. The Cotton Rofe.

This little rofe Cotton weede hath many fuch like woolly leaues, growing as the former from the roote vpon fmall fhort branches, not full an hand breadth high, in fafhion fomewhat like vnto Dayfie leaues, but leffer, and round pointed : at the toppe of euery ftalke or branch, ftandeth one flower, compofed of two rowes of fmall white leaues, layd open like a Starre or a Rofe, as it beareth the name, hauing a round head in the middle made of many yellow threds or thrumes, which falling away, there rifeth vp a fmall round head, full of fmall feedes : the root is fmall, long and threddy.

The Place.

The foure firft plants doe grow naturally in many of the hot Countries of Europe, as Spaine, Italie, and Prouince in France ; as alfo in Candy, Barbary, and other places, and muft be carefully kept with vs in the winter time. The Liuelong was brought out of the Weft Indies, and groweth plentifully in our gardens. The two laft doe grow as well in the colder Countries of Germany, as in France and other places.

The Time.

They all flower in the end of September, if they will fhew out their beauty at all with vs, for fometimes it is fo late, that they haue no faire colour at all, efpecially the foure firft forts.

The Names.

Variable and many are the names that feuerall Writers do call thefe foure firft forts of plants, as *Helichryfum, Heliochryfum,* or *Elichrylum. Eliochryfum, Chryfocome, Coma aurea, Amaranthus luteus, Stoechas Citrina,* and *Aurelia,* with others, needleffe here to be recited : it is fufficient for this worke, to giue you knowledge that their names are fufficient as they are expreffed in their titles : The fift is called *Gnaphalium* by Carolus Clufius, from the likeneffe of the vmbels or tufts of heads, though greater and white : for as I faid before, the Cotton weedes are of kindred with the golden tufts : It hath been called by our Englifh Gentlewomen, Liue long, and Life euerlafting, becaufe of the durabilitie of the flowers in their beautie. The two laft are cal-

I i 2   led

led *Gnaphalium*, according to their titles ; and in Englifh they may paffe vnder thofe names are fet downe with them.

The Vertues.

The foure firft are accounted to bee hot and dry, and the three laft to bee cold and dry : yet all of them may to fome good purpofe bee applyed to rheumaticke heads. The former foure are likewife vfed to caufe vrine, and in baths to comfort and heate cold parts. They are alfo layd in chefts and wardrobes, to keepe garments from moths ; and are worne in the heads and armes of Gentiles and others, for their beautifull afpect.

Chap. XCIIII.

Canna Indica. The Indian flowring Reede.

THere are two kindes or forts of this beautifull plant, the one with a red flower, the other with a yellow, fpotted with reddifh fpots, both which in fome kind ly yeares haue borne their braue flowers, but neuer any ripe feede, and doth not abide the extremities of our winters, eyther abroade or vnder couert, vnleffe it meete with a ftoue or hot-houfe, fuch as are vfed in Germany, or fuch other like place: For neyther houfe nor cellar will preferue it, for want of heate.

Canna Indica flore rubro. Red flowred Indian Reede.

This beautifull plant rifeth vp with faire greene, large, broade leaues, euery one rifing out of the middle of the other, and are folded together, or writhed like vnto a paper Coffin (as they call it) fuch as Comfitmakers and Grocers vfe, to put in their Comfits and Spices, and being fpread open, another rifeth from the bottome thereof, folded in the fame manner, which are fet at the ioynts of the ftalke when it is rifen vp, like vnto our water Reede, and growing (if it runne vp for flower) to be three or foure foote high, as I haue obferued in mine owne garden : the flowers grow at the toppe of the ftalke one aboue another, which before their opening are long, fmall, round, and pointed at the end, very like vnto the claw of a Crauife or Sea-Crab, and of the fame red or crimfon colour, but being open, are very like vnto the flower of *Gladiolus* or Corne-flagge, but of a more orient colour then at the firft, and ftanding in a rough huske, wherein afterwards ftandeth a three fquare head, containing therein round blacke feede, of the bigneffe of a peafe : the roote is white and tuberous, growing into many knobs, from whence arife fuch other leaues and ftalkes, whereby it encreafeth very much, if it be righrly kept and defended.

Canna Indica flore flauo punctato,
Yellow fpotted Indian Reede.

This Reede groweth vp with leaues and flowers, in all points fo like vnto the former, that it cannot bee knowne from it, vntill it come to flower, which is of a yellow colour, fpotted with reddifh fpots, without any other difference.

The Place.

Thefe plants grow naturally in the Weft Indies, from whence they were firft fent into Spaine, and Portugall, where Clufius faith he faw them planted by the houfes fides, flowring in winter, which might be in thofe warme Countreyes. We preferue them with great care in our gardens, for the beautifull afpect of their flowers.

The

The Time.

They flower not with vs vntill the end, or middle of Auguſt, at the ſooneſt.

The Names.

They are called of ſome *Canna Indica*, and *Arundo Indica*, of others *Cannacorus*, and of ſome *Flos Cancri*, becauſe the colour of the flowers, as well as the forme of the buds, are ſo like vnto a Sea-Crabs cle, or claw.

The Vertues.

There is not any vſe of theſe in Phyſicke that I know.

Chap. XCV.

Mandragoras. Mandrake.

THe Mandrake is diſtinguiſhed into two kindes, the male and the female ; the male hath two ſorts, the one differing from the other, as ſhall be ſhewed; but of the female I know but one : the male is frequent in many gardens, but the female, in that it is more tender and rare, is nourſed vp but in a few.

Manadrgoras mas. The male Mandrake.

The male Mandrake thruſteth vp many leaues together out of the ground, which being full growne, are faire, large and greene, lying round about the roote, and are larger and longer then the greateſt leaues of any Lettice, whereunto it is likened by Dioſcorides and others : from the middle, among theſe leaues, riſe vp many flowers, euery one vpon a long ſlender ſtalke, ſtanding in a whitiſh greene huske, conſiſting of fiue pretty large round pointed leaues, of a greeniſh white colour, which turne into ſmall round apples, greene at the firſt, and of a pale red colour when they are ripe, very ſmooth and ſhining on the outſide, and of a heady or ſtrong ſtuffing ſmell, wherein is contained round whitiſh flat ſeede : the roote is long and thicke, blackiſh on the outſide, and white within, conſiſting many times but of one long roote, and ſometimes diuided into two branches a little below the head, and ſometimes into three or more, as nature liſteth to beſtow vpon it, as my ſelfe haue often ſeene, by the tranſplanting of many, as alſo by breaking and cutting off of many parts of the rootes, but neuer found harme by ſo doing, as many idle tales haue beene ſet downe in writing, and deliuered alſo by report, of much danger to happen to ſuch, as ſhould digge them vp or breake them ; neyther haue I euer ſeene any forme of man like or woman-like parts, in the rootes of any : but as I ſaid, it hath oftentimes two maine rootes running down-right into the ground, and ſometimes three, and ſometimes but one, as it likewiſe often happeneth to Parſneps, Carrots, or the like. But many cunning counterfeit rootes haue bin ſhaped to ſuch formes, and publickly expoſed to the view of all that would ſee them, and haue been tolerated by the chiefe Magiſtrates of the Citie, notwithſtanding that they haue beene informed that ſuch practices were meere deceit, and vnſufferable ; whether this happened through their ouer-credulitie of the thing, or of the perſons, or through an opinion that the information of the truth roſe vpon enuy, I know not, I leaue that to the ſearcher of all hearts : But this you may bee bold to reſt vpon, and aſſure your ſelues, that ſuch formes as haue bin publickly expoſed to be ſeene, were neuer ſo formed by nature, but only by the art and cunning of knaues and deceiuers. and let this be your *Galeatum* againſt all ſuch vaine, idle and ridiculous toyes of mens inuentions.

There

There is likewise another fort of thefe male Mandrakes, which I firft faw at Canterbury, with my very louing and kinde friende Iohn Tradefcante, in the garden of the Lord Wotton, whofe gardiner he was at that time ; the leaues whereof were of a more grayifh greene colour, and fomewhat folded together, when as the former kind that grew hard by it, was of the fame forme that is before defcribed, and ordinary in all others : but whether the apples were differing from the other, I know not, nor did they remember that euer it had borne any.

Mandragoras fœmina. The female Mandrake.

The female Mandrake doth likewife put vp many leaues together, from the head of the roote, but they are nothing fo large, and are of a darker greene colour, narrower alfo and fhining, more crumpled, and of a ftronger fent : the flowers are many, rifing vp in the middle of the leaues, vpon flender ftalkes, as in the male kind, but of a blewifh purple colour, which turne into fmall round fruite or apples, and not long like a peare (as Clufius reporteth that faw them naturally growing in Spaine) greene at the firft, and of a pale yellowifh colour, when they are full ripe; of a more pleafing, or if you will, of a leffe heady fent then the apples of the male, wherein is contained fuch like feede, but fmaller and blacker : the rootes are like the former, blacke without and white within, and diuided in the fame manner as the male is, fometimes with more, and fometimes with fewer parts or branches.

The Place.

They grow in many places of Italie, as Matthiolus reporteth, but efpecially on Mount Garganus in Apulia. Clufius faith hee found the female in many wet grounds of Spaine, as alfo in the borders of thofe medowes that lye neere vnto riuers and water courfes. The male is cherifhed in many Gardens, for pleafure as well as for vfe : but the female as is faid, is both very rare, and farre more tender.

The Time.

The Male flowreth in March, and the fruit is ripe in Iuly. The Female, if it be well preferued, flowreth not vntill Auguft, or September; fo that without extraordinary care, we neuer fee the fruite thereof in our gardens.

The Names.

Mandragoras mas is called *albus*, as the *Fœmina* is called *niger*, which titles of blacke and white, are referred vnto the colour of the leaues : the female is called alfo *Thridacias*, from the likeneffe of Lettice, whereunto they fay in forme it doth carry fome fimilitude. Diofcorides faith, that in his time the male was called *Morion*, and both of them *Antimelum*, and *Circæa*. Wee call them in Englifh, The male, and the female Mandrake.

The Vertues.

The leaues haue a cooling and drying qualitie, fit for the oyntment *Populeon*, wherein it is put. But the Apples haue a foporiferous propertie, as Leuinus Lemnius maketh mention in his Herball to the Bible, of an experiment of his owne. Befides, as Diofcorides firft, and then Serapio, Auicen, Paulus Ægineta, and others alfo do declare, they conduce much to the cooling and cleanfing of an hot *matrix*. And it is probable, that Rachel knowing that they might be profitable for her hot and dry body, was the more earneft with Leah for her Sonne Rubens Apples, as it is fet downe *Genefis* 30. *verfe* 14. The ftrong fent of thefe apples is remembred alfo, *Cant.* 7.13. although fome would diuert the fignification of the Hebrew word, דודאים,

vnto

vnto Violets, or some other sweet flowers, in the former place of *Genesis*, and the fruit of *Musa*, or Adams Apples in this place of the *Canticles*. Hamilcar the Carthaginian Captaine is said to haue infected the wine of the Lybians (his enemies against whom he fought) with the apples of Mandrake, whereby they being made exceeding drowsie, he obtained a famous victory ouer them.

Chap. XCVI.

Pomum Amoris. Loue Apples.

ALthough the beautie of this plant consisteth not in the flower, but fruit, yet giue me leaue to insert it here, lest otherwise it haue no place : whereof there are two especiall sorts, which wee comprehend in one Chapter, and distinguish them by *maius* and *minus*, greater and smaller : yet of the greater kinde, we haue noursed vp in our Gardens two sorts, that differ only in the colour of the fruite, and in nothing else.

Pomum Amoris maius fructu rubro.
Great Apple of Loue the ordinary red sort.

This greater kinde of Loue Apples, which hath beene most frequently cherished with vs, hath diuers long and trayling branches, leaning or spreading vpon the ground, not able to sustaine themselues, whereon doe grow many long winged leaues, that is, many leaues set on both sides, and all along a middle ribbe, some being greater, and others lesse, iagged also and dented about the edges, of a grayish ouer-worne greene colour, somewhat rough or hairy in handling, from among the leaues and the branches come forth long stalkes, with diuers flowers set thereon, vpon seuerall short footstalks, consisting of sixe, and sometimes of eight small long yellow leaues, with a middle pricke or vmbone, which after the flowers are fallen, riseth to be the fruite, which are of the bignesse of a small or meane Pippin, vneuenly bunched out in diuers places, and scarce any full round without bunches, of a faire pale reddish colour, or somewhat deeper, like vnto an Orenge, full of a slimie iuice and watery pulpe, wherein the seede lyeth, which is white, flat and somewhat rough : the roote shooteth with many small strings and bigger branches vnder ground, but perisheth at the first feeling of our winter weather. The fruite hereof by often sowing it in our Land, is become much smaller then I haue here described it : but was at the first, and so for two or three yeares after, as bigge as I haue related it.

Pomum Amoris maius fructu luteo. Yellow Amorous Apples.

Of the same kinde is this other sort of Amorous Apples, differing in nothing but the colour of the fruite, which is of a pale yellow colour, hauing bunches or lobes in the same manner, and seede also like the former.

Pomum Amoris minus, siue Mala Ethiopica parua.
Small Loue Apples.

The small Apples of Loue in the very like manner, haue long weake trayling branches, beset with such like leaues as the greater kinde hath, but smaller in euery part : the flowers also stand many together on a long stalke, and yellow as the former, but much smaller : the fruite are small, round, yellowish red berries, not much bigger then great grapes, wherein are contained white flat seede, like the other, but smaller : the roote perisheth in like manner euery yeare, and therefore must bee new sowen euery spring, if you will haue the pleasure of their sight in the garden ; yet some yeares I haue known them rise of their owne sowing in my garden.

The

The Place.

They growe naturally in the hot Countries of Barbary, and Ethiopia; yet some report them to be first brought from Peru, a Prouince of the West Indies. Wee onely haue them for curiosity in our Gardens, and for the amorous aspect or beauty of the fruit.

The Time.

¶They flower in Iuly and August, and their fruit is ripe in the middle or end of September for the most part.

The Names.

The first is named diuersly by diuers Authors; for Lobel, Camerarius, and others, call them *Poma amoris.* Dodonæus *Aurea Mala.* Gesnerus first, and Bauhinus after him, make it to be a kinde of *Solanum Pomiferum.* Anguillara taketh it to be *Lycopersicum* of Galen. Others thinke it to bee *Glaucium* of Dioscorides. The last is called *Mala Æthiopica parua,* and by that title was first sent vnto vs, as if the former were of the same kinde and country. We call them in English, Apples of Loue, Loue-Apples, Golden Apples, or Amorous Apples, and all as much to one purpose as another, more then for their beautifull aspect.

The Vertues.

In the hot Countries where they naturally growe, they are much eaten of the people, to coole and quench the heate and thirst of their hot stomaches. The Apples also boyled, or infused in oyle in the sunne, is thought to be good to cure the itch, assuredly it will allay the heate thereof.

Chap. XCVII.

Digitalis. Foxegloue.

There are three principall sorts of Foxegloues, a greater, a middle or meane sort, and a lesser, and of them, three especiall colours, that is, purple, white, and yellow; the common purple kinde that groweth abroad in the fields, I leaue to his wilde habitation: and of the rest as followeth.

1. *Digitalis maxima ferruginea.* Dun coloured Foxegloues.

The leaues of this Foxegloue are long and large, of a grayish green colour, finely cut or dented about the edges, like the teeth of a fine sawe; among which commeth vp a strong tall stalke, which when it was full growne, and with ripe seede thereon, I haue measured to be seuen foot high at the least, wheron grow an innumerable company (as I may so say, in respect of the aboundance) of flowers, nothing so large as the common purple kinde, that groweth wilde euery where in our owne Countrey, and of a kinde of browne or yellowish dunne colour, with a long lippe at euery flower; after them come seede, like the common kinde, but in smaller heads: the rootes are stringie like the ordinary, but doe vsually perish, or seldome abide after it hath giuen seed.

2. *Digitalis maior flore carneo.* Blush coloured Foxegloues.

This kinde of Foxegloues hath reasonable large leaues, yet not altogether so large

as

1 *Canna Indica.* The Indian Reed. 2 *Mandragoras mas.* The male Mandrake. 3 *Pomum amoris maius.* Great Apples of loue. 4 *Digitalis maior flore luteo amplo.* The great yellow Foxegloue. 5 *Digitalis media flore luteo rubente.* Orenge tawny Foxegloues. 6 *Digitalis maxima ferruginea.* Dun coloured Foxegloues.

as the common field kinde : the flowers are also smaller then the common sort, but of a blush colour.

3. *Digitalis media flore luteo rubente.* Orenge tawnie Foxegloue.

As this Foxegloue is none of the greatest, so also is it none of the smallest ; but a sort betweene both, hauing leaues in some proportion correspondent to the lesser yellow Foxegloue, but not so large as the lesser white : the flowers are long and narrow, almost as large as the last white, but nothing so large as the first white, of a faire yellowish browne colour, as if the yellow were ouershadowed with a reddish colour, and is that colour wee vsually call an Orenge tawnie colour : the seede is like the former : the rootes perish euery yeare that they beare seede, which is vsually the second yeare of the springing.

4. *Digitalis maior alba.* The greater white Foxegloue.

This white Foxegloue is in all things so like vnto the purple wilde kinde, that it can hardly be distinguished from it, vnlesse it be in the fresher greennesse and largenesse of the leaues : the flowers are as great in a manner as the purple, but wholly white, without any spot in them : the seed and other things agree in all points.

5. *Digitalis alba altera seu minor.* The lesser white Foxegloue.

We haue in our Gardens another sort of white Foxegloue, whose leaues are like vnto the last described, but not altogether so long or large, and of a darker greene colour : the stalke groweth not so high, as not full three foote : the flowers are pure white, fashioned like vnto the former, but not so great or large, in all other things alike : the rootes hereof did abide sometime in our Gardens, but since perished, and the seede also, since when we neuer could obtaine from any our friends of that kinde againe.

6. *Digitalis maior lutea flore amplo.*
The great yellow Foxegloue.

The leaues of this greater yellow Foxegloue, are in forme somewhat like vnto the common purple kinde, but not altogether so large : the stalke groweth to bee three or foure foote high, whereon stand many long hollow pendulous flowers, in shape like the ordinary purple : but somewhat shorter, and more large and open at the brimmes, of a faire yellow colour, wherein are long threads, like as in the others : the roote hereof is greater at the head, and more wooddy then any of the rest, with many smaller fibres, spreading themselues in the ground, and abideth almost as well as our common purple kinde.

7. *Digitalis minor lutea siue pallida.*
The small pale yellow Foxegloue.

This small pale yellow Foxegloue hath somewhat short, broad, smooth and darke greene leaues, snipt or dented about the edges very finely : the stalke is two foot high, beset with such like leaues, but lesser : the flowers are moe in number then in any of the rest, except the first and greatest, and growe along the vpper part of the stalke, being long and hollow, like the other, but very small, and of a pale yellow colour almost white : the seede vessels are small like the former, wherein are contained seede like the rest, but smaller : the rootes are stringy, but durable, and seldome perish with any iniury of the extreamest frosts.

The Place.

The great white kinde hath been often, and in many places found wilde in our owne Country, among or hard by the common purple kinde. All the rest are strangers, but cherished in our Gardens.

The

The Time.

They flower in Iune and Iuly, and some in August, their seede becomming ripe quickly after.

The Names.

Onely the name *Digitalis*, is of all Writers giuen vnto these plants ; for it is not knowne to bee remembred of any of the old Authors. Wee call them generally in English, Foxegloue ; but some (as thinking it to bee too foolish a name) doe call them Finger-flowers, because they are like vnto the fingers of a gloue, the ends cut off.

The Vertues.

Foxegloues are not vsed in Physicke by any iudicious man that I know ; yet some Italians of Bononia, as Camerarius saith, in his time vsed it as a wound herbe.

Chap. XCVIII.

Verbascum. Mullein.

THere be diuers kindes of Mullein, as white Mullein, blacke Mullein, wooddy Mullein, base Mullein, Moth Mullein, and Ethiopian Mullein, all which to distinguish or to describe, is neither my purpose, nor the intent of this worke, which is to store a Garden with flowers of delight, and sequester others not worthy of that honour. Those that are fit to bee brought to your consideration in this place, are first, the *Blattarias*, or Moth Mulleins, and then the wooddy Mullein, which otherwise is called French Sage, and lastly, the Ethiopian Mullein, whose beauty consisteth not in the flower, but in the whole plant ; yet if it please you not, take it according to his Country for a Moore, an Infidell, a Slaue, and so vse it.

1. *Blattaria lutea odorata.* Sweete yellow Moth Mullein.

The yellow Moth Mullein whose flower is sweete, hath many hard grayish greene leaues lying on the ground, somewhat long and broad, and pointed at the end : the stalks are two or three foot high, with some leaues on them, & branching out from the middle vpwards into many long branches, stored with many small pale yellow flowers, of a pretty sweete sent, somewhat stronger then in the other sorts, which seldome giueth seede, but abideth in the roote, liuing many yeares, which few or none of the others doe.

2. *Blattaria lutea maior siue Hispanica.* The great yellow Moth Mullein.

This Spanish kinde hath larger and greener leaues then the former, and rounder and larger then the next that followeth : the stalke is higher then in any of the Moth Mulleins, being for the most part foure or fiue foote high, whereon toward the toppe growe many goodly yellow flowers, consisting of fiue leaues, as all the rest doe, not so thicke set as the former, but much larger, with some small purplish threads in the middle : the ends whereof are fashioned somewhat like as if a Flie were creeping vp the flower, which turne into round heads, sometimes two or three or more standing together, but vsually one, wherein lye small duskie seed : the roote is not great nor full of threads, and doth perish most vsually hauing giuen seede, except the Winter bee very milde.

3. Blattaria

3. *Blattaria lutea altera vulgatior.* The ordinary yellow Moth Mullein.

This yellow Moth Mullein (which is the moſt frequent in our Gardens) hath longer, and narrower leaues then any of the former, and roundly notched or dented on the edges, of a darke greene colour : the ſtalke is ſometimes branched, but moſt viſually ſingle, whereon ſtand many gold yellow flowers, not fully ſo large as the Spaniſh kinde, but with the like purple threads in the middle : the ſeede is ſmall, and contained in the like round heads, but alwaies euery one ſingle by it ſelfe : the roote periſheth euery yeare that it beareth ſeede.

4. *Blattaria flore luteo purpuraſcente.* Cloth of gold Moth Mullein.

The greateſt point of difference betweene this and the laſt deſcribed, conſiſteth chiefly in the colour of the flower, which in this is of the colour of cloth of gold, that is, the ground yellow, and ouerſhadowed with a bright crimſon colour, which is a fine colour of much delight : the threads in the middle are nor ſo purple red as in the former, but much about the colour of the flower : this is not ſo willing to giue ſeede, and will as hardly abide in the roote, and hath out of queſtion riſen from the ſeede of the former.

5. *Blattaria flore albo.* White Moth Mullein.

The leaues of the white Moth Mullein are ſomewhat like vnto the yellow, yet not altogether ſo much roundly notched about the edges, but rather a little dented, with ſharper notches : the ſtalke riſeth as high as the yellow, and hath now and then ſome branches about it : the flowers hereof are pure white, as large and great as the ordinary yellow, or ſomewhat larger, with the like purple threads in the middle, as are in the yellow : the ſeed is like the other, the root periſheth in like maner, and will not endure.

6. *Blattaria flore purpureo.* Purple Moth Mullein.

The Purple Moth Mullein hath his leaues lying on the ground, broader and ſhorter then any of the other, of a more grayiſh greene colour, and without any denting for the moſt part about the edges, ſharpe pointed alſo at the end of the leafe ; among the leaues riſeth vp the ſtalke, not ſo high as either the white or the yellow, and many times branched, bearing many flowers thereon, of the ſame faſhion, and no whit ſmaller, of a faire deepe blewiſh colour tending to redneſſe, the threads in the middle of the flowers being yellow : the ſeede veſſels hereof are ſomewhat ſmaller then any of the former, except the firſt ſweete yellow kinde : the roote hereof is long, thicke, and blackiſh on the outſide, abiding very well from yeare to yeare, and riſeth well alſo from the ſowing of the ſeede.

7. *Blattaria flore caruleo.* Blew Moth Mullein.

This blew Moth Mullein is in all reſpects like vnto the former purple kinde, ſauing onely in the colour of the flower, which is of a blewiſh violet colour, and is not much inferiour either in greatneſſe of the plant, or in the largeneſſe of the flower, vnto the former purple kinde, and endureth many yeares in the like manner. And theſe be all the ſorts of this kinde of Moth Mullein, that I haue ſeene and nourſed vp for this my Garden, without interpoſing any vnknowne, not ſeene, or vnworthy.

8. *Verbaſcum ſilueſtre ſiue quartum Matthioli.* Wooddy Mullein or French Sage.

Wooddy Mullein or French Sage, hath diuers wooddy branches two or three foot high, very hoary or white, whereon at ſeuerall ioynts ſtand diuers thicke leaues, white alſo and hoary, long, ſomewhat broad, round pointed, and rough, ſomewhat reſembling the leaues of Sage in the forme and roughneſſe, out not in the ſent, whereof our
people

people gaue it the name of Sage, calling it French Sage (when as it is as great a ſtranger in France as in England, yet they doe with this as with many other things, calling them French, which come from beyond the Seas; as for example, all or moſt of
our bulbous flowers, they call French flowers, &c.) at the toppes of the ſtalkes and
branches, at certaine diſtances, are placed round about them many gaping flowers,
like vnto the flowers of Sage, but yellow : after which now and then come ſeede,
ſomewhat bigger then the Moth Mulleins, and leſſe then the next Mullein of Ethiopia:
the roote is wooddy at the toppe, with diuers blackiſh ſtrings growing from it, and endureth as well aboue ground with his leaues, as vnder it with his rootes.

9. *Æthiopis.* Ethiopian Mullein.

This Mullein of Ethiopia hath many great, broad, and large leaues lying on the
ground, rent or torne in diuers of them very much on the ſides, of ſo hoary a white
greene colour, that it farre paſſeth any of the white Mulleins, that growe wilde abroad
in our owne Country; for they are of a yellowiſh white hoarineſſe, nothing ſo pleaſant to looke on as this : in the middle of theſe leaues riſeth vp a ſquare ſtrong ſtalke,
foure or fiue foote high, ſet full of ſuch like leaues as growe belowe, but much leſſer,
and leſſer ſtill vp to the toppe, all hoary and woolly, as the reſt, and diuided into manie branches, ſpreading farre, and taking vp a great compaſſe of ground, more then
any one roote of Garden Clary, or other ſuch like plant : at each of the ſtalkes and
branches are ſet two ſmall leaues, and with them, round about the ſtalkes, ſtand many
ſmall gaping flowers, of a pale bleake blew colour : the ſeede is almoſt as large as Garden Clary ſeede, and of the ſame forme and colour : the roote is wooddy, and periſheth as ſoone as it hath borne ſeede, which is vſually the ſecond yeare after the ſowing; for the firſt yeare it ſeldome runneth vp to flower.

10. *Lamium Pannonicum ſiue Galeopſis Pannonica.* Hungary dead Nettle or the Dragon flower.

Let mee thruſt this plant into this place, rather then make a peculiar Chapter, becauſe I haue no other of the ſame ſtocke or kindred to be ioyned with it, and is a pretty
ornament in a Garden. The leaues whereof are very large, round, and great, rough
or full of veines, which make it ſeeme crumpled, dented or deepely notched about the
edges, and of a very darke greene colour, and ſometimes browniſh, or of a darke
reddiſh colour withall, euery one ſtanding on a long foote-ſtalke, very like in forme
vnto the great white Arch-Angell leaues, but farre larger and blacker : the ſtalkes are
great and foure ſquare, hauing leaues and flowers ſtanding round about them at the
ioynts like coronets, which flowers are very great, long, and wide gaping open, of a
darke red or purple colour, with ſome whiteneſſe or ſpots in the iawes, and ſome hai-
rineſſe alſo on the ſides, which ſtand in full flower two or three moneths moſt vſually,
and ſometimes longer, after which come browniſh ſeede : the roote is a great tuft or
buſh of long whitiſh ſtrings, and encreaſeth euery yeare, not fearing the greateſt
iniuries of our coldeſt and extreameſt Winters.

The Place.

All theſe plants are ſtrangers in our Countrey, and onely preſerued in
Gardens, to furniſh them with variety; but (as I ſaid) the cloth of gold
Moth Mullein hath been raiſed from ſeed in our owne Country.

The Time.

The laſt flowreth firſt, before all the reſt, beginning in Aprill. The
Moth Mulleins in May and Iune. The French Sage in Iuly.

The Names.

All the ſorts of *Blattaria* may bee comprehended vnder the kindes of
Kk　　　　　*Verbaſcum*

Verbascum nigrum, as any one but meanely exercised in the knowledge of plants, may discerne. And although Plinie saith, that Moths doe most frequently haunt where *Blattaria* either groweth, or is laid, yet it is not obserued sufficiently in our Country so to doe, notwithstanding the name of Moth Mullein is generally giuen them. The last is generally called with vs *Lamium Pannonicum*, but certainely it is the *Galeopsis maxima Pannonica* of Clusius.

The Vertues.

Other qualities I haue not found hath been allotted vnto the *Blattaria* or Moth Mullein, then those of Plinie, to engender Moths. Wee vse none of these plants in Physicke in these daies.

CHAP. XCIX.

Valeriana. Valerian.

THe many sorts of Valerian (or Set-wall as many doe call them) are fitter for a generall worke, or a generall Physicall Garden of Simples, then this of delightfull flowers. I will therefore select out a few, worthy of the place, and offer them to your considerations.

1. *Valeriana rubra Dodonæi.* Red Valerian.

This Valerian hath diuers hard, but brittle whitish greene stalkes, rising from the roote, full of tuberous or swelling ioynts, whereat stand two leaues, on each side one, and now and then some small leaues from betweene them, which are somewhat long and narrow, broadest in the middle, and small at both ends, without either diuision or incisure on the edges, of a pale greene colour : the stalkes are branched at the top into diuers parts, at the ends whereof stand many flowers together, as it were in an vmbell or tuft, somewhat like vnto the flowers of our ordinary Valerian, but with longer neckes, and of a fine red colour, very pleasant to behold, but of no sent of any Valerian : after these flowers haue stood blowne a very great while, they sodainely fall away, and the seede is ripe very quickly after, which is whitish, standing vpon the branches naked, as the Valerians doe, and very like vnto them, with a little white doune at the end of euery one of them, whereby they are soone carried away with the winde : the roote is great, thicke, and white, continuing long, and shooting out new branches euery yeare, and smelling somewhat like a Valerian.

2. *Nardus Montana tuberosa.* Knobbed Mountaine Valerian.

This kinde of Valerian or Spiknard, if you will so call it, hath his first leaues lying on the ground, without any diuision in them at all, being smooth, and of a dark greene colour, which so abide all the winter ; but those that spring vp after, and when it runneth vp to flower, are cut in on the edges, very like vnto the iagged leaues of the great garden Valerian, and so the elder they grow, the more cut and iagged they are : the stalke and flowers are very like the stalke with flowers of the garden Valerian, but of a darke or deepe red colour, and more store of them thrust together, by double the number almost : the seede is like the seede of the great Valerian : the root is tuberous, or knobbed in many parts, round about, aboue and below also, with some fibres shooting from them, whereby it is encreased, and smelleth very like the roote of the garden Setwall, or not altogether so strong.

1 *Blattaria flore albo.* Moth Mullein with a white flower. 2 *Blattaria flore purpureo* Moth Mullein with a purple flower. 3 *Verbasium quartum Matthioli* French Sage. 4 *Æthiopis.* Ethiopian Mullein. 5 *Valeriana rubra Dodonæi,* Red Valerian. 6 *Valeriana Graca.* Greek Valerian. 7 *Lamium Pannonicum,* Hungary dead Nettle. 8 *Cardamine flore pleno,* Double Cuckowe flower or Ladies smocks.

3. *Valeriana Græca.* Greeke Valerian.

The Greek Valerian hath many winged leaues lying vpon the ground, that is, many small leaues set on both sides of a middle ribbe, very like vnto the wilde Valerian, that groweth by the ditch sides, but much smaller and tenderer, among which rise vp one or two round brittle stalkes, two foote high or thereabouts, whereon are set at the ioynts, such like leaues as grow below, but smaller : the toppes of the stalkes are diuided into many small branches, thicke set together, full with flowers, consisting of fiue small round leaues a peece, layd open like vnto the Cinquefoile flower, with some white threds in the middle, tipt with yellow pendents : the colour of these flowers in some plants, is of a faire bleake blew colour, and in others pure white : And I doe heare of one beyond the Seas (if the report bee true, for I haue not seene such a one) which should beare red flowers : after the flowers are past, there come vp in their pla-ces small hard huskes or heads, containing small blackish seedes : the roote is com-posed of a number of small long blackish threds, fastened together at the head, with-out any sent at all of a Valerian, eyther in roote or leafe ; and why it should bee called a Valerian I see no great reason, for it agreeth with none of them, in flower or seede, and but onely with the wilde Valerian in leafe, as I said before : but as it is, we so giue it you, and for the flowers sake is receiued into our gardens, to helpe to fill vp the num-ber of natures rarities and varieties.

The Place.

All these Valerians are strangers, but endenizond for their beauties sake in our Gardens. The Mountaine Valerian I had of the liberalitie of my louing friend Iohn Tradescante, who in his trauaile, and search of natures varieties, met with it, and imparted thereof vnto me.

The Time.

They flower in the Summer moneths, and seed quickly after.

The Names.

The first is generally called of most, *Valeriana rubra Dodonei,* who saith also that some would haue it to be *Behen rubrum.* Some call it *Valerianthon,* others make it a kinde of *Osimastrum,* and some *Saponaria altera,* with other names, which are to no great purpose to set downe in this place, it beeing fitter for a generall worke to discusse of names, wherein both reading, know-ledge and iudgement must bee shewen, to correct errours, and set downe the truth, that one may rest thereon. The others haue their names in their titles sufficient to distinguish them.

The Vertues.

The Mountaine Valerian is of all the the rest here set downe of most vse in Physicke, the rest hauing little or none that I know, although it be much weaker then the great garden kinde, or the Indian Nardus, in whose steed anciently it was vsed, in oyles, oyntments, &c.

Chap. C.

Cardamine. Cuckow flowers, or Ladies smockes.

OF the common sorts of Cuckow flowers that grow by ditch-sides, or in moist medowes, & wet grounds, it is not my purpose here to write, but of one or two other, the most specious or faire of all the tribe, that doe best befit this garden.

1. Cardamine

1. *Cardamine flore pleno.* Double Cuckow flowers.

The double *Cardamine* hath a few winged leaues, weake and tender, lying on the ground, very like vnto the single medow kinde ; from among which riseth vp a round greene stalke, set here and there, with the like leaues that grow below, the top wherof hath a few branches, whereon stand diuers flowers, euery one vpon a small footestalk, consisting of many small whitish round leaues, a little dasht ouer with a shew of blush, set round together, which make a double flower : the roote creepeth vnder ground, sending forth small white fibres, and shooteth vp in diuers places.

2. *Cardamine trifolia.* Trefoile Ladies smockes.

This small plant hath diuers hard, darke round greene leaues, somewhat vneuen about the edges, alwayes three set together on a blackish small footstalke, among which rise vp small round blackish stalkes, halfe a foote high, with three small leaues at the ioynts, where they branch forth; at the toppes whereof stand many flowers, consisting of foure leaues a peece, of a whitish or blush colour very pale : after which come vp small, thicke and long pods, wherein is contained small round seede : the root is composed of many white threds, from the heads whereof runne out small strings, of a dark purple colour, whereby it encreaseth.

The Place.

The first with the double flower is found in diuers places of our owne Countrey, as neere Micham about eight miles from London ; also in Lancashire, from whence I receiued a plant, which perished, but was found by the industrie of a worthy Gentlewoman, dwelling in those parts heretofore remembred, called Mistresse Thomasin Tunstall, a great louer of these delights. The other was sent me by my especial good friend Iohn Tradescante, who brought it among other dainty plants from beyond the Seas, and imparted thereof a roote to me.

The Time.

The last most vsually flowreth before the former, yet not much differing, that is, in the end of Aprill or in May.

The Names.

The first is a double kinde of that plant, that growing wilde abroad, is vsually called *Cardamine altera*, and *Sisymbrium alterum* of Dioscorides, and of some *Flos cuculi*, but not fitly ; for that name is more vsually giuen vnto the wilde featherd Campions, both single and double, as is before expressed : yet for want of a fitter name, wee may call it in English, eyther Cuckowe flower, or Ladyes smockes, which you will. The second hath beene sent vnder the name of *Sanicula trifolia*, but the most frequent name now receiued, is *Cardamine trifolia*, and in English Trefoile Ladies spockes.

The Vertues.

The double Ladies smockes are of the same qualitie with the single, and is thought to be as effectuall as Watercresses. The propertie of the other I thinke is not much knowne, although some would make it a wound herbe.

CHAP. CI.

Thlaspi Creticum. Candy Tufts.

OF the many forts of *Thlaspi* it is not the fcope of this worke to relate, I will fe-lect but onely two or three, which for their beautie are fit to bee inferted into this garden.

Thlaspi Creticum vmbellatum flore albo & purpureo.
Candy Tufts white and purple.

This fmall plant rifeth feldome aboue a foote and a halfe high, hauing fmall, narrow, long and whitifh greene leaues, notched or dented with three or foure notches on each fide, from the middle to the point-wards ; from among which rife vp the ftalkes, branched from the bottome almoft into diuers fmall branches, at the toppes whereof ftand many fmall flowers, thick thruft together in an vmbell or tuft, making them feeme to be fmall, round, double flowers of many leaues, when as euery flower is fingle, and ftandeth a part by it felfe, of a faire white colour in fome plants, without any fpot, and in others with a purplifh fpot in the centre or middle, as if fome of the middle leaues were purple; in others againe the whole flower is purplifh all ouer, which make a pretty fhew in a garden : the feede is contained in many fmall and flat feed veffels, which ftand together in an vmbell, as the flowers did, in which are contained fomewhat reddifh feede, like vnto fome other forts of *Thlaspi*, called Treakle Muftards : the roote is fmall and hard, and perifheth euery yeare hauing giuen feede.

Thlaspi Mari- We haue another fort, whofe leaues before it fendeth forth any ftalke, are a little
num Baticum. toothed, or finely dented about the edges, and brancheth not fo much out, but carryeth an vmbell of purplifh flowers like vnto the former, and paler yellow feede.

The Place.

Thefe doe grow in Spaine and Candie, not farre from the Sea fide.

The Time.

Thefe *Thlaspi* giue not their flowers vntill the end of Iune, or beginning of Iuly, and the feed is ripe foone after.

The Names.

. The firft is named by fome, *Draba*, or *Arabis*, as Dodonæus, but *Draba* is another plant differing much from this. Weq call one fort, *Thlaspi Creti-cum*, and the other *Thlaspi Baticum marinum*, becaufe the one came from Spaine, and the other from Candy ; we giue it in Englifh, the name of Tufts, becaufe it doth fit the forme of the flowers beft, although ordinarily all the *Thlaspi* are Englifhed Wilde Muftardes.

The Vertues.

Candy, or Spanifh Tufts, is not fo fharpe biting in tafte, as fome other of the Thlafpies are, and therefore is not to be vfed in medicines, where *Thlaspi* fhould be in the ftead thereof.

Chap. CII.

Clematis. Clamberers, or Creepers.

Auing shewed you all my store of herbes bearing fine flowers, let mee now bring to your consideration the rest of those plants, be they Shrubs or Trees, that are cherished in our garden, for the beauty of their flowers chiefly, or for some other beautifull respect: and first I will begin with such as creepe on the ground, without climing, and then such as clime vp by poles, or other things, that are set or grow neere them, fit to make Bowers, and Arbours, or else are like them in forme, in name, or some other such qualitie or propertie.

1. *Clematis Daphnoides, siue Vinca peruinca simplex minor diuersorum colorum.*
Single Perwinkle of diuers colours.

The smaller Perwinkle which not onely groweth wilde in many places, but is most vsuall in our Gardens, hath diuers creeping branches, trayling or running vpon the ground, shooting out small fibres at the ioynts, as it creepeth, taking thereby hold in the ground, and rooteth in diuers places : at the ioynts of these branches stand two small darke greene shining leaues, somewhat like vnto small Baye leaues, but smaller, and at the ioynts likewise with the leaues, come forth the flowers, one at a ioynt, standing vpon a tender footestalke, being somewhat long and hollow, parted at the brims, sometimes into foure leaues, and sometimes into fiue, the most ordinary sort is of a pale or bleake blew colour, but some are pure white, and some of a darke reddish purple colour: the root is in the body of it, little bigger then a rush, bushing in the ground, and creeping with his branches farre about, taking roote in many places, whereby it quickely possesseth a great compasse ; and is therefore most vsually planted vnder hedges, or where it may haue roome to runne.

2. *Vinca peruinca flore duplici purpureo.*
Double purple Perwinkle.

The double Perwinkle is like vnto the former single kinde, in all things except in the flower, which is of that darke reddish purple colour that is in one of the single kindes ; but this hath another row of leaues within the flower, so that the two rowes of leaues causeth it to be called double, but the leaues of these are lesser then the single. I haue heard of one with a double white flower, but I haue not yet seene it.

3. *Clematis Daphnoides siue Peruinca maior.*
The greater Perwinkle.

This greater Perwinkle is somewhat like the former, but greater, yet his branches creepe not in that manner, but stand more vpright, or lesse creeping at the least : the leaues also hereof stand by couples at the ioynts, but they are broader and larger by the halfe : the flowers are larger, consisting of fiue leaues that are blew, a little deeper then the former blew : this plant is farre tenderer to keepe then the other, and therefore would stand warme, as well as in a moist shadowie place.

4. *Clematis altera siue vrens flore albo.*
Burning Clamberer, or Virgins Bower.

This Causticke or burning Climer, hath very long and climing tender branches, yet somewhat woody below, which winde about those things that stand neere it, couered with a brownish greene barke, from the ioynts whereof shoote forth many winged leaues, consisting for the most part of fiue single leaues, that is, two and two together, and one at the end, which are a little cut in or notched on the edges here and
there,

there, but euery part of them is leſſer then the leaues of the next following Climer, without any claſping tendrels to winde about any thing at all : towards the vpper part of the branches, with the ſaid leaues, come forth long ſtalks, wheron ſtand many white flowers cluſtering together, opening the brims into ſixe or eight ſmall leaues, ſpreading like a ſtarre, very ſweet of ſmell, or rather of a ſtrong heady ſent, which after turne into flattiſh and blackiſh ſeede, plumed at the head, which plume or feather flyeth away with the winde after it hath ſtood long, and leaueth the ſeede naked or bare : the roote is white and thicke, fleſhie and tender, or eaſie to be broken, as my ſelfe can well teſtifie, in that deſiring to take a ſucker from the roote, I could not handle it ſo tenderly, but that it broke notwithſtanding all my care. Maſter Gerard in his Herball maketh mention of one of this kinde with double white flowers, which hee ſaith he recouered from the ſeede was ſent him from Argentine, that is Strasborough, whereof hee ſetteth forth the figure with double flowers : but I neuer ſaw any ſuch with him, neither did I euer heare of any of this kinde with double flowers. Cluſius indeed ſaith, that hee receiued from a friend ſome ſeede vnder the name of *Clematis flore albo pleno* : but hee doubteth whether there bee any ſuch : the plants that ſprang with him from that ſeede, were like vnto the vpright kinde called *Flammula Matthioli*, or *Iouis creſta*, as he there ſaith : but aſſuredly I haue beene informed from ſome of my eſpeciall friends beyond Sea, that they haue a double white *Clematis*, and haue promiſed to ſend it, but whether it will be of the climing or vpright ſort, I cannot tell vntill I ſee it : but ſurely I doe much doubt whether the double will giue any good ſeede.

5. *Clematis altera ſiue peregrina flore rubro.* Red Ladies Bower.

This Climer hath many limber and weake climing branches like the former, couered with a browne thin outer barke, and greene vnderneath : the leaues ſtand at the ioynts, conſiſting but of three leaues or parts, whereof ſome are notched on one ſide, and ſome on both, without any claſping tendrels alſo, but winding with his branches about any thing ſtandeth next vnto it : the flowers in like manner come from the ſame ioynts with the leaues, but not ſo many together as the former vpon long footſtalkes, conſiſting of foure leaues a peece, ſtanding like a croſſe, of a darke red colour ; the ſeed is flat and round, and pointed at the end, three or foure or more ſtanding cloſe together vpon one ſtalk, without any doune vpon them at all, as in the former : the roots are a bundell of browniſh yellow ſtrong ſtrings, running down deep into the ground, from a bigge head aboue.

6. *Clematis peregrina flore purpureo ſimplici.* Single purple Ladies Bower.

This Ladies Bower differeth in nothing from the laſt deſcribed, but onely in the colour of the flower, which is of a ſad blewiſh purple colour ; ſo that the one is not poſſible to be known from the other, vntill they be in flower.

7. *Clematis peregrina flore purpureo pleno.* Double flowred purple Ladies Bower.

This double *Clematis* hath branches and leaues ſo neere reſembling the ſingle kinds, that there can be knowne no difference, vnleſſe it be, that this groweth more goale and great, and yeeldeth both more ſtore of branches from the ground, and more ſpreading aboue : the chiefeſt marke to diſtinguiſh it is the flower, which in this is very thicke and double, conſiſting of a number of ſmaller leaues, ſet cloſe together in order in the middle, the foure outermoſt leaues that encompaſſe them, being much broader and larger then any of the inward, but all of a dull or ſad blewiſh purple colour, the points or ends of the leaues ſeeming a little darker then the middle of them : this beareth no ſeede that euer I could ſee, heare of, or learne by any of credit, that haue nourſed it a great while ; and therefore the tales of falſe deceitfull gardiners, and others, that diliuer ſuch for truth, to deceiue perſons ignorant thereof, muſt not bee credulouſly entertained.

In

In the great booke of the Garden of the Bishop of Eystot (which place is neere ~vnto Noremberg) in Germany, I reade of a *Clematis* of this former kinde, whose figure is thereto also annexed, with double flowers of an incarnate, or pale purple tending to a blush colour, whereof I haue not heard from any other place.

Clematis peregrina flore carneccarueo.

8. *Flammula Iouis erecta.* Vpright Virgins Bower.

This kinde of *Clematis* hath diuers more vpright stalkes then any of the foure last described, sometimes foure or fiue foote high, or more; yet leaning or bending a little, so that it had some neede of sustaining, couered with a brownish barke; from whence come forth on all sides diuers winged leaues, consisting of fiue or seuen leaues, set on both sides of a middle ribbe, whereof one is at the end: the tops of the stalkes are diuided into many branches, bearing many white sweet smelling flowers on them, like in fashion vnto the white Virgins Bower; after which come such like feather topt seede, which remaine and shew themselues, being flat like the other, when the plumes are blowne abroad: the roote spreadeth in the ground from a thicke head, into many long strings, and fasteneth it selfe strongly in the earth; but all the stalkes dye downe euery yeare, and spring afresh in the beginning of the next.

9. *Clematis carulea Pannonica.* The Hungarian Climer.

The stalks of this plant stand vpright, & are foure square, bearing at euery ioynt two leaues, which at the first are closed together, and after they are open, are somewhat like vnto the leaues of *Asclepias*, or Swallow-wort: from the tops of the stalks, and sometimes also from the sides by the leaues commeth forth one flower, bending the head downward, consisting of foure leaues, somewhat long & narrow, standing like a crosse, and turning vp their ends a little againe, of a faire blew or skie colour, with a thicke pale yellow short thrumme, made like a head in the middle: after the flower is past, the head turneth into such a like round feather topt ball, as is to be seene in the Trauellers ioy, or *Viorna* (as it is called) that groweth plentifully in Kent, and in other places by the way sides, and in the hedges, wherein is included such like flat seede. These stalkes (like as the last) dye downe to the ground euery yeare, and rise againe in the Spring following, shooting out new branches, and therby encreaseth in the root.

10. *Maracoc siue Clematis Virginiana.* The Virginia Climer.

Because this braue and too much desired plant doth in some things resemble the former Climers, so that vnto what other family or kindred I might better conioyne it I know not; let me I pray insert it in the end of their Chapter, with this description. It riseth out of the ground (very late in the yeare, about the beginning of May, if it be a plant hath risen from the seed of our owne sowing, and if it be an old one, such as hath been brought to vs from Virginia, not till the end thereof) with a round stalke, not aboue a yard and a halfe high (in any that I haue seene) but in hotter Countries, as some Authors haue set it downe, much higher, bearing one leafe at euery ioynt, which from the ground to the middle thereof hath no claspers, but from thence vpwards hath at the same ioynt with the leafe both a small twining clasper, like vnto a Vine, and a flower also: euery leafe is broad at the stalke thereof, and diuided about the middle on both sides, making it somewhat resemble a Figge leafe, ending in three points, whereof the middlemost is longest: the bud of the flower, before it doe open, is very like vnto the head or seede vessell of the ordinary single *Nigella*, hauing at the head or top fiue small crooked hornes, which when this bud openeth, are the ends or points of fiue leaues, that are white on the inside, and lay themselues flat, like vnto an Anemone, and are a little hollow like a scoope at the end, with fiue other smaller leaues, and whiter then they lying betweene them, which were hid in the bud before it opened, so that this flower being full blowne open, consisteth often white leaues, laide in order round one by another: from the bottome of these leaues on the inside, rise diuers twined threads, which spread and lay themselues all ouer these white leaues, reaching beyond the points of them a little, and are of a reddish peach colour: towards the bottomes

tomes likewife of thefe white leaues there are two red circles, about the breadth of an Oten ftrawe, one diftant from another (and in fome flowers there is but one circle feen) which adde a great grace vnto the flower, for the white leaues fhew their colour through the peach coloured threads, and thefe red circles or rings vpon them being alfo perfpicuous, make a tripartite fhew of colours moft delightfull: the middle part of this flower is hollow, and yellow-ifh; in the bottome whereof rifeth vp an vmbone, or round ftile, fomewhat bigge, of a whitifh greene colour, fpotted with reddifh fpots like the ftalkes of Dragons, with fiue round threads or chiues, fpot-ted in the like manner, and tipt at the ends with yellow pendents, ftanding a-bout the middle part of the faid vm-bone, and from thence rifing higher, en-deth in three long crooked hornes moft vfually (but fometimes in foure, as hath beene obferued in Rome by D^r. Aldine, that fet forth fome principall things of Cardinall Farnefius his Garden) fpotted like the reft, hauing three round greene buttons at their ends: thefe flowers are of a comfortable fweete fent, very ac-ceptable, which perifh without yeelding fruit with vs, becaufe it flowreth fo late: but in the naturall place; and in hot Countries, it beareth a fmall round whi-tifh fruit, with a crowne at the toppe thereof, wherein is contained (while it is frefh, and before it be ouer dried) a fweet

The Iefuites Figure of the Maracoc.

GRANADILLVS FRVTEX INDICVS
CHRISTI PASSIONIS IMAGO.

liquor, but when it is dry, the feede within it, which is fmall, flat, fomewhat rough and blacke, will make a ratling noife: the rootes are compofed of a number of excee-ding long and round yellowifh browne ftrings, fpreading farre abroad vnder the ground (I haue feene fome rootes that haue beene brought ouer, that were as long as any rootes of *Sarfa parilla*, and a great deale bigger, which to be handfomely laid into the ground, were faine to be coyled like a cable) and fhooting vp in feuerall places a good diftance one from another, whereby it may be well encreafed.

The Place.

The firft blew Perwinkle groweth in many Woods and Orchards, by the hedge fides in England, and fo doth the white here and there, but the other fingle and double purple are in our Gardens onely. The great Per-winkle groweth in Prouence of France, in Spaine, and Italy, and other hot Countries, where alfo growe all the twining Clamberers, as well fingle as double: but both the vpright ones doe growe in Hungary and therea-bouts. The furpaffing delight of all flowers came from Virginia. Wee preferue them all in our Gardens.

The Time.

The Perwinkles doe flower in March and Aprill. The Climers not vn-till the end of Iune, or in Iuly, and fometimes in Auguft. The Virginian fomewhat later in Auguft; yet fometimes I haue knowne the flower to fhew it felfe in Iuly.

The Names.

The firft is out of queftion the firft *Clematis* of Diofcorides, and called of many

many *Clematis Daphnoides* (but not that plant that is ſimply called *Daphnoides*, for that is *Laureola*) and is vſually called *Vinca pervinca*: but it is not *Chamædaphne*, for that is another plant, as ſhall be ſhewed in his place; ſome call it *Centunculus*: In Engliſh wee call it Perwinkle. The other is *Clematis altera* of Dioſcorides, and is called alſo *Clematis peregrina*, whoſe diſtinctions are ſet downe in their titles: In Engliſh, Ladies Bower, or Virgins Bower, becauſe they are fit to growe by Arbours, to couer them. The firſt vpright Clamberer is called, and that rightly of ſome, *Clematis erecta*, or *ſurrecta*. Of others, *Flammula frutex*, and *Flammula Iouis*, or *ſurrecta*: In Engliſh, Vpright Virgins Bower. The next is called by Cluſius, *Clematis Pannonica cærulea*, who thought it to be *Climeni ſpecies*, by the relation of others, at the firſt, but after entituled it, *Clematis*: In Engliſh, the Hungarian Climer. The laſt may be called in Latine, *Clematis Virginiana*: In Engliſh, The Virgin or Virginian Climer; of the Virginians, *Maracoc*: of the Spaniards in the Weſt Indies *Granadillo*, becauſe the fruit (as is before ſaid) is in ſome faſhion like a ſmall Pomegranate on the outſide; yet the ſeede within is flattiſh, round, and blackiſh. Some ſuperſtitious Ieſu-ite would faine make men beleeue, that in the flower of this plant are to be ſeene all the markes of our Sauiours Paſſion; and therefore call it *Flos Paſſionis*: and to that end haue cauſed figures to be drawne, and printed, with all the parts proportioned out, as thornes, nailes, ſpeare, whippe, pillar, &c. in it, and all as true as the Sea burnes, which you may well perceiue by the true figure, taken to the life of the plant, compared with the figures ſet forth by the Ieſuites, which I haue placed here likewiſe for euery one to ſee: but theſe bee their aduantagious lies (which with them are tolerable, or rather pious and meritorious) wherewith they vſe to inſtruct their people; but I dare ſay, God neuer willed his Prieſts to inſtruct his people with lyes: for they come from the Diuell, the author of them. But you may ſay I am beſide my Text, and I am in doubt you will thinke, I am in this beſides my ſelfe, and ſo nothing to be beleeued herein that I ſay. For, for the moſt part, it is an inherent errour in all of that ſide, to beleeue nothing, be it neuer ſo true, that any of our ſide ſhall affirme, that contrarieth the aſſertions of any of their Fathers, as they call them: but I muſt referre them to God, and hee knoweth the truth, and will reforme or deforme them in his time. In regard whereof I could not but ſpeake (the occaſion being thus offered) againſt ſuch an erroneous opinion (which euen Dr. Aldine at Rome, before remembred, diſproued, and contraried both the ſaid figures and name) and ſeek to diſproue it, as doth (I ſay not almoſt, but I am affraid altogether) leade many to adore the very picture of ſuch things, as are but the fictions of ſuperſtitious brains: for the flower it ſelfe is farre differing from their figure, as both Aldine in the aforeſaid booke, and Robinus at Paris in his *Theatrum Flora*, doe ſet forth; the flowers and leaues being drawne to the life, and there exhibited, which I hope may ſatisfie all men, that will not be perpetually obſtinate and contentious.

The Vertues.

Coſtæus ſaith hee hath often ſeene, that the leaues of Perwinkle held in the mouth, hath ſtayed the bleeding at the noſe. The French doe vſe it to ſtay the menſtruall fluxes. The other are cauſticke plants, that is, fiery hot, and bliſtering the ſkinne; and therefore (as Dioſcorides ſaith) is profitable to take away the ſcurfe, lepry, or ſuch like deformities of the skin. What property that of Virginia hath, is not knowne to any with vs I thinke, more then that the liquor in the greene fruit is pleaſant in taſte; but aſſuredly it cannot be without ſome ſpeciall properties, if they were knowne.

Chap. CIII.

Chamælæ. Dwarfe Spurge Oliue, or Dwarfe Baye.

I Haue three forts of *Chamælæa* to bring to your confideration, euery one differing notably from other; two of them of great beauty in their flowers, as well as in the whole plant: the third abiding with greene leaues, although it haue no beauty in the flower, yet worthy of the place it holds. And vnto thefe I muft adioyne another plant, as comming neareft vnto them in the brauery of the flowers.

1. *Chamælæa Germanica fiue Mezereon floribus dilutioris coloris & faturatioris.*
Dwarfe Baye, or flowring Spurge Oliue.

We haue two forts of this Spurge Oliue or Dwafe Baye, differing onely in the co-lour of the flowers. They both rife vp with a thicke wooddy ftemme, fiue or fix foot high fometimes, or more, and of the thickneffe (if they be very old) of a mans wreft at the ground, fpreading into many flexible long branches, couered with a tough grayifh barke, befet with fmall long leaues, fomewhat like vnto Priuet leaues, but fmaller and paler, and in a manner round pointed: the flowers are fmall, confifting of foure leaues, many growing together fometimes, and breaking out of the branches by themfelues: in the one fort of a pale red at the firft blowing, and more white afterwards; the other of a deeper red in the bloffome, and continuing of a deeper red colour all the time of the flowring, both of them very fweete in fmell: after the flowers are paft, come the berries, which are greene at the firft, and very red afterwards, turning blackifh red, if they ftand too long vpon the branches: the rootes fpread into many tough long bran-ches, couered with a yellowifh barke.

2. *Chamælæa Alpina.* Mountaine Spurge Oliue.

This Mountaine Laurell rifeth vp with a fmall wooddy ftemme, three or foure foot high, or more, branching forth towards the vpper parts into many flender and tough branches, couered with a rough hoary greene barke, befet at the ends thereof with flatter, fuller, and fmaller round pointed leaues then the former, of a grayifh greene colour on the vpperfide, and hoary vnderneath, which abide on the branches in Winter, and fall not away as the former: the flowers are many fet together at the ends of the branches, greater then the former, and confifting of foure leaues a peece, of a light blufh colour, ftanding in fmall grayifh huskes, of little or no fent at all: the fruit followeth, which are fmall long graines or berries, of an excellent red colour, which afterwards turne blacke: the roote is long, and fpreadeth about vnder the vpper part of the earth.

3. *Chamælæa tricoccos.* Widowe Wayle.

This three berried Spurge Oliue hath no great ftemme at all, but the whole plant fpreadeth from the ground into many flexible tough greene branches, whereon are fet diuers narrow, long, darke greene leaues all along the branches, which abide greene all the Winter: the flowers are very fmall, fcarce to be feene, and come forth between the leaues and the ftalke, of a pale yellow colour, made of three leaues; after which come fmall blackifh berries, three vfually fet together: the roote fpreadeth it felfe in the ground not very farre, being hard and wooddy, and often dyeth, if it bee not well defended from the extremity of our fharpe Winters.

4. *Cneorum Matthioli.* Small Rocke Rofes.

I was long in doubt in what place I fhould difpofe of this plant, whether among the Campions, as Bauhinus, or among thefe, as Clufius doth; but left my Gorden fhould want it wholly, let it take vp roome for this time here. This gallant plant hath diuers long, weake, slender, but yet tough branches lying vpon the ground, diuided vsually into other smaller branches, whereon growe many, small, long, and somewhat thicke leaues, somewhat like vnto the leaues of the former *Mezereon,* set without any order to the very tops, from whence doe come forth a tuft of many small flowers together, made or consisting of foure leaues a peece, of a bright red or carnation colour, and very sweete withall, which turne into small round whitish berries, wherein is contained small round seede, couered with a grayish coate or skinne: the roote is long and yellowish, spreading diuers wayes vnder the ground, and abideth many yeares shooting forth new branches.

Flore albo.   It hath beene obserued in some of these plants, to bring forth white flowers, not differing in any thing else.

The Place.

The first sorts growe plentifully in many places of Germany. The second in the mountaines by Sauoye. The third in Prouence and Spaine. The last in diuers parts of Germany, Bohemia, and Austria, and about Franckford.

The Time.

The two first sorts are most vsually in flower about Christmas, or in Ianuary, if the weather be not violent, and sometimes not vntill February. The second flowreth not vntill Aprill. The third in May. The berries of them ripen some in Iune and Iuly; some in August and September, as their flowring is earlier or later. The last flowreth as well in the Spring as in Autumne, so apt and plentifull it is in bearing, and the seede at both times doth ripen soone after.

The Names.

The first is called of some *Chamælæa,* with this addition *Germanica,* that it may differ from the third, which is the true *Chamælæa* of Dioscorides, as all the best Authors doe agree, and is also called *Piper montanum* of the Italians. It is generally called *Mezereon,* and is indeede the true *Mezereon* of the Arabians, and so vsed in our Apothecaries shops, wheresoeuer the Arabians *Mezereon* is appointed, although the Arabians are so intricate and vncertaine in the descriptions of their plants, confounding *Chamælæa* and *Thymælæa* together. Matthiolus maketh it to be *Daphnoides* of Dioscorides; but in my opinion he is therein mistaken: for all our best moderne Writers doe account our *Laureola,* which hath blacke berries, to bee the true *Daphnoides:* the errour of his Countrey might peraduenture drawe him thereunto; but if hee had better considered the text of Dioscorides, that giueth black berries to *Daphnoides,* and red to *Chamædaphne,* he would not so haue written; and truly, I should thinke (as Lobel doth) with better reason, that this *Chamælæa* were Dioscorides *Chamædaphne,* then hee to say it were *Daphnoides:* for the description of *Chamædaphne,* may in all parts be very fitly applyed to this *Chamælæa:* and euen these words, *Semen annexum folijs,* wherein may be the greatest doubt in the description, may not vnfitly bee construed, that as is seene in the plant, the berries growe at the foote of the leaues, about the branches: the faculties indeede that Dioscorides giueth to *Chamædaphne,* are (if any repugnancie be) the greatest let or hinderance, that this *Chamælæa* should not be it: but I leaue the discussing of these and others of the like nature, to our learned Physitians; for I deale not so much with vertues as with descriptions. The second is called of Lobel *Chamælæa Alpina incana,* of Clusius *Chamælæa secunda,* and saith hee had it out of Italy. Wee may call it in English, Mountaine Spurge Oliue, as it is in the description, or Mountaine Laurell, which you will. The last hath the name of *Cneorum,* first giuen it by Matthiolus, which since is continued by all others. Bauhinus (as I said) referreth it to the Mountaine Campions, but Clusius

(as

1 *Chamalæa Germanica seu Mezereon.* Mezereon or Dwarfe Bay. 2 *Chamalæa Alpina.* Mountain Spurge Oliue. 3 *Cneorum Matthioli.* Small Rocke Roles. 4 *Laurus Tinus sine siluestris.* The wild Bay tree. 5 *Oleander sine Laurus Rosea.* The Rose Bay tree. 6 *Laurocerasus.* The Bay Cherrie tree.

(as I doe) to the kindes of *Chamalaa* or *Thymalaa*. For want of an Englifh name I haue (as you fee, and that is according to the name the Germane women, as Clufius faith, doe call it) entituled it the Small Rocke Rofe; which may abide vntill a fitter may be conferred vpon it.

The Vertues.

All thefe plants except the laft, as well leaues as berries, are violent purgers, and therefore great caution is to bee had in the vfe of them. The laft hath not beene applyed for any difeafe that I know.

Chap. CIII.

Laurus. The Bay Tree.

MY meaning is not to make any defcription of our ordinary Bayes in this place (for as all may very well know, they may be for an Orchard or Courtyard, and not for this Garden) but of two or three other kindes, whofe beautifull afpect haue caufed them to be worthy of a place therein : the one is called *Laurus Tinus,* The wilde Baye : the other *Laurus Rofea* or *Oleander,* The Rofe Bay : and a third is *Laurocerafus,* The Cherry Bay ; which may haue not onely fome refpect for his long bufh of fweet fmelling flowers, but efpecially for the comely ftatelineffe of his gallant euer frefh greene leaues ; and the rather, becaufe with vs in moft places, it doth but *frutefcere*, vfe to bee Shrub high, not *arborefcere*, Tree high, which is the more fit for this Garden.

1. *Laurus Tinus fiue filueftris.* The wilde Bay tree.

This wilde Baye groweth feldome to bee a tree of any height, but abideth for the moft part low, fhooting forth diuers flender branches, whereon at euery ioynt ftand two leaues, long, fmooth, and of a darke greene colour, fomewhat like vnto the leaues of the Female Cornell tree, or between that and Baye leaues : at the toppes of the branches ftand many fmall white fweete fmelling flowers, thrufting together, as it were in an vmbell or tuft, confifting of fiue leaues a peece, the edges whereof haue a fhew of a wafh purple, or light blufh in them, which for the moft part fall away without bearing any perfect ripe fruit in our Countrey : Yet fometimes it hath fmall black berries, as if they were good, but are not. In his naturall place it beareth fmall, round, hard and pointed berries, of a fhining blacke colour, for fuch haue come often to my hands (yet Clufius writeth they are blew) ; but I could neuer fee any fpring that I put into the ground. This that I here defcribe, feemeth to me to be neither of both thofe that Clufius faw growing in Spain and Potugall, but that other, that (as he faith) fprang in the low Countreyes of Italian feede.

2. *Laurus Rofea fiue Oleander.* The Rofe Bay.

Of the Rofe Bay there are two forts, one bearing crimfon coloured flowers, which is more frequent, and the other white, which is more rare. They are fo like in all other things, that they neede but one defcription for both. The ftemme or trunke is many times with vs as bigge at the bottome as a good mans thumbe, but growing vp fmaller, it diuideth it felfe into branches, three for the moft part comming from one ioynt or place, and thofe branches againe doe likewife diuide themfelues into three other, and fo by degrees from three to three, as long as it groweth : the loweft of thefe are bare of leaues, hauing fhed or loft them by the cold of winters, keeping onely leaues on the vppermoft branches, which are long, and fomewhat narrow, like in forme vnto Peach leaues, but thicker, harder, and of a darke greene colour on the vpperfide, and

yellowifh

yellowish greene vnderneath : at the tops of the young branches come forth the flowers, which in the one fort before they are open, are of an excellent bright crimson colour, and being blowen, confift of foure long and narrow leaues, round pointed, fomewhat twining themfelues, of a paler red colour, almoft tending to blufh, and in the other are white, the greene leaues alfo being of a little frefher colour : after the flowers are paft, in the hot countries, but neuer in ours, there come vp long bending or crooked flat pods, whofe outward fhell is hard, almoft woody, and of a browne colour, wherein is contained fmall flat brownifh feede, wrapped in a great deale of a brownifh yellow doune, as fine almoft as filke, fomewhat like vnto the huskes of *Afclepias*, or *Periploca*, but larger, flatter and harder ; as my felfe can teftifie, who had fome of the pods of this Rofe bay, brought mee out of Spaine, by Mafter Doctor Iohn More, the feedes whereof I fowed, and had diuers plants that I raifed vp vnto a reafonable height, but they require, as well old as young, to bee defended from the colde of our winters.

3. *Laurocerafus.* The Bay Cherry.

This beautifull Bay in his naturall place of growing, groweth to bee a tree of a reafonable bigneffe and height, and oftentimes with vs alfo if it bee pruined from the lower branches ; but more vfually in thefe colder Countries, it groweth as a fhrub or hedge bufh, fhooting forth many branches, whereof the greater and lower are couered with a darke grayifh greene barke, but the young ones are very greene, whereon are fet many goodly, faire, large, thicke and long leaues, a little dented about the edges, of a more excellent frefh fhining greene colour, and farre larger then any Bay leafe, and compared by many to the leaues of the *Pomeritron* tree (which becaufe wee haue none in our Countrey, cannot be fo well known) both for colour and largeneffe, which yeeld a moft gracefull afpect : it beareth long ftalkes of whitifh flowers, at the ioynts of the leaues both along the branches and towards the ends of them alfo, like vnto the Birds Cherry or *Padus Theophrafti*, which the French men call *Putier & Cerifier blanc*, but larger and greater, confifting of fiue leaues with many threds in the middle: after which commeth the fruite or berries, as large or great as Flanders Cherries, many growing together one by another on a long ftalke, as the flowers did, which are very blacke and fhining on the outfide, with a little point at the end, and reafonable fweete in tafte, wherein is contained a hard round ftone, very like vnto a Cherry ftone, as I haue obferued as well by thofe I receiued out of Italie, as by them I had of Mafter Iames Cole a Merchant of London lately deceafed, which grew at his houfe in Highgate, where there is a faire tree which hee defended from the bitterneffe of the weather in winter by cafting a blanket ouer the toppe thereof euery yeare, thereby the better to preferue it.

The Place.

The firft is not certainly knowne from whence it came, and is communicated by the fuckers it yeeldeth. The fecond groweth in Spaine, Italie, Grece, and many other places : that with white flowers is recorded by Bellonius, to grow in Candy. The laft, as Matthiolus, and after him Clufius report, came firft from Conftantinople : I had a plant hereof by the friendly gift of Mafter Iames Cole, the Merchant before remembred, a great louer of all rarities, who had it growing with him at his countrey houfe in Highgate aforefaid, where it hath flowred diuers times, and borne ripe fruit alfo.

The Time.

The firft flowreth many times in the end of the yeare before Chriftmas, and often alfo in Ianuary, but the moft kindly time is in March and Aprill, when the flowers are fweeteft. The fecond flowreth not vntill Iuly. The laft in May, and the fruit is ripe in Auguft and September.

The

The Names.

The firſt is called *Laurus ſilueſtris*, and *Laurus Tinus* : in Engliſh Wilde Bay, or Sweete flowring Bay. The ſecond is called *Laurus Roſea, Oleander, Nerium*, and *Rhododendros* : in Engliſh The Roſe Bay, and Oleander. The laſt was ſent by the name of *Trebezon Curmaſi*, that is to ſay, *Dactylus Trapezuntina*, but not hauing any affinitie with any kinde of Date, Bellonius as I thinke firſt named it *Laurocera/us*, and *Ceraſus Trapezuntina*. Dalechampius thinketh it to bee *Lotus Aphricana*, but Cluſius refuteth it. Thoſe ſtones or kernels that were ſent me out of Italie, came by the name of *Laurus Regia*, The Kings Bay. Wee may moſt properly call it according to the Latine name in the title, The Cherry bay, or Bay Cherry, becauſe his leaues are like vnto Bay-leaues, and both flowers and fruit like vnto the Birdes Cherry or Cluſter Cherry, for the manner of the growing ; and therfore I might more fitly I confeſſe haue placed it in my Orchard among the ſorts of Cherries : but the beautifulneſſe of the plant cauſed mee rather to inſert it here.

The Vertues.

The wilde Bay hath no propertie allotted vnto it in Phyſicke, for that it is not to be endured, the berries being chewed declare it to be ſo violent hot and choking. The Roſe Bay is ſaid by Dioſcorides, to be death to all foure footed beaſts, but contrariwiſe to man it is a remedie againſt the poiſon of Serpents, but eſpecially if Rue bee added vnto it. The Cherry Bay is not knowne with vs to what phyſicke vſe it may be applyed.

CHAP. CIIII.

Ceraſus flore multiplici. The Roſe or double bloſſomd Cherry.
Malus flore multiplici. The double bloſſomd Apple tree. And
Malus Perſica flore multiplici. The double bloſſomd Peach tree.

THe beautifull ſhew of theſe three ſorts of flowers, hath made me to inſert them into this garden, in that for their worthineſſe I am vnwilling to bee without them, although the reſt of their kindes I haue tranſferred into the Orchard, where among other fruit trees, they ſhall be remembred : for all theſe here ſet downe, ſeldome or neuer beare any fruite, and therefore more fit for a Garden of flowers, then an Orchard of fruite.

Ceraſus flore pleno vel multiplici.
The Roſe Cherry, or double bloſſomd Cherry.

The double bloſſomed Cherry tree is of two ſorts for the flower, but not differing in any other part, from the ordinary Engliſh or Flanders Cherry tree, growing in the very like manner : the difference conſiſteth in this, that the one of theſe two ſorts hath white flowers leſſe double, that is, of two rowes or more of leaues, and the other more double, or with more rowes of leaues, and beſides I haue obſerued in this greater double bloſſomd Cherry, that ſome yeares moſt of the flowers haue had another ſmaller and double flower, riſing vp out of the middle of the other, like as is to bee ſeene in the double Engliſh Crow-foote, and double redde *Ranunculus* or Crowfoote, before deſcribed : this I ſay doth not happen euery yeare, but ſometimes. Sometimes alſo theſe trees will giue a few berries, here and there ſcattered, and that with leſſe double flowers more often, which are like vnto our Engliſh Cherries both for taſte and bigneſſe. Theſe be very fit to be ſet by Arbours.

Malus

Malus flore multiplici. The double bloſſomd Apple tree.

This double bloſſomd Apple tree is altogether like vnto our ordinary Pippin tree in body, branch and leafe, the only difference is in the flower, which is altogether whitiſh, ſauing that the inner leaues towards the middle are more reddiſh, but as double and thicke as our double Damaske Roſes, which fall away without bearing fruit.

Malus Perſica flore multiplici. The double bloſſomd Peach tree.

This Peach tree for the manner of growing, is ſo like vnto an ordinary Peach tree, that vntill you ſee it in bloſſome you can perceiue no difference : the flower is of the ſame colour with the bloſſomes of the Peach, but conſiſting of three or foure, or more rowes of leaues, which fall often away likewiſe without bearing any fruite ; but after it hath abiden ſome yeares in a place doth forme into fruite, eſpecially being planted againſt a wall.

The Place.

Both the Cherry trees are frequent in many places of England, nourſed for their pleaſant flowers. The Apple is as yet a ſtranger. And the Peach hath not been ſeen or knowne, long before the writing hereof.

The Time.

They all flower in April & May, which are the times of their other kinds.

The Names.

Their names are alſo ſufficiently expreſſed to know them by.

The Vertues.

Cherries, Peaches and Apples, are recorded in our Orchard, and there you ſhall finde the properties of their fruit: for in that theſe beare none or very few, their bloſſomes are of moſt vſe to grace and decke the perſons of thoſe that will weare or beare them.

Chap. CV.

Periclymenum. Honyſuckles.

THe Honiſuckle that groweth wilde in euery hedge, although it be very ſweete, yet doe I not bring into my garden, but let it reſt in his owne place, to ſerue their ſenſes that trauell by it, or haue no garden. I haue three other that furniſh my Garden, one that is called double, whoſe branches ſpreade far, and being very fit for an arbour will ſoone couer it : the other two ſtand vpright, and ſpreade not any way far, yet their flowers declaring them to be Honiſuckles, but of leſſe delight, I conſort them with the other.

Periclymenum perfoliatum ſiue Italicum. The double Honiſuckle.

The truncke or body of the double Honiſuckle, is oftentimes of the bigneſſe of a good ſtaffe, running out into many long ſpreading branches, couered with a whitiſh barke, which had neede of ſome thing to ſuſtaine them, or elſe they will fall down to the ground (and therefore it is vſually planted at an arbour, that it may run thereon,

or

or againſt a houſe wall, and faſtened thereto in diuers places with nailes) from whence ſpring forth at ſeuerall diſtances, and at the ioynts, two leaues, being like in forme vnto the wilde Honiſuckles, and round pointed for the moſt part; theſe branches diuiding themſelues diuers wayes, haue at the toppes of them many flowers, ſet at certaine diſtances one aboue another, with two greene leaues at euery place, where the flowers doe ſtand, ioyned ſo cloſe at the bottome, and ſo round and hollow in the middle, that it ſeemeth like a hollow cuppe or ſawcer of flowers : the flowers ſtand round about the middle of theſe cuppes or ſawcers, being long, hollow, and of a whitiſh yellow colour, with open mouthes daſht ouer with a light ſhew of purple, and ſome threds within them, very ſweet in ſmell, like both in forme and colour vnto the common Honiſuckles, but that theſe cuppes with the flowers in them are two or three ſtanding one aboue another (which make a far better ſhew then the common, which come forth all at the heade of the branches, without any greene leaues or cuppes vnder them) and therefore theſe were called double Honiſuckles.

Periclymenum rectum fructu rubro. Red Honiſuckles.

This vpright Woodbinde hath a ſtraight woody ſtemme, diuided into ſeuerall branches, about three or foure foote high, couered with a very thinne whitiſh barke, whereon ſtand two leaues together at the ioynts, being leſſer then the former, ſmooth and plaine, and a little pointed : the flowers come forth vpon ſlender long footſtalks at the ioynts where the leaues ſtand, alwayes two ſet together, and neuer more, but ſeldome one alone, which are much ſmaller then the former, but of the ſame faſhion, with a little button at the foote of the flower; the buds of the flowers before they are open are very reddiſh, but being open are not ſo red, but tending to a kinde of yellowiſh bluſh colour : after which come in their places two ſmall red berries, the one withered for the moſt part, or at leaſt ſmaller then the other, but (as Cluſius ſaith) in their naturall places they are both full and of one bigneſſe.

Periclymenum rectum fructu caruleo. Blew berried Honiſuckles.

This other vpright Woodbinde groweth vp as high as the former, or rather ſomewhat higher, couered with a blackiſh rugged barke, chapping in diuers places, the younger branches whereof are ſomewhat reddiſh, and couered with an hoary doune : the leaues ſtand two together at the ioints, ſomewhat larger then the former, and more whitiſh vnderneath : the flowers are likewiſe two ſtanding together, at the end of a ſlender footeſtalke, of a pale yellowiſh colour when they are blowne, but more reddiſh in the bud : the berries ſtand two together as the former, of a darke blewiſh colour when they are fully ripe, and full of a red liquour or iuice, of a pleaſant taſte, which doth not only dye the hands of them that gather them, but ſerueth for a dying colour to the inhabitants where they grow plentifully, wherein are contained many flat ſeede : The roote is woody as the former is.

The Place.

The firw groweth in Italie, Spaine, and Prouence of France, but not in the colder countreyes, vnleſſe it be there planted, as is moſt frequent in our countrey. The others grow in Auſtria, and Stiria, as Cluſius ſaith, and are entertained into their gardens onely that are curious.

The Time.

The firſt flowreth vſually in Aprill, the reſt in May.

The Names.

The firſt is called *Periclymenum, Caprifolium perfoliatum,* and *Italicum,* as a difference from the common kinde : In Engliſh Double Woodbinde,
or

or double Honifuckles. The others, as they are rare, and little knowne, fo are their names alfo : yet according to their Latine, I haue giuen them English names.

The Vertues.

The double Honifuckle is as effectuall in all things, as the fingle wilde kinde, and befides, is an efpeciall good wound herbe for the head or other parts. I haue not knowne the vpright kindes vfed in Phyficke.

CHAP. CVI.

Iafminum fiue Gelfeminum. Iafmine or Gefmine.

WE haue but one fort of true Iafmine ordinarily in our Gardens through-out the whole Land ; but there is another greater fort, which is farre more tender, brought out of Spaine, and will hardly endure any long time with vs, vnleffe it be very carefully preferued. Wee haue a third kinde called a yellow Iafmine, but differeth much from their tribe in many notable points : but be-caufe the flowers haue fome likeneffe with the flowers of the true Iafmine, it hath been vfually called a Iafmine ; and therefore I am content for this Garden to conioyne them in one Chapter.

1. *Iafminum album.* The white Iafmine.

The white Iafmine hath many twiggy flexible greene branches, comming forth of the fundry bigger boughes or ftems, that rife from the roote, which are couered with a grayifh darke coloured barke, hauing a white pith within it like the Elder, but not fo much : the winged leaues ftand alwaies two together at the ioynts, being made of ma-nie fmall and pointed leaues, fet on each fide of a middle ribbe, fix moft vfually on both fides, with one at the end, which is larger, more pointed then any of the reft, and of a darke greene colour : at the toppes of the young branches ftand diuers flow-ers together, as it were in an vmbell or tuft, each whereof ftandeth on a long greene ftalke, comming out of a fmall huske, being fmall, long, and hollow belowe, ope-ning into fiue white fmall, pointed leaues, of a very ftrong fweete fmell, which fall away without bearing any fruit at all, that euer I could learne in our Country ; but in the hot Countries where it is naturall, it is faid to beare flat fruit, like Lupines : the rootes fpread farre and deepe, and are long and hard to growe, vntill they haue taken ftrong hold in the ground.

2. *Iafminum Catalonicum.* The Spanifh Iafmine.

This Catalonia Iafmine groweth lower then the former, neuer rifing halfe fo high, and hath flender long greene branches, rifing from the toppe of the wooddy ftemme, with fuch like leaues fet on them as the former, but fomewhat fhorter and larger : the flowers alfo are like vnto the former, and ftand in the fame manner at the end of the branches, but are much larger, being of a blufh colour before they are blowne, and white with blufh edges when they are open, exceeding fweete of fmell, more ftrong then the former.

3. *Iafminum luteum, fiue Trifolium fruticans alÿs Polemonium.* The yellow Iafmine.

This that is called the yellow Iafmine, hath many long flender twiggy branches ri-fing from the roote, greene at the firft, and couered with a darke grayifh barke after-wards, whereon are fet at certaine diftances, three fmall darke greene leaues together, the end leafe being alwaies the biggeft : at the ioynts where the leaues come forth,
ftand

stand long stalkes, bearing long hollow flowers, ending in fiue, and some in six leaues, very like vnto the flowers of the first Iasmine, but yellow, whereupon it is vsually called the Yellow Iasmine : after the flowers are past, there come in their places round blacke shining berries, of the bignesse of a great Pease, or bigger, full of a purplish iuyce, which will dye ones fingers that bruise them but a little : the roote is tough, and white, creeping farre about vnder the ground, shooting forth plentifully, whereby it greatly encreaseth.

The Place.

The first is verily thought to haue been first brought to Spaine out of Syria, or thereabouts, and from Spaine to vs, and is to be seene very often, and in many of our Country Gardens. The second hath his breeding in Spaine also, but whether it be his originall place we know not, and is scarce yet made well acquainted with our English ayre. The third groweth plentifully about Mompelier, and will well abide in our London Gardens, and any where else.

The Time.

The first flowreth not vntill the end of Iuly. The second somewhat earlier. The third in Iuly also.

The Names.

The first is generally called *Iasminum album*, and *Gelseminum album* : In English, The white Iasmine. The second hath his name in his title, as much as may be said of it. The third hath been taken of some to be a *Cytisu*, others iudge it to be *Polemonium*, but the truest name is *Trifolium fruticans*, although many call it *Iasminum luteum* : In English most vsually, The yellow Iasmine, for the reasons aforesaid ; or else after the Latine name, Shrubbie Trefoile, or Make-bate.

The Vertues.

The white Iasmines haue beene in all times accepted into outward medicines, eyther for the pleasure of the sweete sent, or profit of the warming properties. And is in these dayes onely vsed as an ornament in Gardens, or for sent of the flowers in the house, &c. The yellow Iasmine, although some haue adiudged it to be the *Polemonium* of Dioscorides, yet it is not vsed to those purposes by any that I know.

CHAP. CVII.

Syringa. The Pipe tree.

VNder the name of *Syringa*, is contained two speciall kinds of Shrubs or Trees, differing one from another ; namely, the *Lilac* of Matthiolus, which is called *Syringa cærulea*, and is of two or three sorts : And the *Syringa alba*, which also is of two sorts, as shall bee declared.

1. *Lilac siue Syringa cærulea.* The blew Pipe tree.

The blew Pipe tree riseth sometimes to be a great tree, as high and bigge in the bodie as a reasonable Apple tree (as I haue in some places seene and obserued) but most vsually groweth lower, with many twigs or branches rising from the roote, hauing as much pith in the middle of them as the Elder hath, couered with a grayish greene barke,

barke, but darker in the elder branches, with ioynts set at a good diſtance one from another, and two leaues at euery ioynt, which are large, broad, and pointed at the ends, many of them turning or folding both the ſides inward, and ſtanding on long foote ſtalkes : at the toppes of the branches come forth many flowers, growing ſpike-faſhion, that is, a long branch of flowers vpon a ſtalke, each of theſe flowers are ſmall, long, and hollow belowe, ending aboue in a pale blewiſh flower, conſiſting of foure ſmall leaues, of a pretty ſmall ſent : after the flowers are paſt, there come ſometimes (but it is not often in our Country, vnleſſe the tree haue ſtood long, and is grown great, the ſuckers being continually taken away, that it may growe the better) long and flat cods, conſiſting as it were of two ſides, a thin ſkinne being in the midſt, wherein are contained two long flattiſh red ſeede : the rootes are ſtrong, and growe deepe in the ground.

2. *Syringa flore lacteo ſiue argenteo.*
The ſiluer coloured Pipe tree.

This Pipe tree differeth not from the former blew Pipe tree, either in ſtemme or branches, either in leaues or flowers, or manner of growing, but onely in the colour of the flower, which in this is of a milke, or ſiluer colour, which is a kinde of white, wherein there is a thinne waſh, or light ſhew of blew ſhed therein, comming ſomewhat neare vnto an aſh-colour.

3. *Lilac lacimatis folijs.* The blew Pipe tree with cut leaues.

This Pipe tree ſhould not differ from the firſt in any other thing then in the leaues, which are ſaid to be cut in on the edges into ſeuerall parts, as the relation is giuen *à viris fide dignis*, for as yet I neuer ſaw any ſuch, but I here am bold to ſet it downe, to induce and prouoke ſome louer of plants to obtaine it for his pleaſure, and others alſo.

4. *Syringa flore albo ſimplici.* The ſingle white Pipe tree.

The ſingle white Pipe tree or buſh, neuer commeth to that height of the former, but abideth alwaies like a hedge tree or buſh, full of ſhootes or ſuckers from the roote, much more then the former : the young ſhootes hereof are reddiſh on the outſide, and afterward reddiſh at the ioynts, and grayiſh all the reſt ouer : the young as well as the old branches, haue ſome pith in the middle of them, like as the Elder hath : the leaues ſtand two at a ioynt, ſomewhat like the former, but more rugged or crumpled, as alſo a little pointed, and dented about the edges : the flowers growe at the toppes of the branches, diuers ſtanding together, conſiſting of foure white leaues, like vnto ſmall Muske Roſes, and of the ſame creame colour, as I may call it, with many ſmall yellowiſh threads in the middle, and are of a ſtrong, full, or heady ſent, not pleaſing to a great many, by reaſon of the ſtrange quickneſſe of the ſent : the fruit followeth, being flat at the head, with many leaſie ſhels or ſcales compaſſing it, wherein is encloſed ſmall long ſeede : the rootes runne not deepe, but ſpread vnder the ground, with many fibres annexed vnto them.

5. *Syringa Arabica flore albo duplici.*
The double white Pipe tree.

This Pipe tree hath diuers long and ſlender branches, whereon growe large leaues, ſomewhat like vnto the leaues of the former ſingle white kinde, but not ſo rough or hard, and not at all dented about the edges, two alwaies ſtanding one againſt another at euery ioynt of the ſtalke, but ſet or diſpoſed on contrary ſides, and not all vpon one ſide, at the ends whereof come forth diuers flowers, euery one ſtanding on his owne foote-ſtalke, the hoſe or huſke being long and hollow, like vnto the white Iaſmine, and the flowers therin conſiſting of a double rowe of white and round pointed leaues, fiue or ſix in a rowe, with ſome yellowneſſe in the middle, which is hollow, of a very ſtrong and heady ſweet ſent, and abiding a long time flowring, eſpecially in the hotter Countries, but is very tender, and not able to abide any the leaſt cold weather with vs,

for

1 *Iasminum album vulgare.* The ordinary white Iasmine. 2 *Iasminum Americanum siue Conuoluulus Americanus.* The Iasmine or Bindweed of America.
3 *Iasminum luteum vulgare.* The yellow Iasmine. 4 *Lilac seu Syringa caerulea.* The blew Pipe tree. 5 *Syringa alba vulgaris.* The single white Syringa or
Pipe tree. 6 *Syringa flore albo duplici.* The double white Syringa. 7 *Sambucus rosea.* The Elder or Gelder Rose.

for the cold windes will (as I vnderſtand) greatly moleſt it : and therefore muſt as charily be kept as Orenge trees with vs, if wee will haue it to abide.

The Place.

The firſt groweth in Arabia (as Matthiolus thinketh, that had it from Conſtantinople.) We haue it plentifully in our Gardens. The ſecond and third are ſtrangers with vs as yet. The fourth is as frequent as the firſt, or rather more, but his originall is not knowne. The laſt hath his originall from Arabia, as his name importeth.

The Time.

The firſt, ſecond, and third flower in Aprill, the other two not vntill May.

The Names.

The firſt is called of Matthiolus *Lilac*, and by that name is moſt vſually called in all parts. It is alſo called *Syringa cærulea*, becauſe it commeth neareſt vnto thoſe woods, which for their pithy ſubſtance, were made hollow into pipes. It is called of all in Engliſh, The blew Pipe tree. It ſeemeth likely, that Petrus Bellonius in his third Booke and fiftieth Chapter of his obſeruations (making mention of a ſhrubbe that the Turkes haue, with Iuie leaues alwaies greene, bearing blew or violet coloured flowers on a long ſtalke, of the bigneſſe and faſhion of a Foxe taile, and thereupon called in their language a Foxe taile) doth vnderſtand this plant here expreſſed. The certainty whereof might eaſily be knowne, if any of our Merchants there reſiding, would but call for ſuch a ſhrubbe, by the name of a Foxe taile in the Turkiſh tongue, and take care to ſend a young roote, in a ſmall tubbe or basket with earth by Sea, vnto vs here at London, which would be performed with a very little paines and coſt. The ſecond and third, as kindes thereof, haue their names in their titles. The fourth is called by Cluſius and others, *Frutex Coronarius*; ſome doe call it *Lilac flore albo*, but that name is not proper, in that it doth confound both kindes together. Lobel calleth it *Syringa Italica*. It is now generally called of all *Syringa alba*, that is in Engliſh, The white Pipe tree. Some would haue it to bee *Oſtrys* of Theophraſtus, but Cluſius hath ſufficiently cleared that doubt. Of others *Liguſtrum Orientale*, which it cannot be neither ; for the *Cyprus* of Plinie is Dioſcorides his *Liguſtrum*, which may be called *Orientale*, in that it is moſt proper to the Eaſterne Countries, and is very ſweete, whoſe ſeede is like vnto Coriander ſeede. The laſt is called by diuers *Syringa Arabica flore albo duplici*, as moſt fitly agreeing thereunto. Of Baſilius Beſlerus that ſet forth the great booke of the Biſhop of Eyſtot in Germany his Garden, *Syringa Italica flore albo pleno*, becauſe, as it is likely, hee had it from Italy. It is very likely, that Proſper Alpinus in his booke of Egyptian plants, doth meane this plant, which hee there calleth *Sambach, ſiue Iaſminum Arabicum*. Matthæus Caccini of Florence in his letter to Cluſius entituleth it *Syringa Arabica, ſiue Iaſminum Arabicum, ſiue Iaſminum ex Gine*, whereby hee declareth that it may not vnfitly be referred to either of them both. We may call it in Engliſh as it is in the title, The double white Pipe tree.

The Vertues.

We haue no vſe of theſe in Phyſicke that I know, although Proſper Alpinus ſaith, the double white Pipe tree is much vſed in Egypt, to help women in their trauailes of childbirth.

CHAP.

Chap. CVIII.

Sambucus Rosea. The Elder or Gelder Rose.

Lthough there be diuers kindes of Elders, yet there is but one kinde of Elder Rose, whereof I meane to intreate in this Chapter, being of neare affinity in some things vnto the former Pipe trees, and which for the beauty of it deserueth to be remembred among the delights of a Garden.

Sambucus Rosea. The Gelder Rose.

The Gelder Rose (as it is called) groweth to a reasonable height, standing like a tree, with a trunke as bigge as any mans arme, couered with a darke grayish barke, somewhat rugged and very knotty : the younger branches are smooth and white, with a pithy substance in the middle, as the Elders haue, to shew that it is a kind thereof, whereon are set broad leaues, diuided into three parts or diuisions, somewhat like vnto a Vine leafe, but smaller, and more rugged or crumpled, iagged or cut also about the edges : at the toppes of euery one of the young branches, most vsually commeth forth a great tuft, or ball as it were, of many white flowers, set so close together, that there can be no distinction of any seuerall flower seene, nor doth it seeme like the double flower of any other plant, that hath many rowes of leaues set together, but is a cluster of white leaued flowers set together vpon the stalke that vpholdeth them, of a small sent, which fall away without bearing any fruit in our Country, that euer I could obserue or learne : The roote spreadeth neither farre nor deepe, but shooteth many small rootes and fibres, whereby it is fastened in the ground, and draweth nourishment to it, and sometimes yeeldeth suckers from it.

The Place.

It should seeme, that the naturall place of this Elder is wet and moist grounds, because it is so like vnto the Marsh Elder, which is the single kind hereof. It is onely noursed vp in Gardens in all our Country.

The Time.

It flowreth in May, much about the time of the double Peony flower, both which being set together, make a pleasant variety, to decke vp the windowes of a house.

The Names.

It is generally called Sambucus Rosea : In English, The Elder Rose, and more commonly after the Dutch name, the Gelder Rose. Dalechampius seemeth to make it Thraupalus of Theophrastus, or rather the single Marsh Elder ; for I thinke this double kinde was not knowne in Theophrastus his time.

The Vertues.

It is not applyed to any Physicall vse that I know.

Chap.

Rosa. The Rose tree or bush.

THe great varietie of Roses is much to be admired, beeing more then is to bee seene in any other shrubby plant that I know, both for colour, forme and smell. I haue to furnish this garden thirty sorts at the least, euery one notably differing from the other, and all fit to be here entertained : for there are some other, that being wilde and of no beautie or smell, we forbeare, and leaue to their wilde habitations. To distinguish them by their colours, as white, red, incarnate, and yellow, were a way that many might take, but I hold it not so conuenient for diuers respects : for so I should confound those of diuers sorts one among another, and I should not keepe that methode which to me seemeth most conuenient, which is, to place and ranke euery kinde, whether single or double, one next vnto the other, that so you may the better vnderstand their varieties and differences : I will therefore beginne with the most ancient, and knowne Roses to our Countrey, whether naturall or no I know not, but assumed by our precedent Kings of all others, to bee cognisances of their dignitie, the white Rose and the red, whom shall follow the damaske, of the finest sent, and most vse of all the other sorts, and the rest in their order.

1. *Rosa Anglica alba.* The English white Rose.

The white Rose is of two kindes, the one more thicke and double then the other : The one riseth vp in some shadowie places, vnto eight or ten foote high, with a stocke of a great bignesse for a Rose. The other growing seldome higher then a Damaske Rose. Some doe iudge both these to be but one kinde, the diuersitie happening by the ayre, or ground, or both. Both these Roses haue somewhat smaller and whiter greene leaues then in many other Roses, fiue most vsually set on a stalke, and more white vnderneath, as also a whiter greene barke, armed with sharpe thornes or prickles, whereby they are soone known from other Roses, although the one not so easily from the other: the flowers in the one are whitish, with an eye or shew of a blush, especially towards the ground or bottome of the flower, very thicke double, and close set together, and for the most part not opening it selfe so largely and fully as eyther the Red or Damaske Rose. The other more white, lesse thicke and double, and opening it selfe more, and some so little double as but of two or three rowes, that they might be held to be single, yet all of little or no smell at all. To describe you all the seuerall parts of the Rose, as the bud, the beards, the threds &c. were needlesse, they are so conuersant in euery ones hand, that I shall not neede but to touch the most speciall parts of the varieties of them, and leaue a more exact relation of all things incident vnto them, vnto a generall worke.

2. *Rosa Incarnata.* The Carnation Rose.

The Carnation Rose is in most things like vnto the lesser white rose, both for the growing of the stocke, and bignesse of the flower, but that it is more spreade abroade when it is blown then the white is, and is of a pale blush colour all the flower throughout, of as small a sent as the white one is almost.

Rosa belgica siue Vitrea. This kinde of Rose is not very great, but very thicke and double, and is very variable in the flowers, in that they will be so different one from another : some being paler then others, and some as it were blasted, which commeth not casually, but naturally to this rose : but the best flowers (whereof there will bee still some) will be of a bright pale murrey colour, neere vnto the Veluet rose, but nothing so darke a colour.

3. *Rosa Anglica rubra.* The English red Rose.

The red Rose (which I call English, not only for the reason before expressed, but because
cause

cause (as I take it) this Rose is more frequent and vsed in England, then in other places) neuer groweth so high as the damaske Rose bush, but most vsually abideth low, and shooteth forth many branches from the roote (and is but seldome suffered to grow vp as the damaske Rose into standards) with a greene barke, thinner set with prickles, and larger and greener leaues on the vpperside then in the white, yet with an eye of white vpon them, fiue likewise most vsually set vpon a stalke, and grayish or whitish vnderneath. The Roses or Flowers doe very much vary, according to their site and abiding; for some are of an orient, red or deepe crimson colour, and very double (although neuer so double as the white) which when it is full blowne hath the largest leaues of any other Rose; some of them againe are paler, tending somewhat to a damaske; and some are of so pale a red, as that it is rather of the colour of the canker Rose, yet all for the most part with larger leaues then the damaske, and with many more yellow threds in the middle: the sent hereof is much better then in the white, but not comparable to the excellencie of the damaske Rose, yet this Rose being well dryed and well kept, will hold both colour and sent longer then the damaske, bee it neuer so well kept.

4. *Rosa Damascena.* The Damaske Rose.

The Damaske Rose bush is more vsually nourced vp to a competent height to stand alone, (which we call Standards) then any other Rose: the barke both of the stocke and branches, is not fully so greene as the red or white Rose: the leaues are greene with an eye of white vpon them, so like vnto the red Rose, that there is no great difference betweene them, but that the leaues of the red Rose seeme to bee of a darker greene. The flowers are of a fine deepe blush colour, as all know, with some pale yellow threds in the middle, and are not so thicke and double as the white, nor being blowne, with so large and great leaues as the red, but of the most excellent sweet pleasant sent, far surpassing all other Roses or Flowers, being neyther heady nor too strong, nor stuffing or vnpleasant sweet, as many other flowers.

5. *Rosa Prouincialis siue Hollandica Damascena.*
The great double Damaske Prouince or Holland Rose.

This Rose (that some call *Centifolia Batauica incarnata*) hath his barke of a reddish or browne colour, whereby it is soone discerned from other Roses. The leaues are likewise more reddish then in others, and somewhat larger, it vsually groweth very like the Damaske rose, and much to the same height: the flowers or roses are of the same deepe blush colour that the damaske roses are, or rather somewhat deeper, but much thicker, broader, and more double, or fuller of leaues by three parts almost, the outer leaues turning themselues backe, when the flower hath stood long blowne, the middle part it selfe (which in all other roses almost haue some yellow threds in them to be seene) being folded hard with small leaues, without any yellow almost at all to be seene, the sent whereof commeth neerest vnto the damaske rose, but yet is short of it by much, howsoeuer many doe thinke it as good as the damask, and to that end I haue known some Gentlewomen haue caused all their damaske stockes to bee grafted with prouince Roses, hoping to haue as good water, and more store of them then of damask Roses; but in my opinion it is not of halfe so good a sent as the water of damaske Roses: let euery one follow their own fancie.

6. *Rosa Prouincialis rubra.* The red Prouince Rose.

As the former was called *incarnata*, so this is called *Batauica centifolia rubra*, the difference being not very great: the stemme or stocke, and the branches also in this, seeming not to be so great but greener, the barke being not so red; the leaues of the same largenesse with the former damaske Prouince. The flowers are not altogether so large, thicke and double, and of a little deeper damaske or blush colour, turning to a red Rose, but not comming neere the full colour of the best red Rose, of a sent not so sweete as the damaske Prouince, but comming somewhat neere the sent of the ordinary

dinary

dinary red rose, yet exceeding it. This rose is not so plentifull in bearing as the damaske Prouince.

7. *Rosa Prouincialis alba.* The white Prouince Rose.

It is said of diuers, that there is a white Prouince Rose, whereof I am not *oculatus testis,* and therfore I dare not giue it you for a certaintie, and indeed I haue some doubt, that it is the greater and more double white rose, whereof I gaue you the knowledge in the beginning: when I am my selfe better satisfied, I shall bee ready to satisfie others.

8. *Rosa versicolor.* The party coloured Rose, of some Yorke and Lancaster.

This Rose in the forme and order of the growing, is neerest vnto the ordinary damaske rose, both for stemme, branch, leafe and flower : the difference consisting in this, that the flower (being of the same largenesse and doublenesse as the damask rose) hath the one halfe of it, sometimes of a pale whitish colour, and the other halfe, of a paler damaske colour then the ordinary; this happeneth so many times, and sometimes also the flower hath diuers stripes, and markes in it, as one leafe white, or striped with white, and the other halfe blush, or striped with blush, sometimes also all striped, or spotted ouer, and other times little or no stripes or markes at all, as nature listeth to play with varieties, in this as in other flowers : yet this I haue obserued, that the longer it abideth blowen open in the sun, the paler and the fewer stripes, markes or spots will be seene in it : the smell whereof is of a weake damaske rose sent.

9. *Rosea Chrystallina.* The Chrystall Rose.

This Rose is very like vnto the last described, both for stocke, branch and leafe : the flower hereof is not much different from it, being no great large or double Rose, but of a meane size, striped and marked with a deeper blush or red, vpon the pale coloured leafe, that it seemeth in the marking and beauty thereof, to bee of as much delight as the Chrystall Gilloflower : this, euen like the former, soone fadeth and passeth away, not yeelding any great store of flowers any yeare.

10. *Rosa rubra humilis siue pumilio.* The dwarfe red Rose, or Gilloflower Rose.

This Rose groweth alwayes low and small, otherwise in most respects like vnto the ordinary redde Rose, and with few or no thornes vpon it : the Flowers or Roses are double, thicke, small and close, not so much spread open as the ordinary red, but somewhat like vnto the first double white Rose before expressed ; yet in some places I haue seene them more layde open then these, as they grew in my garden, being so euen at the toppes of the leaues, as if they had been clipt off with a paire of sheeres, and are not fully of so red a colour as the red Prouince Rose, and of as small or weak sent as the ordinary red Rose, or not so much.

11. *Rosa Francafurtensis.* The Franckford Rose.

The young shootes of this Rose are couered with a pale purplish barke, set with a number of small prickes like haires, and the elder haue but very few thornes : the flower or rose it selfe hath a very great bud or button vnder it, more then in any other rose, and is thicke and double as a red rose, but so strongly swelling in the bud, that many of them breake before they can be full blowen, and then they are of a pale red rose colour, that is, betweene a red and a damaske, with a very thicke broade and hard vmbone of short yellow threds or thrumes in the middle, the huske of the flower hauing long ends, which are called the beards of the rose, which in all other are iagged in some of them, in this hath no iagge at all : the smell is neerest vnto a red Rose.

1 *Rosa Damascena.* The Damaske Rose. 2 *Rosa Prouincialis siue Hollandica.* The great Prouince Rose. 3 *Rosa Francofortensis.* The Franckford Rose. 4 *Rosa rubra humilis.* The dwarfe red Rose. 5 *Rosa Hungarica.* The Hungarian Rose. 6 *Rosa lutea multiplex.* The great double yellow Rose.

12. *Rosa Hungarica.* The Hungarian Rose.

The Hungarian Rose hath greene shootes slenderly set with prickes, and seldome groweth higher then ordinarily the red Rose doth; the stemme or stocke being much about that bignesse : the flower or rose is as great, thicke and double, as the ordinary red Rose, and of the same fashion, of a paler red colour, and beeing neerely looked vpon is finely spotted with faint spots, as it were spreade ouer the red; the smell wherof is somewhat better then the smell of the ordinary red Rose of the best kinde.

13. *Rosa Holoserica simplex & multiplex.* The Veluet Rose single and double.

The old stemme or stock of the veluet Rose is couered with a dark coloured barke, and the young shootes of a sad greene with very few or no thornes at all vpon them : the leaues are of a sadder greene colour then in most sorts of Roses, and very often seuen on a stalke, many of the rest hauing but fiue: the Rose is eyther single or double : the single is a broade spread flower, consisting of fiue or sixe broade leaues with many yellow threds in the middle : the double hath two rowes of leaues, the one large, which are outermost, the other smaller within, of a very deepe red crimson colour like vnto crimson veluet, with many yellow threds also in the middle; and yet for all the double rowe of leaues, these Roses stand but like single flowers : but there is another double kinde that is more double then this last, consisting oftentimes of sixteene leaues or more in a flower, and most of them of an equall bignesse, of the colour of the first single rose of this kinde, or somewhat fresher; but all of them of a smaller sent then the ordinary red Rose.

14. *Rosa sine spinis simplex & multiplex.* The Rose without thornes single and double.

The Rose without thornes hath diuers greene smooth shootes, rising from the root, without any pricke or thorne at all vpon them, eyther young or old: the leaues are not fully so large as of the red rose : the flowers or roses are not much bigger then those of the double Cinamon Rose, thicke set together and short, of a pale red Rose colour, with diuers pale coloured veines through euery leafe of the flower, which hath caused some to call it The marbled Rose, and is of a small sent, not fully equall to the red Rose. The single of this kinde differeth not in any other thing from the former, then in the doublenesse or singlenesse of the flowers, which in this are not halfe so double, nor yet fully single, and are of a paler red colour.

Rosa sine spina flore albo. I haue heard likewise of a white Rose of this kinde, but I haue seene none such as yet, and therefore I can say no more thereof.

15. *Rosa Cinamomea simplex & multiplex.* The Cinamon Rose single and double.

The single Cinamon Rose hath his shootes somewhat red, yet not so red as the double kinde, armed with great thornes, like almost vnto the Eglantine bush, thereby showing, as well by the multiplicitie of his shootes, as the quicknesse and height of his shooting, his wilde nature : On the stemme and branches stand winged leaues, sometimes seuen or more together, which are small and greene, yet like vnto other Roses. The Roses are single, of fiue leaues a peece, somewhat large, and of a pale red colour, like vnto the double kinde, which is in shootes redder, and in all other things like vnto the single, but bearing small, short, thicke and double Roses, somewhat like vnto the Rose without thornes, but a little lesser, of a paler red colour at the end of the leaues, and somewhat redder and brighter toward the middle of them, with many yellow short thrumes; the small sent of Cinamon that is found in the flowers hath caused it to beare the name.

16. *Rosa lutea simplex.* The single yellow Rose.

This single yellow Rose is planted rather for variety then any other good vse. It often groweth to a good height, his stemme being great and wooddy, with few or no prickes vpon the old wood, but with a number of small prickes like haires, thicke set, vpon the younger branches, of a darke colour somewhat reddish, the barke of the young shootes being of a sad greene reddish colour: the leaues of this Rose bush are smaller, rounder pointed, of a paler greene colour, yet finely sniptabout the edges, and more in number, that is, seuen or nine on a stalke or ribbe, then in any other Garden kinde, except the double of the same kinde that followeth next: the flower is a small single Rose, consisting of fiue leaues, not so large as the single Spanish Muske Rose, but somewhat bigger then the Eglantine or sweete Briar Rose, of a fine pale yellow colour, without any great sent at all while it is fresh, but a little more, yet small and weake when it is dryed.

17. *Rosa lutea multiplex siue flore pleno.* The double yellow Rose.

The double yellow Rose is of great account, both for the rarity, and doublenesse of the flower, and had it sent to the rest, would of all other be of highest esteeme. The stemme or stocke, the young shoots or branches, the small hairy prickes, and the small winged leaues, are in all parts like vnto the former single kinde; the chiefest difference consisteth in the doublenesse of the flower or Rose, which is so thicke and double, that very often it breaketh out on one side or another, and but a few of them abiding whole and faire in our Countrey, the cause whereof wee doe imagine to bee the much moisture of our Countrey, and the time of flowring being subiect to much raine and showers; many therefore doe either plant it against a wall, or other wayes defend it by couering: againe, it is so plentifull in young shootes or branches, as also in flowers at the toppe of euery branch, which are small and weake for the most part, that they are not able to bring all the flowers to ripenesse; and therefore most of them fall or wither away without comming to perfection (the remedy that many doe vse for this inconuenience last recited is, that they nippe away most of the buds, leauing but some few vpon it, that so the vigour of the plant may be collected into a few flowers, whereby they may the better come to perfection, and yet euen thus it is hardly effected) which are of a yellowish greene colour in the bud, and before they be blowne open, but then are of a faire yellow colour, very full of leaues, with many short haires rather then leaues in the middle, and hauing short, round, greene, smooth buttons, almost flat vnder them: the flower being faire blowne open, doth scarce giue place for largenesse, thicknesse, and doublenesse, vnto the great Prouence or Holland Rose. This Rose bush or plant is very tender with vs here about London, and will require some more care and keeping then the single of this kinde, which is hardy enough; for I haue lost many my selfe, and I know but a few about this towne that can nourse it vp kindly, to beare or scarce to abide without perishing; but abideth well in euery free aire of all or the most parts of this Kingdome: but (as I heare) not so well in the North.

18. *Rosa Moschata simplex & multiplex.* The Muske Rose single and double.

The Muske Rose both single and double, rise vp oftentimes to a very great height, that it ouergroweth any arbour in a Garden, or being set by an house side, to bee ten or twelue foote high, or more, but more especially the single kinde, with many green farre spread branches, armed with a few sharpe great thornes, as the wilder sorts of Roses are, whereof these are accounted to be kindes, hauing small darke greene leaues on them, not much bigger then the leaues of Eglantine: the flowers come forth at the toppes of the branches, many together as it were in an vmbell or tuft, which for the most part doe flower all at a time, or not long one after another, euery one standing on a pretty long stalke, and are of a pale whitish or creame colour, both the single and
the

the double; the single being small flowers, consisting of fiue leaues, with many yellow threads in the middle: and the double bearing more double flowers, as if they were once or twice more double then the single, with yellow thrummes also in the middle, both of them of a very sweete and pleasing smell, resembling Muske: some there be that haue auouched, that the chiefest sent of these Roses consisteth not in the leaues, but in the threads of the flowers.

19. *Rosa Moschata multiplex altera: alijs Damascena alba, vel verisimilior Cinamomea flore pleno albo.* The double white Damaske Muske Rose.

This other kinde of Muske Rose (which with some is called the white Damaske Muske, but more truely the double white Cinamon Rose) hath his stemme and branches also shorter then the former, but as greene: the leaues are somewhat larger, and of a whiter greene colour; the flowers also are somewhat larger then the former double kinde, but standing in vmbels after the same manner, or somewhat thicker, and of the same whitish colour, or a little whiter, and somewhat, although but a little, neare the smell of the other, but nothing so strong. This flowreth at the time of other Roses, or somewhat later, yet much before the former two sorts of Muske Roses, which flower not vntill the end of Summer, and in Autumne; both which things, that is, the time of the flowring, and the sent being both different, shew plainly it cannot be of the tribe of Muske Roses.

20. *Rosa Hispanica Moschata simplex.* The Spanish Muske Rose.

This Spanish Rose riseth to the height of the Eglantine, and sometimes higher, with diuers great greene branches, the leaues whereof are larger and greener then of the former kindes: the flowers are single Roses, consisting of fiue whiter leaues then in any of the former Muske Roses, and much larger, hauing sometimes an eye of a blush in the white, of a very sweete smell, comming nearest vnto the last recited Muske Rose, as also for the time of the flowring.

21. *Rosa Pomifera maior.* The great Apple Rose.

The stemme or stocke of this Rose is great, couered with a darke grayish barke, but the younger branches are somewhat reddish, armed here and there with great and sharpe thornes, but nothing so great or plentifull as in the Eglantine, although it bea wilde kinde: the leaues are of a whitish greene colour, almost like vnto the first white Rose, and fiue alwaies set together, but seldome seuen: the flowers are small and single, consisting of fiue leaues, without any sent, or very little, and little bigger then those of the Eglantine bush, and of the very same deepe blush colour, euery one standing vpon a rough or prickly button, bearded in the manner of other Roses, which when the flowers are fallen growe great, somewhat long and round, peare-fashion, bearing the beards on the tops of them; and being full ripe are very red, keeping the small prickles still on them, wherein are many white, hard, and roundish seedes, very like vnto the seede of the Heppes or Eglantine berries, lying in a soft pulpe, like vnto the Hawthorne berries or Hawes: the whole beauty of this plant consisteth more in the gracefull aspect of the red apples or fruit hanging vpon the bushes, then in the flowers, or any other thing. It seemeth to be the same that Clusius calleth *Rosa Pamila*, but that with me it groweth much higher and greater then he saith his doth.

22. *Rosa siluestris odora siue Eglenteria simplex.* The single Eglantine or sweete Briar bush.

The sweete Briar or Eglantine Rose is so well knowne, being not onely planted in Gardens, for the sweetenesse of the leaues, but growing wilde in many woods and hedges, that I thinke it lost time to describe it; for that all know it hath exceeding long greene shootes, armed with the cruellest sharpe and strong thornes, and thicker set then

1 *Rosa sine spinis multiplex.* The double Rose without thorns. 2 *Rosa Cinamomea flore p.eno.* The double Cinamon Rose. 3 *Rosa Holoserica simplex.* The single Veluet Rose. 4 *Rosa Holoserica duplex.* The double Veluet Rose. 5 *Rosa Moschata multiplex.* The double Muske Rose. 6 *Rosa Moschata Hispanica simplex.* The single Spanish Muske Rose. 7 *Rosa Pomifera maior.* The great Apple Rose. 8 *Rosa siluestris sine Eglanteria duplex.* The double Eglantine Rose.

then is in any Rose either wilde or tame : the leaues are smaller then in most of those that are noursed vp in Gardens, seuen or niue most vsually set together on a ribbe or stalke, very greene and sweete in smell, aboue the leaues of any other kinde of Rose : the flowers are small single blush Roses, of little or no sent at all, which turne into reddish berries, stuffed within with a dounie or flocky matter or substance, wherein doth lye white hard seede.

23. *Rosa siluestris odora siue Eglenteria flore duplici.* The double Eglantine.

The double Eglantine is in all the places that I haue seene it a grafted Rose, (but I doubt not, but that his originall was naturall, and that it may be made naturall againe, as diuers other Roses are.) It groweth and spreadeth very well, and with a great head of branches, whereon stand such like leaues as are in the single kinde, but a little larger, not smelling fully so sweete as it : the flowers are somewhat bigger then the single, but not much, hauing but one other rowe of leaues onely more then the former, which are smaller, and the outer leaues larger, but of the same pale reddish purple colour, and smelleth somewhat better then the single.

24. *Rosa semper virens.* The euer greene Rose bush.

This Rose or bush is very like vnto a wilde single Eglantine bush in many respects, hauing many very long greene branches, but more slender and weake, so that many times they bend downe againe, not able to sustaine themselues without some helpe, and armed with hooked thornes as other Roses be ; the winged leaues consist of seuen for the most part, whereof those two that are lowest and opposite, are smallest, the next two bigger then they, the third couple bigger then any of the rest belowe, and the end leafe biggest of all : this proportion generally it holdeth in euery winged leafe through the whole plant, which at the first comming forth are somewhat reddish, with the young branch that shooteth out with them, but being full growne, are of a deepe greene colour, and somewhat shining, dented about the edges, and fall not away from the branches as other Roses doe, but abide thereon for the most part all the Winter : the flowers stand foure or fiue together at the tops of the branches, being single Roses, made of fiue leaues a peece, of a pure white colour, much larger then the ordinary Muske Rose, and of a fine sent, comming nearest thereunto, with many yellow chiues or threads in the middle.

The Place.

Some of these Roses had their originall, as is thought in England, as the first and second ; for these dryed red Roses that come ouer to vs from beyond the Seas, are not of the kinde of our red Rose, as may well be perceiued by them that will compare our English dryed leaues with those. Some in Germany, Spaine, and Italy. Some againe in Turkie, as the double yellow Rose, which first was procured to be brought into England, by Master Nicholas Lete, a worthy Merchant of London, and a great louer of flowers, from Constantinople, which (as wee heare) was first brought thither from Syria ; but perished quickly both with him, and with all other to whom hee imparted it : yet afterwards it was sent to Master Iohn de Franqueuille, a Merchant also of London, and a great louer of all rare plants, as well as flowers, from which issprung the greatest store, that is now flourishing in this Kingdome.

The Time.

The Cinamon Rose is the earliest for the most part, which flowreth with vs about the middle of May, and sometimes in the beginning. The ordinary Muske Roses both single and double flower latest, as is said. All the other flower much about one time, in the beginning of Iune, or thereabouts, and continue flowring all that moneth, and the next throughout for the most part, and the red vntill August be halfe past.

The

The Names.

The feuerall names, whereby they are moft commonly knowne vnto vs in this Countrey, are expreffed in their titles; but they are much differing from what they are called in other Countries neare vnto vs, which to compare, conferre, and agree together, were a worke of more paines then vfe: But to proportion them vnto the names fet downe by Theophraftus, Pliny, and the reft of the ancient Authors, were a worke, wherein I might be fure not to efcape without falling into errour, as I verily beleeue many others haue done, that haue vndertaken to doe it: I will therefore for this worke defire that you will reft contented, with fo much as hath already been deliuered, and expect an exact definition and complete fatisfaction by fuch a methodicall courfe as a generall Hiftory will require, to be performed by them that fhall publifh it.

The Vertues.

The Rofe is of exceeding great vfe with vs; for the Damaske Rofe (befides the fuperexcellent fweete water it yeeldeth being diftilled, or the perfume of the leaues being dryed, feruing to fill fweete bags) ferueth to caufe folublenefle of the body, made into a Syrupe, or preferued with Sugar moift or dry candid. The Damaske Prouince Rofe, is not onely for fent neareft of all other Rofes vnto the Damaske, but in the operation of folubility alfo. The red Rofe hath many Phyficall vfes much more then any other, feruing for many forts of compofitions, both cordiall and cooling, both binding and loofing. The white Rofe is much vfed for the cooling of heate in the eyes: diuers doe make an excellent yellow colour of the iuyce of white Rofes, wherein fome Allome is diffolued, to paint or colour flowers or pictures, or any other fuch things. There is little vfe of any other fort of Rofes; yet fome affirme, that the Muske Rofes are as ftrong in operation to open or loofen the belly as the Damaske Rofe or Prouince.

Chap. CXI.

Ciftus. The Holly Rofe or Sage Rofe.

THere are three principall kindes of *Ciftus*, the male, the female, and the gumme or fweete fmelling *Ciftus* bearing *Ladanum*, called *Ledon*. Of each of thefe three there are alfo diuers forts: Of them all to intreate in this worke is not my minde, I will onely felect out of the multitude fome few that are fit for this our Garden, and leaue the reft to a greater.

1. *Ciftus mas.* The male Holly Rofe or Sage Rofe.

The male *Ciftus* that is moft familiar vnto our Countrey, I meane that will beft abide, is a fmall fhrubby plant, growing feldome aboue three or foure foote high with vs, hauing many flender brittle wooddy branches, couered with a whitifh barke, whereon are fet many whitifh greene leaues, long and fomewhat narrow, crumpled or wrinckled as it were with veines, and fomewhat hard in handling, efpecially the old ones; for the young ones are fofter, fomewhat like vnto Sage leaues for the forme and colour, but much fmaller, two alwaies fet together at a ioynt: the flowers ftand at the toppe of the branches, three or foure together vpon feuerall flender footftalkes, confifting of fiue fmall round leaues a peece, fomewhat like vnto a fmall fingle Rofe, of a fine reddifh purple colour, with many yellow threads in the middle, with-

out

out iny fent at all, and quickly fading or falling away, abiding feldome one whole day blowne at the moft : after the flowers are paft, there come vp round hard hairie heads in their places, containing fmall brownifh feede : the roote is wooddy, and will abide fome yeares with vs, if there be fome care had to keepe it from the extreamity of our Winters froftes, which both this, and many of the other forts and kinds, will not abide doe what we can.

2. *Ciftus femina*. The female Holly Rofe.

The female Holly Rofe groweth lower, and fmaller then the former male kinde, hauing blackifh branches, leffe woody, but not leffe brittle then it : the leaues are fomewhat rounder and greener, but a little hard or rough withall, growing in the fame manner vpon the branches by couples : the flowers grow at the toppes of the branches, like vnto the former, confifting of fiue leaues, but fomewhat leffer, and wholly white, with yellow threds in the middle, as quickly fading, and of as little fent as the former : the heads and feede are fomewhat bigger then in the former.

3. *Chamæciftus Frificus*. The dwarfe Holly Rofe of Frifeland.

This dwarfe Ciftus is a fmall low plant, hauing diuers fhootes from the rootes, full of leaues that are long and narrow, very like vnto the leaues of the French Spikenard or *Spica Celtica* ; from among which leaues fhoote forth fhort ftalkes, not aboue a fpan high, with a few fmaller leaues thereon ; and at the toppes diuers fmall flowers one aboue another, confifting of fix fmall round leaues, of a yellow colour, hauing two circles of reddifh fpots round about the bottome of the leaues, a little diftant one from another, which adde much grace to the flower : after the flowers are paft, there come in their places fmall round heads, being two forked at the end, containing within them fmall brownifh chaffie feede : the roote is fmall and flender, with many fibres thereat creeping vnder ground, and fhooting forth in diuers places, whereby it much encreafeth : the whole plant, and euery part of it, fmelleth ftrong without any pleafant fent.

4. *Ciftus annuus*. The Holly Rofe of a yeare.

This fmall Ciftus that endureth but a year (and will require to be fowne euery year, if ye will haue it) rifeth vp with ftraight, but flender hard ftalkes, fet here and there confufedly with long and narrow greenifh leaues, very like vnto the leaues of the Gum Ciftus or Ledon, being a little clammy withall : at the toppe of the ftalkes, and at the ioynts with the leaues, ftand two or three pale yellow flowers, confifting of fiue leaues a peece, with a reddifh fpot neere the bottome of euery leafe of the flower, as quickely fading as any of the former : after which follow fmall three fquare heades, containing fmall feede, like vnto the firft female kinde, but fomewhat paler or yellower : the root is fmall and woody, and perifheth as foone as it hath borne feede.

5. *Ciftus Ledon*. The Gum Ciftus, or Sweete Holly Rofe.

This fweete Holly Rofe or Gum Ciftus, rifeth higher, and fpreadeth larger then the former male kind doth, with many blackifh woody branches, whereon are fet diuers long and narrow darke greene leaues, but whitifh vnderneath, two alwayes ftanding together at a ioint, both ftalks and leaues bedeawed as it were continually with a clammy fweete moifture (which in the hot Countries is both more plentifull, and more fweet then in ours) almoft tranfparent, and which being gathered by the inhabitants, with certaine inftruments for that purpofe (which in fome places are leather thongs, drawne ouer the bufhes, and after fcraped off from the thongs againe, and put together) is that kind of blacke fweet gum, which is called *Ladanum* in the Apothecaries fhops : at the tops of the branches ftand fingle white flowers, like vnto fingle Rofes, being larger then in any of the former kindes, confifting of fiue leaues, whereof euery one hath at the bottome a dark purplifh fpot, broad below, and fmall pointed vpwards, with fome yellow threds in the middle : after which are paft, there arife cornered heads,

1 *Ciſtus mas* The male Holly Roſe. 2 *Chamaciſtus Friſicus.* The dwarfe Holly Roſe of Friſia. 3 *Ciſtus Ledon.* The ſweet Holly Roſe or gumme Ciſtus. 4 *Ledum Alpinum.* The mountaine Holly Roſe. 5 *Ledum Sileſiacum.* The ſweet Mary Roſe of Sileſia. 6 *Roſmarinum aureum,* Gilded Roſemary.

heads, containing such small brownish seede as is in the former male kinde : the roote is woody, and spreadeth vnder ground, abiding some yeares, if it be placed vnder a wall, where it may bee defended from the windes that often breake it, and from the extremitie of our winters, and especially the snow, if it lye vpon it, which quickly causeth it to perish.

6. *Ledum Alpinum seu Rosa Alpina.* The Mountaine sweet Holly Rose.

The fragrant smell with properties correspondent of two other plants, causeth me to insert them in this Chapter, and to bring them to your knowledge, as well worthy a fit place in our Garden. The first of them hath diuers slender woody branches, two foote high or thereabouts, couered with a grayish coloured barke, and many times leaning downe to the ground, whereby it taketh roote againe : vpon these branches grow many thicke, short, hard greene leaues, thicke set together, confusedly without order, sometimes whitish vnderneath, and sometimes yellowish : the toppes of the branches are loden with many flowers, which cause them to bend downwards, being long, hollow and reddish, opening into fiue corners, spotted on the outside with many white spots, and of a paler red colour on the inside, of a fine sweet sent : after the flowers are past, there follow small heads, containing small brownish seede : the root is long, hard and woody, abiding better if it comprehend in the ground, then some of the former, because his originall is out of a colder country.

7. *Ledum Silesiacum.* The sweete Mary Rose, or Rosemary of Silesia.

This other sweete plante riseth vp with woody ash-coloured branches two foote high or more, which shoote forth other branches, of a reddish or purplish colour, couered with a brownish yellow hoarinesse, on which are set many narrow long greene leaues, like vnto Rosemary leaues, but couered with the like hoarinesse as the stalks are (especially in the naturall places, but not so much being transplanted) and folding the sides of the leaues so close together, that they seeme nothing but ribbes, or stalkes, of an excellent sweet and pleasant sent ; at the ends of the branches there grow certaine brownish scaly heads, made of many small leaues set thicke together, out of which breake forth many flowers, standing in a tuft together, yet seuerally euery one vpon his owne footstalke, consisting of fiue white leaues, with certaine white threds in the middle, smelling very sweete : after which rise small greene heads, spotted with brownish spots, wherein is contained very small, long, yellowish seede : the roote is hard and woodie.

The Place.

The first, second, fourth and fifth, grow in the hot Countries, as Italie, Spaine, &c. The third, and the two last in the colder Countries, as Frise-land, Germanie, Bohemia.

The Time.

They do all flower in the Summer moneths of Iune, Iuly and August, and their seede is ripe quickly after.

The Names.

The first, second, fourth and fift, haue their names sufficiently expressed in their descriptions. The third was sent vnto Clusius, vnder the name of *Herculus Frisicus*, because of the strong sent : but he referreth it to the kinds of *Chamæcistus*, that is, dwarfe or low *Cistus*, both for the low growth, and for the flowers and seede sake. The fixt is diuersly called; for Clusius calleth it *Ledum Alpinum* : others, *Nerium alpinum*, making it to bee a Rose Bay. Gesner

Gesner according to the Countrey peoples name, *Rosa Alpina*, and *Rosa Montana*. Lobel calleth it *Balsamum Alpinum*, of the fragrant smell it hath, and *Chamærhododendros Chamælexfolio*. And some haue called it *Euonymus*, without all manner of iudgement. In English wee may call it, The Mountaine Rose, vntill a fitter name be giuen it. The last is called of Matthiolus, *Rosmarinum siluestre*, but of Clusius *Ledum*, referring it to their kindred; and *Silesiacum*, because he found it in that Countrey; or for distinction sake, as he saith, it may bee called, *Ledum folijs Rosmarini*, or *Ledum Bohemicum*. Cordus, as it seemeth in his History of Plants, calleth it *Chamæpeuce*, as though he did account it a kinde of low Pine, or Pitch tree.

The Vertues.

The first, second, and fift, are very astringent, effectuall for all sorts of fluxes of humours. The sweet Gum called *Ladanum*, made artificially into oyle, is of singular vse for *Alopecia*, or falling of the haire. The seed of the fourth is much commended against the stone of the Kidneyes. The sweete Rosemary of Silesia is vsed of the inhabitants, where it naturally groweth, against the shrinking of sinewes, crampes, or other such like diseases, wherof their daily experience makes it familiar, being vsed in bathing or otherwise.

Chap. CXII.

Rosmarinum. Rosemary.

THere hath beene vsually knowne but one sort of Rosemary, which is frequent through all this Country: but there are some other sorts not so well knowne; the one is called Gilded Rosemary; the other broade leafed Rosemary; a third I will adioyne, as more rare then all the other, called Double flowred Rosmary, because few haue heard thereof, much lesse seene it, and my selfe am not well acquainted with it, but am bold to deliuer it vpon credit.

1. *Libanotis Coronaria siue Rosmarinum vulgare.*
Our Common Rosmary.

This common Rosemary is so well knowne through all our Land, being in euery womans garden, that it were sufficient but to name it as an ornament among other sweete herbes and flowers in our Garden, seeing euery one can describe it : but that I may say something of it, It is well obserued, as well in this our Land (where it hath been planted in Noblemens, and great mens gardens against bricke wals, and there continued long) as beyond the Seas, in the naturall places where it groweth, that it riseth vp in time vnto a very great height, with a great and woody stemme (of that compasse, that (being clouen out into thin boards) it hath serued to make lutes, or such like instruments, and here with vs Carpenters rules, and to diuers other purposes) branching out into diuers and sundry armes that extend a great way, and from them againe into many other smaller branches, wheron are set at seuerall distances, at the ioynts, many very narrow long leaues, greene aboue, and whitish vnderneath ; among which come forth towards the toppes of the stalkes, diuers sweet gaping flowers, of a pale or bleake blewish colour, many set together, standing in whitish huskes ; the seed is small and red, but thereof seldome doth any plants arise that will abide without extraordinary care ; for although it will spring of the seede reasonable well, yet it is so small and tender the first yeare, that a sharpe winter killeth it quickly, vnlesse it be very well defended : the whole plant as well leaues as flowers, smelleth exceeding sweete.

2. *Rosmarinum striatum, siue aureum.* Gilded Rosemary.

This Rosemary differeth not from the former, in forme or manner of growing, nor

in

in the forme or colour of the flower, but only in the leaues, which are edged, or stri-ped, or pointed with a faire gold yellow colour, which so continueth all the yeare throughout, yet fresher and fairer in Summer then in Winter; for then it will looke of a deader colour, yet so, that it may be discerned to be of two colours, green & yellow.

3. *Rosmarinum latifolium.* Broade leafed Rosemary.

This broad leafed Rosemary groweth in the same manner that the former doth, but that we haue not seene it in our Countrey since we had it to grow so great, or with such woody stemmes : the leaues stand together vpon the long branches after the same fa-shion, but larger, broader and greener then the other, and little or nothing whitish vnderneath : the flowers likewise are of the same forme and colour with the ordinary, but larger, and herein consisteth the difference.

4. *Rosmarinum flore duplici.* Double flowred Rosmary.

The double flowred Rosmary thus far differeth from the former, that it hath stron-ger stalkes, not so easie to breake, fairer, bigger and larger leaues, of a faire greene co-lour, and the flowers are double, as the Larkes heele or spurre : This I haue onely by relation, which I pray you accept, vntill I may by sight better enforme you.

The Place.

Our ordinary Rosmary groweth in Spaine, and Prouence of France, and in others of those hot Countryes, neere the Sea side. It will not abide (vnlesse kept in stoues) in many places of Germany, Denmarke, and those colder Countries. And in some extreame hard winters, it hath well neere perished here in England with vs, at the least in many places: but by slipping it is vsually, and yearly encreased, to replenish any garden.

The Time.

It flowreth oftentimes twice in the yeare ; in the Spring first, from April vntill the end of May or Iune, and in August and September after, if the yeare before haue been temperate.

The Names.

Rosmary is called of the ancient Writers, *Libanotis,* but with this distin-ction, *Stephanomatica,* that is, *Coronaria,* because there were other plants cal-led *Libanotis,* that were for other vses, as this for garlands, where flowers and sweete herbes were put together. The Latines call it *Rosmarinum.* Some would make it to be *Cneorum nigrum* of Theophrastus, as they would make Lauander to bee his *Cneorum album,* but Matthiolus hath sufficiently confu-ted that errour.

The Vertues.

Rosmary is almost of as great vse as Bayes, or any other herbe both for inward and outward remedies, and as well for ciuill as physicall purposes. Inwardly for the head and heart ; outwardly for the sinewes and ioynts : for ciuill vses, as all doe know, at weddings, funerals, &c. to bestow among friends : and the physicall are so many, that you might bee as well tyred in the reading, as I in the writing, if I should set down all that might be said of it. I will therefore onely giue you a taste of some, desiring you will be con-tent therewith. There is an excellent oyle drawne from the flowers alone by the heate of the Sunne, auaileable for many diseases both inward and outward, and accounted a soueraigne Balsame: it is also good to helpe dim-

nesse

neffe of fight, and to take away fpots, markes and fcarres from the skin ; and is made in this manner. Take a quantitie of the flowers of Rofemary, according to your owne will eyther more or leffe, put them into a ftrong glaffe clofe ftopped, fet them in hot horfe dung to digeft for fourteene dayes, which then being taken forth of the dung, and vnftopped, tye a fine linnen cloth ouer the mouth, and turne downe the mouth thereof into the mouth of another ftrong glaffe, which being fet in the hot Sun, an oyle will diftill downe into the lower glaffe ; which preferue as precious for the vfes before recited, and many more, as experience by practice may enforme diuers.

There is another oyle Chymically drawne, auaileable in the like manner for many the fame inward and outward difeafes, *viz.* for the heart, rheumaticke braines, and to ftrengthen the memory, outwardly to warme and comfort cold benummed finewes, whereof many of good iudgement haue had much experience.

Chap. CXIII.

Myrtus. The Mirtle tree or bufh.

IN the hot Countreyes, there haue been many forts of Mirtles found out, naturally growing there, which will not fructifie in this of ours, nor yet abide without extraordinary care, and conueniencie withall, to preferue them from the fharpeneffe of our winters. I fhall only bring you to view three forts in this my Garden, the one with a greater, the other two with leffer leaues, as the remainder of others which wee haue had, and which are preferued from time to time, not without much paine and trouble.

1. *Myrtus latifolia.* The greater leafed Mirtle.

The broader leafed Mirtle rifeth vp to the height of foure or fiue foote at the moft with vs, full of branches and leaues growing like a fmall bufh, the ftemme and elder branches whereof are couered with a dark coloured bark, but the young with a green, and fome with a red, efpecially vpon the firft fhooting forth, whereon are fet many frefh greene leaues, very fweet in fmell, and very pleafant to behold, fo neer refembling the leaues of the Pomegranate tree that groweth with vs, that they foone deceiue many that are not expert therein, being fomewhat broade and long, and pointed at the ends, abiding alwaies greene: at the ioynts of the branches where the leaues ftand, come forth the flowers vpon fmall footeftalkes, euery one by it felfe confifting of fiue fmall white leaues, with white threds in the middle, fmelling alfo very fweet: after the flowers are paft, there doe arife in the hot Countries, where they are naturall, round blacke berries, when they are ripe, wherein are contained many hard white crooked feedes, but neuer in this Countrey, as I faid before : the roote difperfeth it felfe into many branches, with many fibres annexed thereto.

2. *Myrtus minor, feu minore folio.* The fmaller leafed Mirtle.

The fmaller leafed Mirtle is a low fhrub or bufh, like vnto the former, but fcarce rifing fo high, with branches fpreading about the ftemme, much thicker fet with leaues then the former, fmaller alfo, and pointed at the ends, of a little deeper greene colour, abiding greene alfo winter and fummer, and very fweete likewife : the flowers are white like vnto the former, and as fweete, but fhew not themfelues fo plentifull on the branches : the fruit is blacke in his naturall places, with feedes therein as the former.

3. *Myrtus minor rotundiore folio.* Boxe Mirtle.

Wee haue another fort of this fmall kinde of Mirtle, fo like vnto the former both for fmalneffe, deepe greene colour of the leaues, and thicke growing of the branches,

that

that it will be thought of most, without good heede, and comparing the one with the other, to be the very same with the former: but if it bee well viewed, it will shew, by the roundnesse at the ends of the leaues very like vnto the small Boxe leaues, to be another differing kinde, although in nothing else. Wee nourse them with great care, for the beautifull aspect, sweete sent and raritie, as delights and ornaments for a garden of pleasure, wherein nothing should be wanting that art, care and cost might produce and preserue: as also to set among other euer greene plants to sort with them.

The Place.

These, and many other sorts of Mirtles grow in Spaine, Portugall, Italie, and other hot Countries in great aboundance, where they make their hedges of them : wee (as I said) keepe them in this Countrey, with very great care and diligence.

The Time.

The Mirtles doe flower very late with vs, not vntill August at the soonest, which is the cause of their not fructifying.

The Names.

They are called in Latine *Myrtus*, and in English Mirtle tree, without any other diuersitie of names, for the generall title. Yet the seuerall kindes haue had seuerall denominations, in Plinies time, and others, as *Romana, Coniugala, Terentina, Egyptia, alba, nigra, &c.* which haue noted the differences, euen then well obserued.

The Vertues.

The Mirtle is of an astringent qualitie, and wholly vsed for such purposes.

Chap. CXIIII.

Malus Punica siue Granata. The Pomegranet tree.

THere are two kindes of Pomegranet trees, The one tame or manured, bearing fruit, which is distinguished of some into two sorts, of others into three, that is, into sower, and sweet, and into sower sweete. The other wilde, which beareth no fruite, because it beareth double flowers, like as the Cherry, Apple, and Peach tree with double blossomes, before described, and is also distinguished into two sorts, the one bearing larger, the other lesser flowers. Of the manured kinde wee haue onely one sort (so farre as we know) for it neuer beareth ripe fruit in this our Countrey) which for the beautifull aspect, both of the greene verdure of the leaues, and faire proportion and colour of the flowers, as also for the raritie, are noursed in some few of their gardens that delight in such rarities : for in regard of the tendernesse, there is neede of diligent care, that is, to plant it against a brick wall, and defend it conueniently from the sharpenesse of our winters, to giue his Master some pleasure in seeing it beare flowers : And of the double kinde we haue as yet obtained but one sort, although I shall giue you the knowledge and description of another.

1. *Malus Punica satiua.* The tame Pomegranet tree.

This Pomegranet tree groweth not very high in his naturall places, and with vs sometimes it shooteth forth from the roote many brownish twigges or branches, or if it bee pruned from them, and suffered to grow vp, it riseth to bee seuen or eight foote high,

spreading

1 *Myrtus latifolia maior.* The Great leafed Myrtle. 2 *Myrtus angustifolia minor.* The small leafed Myrtle. 3 *Myrtus buxifolia minor.* The Boxe leafed Myrtle. 4 *Malus Granata flos plenus flore.* The ordinary Pomegranet tree. 5 *Balaustium Romanum seu minus.* The lesser doublest wred Pomegranet tree. 6 *Balaustium maius siue Cyprium.* The greatest double flowred Pomegranet. 7 *Pseudocapsicum seu Amomum Plinij.* The Winter Cherry tree. 8 *Ficus Indica cum suo fructu.* The Indian Figge tree and his fruit.

spreading into many small and slender branches, here and there set with thornes, and with many very faire greene shining leaues, like in forme and bignesse vnto the leaues of the larger Myrtle before described, euery one hauing a small reddish foote-stalke vpon these branches: among the leaues come forth here and there, long, hard, and hollow reddish cups, diuided at the brimmes, wherein doe stand large single flowers, euery one consisting of one whole leafe, smaller at the bottome then at the brimme, like bels, diuided as it were at the edges into fiue or six parts, of an orient red or crimson colour in the hotter Countries; but in this it is much more delayed, and tendeth neare vnto a blush, with diuers threads in the middle. The fruit is great and round, hauing as it were a crowne on the head of it, with a thicke tough hard skinne or rinde, of a brownish red colour on the outside, and yellow within, stuffed or packt full of small graines, euery one encompast with a thin skin, wherein is contained a cleare red iuyce or liquor, either of a sweet (as I said before) or sower taste, or betweene them both of a winie taste: the roote disperseth it selfe very much vnder ground.

2. *Balaustium maius siue Malus Punica siluestris maior.* The greater wilde or double blossomd Pomegranet tree.

The wilde Pomegranet is like vnto the tame in the number of purplish branches, hauing thornes, and shining faire greene leaues, somewhat larger then the former: from the branches likewise shoote forth flowers, farre more beautifull then those of the tame or manured sort, becaufe they are double, and as large as a double Prouince Rose, or rather more double, of an excellent bright crimson colour, tending to a silken carnation, standing in brownish cups or huskes, diuided at the brims vsually into foure or fiue seuerall points, like vnto the former, but that in this kinde there neuer followeth any fruit, no not in the Country, where it is naturally wilde.

3. *Balaustium minus.* The smaller wilde Pomegranet tree.

This smaller kinde differeth from the former in his leaues, being of a darker greene colour, but not in the height of the stemme, or purplishnesse of his branches, or thorns vpon them; for this doth shew it selfe more like vnto a wilde kind then it: the flowers hereof are much smaller, and not so thicke and double, of a deeper or sadder red Orenge tawny colour, set also in such like cups or huskes.

The Place.

The tame or manured kinde groweth plentifully in Spaine, Portugall, and Italy, and other in other warme and hot countries. Wee (as I said before) preserue it with great care. The wilde I thinke was neuer seene in England, before Iohn Tradescante my very louing good friend brought it from the parts beyond the Seas, and planted it in his Lords Garden at Canterbury.

The Time.

They flower very late with vs, that is, not vntill the middle or end of August, and the cold euenings or frosts comming so soone vpon it, doth not onely hinder it from bearing, but many times the sharpe winters so pinch it, that it withereth it downe to the ground, so that oftentimes it hardly springeth againe.

The Names.

The name *Malus Punica* for the tree, and *Malum Punicum* for the fruit, or *Malus Granata*, and *Malum Granatum*, is the common name giuen vnto this tree, which is called in English the Pomegarnet or Pomegranet tree. The flowers of the tame kinde are called *Cytini*, as Dioscorides saith, although Plinie seemeth either to make *Cytinus* to be the flower of the wilde kinde, or

Balaustium

Balaustium to be the flower of both tame and wilde kinde : but properly, as I take it, *Cytinus* is the cup wherein the flower as well of the tame as wilde kinde doth stand; for vnto the similitude of them, both the flowers of *Asarum*, and the seede vessels of *Hyosciamus* are compared and resembled, and not vnto the whole flower : the barke or rinde of the fruit is called of diuers *Sidion*, and in the Apothecaries shops *Psidium*, and *cortex Granatorum*. The wilde kinde is called *Malus Punica siluestris* : In English, The wilde Pomegranet tree; the flower thereof is properly called *Balaustium*. The lesser kind is vsually called *Balaustium Romanum*, as the greater is called *Creticum* and *Cyprinum*, because they growe in Candy and Cyprus.

The Vertues.

The vse of all these Pomegranets is very much in Physicke, to coole and binde all fluxibility both of body and humours : they are also of singular effect in all vlcers of the mouth, and other parts of the body, both of man and woman. There is no part of them but is applyed for some of these respects. The rinde also of the Pomegranet is vsed of diuers in stead of Gaules, to make the best sort of writing Inke, which is durable to the worlds end.

Chap. CXV.

Amomum Plinÿ seu Pseudocapsicum.
Tree Night shade or the Winter Cherry tree.

I Haue adioyned this plant, for the pleasurable beauty of the greene leaues, and red berries. It groweth vp to be a yard or foure foote high at the most, hauing a small wooddy stemme or stocke, as bigge as ones finger or thumbe, couered with a whitish greene barke, set full of greene branches, and faire greene leaues, somewhat vneuen sometimes on the edges, narrower then any Night shade leaues, and very neare resembling the leaues of the *Capsicum*, or Ginny pepper, but smaller and narrower, falling away in the Winter, and shooting fresh in the Spring of the yeare : the flowers growe often two or three together, at the ioynts of the branches with the leaues, being white, opening starre-fashion, and sometimes turning themselues backe, with a yellow pointell in the middle, very like vnto the flowers of Night shade: after the flowers are past, come forth in their stead small greene buttons, which after turne to be pleasant round red berries, of the bignesse of small Cherries when they are ripe, which with vs vsually ripen not vntill the Winter, or about Christmas, wherein are contained many small whitish seede that are flat : all the whole plant, as well leaues and flowers as seede, are without either smell or taste : the roote hath many yellowish strings and fibres annexed vnto it.

The Place.

The originall place hereof is not well knowne, but is thought to bee the West Indies. It hath been planted of long time in most of these Countries, where it abideth reasonable well, so that some care bee had thereof in the extreamity of the Winter.

The Time.

It flowreth sometimes in Iune, but vsually in Iuly and August, and the fruit is not ripe (as is said) vntill the Winter.

The

The Names.

This plant hath diuers names ; for it is thought to be that kinde of *Amomum* that Plinie setteth downe. Dodonæus calleth it *Pseudocapsicum*, for some likenesse in the leafe and fruit vnto the small *Capsicum* or Ginnie Pepper, although much vnlike in the taste and property. Others doe call it *Strichnodendron*, that is, *Solanum arborescens*, and wee in English according thereunto, Tree Night shade. But some Latine asses corrupting the Latine word *Amomum*, doe call it the Mumme tree. Dalechampius calleth it *Solanum Americum, seu Indicum*, and saith the Spaniards call it in their tongue, *Guindas de las Indias*, that is, *Cerasa Indiana*, Indian Cherries, which if any would follow, I would not bee much againſt it : but many Gentlewomen doe call them Winter Cherries, becauſe the fruit is not throughly ripe vntill Winter.

The Vertues.

I finde no Phyſicall property allotted vnto it, more then that by reaſon of the inſipidity, it is held to be cooling.

Chap. CXVI.

Ficus Indica minor. The ſmaller Indian Figge tree.

THis Indian Figge tree, if you will call it a tree (becauſe in our Country it is not ſo, although it groweth in the naturall hot Countries from a wooddy ſtemme or body into leaues) is a plant conſiſting only of leaues, one ſpringing out of another, into many branches of leaues, and all of them growing out of one leafe, put into the ground halfe way, which taking roote, all the reſt riſe out thereof, thoſe belowe for the moſt part being larger then thoſe aboue ; yet all of them ſomewhat long, flat, and round pointed, of the thickneſſe of a finger vſually, and ſmalleſt at the lower end, where they are ioyned or ſpring out of the other leaues, hauing at their firſt breaking out a ſhew of ſmall, red, or browne prickes, thicke ſet ouer all the vpper ſide of the leaues, but with vs falling away quickly, leauing onely the markes where they ſtood : but they haue beſides this ſhew of great prickes, a few very fine, and ſmall, hard, white, and ſharpe, almoſt inſenſible prickes, being not ſo bigge as haires on the vnderſide, which will often ſticke in their fingers that handle them vnaduiſedly, neither are they to be diſcerned vnleſſe one look preciſely for them: the leaues on the vnderſide hauing none of thoſe other great pricks or marks at all, being of a faire freſh pale green colour: out of the vppermoſt leaues breake forth certaine greene heads, very like vnto leaues (ſo that many are deceiued, thinking them to be leaues, vntill they marke them better, and be better experienced in them) but that they growe round and not flat, and are broad at the toppe ; for that out of the tops of euery of them ſhooteth out a pale yellow flower, conſiſting of two rowes of leaues, each containing fiue leaues a peece, laid open with certaine yellow threads, tipt with red in the middle : this greene head, vntill the flower be paſt, is not of halfe that bigneſſe that it attaineth vnto after, yet ſeldome or neuer commeth vnto perfection with vs, being long and round, like vnto a Figge, ſmall belowe, and greater aboue, bearing vpon the flat or broad head the marke of the flower ; ſome holding ſtill on them the dryed leaues, and others hauing loſt them, ſhew the hollowneſſe which they haue in the toppe or middle of the head, the ſides round about being raiſed or ſtanding vp higher : this head or figge in our Country abideth greene on the outſide, and little or nothing reddiſh within (although it abide all the Winter, and the Summer following, as ſometimes it doth) for want of that heate and comfort of the Sunne it hath in his naturall place, where it groweth

reddiſh

reddish on the outside, and containing within it a bloudy red clammy iuyce, making the vrine of them that eate of them as red as bloud, which many seeing, were in doubt of themselues, left their vrine were not very bloud ; of what sweetnesse, like a figge, in the naturall places, I am not well assured , yet affirmed : but those that haue beene brought vnto me, whose colour on the outside was greenish, were of a reddish purple within, and contained within them round, small, hard seede, the taste was flat, wate-rish, or insipide : the roote is neither great, nor disperseth it selfe very deepe or farre, but shooteth many small rootes vnder the vpper cruft of the earth.

There is a greater kinde hereof, whose leaues are twice or thrice as bigge, which ha-uing been often brought vs , will seldome abide more then one Summer with vs , our Winters alwaies rotting the leaues, that it could not be longer kept.

The Place.

This Indian Figge tree groweth disperfedly in many places of Ame-rica, generally called the West Indies : The greater kinde in the more re-mote and hot Countries , as Mexico, Florida, &c. and in the Bermudas or Summer Islands, from whence wee haue often had it. The lesser in Virgi-nia, and those other Countries that are nearer vnto vs, which better endu-reth with vs.

The Time.

It floureth with vs fometimes in May, or Iune ; but (as I said) the fruit ne-uer commeth to perfection in this Country.

The Names.

Diuers doe take it to bee *Opuntia Pliny*, whereof hee speaketh in the 21. Booke and 17. Chapter of his Naturall History : but he there faith , *Opun-tia* is an herbe, sweete and pleasant to be eaten, and that it is a wonder that the roote should come from the leafe , and so to growe ; which words al-though they descipher out the manner of the growing of this plant, yet be-cause this is a kinde of tree, and not an herbe, nor to be eaten , it cannot bee the same : but especially because there is an herbe which groweth in the fame manner, or very neare vnto it , one leafe standing on the toppe or side of another, being a Sea plant, fit to be eaten with vinegar and oyle (as many other herbes are that growe in the falt marshes , or neare the Sea , whereof Sea Purflane is one) which Clufius calleth *Lychen Marinus*, and (as Clufius faith) Cortufus very fitly called *Opuntia marina*, and out of doubt is the ve-rie fame *Opuntia* that Theophraftus maketh mention of, and Plinie out of him. Our English people in Virginia , and the Bermuda Island, where it groweth plentifully, because of the form of the fruit, which is somewhat like to a Peare, & not being fo familiarly acquainted with the growing of Figs, fent it vnto vs by the name of the prickly Peare, from which name many haue fuppofed it to be a Peare indeede, but were therein deceiued.

The Vertues.

There is no other especiall property giuen hereunto, by any that haue written of the West Indies, then of the colouring of the vrine , as is be-fore said.

CHAP. CXVII.

Yuca siue Iucca. The suppofed Indian Iucca.

THis rare Indian plant hath a great thicke tuberous roote (fpreading in time into many tuberous heads) from the head whereof fhooteth forth many long, hard, and guttured leaues, very fharpe pointed, compaffing as it were one another at the bottome, of a grayifh greene colour, which doe not fall away, but abide euer greene on the plant; from the middle whereof fpringeth forth (now and then, but not euery yeare) a ftrong round ftalke, diuided into diuers branches, whereon ftand diuers white, and fomewhat large flowers, hanging downe their heads, confifting of fix leaues, with diuers veines, of a weake reddifh or blufh colour, fpread on the backe of the three outer leaues, efpecially from the middle of the leaues to the bottome, and not rifing to the edge of the leafe of any flower, which fall away without bearing any feede in our Country, as farre as euer could be obferued either in the plant that Mafter Gerard kept a long time by him, or by Robinus at Paris his plant, which Mafter Gerard fent vnto him, or yet by that plant, that Vefpafian Robin the fonne of old Robin feat vnto Mafter Iohn de Franqueuille, and now abideth and flourifheth in my Garden.

The Place.

It was firft brought into England (as Mafter Gerard faith) from the Weft Indies, by a feruant of Mafter Thomas Edwards, an Apothecary of Exeter, and imparted to him, who kept it vnto his death: but perifhed with him that got it from his widow, intending to fend it to his Country houfe.

The Time.

It flowreth not vntill Iuly, and the flowers fall away fodainely, after they haue beene blowne open a while.

The Names.

Mafter Gerard firft as I thinke called it *Iucca*, fuppofing it to bee the true *Yuca* of *Theuet*, wherewith the Indians make bread, called *Caffaua*: but the true *Iucca* is defcribed to haue a leafe diuided into feuen or nine parts, which this hath not: Yet not knowing by what better name to call it, let it hold ftill his firft impofition, vntill a fitter may be giuen it.

The Vertues.

Wee haue not heard of any, that hath either read, heard, or experimented the faculties hereof, nor yet whether it hath good or euill tafte; for being rare, and poffeffed but by a few, they that haue it are loth to cut any thereof, for feare of fpoiling and lofing the whole roote.

Some haue affirmed, that in fome parts of Turkie, where as they fay this plant groweth, they make a kinde of cloth from the threads are found running through the leaues; but I finde the threads are fo ftrong and hard, that this cannot be that plant the relators meane is vfed in that manner.

1 Yuca sive Iucca. The Indian Iucca.   2 Arbor vitæ. The tree of life.   3 Arbor Iudæ. Iudas tree.   4 Laburnum Beane Trefoile.
5 Cytisus. Tree Trefoile.

Chap. CXVIII.

Arbor vitæ. The tree of life.

THe tree of life riseth vp in some places where it hath stood long, to be a tree of a reasonable great bignesse and height, couered with a redder barke then any other tree in our Country that I know, the wood whereof is firme and hard, and spreadeth abroad many armes and branches, which againe send forth many smaller twigges, bending downewards; from which twiggy or slender branches, being flat themselues like the leaues, come forth on both sides many flat winged leaues, somewhat like vnto Sauine, being short and small, but not pricking, seeming as if they were brayded or folded like vnto a lace or point, of a darke yellowish greene colour, abiding greene on the branches Winter and Summer, of a strong resinous taste, not pleasing to most, but in some ready to procure casting, yet very cordiall and pectorall also to them that can endure it : at the toppes of the branches stand small yellowish dounie flowers, set in small scaly heads, wherein lye small, long, brownish seede, which ripen well in many places, and being sowne, doe spring and bring forth plants, which with some small care will abide the extreamest Winters we haue.

The Place.

The first or originall place where it naturally groweth, as farre as I can learne or vnderstand, is that part of America which the French doe inhabite, about the riuer of Canada, which is at the backe of Virginia Northward, and as it seemeth, first brought by them from thence into Europe, in the time of Francis the first French King, where it hath so plentifully encreased, and so largely beene distributed, that now few Gardens of respect, either in France, Germany, the Lowe-Countries, or England, are without it.

The Time.

It flowreth in the end of May, and in Iune; the fruit is ripe in the end of August and Sptember.

The Names.

All the Writers that haue written of it, since it was first knowne, haue made it to be *Thuyæ genus,* a kinde of Thuya, which Theophrastus compareth vnto a Cypresse tree, in his fifth Book and fifth Chapter : but *Omne simile non est idem,* and although it haue some likenesse, yet I verily beleeue it is *proprium sui genus,* a proper kinde of it owne, not to bee paralleld with any other. For wee finde but very few trees, herbes, or plants in America, like vnto those that growe in Europe, the hither part of Africa, or in the lesser Asia, as experience testifieth. Some would make it to be *Cedrus Lycia,* but so it cannot be. The French that first brought it, called it *Arbor vitæ,* with what reason or vpon what ground I know not : but euer since it hath continued vnder the title of the Tree of life.

The Vertues.

It hath beene found by often experience, that the leaues hereof chewed in the morning fasting, for some few dayes together, haue done much good to diuers, that haue beene troubled with shortnesse of breath, and to helpe to expectorate thinne purulentous matter stuffing the lungs. Other properties I haue not heard that it hath ; but doubtlesse, the hot resinous smell and
taste

tafte it hath, both while it is frefh, and after it hath beene long kept dry, doth euidently declare his tenuity of parts, a digefting and cleaning quality it is poffeffed with, which if any induftrious would make tryall, hee fhould finde the effects.

Chap. CXIX.

Arbor Iuda. Iudas tree.

IVdas tree rifeth vp in fome places, where it ftandeth open from a wall, and alone free from other trees (as in a Garden at Battherfey, which fometimes agoe belonged to Mafter Morgan, Apothecary to the late Queene Elizabeth of famous memory) to be a very great and tall tree, exceeding any Apple tree in height, and equall in bigneffe of body thereunto (as my felfe can teftifie, being an eye witneffe thereof) when as it had many ftalkes of flowers, being in the bud, breaking out of the body of the tree through the barke in diuers places, when as there was no bough or branch neare them by a yard at the leaft, or yet any leafe vpon the tree, which they gathered to put among other flowers, for Nofegayes) and in other places it groweth to bee but an hedge bufh, or plant, with many fuckers and fhootes from belowe, couered with a darke reddifh barke, the young branches being more red or purplifh: the flowers on the branches come forth before any fhew or budding of leaues, three or foure ftanding together vpon a fmall foote-ftalke, which are in fafhion like vnto Peafe bloffomes, but of an excellent deepe purplifh crimfon colour: after which come in their places fo many long, flat, large, and thinne cods, of a brownifh colour, wherein are contained fmall, blackifh browne, flat, and hard feede: the roote is great, and runneth both deepe, and farre fpreading in the earth: the leaues come forth by themfelues, euery one ftanding on a long ftalke, being hard & very round, like vnto the leafe of the largeft *Afarum*, but not fo thick, of a whitifh green on the vpper fide, and grayifh vnderneath, which fall away euery yeare, and fpring afrefh after the Spring is well come in, and the buds of flowers are fprung.

There is another of this kinde, growing in fome places very high, fomewhat like *Flore albo* the former, and in other places alfo full of twiggy branches, which are greener then the former, as the leaues are likewife: the flowers of this kinde are wholly white, and the cods nothing fo red or browne, in all other things agreeing together.

The Place.

The former groweth plentifully in many places of Spaine, Italy, Prouence in France, and in many other places. The other hath beene fent vs out of Italy many times, and the feede hath fprung very well with vs, but it is fomewhat tender to keepe in the Winter.

The Time.

The flowers (as I faid) appeare before the leaues, and come forth in Aprill and May, and often fooner alfo, the leaues following fhortly after, but neither of them beareth perfect feede in our Country, that euer I could learne, or know by mine owne or others experience.

The Names.

Some would referre this to *Cercis*, whereof Theophraftus maketh mention in his firft Booke and eighteenth Chapter, among thofe trees that beare their fruit in cods, like as Pulfe doe: and hee remembreth it againe in the fourteenth Chapter of his third Booke, and maketh it not vnlike the white

Poplar

Poplar tree, both in greatnesse and whitenesse of the branches, with the leafe of an Iuie, without corners on the one part, cornered on the other, and sharpe pointed, greene on both sides almost alike, hauing so slender long footestalkes that the leaues cannot stand forthright, but bend downwards, with a more rugged barke then the white Poplar tree. Clusius thinketh this large description is but an ample description of the third kinde of Poplar, called *Lybica*, the Aspen tree, which Gaza translateth *Alpina* : but who so will well consider it, shall finde it neyther answerable to any Poplar tree, in that it beareth not cods as *Cercis* doth ; nor vnto this *Arbor Iuda*, because it beareth not white branches. Clusius saith also, that the learned of Mompelier in his time, referred it to *Colytea* of Theophrastus in his third booke and seuenteenth chapter, where he doth liken it to the leaues of the broadest leafed Bay tree, but larger and rounder, green on the vpperside, and whitish vnderneath, and whereunto (as he saith) Theophrastus giueth cods in the fourteenth chapter of the same third booke: and by the contracting of their descriptions both together, saith, they agree vnto this Iudas tree. But I find some doubts and differences in these places : for the *Colutea* that Dioscorides mentioneth in the said fourteenth chapter of his third booke, hath (as he saith there) a leafe like vnto the Willow, and therefore cannot bee the same *Colutea* mentioned in the seuenteenth chapter of the same third book, which hath a broade Bay leafe : indeede hee giueth seede in cods : but that with broade Bay leaues is (as he saith) without eyther flower or fruite ; and besides all this, he saith the rootes are very yellow, which is not to bee found in this *Arbor Iuda*, or Iudas tree: let others now iudge if these things can bee well reconciled together. Some haue for the likenesse of the cods vnto Beane cods, called it *Fabago*. And Clusius called it *Siliqua siluestris*. It is generally in these dayes called *Arbor Iuda*, and in English after the Latine name, vntill a fitter may be had, Iudas tree.

The Vertues.

There is nothing extant in any Author of any Physicall vse it hath, neyther hath any later experience found out any.

Chap. CXX.

Laburnum. Beane Trefoile.

THere be three sorts of these codded trees or plants, one neere resembling another, whereof *Anagyris* of Dioscorides is one. The other two are called *Laburnum* ; the larger whereof Matthiolus calleth *Anagyris altera*, and so doe some others also : the third is of the same kinde with the second, but smaller. I shall not for this our Garden trouble you or my selfe with any more of them then one, which is the lesser of the two *Liburnum*, in that it is more frequent, and that it will far better abide then the *Anagyris*, which is so tender, that it will hardly endure the winters of our Countrey : and the greater *Laburnum* is not so easily to be had.

Laburnum. Beane Trefoile.

This codded tree riseth vp with vs like vnto a tall tree, with a reasonable great body, if it abide any long time in a place, couered with a smooth greene barke; the branches are very long, greene, pliant, and bending any way, whereon are set here and there diuers leaues, three alwaies standing together vpon a long stalk, being somwhat long, and not very narrow, pointed at the ends, greene on the vpperside, and of a siluer shining colour vnderneath, without any smell at all : at the ioynts of these branches, where the leaues stand, come forth many flowers, much like vnto broome flowers, but not so

large

large or open, growing about a very long branch or ftalke, fometimes a good fpan or more in length, and of a faire yellow colour, but not very deepe ; after which come flat thin cods, not very long or broade, but as tough and hard as the cods of Broome; wherein are contained blackifh feede, like, but much leffe then the feede of *Anagyris vera* (which are as big as a kidney beane, purplifh and fpotted) : the roote thrufteth down deepe into the ground, fpreading alfo farre, and is of a yellowifh colour.

The Place.

This tree groweth naturally in many of the woods of Italie, and vpon the Alpes alfo, and is therefore ftill accounted to be that *Laburnum* that Plinie calleth *Arbor Alpina*. It groweth in many gardens with vs.

The Time.

It floweth in May, the fruit or cods, and the feedes therein are ripe in the end of Auguft, or in September.

The Names.

This tree (as I faid before) is called of Matthiolus *Anagyris altera, fiue fecunda*, of Cordus, Gefner and others, efpecially of moft now adayes, *Laburnum*. It is probable in my opinion, that this fhould bee that *Colutea* of Theophraftus, mentioned in the fourteenth Chapter of his third book with the leafe of a Willow ; for if you take any one leafe by it felfe, it may well refemble a Willow leafe both for forme and colour, and beareth fmall feed in cods like vnto pulfe as that doth. Of fome it hath beene taken for a kinde of *Cytifus*, but not truely. We call it in Englifh, Beane Trefoile, in regard of his cods and feede therein, fomewhat like vnto Kidney Beanes, and of the leaues, three alwayes ftanding together, vntill a more proper name may bee giuen it.

The Vertues.

There is no vfe hereof in Phyficke with vs, nor in the naturall place of the growing, faue only to prouoke a vomit, which it will doe very ftrongly,

---

CHAP. CXXI.

Cytifus. Tree Trefoile.

THere are fo many forts of *Cytifus* or Tree trefoiles, that if I fhould relate them all, I fhould weary the Reader to ouerlooke them, whereof the moft part pertaine rather to a generall worke then to this abftract. I fhall not therefore trouble you with any fuperfluous, but only with two, which we haue nourfed vp to furnifh wafte places in a garden.

Cytifus Maranthæ. Horned Tree Trefoile.

This Tree Trefoile which is held of moft Herbarifts to bee the true *Cytifus* of Diofcorides, rifeth vp to the height of a man at the moft, with a body of the bigneffe of a mans thumbe, couered with a whitifh bark, breaking forth into many whitifh branches fpreading farre, befet in many places with fmall leaues, three alwayes fet together vpon a fmall fhort footeftalke, which are rounder, and whiter then the leaues of Beane Trefoile : at the ends of the branches for the moft part, come forth the flowers three or foure togethers, of a fine gold colour, and of the fafhion of Broome flowers, but

not

not to large : after the flowers are paft, there come in their places crooked flat thinne cods, of the fafhion of a halfe moone, or crooked horne, whitifh when they are ripe, wherein are contained blackifh feede : the roote is hard and woody, fpreading diuers wayes vnder the ground : the whole plant hath a pretty fmall hot fent.

Cytifus vulgatior. The common Tree Trefoile.

This *Cytifus* is the moft common in this Land, of any the other forts of tree trefoiles, hauing a blackifh colourd barke, the ftemme or body whereof is larger then the former, both for height and fpreading, bearing alfo three leaues together, but fmaller and greener then the former : the flowers are fmaller, but of the fame fafhion and colour: the cods blackifh and thin, and not very long, or great, but leffer then Broome cods, wherein there lyeth fmall blackifh hard feede : the roote is diuerfly difperfed in the ground.

The Place.

The firft groweth in the kingdome of Naples, and no doubt in many other places of Italie, as Matthiolus faith. The other groweth in diuers places of France.

The Time.

They flower for the moft part in May or Iune : the feede is ripe in Auguft or September.

The Names.

The firft (as I faid) is thought of moft to be the true *Cytifus* of Diofcorides, and as is thought, was in thefe later dayes firft found by Bartholomæus Maranta of Naples, who fent it firft to Matthiolus, and thereupon hath euer fince beene called after his name, *Cytifus Marantha.* Some doe call it *Cytifus Lunatus,* becaufe the cods are made fomewhat like vnto an halfe Moone. We call it in Englifh, Horned Tree Trefoile. The other is called *Cytifus vulgaris,* or *vulgatior,* in Englifh, The common Tree Trefoile, becaufe we haue not any other fo common.

The Vertues.

The chiefeft vertues that are appropriate to thefe plants, are to procure milke in womens breafts, to fatten pullen, fheep &c. and to be good for bees.

CHAP. CXXII.

Colutea. The Baftard Sena Tree.

WEe haue in our Gardens two or three forts of the Baftard Sena tree ; a greater as I may fo call it, and two leffer : the one with round thin tranfparent skins like bladders, wherein are the feede : the others with long round cods, the one bunched out or fwelling in diuers places, like vnto a Scorpions tale, wherein is the feede, and the other very like vnto it, but fmaller.

1. *Colutea Veficaria.* The greater Baftard Sena with bladders.

This fhrub or tree, or fhrubby tree, which you pleafe to call it, rifeth vp to the height of a pretty tree, the ftemme or ftock being fometimes of the bigneffe of a mans arme, couered with a blackifh greene rugged barke, the wood whereof is harder then of an

Elder,

Elder, but with an hollowneſſe like a pith in the heart or middle of the branches, which are diuided many wayes, and whereon are ſet at ſeuerall diſtances, diuers winged leaues, compoſed of many ſmall round pointed, or rather flat pointed leaues, one ſet againſt another, like vnto Licoris, or the Hatchet Fitch ; among theſe leaues come forth the flowers, in faſhion like vnto Broome flowers, and as large, of a very yellow colour : after which appeare cleare thinne ſwelling cods like vnto thinne tranſparent bladders, wherein are contained blacke ſeede, ſet vpon a middle ribbe or ſinew in the middle of the bladder, which if it be a little cruſhed betweene the fingers, will giue a cracke, like as a bladder full of winde. The roote groweth branched and woody.

2. *Colutæa Scorpioides maior.* The greater Scorpion podded Baſtard Sena.

This Baſtard Sena groweth nothing ſo great or tall, but ſhooteth out diuerſly, like vnto a ſhrub, with many ſhoots ſpringing from the root : the branches are greener, but more rugged, hauing a white barke on the beſt part of the elder growne branches ; for the young are greene, and haue ſuch like winged leaues ſet on them as are to be ſeen in the former, but ſmaller, greener, and more pointed : the flowers are yellow, but much ſmaller, faſhioned ſomewhat like vnto the former, with a reddiſh ſtripe downe the backe of the vppermoſt leafe: the long cods that follow are ſmall, long and round, diſtinguiſhed into many diuiſions or dents, like vnto a Scorpions tayle, from whence hath riſen the name : in theſe ſeuerall diuiſions lye ſeuerall blacke ſeede, like vnto the ſeede of Fenigrecke : the roote is white and long, but not ſo woody as the former.

3. *Colutæa Scorpioides minor.* The leſſer Scorpion Baſtard Sena.

This leſſer Baſtard Sena is in all things like the former, but ſomewhat lower, and ſmaller both in leafe, flower, and cods of ſeede, which haue not ſuch eminent bunches on the cods to be ſeene as the former.

The Place.

They grow as Matthiolus ſaith about Trent in Italie, and in other places : the former is frequent enough through all our Countrey, but the others are more rare.

The Time.

They flower about the middle or end of May, and their ſeede is ripe in Auguſt. The bladders of the firſt will abide a great while on the tree, if they be ſuffered, and vntill the winde cauſe them to rattle, and afterwards the skins opening, the ſeed will fall away.

The Names.

The name *Colutæa* is impoſed on them, and by the iudgement of moſt writers, the firſt is taken to bee that *Colutæa* of *Lipara* that Theophraſtus maketh mention of, in the ſeuenteenth chapter of his third booke. But I ſhould rather thinke that the *Scorpioides* were the truer *Colutæa* of Theophraſtus, becauſe the long pods thereof are more properly to bee accounted *ſiliquæ*, then the former which are *veſicæ tumentes*, windy bladders, and not *ſiliquæ* : and no doubt but Theophraſtus would haue giuen ſome peculiar note of difference if he had meant thoſe bladders, and not theſe cods. Let others of iudgement be vmpeeres in this caſe ; although I know the currant of writers ſince Matthiolus, doe all hold the former *Colutæa veſicaria* to be the true *Colutæa Liparæ* of Theophraſtus. Wee call it in Engliſh, Baſtard Sena, from Ruellius, who as I thinke firſt called it Sena, from the forme of the leaues. The ſecond and third (as I ſaid before) from the forme of the cods receiued their names, as it is in the titles and deſcriptions ; yet they may as properly be called *Siliquoſæ*, for that their fruite are long cods.

The

The Vertues.

Theophraſtus ſaith it doth wonderfully helpe to fatten ſheepe : But ſure it is found by experience, that if it be giuen to man it cauſeth ſtrong caſtings both vpwards and downwards ; and therefore let euery one beware that they vſe not this in ſteede of good Sena, leſt they feele to their coſt the force thereof.

CHAP. CXII.

Spartum Hiſpanicum frutex. Spaniſh Broome.

ALthough Cluſius and others haue found diuers ſorts of this ſhrubby Spartum or Spaniſh Broome, yet becauſe our Climate will nouriſe vp none of them, and euen this very hardly, I ſhall leaue all others, and deſcribe vnto you this one only in this manner : Spaniſh Broome groweth to bee fiue or ſixe foote high, with a woody ſtemme below, couered with a darke gray, or aſh-coloured barke, and hauing aboue many pliant, long and ſlender greene twigs, whereon in the beginning of the yeare are ſet many ſmall long greene leaues, which fall away quickly, not abiding long on ; towards the tops of theſe branches grow the flowers, faſhioned like vnto Broom flowers, but larger, as yellow as they, and ſmelling very well ; after which come ſmall long cods, creſted at the backe, wherein is contained blackiſh flat ſeede, faſhioned very like vnto the Kidney beanes: the roote is woody, diſperſing it ſelfe diuers waies.

The Place.

This groweth naturally in many places of France, Spaine and Italie, wee haue it as an ornament in our Gardens, among other delightfull plants, to pleaſe the ſenſes of ſight and ſmelling.

The Time.

It flowreth in the end of May, or beginning of Iune, and beareth ſeede, which ripeneth not with vs vntill it be late.

The Names.

It is called *Spartium Græcorum,* and *Spartum frutex,* to diſtinguiſh it from the ſedge or ruſh, that is ſo called alſo. Of ſome it is called *Geniſta,* and thought not to differ from the other *Geniſta,* but they are much deceiued ; for euen in Spaine and Italie, the ordinary *Geniſta* or Broome groweth with it, which is not pliant, and fit to binde Vines, or ſuch like things withall as this is.

The Vertues.

There is little vſe hereof in Phyſicke, by reaſon of the dangerous qualitie of vomiting, which it doth procure to them that take it inwardly : but being applyed outwardly, it is found to helpe the *Sciatica,* or paine of the hippes.

1 *Colutea vulgaris.* Ordinary baſtard Sene. 2 *Periploca rella Virginiana.* Virginian Silke. 3 *Colutea Scorpioides.* Scorpion baſtard Sene. 4 *Spartum Hiſpanicum.* Spaniſh Broome. 5 *Liguſtrum.* Priuet. 6 *Saluia variegata.* Party coloured Sage. 7 *Maiorana aurea.* Guilded Maierome.

CHAP. CXXIIII.

Periploca recta Virginiana. Virginian Silke.

LEst this stranger should finde no hospitality with vs, being so beautifull a plant, or not finde place in this Garden, let him be here receiued, although with the last, rather then not at all. It riseth vp with one or more strong and round stalkes, three or foure foote high, whereon are set at the seuerall ioynts thereof two faire, long, and broad leaues, round pointed, with many veines therein, growing close to the stemme, without any foote-stalke : at the tops of the stalkes, and sometimes at the ioynts of the leaues, groweth forth a great bush of flowers out of a thinne skinne, to the number of twenty, and sometimes thirty or forty, euery one with a long foote-stalke, hanging downe their heads for the most part, especially those that are outermost, euery one standing within a small huske of greene leaues, turned to the stalkeward, like vnto the Lysimachia flower of Virginia before described, and each of them consisting of fiue small leaues a peece, of a pale purplish colour on the vpperside, and of a pale yellowish purple vnderneath, both sides of each leafe being as it were folded together, making them seeme hollow and pointed, with a few short chiues in the middle : after which come long and crooked pointed cods standing vpright, wherein are contained flat brownish seede, dispersedly lying within a great deale of fine, soft, and whitish browne silke, very like vnto the cods, seede, and silke of *Asclepias*, or Swallow-wort, but that the cods are greater and more crooked, and harder also in the outer shell : the roote is long and white, of the bignesse of a mans thumbe, running vnder ground very far, and shooting vp in diuers places, the heads being set full of small white grumes or knots, yeelding forth many branches, if it stand any time in a place: the whole plant, as well leaues as stalkes, being broken, yeeld a pale milke.

The Place.

It came to me from Virginia, where it groweth aboundantly, being raised vp from the seede I receiued.

The Time.

It flowreth in Iuly, and the seede is ripe in August.

The Names.

It may seeme very probable to many, that this plant is the same that Prosper Alpinus in the twenty fift Chapter of his Booke of Egyptian plants, nameth *Beidelsar*; and Honorius Bellus in his third and fourth Epistles vnto Clusius (which are at the end of his History of plants) calleth *Ossar frutex* : And Clusius himselfe in the same Booke calleth *Apocynum Syriacum, Palaestinum,* and *Aegyptiacum,* because this agreeth with theirs in very many and notable parts ; yet verily I thinke this plant is not the same, but rather another kinde of it selfe : First, because it is not *frutex,* a shrub or wooddy plant, nor keepeth his leaues all the yeare, but loseth both leaues and stalks, dying down to the ground euery yeare: Secondly, the milke is not causticke or violent, as Alpinus and Bellus say *Ossar* is : Thirdly, the cods are more crooked then those of Clusius, or of Alpinus, which Honorius Bellus acknowledgeth to be right, although greater then those he had out of Egypt : And lastly, the rootes of these doe runne, whereof none of them make any mention. Gerard in his Herball giueth a rude figure of the plant, but a very true figure of the cods with seede, and saith the Virginians call it *wisanck,* and referreth it to the *Asclepias,* for the likenesse of the cods stuffed with

silken

silken doune. But what reason Caspar Bauhinus in his *Pinax Theatri Bota-nici* had, to call it (for it is Clusius his *Apocynum Syriacum*) by the name of *Lapathum Ægyptiacum lactescens siliqua Asclepiadis*, I know none in the world: for but that he would shew an extreame singularity in giuing names to plants, contrary to all others (which is very frequent with him) how could he thinke, that this plant could haue any likenesse or correspondencie, with any of the kindes of Dockes, that euer he had seene, read, or heard of, in face, or shew of leaues, flowers, or seede; but especially in giuing milke. I haue you see (and that not without iust and euident cause) giuen it a differing Latine name from Gerard, because the *Asclepias* giueth no milke, but the *Periploca* or *Apocynum* doth; and therefore fitter to be referred to this then to that. And because it should not want an English name answerable to some peculiar property thereof, I haue from the silken doune called it Virginian Silke: but I know there is another plant growing in Virginia, called Silke Grasse, which is much differing from this.

The Vertues.

I know not of any in our Land hath made any tryall of the properties hereof. Captaine Iohn Smith in his booke of the discouery and description of Virginia, saith, that the Virginians vse the rootes hereof (if his be the same with this) being bruised and applyed to cure their hurts & diseases.

Chap. CXXV.

Ligustrum. Primme or Priuet.

Ecause the vse of this plant is so much, and so frequent throughout all this Land, although for no other purpose but to make hedges or arbours in Gardens, &c. whereunto it is so apt, that no other can be like vnto it, to bee cut, lead, and drawne into what forme one will, either of beasts, birds, or men armed, or otherwise: I could not forget it, although it be so well knowne vnto all, to be an hedge bush growing from a woodly white roote, spreading much within the ground, and bearing manie long, tough, and plyant sprigs and branches, whereon are set long, narrow, and pointed sad greene leaues by couples at euery ioynt: at the tops whereof brake forth great tufts of sweete smelling white flowers, which when they are fallen, turne into small blacke berries, hauing a purple iuyce within them, and small seede, flat on the one side, with an hole or dent therein: this is seene in those branches that are not cut, but suffered to beare out their flowers and fruit.

The Place.

This bush groweth as plentifully in the Woods of our owne Couutrey, as in any other beyond the Seas.

The Time.

It flowreth sometimes in Iune, and in Iuly; the fruit is ripe in August and September.

The Names.

There is great controuersie among the moderne Writers concerning this plant, some taking it to be κύπρος of Dioscorides, other to be *Phillyrea* of Dioscorides, which followeth next after *Cyprus*. Plinie maketh mention of *Cyprus* in two places; in the one he saith, *Cyprus* hath the leafe of *Ziziphus*,

P p us

or the Iuiube tree: in the other he saith, that certain do affirme, that the *Cyprus* of the East Country, and the *Ligustrum* of Italy is one and the same plant : whereby you may plainly see, that our Priuet which is *Ligustrum*, cannot be that *Cyprus* of Plinie with Iuiube leaues: Besides, both Dioscorides & Plinie say, that *Cyprus* is a tree ; but all know that *Ligustrum*, Priuet, is but an hedge bush : Againe, Dioscorides saith, that the leaues of *Cyprus* giue a red colour, but Priuet giueth none. Bellonius and Prosper Alpinus haue both recorded, that the true *Cyprus* of Dioscorides groweth plentifully in Egypt, Syria, and those Easterne Countries, and nourced vp also in Constantinople, and other parts of Greece, being a merchandise of much worth, in that they transport the leaues, and young branches dryed, which laid in water giue a yellow colour, wherewith the Turkish women colour the nailes of their hands, and some other parts of their bodies likewise, delighting much therein : and that it is not our *Ligustrum*, or Priuet, because *Cyprus* beareth round white seede, like Coriander seede, and the leaues abide greene alwaies vpon the tree, which groweth (if it bee not cut or pruined) to the height of the Pomegranet tree. I haue (I confesse) beyond the limits I set for this worke spoken concerning our Priuet, because I haue had the seede of the true *Cyprus* of Dioscorides sent mee, which was much differing from our Priuet, and although it sprang vp, yet would not abide any time, whereas if it had beene our Priuet, it would haue beene familiar enough to our Countrey.

The Vertues.

It is of small vse in physicke, yet some doe vse the leaues in Lotions, that serue to coole and dry fluxes or sores in diuers parts.

CHAP. CXXVI.

Saluia variegata. Party coloured Sage. And
Maiorana versicolor siue aurea. Yellow or golden Marierome.

VNto all these flowers of beauty and rarity, I must adioyne two other plants, whose beauty consisteth in their leaues, and not in their flowers : as also to separate them from the others of their tribe, to place them here in one Chapter, before the sweete herbes that shall follow, as is fittest to furnish this our Garden of pleasure. This kinde of Sage groweth with branches and leaues, very like the ordinary Sage, but somewhat smaller, the chiefest difference consisteth in the colour of the leaues, being diuersly marked and spotted with white and red among the greene: for vpon one branch you shall haue the leaues seuerally marked one from another, as the one halfe of the leafe white, and the other halfe greene, with red shadowed ouer them both, or more white then greene, with some red in it, either parted or shadowed, or dasht here and there, or more greene then white, and red therein, eyther in the middle or end of the leafe, or more or lesse parted or striped with white and red in the greene, or else sometimes wholly greene the whole branch together, as nature lifteth to play with such varieties : which manner of growing rising from one and the same plant, because it is the more variable, is the more delightfull and much respected.

There is another speckled Sage parted with white and greene, but it is nothing of that beauty to this, because this hath three colours euidently to bee discerned in euery leafe almost, the red adding a superaboundant grace to the rest.

Maiorana aurea siue versicolor. Yellow or golden Marierome.

This kinde of Marierome belongeth to that sort is called in Latine *Maiorana latifolia,*

lia, which Lobel setteth forth for *Hyssopus Græcorum genuina*: In English Winter Mar-ierome, or po. Marierome : for it hath broader and greater leaues then the sweete Marierome, and a different vmbell or tuft of flowers. The difference of this from that set forth in the Kitchin Garden, consisteth chiefly in the leaues, which are in Summer wholly yellow in some, or but a little greene, or parted with yellow and greene more or lesse, as nature listeth to play : but in Winter they are of a darke or dead greene co-lour, yet recouering it selfe againe : the sent hereof is all one with the pot Marierome.

Wee haue another parted with white and greene, much after the manner with the former.

The Place, Time, Names, and Vertues of both these plants, shall be de-clared where the others of their kindes are specified hereafter, and in the Kitchen Garden ; for they differ not in properties.

Chap. CXXVII.

Lauendula. Lauender Spike.

AFter all these faire and sweete flowers before specified, I must needes adde a few sweete herbes, both to accomplish this Garden, and to please your senses, by placing them in your Nosegayes, or else where, as you list. And although I bring them in the end or last place, yet are they not of the least account.

1. *Lauendula maior.* Garden Lauender.

Our ordinary Garden Lauender riseth vp with a hard wooddy stemme aboue the ground, parted into many small branches, whereon are set whitish, long, and narrow leaues, by couples one against another ; from among which riseth vp naked square stalkes, with two leaues at a ioynt, and at the toppe diuers small huskes standing round about them, formed in long and round heads or spikes, with purple gaping flowers springing out of each of them : the roote is wooddy, and spreadeth in the ground : The whole plant is of a strong sweete sent, but the heads of flowers much more, and more piercing the senses, which are much vsed to bee put among linnen and ap-parrell.

There is a kinde hereof that beareth white flowers, and somewhat broader leaues, *Flore albis.* but it is very rare, and seene but in few places with vs, because it is more tender, and will not so well endure our cold Winters.

2. *Lauendula minor seu Spica.* Small Lauender or Spike.

The Spike or small Lauender is very like vnto the former, but groweth not so high, neither is the head or spike so great and long, but shorter and smaller, and of a more purplish colour in the flower : the leaues also are a little harder, whiter, and shorter then the former ; the sent also is somewhat sharper and stronger. This is not so fre-quent as the first, and is nourished but in some places that are warme, and where they delight in rare herbes and plants.

The Place.

Lauender groweth in Spaine aboundantly, in many places so wilde, and little regarded, that many haue gone, and abiden there to distill the oyle thereof whereof great quantity now commeth ouer from thence vnto vs : and also in Lanquedocke, and Prouence in France.

The Time.

It flowreth early in those hot Countries, but with vs not vntill Iune and Iuly.

The

The Names.

It is called of some *Nardus Italica*, and *Lauendula*, the greater is called *Fæmina*, and the lesser *Mas*. We doe call them generally Lauender, or Lauender Spike, and the lesser Spike, without any other addition.

The Vertues.

Lauender is little vsed in inward physicke, but outwardly; the oyle for cold and benummed parts, and is almost wholly spent with vs, for to perfume linnen, apparrell, gloues, leather, &c. and the dryed flowers to comfort and dry vp the moisture of a cold braine.

Chap. CXXVIII.

Stœchas. Sticadoue, Cassidony, or French Lauender.

CAssidony that groweth in the Gardens of our Countrey, may peraduenture somewhat differ in colour, as well as in strength, from that which groweth in hotter Countries; but as it is with vs, it is more tender a great deale then Lauender, and groweth rather like an herbe then a bush or shrub, not aboue a foote and a halfe high, or thereabouts, hauing many narrow long greene leaues like Lauender, but softer and smaller, set at seuerall distances together about the stalkes, which spread abroad into branches: at the tops whereof stand long and round, and sometimes foure square heads, of a darke greenish purple colour, compact of many scales set together; from among which come forth the flowers, of a blewish purple colour, after which follow seede vessels, which are somewhat whitish when they are ripe, containing blackish browne seede within them: the roote is somewhat wooddy, and will hardly abide the iniuries of our cold Winters, except in some places onely, or before it haue flowred: The whole plant is somewhat sweete, but nothing so much as Lauender.

The Place.

Cassidony groweth in the Islands Stœchades, which are ouer against Marselles, and in Arabia also: we keep it with great care in our Gardens.

The Time.

It floweth the next yeare after it is sowne, in the end of May, which is a moneth before any Lauender.

The Names.

It is called of some *Lauendula siluestris*, but most vsually *Stœchas*: in English, of some Stickadoue, or French Lauender; and in many parts of England, Cassidony.

The Vertues.

It is of much more vse in physicke then Lauen'er, and is much vsed for old paines in the head. It is also held to be good for to open obstructions, to expell melancholy, to cleanse and strengthen the liuer, and other inward parts, and to be a Pectorall also.

Chap. CXXIX.

Abrotanum femina siue Santolina. Lauender Cotton.

THis Lauender Cotton hath many wooddy, but brittle branches, hoary or of a whitish colour, whereon are set many leaues, which are little, long, and foure square, dented or notched on all edges, and whitish also: at the tops of these branches stand naked stalkes, bearing on euery one of them a larger yellow head or flower, then eyther Tansie or Maudeline, whereunto they are somewhat like, wherein is contained small darke coloured seede: the roote is hard, and spreadeth abroad with many fibres: the whole plant is of a strong sweete sent, but not vnpleasant, and is in many places planted in Gardens, to border knots with, for which it will abide to be cut into what forme you thinke best; for it groweth thicke and bushy, very fit for such workes, besides the comely shew the plant it selfe thus wrought doth yeeld, being alwayes greene, and of a sweet sent; but because it quickly groweth great, and will soon runne out of forme, it must be euery second or third yeare taken vp, and new planted.

The Place.

It is onely planted in Gardens with vs, for the vses aforesaid especially.

The Time.

It flowreth in Iuly, and standeth long in the hot time of the yeare in his colour, and so will doe, if it be gathered before it haue stood ouer long.

The Names.

Diuers doe call it as Matthiolus doth, *Abrotanum femina*, and *Santolina*; and some call it *Chamaecyparissus*, because the leaues thereof, are somewhat like the leaues of the Cypresse tree: Wee call it in English generally Lauender Cotton.

The Vertues.

This is vsually put among other hot herbes, eyther into bathes, ointments, or other things, that are vsed for cold causes. The seede also is much vsed for the wormes.

Chap. CXXX.

Ocimum. Bassill.

BAssill is of two sorts (besides other kindes) for this our Garden, the one whereof is greater, the other lesse in euery part thereof, as shall be shewed.

1. *Ocimum Citratum.* Common Bassill.

Our ordinary Garden Bassill hath one stalke rising from the root, diuersly branched out, whereon are set two leaues alwayes at a ioynt, which are broad, somewhat round, and pointed, of a pale greene colour, but fresh, a little snipt or dented about the edges, and of a strong or heady sent, somewhat like a Pomecitron, as many haue compared it, and thereof call it *Citratum*: the flowers are small and white, standing at the tops of the branches, with two small leaues at euery ioynt vnder them, in some plants green, in others

thers

thers browne vnder them : after which commeth blackish seede : the roote perisheth at the first approach of winter weather, and is to be new sowen euery yeare.

2. *Ocimum minimum siue Gariophyllatum.* Bush Basill.

The bush Basill groweth not altogether so high, but is thicker spreade out into branches, whereon grow smaller leaues, and thicker set then the former, but of a more excellent and pleasant smell by much : the flowers are white like the former, and the seede blacke also like it, and perisheth as suddenly, or rather sooner then it, so that it requireth more paines to get it, and more care to nourse it, because we seldome or neuer haue any seede of it.

Ocimum Indicum. Indian Basill.

The Indian Basill hath a square reddish greene stalke, a foote high or better, from the ioynts whereof spreade out many branches, with broade fat leaues set thereon, two alwayes together at the ioynt, one against another, as other Basils haue, but somewhat deepely cut in on the edges, and oftentimes a little crumpled, standing vpon long reddish footestalkes, of a darke purple colour, spotted with deeper purple spots, in some greater, in others lesser: the flowers stand at the tops of the stalkes spike-fashion, which are of a white colour, with reddish stripes and veines running through them, set or placed in darke purple coloured huskes : the seede is greater and rounder then the former, and somewhat long withall : the roote perisheth in like manner as the other former doe. The whole plant smelleth strong, like vnto the other Basils.

The Place.

The two last sorts of Basils are greater strangers in our Country then the first which is frequent, and only sowen and planted in curious gardens. The last came first out of the West Indies.

The Time.

They all flower in August, or Iuly at the soonest, and that but by degrees, and not all at once.

The Names.

The first is vsually called *Ocimum vulgare,* or *vulgatius,* and *Ocimum Citratum.* In English, Common or Garden Basill. The other is called *Ocimum minimum,* or *Gariophyllatum,* Cloue Basill, or Bush Basill. The last eyther of his place, or forme of his leaues, being spotted and curled, or all, is called *Ocimum Indicum maculatum, latifolium & crispum.* In English according to the Latine, Indian Basill, broade leafed Basill, spotted or curled Basill, which you pleafe.

The Vertues.

The ordinary Basill is in a manner wholly spent to make sweet, or washing waters, among other sweet herbes, yet sometimes it is put into nosegayes. The Physicall properties are, to procure a cheerefull and merry heart, whervnto the feede is chiefly vsed in pouder, &c. and is most vsed to that, and to no other purpose.

CAP.

1 *Santolina.* Lauender Cotton. 2 *Lauendula.* Lauender Spike. 3 *Stœbas.* Caſſidony. 4 *chamædrys* Germander. 5 *Ocimum minus* Fine Baſſill. 6 *Marum.* Herbe Maſticke. 7 *Maio ana.* Sweete Marierome.

CHAP. CXXXI.

Maiorana. Sweete Marierome.

Ee haue many forts of Marierome ; fome that are fweete, and but Summer plants ; others that are greater and not fo fweet ; and fome alfo that are wilde. Of all thefe I will onely felect fome of the choifeft that are fit for this place, and leaue the other for the next garden, and the garden of fimples, or a generall worke : yet hereunto I will adioyne another fweete plant called Mafticke, as participating neerer with them then with Time, whereunto many doe referre it.

1. *Maiorana maior æftiua.* Common fweet Marierome.

The fweet Marierome that is moft frequently fowen in our Country, is a low herbe little aboue a foote high when it is at the higheft, full of branches, and fmall whitifh foft roundifh leaues, fmelling very fweet : at the toppes of the branches ftand diuers fmall fcaly heads, like vnto knots, (and therefore of fome called knotted Marierome) of a whitifh greene colour, out of which come here and there fmall white flowers, and afterwards fmall reddifh feede : the roote is compofed of many fmall threds or ftrings, which perifh with the whole plant euery yeare.

2. *Maiorana tenuifolia.* Marierome gentle.

This Marierome hath likewife diuers fmall branches, growing low, and not higher then the former, but hauing finer and fmaller leaues, hoary and foft, but much fweeter : the heads are like vnto the former, and fo are the flowers and feede, and the whole plant abiding but a Summer in the like manner.

3. *Marum.* Herbe Mafticke.

The neerer refemblance that this herbe hath with Marierome then with Tyme (as I faid before) hath made me place it next vnto the fmall fweet Marierome. It rifeth vp with a greater, and a more woody ftalke then Marierome, two foote high or better in fome places, where it liketh the ground and ayre, branching out on all fides towards the vpper part, leauing the ftemme bare below, if it bee old, otherwife being young, thinly furnifhing the branches from the bottome with fmall greene leaues, bigger then the leaues of any Tyme, and comming neere vnto the bigneffe and forme of the laft recited finer Marierome, but of a greener colour : at the toppes of the branches ftand fmall white flowers on a head, which afterwards turne into a loofe tuft of a long white hoary matter, like vnto foft doune, with fome leaues vnderneath and about it, which abide not long on the ftalkes, but are blowne away with the winde : the feede is fo fmall if it haue any, that I haue not obferued it : the roote is thready : the whole plant is of a fweete refinous fent, ftronger then the Marierome, and abideth our winters, if it be carefully planted and regarded.

The Place.

The fweete Marieromes grow naturally in hot Countreyes : the firft in Spaine &c. the fecond is thought to come out of Syria, or Perfia firft into Italie, where they much efteeme it, and plant it curioufly and carefully in pots, and fet them in their windowes, beeing much delighted therewith for the fweet fent it hath. The firft is vfually fowen euery yeare in moft gardens with vs : but the fecond is very rare and daintie, and muft as daintely be preferued, being more tender then the former. The herbe Mafticke is thought to be firft brought out of Candie, Clufius faith he found it in Spaine : It is planted by flippes, (and not fowen) in many gardens, and is much replanted

for

for increase, but profpereth onely, or more frequently, in loamie or clay grounds then in any other foyle.

The Time.

The fweete Marieromes beare their knots or fcaly heads in the end of Iuly, or in Auguſt. Herbe Maſticke in Iune many times, or in the beginning of Iuly.

The Names.

The firſt of the two fweet Marieromes called *Maiorana* in Latine *à maiore cura*, is taken of moſt writers to be the *Amaracus* or *Sampſuchum* of Dioſcorides, Theophraſtus and Plinie, although Galen doth feem a little to diſſent therefrom. The other fweet Marierome hath his name in his title as much as can be faid of it. The next is thought by the beſt of the moderne Writers to be the true *Marum* that Galen preferreth for the excellent fweetneſſe, before the former Marierome in making the *Oleum*, or *vnguentum Amaricinum*, and feemeth to incline to their opinion that thought *Amaracus* was deriued from *Marum*. It is the fame alſo that Galen and others of the ancient Writers make mention of, to go into the compoſition of the *Trochiſci Hedychroi*, as well as *Amaracus* among the ingredients of the *Theriaca Andromachi*. In Engliſh we call it Maſticke fimply, or Herbe Maſtick, both to diſtinguiſh it from that Tyme that is called Maſticke Tyme, and from the Maſticke Tree, or Gum, fo called. Some of later times, and Cluſius with them, haue thought this to be Dioſcorides his *Tragoriganum*, which doth fomewhat reſemble it : but there is another plant that Matthiolus fetteth forth for *Marum*, that in Lobels opinion and mine is the trueſt *Tragoriganum*, and this the trueſt *Marum*.

The Vertues.

The fweete Marieromes are not onely much vfed to pleafe the outward fenſes in nofegayes, and in the windowes of houfes, as alfo in fweete pouders, fweete bags, and fweete waſhing waters, but are alſo of much vſe in Phyficke, both to comfort the outward members, or parts of the body, and the inward alfo : to prouoke vrine being ſtopped, and to eafe the paines thereof, and to caufe the feminine courfes. Herbe Maſticke is of greater force to helpe the ſtopping of vrine, then the Marierome, and is put into Antidotes, as a remedie againſt the poyſon of venemous Beaſts.

Chap. CXXXI.

Thymum. Tyme.

THere are many kindes of Tyme, as they are vfually called with vs, fome are called of the garden, and others wilde, which yet for their fweetneſſe are brought into gardens, as Muske Tyme, and Lemon Tyme ; and fome for their beauty, as embroidered or gold yellow Tyme, and white Tyme. But the true Tyme of the ancient Writers, called *Capitatum*, as a fpeciall note of diſtinction from all other kindes of Tyme, is very rare to be feene with vs here in England, by reaſon of the tenderneſſe, that it will not abide our Winters. And all the other forts that with vs are called garden Tymes, are indeede but kindes of wilde Tyme, although in the defect or want of the true Tyme, they are vfed in the ſtead of it. With the Tymes I muſt doe as I did with the Marieromes in the Chapter before, that is, reſerue the moſt common in vſe, for the common vſe of the Kitchen, and ſhew you only thoſe here, that are not put to that vſe : and firſt with the true Tyme, becauſe it is knowne but to a few.

1, *Thymum*

1. *Thymum legitimum capitatum.* The true Tyme.

The true Tyme is a very tender plant, hauing hard and hoary brittle branches, spreading from a small wooddy stemme, about a foote and a halfe high, whereon are set at seuerall ioynts, and by spaces, many small, long, whitish, or hoary greene leaues, of a quicke sent and taste: at the tops of the branches stand small long whitish greene heads, somewhat like vnto the heads of *Stœchas*, made as it were of many leaues or scales, out of which start forth small purplish flowers (and in some white, as Bellonius saith) after which commeth small seede, that soone falleth out, and if it be not carefully gathered, is soone lost, which made (I thinke) Theophrastus to write, that this Tyme was to be sowne of the flowers, as not hauing any other seede: the root is small and wooddy. This holdeth not his leaues in Winter, no not about Seuill in Spaine, where it groweth aboundantly, as Clusius recordeth, finding it there naked or spoiled of leaues. And will not abide our Winters, but perisheth wholly, roote and all.

2. *Serpillum hortense siue maius.* Garden wilde Tyme.

The wilde Tyme that is cherished in gardens groweth vpright, but yet is lowe, with diuers slender branches, and small round greene leaues, somewhat like vnto small fine Marierome, and smelling somewhat like vnto it: the flowers growe in roundels at the toppes of the branches, of a purplish colour: And in another of this kinde they are of a pure white colour.

There is another also like hereunto, that smelleth somewhat like vnto Muske; and therefore called Muske Tyme, whose greene leaues are not so small as the former, but larger and longer.

3. *Serpillum Citratum.* Lemon Tyme.

The wilde Tyme that smelleth like vnto a Pomecitron or Lemon, hath many weake branches trayling on the ground, like vnto the first described wilde Tyme, with small darke greene leaues, thinly or sparsedly set on them, and smelling like vnto a Lemon, with whitish flowers at the toppes in roundels or spikes.

4. *Serpillum aureum siue versicolor.* Guilded or embroidered Tyme.

This kinde of wilde Tyme hath small hard branches lying or leaning to the ground, with small party coloured leaues vpon them, diuided into stripes or edges, of a gold yellow colour, the rest of the leafe abiding greene, which for the variable mixture or placing of the yellow, hath caused it to be called embroidered or guilded Tyme.

The Place.

The first groweth as is said before, about Seuill in Spaine, in very great aboundance as Clusius saith; and as Bellonius saith, very plentifully on the mountaines through all Greece. The others growe some in this Country, and some in others: but wee preserue them with all the care wee can in our gardens, for the sweete and pleasant sents and varieties they yeeld.

The Time.

The first flowreth not vntill August; the rest in Iune and Iuly.

The Names.

Their names are seuerally set downe in their titles, as is sufficient to distinguish them; and therefore I shall not neede to trouble you any further with them.

The

The Vertues.

The true Tyme is a speciall helpe to melancholicke and spleneticke diseases, as also to flatulent humours, either in the vpper or lower parts of the body. The oyle that is Chimically drawne out of ordinary Tyme, is vsed (as the whole herbe is, in the stead of the true) in pils for the head and stomach. It is also much vsed for the toothach, as many other such like hot oyles are.

Chap. CXXXII.

Hyssopus. Hyssope.

THere are many varieties of Hyssope, beside the common or ordinary, which I reserue for the Kitchen garden, and intend onely in this place to giue you the knowledge of some more rare: *viz.* of such as are nourced vp by those that are curious, and fit for this garden: for there are some other, that must be remembred in the Physicke garden, or garden of Simples, or else in a generall worke.

1. *Hyssopus folijs niueis.* White Hyssope.

This white Hyssope is of the same kinde and smell with the common Hyssope; but differeth, in that this many times hath diuers leaues, that are wholly of a white colour, with part of the stalke also: others are parted, the one halfe white, the other halfe greene, and some are wholly greene, or with some spots or stripes of white within the greene, which makes it delightfull to most Gentlewomen.

2. *Hyssopus folijs cinereis.* Russet Hyssope.

As the last hath party coloured leaues, white and greene, so this hath his leaues of an ash-colour, which of some is called russet; and hath no other difference either in forme or smell.

3. *Hyssopus aureus.* Yellow or golden Hyssope.

All the leaues of this Hyssope are wholly yellow, or but a little greene in them, and are of so pleasant a colour, especially in Summer, that they prouoke many Gentlewomen to weare them in their heads, and on their armes, with as much delight as many fine flowers can giue: but in Winter their beautifull colour is much decayed, being of a whitish greene, yet recouer themselues againe the next Summer.

4. *Hyssopus surculis densis.* Double Hyssope.

As this kinde of Hyssope groweth lower then the former or ordinary kinde, so it hath more branches, slenderer, and not so wooddy, leaning somewhat downe toward the ground, so wonderfully thicke set with leaues, that are like vnto the other, but of a darker greene colour, and somewhat thicker withall, that it is the onely fine sweete herbe, that I know fittest (if any be minded to plant herbes) to set or border a knot of herbes or flowers, because it will well abide, and not growe too wooddy or great, nor be thinne of leaues in one part, when it is thicke in another, so that it may be kept with cutting as smooth and plaine as a table. If it be suffered to growe vp of it selfe alone, it riseth with leaues as before is specified, and flowreth as the common doth, and of the same sent also, not differing in any thing, but in the thicknesse of the leaues on the stalkes and branches, and the aptnesse to be ordered as the keeper pleaseth.

5. *Chamædrys*

Chamædrys. Germander.

Left Germander fhould be vtterly forgotten, as not worthy of our Garden, feeing many (as I faid in my treatife or introduction to this Garden) doe border knots therewith: let me at the leaft giue it a place, although the laft, being more vfed as a ftrewing herbe for the houfe, then for any other vfe. It is (I thinke) fufficiently knowne to haue many branches, with fmall and fomewhat round endented leaues on them, and purplifh gaping flowers: the rootes fpreading far abroad, and rifing vp againe in many places.

The Place.

Thefe Hyffopes haue beene moft of them nourfed vp of long time in our Englifh Gardens, but from whence their firft originall fhould be, is not well knowne. The Germander alfo is onely in Gardens, and not wilde.

The Time.

They flower in Iune and Iuly.

The Names.

The feuerall names whereby they are knowne to vs, are fet forth in their titles; and therefore I neede not here fay more of them then onely this, that neyther they here fet downe, nor the common or ordinary fort, nor any of the reft not here expreffed, are any of them the true Hyffope of the ancient Greeke Writers, but *fuppofititia*, vfed in the ftead thereof. The Germander, from the forme of the leaues like vnto fmall oaken leaues, had the name *Chamædrys* giuen it, which fignifieth a dwarfe Oake.

The Vertues.

The common Hyffope is much vfed in all pectorall medicines, to cut fleagme, and to caufe it eafily to be auoided. It is vfed of many people in the Country, to be laid vnto cuts or frefh wounds, being bruifed, and applyed eyther alone, or with a little Sugar. It is much vfed as a fweet herbe, to be in the windowes of an houfe. I finde it much commended againft the Falling Sickneffe, efpecially being made into Pils after this manner: Of Hyffope, Horhound, and Caftor, of each halfe a dramme, of Peony rootes (the male kinde is onely fit to be vfed for this purpofe) two drams, of *Affa fætida* one fcruple: Let them be beaten, and made into pils with the iuyce of Hyffope; which being taken for feuen dayes together at night going to bed, is held to be effectuall to giue much eafe, if not thoroughly to cure thofe that are troubled with that difeafe. The vfe of Germander ordinarily is as Tyme, Hyffope, and other fuch herbes, to border a knot, whereunto it is often appropriate, and the rather, that it might be cut to ferue (as I faid) for a ftrewing herbe for the houfe among others. For the phyficall vfe it ferueth in difeafes of the fplene, and the ftopping of vrine, and to procure womens courfes.

Thus haue I led you through all my Garden of Pleafure, and fhewed you all the varieties of nature nourfed therein, pointing vnto them, and defcribing them one after another. And now laftly (according to the vfe of our old ancient Fathers) I bring you to reft on the Graffe, which yet fhall not be without fome delight, and that not the leaft of all the reft.

1 *Thymum legitimum.* The true Tyme. 2 *Serpillum maius hortense.* Garten wilde Tyme. 3 *Serpillum Citratum.* Lemon Tyme. 4 *Hyssopus versicolor sine aureus.* Golden Hyssope. 5 *Chamædrys.* Germander. 6 *Spartum Austriacum siue Gramen plumarium minus.* The lesser feather Grasse. 7 *Gramen striatum vel falcatum.* Painted Grasse or Ladies Laces.

CHAP. CXXXIII.

Gramina. Graſſes.

THere are among an infinite number (as I may ſo ſay) of Graſſes, a few onely which I thinke fit to be planted in this Garden, both for the rarity of them, and alſo for your delight, and the excellent beauty that is in them aboue ma-ny other plants. One of them hath long agoe bin reſpected, and cheriſhed in the coun-try gardens of many Gentlewomen, and others. The others are knowne but vnto a few.

1. *Gramen ſtriatum.* Painted Graſſe or Ladies laces.

This kinde of Graſſe hath many ſtiffe, hard, round ſtalkes, full of ioynts, whereon are ſet at euery ioynt one long leafe, ſomewhat broad at the bottome, where it com-paſſeth the ſtalke, and ſmaller to the end, where it is ſharpe pointed, hard or rough in handling, and ſtriped all the length of the leafe with white ſtreakes or lines, that they ſeeme party coloured laces of white and greene: the tops of the ſtalkes are furniſhed with long ſpikie tufts, like vnto the tufts of Couch Graſſe: the rootes are ſmall, white, and thredday, like the rootes of other Graſſes.

2. *Gramen Plumarium minus.* The leſſer Feather-Graſſe.

This leſſer Feather-Graſſe hath many ſmall, round, and very long leaues or blades, growing in tufts, much finer and ſmaller then any other Graſſe that I know, being al-moſt like vnto haires, and of a freſh greene colour in Summer, but changing into gray, like old hay in Winter, being indeede all dead, and neuer reuiuing; yet hardly to be plucked away vntill the Spring, and then other greene leaues or ruſhes riſe vp by them, and in their ſtead, and are aboue a foote in length: from the middle of theſe tufts come forth rounder and bigger ruſhes, which are the ſtalkes, and which haue a chaffie round eare about the middle thereof, which when it is full growne, is ſomewhat higher then the toppes of the leaues or ruſhes, opening it ſelfe (being before cloſe) at the top, and ſhewing forth three or foure long ayles or beards, one aboue another, which bend themſelues a little downewards (if they ſtand ouer long before they are gathered, and will fall off, and be blowne away with the winde) being ſo finely feathered on both ſides, all the length of the beard, and of a pale or grayiſh colour, that no feather in the taile of the Bird of Paradiſe can be finer, or to be compared with them, hauing ſticking at the end of euery one of them, within the eare, a ſmall, long, whitiſh, round, hard, and very ſharpe pointed graine, like vnto an oaten graine, that part of the ſtalke of the feather that is next vnder it, and aboue the ſeede for ſome two or three inches, being ſtiffe and hard, and twining or curling it ſelfe, if it be ſuffered to ſtand too long, or to fall away, otherwiſe being ſtraight as the feather it ſelfe: the roote is compoſed of many long, hard, ſmall thredday ſtrings, which runne deepe and far, and will not wil-lingly be remoued, in that it gaineth ſtrength euery yeare by ſtanding.

3. *Gramen Plumarium maius.* The greater Feather-Graſſe.

The greater Feather-Graſſe is like vnto the leſſer, but that both the leaues and the feathers are greater, and nothing ſo fine, groſſer alſo, and of leſſe beauty and reſpect, though whiter then it; and therefore is not ſo much regarded: for I haue knowne, that many Gentlewomen haue vſed the former leſſer kinde, being tyed in tufts, to ſet them in ſtead of feathers about their beds, where they haue lyen after childe-bearing, and at other times alſo, when as they haue been much admired of the Ladies and Gentles that haue come to viſit them.

The Place.

The firſt of theſe Graſſes, as Lobel ſaith, groweth naturally in the woods and hils of Sauoy. It hath long agoe beene receiued into our Engliſh gardens.

dens. The second, as Clusius saith, in Austria, from whence also (as I take it) the greater came, and are both in the gardens of those, that are curious obseruers of these delights.

The Time.

The first is in its pride for the leaues all the Spring and Summer, yeelding his bush in Iune. The other giue their feather-like sprigs in Iuly and August, and quickly (as I said) are shed, if they be not carefully gathered.

The Names.

The first is called by Lobel *Gramen sulcatum*, or *striatum album*, of others *Gramen pictum*. The French call it *Aiguellettes d'armes*, of the fashion that their Ensignes, Pennons, or Streamers vsed in wars were of, that is, like vnto a party coloured curtaine. In English vsually Ladies laces, and Painted Grasse. The first of the other two is called *Gramen plumarium* or *plumosum*, and *minus* is added for the distinction of it. Clusius calleth it *Spartum Austriacum*, of the likenesse and place where he found it. The last is called *Gramen plumarium*, or *plumosum maius*, The greater Feather-Grasse.

The Vertues.

These kindes of Grasses are not in any time or place that I doe heare of applyed to any Physicall vse ; and therefore of them I will say no more : but here I will end the prime part of this worke.

THE ORDERING
OF THE KITCHEN
GARDEN.

Chap. I.

The situation of a Kitchen Garden, or Garden of Herbes, and what sort of manure is fittest to helpe the decaying of the soyle thereof.

Auing giuen you the best rules and instructions that I can for your flower Garden, and all the flowers that are fit to furnish it, I now proceede to your herbe garden, which is not of the least respect belonging to any mans house, nor vtterly to bee neglected for the many vtilities are to be had from it, both for the Masters profit and pleasure, and the meynies content and nourishment : all which if I should here set down, I had a large field to wander in, and matter sufficient to entreat of, but this worke permitteth not that libertie : and I thinke there are but few but eyther know it already, or conceiue it sufficiently in their minds. Passing therefore no further in such discourses, I come to the matter in hand, which is to shew you where the fittest place is for an herbe garden. As before I shewed you that the beautie of any worthy house is much the more commended for the pleasant situation of the garden of flowers, or of pleasure, to be in the sight and full prospect of all the chiefe and choisest roomes of the house; so contrariwise, your herbe garden should bee on the one or other side of the house, and those best and choyse roomes : for the many different sents that arise from the herbes, as Cabbages, Onions, &c. are scarce well pleasing to perfume the lodgings of any house; and the many ouertures and breaches as it were of many of the beds thereof, which must necessarily bee, are also as little pleasant to the sight. But for priuate mens houses, who must like their habitations as they fall vnto them, and cannot haue time or meanes to alter them, they must make a vertue of necessity, and conuert their places to their best aduantage, by making their profit their chiefest pleasure, and making one place serue for all vses. The choyce of ground for this Garden, is (as I said before) where it is fat, fertill and good, there needeth the lesse labour and cost : and contrariwise, where it is cold, wet, dry or barren, there must bee the more helpes still added to keepe it in heart. For this Garden by reason of the much and continuall stirring therein, the herbes and rootes drawing out the substance of the fertilitie thereof more aboundantly then in the former, must be continually holpen with soyle, or else few things of goodnesse or worth will come forward therein. The stable soyle of horses is best and more proper for any colde grounds, for being the hottest, it will cause any the seedes for this Garden to prosper well, and be more forward then in any other ground that is not so holpen. The stable soyle of Cattell is of a colder and moister nature, and is therefore more proper for

the

the hot fandy or grauelly grounds, and although it bee longer before it bee brought to mould then that of horſes, yet it will outlaſt it more then twice ſo long. Let euery one therefore take according to the nature of the ground ſuch helpes as are moſt fit and conuenient, as I haue here and before ſhewed. But I doe here ingenuouſly confeſſe my opinion of theſe forcings and helpings of ground, that howſoeuer it doth much good to ſome particular things, which becauſe they delight in heate, and cannot be brought to perfection without it in this our Countrey, which is colder then their naturall from whence they are brought, muſt therfore haue artificiall helpes to forward them ; yet for many other things the compoſt doth much alter and abate the naturall vigour, and quickeneſſe of taſte, that is perceiued in them that grow in a naturall fat or ſandy ſoile that is not ſo holpen.

CHAP. II.

The forme of a Garden of herbes for neceſſary vſes, with the ordering thereof.

AS our former Garden of pleaſure is wholly formable in euery part with ſquares, trayles, and knots, and to bee ſtill maintained in their due forme and beautie : ſo on the contrary ſide this Garden cannot long conſerue any forme, for that euery part thereof is ſubiect to mutation and alteration. For although it is conuenient that many herbes doe grow by themſelues on beds, caſt out into ſome proportion fit for them, as Tyme, Hiſſope, Sage, &c. yet many others may bee ſowen together on a plot of ground of that largeneſſe that may ſerue euery mans particular vſe as he ſhall haue occaſion to employ it, as Reddiſh, Lettice and Onions, which after they are growne vp together may be drawne vp and taken away, as there is occaſion to ſpend them : but Carrots or Parſneps being ſowen with others muſt bee ſuffered to grow laſt, becauſe they require a longer time before they be fit to be taken vp. Other herbes require ſome large compaſſe of ground whereon they may grow of themſelues without any other herbes growing among them, as Artichokes, Cowcumbers, Melons, Pompions. And ſome will doe ſo with their Cabbages alſo, but the beſt and moſt frugall way now vſed, is to plant them round about the border of your plot or ground whereon you plant Cowcumbers, Pompions, or other things, in that by this meanes ſo much ground will be well ſaued, and the other things be no whit hindered thereby, which elſe a great deale of ground muſt be employed for them apart. So that by this that I haue here ſaid, you may perceiue the forme of this Garden is for the moſt part, to bee ſtill out of forme and order, in that the continuall taking vp of the herbes and rootes that are ſowen and planted, cauſeth the beds or parts of this Garden to lye broken, diſmembered, and out of the order that at the firſt it was put into. Remember herewithall that (as I ſaid before) this Garden requireth the continuall helpe of ſoyle to be brought into it, in that the plenty of theſe manner of herbes and rootes doe ſo much waſte the fertilitie and fatneſſe of the ground, that without continuall refreſhing it would quickly become ſo poore and barren, that it would not yeelde the worth of the ſeede. The ordinary time to ſoyle a Garden, is to bring in manure or dung before Chriſtmas, and eyther bury it ſome ſmall depth, not too deepe, or elſe to lay it vpon the ground that the winter froſtes may pierce it, and then turne it ſhallow into the ground to ſow your ſeeds in the Spring.

*How to order diuers Garden herbes, both for their sowing,
spending, and gathering of the seede.*

OVr chiefest and greatest Gardiners now adaies, doe fo prouide for themfelues
:uery yeare, that from their owne grounds they gather the feede of many
herbes that they fowe againe: for hauing gained the beft kind of diuers herbes,
they will be ftill furnifhed with the fame, and be not to feeke euery yeare for new that
oftentimes will not yeelde them halfe the profit that their choyce feede will: I fay of
many herbes, but not of all ; for the beft of them all hath not ground fufficient for all
forts, nor will our climate bring fome to that perfection that other forraine doth, and
therefore the feede of fome things are continually brought from beyond Sea vnto vs.
And againe although our chiefe Gardiners doe ftill prouide their owne feede of diuers
things from their owne ground, becaufe as I faid it is of the beft kinde, yet you muft vnderftand alfo, that good ftore of the fame fortes of feeds are brought from beyond the
Seas, for that which is gathered in this Land is not fufficient to ferue euery mans vfe in
the whole Kingdome by many parts ; yet ftill it is true, that our Englifh feede of many
things is better then any that commeth from beyond the Seas : as for example, Reddifh, Lettice, Carrots, Parfneps, Turneps, Cabbages, and Leekes, of all which I intend to write in this place ; for thefe are by them fo husbanded, that they doe not fow
their owne grounds with any other feede of thefe forts but their owne : which that
you may know the manner how to doe, I will here fet it downe, that euery one may
haue the beft directions if they will follow them. Of Reddifh there are two forts, one
more early then the other : they vfe therefore to fow their early Reddifh firft, that they
may haue the earlieft profit of them, which is more worth in one fortnight, then in
a moneth after. And to effect this they haue fome artificiall helps alfo, which are thefe:
They vfe to digge vp a large plot of ground where they intend to fow their feede a little before or after Chriftmas, cafting it into high balkes or ridges fiue or fixe foote a-
funder, which they fuffer to lye and take all the extreame frofts in Ianuary to mellow
the earth, and when the froftes are paft, they then beginne to bring into it good ftore
of frefh ftable dung, which they laye neyther too deepe nor too thicke, and couer it
with the mould a hand breadth thickneffe aboue the dung, which doth giue fuch a
warmth and comfort to whatfoeuer is fowen thereon, that it forceth it forward much
fooner then any other way can doe : And to preuent both the froftes, and the cold
bitter windes which often fpoyle their feede new fprung vp, they vfe to fet great high
and large mattes made of reedes, tyed together, and faftened vnto ftrong ftakes, thruft
into the ground to keepe them vp from falling, or being blowne down with the winde ;
which mattes they place on the North and Eaft fide to breake the force of thefe winds,
and are fo fure and fafe a defence, that a bricke wall cannot better defend any thing
vnder it, then this fence will. In this manner they doe euery yeare to bring forward
their feede to gaine the more by them, and they that will haue Reddifh early, muft
take the fame courfe. The other fort of Reddifh for the moft part is fowen in Februa-
rie, a fortnight after the other at the leaft, and likewife euery moneth after vnto September, that they may haue young continually. For the blacke Reddifh, although many in many places doe fowe it in the fame time, and in the fame manner that the ordinary is fowen, yet the nature thereof is to runne vp to feede more fpeedily then the
other, if it haue fo rich ground to grow vpon, and therefore the beft time to fow it is
in Auguft, that fo it may abide all winter, wherein is the chiefeft time for the fpending
thereof, and to keepe it vntill the beginning of the next yeare from running vp to feed:
the gathering whereof, as alfo of the other fort, is all after one manner, that is, to be
pulled vp when the pods change whitifh, and then hanged vpon bufhes, pales, or fuch
other thing, vntill they bee thorough dry, and then beaten or thrafhed out vpon a
fmooth plancher, or vpon clothes, as euery ones ftore is, and their conueniencie. Let-
tic : is fowen oftentimes with the early Reddifh, in the fame manner before faid, tl
they may haue Lettice likewife as early as the time of the year will permit them, w

they pull vp where they grow too thicke, spending them first, and so taking vp from time to time, vntill they stand two foote in sunder one from another, and beginne to spindle and shoote vp for seede. In this is vsed some arte to make the plants strong to giue the better seede without danger of rotting or spoyling with the wet, which often happeneth to those about whom this caution is not obserued : Before your Lettice is shot vp, marke out the choysest and strongest plantes which are fittest to grow for seede, and from those when they are a foote high, strippe away with your hand the leaues that grow lowest vpon the stalke next the ground, which might rot, spoyle or hinder them from bearing so good seede; which when it is neere to be ripe, the stalkes must be cut off about the middle, and layde vpon mats or clothes in the Sunne, that it may there fully ripen and be gathered ; for it would be blowne away with the winde if it should be suffered to abide on the stalkes long. Parsneps must be sowen on a deep trenched mellow ground, otherwise they may run to seede the first yeare, which then are nothing worth : or else the rootes will be small staruelings and short, and runne into many spires or branches, whereby they will not bee of halfe the worth. Some vse to sow them in August and September, that so they may bee well growne to serue to spend in Lent following, but their best time is in February, that the Summers growth may make them the fairer and greater. When they runne vp to seede, you shall take the principall or middle heades, for those carry the Master seede, which is the best, and will produce the fairest rootes againe. You shall hardly haue all the seede ripe at one instant, for vsually the chiefest heads will be fallen before the other are ripe : you must therefore still looke them ouer, and cut them as they ripen. Carrots are vsually sowen in March and Aprill, and if it chance that some of them doe runne vp for seede the same year, they are to be weeded out, for neyther the seed nor roots of them are good : You must likewise pull them vp when they are too thicke, if you will haue them grow fair, or for seed, that they may grow at the least three or foure foot in sunder : the stalkes of Carrots are limber, and fall downe to the ground ; they must therefore be sustained by poles layde acrosse on stalkes thrust into the ground, and tyed to the poles and stalkes to keepe them vp from rotting or spoyling vpon the ground : the seed hereof is not all ripe at once, but must be tended and gathered as it ripeneth, and layd to dry in some dry chamber or floore, and then beaten out with a stick, and winnowed from the refuse. Turneps are sowne by themselues vpon a good ground in the end of Iuly, and beginning of August, to haue their rootes best to spend in winter ; for it often happeneth that those seedes of Turneps that are sowen in the Spring, runne vp to seede the same yeare, and then it is not accounted good. Many doe vse to sow Turneps on those grounds from whence the same yeare they haue taken off Reddish and Lettice, to make the greater profit of the ground, by hauing two crops of increase in one yeare. The stalkes of Turneps will bend downe to the ground, as Carrots doe, but yet must not be bound or ordered in that manner, but suffered to grow without staking or binding, so as they grow of some good distance in sunder : when the seede beginneth to grow ripe, be very carefull to preserue it from the birds, which will be most busie to deuour them. You shall vnderstand likewise that many doe account the best way to haue the fairest and most principall seede from all these fore-recited herbes, that after they are sowen, and risen to a reasonable growth, they be transplanted into fresh ground. Cabbages also are not only sowen for the vse of their heads to spend for meat, but to gather their seede likewise, which howsoeuer some haue endeauoured to doe, yet few haue gained good seede, because our sharpe hard frostes in winter haue spoyled and rotted their stockes they preserued for the purpose ; but others haue found out a better and a more sure way, which is, to take vp your stocks that are fittest to be preserued, and bring them into the house, and there wrap them eyther in clothes, or other things to defend them from the cold, and hang them vp in a dry place, vntill the beginning of March following, then planting them in the ground, and a little defend them at the first with straw cast ouer them from the cold nights, thereby you may be sure to haue perfect good seede, if your kinde be of the best : Sowe your seed in the moneths of February or March, and transplant them in May where they may stand to grow for your vse, but be carefull to kill the wormes or Caterpillers that else will deuoure all your leaues, and be carefull also that none of the leaues bee broken in the planting, or otherwise rubbed, for that oftentimes hindereth the well closing of them. Leekes are

for

for the moſt part wholly nourſed vp from the ſeede that is here gathered; and becauſe there is not ſo much ſtore of them either ſowne or ſpent, as there is of Onions by the twentieth part, we are ſtill the more carefull to be prouided from our owne labours; yet there be diuers Gardiners in this Kingdome, that doe gather ſome ſmall quantity of Onion ſeede alſo for their owne or their priuate friends ſpending. The ſowing of them both is much about one time and manner, yet moſt vſually Leeks are ſowne later then Onions, and both before the end of March at the furtheſt; yet ſome ſowe Onions from the end of Iuly to the beginning of September, for their Winter prouiſion. Thoſe that are ſowne in the Spring, are to be taken vp and tranſplanted on a freſh bed prepared for the purpoſe, or elſe they will hardly abide a Winter; but hauing taken roote before Winter, they will beare good ſeede in the Summer following : You muſt ſtake both your Leekes and your Onion beds, and with poles laid a croſſe, binde your lopple headed ſtalkes vnto them, on high as well as belowe, or elſe the winde and their owne weight will beare them downe to the ground, and ſpoile your ſeede. You muſt thinne them, that is, pull vp continually after they are firſt ſprung vp thoſe that growe too thicke, as you doe with all the other herbes before ſpoken of, that they may haue the more roome to thriue. Of all theſe herbes and rootes before ſpoken of, you muſt take the likelieſt and faireſt to keepe for your ſeede; for if you ſhould not take the beſt, what hope of good ſeede can you expect ? The time for the ſpending of theſe herbes and rootes, not particularly mentioned, is vntill they begin to runne vp for ſeede, or vntill they are to be tranſplanted for ſeede, or elſe vntill Winter, while they are good, as euery one ſhall ſee cauſe.

Chap. IIII.

How to order Artichokes, Melons, Cowcumbers, and Pompions.

THere are certaine other herbes to be ſpoken of, which are wholly nourſed vp for their fruit ſake, of whom I ſhall not need to ſay much, being they are ſo frequent in euery place. Artichokes being planted of faire and large ſlips, taken from the roote in September and October (yet not too late) will moſt of them beare fruit the next yeare, ſo that they be planted in well dunged ground, and the earth raiſed vp like vnto an Anthill round about each roote, to defend them the better from the extreame froſts in Winter. Others plant ſlips in March and Aprill, or ſooner, but although ſome of them will beare fruit the ſame yeare, yet all will not. And indeede many doe rather chooſe to plant in the ſpring then in the fall, for that oftentimes an extreame hard Winter following the new ſetting of ſlips, when they haue not taken ſufficient heart and roote in the ground, doth vtterly pierce and periſh them, when as they that are ſet in the Spring haue the whole Summers growth, to make them ſtrong before they feele any ſharpe froſts, which by that time they are the better able to beare. Muske Melons haue beene begun to bee nourſed vp but of late dayes in this Land, wherein although many haue tryed and endeauoured to bring them to perfection, yet few haue attained vnto it : but thoſe rules and orders which the beſt and skilfulleſt haue vſed, I will here ſet downe, that who ſo will, may haue as good and ripe Melons as any other in this Land. The firſt thing you are to looke vnto, is to prouide you a peece of ground fit for the purpoſe, which is either a ſloping or ſheluing banke, lying open and oppoſite to the South Sunne, or ſome other fit place not ſheluing, and this ground alſo you muſt ſo prepare, that all the art you can vſe about it to make it rich is little enough ; and therefore you muſt raiſe it with meere ſtable ſoyle, thorough rotten & well turned vp, that it may be at the leaſt three foote deepe thereof, which you muſt caſt alſo into high beds or balkes, with deepe trenches or furrowes betweene, ſo as the ridges may be at the leaſt a foot and a halfe higher then the furrowes; for otherwiſe it is not poſſible to haue good Melons growe ripe. The choiſe of your ſeede alſo is another thing of eſpeciall regard, and the beſt is held to be Spaniſh, and not French, which hauing once gained, be ſure to haue ſtill of the ſame while they laſt
good,

good, that you may haue the feede of your owne ripe Melons from them that haue eaten them, or faue fome of the beft your felfe for the purpofe. I fay while they laft good; for many are of opinion, that no feede of Muske Melons gathered in England, will endure good to fowe againe here aboue the third yeare, but ftill they muft be renewed from whence you had your choifeft before. Then hauing prepared a hot bed of dung in Aprill, fet your feedes therein to raife them vp, and couer them, and order them with as great care or greater then Cowcumbers, &c. are vfed, that when they are ready, they may be tranfplanted vpon the beds or balkes of that ground you had before prepared for them, and fet them at the leaft two yards in funder, euery one as it were in a hole, with a circle of dung about them, which vpon the fetting being watered with water that hath ftood in the Sunne a day or two, and fo as often as neede is to water, couer them with ftrawe (fome vfe great hollow glaffes like vnto bell heads) or fome fuch other things, to defend them both from the cold euenings or dayes, and the heate of the Sunne, while they are young and new planted. There are fome that take vpon them great skill, that miflike of the raifing vp of Melons, as they doe alfo of Cowcumbers, on a hot bed of horfe dung, but will put two or three feedes in a place in the very ground where they fhall ftand and growe, and thinke without that former manner of forcing them forwards, that this their manner of planting will bring them on faft and fure enough, in that they will plucke away fome of the worft and weakeft, if too many rife vp together in a place; but let them know for certaine, that howfoeuer for Cowcumbers their purpofe and order may doe reafonable well, where the ground is rich and good, and where they ftriue not to haue them fo early, as they that vfe the other way, for Muske Melons, which are a more tender fruit, requiring greater care and trouble in the nourfing, and greater and ftronger heate for the ripening, they muft in our cold climate haue all the art vfed vnto them that may be, to bring them on the more early, and haue the more comfort of the Sunne to ripen them kindly, or elfe they will not bee worth the labour and ground. After you haue planted them as aforefaid, fome of good skill doe aduife, that you be carefull in any dry feafon, to giue them water twice or thrice euery weeke while they are young, but more afterward when they are more growne, and that in the morning efpecially, yea and when the fruit is growne fomewhat great, to water the fruit it felfe with a watering pot in the heate of the day, is of fo good effect, that it ripeneth them much fafter, and will giue them the better tafte and fmell, as they fay. To take likewife the fruit, and gather it at the full time of his ripeneffe is no fmall art; for if it be gathered before his due time to be prefently eaten, it will be hard and greene, and not eate kindly; and likewife if it be fuffered too long, the whole goodneffe will be loft: You fhall therefore know, that it is full time to gather them to fpend prefently, when they begin to looke a little yellowifh on the outfide, and doe fmell full and ftrong; but if you be to fend them farre off, or keepe them long vpon any occafion, you fhall then gather them fo much the earlier, that according to the time of the carriage and fpending, they may ripen in the lying, being kept dry, and couered with woollen clothes: When you cut one to eate, you fhall know it to be ripe and good, if the feede and pulpe about them in the middle be very waterifh, and will eafily be feparated from the meate, and likewife if the meate looke yellow, and be mellow, and not hard or greene, and tafte full and pleafant, and not waterifh: The vfuall manner to eate them is with pepper and falt, being pared and fliced, and to drowne them in wine, for feare of doing more harme. Cowcumbers and Pompions, after they are nourfed vp in the bed of hot dung, are to be feuerally tranfplanted, each of them on a large plot of ground, a good diftance in funder: but the Pompions more, becaufe their branches take vp a great deale more ground, & befides, will require a great deale more watering, becaufe the fruit is greater. And thus haue you the ordering of thofe fruits which are of much efteeme, efpecially the two former, with all the better fort of perfons; and the third kinde is not wholly refufed, of any, although it ferueth moft vually for the meaner and poorer fort of people, after the firft early ripe are fpent.

CHAP. V.

The ordering of diuers forts of herbes for the pot, for meate, and for the table.

Yme, Sauory, and Hyſſope, are vſually ſowne in the Spring on beds by them-ſelues, euerie one a part ; but they that make a gaine by ſelling to others the young rootes, to ſet the knots or borders of Gardens, doe for the moſt part ſowe them in Iuly and Auguſt, that ſo being ſprung vp before Winter, they will be the fitter to be taken vp in the Spring following, to ſerue any mans vſe that wonld haue them. Sage, Lauender, and Roſemary, are altogether ſet in the Spring, by ſlipping the old ſtalkes, and taking the youngeſt and likelieſt of them, thruſting them either twined or otherwiſe halfe a foote deepe into the ground, and well watered vpon the ſetting ; if any ſeaſonable weather doe follow, there is no doubt of their well thri-uing : the hot Sunne and piercing drying Windes are the greateſt hinderances to them; and therefore I doe aduiſe none to ſet too ſoone in the Spring, nor yet in Autumne, as many doe practiſe : for I could neuer ſee ſuch come to good, for the extremity of the Winter comming vpon them ſo ſoone after their ſetting, will not ſuffer their young ſhootes to abide, not hauing taken ſufficient ſtrength in the ground, to maintain them-ſelues againſt ſuch violence, which doth often pierce the ſtrongeſt plants. Marierome and Baſſill are ſowne in the Spring, yet not too early ; for they are tender plants, and doe not ſpring vntill the weather bee ſomewhat warme : but Baſſill would bee ſowne dry, and not haue any water of two or three daies after the ſowing, elſe the ſeede will turne to a gelly in the ground. Some vſe to ſowe the ſeed of Roſemary, but it ſeldome abideth the firſt Winter, becauſe the young plants being ſmall, and not of ſufficient ſtrength, cannot abide the ſharpneſſe of ſome Winters, notwithſtanding the couering of them, which killeth many old plants; but the vſuall way is to ſlippe and ſet, and ſo they thriue well. Many doe vſe to ſowe all or the moſt ſorts of Pot-herbes together on one plot of ground, that they neede not to goe farre to gather all the ſorts they would vſe. There are many ſorts of them well knowne vnto all, yet few or none doe vſe all ſorts, but as euery one liketh; ſome vſe thoſe that others refuſe, and ſome eſteem thoſe not to bee wholeſome and of a good relliſh, which others make no ſcruple of. The names of them are as followeth, and a ſhort relation of their ſowing or planting.

Roſemary, Tyme, and Sauorie are ſpoken of before, and Onions and Leekes.

Mints are to bee ſet with their rootes in ſome by-place, for that their rootes doe creepe ſo farre vnder ground, that they quickly fill vp the places neare adioyning, if they be not puld vp.

Clarie is to be ſowne, and ſeedeth and dyeth the next yeare, the herbe is ſtrong, and therefore a little thereof is ſufficient.

Nep is ſowne, and dyeth often after ſeeding, few doe vſe it, and that but a little at a time : both it and Clarie are more vſed in Tanſies then in Broths.

Coſtmarie is to be ſet of rootes, the leaues are vſed with ſome in their Broths, but with more in their Ale.

Pot Marierome is ſet of rootes, being ſeparated in ſunder.

Penniroyall is to be ſet of the ſmall heads that haue rootes, it creepeth and ſprea-deth quickly.

Alliſanders are to be ſowne of ſeede, the tops of the rootes with the greene leaues are vſed in Lent eſpecially.

Parſley is a common herbe, and is ſowne of ſeede, it ſeedeth the next yeare and dyeth : the rootes are more vſed in broths then the leaues, and the leaues almoſt with all ſorts of meates.

Fennell is ſowne of ſeede, and abideth many yeares yeelding ſeede : the rootes al-ſo are vſed in broths, and the leaues more ſeldome, yet ſerue to trimme vp many fiſh meates.

Borage is ſowne of ſeede, and dyeth the next yeare after, yet once being ſuffered to ſeede in a Garden, will ſtill come of it owne ſhedding.

Bugloſſe

Buglosse commeth of seede, but abideth many yeares after it hath giuen seede, if it stand not in the coldest place of the Garden.

Marigolds are sowne of seede, and may be after transplanted, they abide two or three yeares, if they be not set in too cold a place : the leaues and flowers are both vsed.

Langedebeefe is sowne of seede, which shedding it selfe will hardly be destroyed in a Garden.

Arrach is to be sowne of seede, this likewise will rise euery yeare of it owne seed, if it be suffered to shed it selfe.

Beetes are sowne of seede, and abideth some yeares after, still giuing seede.

Blites are vsed but in some places ; for there is a generall opinion held of them, that they are naught for the eyes : they are sowne euery yeare of seede.

Bloodwort once sowne abideth many yeares, if the extremity of the frosts kill it not, and seedeth plentifully.

Patience is of the same nature, and vsed in the same manner.

French Mallowes are to be sowne of seede, and will come of it owne sowing, if it be suffered to shed it selfe.

Ciues are planted onely by parting the rootes ; for it neuer giueth any seede at all.

Garlicke is ordered in the same manner, by parting and planting the rootes euerie yeare.

These be all the sorts are vsed with vs for that purpose, whereas I said before, none vseth all, but euery one will vse those they like best : and so much shall suffice for pot-herbes.

CHAP. VI.

The manner and ordering of many sorts of herbes and rootes for Sallets.

IF I should set downe all the sorts of herbes that are vsually gathered for Sallets, I should not onely speake of Garden herbes, but of many herbes, &c. that growe wilde in the fields, or elfe be but weedes in a Garden ; for the vsuall manner with many, is to take the young buds and leaues of euery thing almost that groweth, as well in the Garden as in the Fields, and put them all together, that the taste of the one may amend the rellish of the other: But I will only shew you those that are sown or planted in gardens for that purpose. Asparagus is a principall & delectable Sallet herbe, whose young shootes when they are a good handfull high aboue the ground, are cut an inch within the ground, which being boyled, are eaten with a little vinegar and butter, as a Sallet of great delight. Their ordering with the best Gardiners is on this wise: When you haue prouided seede of the best kinde, you must sowe it either before Christmas, as most doe, or before the end of February ; the later you sowe, the later and the more hardly will they spring : after they are growne vp, they are to be transplanted in Autumne on a bed well trenched in with dung; for elfe they will not bee worth your labour, and set about a foote distance in sunder, and looke that the more carefull you are in the replanting of them, the better they will thriue, and the sooner growe great : after fiue or six yeares standing they vsually doe decay ; and therefore they that striue to haue continually faire and great heads, doe from seede raise vp young for their store. You must likewise see that you cut not your heads or young shoote too nigh, or too much, that is, to take away too many heads from a roote, but to leaue a sufficient number vncut, otherwise it will kill the heart of your rootes the sooner, causing them to dye, or to giue very small heads or shootes ; for you may well consider with your selfe, that if the roote haue not head enough left it aboue the ground to shoote greene this yeare, it will not, nor cannot prosper vnder ground to giue encrease the next yeare. The ordering of Lettice I haue spoken of before, and shall not neede here to repeate what hath beene already said, but referre you thereunto for the sowing, planting, &c. onely I will here shew you the manner of ordering them for Sallets. There are some forts of Lettice that growe very great, and close their heads, which are called Cab-bage

bage Lettice, both ordinary and extraordinary, and there are other forts of great Lettice that are open, and clofe not, or cabbage not at all, which yet are of an excellent kinde, if they be vfed after that efpeciall manner is fit for them, which is, That when they are planted (for after they are fowne, they muft be tranfplanted) of a reafonable diftance in funder, and growne to be of fome bigneffe, euery one of them muft bee tyed together with baft or thread toward the toppes of the leaues, that by this meanes all the inner leaues may growe whitifh, which then are to be cut vp and vfed: for the keeping of the leaues clofe doth make them tafte delicately, and to bee very tender. And thefe forts of Lettice for the moft part are fpent after Summer is paft, when other Lettice are not to be had. Lambes Lettice or Corne Sallet is an herbe, which abiding all Winter, is the firft Sallet herbe of the yeare that is vfed before any ordinarie Lettice is ready; it is therefore vfually fowne in Auguft, when the feede thereof is ripe. Purflane is a Summer Sallet herbe, and is to be fowne in the Spring, yet fomewhat late, becaufe it is tender, and ioyeth in warmth; and therefore diuers haue fowne it vpon thofe beddes of dung, whereon they nourfed vp their Cowcumbers, &c. after they are taken away, which being well and often watered, hath yeelded Sallet vntill the end of the yeare. Spinach is fowne in the Spring, of all for the moft part that vfe it, but yet if it be fowne in Summer it will abide greene all the Winter, and then feedeth quickly: it is a Sallet that hath little or no tafte at all therein, like as Lettice and Purflane; and therefore Cookes know how to make many a good difh of meate with it, by putting Sugar and Spice thereto. Coleworts are of diuers kinds, and although fome of them are wholly fpent among the poorer fort of people, yet fome kindes of them may be dreffed and ordered as may delight a curious palate, which is, that being boyled tender, the middle ribs are taken cold, and laid in difhes, and vinegar and oyle poured thereon, and fo eaten. Coleflowers are to be had in this Country but very feldome, for that it is hard to meete with good feede: it muft bee fowne on beds of dung to force it forward, or elfe it would perifh with the froft before it had giuen his head of flowers, and tranfplanted into verie good and rich ground, left you lofe the benefit of your labours. Endiue is of two forts, the ordinary, and another that hath the edges of the leaues curld or crumpled; it is to be whited, to make it the more dainty Sallet, which is vfually done in this manner: After they are grown to fome reafonable greatneffe (but in any cafe before they fhoote forth a ftalke in the midft for feede) they are to be taken vp, and the rootes being cut away, lay them to dry or wither for three or foure houres, and then bury them in fand, fo as none of them lye one vpon another, or if you can, one to touch another, which by this meanes will change whitifh, and thereby become verie tender, and is a Sallet both for Autumne and Winter. Succorie is vfed by fome in the fame manner, but becaufe it is more bitter then Endiue, it is not fo generally vfed, or rather vfed but of a verie few: and whereas Endiue will feede the fame yeare it is fowne, and then dye, Succorie abideth manie yeares, the bitterneffe thereof caufing it to be more Phyficall to open obftructions; and therefore the flowers pickled vp, as diuers other flowers are vfed to be now adaies, make a delicate Sallet at all times when there is occafion to vfe them. Of red Beetes, the rootes are onely vfed both boyled and eaten cold with vinegar and oyle, and is alfo vfed to trimme vp or garnifh forth manie forts of difhes of meate: the feede of the beft kinde will not abide good with vs aboue three yeares, but will degenerate and growe worfe; and therefore thofe that delight therein muft be curious, to be prouided from beyond Sea, that they may haue fuch as will giue delight. Sorrell is an herbe fo common, and the vfe fo well knowne, both for fawce, and to feafon broths and meates for the found as well as ficke perfons, that I fhall not neede to fay anie more thereof. Cheruill is a Sallet herbe of much vfe, both with French and Dutch, who doe much more delight in herbes of ftronger tafte then the Englifh doe: it is fowne early, and vfed but a while, becaufe it quickly runneth vp to feede. Sweete Cheruill, or as fome call it, Sweete Cis, is fo like in tafte vnto Anife feede, that it much delighteth the tafte among other herbes in a Sallet: the feede is long, thicke, blacke, and cornered, and muft be fowne in the end of Autumne, that it may lye in the ground all the Winter, and then it will fhoote out in the Spring, or elfe if it be fowne in the Spring, it will not fpring vp that yeare vntill the next: the leaues (as I faid before) are vfed among other herbes: the rootes likewife are not onely cordiall, but alfo held to be preferuatiue againft the Plague, either greene, dryed, or preferued

with

with fugar. Rampion rootes are a kinde of Sallet with a great many, being boyled tender, and eaten cold with vinegar and pepper. Cresses is an herbe of easie and quick growth, and while it is young eaten eyther alone, or with parsley and other herbes : it is of a strong tafte to them that are not accuftomed thereunto, but it is much vfed of ftrangers. Rocket is of the same nature and qualitie, but fomewhat ftronger in tafte : they are both fowen in the Spring, and rife, feede and dye the fame yeare. Tarragon is an herbe of as ftrong a tafte as eyther Rocket or Cresses, it abideth and dyeth not euery yeare, nor yet giueth ripe feede (as far as euer could bee found with vs) any yeare, but maketh fufficient increafe within the ground, fpreading his roots all abroad a great way off. Muftard is a common fawce both with fish and flesh, and the feed thereof (and no part of the plant befides) is well knowne how to be vfed being grownded, as euery one I thinke knoweth. The rootes of horse Radish likewife beeing grownd like Muftard, is vfed both of ftrangers and our owne nation, as fawce for fish. Tanfie is of great vfe, almoft with all manner of perfons in the Spring of the yeare : it is more vfually planted of the rootes then otherwife ; for in that the rootes fpread far and neere they may be eafily taken away, without any hurt to the reft of the rootes. Burnet, although it be more vfed in wine in the Summer time then any way elfe, yet it is likewife made a fallet herbe with many, to amend the harsh or weak rellish of fome other herbs. Skirrets are better to be fowen of the feed then planted from the roots, and will come on more fpeedily, and be fairer rootes : they are as often eaten cold as a Sallet, being boyled and the pith taken out, as ftewed with butter and eaten warme. Let not Parfley and Fenell be forgotten among your other Sallet herbes, wherof I haue fpoken before, and therefore need fay no more of them. The flowers of Marigolds pickt cleane from the heads, and pickled vp againft winter, make an excellent Sallet when no flowers are to be had in a garden. Cloue Gilloflowers likewife preferued or pickled vp in the fame manner (which is *ftratum fuper ftratum*, a lay of flowers, and then ftrawed ouer with fine dry and poudered Sugar, and fo lay after lay ftrawed ouer, vntill the pot bee full you meane to keepe them in, and after filled vp or couered ouer with vinegar) make a Sallet now adayes in the higheft efteeme with Gentles and Ladies of the greateft note : the planting and ordering of them both is fpoken of feuerally in their proper places. Goates beard that groweth in Gardens only, as well as that which groweth wilde in Medowes, &c. bearing a yellow flower, are vfed as a Sallet, the rootes beeing boyled and pared are eaten cold with vinegar, oyle and pepper; or elfe ftewed with butter and eaten warme as Skirrets, Parfneps &c. And thus haue you here fet downe all thofe moft vfuall Sallets are vfed in this Kingdome : I fay the moft vfuall, or that are nourfed vp in Gardens; for I know there are fome other wilde herbes and rootes, as Dandelion &c. but they are vfed onely of ftrangers, and of thofe whofe curiofitie fearcheth out the whole worke of nature to fatisfie their defires.

Chap. VII.

Of diuers Phyficall herbes fit to be planted in Gardens, to ferue for the efpeciall vfes of a familie.

Hauing thus fhewed you all the herbes that are moft vfually planted in Kitchen Gardens for ordinary vfes, let mee alfo adde a few other that are alfo nourfed vp by many in their Gardens, to preferue health, and helpe to cure fuch fmall difeafes as are often within the compaffe of the Gentlewomens skils, who, to helpe their owne family, and their poore neighbours that are farre remote from Phyfitians and Chirurgions, take much paines both to doe good vnto them, and to plant thofe herbes that are conducing to their defires. And although I doe recite fome that are mentioned in other places, yet I thought it meete to remember them altogether in one place. Angelica, the garden kinde, is fo good an herbe, that there is no part thereof but is of much vfe, and all cordiall and preferuatiue from infectious or contagious difeafes, whether you will diftill the water of the herbe, or preferue or candie the rootes or the greene ftalkes, or vfe the feede in pouder or in diftillations, or decoctions with other things : it is fowen of feede, and will abide vntill it giue feede, and then dyeth. Rue or Herbe grace is a ftrong herbe, yet vfed inwardly againft the plague as an Antidote with Figs and Wall-nuts, and helpeth much againft windy bodies: outwardly it is vfed to bee layde to the wreftes of the hands, to driue away agues: it is more vfually planted of flips then raifed from feede, and abideth long if fharpe froftes kill it not. Dragons being diftilled are held to be good to expell any euill thing from the heart: they are altogether planted of the rootes. Setwall, Valerian, or Capons tayle, the herbe often, but the roote much better, is vfed to prouoke fweating, thereby to expell euill vapours that might annoy the heart: it is only planted of the rootes when they are taken vp, and the young replanted. Afarabacca, the leaues are often vfed to procure vomiting being ftamped, and the ftrained iuice to a little quantitie, put into a draught of ale and drunke, thereby to eafe the ftomacke of many euill and groffe humours that there lye and offend it; diuers alfo take the leaues and rootes a little boyled in wine, with a little fpice added thereunto, to expell both tertian and quartan agues: the rootes of our Englifh growing is more auaileable for thefe purpofes then any outlandifh: it is planted by the roote; for I could neuer fee it fpring of feede. Mafterwort commeth fomewhat neere in propertie vnto Angelica, and befides very effectuall to difperfe winde in the bodie, whether of the collicke or otherwife; as alfo very profitable to comfort in all cold caufes: it yeeldeth feede, but yet is more vfually planted from the rootes being parted. Balme is a cordiall herbe both in fmell and tafte, and is wholly vfed for thofe purpofes, that is, to comfort the heart being diftilled into water either fimple or compound, or the herbe dryed and vfed: it is fet of the rootes being parted, becaufe it giueth no feede that euer I could obferue. Camomill is a common herbe well knowne, and is planted of the rootes in alleyes, in walkes, and on bankes to fit on, for that the more it is troden on, and preffed downe in dry weather, the clofer it groweth, and the better it will thriue: the vfe thereof is very much, both to warme and comfort, and to eafe paines being applyed outwardly after many fafhions: the decoction alfo of the flowers prouoketh fweat, and they are much vfed againft agues. Featherfew is an herbe of greater vfe for women then for men, to diffolue flatulent or windy humours, which caufeth the paines of the mother: fome vfe to take the iuice thereof in drinke for agues: it is as well fowen of the feede as planted of the rootes. Coftmary is vfed among thofe herbes that are put ino ale to caufe it haue a good rellifh, and to be fomewhat phyficall in the moneth of May, and doth helpe to prouoke vrine: it is fet of the rootes being parted. Maudlin is held to be a principall good herbe to open and cleanfe the liuer, and for that purpofe is vfed many wayes, as in ale, in tanfies, and in broths &c. the feed alfo is vfed, and fo is the herbe alfo fometimes, to kill the wormes in children: it is fowen of the feede, and planted alfo of the feparated rootes. Caffidonie is a fmall kinde of Lauender, but differing both in forme and qualitie: it is much vfed for the head to eafe paines thereof, as alfo put among other things to purge melancholicke difeafes: it is fowen of feede, and abideth not a winter vnleffe it bee well defended, and yet hardly giueth ripe feede againe with vs. Smallage is a great opening herbe, and much more then eyther Parfley or Fenell, and the rootes of them all are often vfed together in medicines: it is fowen of feede, and will not bee wanting in a Garden if once you fuffer it to fow it felfe. Cardus Benedictus, or the Bleffed Thiftle, is much vfed in the time of any infection or plague, as alfo to expell any euill fymptome from the heart at all other times. It is vfed likewife to be boyled in poffet drink, & giuen to them that haue an ague, to help to cure it by fweating or otherwife. It is vfually fowen of feed, and dyeth when it hath giuen feed. Winter Cherries are likewife nurfed vp in diuers gardens, for that their propertie is to giue helpe to them that are troubled eyther with the ftopping or heate of their vrine: the herbe and berries are often diftilled, but the berries alone are more often vfed: after it is once planted in a garden it will runne vnder ground, & abide well enough. Celondine is held to bee good for the iaundife, it is much vfed for to cleere dim eyes, eyther the iuice or the water dropped into them: it is fowen of feede, and being once brought into a garden, will hardly be weeded out; the feede that fheddeth will fo fow it felfe, and therefore fome cornerin a garden is the fitteft place for it. Tabacco is of two forts, and both vfed to be planted in Gardens, yet the Englifh kinde (as it is called) is more to be found in our Countrey Gardens then the Indian fort: the leaues of both forts indifferently, that is, of eyther of which is next at hand, being ftamped and boy-

led

led eyther by it felfe, or with other herbes in oyle or hogs fuet, doe make an excellent falue for greene wounds, and alfo to clenfe old vlcers or fores; the iuice of the greene leaues drunke in ale, or a dryed leafe fteeped in wine or ale for a night, and the wine or ale drunke in the morning, prouoketh to caft, but the dryed leafe much ftronger then the greene: they are fo ven of feede, but the Indian kinde is more tender, and will not abide a winter with vs abroade. Spurge that vfually groweth in Gardens, is a violent purger, and therefore it is needfull to be very carefull how it is vfed: the feede is more ordinarily vfed then any other part of the plant, which purgeth by vomiting in fome, and both vpwards and downwards in many; the iuice of the herbe, but efpecially the milke thereof, is vfed to kill wartes: it is fowen of feede, and when it doth once fhed it felfe, it will ftill continue fpringing of the fallen feede. Bearefoote is fowen of feed, and will hardly abide tranfplanting vnleffe it bee while it is young; yet abideth diuers yeares, if it ftand not in too cold a place. This I fpeake of the greater kinde; for the lower fmall wilde kind (which is the moft ordinary in this land) will neuer decay: the leaues are fometimes vfed greene, but moft vfually dryed and poudered, and giuen in drinke to them that haue the wormes: it purgeth melancholy, but efpecially the roots. In many Countries of this Land, and elfewhere, they vfe to thruft the ftalk of the great kinde through the eare or dewlap of Kine and Cattell, to cure them of many difeafes. Salomons Seale, or (as fome call it) Ladder to heauen, although it doth grow wilde in many places of this Land, yet is planted in Gardens: it is accounted an excellent wound herbe to confolidate, and binde, infomuch that many vfe it with good fucceffe to cure ruptures, and to ftay both the white and the red fluxe in women: it is planted altogether of the rootes, for I could neuer finde it fpring from the feede, it is fo ftrong. Comfry likewife is found growing wilde in many places by ditch fides, and in moift places, and therefore requireth fome moift places of the garden: it is wholly vfed for knitting, binding, and confolidating fluxes and wounds, to be applyed either inwardly or outwardly: The rootes are ftronger for thofe purpofes then any other parts of the plant. Licoris is much vfed now adaies to bee planted in great quantitie, euen to fill many acres of ground, whereof rifeth a great deale of profit to thofe that know how to order it, and haue fit grounds for it to thriue in; for euery ground will not be aduantagious: It will require a very rich, deepe and mellow ground, eyther naturall or artificiall; but for a priuate houfe where a fmall quantitie will ferue, there needeth not fo much curiofitie: it is vfually planted of the top heads, when the lower rootes (which are the Licoris that is vfed) and the runners are cut from them. Some vfe to make an ordinary drinke or beuerage of Licoris, boyled in water as our vfuall ale or beere is with malt, which fermented with barme in the fame manner, and tunned vp, ferueth in ftead thereof, as I am credibly informed: It is otherwife in a manner wholly fpent for colds, coughes and rheumes, to expectorate flegme, but vfed in diuers formes, as in iuice, in decoctions, fyrrups, roules, trochifces, and the greene or dryed roote of it felfe.

And thefe are the moft ordinary Phyficall herbes that are vfed to be planted in gardens for the vfe of any Country familie, that is (as I faid before) farre remote from Phyfitians or Chirurgions abidings, that they may vfe as occafion ferueth for themfelues or their neighbours, and by a little care and paines in the applying may doe a great deale of good, and fometimes to them that haue not wherewith to fpend on themfelues, much leffe on Phyfitians or Chirurgions, or if they haue, may oftentimes receiue leffe good at their hands then at others that are taught by experience in their owne families, to be the more able to giue helpe to others.

The

THE
KITCHEN
GARDEN.

THE SECOND PART,

Ontaining as well all forts of herbes, as rootes and fruits, that are vfually planted in Gardens, to ferue for the vfe of the Table whether of the poore or rich of our Countrey : but herein I intend not to bring any fruite bearing trees, fhrubbes, or bufhes ; for I referue them for my Orchard, wherin they fhal be fet forth. So that in thefe three parts, I fuppofe the exquifite ornament of any worthy houfe is confummate for the exteriour bounds, the benefit of their riches extending alfo to the furnifhing of the moft worthy inward parts thereof : but becaufe many take pleafure in the fight and knowledge of other herbes that are Phyficall, and much more in their properties and vertues, if vnto thefe three I fhould adde a Phyficke Garden, or Garden of Simples, there would be a quadripartite complement, of whatfoeuer arte or nature, neceffitie or delight could affect : which to effect (as many my friends haue intreated it at my hands) will require more paines and time then all this worke together : yet to fatisfie their defires and all others herein, that would bee enformed in the truth, and reformed of the many errours and flips fet forth and publifhed heretofore of plants by diuers, I fhall (God affifting and granting life) labour to performe, that it may fhew it felfe to the light in due conueniencie, if thefe bee well and gratefully accepted. And becaufe I ended with fome fweete herbes in the former part, I will in this part beginne with the reft, which I referued for this place, as fitter for the pot and kitchen then for the hand or bofome, and fo defcend to other herbes that are for meat or fallets : and after them to thofe rootes that are to be eaten, as meate or as fallets : and laftly the fruits that grow neere, or vpon the ground, or not much aboue it ; as the Artichoke, &c. in which I make a fhorter defcription then I did in the former, rather endeauouring to fhew what they are, and whereunto they are vfed, then the whole varietie or any exact declaration : which methode, although in fome fort it may bee fitting for this purpofe, yet it is not for an hiftory or herball : I fhall therefore require their good acceptance for whofe fake I doe it, not doubting, but that I, or others, if they write againe of this fubiect, may polifh and amende what formerly hath beene eyther mif fet, or not fo thoroughly expreffed, befides fome additions of new conceits ; feeing I treade out a new path, and therefore thofe that follow may the eafilier fee the Meanders, and fo goe on in a direct line.

CHAP.

Chap. I.

Maierana latifolia, sine maior Anglica. Winter, or pot Marierome.

Winter Marierome is a small bushie herbe like vnto sweete Marierome, being parted or diuided into many branches, whereon doe grow broader and greener leaues, set by couples, with some small leaues likewise at the seuerall ioynts all along the branches: at the tops whereof grow a number of small purplish white flowers set together in a tuft, which turne into small and round seed, bigger then sweet Marierome seede: the whole plant is of a small and fine sent, but much inferiour to the other, and is nothing so bitter as the sweete Marierome, and thereby both the fitter and more willingly vsed for meates: the roote is white and threddy, and perisheth not as the former, but abideth many yeares.

The Vse of winter Marierome.

The vse of this Marierome is more frequent in our Land then in others, being put among other pot-herbes and farsing (or faseting herbes as they are called) and may to good profit bee applyed in inward as well as outward griefes for to comfort the parts, although weaker in effect then sweete Marieromes.

Chap. II.

Thymum vulgatius siue durius. Ordinary Garden Tyme.

The ordinary Garden Tyme is a small low wooddy plant with brittle branches, and small hard greene leaues, as euery one knoweth, hauing small white purplish flowers, standing round about the tops of the stalkes: the seed is small and browne, darker then Marierome seed: the root is woody, and abideth well diuers Winters.

Thymum latifolium. Masticke Tyme.

This Tyme hath neyther so wooddy branches, nor so hard leaues, but groweth lower, more spreading, and with somewhat broader leaues: the flowers are of a purplish white colour, standing in roundles round about the stalkes, at the ioynts with leaues at them likewise. This Tyme endureth better and longer then the former, and by spreading it selfe more then the former, is the more apt to bee propagated by slipping, because it hath beene seldome seene to giue seede: It is not so quicke in sent or taste as the former, but is fitter to set any border or knot in a garden, and is for the most part wholly employed to such vses.

The Vse of Tyme.

To set downe all the particular vses whereunto Tyme is applyed, were to weary both the Writer and Reader; I will but only note out a few: for besides the physicall vses to many purposes, for the head, stomacke, spleene, &c. there is no herbe almost of more vse, in the houses both of high and low, rich and poore, both for inward and outward occasions; outwardly for bathings among other hot herbes, and among other sweete herbes for strewings: inwardly in most sorts of broths, with Rosmary, as also with other faseting (or rather farsing) herbes, and to make sawce for diuers sorts both fish and flesh, as to stuffe the belly of a Goose to bee rosted, and after put into the sawce, and the pouder with breade to strew on meate when it

is

1 *Maiorana maior Anglica*. Pot Marieroine. 2 *Thymum vulgatius*. Garden Tyme. 3 *Satureia*. Sauorie. 4 *Hysso-pus*. Hyssope. 5 *Pulegium*. Penniroyall. 6 *Saluia maior*. Common Sage. 7 *Saluia minor primata*. Sage of vertue.

is rofted, and fo likewife on rofted or fryed fifh. It is held by diuers to bee a fpeedy remedy againft the fting of a Bee, being bruifed and layd thereon.

Chap. III.

Satureia fiue Thymbra. Sauorie.

THere are two forts of of Sauory, the one called Summer, and the other Winter Sauorie : The Summer Sauory is a fmall tender herbe, growing not aboue a foote and a halfe high, or thereabouts, rifing vp with diuers brittle branches, flenderly or fparfedly fet with fmall long leaues, foft in handling, at euery ioynt a couple, one againft another, of a pleafant ftrong and quicke fent and tafte : the flowers are fmall and purplifh, growing at the toppes of the ftalkes, with two fmall long leaues at the ioynts vnder them: the feede is fmall, and of a darke colour, bigger then Tyme feede by the halfe : the roote is wooddy, and hath many ftrings, perifhing euery yeare wholly, and muft bee new fowen againe, if any will haue it.

The Winter Sauorie is a fmall low bufhie herbe, very like vnto Hyffope, but not aboue a foote high, with diuers fmall hard branches, and hard darke green leaues thereon, thicker fet together then the former by much, and as thicke as common Hyffope, fometimes with foure leaues or more at a ioynt, of a reafonable ftrong fent, yet not fo ftrong or quicke as the former : the flowers are of a pale purplifh colour, fet at feuerall diftances at the toppes of the ftalkes, with leaues at the ioynts alfo with them, like the former : the roote is woody, with diuers fmall ftrings thereat, and abideth all the winter with his greene leaues : it is more vfually encreafed by flipping or diuiding the roote, and new fetting it feuerally againe in the Spring, then by fowing the feed.

The Vfe of Sauorie.

The Summer Sauorie is vfed in other Countryes much more then with vs in their ordinary diets, as condiment or fawce to their meates, fometimes of it felfe, and fometimes with other herbes, and fometimes ftrewed or layde vpon the difhes as we doe Parfley, as alfo with beanes and peafe, rife and wheate ; and fometimes the dryed herbe boyled among peafe to make pottage.

The Winter Sauorie is one of the (farfing) fafeting herbes as they call them, and fo is the Summer Sauorie alfo fometimes. This is vfed alfo in the fame manner that the Summer Sauorie is, fet downe before, and to the fame purpofes : as alfo to put into puddings, fawfages, and fuch like kindes of meates. Some doe vfe the pouder of the herbe dryed (as I fayd before of Tyme) to mixe with grated bread, to breade their meate, be it fifh or flefh, to giue it the quicker rellifh. They are both effectuall to expell winde.

Chap. IIII.

Hyffopus. Hyffope.

GArden Hyffope is fo well knowne to all that haue beene in a Garden, that I fhall but *aclum agere*, to beftow any time thereon, being a fmall bufhie plant, not rifing aboue two foote high, with many branches, woody below, and tender aboue, whereon are fet at certaine diftances, fundry fmall, long and narrow greene leaues : at the toppe of euery ftalke ftand blewifh purple gaping flowers, one aboue another in a long fpike or eare : after which followeth the feede, which is fmall and blackifh : the rootes are compofed of many threddy ftrings ; the whole plant is of a ftrong fweet fent.

The

The Vfe of Hyffope.

Hyffope is much vfed in Ptifans and other drinkes, to help to expectorate flegme. It is many Countrey peoples medicine for a cut or greene wound, being bruifed with fugar and applyed. I finde it is alfo much commended againft the falling fickeneffe, efpecially being made into pils after the manner before rehearfed. It is accounted a fpeciall remedy againft the fting or biting of an Adder, if the place be rubbed with Hyffope, bruifed and mixed with honey, falt and cummin feede. A decoction thereof with oyle, and annointed, taketh away the itching and tingling of the head, and vermine alfo breeding therein. An oyle made of the herbe and flowers, being annointed, doth comfort benummed finewes and ioynts.

CHAP. V.

Pulegium. Pennyroyall.

PEnnyroyall alfo is an herbe fo well knowne, that I fhall not neede to fpend much time in the defcription of it : hauing many weake round ftalkes, diuided into fundry branches, rather leaning or lying vpon the ground then ftanding vpright, whereon are fet at feuerall ioynts, fmall roundifh darke greene leaues : the flowers are purplifh that grow in gardens, yet fome that grow wilde are white, or more white then purple, fet in roundles about the tops of the branches ; the ftalkes fhoote forth fmall fibres or rootes at the ioynts, as it lyeth vpon the ground, thereby faftening it felfe therein, and quickly increafeth, and ouer-runneth any ground, efpecially in the fhade or any moift place, and is replanted by breaking the fprouted ftalkes, and fo quickely groweth.

Other forts of Pennyroyall are fit for the Phyficke Garden, or Garden of Simples.

The Vfe of Pennyroyall.

It is very good and wholefome for the lunges, to expell cold thin flegme, and afterwards to warme and dry it vp : and is alfo of the like properties as Mintes, to comfort the ftomacke, and ftay vomiting. It is alfo vfed in womens baths and wafhings : and in mens alfo to comfort the finewes. It is yet to this day, as it hath beene in former times, vfed to bee put into puddings, and fuch like meates of all forts, and therefore in diuers places they know it by no other name then Pudding-graffe.

The former age of our great Grandfathers, had all thefe hot herbes in much and familiar vfe, both for their meates and medicines, and therewith preferued themfelues in long life and much health : but this delicate age of ours, which is not pleafed with any thing almoft, be it meat or medicine, that is not pleafant to the palate, doth wholly refufe thefe almoft, and therefore cannot be partaker of the benefit of them.

CHAP. VI.

Salvia. Sage.

THere are two efpeciall kindes of Sage nourfed vp in our Gardens, for our ordinary vfe, whereof I intend to write in this place, leauing the reft to his fitter place. Our ordinary Sage is reckoned to bee of two forts, white and red,

both

both of them bearing many foure square wooddy ſtalkes, in ſome whiter, in others redder, as the leaues are alſo, ſtanding by couples at the ioynts, being long, rough, and wrinkled, of a ſtrong ſweete ſent : at the tops of the ſtalkes come forth the flowers, ſet at certaine ſpaces one aboue another, which are long and gaping, like vnto the flowers of Clary, or dead Nettles, but of a blewiſh purple colour, after which come ſmall round ſeede in the huske that bore the flower : the roote is wooddy, with diuers ſtrings at it : It is more vſually planted of the ſlips, pricked in the Spring time into the ground, then of the ſeed.

Saluia minor ſiue pinnata. Small Sage or Sage of vertue.

The leſſer Sage is in all things like vnto the former white Sage, but that his branches are long and ſlender, and the leaues much ſmaller, hauing for the moſt part at the bottome of each ſide of the leafe a peece of a leafe, which maketh it ſhew like finns or eares : the flowers alſo are of a blewiſh purple colour, but leſſer. Of this kinde there is one that beareth white flowers.

The Vſe of Sage.

Sage is much vſed of many in the moneth of May faſting, with butter and Parſley, and is held of moſt much to conduce to the health of mans body.

It is alſo much vſed among other good herbes to bee tund vp with Ale, which thereupon is termed Sage Ale, whereof many barrels full are made, and drunke in the ſaid moneth chiefly for the purpoſe afore recited: and alſo for teeming women, to helpe them the better forward in their childebearing, if there be feare of abortion or miſcarrying.

It is alſo vſed to be boyled among other herbes, to make Gargles or waters to waſh ſore mouths and throates : As alſo among other herbes, that ſerue as bathings, to waſh mens legs or bodies in the Summer time, to comfort nature, and warme and ſtrengthen aged cold ſinewes, and lengthen the ſtrength of the younger.

The Kitchen vſe is either to boyle it with a Calues head, and being minced, to be put with the braines, vinegar and pepper, to ſerue as an ordinary ſawce thereunto : Or being beaten and iuyced (rather then minced as manie doe) is put to a roſted Pigges braines, with Currans for ſawce thereunto. It is in ſmall quantity (in regard of the ſtrong taſte thereof) put among other faſting herbes, to ſerue as ſawce for peeces of Veale, when they are farſed or ſtuffed therewith, and roſted, which they call Olliues.

For all the purpoſes aforeſaid, the ſmall Sage is accounted to be of the more force and vertue.

Chap. VII.

Horminum ſatiuum. Garden Clary.

THere is but one ſort of Garden Clary, though many wilde, which hath foure ſquares ſtalks, with broad rough wrinkled whitiſh leaues, ſomewhat vneuenly cut in on the edges, and of a ſtrong ſweete ſent, growing ſome next the ground, & ſome by couples vpon the ſtalkes: the flowers growe at certaine diſtances, with two ſmall leaues at the ioynts vnder them, ſomewhat like vnto the flowers of Sage, but leſſer, and of a very whitiſh or bleake blew colour : the ſeede is of a blackiſh browne colour, ſomewhat flat, and not ſo round as the wilde : the rootes ſpread not farre, and periſh euery yeare that they beare flowers and ſeede. It is altogether to bee ſowne of ſeed in the Spring time, yet ſometimes it will riſe of it owne ſowing.

The

The Vſe of Clary.

The moſt frequent and common vſe of Clary, is for men or women that haue weake backes, to helpe to comfort and ſtrengthen the raines, being made into Tanſies and eaten, or otherwiſe. The ſeede is vſed of ſome to be put into the corner of the eye, if any mote or other thing haue happened into it : but aſſuredly although this may peraduenture doe ſome good, yet the ſeede of the wilde will due much more. The leaues taken dry, and dipped into a batter made of the yolkes of egges, flower, and a little milke, and then fryed with butter vntill they be criſpe, ſerue for a diſh of meate accepted with manie, vnpleaſant to none.

CHAP. VIII.

Nepeta. Nep.

ALthough thoſe that are Herbariſts do know three ſorts of Nep, a greater & two leſſer, yet becauſe the leſſer are not vſuall, but in the Gardens of thoſe that delight in natures varieties, I do not here ſhew you them. That which is vſuall (and called of manie Cat Mint) beareth ſquare ſtalkes, but not ſo great as Clarie, hauing two leaues at euery ioynt, ſomewhat like vnto Balme or Speare Mintes, but whiter, ſofter, and longer, and nicked about the edges, of a ſtrong ſent, but nothing ſo ſtrong as Clary : the flowers growe at the toppes of the ſtalkes, as it were in long ſpikes or heads, ſomewhat cloſe together, yet compaſſing the ſtalkes at certaine ioynts, of a whitiſh colour, for forme and bigneſſe like vnto Balme, or ſomewhat bigger : the rootes are compoſed of a number of ſtrings, which dye not, but keepe greene leaues vpon them all the Winter, and ſhoote anew in the Spring. It is propagated both by the ſeede, and by ſlipping the rootes.

The Vſe of Nep.

Nep is much vſed of women either in baths or drinkes to procure their feminine courſes : as alſo with Clarie, being fryed into Tanſies, to ſtrengthen their backes. It is much commended of ſome, if the iuyce thereof be drunke with wine, to helpe thoſe that are bruiſed by ſome fall, or other accident. A decoction of Nep is auaileable to cure the ſcabbe in the head, or other places of the body.

CHAP. IX.

Meliſſa. Baulme.

THe Garden Baulme which is of common knowne vſe, hath diuers ſquare blackiſh greene ſtalkes, and round, hard, darke, greene pointed leaues, growing thereon by couples, a little notched about the edges, of a pleaſant ſweete ſent, drawing neareſt to the ſent of a Lemon or Citron ; and therefore of ſome called *Citrago* : the flowers growe about the toppes of the ſtalkes at certaine diſtances, being ſmall and gaping, of a pale carnation colour, almoſt white : the rootes faſten themſelues ſtrongly in the ground, and endure many yeares, and is encreaſed by diuiding the rootes ; for the leues dye downe to the ground euery yeare, leauing no ſhew of leafe or ſtalke in the Winter.

The

The Vfe of Baulme.

Baulme is often vfed among other hot and fweete herbes, to make baths and wafhings for mens bodies or legges, in the Summer time, to warme and comfort the veines and finewes, to very good purpofe and effect, and hath in former ages beene of much more vfe then now adaies. It is alfo vfed by diuers to be ftilled, being fteeped in Ale, to make a Baulme water, after the manner they haue beene taught, which they keepe by them, to vfe in the ftead of *Aqua vitæ,* when they haue any occafion for their owne or their neighbours Families, in fuddaine qualmes or paffions of the heart : but if they had a little better direction (for this is fomewhat too rude) it would doe them more good that take it : For the herbe without all queftion is an excellent helpe to comfort the heart, as the very fmell may induce any fo to beleeue. It is alfo good to heale greene wounds, being made into falues : and I verily thinke, that our forefathers hearing of the healing and comfortable properties of the true naturall Baulme, and finding this herbe to be fo effectuall, gaue it the name of Baulme, in imitation of his properties and vertues. It is alfo an herbe wherein Bees doe much delight, as hath beene found by experience of thofe that haue kept great ftore ; if the Hiues bee rubbed on the infide with fome thereof, and as they thinke it draweth others by the fmell thereof to refort thither. Plinie faith, it is a prefent remedy againft the ftinging of Bees.

Chap. X.

Mentha. Mintes.

THere are diuers forts of Mints, both of the garden, and wilde, of the woods, mountaines, and ftanding pooles or waters: but I will onely in this place bring to your remembrance two or three forts of the moft vfuall that are kept in gardens, for the vfes whereunto they are proper.

Red Mint or browne Mint hath fquare brownifh ftalkes, with fomewhat long and round pointed leaues, nicked about the edges, of a darke greene colour, fet by couples at euery ioynt, and of a reafonable good fent : the flowers of this kinde are reddifh, ftanding about the toppes of the ftalkes at diftances : the rootes runne creeping in the ground, and as the reft, will hardly be cleared out of a garden, being once therein, in that the fmalleft peece thereof will growe and encreafe apace.

Speare Mint hath a fquare greene ftalke, with longer and greener leaues then the former, fet by couples, of a better and more comfortable fent, and therefore of much more vfe then any other : the flowers hereof growe in long eares or fpikes, of a pale red or blufh colour : the rootes creepe in the ground like the other.

Party coloured or white Mint hath fquare greene ftalkes and leaues, fomewhat larger then Speare Mint, and more nicked in the edges, whereof many are parted, halfe white and halfe greene, and fome more white then greene, or more green then white, as nature lifteth : the flowers ftand in long heads clofe fet together, of a blufh colour : the rootes creepe as the reft doe.

The Vfe of Mintes.

Mintes are oftentimes vfed in baths, with Baulme and other herbes, as a helpe to comfort and ftrengthen the nerues and finewes.

It is much vfed either outwardly applyed, or inwardly drunke, to ftrengthen and comfort weake ftomackes, that are much giuen to cafting : as alfo for feminine fluxes. It is boyled in milke for thofe whofe ftomackes are apt

1 *Horminum sativum.* Garden Clary. 2 *Nepeta.* Nep. 3 *Melissa.* Baulme. 4 *Mentha sativa.* Garden Mintes.
5 *Balsamita mas, seu Costus hortorum.* Costmary. 6 *Ageratum.* Maudeline.

apt to cauſe it to curdle. And applyed with ſalt, is a good helpe for the biting of a mad dogge.

It is vſed to be boyled with Mackarell, and other fiſh.

Being dryed, is often and much vſed with Penniroyall, to bee put into
puddings : as alſo among peaſe that are boyled for pottage.

Where Dockes are not ready at hand, they vſe to bruiſe Mintes, and
lay them vpon any place that is ſtung with Bees, Waſpes, or ſuch like, and
that to good purpoſe.

CHAP. XI.

Balſamita mas & famina, ſeu Coſtus hortorum maior & minor.
Coſtmary and Maudeline.

COſtmary or Alecoaſt is a ſweet herbe, bearing many broad and long pale green
leaues, ſnipped about the edges, euery one vpon a long foote-ſtalke ; among
which riſe vp many round greene ſtalkes, with ſuch like leaues on them, but
leſſer vp to the toppe, where it ſpreadeth it ſelfe into three or foure branches, euery
one bearing an vmbell or tuft of gold yellow flowers, ſomewhat like vnto Tanſie
flowers, but leſſer, which turne into ſmall heads, containing ſmall flat long ſeede : the
roote is ſomewhat hard and ſtringy, and being diuided, is replanted in the Spring of
the yeare for increaſe.

Maudeline hath ſomewhat long and narrow leaues, ſnipt about the edges : the ſtalks
are two foot high, bearing many yellow flowers on the tops of the branches, in an vmbell or tuft like vnto Tanſie : the whole herbe is ſweete, and ſomewhat bitter, and is
replanted by ſlipping.

The Vſe of Coſtmary and Maudeline.

Coſtmary is of eſpeciall vſe in the Spring of the yeare, among other ſuch
like herbes, to make Sage Ale, and thereupon I thinke it tooke the name of
Alecoaſt.

It is alſo vſed to be put among other ſweete herbes, to make ſweete waſhing water, whereof there is great ſtore ſpent.

The leaues haue an eſpeciall vertue to comfort both the ſtomack and
heart, and to warme and dry a moiſt braine. The ſeede is much vſed in the
Country, to be giuen to children for the wormes, in the ſtead of wormſeed,
and ſo is the ſeede of Maudeline alſo.

Maudeline is much vſed with Coſtmary and other ſweet herbes, to make
ſweete waſhing water : the flowers alſo are tyed vp with ſmall bundels of
Lauender toppes, theſe being put in the middle of them, to lye vpon the
toppes of beds, preſſes, &c. for the ſweete ſent and ſauour it caſteth. It is
generally accounted of our Apothecaries to be the true *Eupatorium* of Auicen, and the true *Ageratum* of Dioſcorides ; but Dodonæus ſeemeth to contradict both.

CHAP. XII.

Tanacetum vulgare & criſpum. Tanſie.

OVr Garden Tanſie hath many hard greene leaues, or rather wings of leaues ;
for they are many ſmall ones, ſet one againſt another all along a middle ribbe
or ſtalke, and ſnipt about the edges : in ſome the leaues ſtand cloſer and thicker, and ſomewhat crumpled, which hath cauſed it to be called double or curld Tanſie,

fie, in others thinner and more fparfedly : It rifeth vp with many hard ftalks, whereon growe at the tops vpon the feuerall fmall branches gold yellow flowers like buttons, which being gathered in their prime, will hold the colour frefh a long time : the feede is fmall, and as it were chaffie : the roote creepeth vnderground, and fhooteth vp againe in diuers places : the whole herbe, both leaues and flowers, are of a fharpe, ftrong, bitter fmell and tafte, but yet pleafant, and well to be endured.

The Vfe of Tanfie.

The leaues of Tanfie are vfed while they are young, either fhred fmall with other herbes, or elfe the iuyce of it and other herbes fit for the pur-pofe, beaten with egges, and fryed into cakes (in Lent and the Spring of the yeare) which are vfually called Tanfies, and are often eaten, being taken to be very good for the ftomack, to helpe to digeft from thence bad humours that cleaue thereunto : As alfo for weak raines and kidneyes, when the vrine paffeth away by drops : This is thought to be of more vfe for men then for women. The feed is much commended againft all forts of wormes in chil-dren.

Chap. XIII.

Pimpinella fiue Sanguiforba. Burnet.

Vrnet hath many winged leaues lying vpon the ground, made of many fmall, round, yet pointed greene leaues, finely nicked on the edges, one fet againft an-other all along a middle ribbe, and one at the end thereof ; from among which rife vp diuers round, and fometimes crefted browne ftalkes, with fome few fuch like leaues on them as growe belowe, but fmaller : at the toppes of the ftalkes growe fmall browne heads or knaps, which fhoote forth fmall purplifh flowers, turning into long and brownifh, but a little cornered feede : the roote groweth downe deepe, being fmall and brownifh : the whole plant is of a ftipticke or binding tafte or quality, but of a fine quicke fent, almoft like Baulme.

The Vfe of Burnet.

The greateft vfe that Burnet is commonly put vnto, is to put a few leaues into a cup with Claret wine, which is prefently to be drunke, and giueth a pleafant quicke tafte thereunto, very delightfull to the palate, and is ac-counted a helpe to make the heart merrie. It is fometimes alfo while it is young, put among other Sallet herbes, to giue a finer rellifh thereunto. It is alfo vfed in vulnerary drinkes, and to ftay fluxes and bleedings, for which purpofes it is much commended. It hath beene alfo much commended in contagious and peftilentiall agues.

Chap. XIIII.

Hippolapathum fatiuum, fiue Rhabarbarum Monachorum.
Monkes Rubarbe or Patience.

Arden Patience is a kinde of Docke in all the parts thereof, but that it is larger and taller then many others, with large and long greene leaues, a great, ftrong, and high ftalke, with reddifh or purplifh flowers, and three fquare feede, like as all other Dockes haue : the roote is great and yellow, not hauing any fhew of flefh coloured veines therein, no more then the other kinde with great round thin leaues,

commonly

commonly called *Hippolapathum rotundifolium*, Baſtard Rubarbe, or Monkes Rubarbe, the properties of both which are of very weake effect : but I haue a kinde of round leafed Dock growing in my Garden, which was ſent me from beyond Sea by a worthy Gentleman, Mr. Dr. Matth. Liſter, one of the Kings Phyſitians, with this title, *Rhaponticum verum*, and firſt grew with me, before it was euer ſeen or known elſewhere in England, which by proof I haue found to be ſo like vnto the true Rubarbe, or the Rha of Pontus, both for forme and colour, that I dare ſay it is the very true Rubarbe, our climate only making it leſſe ſtrong in working, leſſe heauy, and leſſe bitter in taſte: For this hath great and thicke rootes, as diuerſly diſcoloured with fleſh coloured veines as the true Rubarbe, as I haue to ſhew to any that are deſirous to ſee and know it ; and alſo other ſmaller ſprayes or branches of rootes, ſpreading from the maine great roote, which ſmaller branches may well be compared to the *Rhaponticum* which the Merchants haue brought vs, which we haue ſeene to be longer and ſlenderer then Rubarbe, but of the very ſame colour : this beareth ſo goodly large leaues, that it is a great beauty in a garden to behold them : for I haue meaſured the ſtalke of the leafe at the bottome next the roote to bee of the bigneſſe of any mans thumbe ; and from the roote to the leafe it ſelfe, to bee two foote in length, and ſometimes more; and likewiſe the leafe it ſelfe, from the lower end where it is ioyned to the ſtalke, to the end or point thereof, to bee alſo two foote in length, and ſometimes more ; and alſo in the broadeſt part of the leafe, to be two foote or more ouer in breadth : it beareth whitiſh flowers, contrary to all other Dockes, and three ſquare browniſh ſeede as other Dockes doe, but bigger, and therefore aſſuredly it is a Docke, and the true Rubarbe of the Arabians, or at the leaſt the true *Rhaponticum* of the Ancients. The figure of the whole plant I haue cauſed to be cut, with a dryed roote as it grew in my garden by it ſelfe, and haue inſerted it here, both becauſe Matthiolus giueth a falſe figure of the true Rubarbe, and that this hath not been expreſſed and ſet forth by any before.

The Vſe of Patience, and of the Rubarbe.

The leaues of Patience are often, and of many vſed for a pot-herbe, and ſeldome to any other purpoſe : the roote is often vſed in Diet-beere, or ale, or in other drinkes made by decoction, to helpe to purge the liuer, and clenſe the blood. The other Rubarbe or *Rhaponticum*, wherof I make mention, and giue you here the figure, I haue tryed, and found by experience to purge gently, without that aſtriction that is in the true Rubarbe is brought vs from the Eaſt Indies, or China, and is alſo leſſe bitter in taſte ; whereby I coniecture it may bee vſed in hot and feaueriſh bodies more effectually, becauſe it doth not binde after the purging, as the Eaſt India Rubarbe doth : but this muſt bee giuen in double quantitie to the other, and then no doubt it will doe as well : The leaues haue a fine acide taſte : A ſyrrupe therefore made with the iuice and ſugar, cannot but be very effectuall in deiected appetites, and hot fits of agues; as alſo to helpe to open obſtructions of the liuer, as diuers haue often tryed, and found auaileable by experience.

Chap. XV.

Lapathum ſanguineum. Blood-wort.

Among the ſorts of pot-herbes Blood-worte hath alwayes beene accounted a principall one, although I doe not ſee any great reaſon therein, eſpecially ſeeing there is a greater efficacie of binding in this Docke, then in any of the other : but as common vſe hath receiued it, ſo I here ſet it downe. Blood-worte is one of the ſorts of Dockes, and hath long leaues like vnto the ſmaller yellow Docke, but ſtriped with red veines, and ouer-ſhadowed with red vpon the greene leafe, that it ſeemeth almoſt wholly red ſometimes : the ſtalke is reddiſh, bearing ſuch like leaues, but

ſmaller

1 *Tanacetum,* Tanfie. 2 *Pimpinella.* Burnet. 3 *Rhaponticum verum feu potius Rhabarbarum verum.* True Raponticke or rather true Rubarbe. 4 *Lapathum fativum feu Patientia.* Monkes Rubarbe or Patience. 5 *Lapathum fanguineum.* Bloudwort. 6 *Acet.fa.* Sorrell.

smaller vp to the toppe, where it is diuided into diuers small branches, whereon grow purplish flowers, and three square darke red seede, like vnto others : the roots are not great, but somewhat long, and very red, abiding many yeares, yet sometimes spoiled with the extremitie of winter.

The Vse of Blood-worte.

The whole and onely vse of the herbe almost, serueth for the pot, among other herbes, and, as I said before, is accounted a most especiall one for that purpose. The seede therof is much commended for any fluxe in man or wo-man, to be inwardly taken, and so no doubt is the roote, being of a stipticke qualitie.

Chap. XVI.

Oxalis siue Acetosa. Sorrell.

SOrrell must needes bee reckoned with the Dockes, for that it is so like vnto them in all things, and is of many called the sower Docke. Of Sorrels there are ma-ny sorts, but I shall not trouble you with any other in this place, then the com-mon Garden Sorrell, which is most knowne, and of greatest vse with vs ; which hath tender greene long leaues full of iuice, broade, and bicorned as it were, next vnto the stalke, like as Arrach, Spinach, and our English Mercurie haue, of a sharpe sower taste : the stalkes are slender, bearing purplish long heads, wherein lye three square shining browne seede, like, but lesser then the other : the root is smaller then any of the other Dockes, but browne, and full of strings, and abideth without decaying, hauing greene leaues all the winter, except in the very extremitie thereof, which often taketh away all or most of his leaues.

The Vse of Sorrell.

Sorrell is much vsed in sawces, both for the whole, and the sicke, cooling the hot liuers, and stomackes of the sicke, and procuring vnto them an ap-petite vnto meate, when their spirits are almost spent with the violence of their furious or fierie fits ; and is also of a pleasant rellish for the whole, in quickning vp a dull stomacke that is ouer-loaden with euery daies plenty of dishes. It is diuers waies dressed by Cooks, to pleafe their Masters stomacks.

Chap. XVII.

Buglossum luteum, siue Lingua Bouis. Langdebeefe.

VNto this place may well bee referred our ordinary Borage and Buglosse, set forth in the former Booke, in regard of the properties whereunto they are much employed, that is, to serue the pot among other herbes, as is sufficiently knowne vnto all. And yet I confesse, that this herbe (although it bee called *Buglossum luteum*, as if it were a kind of Buglosse) hath no correspondency with Buglosse or Bo-rage in any part, sauing only a little in the leafe; & our Borage or Buglosse might more fitly, according to the Greeke name, bee called Oxe tongue or Langdebeefe; and this might in my iudgement more aptly be referred to the kinds of *Hieratium* Hawkeweed, whereunto it neerest approacheth : but as it is commonly receiued, so take it in this place, vntill it come to receiue the place is proper for it. It hath diuers broad and long darke green leaues, lying vpon the ground, very rough in handling, full of small haires or prickes, ready to enter into the hands of any that handle it ; among which riseth

vp

vp a round greene hairy or prickly ftalk, bearing at the toppe, among a few fmall green leaues, diuers fmall yellow flowers in rough heads, which turne into doune, containing within them browne yellowifh fmall long feedes, fomewhat like vnto the feede of Hawkeweede: the roote is woodly, which perifheth quickly after it hath borne feed, but is tender while it is young.

The Vfe of Langdebeefe.

The leaues are onely vfed in all places that I know, or euer could learne, for an herbe for the pot among others, and is thought to bee good to loo-fen the belly.

Chap. XVIII.

Atriplex fiue Olus Aureum. Arrach.

THere be diuers kindes of Arrach, or Orach, as fome doe call them ; fome of the Garden, whereof I meane to entreate in this place; others wilde of the Fieldes, &c. and others of the Sea, which are not to bee fpoken of in this worke, but referred to a generall hiftorie. The white garden Arrach, or Orach, hath diuers leaues, ftanding vpon their feuerall footeftalkes, broade at the bottome, ending in two points like an arrow, with two feathers at the head, and fmall pointed at the end of the leafe, of a whitifh yellow greene colour, and as it were ftrewed ouer with flower or meale, efpecially while they are young : the ftalke likewife is mealy, bearing many branches with fmall yellow flowers on them, which turne into fmall leafie feeds: the roote groweth fomewhat deepe in the ground, with many fmall threds faftened thereto : it quickly fpringeth vp of the feede, groweth great, and fadeth away as foon as it hath borne feede.

The purple Arrach is in all things like vnto the white, fauing onely in the colour of the leafe, ftalke, feede, &c. which are all of a mealy dufty purplifh colour.

The Vfe of Arrach.

Arrach is cold and moift, and of a lubricke or flippery qualitie, whereby it quickely paffeth through the ftomacke and belly, and maketh it foluble, and is of many vfed for that purpofe, being boyled and buttered, or put a-mong other herbes into the pot to make pottage.

There are many difhes of meate made with them while they are young, for being almoft without fauour of themfelues, they are the more conuer-tible into what rellifh any one will make them with Sugar, Spice &c.

Chap. XIX.

Blitum. Blites.

THere be diuers forts of Blites, fome whereof I haue entreated in the former part of this worke, vnder the title of *Amaranthus,* Flower gentle : others that are nourfed vp in Gardens, I will fet forth in this place, which are onely two, that haue come to my knowledge, that is, the white and the red, and are of a qualitie as neere vnto Arrach as vnto Beetes, participating of both, and therefore I haue pla-ced them betwixt them. The white Blite hath leaues fomewhat like vnto Beetes, but fmaller, rounder, and of a whitifh greene colour, euery one ftanding vpon a fmall long footeftalke : the ftalke rifeth vp two or three foote high, with many fuch like leaues thereon : the flowers grow at the top in long round tufts or clufters, wherein are con-

tained

tained small round seede : the roote is very full of threds or strings.

The red Blite is in all things like the white, but that his leaues and tufted heades are exceeding red at the first, and after turne more purplish.

The Vſe of Blites.

Blites are vſed as Arrach, eyther boyled of it ſelfe or ſtewed, which they call Loblolly, or among other herbes to bee put into the pot ; and yet ſome doe vtterly refuſe it, becauſe in diuers it prouoketh caſtings. It is altogether inſipide or without taſte, but yet by reaſon of the moiſt ſlipperie qualitie it hath, it helpeth to looſen the belly. The vnſauorineſſe whereof hath in many Countries growne into a prouerbe, or by-word, to call dull, ſlow, or lazie perſons by that name : They are accounted more hurtfull to the ſtomacke, and ſo to the head and eyes, then other herbes, and therefore they are the leſſe vſed.

Chap. XX.

Beta. Beetes.

THere are many diuerſities of Beetes, ſome growing naturally in our own Coun-try, others brought from beyond Sea ; whereof ſome are white, ſome greene, ſome yellow, ſome red : the leaues of ſome are of vſe only, and the root not vſed : others the roote is only vſed, and not the leaues : and ſome againe, both roote and leafe. The ancient Authors, as by their workes appeare, knew but two ſorts, the white and the blacke Beete, whereof the white is ſufficiently known, and was of them termed *Sicula*, of the later Phyſitians *Sicla*, becauſe it was thought firſt to be brought from Sicilie : the blacke abideth ſome controuerſie ; ſome thinking that our common greene Beete, becauſe it is of a darke greene colour, was that they called the blacke Beete ; others that our ſmall red Beete, which is of a darke red colour, was their black Beete, which in my opinion is the more likely : But to come to the matter in hand, and giue you the deſcriptions of them which are in vſe with vs, and leaue controuerſies to ſuch a worke as is fit for them, wherein all ſuch matters may be diſcuſſed at large.

The common white Beete hath many great leaues next the ground (in ſome hot Countries growing to be three foote long, and very broade, in our Countrey they are very large, but nothing neere that proportion) of a whitiſh greene colour ; the ſtalke is great, ſtrong, and ribbed or creſted, bearing great ſtore of leaues vpon it vp to the very toppe almoſt : the flowers grow in very long tufts, ſmall at the ends, and turning down their heads, which are ſmall pale greeniſh yellow burres, giuing cornered prick-ly ſeede : the roote is great, long and hard, when it hath giuen ſeede, of no vſe at all, but abideth a former winter with his leaues vpon it, as all other ſorts following doe.

The common red Beet differeth not from the white Beete, but only that it is not ſo great, and both the leaues and rootes are ſomewhat red : the leaues bee in ſome more red then in others, which haue but red veines or ſtrakes in them, in ſome alſo of a freſh red, in others very darke red : the roote hereof is red, ſpongy, and not vſed to bee eaten.

The common greene Beete is alſo like vnto the white Beete, but of a darke greene colour. This hath beene found neere the ſalt Marſhes by Rocheſter, in the foote-way going from the Lady Leueſons houſe thither, by a worthy, diligent and painefull ob-ſeruer and preſeruer both of plants and all other natures varieties, often remembred before in this worke, called Iohn Tradeſcante, who there finding it, gaue me the know-ledge thereof, and I haue vpon his report ſet it here down in this manner :

The Romane red Beete, called *Beta rapoſa*, is both for leafe and roote the moſt excel-lent Beete of all others : his rootes bee as great as the greateſt Carrot, exceeding red both within and without, very ſweete and good, fit to bee eaten : this Beete groweth higher then the laſt red Beete, whoſe rootes are not vſed to bee eaten : the leaues like-

wiſe

wife are better of tafte, and of as red a colour as the former red Beete : the roote is fometimes fhort like a Turnep, whereof it took the name of *Rapa* or *rapofa* ; and fometimes as I faid before, like a Carrot and long : the feede is all one with the leffer red Beete.

The Italian Beete is of much refpect, whofe faire greene leaues are very large and great, with great white ribbes and veines therein: the ftalke in the Summer time, when it is growen vp to any height, is fix fquare in fhew, and yellowifh withall, as the heades with feede vpon them feeme likewife.

The great red Beete that Mafter Lete a Merchant of London gaue vnto Mafter Gerrard, as he fetteth it downe in his Herball, feemeth to bee the red kinde of the laft remembred Beete, whofe great ribbes as he faith, are as great as the middle ribbe of the Cabbage leafe, and as good to bee eaten; whofe ftalke rofe with him to the height of eight cubits, and bore plenty of feede.

The Vfe of Beetes.

Beetes, both white, greene and red, are put into the pot among other herbes, to make pottage, as is commonly known vnto all, and are alfo boyled whole, both in France vfually with moft of their boyled meates, and in our Countrey, with diuers that delight in eating of herbes.

The Italian Beete, and fo likewife the laft red Beete with great ribbes, are boyled, and the ribbes eaten in fallets with oyle, vinegar and pepper, and is accounted a rare kinde of fallet, and very delicate.

The roote of the common red Beete with fome, but more efpecially the Romane red Beete, is of much vfe among Cookes to trimme or fet out their difhes of meate, being cut out into diuers formes and fafhions, and is grown of late dayes into a great cuftome of feruice, both for fifh and flefh.

The rootes of the Romane red Beete being boyled, are eaten of diuers while they are hot with a little oyle and vinegar, and is accounted a delicate fallet for the winter ; and being cold they are fo vfed and eaten likewife.

The leaues are much vfed to mollifie and open the belly, being vfed in the decoction of Glifters. The roote of the white kinde fcraped, and made vp with a little honey and falt, rubbed on and layd on the belly, prouoketh to the ftoole. The vfe of eating Beetes is likewife held to bee helpefull to fpleneticke perfons.

Chap. XXI.

Hippofelinum, fiue Olus atrum. Alifanders.

ALifanders hath beene in former times thought to be the true Macedonian Parfley, and in that errour many doe yet continue : but this place giueth not leaue to difcuffe that doubt : but I muft here only fhew you, what it is, and to what vfe it is put ordinarily for the Kitchen. The leaues of Alifanders are winged or cut into many parts, fomewhat refembling Smallage, but greater, broader, and more cut in about the edges : the ftalkes are round and great, two foote high or better, bearing diuers leaues on them, and at the toppe fpokie roundles of white flowers on feuerall fmall branches, which turne into blacke feede, fomewhat cornered or crefted, of an aromaticall bitter tafte : the roote is blacke without, and white within, and abideth well the firft year of the fowing, perifhing after it hath borne feed.

The Vfe of Alifanders.

The tops of the rootes, with the lower part of the ftalkes of Alifanders, are vfed in Lent efpecially, and Spring of the yeare, to make broth, which although it be a little bitter, yet it is both wholfome, and pleafing to a great

many,

many, by reafon of the aromaticall or fpicie tafte, warming and comforting the ftomack, and helping it digeft the many waterifh and flegmaticke meates are in thofe times much eaten. The rootes alfo either rawe or boyled are often eaten with oyle and vinegar. The feede is more vfed phyfically then the roote, or any other part, and is effectuall to prouoke plenty of vrine in them that piffe by drops, or haue the Strangury : It helpeth womens courfes, and warmeth their benummed bodies or members, that haue endured fierce cold daies and nights, being boyled and drunke.

Chap. XXII.

Selinum dulce. Sweete Parfley or fweete Smallage.

THis kinde of fweete Parfley or Smallage, which foeuer you pleafe to call it, for it refembleth Smallage as well in the largeneffe of the leaues, as in the tafte, yet fweeter and pleafanter, is (as I take it) in this like vnto fweete Fennell (that hath his fweetneffe from his naturall foyle and clymate ; for howfoeuer it bee reafonable fweete the firft yeare it is fowne with vs, yet it quickly doth degenerate, and becommeth no better then our ordinarie Fennell afterwards). The firft yeare it is fowne and planted with vs (and the firft that euer I faw, was in a Venetian Ambaffadours Garden in the Spittle yard, neare Bifhops gate ftreete) is fo fweete and pleafant, efpecially while it is young, as if Sugar had beene mingled with it : but after it is growne vp high and large, it hath a ftronger tafte of Smalladge, and fo likewife much more the next yeare ; that it groweth from the feed was gathered here : the leaues are many, fpreading farre about the roote, broader and of a frefher greene colour then our ordinary Smalladge, and vpon longer ftalkes : the feed is as plentifull as Parfley, being fmall and very like vnto it, but darker of colour.

The Vfe of fweete Parfley.

The Venetians vfe to prepare it for meate many waies, both the herbe and the roote eaten rawe, as many other herbes and rootes are, or boyled or fryed to be eaten with meate, or the dryed herbe poudered and ftrewed vpon meate ; but moft vfually either whited, and fo eaten rawe with pepper and oyle, as a dainty Sallet of it felfe, or a little boyled or ftewed : the tafte of the herbe being a little warming, but the feede much more, helpeth cold windy ftomackes to digeft their meate, and to expell winde.

Chap. XXIII.

Petrofelinum & *Apium.* Parfley and Smalledge.

WE haue three fotts of Parfley in our Gardens, and but one of Smalladge : Our common Parfley, Curld Parfley, and Virginia Parfley ; which laft, although it be but of late knowne, yet it is now almoft growne common, and of as good vfe as the other with diuers. Our common Parfley is fo well knowne, that it is almoft needleffe to defcribe it, hauing diuers frefh greene leaues, three alwaies placed together on a ftalke, and fnipt about the edges, and three ftalkes of leaues for the moft part growing together : the ftalkes growe three or foure foote high or better, bearing fpikie heads of white flowers, which turne into fmall feede, fomewhat fharpe and hot in tafte : the roote is long and white.

Curld Parfley hath his leaues curled or crumpled on the edges, and therein is the onely difference from the former.

Virginia

Virginia Parſley is in his leafe altogether like vnto common Parſley for the forme, conſiſting of three leaues ſet together, but that the leaues are as large as Smallage leaues, but of a pale or whitiſh greene colour, and of the ſame taſte of our common Parſley: the ſeede hereof is as the leaues, twice if not thrice as bigge as the ordinary Parſley, and periſheth when it hath giuen ſeede, abiding vſually the firſt yeare of the ſowing.

Smallage is in forme ſomewhat like vnto Parſley, but greater and greener, and leſſe pleaſant, or rather more bitter in taſte: the ſeede is ſmaller, and the root more ſtringy.

The Vſe of Parſley.

Parſley is much vſed in all ſorts of meates, both boyled, roaſted, fryed, ſtewed, &c. and being greene it ſerueth to lay vpon ſundry meates, as alſo to draw meate withall. It is alſo ſhred and ſtopped into poudered beefe, as alſo into legges of Mutton, with a little beefe ſuet among it, &c.

The rootes are often vſed to be put into broth, to helpe to open obſtructions of the liuer, reines, and other parts, helping much to procure vrine.

The rootes likewiſe boyled or ſtewed with a legge of Mutton, ſtopped with Parſley as aforeſaid, is very good meate, and of very good relliſh, as I haue proued by the taſte; but the rootes muſt bee young, and of the firſt yeares growth, and they will haue their operation to cauſe vrine.

The ſeed alſo is vſed for the ſame cauſe, when any are troubled with the ſtone, or grauell, to open the paſſages of vrine.

Although Smallage groweth in many places wilde in moiſt grounds, yet it is alſo much planted in Gardens, and although his euill taſte and ſauour doth cauſe it not to be accepted into meates as Parſley, yet it is not without many ſpeciall good properties, both for outward and inward diſeaſes, to helpe to open obſtructions, and prouoke vrine. The iuyce cleanſeth vlcers; and the leaues boyled with Hogs greaſe, healeth felons on the ioynts of the fingers.

CHAP. XXIIII.

Fœniculum. Fenell.

THere are three ſorts of Fenell, whereof two are ſweete. The one of them is the ordinary ſweete Fenell, whoſe ſeedes are larger and yellower then the common, and which (as I ſaid before in the Chapter of ſweete Parſley) doth ſoone degenerate in this our Country into the common. The other ſweete Fenell is not much knowne, and called Cardus Fenell by thoſe that ſent it out of Italy, whoſe leaues are more thicke and buſhie then any of the other. Our common Fenell, whereof there is greene and red, hath many faire and large ſpread leaues, finely cut and diuided into many ſmall, long, greene, or reddiſh leaues, yet the thicker tufted the branches be, the ſhorter are the leaues: the ſtalkes are round, with diuers ioynts and leaues at them, growing fiue or ſix foot high, bearing at the top many ſpoakie rundels of yellow flowers: the Common, I meane, doth turne into a darke grayiſh flat ſeede, and the Sweete into larger and yellower: the roote is great, long, and white, and endureth diuers yeares.

The Vſe of Fenell.

Fenell is of great vſe to trimme vp, and ſtrowe vpon fiſh, as alſo to boyle or put among fiſh of diuers ſorts, Cowcumbers pickled, and other fruits, &c. The rootes are vſed with Parſley rootes, to be boyled in broths and drinkes to open obſtructions. The ſeed is of much vſe with other things to expell winde. The ſeede alſo is much vſed to be put into Pippin pies, and diuers

other

1 *Petroſelinum.* Parſley. 2 *Apium.* Smallage. 3 *Fœniculum.* Fenell. 4 *Anethum.* Dill. 5 *Myrrhis ſiue Cereſolium magnum.*
Sweete Cheruill. 6 *Cereſolium vulgare.* Common Cheruill.

other such baked fruits, as also into bread, to giue it the better rellish.

The sweete Cardus Fenell being sent by Sir Henry Wotton to Iohn Tradescante, had likewise a large direction with it how to dresse it; for they vse to white it after it hath been transplanted for their vses, which by reason of the sweetnesse by natnre, and the tendernesse by art, causeth it to be the more delightfull to the taste, especially with them that are accustomed to feede on greene herbes.

Chap. XXV.

Anethum. Dill.

DIll doth much growe wilde, but because in many places it cannot be had, it is therefore sowne in Gardens for the vses whereunto it serueth. It is a smaller herbe then Fenell, but very like, hauing fine cut leaues, not so large, but shorter, smaller, and of a stronger and quicker taste: the stalke is smaller also, and with few ioynts and leaues on them, bearing spoakie tufts of yellow flowers, which turne into thinne, small, and flat seedes: the roote perisheth euery yeare, and riseth againe for the most part of it owne sowing.

The Vse of Dill.

The leaues of Dill are much vsed in some places with Fish, as they doe Fenell; but because it is so strong many doe refuse it.

It is also put among pickled Cowcumbers, wherewith it doth very well agree, giuing vnto the cold fruit a pretty spicie taste or rellish.

It being stronger then Fenell, is of the more force to expell winde in the body. Some vse to eate the seed to stay the Hickocke.

Chap. XXVI.

Myrrhis siue Cerefolium maius & vulgare.
Sweet Cheruill and ordinary Cheruill.

THe great or sweete Cheruill (which of some is called Sweete Cicely) hath diuers great and faire spread winged leaues, consisting of many leaues set together, deeply cut in the edges, and euery one also dented about, very like, and resembling tne leaues of Hemlockes, but of so pleasant a taste, that one would verily thinke, he chewed the leaues or seedes of Aniseedes in his mouth: The stalke is reasonable great, and somewhat cornered or crested about three or foure foote high, at the toppe whereof stand many white spoakie tufts of flowers, which change into browne long cornered great seede, two alwaies ioyned together: the roote is great, blackish on the outside, and white within, with diuers fibres annexed vnto it, and perisheth not, but abideth many yeares, and is of a sweete, pleasant, and spicie hot taste, delightfull vnto many.

The common Cheruill is a small herbe, with slender leaues, finely cut into long peeces, at the first of a pale yellowish greene colour, but when the stalke is growne vp to seede, both stalkes and leaues become of a darke red colour: the flowers are white, standing vpon scattered or thin spread tufts, which turne into small, long, round, and sharpe pointed seedes, of a brownish blacke colour: the roote is small, with diuers long slender white strings, and perisheth euery yeare.

The

The Vfes of thefe Cheruils.

The common Cheruill is much vfed of the French and Dutch people, to bee boyled or ftewed in a pipkin, eyther by it felfe, or with other herbes, whereof they make a Loblolly, and fo eate it. It is vfed as a pot-herbe with vs.

Sweete Cheruill, gathered while it is young, and put among other herbes for a fallet, addeth a meruellous good rellifh to all the reft. Some commend the greene feedes fliced and put in a fallet of herbes, and eaten with vinegar and oyle, to comfort the cold ftomacke of the aged. The roots are vfed by diuers, being boyled, and after eaten with oyle and vinegar, as an excellent fallet for the fame purpofe. The preferued or candid rootes are of fingular good vfe to warme and comfort a cold flegmaticke ftomack, and is thought to be a good preferuatiue in the time of the plague.

Chap. XXVII.

Malua Crifpa. French Mallowes.

THe curld or French Mallow groweth vp with an vpright greene round ftalke, as high vfually as any man, whereon from all fides grow forth round whitifh greene leaues, curld or crumpled about the edges, like a ruffe, elfe very like vnto an ordinary great Mallow leafe : the flowers grow both vpon the ftalke, and on the other branches that fpring from them, being fmall and white ; after which come fmall cafes with blacke feede like the other Mallowes : the roote perifheth when it hath borne feede, but abideth vfually the firft yeare, and the fecond runneth vp to flower and feede.

The Vfe of French Mallowes.

It is much vfed as a pot-herbe, efpecially when there is caufe to moue the belly downward, which by his flippery qualitie it doth helpe forward. It hath beene in times paft, and fo is to this day in fome places, vfed to be boyled or ftewed, eyther by it felfe with butter, or with other herbes, and fo eaten.

Chap. XXVIII.

Intubum. Succorie and Endiue.

I Put both Succorie and Endiue into one chapter and defcription, becaufe they are both of one kindred ; and although they differ a little the one from the other, yet they agree both in this, that they are eaten eyther greene or whited, of many.

Endiue, the fmooth as well as the curld, beareth a longer and a larger leafe then Succorie, and abideth but one yeare, quickely running vp to ftalke and feede, and then perifheth: whereas Succorie abideth many years, and hath long and narrower leaues, fomewhat more cut in, or torne on the edges : both of them haue blew flowers, and the feede of the fmooth or ordinary Endiue is fo like vnto the Succorie, that it is very hard to diftinguifh them afunder by fight ; but the curld Endiue giueth blackifh and flat feede, very like vnto blacke Lettice feede : the rootes of the Endiue perifh, but the Succorie abideth.

The Vfe of Succory and Endiue.

Although Succorie bee fomewhat more bitter in tafte then the Endiues,
Tt 2   yet

yet it is oftentimes, and of many eaten greene, but more vfually being buried a while in fand, that it may grow white, which caufeth it to lofe both fome part of the bitterneffe, as alfo to bee the more tender in the eating ; and Horace fheweth it to be vfed in his time, in the 32.Ode of his firft Book, where he faith,

Me pafcunt Oliue, me Cithorea leuefq; Malue.

Endiue being whited in the fame, or any other manner, is much vfed in winter, as a fallet herbe with great delight ; but the curld Endiue is both farre the fairer, and the tenderer for that purpofe.

Chap. XXIX.

Spinachia, fiue Olus Hifpanicum. Spinach.

SPinach or Spinage is of three forts (yet fome doe reckon of foure, accounting that herbe that beareth no feede to be a fort of it felfe, when it is but an accident of nature, as it falleth out in Hempe, Mercury, and diuers other herbes) two that bear prickly feed, the one much greater then the other: the third that beareth a fmooth feede, which is more daintie, and nourfed vp but in few Gardens : The common Spinach which is the leffer of the two prickly forts, hath long greene leaues, broad at the ftalke, and rent, or torne as it were into foure corners, and fharpe pointed at the ends : it quickly runneth vp to ftalke, if it be fowen in the Spring time ; but elfe, if at the end of Summer, it will abide all the winter green, and then fuddenly in the very beginning of the Spring, runne vp to ftalke, bearing many leaues both below and at the toppe, where there doth appeare many fmal greenifh flowers in clufters, and after them prickly feede : The other greater fort that hath prickly feede, is in all things like the former, but larger both in ftalke, leafe and feede. The fmooth Spinach hath broader, and a little rounder pointed leaues then the firft, efpecially the lower leaues ; for thofe that grow vpwards vpon the ftalke, are more pointed, and as it were three fquare, of as darke a greene colour as the former : at the feuerall ioynts of the ftalkes and branches, ftand cluftering many fmall greenifh flowers, which turne into clufters of round whitifh feede, without any prickles at all vpon them : the roote is long, white and fmall, like vnto the other, with many fibres at it : If it be often cut, it will grow the thicker, or elfe fpindle vp very thinly, and with but few leaues vpon the ftalke.

The Vfe of Spinage.

Spinage is an herbe fit for fallets, and for diuers other purpofes for the table only ; for it is not knowne to bee vfed Phyfically at all. Many Englifh that haue learned it of the Dutch people, doe ftew the herbe in a pot or pipkin, without any other moifture then it owne, and after the moifture is a little preffed from it, they put butter, and a little fpice vnto it, and make therewith a difh that many delight to eate of. It is vfed likewife to be made into Tartes, and many other varieties of difhes, as Gentlewomen and their Cookes can better tell then my felfe ; vnto whom I leaue the further ordering of thefe herbes, and all other fruits and rootes of this Garden : For I intend only to giue you the knowledge of them, with fome briefe notes for their vfe, and no more.

1 *Malua crispa.* French Mallowes. 2 *Endiuia.* Endiue. 3 *Cichorium.* Succory. 4 *Spinachia.* Spinach. 5 *Lactuca crispa.* Curl'd Lettice. 6 *Lactuca patula.* An open Lettice. 7 *Lactuca capitata vulgaris.* Ordinary cabbage Lettice. 8 *Lactuca capitata Romana.* The great Romane cabbage Lettice. 9 *Lactuca vrsina.* Corne Sallet or Lambes Lettice.

Chap. XXX.

Lactuca. Lettice.

THere are so many sorts, and so great diuersitie of Lettice, that I doubt I shall scarce be beleeued of a great many. For I doe in this Chapter reckon vp vnto you eleauen or twelue differing sorts ; some of little vse, others of more, being more common and vulgar ; and some that are of excellent vse and seruice, which are more rare, and require more knowledge and care for the ordering of them, as also for their time of spending, as some in the spring, some in summer, others in autumne, and some being whited for the winter. For all these sorts I shall not neede many descriptions, but only shew you which doe cabbage, and which are loose, which of of them are great or small, white, greene or red, and which of them beare white seeds, and which of them blacke. And lastly I haue thought good to adde another Sallet herbe, which because it is called Lambes Lettice of many, or Corne Sallet of others, is put in only to fill vp a number in this Chapter, and that I must speake of it, and not that I thinke it to be any of the kindes of Lettice.

All sorts of Lettice, after a while that they haue closed themselues, if they bee of the Cabbage kindes, or otherwise being loose, and neuer closing, send forth from among the middle of their leaues a round stalke (in some greater, in others lesser, according to their kinde) full of leaues like vnto the lower, branching at the toppe into sundry parts, whereon grow diuers small star-like flowers, of a pale yellowish colour; after which come seede, eyther white or blackish, as the plant yeeldeth, whereat hangeth some small peece of a cottony doune, wherewith the whole head is stored, and is carried away with the winde, if it be not gathered in time : the roote is somewhat long and white, with some fibres at it, and perisheth quickely after the seede is ripe.

The Romane red Lettice is the best and greatest of all the rest. For Iohn Tradescante that first, as I thinke, brought it into England, and sowed it, did write vnto mee, that after one of them had been bound and whited, when the refuse was cutaway, the rest weighed seuenteene ounces : this hath blacke seede.

The white Romane Lettice is like vnto it, hauing long leaues like a Teasell, it is in goodnesse next vnto the red, but must be whited, that it may eate kindly : the seede hereof is white.

The Virginia Lettice hath single and very broade reddish leaues, and is not of any great regard, and therefore is kept but of a few : it beareth blacke seede.

The common Lumbard Lettice that is loose, and another kinde thereof that doth somewhat cabbage, haue both white seedes.

The Venice Lettice is an excellent Cabbage Lettice, and is best to bee sowen after Midsummer for lateward Lettice ; they be sometimes as great as the crowne of a mans hatt : the seede hereof is white, and groweth to be of a meane height.

Our common Cabbage Lettice is well known, and beareth blacke seede.

The curld Lettice which is open, and differeth but little from Endiue, beareth black seede.

Another sort of curld Lettice doth cabbage, and is called Flanders Cropers, or Cropers of Bruges ; this groweth lowest, and hath the smallest head, but very hard and round, and white while it groweth : the seed is blacke.

A kinde of Romane Lettice is of a darke green colour, growing as low as the Venice Lettice, and is an excellent kinde, bearing blacke seede.

And lastly our winter Lettice is wonderfull hardy to endure our cold : It is but single, and must be sowen at Michaelmas, but will be very good, before any of the other good sorts sowen in the Spring, will be ready to be vsed, and beareth white seed.

To instruct a nouice (for I teach not a Gardiner of knowledge) how to gather his seede that it may be good, is in this manner : Let him marke out those plants that hee meaneth shall run vp for seede, which must be the most likely ; & after they haue begun to shoote forth stalkes, strip away the lower leaues, for two or three hands breadth aboue the ground, that thereby in taking away the lowest leaues, the stalke doe not rot nor the seed be hindered in the ripening.

There

There are two manner of wayes to whiten Lettice to make them eate the more tender : the one is by rayfing vp earth like moale hils, round about the plants while they are growing, which will make them grow white : the other is by tying vp all the loofe leaues round together while it groweth, that fo the clofe tying may make it grow white, and thereby be the more tender.

Lambes Lettice or Corne Sallet is a fmall plant while it is young, growing clofe vpon the ground, with many whitifh greene, long and narrow, round pointed leaues, all the winter, and in the beginning of the fpring (if it bee fowen in autumne, as it is vfuall to ferue for an early fallet) rifeth vp with fmall round ftalkes, with two leaues at euery ioynt, branching forth at the toppe, and bearing tufts of fmall bleake blew flowers, which turne into fmall round whitifh feede : the roote is fmall and long, with fome fmall threds hanging thereat : the whole plant is of a waterifh tafte, almoft infipide.

The Vfe of Lettice.

All forts of Lettice are fpent in fallets, with oyle and vinegar, or as euery one pleafe, for the moft part, while they are frefh and greene, or whited, as is declared of fome of the forts before, to caufe them to eate the more delicate and tender. They are alfo boyled, to ferue for many forts of difhes of meate, as the Cookes know beft.

They all coole a hot and fainting ftomacke.

The iuice of Lettice applyed with oyle of Rofes to the foreheads of the ficke and weake wanting fleepe, procureth reft, and taketh away paines in the head : bound likewife to the cods, it helpeth thofe that are troubled with the Colts euill. If a little camphire be added, it reftraineth immoderate luft : but it is hurtfull to fuch as are troubled with the fhortnetfe of breath.

Lambes Lettice is wholly fpent for fallets, in the beginning of the yeare, as I faid, before any almoft of the other forts of Lettice are to be had.

Chap. XXXI.

Portulaca. Purflane.

PVrflane hath many thicke round fhining red ftalkes, full of iuice, lying vpon the ground for the moft part ; whereon are fet diuers long, thicke, pale green leaues, fometimes alone by themfelues, and fometimes many fmall ones together with them ; among which grow fmall yellow flowers, which ftand in little greene huskes, containing blacke feede: the roote is fmall, and perifheth euery yeare, and muft be new fowen in Aprill, in the alleyes of the Garden betweene the beds, as fome haue heretofore vfed, where it may haue the more moifture, or, as I haue feene in fome Gardens, vpon thofe beds of dung that Gardiners haue vfed to nourfe vp their Cowcumbers, Melons, and Pompions, whereon after they haue beene taken away, they haue fowen Purflane, where if it be much watered, the warmth of the dung, and the water giuen it, the Purflane hath grown great and large, and continued vntill winter.

The Vfe of Purflane.

It is vfed as Lettice in fallets, to coole hot and faint ftomackes in the hot time of the yeare, but afterwards if only for delight, it is not good to bee too prodigall in the vfe thereof.

The feede of Purflane doth coole much any inflammation inward or outward, and doth a little binde withall.

Chap.

Chap. XXXII.

Dracoherba siue Tarchon & Dracunculus hortensis. Tarragon.

Tarragon hath long and narrow darke greene leaues, growing on slender and brittle round stalkes, two or three foote high, at the tops whereof grow forth long slender spikes of small yellowish flowers, which seldome giue any good seede, but a dustie or chaffie matter, which flieth away with the winde: the roote is white, and creepeth about vnder ground, whereby it much encreaseth : the whole herbe is of a hot and biting taste.

The Vse of Tarragon.

It is altogether vsed among other cold herbes, to temper their coldnesse, and they to temper its heate, so to giue the better rellish vnto the Sallet ; but many doe not like the taste thereof, and so refuse it.

There are some Authors that haue held Tarragon not to be an herbe of it owne kinde, but that it was first produced, by putting the seede of Lin or Flaxe into the roote of an Onion, being opened and so set into the ground, which when it hath sprung, hath brought forth this herbe Tarragon, which absurd and idle opinion, Matthiolus by certaine experience saith, hath been found false.

Chap. XXXIII.

Nasturtium hortense. Garden Cresses.

Garden Cresses growe vp to the height of two foote or thereabouts, hauing many small, whitish, broad, endented, torne leaues, set together vpon a middle ribbe next the ground, but those that growe higher vpon the stalkes are smaller and longer : the tops of the stalkes are stored with white flowers, which turne into flat pods or pouches, like vnto Shepheard purse, wherein is contained flat reddish seede : the roote perisheth euery yeare : the taste both of leaues and seedes are somewhat strong, hot, and bitter.

The Vse of Cresses.

The Dutchmen and others vse to eate Cresses familiarly with their butter and bread, as also stewed or boyled, either alone or with other herbes, whereof they make a Hotch potch, and so eate it. Wee doe eate it mixed among Lettice or Purslane, and sometimes with Tarragon or Rocket, with oyle and vinegar and a little salt, and in that manner it is very sauoury to some mens stomackes.

The vse of Cresses physically is, it helpeth to expectorate tough flegme, as also for the paines of the breast ; and as it is thought taketh away spots, being laid to with vinegar. The seede is giuen of many to children for the wormes.

1 *Portulaca*. Purflane.   2 *Dracho herba feu Tarchon*. Tarragon.   3 *Eruca fativa*. Garden Rocket.   4 *Nasturtium fativum*. Garden Creſſes.
5 *Sinapi*. Muſtard.   6 *Aſparagus*. Aſparagus or Sperage.

Chap. XXXIIII.

Eruca sativa. Garden Rocket.

OVr Garden Rocket is but a wilde kinde brought into Gardens ; for the true Romane Rocket hath larger leaues ; this hath many long leaues , much torne or rent on the edges, smaller and narrower then the Romane kinde : the flowers hereof are of a pale yellowish colour , whereas the true is whitish, consisting of foure leaues : the seede of this is reddish, contained in smaller and longer pods then the true, which are shorter and thicker, and the seede of a whitish yellow colour : the rootes of both perish as soone as they haue giuen seede. Some haue taken one sort of the wilde kinde for Mustard, and haue vsed the seede for the same purpose.

The Vse of Rocket.

It is for the most part eaten with Lettice, Purslane, or such cold herbes, and not alone , because of its heate and strength ; but that with the white seede is milder. The seede of Rocket is good to prouoke vrine , and to stirre vp bodily lust.

The seede bruised, and mixed with a little vinegar, and of the gall of an Oxe , cleanseth the face of freckles, spots, and blew markes, that come by beatings, fals, or otherwaies.

Matthiolus saith, that the leaues boyled, and giuen with some Sugar to little children, cureth them of the cough.

The seede is held to be helpfull to spleneticke persons ; as also to kill the wormes of the belly.

Chap. XXXV.

Sinapi sativum. Garden Mustard.

THe Mustard that is most vsuall in this Country, howsoeuer diuers doe for their priuate vses sowe it in their Gardens or Orchards, in some conuenient corner, yet the same is found wilde also abroad in many places. It hath many rough long diuided leaues, of an ouerworne greene colour : the stalke is diuided at the toppe into diuers branches, whereon growe diuers pale yellow flowers , in a great length, which turne into small long pods , wherein is contained blackish seede, inclining to rednesse, of a fiery sharpe taste : the roote is tough and white , running deepe into the ground, with many small fibres at it.

The Vse of Mustard.

The seede hereof grownd between two stones, fitted for the purpose, and called a Querne, with some good vinegar added vnto it , to make it liquid and running , is that kinde of Mustard that is vsually made of all sorts , to serue as sawce both for fish and flesh.

The same liquid Mustard is of good vse, being fresh, for Epilepticke persons, to warme and quicken those dull spirits that are sopite and scarce appeare, if it be applyed both inwardly and outwardly.

It is with good successe also giuen to those that haue short breathes , and troubled with a cough in the lungs.

Chap. XXXVI.

Asparagus. Sperage or Asparagus.

ASparagus riseth vp at the first with diuers whitish greene scaly heads, very brittle or easie to breake while they are young, which afterwards rise vp into very long and slender greene stalkes, of the bignesse of an ordinary riding wand at the bottome of most, or bigger or lesser, as the rootes are of growth, on which are set diuers branches of greene leaues, shorter and smaller then Fennell vp to the toppe, at the ioynts whereof come forth small mossie yellowish flowers, which turne into round berries, greene at the first, and of an excellent red colour when they are ripe, shewing as if they were beades of Corrall, wherein are contained exceeding hard and blacke seede: the rootes are dispersed from a spongious head into many long, thicke, and round strings, whereby it sucketh much nourishment out of the ground, and encreaseth plentifully thereby.

We haue another kinde hereof that is of much greater accou̅nt, because the shootes are larger, whiter, and being dressed taste more sweete and pleasant, without any other difference.

The Vse of Asparagus.

The first shootes or heads of Asparagus are a Sallet of as much esteeme with all sorts of persons, as any other whatsoeuer, being boyled tender, and eaten with butter, vinegar, and pepper, or oyle and vinegar, or as euery ones manner doth please; and are almost wholly spent for the pleasure of the pallate. It is specially good to prouoke vrine, and for those that are troubled with the stone or grauell in the reines or kidneyes, because it doth a little open and cleanse those parts.

Chap. XXXVII.

Brassica. Cabbages and Coleworts.

THere is greater diuersity in the forme and colour of the leaues of this plant, then there is in any other that I know groweth vpon the ground. But this place requireth not the knowledge of all sorts which might be shewen, many of them being of no vse with vs for the table, but for delight, to behold the wonderfull variety of the workes of God herein. I will here therefore shew you onely those sorts that are ordinary in most Gardens, and some that are rare, receiued into some especiall Gardens: And first of Cabbages, and then of Coleworts.

Our ordinary Cabbage that closeth hard and round, hath at the first great large thicke leaues, of a grayish greene colour, with thicke great ribbes, and lye open most part of the Summer without closing, but toward the end of Summer, being growne to haue many leaues, it then beginneth to growe close and round in the middle, and as it closeth, the leaues growe white inward; yet there be some kindes that will neuer be so close as these, but will remaine halfe open, which wee doe not account to be so good as the other: in the middle of this head, the next yeare after the sowing, in other Countries especially, and sometimes in ours, if the Winter be milde, as may be seene in diuers Gardens (but to preuent the danger of our Winter frosts, our Gardiners now doe vse to take vp diuers Cabbages with their rootes, and tying a cloth or some such thing about the rootes, doe hang them vp in their houses, where they may be defended from cold, and then set them againe after the frosts are past) and then there shooteth out a great thicke stalke, diuided at the toppe into many branches, bearing thereon diuers small flowers, sometime white, but most commonly yellow, made of foure leaues, which turne into long, round, and pointed pods, containing therein small
round

round feede, like vnto Turnep feede : the roote fpreadeth not farre nor deepe, and dyeth vfually in any great frofte ; for a fmall froft maketh the Cabbage eate the tenderer.

The red Cabbage is like vnto the white, laft fpoken of, but differing in colour and greatneffe ; for it is feldome found fo great as the white, and the colour of the leaues is very variable, as being in fome ftript with red, in others more red, or very deepe red or purple.

The fugar loafe Cabbage, fo called becaufe it is fmaller at the toppe then it is at the bottome, and is of two forts, the one white, the other greene.

The Sauoy Cabbadge, one is of a deepe greene coloured leafe, and curld when it is to be gathered ; the other is yellowifh : neyther of both thefe doe clofe fo well as the firft, but yet are vfed of fome, and accounted good.

The Cole flower is a kinde of Coleworte, whofe leaues are large, and like the Cabbage leaues, but fomewhat fmaller, and endented about the edges, in the middle wherof, fometimes in the beginning of Autumne, and fometimes much fooner, there appeareth a hard head of whitifh yellow tufts of flowers, clofely thruft together, but neuer open, nor fpreading much with vs, which then is fitteft to be vfed, the green leaues being cut away clofe to the head : this hath a much pleafanter tafte then eyther the Coleworte, or Cabbage of any kinde, and is therefore of the more regard and refpect at good mens tables.

The ordinary Coleworte is fufficiently knowne not to clofe or cabbage, and giueth feede plentifully enough.

The other Colewortes that are nourfed vp with thofe that delight in curiofities, befides the aforefaid ordinary greene, which is much vfed of Dutchmen, and other ftrangers, are thefe : The Curld Coleworte eyther wholly of a greene colour, or of diuers colours in one plant, as white, yellow, red, purple or crimfon, fo variably mixed, the leaues being curld on the edges, like a ruffe band, that it is very beautifull to behold.

There is alfo another curld Colewort of leffe beauty and refpect, being but a little curld on the edges, whofe leaues are white, edged with red, or green edged with white.

Two other there are, the one of a popingaye greene colour : the other of a fine deepe greene, like vnto the Sauoyes.

Then there is the Cole rape, which is alfo a kinde of Coleworte, that beareth a white heade, or headed ftalke aboue the ground, as bigge as a reafonable Turnep, but longer, and from the toppe thereof fpringeth out diuers great leaues, like vnto Colewortes ; among which rife diuers ftalkes that beare yellow flowers, and feede in pods, almoft as fmall as Muftard feede : the roote is fomewhat long, and very bufhie with threds.

The Vfe of Cabbages and Colewortes.

They are moft vfually boyled in poudered beefe broth vntill they be tender, and then eaten with much fat put among them.

The great ribs of the Popingay, and deepe greene Colewortes, beeing boyled and layde into difhes, are ferued to the table with oyle and vinegar in the Lent time for very good fallets.

In the cold Countries of Ruffia and Mufcouia, they pouder vp a number of Cabbages, which ferue them, efpecially the poorer forr, for their moft ordinary foode in winter ; and although they ftinke moft grieuoufly, yet to them they are accounted good meate.

It is thought, that the vfe of them doth hinder the milke in Nurfes breafts, caufing it to dry vp quickely : but many women that haue giuen fucke to my knowledge haue denyed that affertion, affirming that they haue often eaten them, and found no fuch effect. How it might proue in more delicate bodies then theirs that thus faid, I cannot tell : but Matthiolus auerreth it to encreafe milke in Nurfes breaftes ; fo differing are the opinions of many. The feede groffely bruifed and boyled a little in fl fh broth, is a prefent remedie for the Collicke ; the feede and the broth being taken together, eafing them that are troubled therewith of all griping paines : as alfo for the ftone in the kidneyes. A Lohoc or licking Electuary made of the pulpe of

the

1 *Brassica capitata.* Cole Cabbage. 2 *Brassica patula.* Open Cabbage. 3 *Brassica Sabaudica crispa.* Curld Sauoye Colewort. 4 *Caulis florida.* Cole flower. 5 *Caulis crispa.* Curld Colewort. 6 *Caulis crispa varieta,* Changeable curld Colewort. 7 *Rapo caum.* Cole rape.

V

the boyled ftalkes, and a little honey and Almond milke, is very profitable for fhortneffe of breath, and thofe that are entring into a Confumption of the lunges. It hath beene formerly held to be helpefull in all difeafes : for Crifippus, an ancient Phyfitian, wrote a whole Volume of the vertues, applying it to all the parts of the body : which thing neede not feeme wonderfull, in that it is recorded by writers, that the old Romanes hauing expelled Phyfitians out of their Common-wealth, did for many hundred of yeares maintaine their health by the vfe of Cabbages, taking them for euery difeafe.

Chap. XXXVIII.

Sifarum. Skirrets.

AFter all the herbes before rehearfed, fit for fallets, or otherwife to bee eaten, there muft follow fuch rootes as are vfed to the fame purpofe : and firft, Skirrets haue many leaues next the ground, compofed of many fmall fmooth green leaues, fet each againft other vpon a middle ribbe, and euery one fnipt about the edges: the ftalke rifeth vp two or three foote high, fet with the like leaues, hauing at the toppe fpoakie tufts of white flowers, which turne into fmall feede, fomewhat bigger and darker then Parfley feede : the rootes be many growing together at one head, beeing long, flender, & rugged or vneuen, of a whitifh colour on the outfide, and more white within, hauing in the middle of the roote a long fmall hard pith or ftring : thefe heads are vfually taken vp in February and March, or fooner if any fo pleafe, the greater number of them being broken off to bee vfed, the reft are planted againe after the heads are feparated, and hereby they are encreafed euery yeare by many ; but it is now adayes more fowen of the feed, which come forwards well enough if the ground be fat and good.

The Vfe of Skirrets.

The rootes being boyled, peeled and pithed, are ftewed with butter, pepper and falt, and fo eaten ; or as others vfe them, to roule them in flower, and fry them with butter, after they haue beene boyled, peeled and pithed: each way, or any way that men pleafe to vfe them, they may finde their tafte to be very pleafant, far beyond any Parfnep, as all agree that tafte them.

Some doe vfe alfo to eate them as a fallet, colde with vinegar, oyle, &c. being firft boyled and dreffed as before faid. They doe helpe to prouoke vrine, and as is thought, to procure bodily luft, in that they are a little windy.

Chap. XXXIX.

Paftinaca fatiua latifolia. Parfneps.

THe common garden Parfnep hath diuers large winged leaues lying vpon the ground, that is, many leaues fet one by another on both fides of a middle ftalk, fomewhat like as the Skirret hath, but much larger, and clofer fet: the ftalke rifeth vp great and tall, fiue or fix foot high fomtimes, with many fuch leaues thereon at feuerall ioynts ; the top whereof is fpread into diuers branches, whereon ftand fpoakie rundles of yellow flowers, which turne into brownifh flat feede : the root is long, great and white, very pleafant to bee eaten, and the more pleafant if it grow in a fat fandy foyle.

There is another fort of garden Parfnep, called the Pine Parfnep, that is not common in euery Garden, and differeth from the former in three notable parts. The root is not fo long, but thicker at the head and fmaller below ; the ftalke is neither fo bigge,

nor

1 *Sifarum.* Skirrits. 2 *Paſtinaca latifolia.* Parſneps. 3 *Paſtinaca tenuifolia.* Carrets. 4 *Napum.* Turneps. 5 *Napus alvus.* Navewes.
6 *Raphanus niger.* Blacke Raddiſh. 7 *Raphanus vulgaris.* Common-Raddiſh.

not so high; and the seede is smaller: yet as Iohn Tradescante saith (who hath giuen me the relation of this, and many other of these garden plants, to whom euery one is a debtor) the roote hereof is not altogether so pleasant as the other.

Moreouer the wilde kinde, which groweth in many places of England(and wherof in some places there might be gathered a quarter sacke full of the seede) if it be sowen in Gardens, and there well ordered, will proue as good as the former kinde of Garden Parsneps.

The Vse of Parsneps.

The Parsnep root is a great nourisher, and is much more vsed in the time of Lent, being boyled and stewed with butter, then in any other time of the yeare; yet it is very good all the winter long. The seede helpeth to dissolue winde, and to prouoke vrine.

CHAP. XL.

Pastinaca satina tenuifolia. Carrots.

THe Carrot hath many winged leaues, rising from the head of the roote, which are much cut and diuided into many other leaues, and they also cut and diuided into many parts, of a deepe greene colour, some whereof in Autumne will turne to be of a fine red or purple (the beautie whereof allureth many Gentlewomen oftentimes to gather the leaues, and sticke them in their hats or heads, or pin them on their armes in stead of feathers): the stalke riseth vp among the leaues, bearing many likewise vpon it, but nothing so high as the Parsnep, being about three foote high, bearing many sporkie tufts of white flowers, which turne into small rough seede, as if it were hairy, smelling reasonable well if it bee rubbed: the roote is round and long, thicke aboue and small below, eyther red or yellow, eyther shorter or longer, according to his kinde; for there is one kinde, whose roote is wholly red quite throughout; another whose roote is red without for a pretty way inward, but the middle is yellow.

Then there is the yellow, which is of two sorts, both long and short: One of the long yellow sorts, which is of a pale yellow, hath the greatest and longest roote, and likewise the greatest head of greene, and is for the most part the worst, being spongy, and not firme.

The other is of a deepe gold yellow colour, and is the best, hauing a smaller head, or tuft of greene leaues vpon it.

The shorte rootes are likewise distinguished, into pale and deepe yellow colours.

The Vse of Carrots.

All these sorts being boyled in the broth of beefe, eyther fresh or salt, but more vsually of salted beefe, are eaten with great pleasure, becaufe of the sweetenesse of them: but they nourish lesse then Parsneps or Skirrets.

I haue not often knowne the seede of this Garden kinde to bee vsed in Physicke: but the wilde kinde is often and much vsed to expell winde, &c.

CHAP. XLI.

Rapum hortense. Turneps.

THere are diuers sorts of Turneps, as white, yellow, and red: the white are the most common, and they are of two kinds, the one much sweeter then the other. The yellow and the red are more rare, and nourished vp only by those that are curious: as also the Navewe, which is seene but with very few.

The

The ordinary Garden Turnep hath many large, and long rough greene leaues, with deepe and vneuen gashes on both sides of them : the stalke riseth vp among the leaues about two foote high, spread at the toppe into many branches, bearing theron yellow flowers, which turne into long pods, with blackish round seede in them : the roote is round and white, some greater, some smaller, the best kinde is knowne to be flat, with a small pigges tale-like roote vnderneath it ; the worser kinde which is more common in many places of this land, both North and West, is round, and not flat, with a greater pigges tayle-like roote vnderneath.

The yellow kinde doth often grow very great, it is hardly discerned from the ordinary kinde while it groweth, but by the greatnesse and spreading of the leaues beeing boyled, the roote changeth more yellow, somewhat neare the colour of a Carrot.

The red Turnep groweth vsually greater then any of the other, especially in a good ground, being of a faire red colour on the outside, but being pared, as white as any other on the inside. This, as Matthiolus saith, doth grow in the Countrey of Anania, where hee hath seene an infinite number of them that haue weighed fifty pound a peece, and in some places hee saith, a hundred pound a peece, both which we would thinke to be incredible, but that we see the kind is greatly giuen to grow, and in warme Countries they may so thriue, that the bulke or bignesse of the roote may so farre passe the growth of our Countrey, as that it may rise to that quantity aboue specified.

The Navew gentle is of two kindes, a smaller and a greater ; the smaller is vsually called in France, *Naveau de Cane,* the roote is somewhat long with the roundnesse ; this kinde is twice as bigge as a mans thumbe, and many of them lesse : The other is long and great, almost as big as the short Carrot, but for the most part of an vneuen length, and roundnesse vnto the very end, where it spreadeth into diuers small long fibres : neyther of them doth differ much from the Turnep, in leafe, flower or seed.

The Vse of Turneps.

Being boyled in salt broth, they all of them eate most kindly, and by reason of their sweetnesse are much esteemed, and often seene as a dish at good mens tables : but the greater quantitie of them are spent at poore mens feasts. They nourish much, and engender moist and loose flesh, and are very windy. The seede of the Navew gentle is (as I take it) called of Andromachus in the composition of his Treakle, *Bunias dulcis* : for Dioscorides and Plinie doe both say, that the seede of the tame Bunias or Napus is put into Antidotes, and not the seede of the wilde, which is more sharpe and bitter ; neyther the seede of the Turnep, which is called in Greeke γογγύλη in Latine *Rapum*, becaufe the seede is not sweete.

CHAP. XLII.

Raphanus. Raddish.

THere are two principall kindes of Garden Raddish, the one is blackish on the outside, and the other white ; and of both these there is some diuision againe, as shall be shewed. Dittander and horse Raddish be reckoned kinds thereof.

The ordinary Raddish hath long leaues, vneuenly gashed on both sides, the stalke riseth vp to the height of three or foure foote, bearing many purplish flowers at the top, made of foure leaues a peece, which turne into thicke and short pods, wherein are contained round seede, greater then Turnep or Coleworte seede, and of a pale reddish colour : the roote is long, white, and of a reddish purple colour on the outside toward the toppe of it, and of a sharpe biting taste.

There is a small kind of Raddish that commeth earlier then the former, that we haue had out of the low Countries, not differing in any thing else.

The blacke Raddish I haue had brought me out of the lowe Countries, where they sell them in some places by the pound, and is accounted with them a rare winter sallet:

the

the roote of the beſt kinde is blackiſh on the outſide (and yet the ſeede gathered from ſuch an one, hath after the ſowing againe, giuen rootes, whereof ſome haue beene blacke, but the moſt part white on the outſide) and white within, great and round at the head, almoſt like a Turnep, but ending ſhorter then a Raddiſh, and longer then a Turnep, almoſt peare-faſhion, of a firmer and harder ſubſtance then the ordinary Raddiſh, but no leſſe ſharpe and biting, and ſomewhat ſtrong withall; the leaues are ſomewhat ſmaller, and with deeper gaſhes, the flower and ſeede are like the former, but ſmaller.

Another ſort of blacke Raddiſh is like in leafe and ſeede to the former, but the flower is of a lighter purple colour : the roote is longer and ſmaller, and changeth alſo to bee white as the former doth, ſo that I thinke they haue both riſen from one kinde.

The Horſe Raddiſh is a kinde of wilde Raddiſh, but brought into Gardens for the vſe of it, and hath great large and long greene leaues, which are not ſo much diuided, but dented about the edges : the roote is long and great, much ſtronger in taſte then the former, and abideth diuers yeares, ſpreading with branches vnder ground.

Dittander is likewiſe a wilde kinde hereof, hauing long pointed blewiſh greene leaues, and a roote that creepeth much vnder ground : I confeſſe this might haue bin placed among the herbes, becauſe the leaues and not the rootes are vſed; but let it paſſe now with the kindes of Raddiſh.

The Vſe of theſe Raddiſhes.

Raddiſhes doe ſerue vſually as a *ſimulum* before meat, giuing an appetite thereunto ; the poore eate them alone with bread and ſalt. Some that are early ſowen, are eaten in Aprill, or ſooner if the ſeaſon permit ; others come later ; and ſome are ſowen late to ſerue for the end of Summer : but (as of all things elſe) the earlier are the more accepted.

The blacke Raddiſhes are moſt vſed in the winter, (yet ſome in their naturall and not forc'd grounds, haue their rootes good moſt part of the Summer) and therefore muſt bee ſowen after Midſomer ; for if they ſhould bee ſowen earlier, they would preſently runne vp to ſtalke and ſeed, and ſo loſe the benefit of the roote. The Phyſicall propertie is, it is often vſed in medicines that helpe to breake the ſtone, and to auoyde grauell.

The Horſe Raddiſh is vſed Phyſically, very much in Melancholicke, Spleneticke and Scorbuticke diſeaſes. And ſome vſe to make a kinde of Muſtard with the rootes, and eate it with fiſh.

Dittander or Pepperworte is vſed of ſome cold churliſh ſtomackes, as a ſawce or ſallet ſometimes to their meate, but it is too hot, bitter and ſtrong for weake and tender ſtomackes.

Our Gardiners about London vſe great fences of reede tyed together, which ſeemeth to bee a mat ſet vpright, and is as good as a wall to defend the cold from thoſe things that would be defended, and to bring them forwards the earlier.

Chap. XLIII.

Cepa. Onions.

Ee haue diuers ſorts of Onions, both white and red, flat, round and long, as ſhall be preſently ſhewed : but I will doe with theſe as I doe with the reſt, only giue you one deſcription for them all, and afterwards their ſeuerall names and varieties, as they are to be knowen by.

Our common Garden Onion hath diuers long greene hollow leaues, ſeeming halfe flat ; among which riſeth vp a great round hollow ſtalke, bigger in the middle then any where elſe, at the toppe whereof ſtandeth a cloſe round head, couered at the firſt with a thin skinne, which breaketh when the head is growne, and ſheweth forth a great vmbell

1 *Raphanus rusticanus.*|Horse Raddish. 2 *Lepidium siue Piperitis.* Dittander. 3 *Cepa rotunda.* Round Onions. 4 *Cepa longa.* Long Onions. 5 *Porrum.* Leekes 6 *Allium.* Garlicke. 7 *Rapunculus.* Rampions. 8 *Tragopogon.* Goates beard.

bell of white flowers, which turne into blacke feede : but then the head is fo heauie that the ftalke cannot fuftaine it, but muft be vpheld from falling to the ground, left it rot and perifh : the roote as all know is round, in fome greater, in others leffer, or flat, in fome red on the outfide only, in others quite thorough out, in fome white, and very fharpe and ftrong, in others milder, and more pleafant, and fome fo pleafant that they may be eaten as an Apple : All thefe kindes of Onions, contrary to the nature of all other bulbous rootes, haue no off-fet, or other roote growing to it, but are euery one alone fingle by themfelues ; and therefore it feemeth, the Latines, as Columella recordeth, haue giuen it the name *Vnio*, and the French it fhould feeme following the Latine, and the Englifh the French, do call it *Oignon* and *Onion*, as an vnite, or as if they were but one and one, and dye euery yeare after feed bearing.

The red flat kinde is moft vfually with vs the ftrongeft of them all, yet I haue had a great red Onion brought mee from beyond Sea, that was as great almoft as two mens fiftes, flat and red quite thoroughout, and very pleafant both to fmell vnto, and to eate, but did quickly degenerate; fo that we plainly fee, that the foyle and climate doth giue great alteration to plants of all forts.

The long kinde wee call St. Omers Onions, and corruptly among the vulgar, St. Thomas Onions.

The other red kinde we call Strasborough Onions, whofe outfide onely is red, and are very fharpe and fierce.

The white Onions both long and flat, are like vnto Chalke-ftones lying vpon the ground, when they are ripe and fit to be gathered.

And laftly, there is the Spanifh Onion, both long and flat, very fweete, and eaten by many like an apple, but as Iohn Tradefcante faith, who hath beene in Spaine, that the Spaniards themfelues doe not eate them fo familiarly, as they doe thofe white Onions that come out of our owne Countrey, which they haue there more plentifully then their fweete Onions.

The Vfe of Onions.

Onions are vfed many wayes, as fliced and put into pottage, or boyled and peeled and layde in difhes for fallets at fupper, or fliced and put into water, for a fawce for mutton or oyfters, or into meate roafted being ftuffed with Parfly, and fo many waies that I cannot recount them, euery one pleafing themfelues, according to their order, manner or delight.

The iuice of Onions is much vfed to be applyed to any burnings with fire, or with Gun-pouder, or to any fcaldings with water or oyle, and is moft familiar for the Country, where vpon fuch fudden occafions they haue not a more fit or fpeedy remedie at hand : The ftrong fmell of Onions, and fo alfo of Garlicke and Leekes, is quite taken away from offending the head or eyes, by the eating of Parfley leaues after them.

CHAP. XLIIII.

Porrum. Leekes.

THere be likewife fundry forts of Leekes, both great and fmall. Leekes are very like vnto Onions, hauing long green hollow-like leaues, flattifh on the one fide, and with a ridge or creft on the backe fide : if they bee fuffered to grow vncut, then in the fecond or third yeare after the fowing, they will fend forth a round and flender ftalke, euen quite thoroughout, and not fwollen or bigger in the middle like the Onion, bearing at the toppe a head of purplifh flowers, and blacke feede after them, very like vnto Onion feede, that it is hard to diftinguifh them : the root is long and white, with a great bufh of fibres hanging at it, which they call the beards.

The vnfet Leeke hath longer and flenderer roots then the other, which being tranfplanted, groweth thicker and greater.

The

The French Leeke, which is called the Vine Leeke, is the best of all others.
Our common kinde is of two sorts, one greater then another.
Another sort encreaseth altogether by the roote, as Garlicke doth.
And then Ciues, which are the smallest, and encrease aboundantly only by the root.
Some doe account Scalions to be rather a kinde of Onions then Leekes, and call them *Cepa Afcalonica,* or *Afcalonitides,* which will quickly spend it selfe, if it be suffered to be vncut ; but all Authors affirme, that there is no wilde kinde of Onion, vnlesse they would haue it to be *Gethyum,* whereof Theophrastus maketh mention, saying, that it hath a long necke (and so these Scalions haue) and was also of some called *Gethyllides,* which antiquity accounted to be dedicated to Latona, the mother of Apollo, becaufe when she was bigge with childe of Apollo, she longed for these Leekes.

The Vfe of Leekes.

The old World, as wee finde in Scripture, in the time of the children of Israels being in Egypt, and no doubt long before, fed much vpon Leekes, Onions, and Garlicke boyled with flesh ; and the antiquity of the Gentiles relate the same manner of feeding on them, to be in all Countries the like, which howsoeuer our dainty age now refuseth wholly, in all sorts except the poorest ; yet Mufcouia and Ruffia vse them, and the Turkes to this day, (as Bellonius writeth) obferue to haue them among their dishes at their tables, yea although they be *Bafbas, Cades,* or *Vaiuodas,* that is to say, Lords, Iudges, or Gouernours of countries and places. They are vsed with vs also sometimes in Lent to make pottage, and is a great and generall feeding in Wales with the vulgar Gentlemen.
Onions boyled or rosted vnder the embers, and mixed with sugar and butter, are good for those that are troubled with coughes, shortnesse of breath, and wheesing. An Onion made hollow at the bottome, and some good Treakle put into it, with a little iuyce of Citrons (or Lemons in the stead thereof) being well baked together vnder the embers, after the hole is stopped againe, and then strained forth, and giuen to one that hath the plague, is very helpefull, so as hee be laid to sweate vpon it.
Ciues are vsed as well to be shred among other herbes for the pot, as to be put into a Sallet among other herbs, to giue it a quicker rellish.
Leekes are held to free the chest and lungs from much corruption and rotten flegme, that sticketh fast therein, and hard to be auoided, as also for them that through hoarfenesse haue lost their voice, if they be eyther taken rawe, or boyled with broth of barley, or some such other supping, fit and conducing thereunto. And baked vnder hot embers is a remedy against a surfeit of Mushromes.
The greene blades of Leekes being boyled and applyed warme to the *Hemorrhoides* or piles, when they are swolne and painfull, giue a great deale of ease.

Chap. XLV.

Allium. Garlicke.

I Haue spoken of diuers sorts of Garlicke called Moly, in the former booke : I shall neede in this place to shew onely those kindes, that this Garden nourseth vp, and leaue the rest to his fit time and place.
Garlicke hath many long greene leaues, like vnto Onions, but much larger, and not hollow at all as Onions are : the stalke riseth vp to be about three foote high, bearing such a head at the toppe thereof as Onions and Leekes doe, with purplish flowers, and blacke feede like Leekes : the roote is white within, couered ouer with many pur-

plifh skins, and is diuided into many parts or cloues, which ferue both to fet againe for increafe, and alfo to vfe as neede fhall require, and is of a very ftrong fmell and tafte, as euery one knoweth, paffing either Onions or Leekes, but exceeding wholfome withall for them that can take it.

Allium Vrfinum. Ramfons.

Ramfons are another kinde of Garlicke, and hath two or three faire broad leaues, of a frefh or light greene colour, pointed at the end : the ftalke groweth about an hand length high, bearing many fmall and pure white ftarre-like flowers at the toppe, and afterwards fmall, blacke, and fmooth round feede : the roote is alfo diuided into many parts, whereby it is much encreafed, and is much milder then the former, both in fmell and tafte.

The Vfe of Garlicke.

It being well boyled in falt broth, is often eaten of them that haue ftrong ftomackes, but will not brooke in a weake and tender ftomacke.

It is accounted, and fo called in diuers Countries, The poore mans Trea-kle, that is, a remedy for all difeafes. It is neuer eaten rawe of any man that I know, as other of the rootes aforefaid, but fodden alwaies and fo taken.

Ramfons are oftentimes eaten with bread and butter, and otherwife alfo, as euery mans affection and courfe of life leadeth him to vfe.

Chap. XLVI.

Rapunculus fiue Rapuntium. Rampions.

GArden Rampions are of two forts, the one greater, the other leffer : the leaues of Rampions are in the one fomewhat broad like a Beete, in the other fomewhat long and narrow, and a little broader at the end, of a light greene colour, lying flat vpon the ground all the firft winter, or yeare of the fpringing, and the next Spring fhooteth forth ftalkes two or three foote high, bearing at the toppe, in the bigger fort, a long flender fpike of fmall horned or crooked flowers, which open their brimmes into foure leaues ; in the leffer many fmall purplifh bels, ftanding vpon feuerall fmall foote-ftalkes, which turne into heads, bearing fmall blackifh feede : the root is white, branched into two or three rootes, of the bigneffe and length of a mans finger or thumbe.

The Vfe of Rampions.

The rootes of both are vfed for Sallets, being boyled, and then eaten with oyle and vinegar, a little falt and pepper.

Chap. XLVII.

Tragopogon. Goates beard.

GOates beard hath many long and narrow leaues, broader at the bottome, and fharper at the end, with a ridge downe the backe of the leafe, and of a pale greene colour ; among which rifeth vp a ftalke of two or three foote high, fmooth and hollow, bearing thereon many fuch like leaues, but fmaller and fhorter, and at the toppe thereof on euery branch a great double yellow flower, like almoft vnto the flower of a Dandelion, which turneth into a head, ftored with doune, and long whitifh feede therein, hauing on the head of euery one fome part of the doune,

and

and is carried away with the winde: if it bee neglected : the roote is long and round, somewhat like vnto a Parſnep, but farre ſmaller, blackiſh on the outſide, and white within, yeelding a milkie iuyce being broken, as all the reſt of the plant doth, and of a very good and pleaſant taſte. This kinde, as alſo another with narrower leaues, almoſt like graſſe, growe wilde abroad in many places, but are brought into diuers Gardens. The other two kindes formerly deſcribed in the firſt part, the one with a purple flower, and the other with an aſh-coloured, haue ſuch rootes as theſe here deſcribed, and may ſerue alſo to the ſame purpoſe, being of equall goodneſſe, if any will vſe them in the ſame manner ; that is, while they are young, and of the firſt yeares ſowing; elſe they all growe hard, in running vp to ſeede.

The Vſe of Goates beard.

If the rootes of any of theſe kindes being young, be boyled and dreſſed as a Parſnep, they make a pleaſant diſh of meate, farre paſſing the Parſnep in many mens iudgements, and that with yellow flowers to be the beſt.

They are of excellent vſe being in this manner prepared, or after any other fit and conuenient way, to ſtrengthen thoſe that are macilent, or growing into any conſumption.

Chap. XLVIII.

Carum. Carawayes.

CArawayes hath many very fine cut and diuided leaues lying on the ground, being alwaies greene, ſomewhat reſembling the leaues of Carrots, but thinner, and more finely cut, of a quicke, hot, and ſpicie taſte: the ſtalke riſeth not much higher then the Carrot ſtalke, bearing ſome leaues at the ioynts along the ſtalke to the toppe, where it brancheth into three or foure parts, bearing ſpoakie vmbels of white flowers, which turne into ſmall blackiſh ſeede, ſmaller then Aniſeede, and of a hotter and quicker taſte : the roote is whitiſh, like vnto a Parſnep, but much ſmaller, more ſpreading vnder ground, and a little quicke in taſte, as all the reſt of the plant is, and abideth long after it hath giuen ſeede.

The Vſe of Carawayes.

The rootes of Carawayes being boyled may be eaten as Carrots, and by reaſon of the ſpicie taſte doth warme and comfort a cold weake ſtomacke, helping to diſſolue winde (whereas Carrots engender it) and to prouoke vrine, and is a very welcome and delightfull diſh to a great many, yet they are ſomewhat ſtronger in taſte then Parſneps.

The ſeede is much vſed to bee put among baked fruit, or into bread, cakes, &c. to giue them a relliſh, and to helpe to digeſt winde in them are ſubiect thereunto.

It is alſo made into Comfits, and put into *Tragea*, or as we call them in Engliſh, Dredges, that are taken for the cold and winde in the body, as alſo are ſerued to the table with fruit.

Chap. XLIX.

Papas sine Battatas. Potatoes.

THree sorts of Potatoes are well knowne vnto vs, but the fourth I rest doubtfull of, and dare not affirme it vpon such termes as are giuen vnto it, vntill I may be better informed by mine owne sight.

The Spanish kinde hath (in the Ilands where they growe, either naturally, or planted for increase, profit, and vse of the Spaniards that nourse them) many firme and verie sweete rootes, like in shape and forme vnto Asphodill rootes, but much greater and longer, of a pale browne on the outside, and white within, set together at one head; from whence rise vp many long branches, which by reason of their weight and weaknesse, cannot stand of themselues, but traile on the ground a yard and a halfe in length at the least (I relate it, as it hath growne with vs, but in what other forme, for flower or fruit, we know not) whereon are set at seuerall distances, broad and in a manner three square leaues, somewhat like triangled Iuie leaues, of a darke greene colour, the two sides whereof are broad and round, and the middle pointed at the end, standing reasonable close together: thus much we haue seene growe with vs, and no more: the roote rather decaying then increasing in our country.

The Potatoes of Virginia, which some foolishly call the Apples of youth, is another kinde of plant, differing much from the former, sauing in the colour and taste of the roote, hauing many weake and somewhat flexible branches, leaning a little downwards, or easily borne downe with the winde or other thing, beset with many winged leaues, of a darke grayish greene colour, whereof diuers are smaller, and some greater then others: the flowers growe many together vpon a long stalke, comming forth from betweene the leaues and the great stalkes, euery one seuerally vpon a short foot-stalke, somewhat like the flower of Tabacco for the forme, being one whole leafe six cornered at the brimmes, but somewhat larger, and of a pale blewish purple colour, or pale doue colour, and in some almost white, with some red threads in the middle, standing about a thicke gold yellow pointell, tipped with greene at the end: after the flowers are past, there come vp in their places small round fruit, as bigge as a Damson or Bulleis, greene at the first, and somewhat whitish afterwards, with many white seedes therein, like vnto Nightshade: the rootes are rounder and much smaller then the former, and some much greater then others, dispersed vnder ground by many small threads or strings from the rootes, of the same light browne colour on the outside, and white within, as they, and neare of the same taste, but not altogether so pleasant.

The Potatos of Canada, (which hath diuers names giuen it by diuers men, as Bauhinus vpon Matthiolus calleth it, *Solanum tuberosum esculentum*, Pelleterius of Middleborough in his *Plantarum Synonimia, Heliotropium Indicum tuberosum*, Fabius Columna in the second part of his *Phytobasanos, Flos Solis Farnesianus, siue Aster Peruanus tuberosus*: We in England, from some ignorant and idle head, haue called them Artichokes of Ierusalem, only because the roote, being boyled, is in taste like the bottome of an Artichoke head: but they may most fitly be called, Potatos of Canada, because their rootes are in forme, colour and taste, like vnto the Potatos of Virginia, but greater, and the French brought them first from Canada into these parts) riseth vp with diuers stiffe, round stalkes, eight or tenne foote high in our Country, where they haue scarce shewed their flowers, whereas the very head of flowers in other Countries, as Fabius Columna expresseth it, being of a Pyramis or Sugar loafe fashion, broade spreading below, and smaller pointed vpwards towards the toppe, is neere of the same length, whereon are set large and broade rough greene leaues, very like vnto the leaues of the flower of the Sunne, but smaller, yet growing in the very same manner, round about the stalkes: at the very later end of Summer, or the beginning of Autumne, if the roote bee well planted and defended, it will giue a shew of a few small yellow flowers at the top, like vnto the flowers of *Aster* or Starre-worte, and much smaller then any flower of the Sunne, which come to no perfection with vs: the roote, while the plant

is

1 _Carum._ Carawayes. 2 _Battatas Hispanorum._ Spanish Potatoes. 3 _Papas seu Battatas Virginianorum._ Virginia Potatoes. 4 _Battatas de Canada._ Potatoes of Canada, or Artichokes of Ierusalem.

is growing aboue ground, encreaseth not to his full growth, but when the Summer is well spent, and the springing of the stalk is past, which is about the end of August, or in September, then the root is perceiued to be encreased in the earth, and will before Autumne be spent, that is, in October, swell like a mound or hillocke, round about the foote of the stalkes, and will not haue his rootes fit to be taken vp, vntill the stalkes be halfe withered at the soonest ; but after they be withered, and so all the winter long vntill the Spring againe, they are good, and fit to bee taken vp and vsed, which are a number of tuberous round rootes, growing close together ; so that it hath beene obserued, that from one roote, being set in the Spring, there hath been forty or more taken vp againe, and to haue ouer-filled a pecke measure, and are of a pleasant good taste as many haue tryed.

The Vse of all these Potato's.

The Spanish Potato's are roasted vnder the embers, and being pared or peeled and sliced, are put into sacke with a little sugar, or without, and is delicate to be eaten.

They are vsed to be baked with Marrow, Sugar, Spice, and other things in Pyes, which are a daintie and costly dish for the table.

The Comfit-makers preserue them, and candy them as diuers other things, and so ordered, is very delicate, fit to accompany such other banquetting dishes.

The Virginia Potato's being dressed after all these waies before specified, maketh almost as delicate meate as the former.

The Potato's of Canada are by reason of their great increasing, growne to be so common here with vs at London, that euen the most vulgar begin to despise them, whereas when they were first receiued among vs, they were dainties for a Queene.

Being put into seething water they are soone boyled tender, which after they bee peeled, sliced and stewed with butter, and a little wine, was a dish for a Queene, beeing as pleasant as the bottome of an Artichoke : but the too frequent vse, especially being so plentifull and cheape, hath rather bred a loathing then a liking of them.

Chap. L.

Cinara. Artichokes.

THe fruits that grow vpon or neere the ground, are next to be entreated of, and first of Artichokes, whereof there be diuers kindes, some accounted tame and of the Garden, others wilde and of late planted in Gardens, Orchards or Fieldes, of purpose to be meate for men.

The Artichoke hath diuers great, large, and long hollowed leaues, much cut in or torne on both edges, without any great shew of prickles on them, of a kinde of whitish greene, like vnto an ash colour, whereof it tooke the Latine name *Cinara* : the stalke is strong, thicke and round, with some skins as it were downe all the length of them, bearing at the toppe one scaly head, made at the first like a Pine-apple, but after growing greater, the scales are more separate, yet in the best kindes lying close, and not staring, as some other kindes doe, which are eyther of a reddish browne, whitish, or greenish colour, and in some broade at the ends, in others sharpe or prickly : after the head hath stood a great while, if it bee suffered, and the Summer proue hot and kindly, in some there will breake forth at the toppe thereof, a tuft of blewish purple thrumes or threds, vnder which grow the seede, wrapped in a great deale of dounie substance : but that roote that yeeldeth flowers will hardly abide the next winter ; but else being cut off when it is well growne, that dounie matter abideth close in the middle of the head, hauing the bottome thereof flat and round, which is that in vtter or substance that is vsed to be eaten : the roote spreadeth it selfe in the ground reasona-

ble

1 *Cinara satina rubra.* The red Artichoke. 2 *Cinara satina alba.* The white Artichoke. 3 *Cinara spinosa.* The French Artichoke. 4 *Cinara sylvestris.* The Thistle Artichoke. 5 *Carduus esculentus.* The Chardon.

ble well, yeelding diuers heads of leaues or suckers, whereby it is increased.

The white Artichoke is in all things like the red, but that the head is of a whitish ashe colour, like the leaues, whereas the former is reddish.

We haue also another, whose head is greene, and very sharpe vpwards, and is common in many places.

Wee haue had also another kinde in former times that grew as high as any man, and branched into diuers stalkes, euery one bearing a head thereon, almost as bigge as the first.

There is another kinde, called the Muske Artichoke, which groweth like the French kinde, but is much better in spending, although it haue a lesser bottome.

The French Artichoke hath a white head, the scales whereof stand staring far asunder one from another at the ends, which are sharpe: this is well known by this qualitie, that while it is hot after it is boyled, it swelleth so strong, that one would verily thinke it had bin boyled in stinking water, which was brought ouer after a great froste that had well nigh consumed our best kindes, and are now almost cleane cast out again, none being willing to haue it take vp the roome of better.

There is a lowe kinde that groweth much about Paris, which the French esteeme more then any other, and is lower then the former French kinde, the head whereof as well as the leaues, is of a fresher greene colour, almost yellowish.

Then there is the Thistle Artichoke, which is almost a wilde kinde, and groweth smaller, with a more open and prickly head then any of the former.

And lastly, the Chardon as they call it, because it is almost of the forme and nature of a Thistle, or wilde Artichoke. This groweth high, and full of sharpe prickles, of a grayish colour. Iohn Tradescante assured mee, hee saw three acres of Land about Brussels planted with this kinde, which the owner whited like Endiue, and then sold them in the winter: Wee cannot yet finde the true manner of dressing them, that our Countrey may take delight therein.

All these kindes are encreased by slipping the young shootes from the root, which being replanted in February, March, or Aprill, haue the same yeare many times, but the next at the most, borne good heads.

Wee finde by dayly experience, that our English red Artichoke is in our Countrey the most delicate meate of any of the other, and therefore diuers thinking it to bee a seuerall kinde, haue sent them into Italie, France, and the Lowe Countries, where they haue not abode in their goodnesse aboue two yeare, but that they haue degenerated; so that it seemeth, that our soyle and climate hath the preheminence to nourish vp this plant to his highest excellencie.

The Vse of Artichokes.

The manner of preparing them for the Table is well knowne to the youngest Housewife I thinke, to bee boyled in faire water, and a little salt, vntill they bee tender, and afterwardes a little vinegar and pepper, put to the butter, poured vpon them for the sawce, and so are serued to the Table.

They vse likewise to take the boyled bottomes to make Pyes, which is a delicate kinde of baked meate.

The Chardon is eaten rawe of diuers, with vinegar and oyle, pepper and salt, all of them, or some, as euery one liketh for their delight.

CHAP. LI.

Faba & Phaseoli. Garden and French Beanes.

THe Garden Beane is of two colours, red or blacke, and white, yet both rise from one; the small or fielde Beanes I make no mention of in this place; but the French or Kidney Beane is almost of infinite sorts and colours: we doe not for all that intend to trouble you in this place, with the knowledge or relation of any more then is fit for a Garden of that nature, that I haue propounded it in the beginning.

Our ordinary Beanes, seruing for foode for the poorer sort for the most part, are planted as well in fieldes as in gardens, because the quantity of them that are spent taketh vp many acres of land to be planted in, and rise vp with one, two or three stalks, according to the fertilitie of the soyle, being smooth and square, higher then any man oftentimes, whereon are set at certaine distances, from the very bottome almost to the toppe, two long smooth fleshy and thicke leaues almost round, one standing by another at the end of a small footestalke : betweene these leaues and the stalke, come forth diuers flowers, all of them looking one way for the most part, which are close a little turned vp at the brimmes, white and spotted with a blackish spot in the middle of them, and somwhat purplish at the foot or bottome, of the forme almost of Broome or Pease flowers, many of which that grow vpward toward the toppe, doe seldome beare fruit, and therefore are gathered to distill, and the toppes of the stalkes cut off, to cause the rest to thriue the better ; after which grow vp long great smooth greene pods, greater then in any other kinde of Pulse, which grow blacke when they are ripe, and containe within them two, three or foure Beanes, which are somewhat flat and round, eyther white or reddish, which being full ripe grow blackish : the roote hath diuers fibres annexed vnto the maine roote, which dyeth euery yeare.

The French or Kidney Beane riseth vp at the first but with one stalke, which afterwards diuideth it selfe into many armes or branches, euery one of them being so weak, that without they be sustained with stickes or poles, whereon with their winding and claspers they take hold, they would lye fruitlesse vpon the ground : vpon these branches grow forth at seuerall places long footestalkes, with euery of them three broade, round and pointed greene leaues at the end of them, towards the tops whereof come forth diuers flowers, made like vnto Pease blossomes, of the same colour for the most part that the fruit will be of, that is to say, eyther white, or yellow, or red, or blackish, or of a deepe purple &c. but white is most vsuall for our Garden ; after which come long and slender flat pods, some crooked, and some straight, with a string as it were running downe the backe thereof, wherein are contained flattish round fruit, made to the fashion of a kidney : the roote is long, and spreadeth with many fibres annexed vnto it, perishing euery yeare.

The Vse of these Beanes.

The Garden Beanes serue (as I said before) more for the vse of the poore then of the rich : I shall therefore only shew you the order the poore take with them, and leaue curiosity to them that will bestow time vpon them. They are only boyled in faire water and a little salt, and afterwards stewed with some butter, a little vinegar and pepper being put vnto them, and so eaten : or else eaten alone after they are boyled without any other sawce. The water of the blossomes distilled, is vsed to take away spots, and to cleer the skin. The water of the greene huskes or cods is good for the stone.

The Kidney Beanes boyled in water huske and all, onely the ends cut off, and the string taken away, and stewed with butter &c. are esteemed more sauory meate to many mens pallates, then the former, and are a dish more oftentimes at rich mens Tables then at the poore.

CHAP.

Chap. LII.

Pisum. Pease.

THere is a very great variety of manured Pease known to vs, and I think more in our Country then in others, whereof some prosper better in one ground and country, and some in others: I shall giue you the description of one alone for all the rest, and recite vnto you the names of the rest.

Garden Pease are for the most part the greatest and sweetest kinds, and are sustained with stakes or bushes. The Field Pease are not so vsed, but growe without any such adoe. They spring vp with long, weake, hollow, and brittle (while they are young and greene) whitish greene stalkes, branched into diuers parts, and at euery ioynt where it parteth one broad round leafe compassing the stalke about, so that it commeth as it were thorough it: the leaues are winged, made of diuers small leaues set to a middle ribbe, of a whitish greene colour, with claspers at the ends of the leaues, whereby it taketh hold of whatsoeuer standeth next vnto it: betweene the leaues and the stalkes come forth the flowers, standing two or three together, euery one by it selfe on his owne seuerall stalke, which are either wholly white, or purple, or mixed white and purple, or purple and blew: the fruit are long, and somewhat round cods, whereof some are greater, others lesser, some thicke and short, some plaine and smooth, others a little crooked at the ends; wherein also are contained diuers formes of fruit or pease; some being round, others cornered, some small, some great, some white, others gray, and some spotted: the roote is small, and quickly perisheth.

The kindes of Pease are these:

The Rounciuall.	The gray Pease.
The greene Hasting.	The white Hasting.
The Sugar Pease.	The Pease without skins.
The spotted Pease.	

The Scottish or tufted Pease, which some call the Rose Pease, is a good white Pease fit to be eaten.

The early or French Pease, which some call Fulham Pease, because those grounds thereabouts doe bring them soonest forward for any quantity, although sometimes they miscarry by their haste and earlinesse.

Cicer Arietinum. Rams Ciches.

This is a kinde of Pulse, so much vsed in Spaine, that it is vsually one of their daintie dishes at all their feasts: They are of two sorts, white and red; the white is onely vsed for meate, the other for medicine. It beareth many vpright branches with winged leaues, many set together, being small, almost round, and dented about the edges: the flowers are either white or purple, according to the colour of the Pease which follow, and are somewhat round at the head, but cornered and pointed at the end, one or two at the most in a small roundish cod.

The Vse of Pease.

Pease of all or the most of these sorts, are either vsed when they are greene, and be a dish of meate for the table of the rich as well as the poore, yet euery one obseruing his time, and the kinde: the fairest, sweetest, youngest, and earliest for the better sort, the later and meaner kindes for the meaner, who doe not giue the deerest price: Or

Being dry, they serue to boyle into a kinde of broth or pottage, wherein many doe put Tyme, Mints, Sauory, or some other such hot herbes, to giue it the better rellish, and is much vsed in Towne and Countrey in the Lent time,

1 *Faba satiua.*Garden Beanes. 2 *Phaseoli satiui.*French Beanes. 3 *Pisum vulga e.* Garden Pease 4 *Pisum vmbelicatum sue xelestin.*Rose Pease or Scottish Pease. 5 *Pisum saccharatum.* Sugar Pease. 6 *Pisum maculatum.* Spotted Pease. 7 *Cicer Arietinum.* Rams Ciches or Cicers.

time, especially of the poorer sort of people.

It is much vsed likewise at Sea for them that goe long voyages, and is for change, because it is fresh, a welcome diet to most persons therein.

The Rams Ciches the Spaniards call *Granancos*, and *Garauancillos*, and eate them boyled and stewed as the most dainty kinde of Pease that are, they are of a very good rellish, and doe nourish much, but yet are not without that windy quality that all sorts of Pulse are subiect vnto : they increase bodily lust much more then any other sorts, and as it is thought, doth helpe to encrease seede.

CHAP. LIII.

Cucumer. The Cowcumber.

OF Cowcumbers there are diuers sorts, differing chiefly in the forme and colour of the fruit, and not in the forme of the plant; therefore one description shall serue in stead of all the rest.

The Cowcumber bringeth forth many trailing rough greene branches lying on the ground, all along whereof growe seuerall leaues, which are rough, broad, vneuen at the edges, and pointed at the ends, with long crooked tendrels comming forth at the same ioynt with the leafe, but on the other side therof: between the stalks & the leaues at the ioynts come forth the flowers seuerally, euery one standing on a short foot-stalke, opening it selfe into fiue leaues, of a yellowish colour, at the bottome whereof groweth the fruit, long and greene at the first, but when it is thorough ripe, a little yellowish, hauing many furrowes, and vneuen bunches all the length of it, wherein is a white firme substance next vnto the skin, and a cleare pulpe or watery substance, with white flat seede lying dispersed through it : the roote is long and white, with diuers fibres at it.

The kindes.

The first described is called, The long greene Cowcumber.

There is another is called, The short Cowcumber, being short, and of an equall bignesse in the body thereof, and of an vnequall bignesse at both ends.

The long Yellow, which is yellowish from the beginning, and more yellow when it is ripe, and hath beene measured to be thirteene inches long : but this is not that small long Cowcumber, called of the Latines, *Cucumis anguinus.*

Another kinde is early ripe, called The French kinde.

The Dantsicke kinde beareth but small fruit, growing on short branches or runners : the pickled Cowcumbers that are vsually sold are of this kind.

The Muscouie kinde is the smallest of all other, yet knowne, and beareth not aboue foure or fiue at the most on a roote, which are no bigger then small Lemons.

The Vse of Cowcumbers.

Some vse to cast a little salt on their sliced Cowcumbers, and let them stand halfe an houre or more in a dish, and then poure away the water that commeth from them by the salt, and after put vinegar, oyle, &c. thereon, as euery one liketh : this is done, to take away the ouermuch waterishnesse and coldnesse of the Cowcumbers.

In many countries they vse to eate Cowcumbers as wee doe Apples or Peares, paring and giuing slices of them, as we would to our friends of some dainty Apple or Peare.

The pickled Cowcumbers that come from beyond Sea, are much vsed
with

with vs for fawce to meate all the Winter long. Some haue ftriuen to equall them, by pickling vp our Cowcumbers at the later end of the yeare, when they are cheapeft, taking the little ones and fcalding them thoroughly well, which after they put in brine, with fome Dill or Fenell leaues and ftalkes: but thefe are nothing comparable to the former, wee either miffing of the right and orderly pickling of them, or the kinde it felfe differing much from ours (as I faid of the Dantficke kinde) for ours are neither fo tender and firme, nor fo fauoury as the other.

The rawe or greene Cowcumbers are fitteft for the hotter time of the yeare, and for hot ftomackes, and not to be vfed in colder weather or cold ftomackes, by reafon of the coldneffe, whereby many liaue been ouertaken.

The feede is vfed phyfically in many medicines that ferue to coole, and a little to make the paffages of vrine flippery, and to giue eafe to hot difeafes.

Chap. LIIII.

Melo. Milions or Muske Melons.

THere bee diuers forts of Melons found out at this day, differing much in the goodneffe of tafte one from another. This Countrey hath not had vntill of late yeares the skill to nourfe them vp kindly, but now there are many that are fo well experienced therein, and haue their ground fo well prepared, as that they will not miffe any yeare, if it be not too extreme vnkindly, to haue many ripe ones in a reafonable time: yet fome will be later then others alwayes.

The Melon is certainly a kinde of Cowcumber, it doth fo neare refemble it, both in the manner of his growing, hauing rough trailing branches, rough vneuen leaues, and yellow flowers: after which come the fruit, which is rounder, thicker, bigger, more rugged, and fpotted on the outfide then the Cowcumber, of a ruffet colour, and greene vnderneath, which when it groweth full ripe, will change a little yellowifh, being as deepe furrowed and ribbed as they, and befides hauing chaps or rifts in diuers places of the rinde: the inward hard fubftance is yellow, which onely is eaten: the feede which is bigger, and a little yellower then the Cowcumber, lying in the middle onely among the moifter pulpe: the fmell and changing of his colour, fore-fhew their ripeneffe to them that are experienced: the roote is long, with many fibres at it. The fruit requireth much watering in the hot time of the day, to caufe them to ripen the fooner, as I haue obferued by diuers of the beft skill therein.

The Vfe of the kindes of Melons.

The beft Melon feede doe come to vs out of Spaine, fome haue come out of Turkie, but they haue been nothing fo good and kindly.

Some are called Sugar Melons, others Peare Melons, and others Muske Melons.

They haue beene formerly only eaten by great perfonages, becaufe the fruit was not only delicate but rare; and therfore diuers were brought from France, and fince were nourfed vp by the Kings or Noblemens Gardiners onely, to ferue for their Mafters delight: but now diuers others that haue skill and conueniencie of ground for them, doe plant them and make them more common.

They paire away the outer rinde, and cut out the inward pulpe where the feede lyeth, flice the yellow firme inward rinde or fubftance, & fo eate it with falt and pepper (and good ftore of wine, or elfe it will hardl digeft) for this is firmer, & hath not that moifture in it that the Cowcumbers haue. It is alfo more delicate, and of more worth, which recompenfeth the pine.

The feed of thefe Melons are vfed as Cowcumbers phyfically, and together with them moft vfually.

CHAP

CHAP. LV.

Pepo. Pompions.

WE haue but one kinde of Pompion (as I take it) in all our Gardens, notwithstanding the diuersities of bignesse and colour.

The Pompion or great Melon (or as some call it Milion) creepeth vpon the ground (if nothing bee by it whereon it may take hold and climbe) with very great, ribbed, rough, and prickly branches, whereon are set very large rough leaues, cut in on the edges with deepe gashes, and dented besides, with many claspers also, which winde about euery thing they meete withall : the flowers are great and large, hollow and yellow, diuided at the brims into fiue parts, at the bottome of which, as it is in the rest, groweth the fruit, which is very great, sometimes of the bignesse of a mans body, and oftentimes lesse, in some ribbed or bunched, in others plaine, and either long or round, either greene or yellow, or gray, as Nature listeth to shew her selfe ; for it is but waste time, to recite all the formes and colours may be obserued in them : the inner rinde next vnto the outer is yellowish and firme : the seede is great, flat, and white, lying in the middle of the watery pulpe: the roote is of the bignesse of a mans thumbe or greater, dispersed vnder ground with many small fibres ioyned thereunto.

Gourds are kindes of Melons ; but because wee haue no vse of them, wee leaue them vnto their fit place.

The Vse of Pompions.

They are boyled in faire water and salt, or in powdered beefe broth, or sometimes in milke, and so eaten, or else buttered. They vse likewise to take out the inner watery substance with the seedes, and fill vp the place with Pippins, and hauing laid on the couer which they cut off from the toppe, to take out the pulpe, they bake them together, and the poore of the Citie, as well as the Country people, doe eate thereof, as of a dainty dish.

The seede hereof, as well as of Cowcumbers and Melons, are cooling, and serue for emulsions in the like manner for Almond milkes, &c. for those are troubled with the stone.

CHAP. LVI.

Fragaria. Strawberries.

THere be diuers sorts of Strawberries, whereof those that are nourfed vp in Gardens or Orchards I intend to giue you the knowledge in this place, and leaue the other to a fitter ; yet I must needs shew you of one of the wilde sorts, which for his strangenesse is worthy of this Garden : And I must also enforme you, that the wilde Strawberry that groweth in the Woods is our Garden Strawberry, but bettered by the soyle and transplanting.

The Strawberry hath his leaues closed together at the first springing vp, which afterwards spread themselues into three diuided parts or leaues, euery one standing vpon a small long foote-stalke, greene on the vpperside, grayish vnderneath, and snipped or dented about the edges ; among which rise vp diuers small stalkes, bearing foure or fiue flowers at the tops, consisting of fiue white round pointed leaues, somewhat yellowish in the bottome, with some yellow threads therein ; after which come the fruit, made of many small graines set together, like vnto a small Mulberry or Raspis, reddish when it is ripe, and of a pleasant winy taste, wherein is enclosed diuers small blackish seede : the roote is reddish and long, with diuers small threads at it, and sendeth
forth

1 *Cucumis vulgaris.* The ordinary Cowcumber. 2 *Cucumis Hispanicus.* The long yellow Spanish Cowcumber. 3 *Mel: vulgaris.* The ordinary Melon. 4 *Melo maximus optimus.* The greatest Muske Melon. 5 *Pepo.* The Pompion. 6 *Fraga vulgaris.* Common strawberries. 7 *Fraga Bohemica maxima.* The great Bohemia Strawberries. 8 *Fraga aculeata.* The prickly Strawberry.

forth from the head therof long reddish strings running vpon the ground, which shoot forth leaues in many places, whereby it is much encreafed.

The white Strawberry differeth not from the red, but in the colour of the fruite, which is whiter then the former when it is thorough ripe, enclining to redneffe.

The greene Strawberry likewife differeth not, but that the fruit is green on all fides when it is ripe, faue on that fide the Sun lyeth vpon it, and there it is fomewhat red.

The Virginia Strawberry carryeth the greateft leafe of any other, except the Bohemian, but fcarce can one Strawberry be feene ripe among a number of plants, I thinke the reafon thereof to be the want of skill, or induftry to order it aright. For the Bohemia, and all other Strawberries will not beare kindly, if you fuffer them to grow with many ftrings, and therefore they are ftill cut away.

There is another very like vnto this, that Iohn Tradefcante brought with him from Bruffels long agoe, and in feuen yeares could neuer fee one berry ripe on all fides, but ftill the better part rotten, although it would euery yeare flower abundantly, and beare very large leaues.

The Bohemia Strawberry hath beene with vs but of late dayes, but is the goodlieft and greateft, both for leafe next to the Virginian, and for beauty farre furpaffing all; for fome of the berries haue beene meafured to bee neere fiue inches about. Mafter Quefter the Poftmafter firft brought them ouer into our Country, as I vnderftand, but I know no man fo induftrious in the carefull planting and bringing them to perfection in that plentifull maner, as Mafter Vincent Sion who dwelt on the Banck fide, neer the old Paris garden ftaires, who from feuen rootes, as hee affirmed to me, in one yeare and a halfe, planted halfe an acree of ground with the increafe from them, befides thofe he gaue away to his friends, and with him I haue feene fuch, and of that bigneffe before mentioned.

One Strawberry more I promifed to fhew you, which although it be a wilde kinde, and of no vfe for meate, yet I would not let this difcourfe paffe, without giuing you the knowledge of it. It is in leafe much like vnto the ordinary, but differeth in that the flower, if it haue any, is greene, or rather it beareth a fmall head of greene leaues, many fet thicke together like vnto a double ruffe, in the midft whereof ftandeth the fruit, which when it is ripe, fheweth to be foft and fomwhat reddifh, like vnto a Strawberry, but with many fmall harmleffe prickles on them, which may be eaten and chewed in the mouth without any maner of offence, and is fomewhat pleafant like a Strawberry: it is no great bearer, but thofe it doth beare, are fet at the toppes of the ftalks clofe together, pleafant to behold, and fit for a Gentlewoman to weare on her arme, &c. as a raritie in ftead of a flower.

The Vfe of Strawberries.

The leaues of Strawberries are alwaies vfed among other herbes in cooling drinkes, as alfo in lotions, and gargles for the mouth and throate: the rootes are fometimes added to make it the more effectuall, and withall fomwhat the more binding.

The berries themfelues are often brought to the Table as a reare feruice, whereunto claret wine, creame or milke is added with fugar, as euery one liketh; as alfo at other times, both with the better and meaner fort, and are a good cooling and pleafant difh in the hot Summer feafon.

The water diftilled of the berries, is good for the paffions of the heart, caufed by the perturbation of the fpirits, being eyther drunke alone, or in wine; and maketh the heart merry.

Some doe hold that the water helpeth to clenfe the face from fpots, and to adde fome cleereneffe to the skinne.

CHAP.

Chap. LVII.

Angelica. Garden Angelica.

HAuing thus furnished you out a Kitchen Garden with all sorts of herbes, roots & fruits fit for it, and for any mans priuate vse, as I did at the first appropriate it, let me a little transcend, and for the profit & vse of Country Gentlewomen and others, furnish them with some few other herbes, of the most especiall vse for those shall need them, to be planted at hand in their Gardens, to spend as occasion shall serue, and first of Angelica.

Angelica hath great and long winged leaues, made of many broade greene ones, diuided one from another vpon the stalk, which is three foot long or better somtimes, among which rise vp great thicke and hollow stalkes with some few ioynts, whereat doth alwayes stand two long leaues compassing the stalke at the bottome, in some places at the ioynts spring out other stalkes or branches, bearing such like leaues but smaller, and at the tops very large vmbels of white flowers, that turne into whitish seede somewhat thicke : the roote groweth great with many branches at it, but quickly perisheth after it hath borne seede : to preserue the roote therefore the better, they vse to cut it often in the yeare, thereby to hinder the running vp to seede : the whole plant, both leafe, roote and seede, is of an excellent comfortable sent, sauour and taste.

The Vse of Angelica.

The distilled water of Angelica, eyther simple or compound, is of especiall vse *in deliquinm animi, vel cordis tremores & passiones*, that is, (swounings, when the spirits are ouercome and faint, or tremblings and passions of the heart, to expell any windy or noysome vapours from it. The green stalkes or the young rootes being preserued or candied, are very effectuall to comfort and warme a colde and weake stomacke : and in the time of infection is of excellent good vse to preserue the spirits and heart from infection. The dryed roote made into pouder, and taken in wine or other drinke, will abate the rage of lust in young persons, as I haue it related vnto me vpon credit : A Syrupe made thereof in this manner, is very profitable to expectorate flegme out of the chest and lunges, and to procure a sweete breath. Into the greene stalke of Angelica as it standeth growing, make a great gashe or incision, wherein put a quantitie of fine white Sugar, letting it there abide for three dayes, and after take it forth by cutting a hole at the next ioynt vnder the cut, where the Syrupe resteth, or cut off the stalke, and turne it downe, that the Syrupe may drayne forth ; which keepe for a most delicate medicine.

Chap. LVIII.

Dracunculus hortensis siue Serpentaria. Dragons.

DRagons riseth out of the ground with a bare or naked round whitish stalke, spotted very much with purplish spots and strakes, bearing at the toppe therof a few greene leaues very much diuided on all sides, standing vpon long footestalkes, in the middle whereof (if the roote be old enough) commeth forth a great long huske or hose, green on the outside, and of a darke purplish colour on the inside, with a slender long reddish pestell or clapper in the middle : the roote is great, round, flat and whitish on the outside, and whiter within, very like vnto the rootes of *Arum*, or Wakerobin, and tasting somewhat sharpe like it.

The

The Vſe of Dragons.

The chiefe vſe whereunto Dragons are applyed, is, that according to an old receiued cuſtome and tradition (and not the iudgement of any learned Author) the diſtilled water is giuen with Mithridatum or Treakle to expell noyſome and peſtilentiall vapours from the heart.

Chap. LIX.

Ruta. Garden Rue, or Herbe Grace.

Garden Rue or Herbe Grace groweth vp with hard whitiſh wooddy ſtalkes, whereon are ſet diuers branches of leaues, being diuided into many ſmall ones, which are ſomewhat thicke and round pointed, of a blewiſh greene colour : the flowers ſtand at the tops of the ſtalkes conſiſting of foure ſmall yellow leaues, with a greene button in the middle, and diuers ſmall yellow threds about it, which growing ripe, containe within them ſmall blacke ſeede : the roote is white and wooddy, ſpreading farre in the ground.

The Vſe of Rue.

The many good properties whereunto Rue ſerueth, hath I thinke in former times cauſed the Engliſh name of Herbe Grace to be giuen vnto it. For without doubt it is a moſt wholeſome herbe, although bitter and ſtrong, and could our dainty ſtomackes brooke the vſe thereof, it would worke admirable effects being carefully and skilfully applyed, as time and occaſion did require : but not vndiſcreetly or hand ouer head, as many vſe to doe that haue no skill. Some doe rippe vp a beade rowle of the vertues of Rue, as Macer the Poet and others, in whom you ſhall finde them ſet downe, to bee good for the head, eyes, breaſt, liuer, heart, ſpleene, &c. In ſome places they vſe to boyle the leaues of Rue, and keep them in pickle, to eate them as Sampire for the helpe of weake eyes. It is very auaileable in gliſters or drinkes againſt the winde or the collicke, and to procure vrine that is ſtayed by the paines therof. The diſtilled water is often vſed for the ſame purpoſes aforeſaid : but beware of the too frequent or ouermuch vſe thereof, becauſe it heateth exceedingly, and waſteth nature mightily.

Chap. LX.

Carduus Benedictus. The Bleſſed Thiſtle.

Carduus benedictus or the bleſſed Thiſtle, hath many weake tender branches lying for the moſt part on the ground, whereon are ſet long and narrow leaues, much cut in or waued about the edges, hairy or rough in handling, yet without any hard or ſharpe thornes or prickles at all, that the tendereſt hand may touch them without harme : but thoſe that grow toward the toppes of the ſtalkes are ſomewhat more prickly, and the heads which grow on the tops of the ſeuerall branches are ſomewhat ſharpe, ſet with prickles like a Thiſtle : the flower is yellow, and the ſeede lying within the woolly or flocky doune like to all other thiſtles, are blackiſh, long and round, with a few haires on the head of them : the roote is white, and periſheth euery yeare after it hath giuen ſeede.

The Vſe of the bleſſed Thiſtle.

The diſtilled water hereof is much vſed to be drunke againſt agues of all ſortes, eyther peſtilentiall or humorall, of long continuance or of leſſe :

1 *Angellica.* Angellica. 2 *Dracunculus hortensis.* Dragons. 3 *Ruta hortensis.* Garden Rue, or Herbegrace. 4 *Cardnus benedictus.* The bleffed Thiftle. 5 *Alkakengi fiue solanum Halicacabum & Vesicarium.* Winter Cherries. 6 *Asarum.* Asrabacca. 7 *Liquaritia.* Licoris.

but the decoction of the herbe giuen in due time, hath the more forcible operation : it helpeth to expell wormes, because of the bitternesse, and is thereby also a friend to the ftomack ouercharged with chollar, and to clenfe the liuer : it prouoketh fweate and vrine, is helpefull to them are troubled with the ftone, and to eafe paines in the fides.

CHAP. LXI.

Solanum veficarium, siue Alkakengi. Winter Cherries.

THe Winter Cherry hath a running or creeping roote in the ground, of the big-neffe wany times of ones little finger, fhooting forth at feuerall ioynts in feue-rall places, whereby it quickly fpreadeth a great compaffe of ground : the ftalke rifeth not aboue a yard high, whereon are fet many broade and long greene leaues, fomewhat like vnto the leaues of Nightfhade, but larger: at the ioynts where-of come forth whitifh flowers made of fiue leaues a peece, which after turne into green berries, inclofed with thin skins or bladders, which change to bee reddifh when they grow ripe, the berry likewife being reddifh, and as large as a Cherry, wherein are contained many flat and yellowifh feed lying within the pulpe : which being gathered and ftrung vp, are kept all the yeare to be vfed vpon occafion.

The Vfe of Winter Cherries.

The diftilled water of the herbe and fruit together, is often taken of them that are troubled with the fharpneffe or difficultie of vrine, and with the ftone in the kidneyes, or grauel in the bladder: but the berries themfelues ei-ther greene or dryed boyled eyther in broth, in wine, or in water, is much more effectuall : It is likewife conducing to open obftructions of the liuer, &c. and thereby to helpe the yellow Iaundife.

CHAP. LXII.

Afarum. Afarabacca.

A Sarabacca, from a fmall creeping roote fet with many fibres, fhooteth forth di-uers heads, and from euery of them fundry leaues, euery one ftanding vpon a long greene ftalke, which are round, thicke, and of a very fad or darke greene colour, and fhining withall : from the rootes likewife fpring vp fhort ftalkes, not fully foure fingers high, at the toppe of euery one of which ftandeth the flower, in fa-fhion very like the feede veffell of Henbane feede, of a greenifh purple colour, which changeth not his forme, but groweth in time to containe therein fmall cornered feed : the greene leaues abide all the winter many times, but vfually fheddeth them in winter, and recouereth frefh in the fpring.

The Vfe of Afarabacca.

The leaues are much and often vfed to procure vomits, fiue or feuen of them bruifed, and the iuice of them drunke in ale or wine. An extract made of the leaues with wine artificially performed, might bee kept all the yeare thorough, to bee vfed vpon any prefent occafion, the quantitie to bee pro-portioned according to the conftitution of the patient. The roote worketh not fo ftrongly by vomit, as the leaues, yet is often vfed for the fame pur-pofe, and befides is held auaileable to prouoke vrine, to open obftructions in the liuer and fpleene, and is put among diuers other fimples, both into Mithridatum and Andromachus Treakle, which is vfually called Venice Treakle. A dram of the dryed roots in pouder giuen in white wine a little before the fit of an ague, taketh away the fhaking fit, & therby caufe the hot fit to be the more remiffe, and in twice taking expell it quite.

CHAP.

CHAP. LXIII.

Glycyrrhiza siue Liqueritia. Licorice.

ALthough there are two sorts of Licorice set downe by diuers Authors, yet because this Land familiarly is acquainted but with one sort, I shall not neede for this Garden, to make any further relation of that is vnknowne, but onely of that sort which is sufficiently frequent with vs. It riseth vp with diuers wooddy stalks, whereon are set at seuerall distances many winged leaues, that is to say, many narrow long greene leaues set together on both sides of the stalke, and an odde one at the end, very well resembling a young Ashe tree sprung vp from the seede : this by many yeares continuance in a place without remouing, and not else, will bring forth flowers many standing together spike-fashion one aboue another vpon the stalkes, of the forme of Pease blossomes, but of a very pale or bleake blew colour, which turne into long somewhat flat and smooth cods, wherein is contained small round hard seede : the roote runneth downe exceeding deep into the ground, with diuers other smaller roots and fibres growing with them, and shoote out suckers from the maine rootes all about, whereby it is much encreased, of a brownish colour on the outside, and yellow within, of a farre more weake sweete taste, yet far more pleasing to vs then that Licorice that is brought vs from beyond Sea ; because that, being of a stronger sweet taste hath a bitternesse ioyned with it, which maketh it the lesse pleasing and acceptable to most.

The Vse of Licorice.

Our English Licorice is now adaies of more familiar vse (as I said before) then the outlandish, and is wholly spent and vsed to helpe to digest and expectorate flegme out of the chest and lunges, and doth allay the sharpenesse or saltnesse thereof. It is good also for those are troubled with shortnesse of breath, and for all sorts of coughes. The iuice of Licorice artificially made with Hyssope water, serueth very well for all the purposes aforesaid. It being dissolued with Gum Tragacanth in Rose water, is an excellent Lohoc or licking medicine to breake flegme, and to expectorate it, as also to avoyde thin frothy matter, or thin salt flegme, which often fretteth the lunges. It doth also lenifie exulcerated kidneyes, or the bladder, and helpeth to heale them. It is held also good for those that cannot make their water but by drops, or a small deale at a time.

The dryed root finely minced, is a speciall ingredient into all Trageas or Dredges, seruing for the purposes aforesaid, but the vse of them is almost wholly left now adaies with all sorts.

Thus haue I shewed you not only the herbes, rootes and fruites, noursed vp in this Garden, but such herbes as are of most necessary vses for the Country Gentlewomens houses : And now I will shew you the Orchard also.

The

THE
ORDERING OF THE
ORCHARD.

The third part, or ORCHARD.

CHAP. I.

*The situation of an Orchard for fruit-bearing trees, and how to
amend the defects of many grounds.*

S I haue done in the two former parts of this Treatise, so I meane to proceede in this ; first to set downe the situation of an Orchard, and then other things in order : And first, I hold that an Orchard which is, or should bee of some reasonable large extent, should be so placed, that the house should haue the Garden of flowers iust before it open vpon the South, and the Kitchen Garden on the one side thereof, should also haue the Orchard on the other side of the Garden of Pleasure, for many good reasons : First, for that the fruit trees being grown great and tall, will be a great shelter from the North and East windes, which may offend your chiefest Garden, and although that your Orchard stand a little bleake vpon the windes, yet trees rather endure these strong bitter blasts, then other smaller and more tender shrubs and herbes can doe. Secondly, if your Orchard should stand behinde your Garden of flowers more Southward, it would shadow too much of the Garden, and besides, would so binde in the North and East, and North and West windes vpon the Garden, that it would spoile many tender things therein, and so much abate the edge of your pleasure thereof, that you would willingly wish to haue no Orchard, rather then that it should so much annoy you by the so ill standing thereof. Thirdly, the falling leaues being still blowne with the winde so aboundantly into the Garden, would either spoile many things, or haue one daily and continuall attending thereon, to cleanse and sweepe them away. Or else to auoide these great inconueniences, appoint out an Orchard the farther off, and set a greater distance of ground betweene. For the ground or soile of the Orchard, what I haue spoken concerning the former Garden for the bettering of the seuerall grounds, may very well serue and be applyed to this purpose. But obserue this, that whereas your Gardens before spoken of may be turned vp, manured, and bettered with soile if they growe out of heart, your Orchard is not so easily done, but must abide many yeares without altering ; and therefore if the ground be barren, or not good, it had the more neede to bee amended, or wholly made good, before you make an Orchard of it ; yet some there be

that

that doe appoint, that where euery tree fhould bee fet, you onely digge that place to make it good : but you muft know, that the rootes of trees runne further after a little times ftanding, then the firft compaffe they are fet in ; and therefore a little compaffe of ground can maintaine them but a little while, and that when the rootes are runne beyond that fmall compaffe wherein they were firft fet, and that they are come to the barren or bad ground, they can thriue no better then if they had beene fet in that ground at the firft, and if you fhould afterwards digge beyond that compaffe, intending to make the ground better further off, you fhould much hurt the fpreading rootes, and put your trees in danger : the fituation of hils in many places is grauelly or chalky, which is not good for trees, becaufe they are both too ftonie, and lacke mellow earth, wherein a tree doth moft ioy and profper, and want moifture alfo (which is the life of all trees) becaufe of the quicke defcent of raine to the lower grounds : and befides all thefe inconueniences there is one more ; your trees planted either on hils or hill fides, are more fubiect to the fury and force of windes to be ouer-turned, then thofe that growe in the lower grounds ; for the ftrongeft and moft forci-ble windes come not vfually out of the North Eaft parts, where you prouide beft de-fence, but from the South and Weft, whence you looke for the beft comfort of the Sunne. To helpe therefore manie of the inconueniences of the hils fides, it were fit to caufe manie leauels to bee made thereon, by raifing the lower grounds with good earth, and fuftaining them with bricke or ftone wals, which although chargeable, will counteruaile your coft, befide the pleafure of the walkes, and profpect of fo worthy a worke. The plaine or leuell grounds as they are the moft frequent, fo they are the moft commendable for an Orchard, becaufe the moulds or earths are more rich, or may better and fooner be made fo ; and therefore the profits are the more may be rai-fed from them. A ftiffe clay doth nourifh trees well, by reafon it containeth moi-fture ; but in regard of the coldneffe thereof, it killeth for the moft part all tender and early things therein : fea-cole afhes therefore, bucke afhes, ftreete foyle, chaulke after it hath lyen abroad and been broken with many yeares frofts and raine, and fheepes dung, are the moft proper and fitteft manure to helpe this kinde of foyle. The dry fandy foile, and grauelly ground are on the contrary fide as bad, by reafon of too much heate and lacke of moifture : the dung of kine or cattell in good quantity beftowed thereon, will much helpe them. The amending or bettering of other forts of grounds is fet down toward the end of the firft Chapter of the firft part of this worke, where-vnto I will referre you, not willing to repeate againe the fame things there fet downe. The beft way to auoide and amend the inconueniences of high, boifterous, and cold windes, is to plant Walnut trees, Elmes, Oakes or Afhes, a good diftance without the compaffe of your Orchard, which after they are growne great, will bee a great fafe-guard thereunto, by breaking the violence of the windes from it. And if the foyle of your Orchard want moifture, the conueying of the finke of the houfe, as alfo any o-ther draine of water thereinto, if it may be, will much helpe it.

CHAP. II.

The forme of an Orchard, both ordinary, and of more grace and rarity.

Accordingto the fituation of mens grounds, fo muft the plantation of them of neceffitie be alfo ; and if the ground be in forme, you fhall haue a formall Or-chard : if otherwife, it can haue little grace or forme. And indeed in the elder ages there was fmall care or heede taken for the formality ; for euery tree for the moft part was planted without order, euen where the mafter or keeper found a vacant place to plant them in, fo that oftentimes the ill placing of trees without fufficient fpace be-tweene them, and negligence in not looking to vphold them, procured more wafte and fpoile of fruit, then any accident of winde or weather could doe. Orchards in moft places haue not bricke or ftone wals to fecure them, becaufe the extent thereof being larger then of a Garden, would require more coft, which euery one cannot vndergoe; and therefore mud wals, or at the beft a quicke fet hedge, is the ordinary and moft vfuall defence it findeth almoft in all places : but with thofe that are of ability to compaffe it with bricke or ftone wals, the gaining of ground, and profit of the fruit trees planted there againft, will in fhort time recompenfe that charge. If you make a doubt how to be fure that your Orchard wall fhall haue fufficient comfort of the Sunne to ripen the fruits, in regard the trees in the Orchard being fo nigh thereunto, and fo high withall, will fo much fhadow the wall, that nothing will ripen well, becaufe it will want the comfort of the Sunne : you may follow this rule and aduice, to remedy thofe inconueniences. Hauing an Orchard containing one acre of ground, two, three, or more, or leffe, walled about, you may fo order it, by leauing a broad and large walke betweene the wall and it, containing twenty or twenty foure foote (or yards if you will) that the wall fhall not be hindered of the Sun, but haue fufficient comfort for your trees, notwithftanding the height of them, the diftance betweene them and the wall being a fufficient fpace for their fhadow to fall into : and by compaffing your Orchard on the infide with a hedge (wherein may bee planted all forts of low fhrubs or bufhes, as Rofes, Cornellian Cherry trees plafhed lowe, Goofeberries, Curran trees, or the like) you may enclofe your walke, and keepe both it and your Orchard in better forme and manner, then if it lay open. For the placing of your trees in this Orchard, firft for the wals : Thofe fides that lye open to the South & Southweft Sunne, are fitteft to bee planted with your tendereft and earlieft fruits, as Apricockes, Peaches, Nectarius, and May or early Cherries : the Eaft, North and Weft, for Plums and Quinces, as you fhall like beft to place them. And for the Orchard it felfe, the ordinary manner is to place them without regard of meafure or difference, as Peares among Apples, and Plums among Cherries promifcuoufly ; but fome keepe both a diftance and a diuifion for

euery fort, without intermingling : yet the moft gracefull Orchard containeth them all, with fome others, fo as they be placed that one doe not hinder or fpoile another ; and therefore to defcribe you the modell of an Orchard, both rare for comelineffe in the proportion, and pleafing for the profitableneffe in the vfe, and alfo durable for continuance, regard this figure is here placed for your direction, where you muft obferue, that your trees are here fet in fuch an equall diftance one from another euery way, & as is fitteft for them, that when they are grown great, the greater branches fhall not gall or rubbe one againft another; for which purpofe twenty or fixteene foot is the leaft to be allowed for the diftance euery way of your trees, & being fet in rowes euery one in the middle diftance, will be the moft gracefull for the plantation, and befides, giue you way fufficient to paffe through them, to pruine, loppe, or dreffe them, as need fhall require, and may alfo bee brought (if you pleafe) to that gracefull delight, that euery alley or diftance may be formed like an arch, the branches of either fide meeting to be enterlaced together. Now for the feuerall forts of fruit trees that you fhall place in this modell, your beft direction is to fet Damfons, Bulleis, and your taler growing Plums on the outfide, and your lower Plums, Cherries, and Apples on the infide, hauing regard, that you place no Peare tree to the Sunward, of any other tree, left it ouerfhadow shadow them : Let your Peare trees therefore be placed behinde, or on the one side of your lower trees, that they may be as it were a shelter or defence on the North & East side. Thus may you also plant Apples among Plums and Cherries, so as you suffer not one to ouer-growe or ouer-toppe another; for by pruning, lopping, and shredding those that growe too fast for their fellowes, you may still keepe your trees in such a conformity, as may be both most comely for the sight, and most profitable for the yeelding of greater and better store of fruit. Other sorts of fruit trees you may mixe among these, if you please, as Filberds, Cornellian Cherries in standerds, and Medlers : but Seruice trees, Baye trees, and others of that high sort, must be set to guard the rest. Thus haue I giuen you the fairest forme could as yet be deuised; and from this patterne, if you doe not follow it precisely, yet by it you may proportion your Orchard, be it large or little, be it walled or hedged.

Chap. III.

Of a nurfery for trees, both from fowing the kernels, and planting fit stockes to graft vpon.

Lthough I know the greater sort (I meane the Nobility and better part of the Gentrie of this Land) doe not intend to keepe a Nursery, to raise vp those trees that they meane to plant their wals or Orchards withall, but to buy them already grafted to their hands of them that make their liuing of it : yet becaufe many Gentlemen and others are much delighted to beftowe their paines in grafting themfelues, and efteeme their owne labours and handie worke farre aboue other mens: for their incouragement and satisfaction, I will here set downe some conuenient directions, to enable them to raise an Orchard of all forts of fruits quickly, both by fowing the kernels or stones of fruit, and by making choife of the best forts of stockes to graft on : First therefore to begin with Cherries; If you will make a Nursery, wherein you may bee stored with plenty of stockes in a little space, take what quantitie you thinke good of ordinarie wilde blacke Cherrie stones, cleansed from the berries, and fowe them, or pricke them in one by one on a peece of ground well turned vp, and large enough for the quantitie of stones you will beftowe thereon, from the midst of Auguft vnto the end of September, which when they are two or three yeares old, according to their growth, you may remoue them, and set them anew in some orderly rowes, hauing pruned their tops and their rootes, which at the next yeares growth after the new planting in any good ground, or at the second, will be of sufficient bigneffe to graft vpon in the bud what forts of Cherries you thinke beft : and it is fitteft to graft them thus young, that pruning your stockes to raife them high, you may graft them at fiue or fix foote high, or higher, or lower, as you shall see good, and being thus grafted in the bud, will both more speedily and safely bring forward your grafts, and with leffe danger of losing your stockes, then by grafting them in the stocke : for if the bud take not by inoculating the first yeare, yet your tree is not loft, nor put in any hazzard of loffe; but may be grafted anew the yeare following, if you will, in another place thereof, whereas if you graft in the stocke, and it doe not take, it is a great chance if the stocke dye not wholly, or at leaft be not so weakened both in strength and height, that it will not bee fit to bee grafted a yeare or two after. In the fame manner as you doe with the blacke, you may deale with the ordinary Englifh red Cherrie stones, or kernels. but they are not so apt to growe so ftraight and high, nor in so short a time as the blacke Cherrie stones are, and befides are fubiect in time to bring out fuckers from the rootes, to the hinderance of the stockes and grafts, or at the leaft to the deformitie of your Orchard, and more trouble to the Gardiner, to pull or digge them away. Plumme stones may bee ordered in this manner likewife, but you muft make choife of your Plums; for although euery Plumme is not so fit for this purpose, as the white Peare Plumme, becaufe it groweth the goaleft and freeft, the barke being fmooth and apteft to be raifed, that they may be grafted vpon; yet diuers other Plummes may be taken, if they be not at hand, or to be had, as the blacke and red Peare Plumme, the

　　　　　　　　　　　　　　　　　　　　　　　　　　　　　　　　w hite

white and red Wheate Plumme, becaufe they are neareft in goodneffe vnto it. Peach ftones will be foone raifed vp to graft other forts of Peaches or Nectorins vpon, but the nature of the Peach roote being fpongie, is not to abide long. As for Almonds, they will be raifed from their ftones to be trees of themfelues; but they will hardly a-bide the remouing, and leffe to bee grafted vpon. Apricocke ftones are the worft to deale withall of any fort of ftone fruit; for although the Apricocke branches are the fitteft ftockes to graft Nectorins of the beft forts vpon, yet thofe that are raifed from the kernels or ftones will neuer thriue to be brought on for this purpofe; but will ftarue and dye, or hardly grow in a long time to be a ftraight and fit ftocke to be grafted, if it be once remoued. Your Cornellian Cherrie trees are wholly, or for the moft part rai-fed from the ftones or kernels; yet I know diuers doe increafe them, by laying in their loweft branches to take roote: and thus much for ftone fruits. Now for Apples and Peares, to be dealt withall in the fame manner as aforefaid. They vfe to take the pref-fing of Crabs whereas Veriuyce is made, as alfo of Cidar and Perry where they are made, and fowing them, doe raife vp great ftore of ftockes; for although the beating of the fruit doth fpoile many kernels, yet there will bee enough left that were neuer toucht, and that will fpring: the Crabbe ftockes fome preferre for the fitteft, but I am fure, that the better Apple and Peare kernels will growe fairer, ftraighter, quicklier, and better to be grafted on. You muft remember, that after two or three yeares you take vp thefe ftockes, and when you haue pruned both toppe and roote, to fet them a-gaine in a thinner and fitter order, to be afterwards grafted in the bud while they are young, as I fhall fhew you by and by, or in the ftocke if you will fuffer them to growe greater. Now likewife to know which are the fitteft ftockes of all forts to choofe, thereon to graft euery of thefe forts of fruits, is a point of fome skill indeede; and therefore obferue them as I doe here fet them downe: for bee you affured, that they are certaine rules, and knowne experiences, whereunto you may truft without being deceiued. Your blacke Cherrie ftockes (as I faid before) are the fitteft and beft for all forts of Cherries long to abide and profper, and euen May or early Cherry will a-bide or liue longer, being grafted thereon, either in the budde or in the ftocke, then on the ordinary red Cherry ftocke; but the red Cherry ftocke is in a manner the onely tree that moft Nurfery men doe take to graft May Cherries on in the ftocke (for it is but a late experience of many, to graft May Cherries in the bud) many alfo doe graft May Cherries on Gafcoigne Cherry ftockes, which doe not onely thriue well, but en-dure longer then vpon any ordinay Cherry ftocke: For indeede the May Cherries that are grafted vpon ordinary red Cherrie ftockes, will hardly hold aboue a dozen yeares bearing well, although they come forwarder at the firft, that is, doe beare foo-ner then thofe that are grafted on Gafcoigne or blacke Cherry ftockes; but as they are earlier in bearing, fo they are fooner fpent, and the Gafcoigne and blacke Cherry ftockes that are longer in comming forward, will laft twice or thrice their time; but many more grafts will miffe in grafting of thefe, then of thofe red Cherry ftockes, and befides, the natures of the Gafcoigne and blacke Cherry ftockes are to rife higher, and make a goodlier tree then the ordinary red ftocke will, which for the moft part fprea-deth wide, but rifeth not very high. The Englifh red Cherry ftocke will ferue very well to graft any other fort of Cherry vpon, and is vfed in moft places of this Land, and I know no other greater inconuenience in it, then that it fhooteth out many fuckers from the roote, which yet by looking vnto may foone bee remoued from doing any harme, and that it will not laft fo long as the Gafcoigne or blacke Cherry ftocke will. May Cherries thus grafted lowe, doe moft vfually ferue to be planted againft a wall, to bring on the fruit the earlier; yet fome graft them high vpon ftandards, although not many, and it is, I thinke, rather curiofity (if they that doe it haue any wals) then anie o-ther matter that caufeth them thus to doe: for the fruit is naturally fmall, though early, and the ftandard Cherries are alwaies later then the wall Cherries, fo that if they can fpare any roome for them at their wals, they will not plant many in ftandards. Now concerning Plummes (as I faid before) for the fowing or fetting of the ftones, fo I fay here for their choife in grafting of them, either in the budde or ftocke. The white Peare Plumme ftocke, and the other there mentioned, but efpecially the white Peare Plumme is the goodlieft, freeft, and fitteft of all the reft, as well to graft all fort of Plummes vpon, as alfo to graft Apricockes, which can be handfomely, and to any

good

good purpofe grafted vpon no other Plum ftocke, to rife to bee worth the labour and paine. All forts of Plums may be grafted in the ftocke, and fo may they alfo in the bud; for I know none of them that will refufe to be grafted in the bud, if a cunning hand performe it well; that is, to take off your bud cleanely and well, when you haue made choice of a fit cyon: for, as I fhal fhew you anon, it is no fmall peece of cunning to chufe your cyon that it may yeeld fit buds to graft withall, for euery plum is not of a like aptnes to yeeld them: But Apricocks cannot be grafted in the ftock for any thing that euer I could heare or learne, but only in the bud, and therefore let your Plum ftocke bee of a reafonable fize for Apricockes efpecially, and not too fmall, that the graft ouergrow not the ftocke, and that the ftocke bee large enough to nourifh the graft. As your Plum ftockes ferue to graft both Apricockes and Plummes, fo doe they ferue alfo very well to graft Peaches of all forts; and although Peach ftockes will ferue to be grafted with Peaches againe, yet the Peach ftocke (as I faid before) will not endure fo long as the Plumme ftocke, and therefore ferueth but for neceffity if Plum ftocks be not ready, or at hand, or for the prefent time, or that they afterwards may graft that fort of Peach on a Plumme ftocke : for many might lofe a good fruit, if when they meete with it, and haue not Plumme ftockes ready to graft it on, they could not be affured that it would take vpon another Peach ftocke or branch, or on the branch of an Apricocke eyther. Plumme ftockes will ferue likewife very well for fome forts of Nectorins; I fay, for fome forts, and not for all : the greene and the yellow Nectorin will beft thriue to be grafted immediately on a Plumme ftocke; but the other two forts of red Nectorins muft not be immediately grafted on the Plumme ftocke, but vpon a branch of an Apricocke that hath beene formerly grafted on a Plumme ftocke, the nature of thefe Nectorins being found by experience to be fo contrary to the Plum ftocke, that it will fterue it, and both dye within a yeare, two or three at the moft : Diuers haue tryed to graft thefe red Nectorins vpon Peach ftockes, and they haue endured well a while; but feeing the Peach ftocke will not laft long it felfe, being ouerweake, how can it hold fo ftrong a nature as thefe red Nectorins, which will (as I faid before) fterue a Plum ftocke that is fufficient durable for any other Plumme?

Apricocke ftockes from the ftones are hardly nurfed vp, and worfe to be remoued, and if a red Nectorin fhould be grafted on an Apricock rayfed from the ftone, and not remoued, I doubt it might happen with it as it doth with many other trees raifed from ftones or kernels, and not remoued, that they would hardly beare fruit : for the nature of moft trees raifed from ftones or kernels, and not remoued, is to fend great downeright rootes, and not to fpread many forwards; fo that if they be not cut away that others may fpreade abroad, I haue feldome feene or known any of them to beare in any reafonable time; and therefore in remouing, thefe great downe-right rootes are alwayes fhred away, and thereby made fit to fhoote others forwards. Hereby you may perceiue, that thefe red Nectorins will not abide to bee grafted vpon any other ftocke well, then vpon an Apricocke branch, although the green and the yellow (as I faid before) will well endure and thriue vpon Plums. The fuckers or fhootes both of Plums and Cherries that rife from their rootes, eyther neare their ftockes, or farther off, fo that they bee taken with fome fmall rootes to them, will ferue to bee ftockes, and will come forward quickly; but if the fuckers haue no fmall roots whereby they may comprehend in the ground, it is almoft impoffible it fhould hold or abide. There is another way to rayfe vp eyther ftockes to graft on, or trees without grafting, which is, by circumcifing a faire and fit branch in this manner : About Midfomer, when the fappe is thoroughly rifen (or before if the yeare be forward) they vfe to binde a good quantity of clay round about a faire and ftraight branch, of a reafonable good fize or bigneffe, with fome conuenient bands, whether it be ropes of hey, or of any other thing, about an handfull aboue the ioynt, where the branch fpreadeth from the tree, and cutting the barke thereof round about vnder the place where the clay is bound, the fap is hereby hindered from rifing, or defcending further then that place fo circumcifed, whereby it will fhoote out fmall knubs and rootes into the clay, which they fuffer fo to abide vntill the beginning of winter, whenas with a fine Sawe they cut off that branch where it was circumcifed, and afterwardes place it in the ground where they would haue it to grow, and ftake it, and binde it faft, which will fhoote forth rootes, and will become eyther a faire tree to beare fruite without grafting, or elfe a fit ftocke to graft on according

ding

ding to the kinde : but oftentimes this kinde of propagation miffeth,in that it fendeth not forth rootes fufficient to caufe it to abide any long time. Let me yet before I leaue this narration of Plummes, giue you one admonition more, that vpon whatfoeuer Plumme ftocke you doe graft,yet vpon a Damfon ftocke that you neuer ftriue to graft, for it (aboue all other forts of Plumme ftockes) will neuer giue you a tree worth your labour. It remaineth only of ftone fruit, that I fpeake of Cornelles, which as yet I neuer faw grafted vpon any ftocke, being as it fhould feeme vtterly repugnant to the nature thereof, to abide grafting, but is wholly rayfed vp (as I faid before)eyther from the ftones,or from the fuckers or layers. For Peares and Apples your vfuall ftockes to graft on are (as I faid before, fpeaking of the nurfing vp of trees from the kernels;)your Crabbe ftockes, and they bee accepted in euery Countrey of this Land as they may conueniently be had, yet many doe take the ftockes of better fruit, whether they bee fuckers, or ftockes rayfed from the kernels (and the moft common and knowne way of grafting, is in the ftocke for all forts of them, although fome doe vfe whipping, packing on,or incifing,as euery one lift to call it : but now we doe in many places begin to deale with Peares and Apples as with other ftone fruit, that is, graft them all in the bud, which is found the moft compendious and fafeft way both to preferue your ftocke from perifhing,and to bring them the fooner to couer the ftock, as alfo to make the goodlier and ftraighter tree, being grafted at what height you pleafe :) for thofe ftockes that are rayfed from the kernels of good fruit (which are for the moft part eafily knowne from others, in that they want thofe thornes or prickles the wilde kindes are armed withall:) I fay for the moft part , for I know that the kernels of fome good fruite hath giuen ftockes with prickles on them (which, as I thinke, was becaufe that good fruite was taken from a wilde ftocke that had not beene long enough grafted to alter his wilde nature ; for the longer a tree is grafted, the more ftrength the fruite taketh from the graft,and the leffe ftill from the ftocke) being fmoother and fairer then the wilde kinds, muft needes make a goodlier tree, and will not alter any whit the tafte of your fruit that is grafted thereon, but rather adde fome better rellifh thereunto ; for the Crabbe ftockes yeelding harfh fruite,muft giue part of their nature to the grafts are fet thereon, and therefore the tafte or rellifh, as well as fome other naturall properties of moft fruits, are fomewhat altered by the ftocke. Another thing I would willingly giue you to vnderftand concerning your fruits and ftockes, that whereas diuers for curiofity and to try experiments haue grafted Cherries vpon Plumme ftockes, or Plums on Cherry ftockes, Apples vpon Peare ftockes,and Peares vpon Apple ftockes, fome of thefe haue held the graft a yeare, two or three peraduenture, but I neuer knew that euer they held long, or to beare fruite, much leffe to abide or doe well : beftow not therefore your paines and time on fuch contrary natures, vnleffe it be for curiofitie, as others haue done : Yet I know that they that graft peares on a white thorne ftocke haue had their grafts feeme to thriue well, and continue long,but I haue feldome feene the fruite thereof anfwerable to the naturall wilde Peare ftocke ; yet the Medlar is knowne to thriue beft on a white thorne. And laftly, whereas diuers doe affirme that they may haue not only good ftockes to graft vpon,but alfo faire trees to bear ftore of fruit from the kernels of Peares or Apples being prickt into the ground, and fuffered to grow without remouing,and then eyther grafted or fuffered to grow into great trees vngrafted ; and for their bearing of fruite, affigne a dozen or twenty yeares from the firft fetting of the kernels,and abiding vngrafted, I haue not feene or heard that experience to hold certaine, or if it fhould be fo, yet it is too long time loft, and too much fruit alfo,to waite twenty yeares for that profit may be gained in a great deale of leffe time, and with more certainty. Vnto thefe inftructions let mee adde alfo one more, which is not much known and vfed,and that is, to haue fruit within foure or fiue years from the firft fowing of your ftones or kernels in this manner : After your ftones or kernels are two or three yeares old, take the faireft toppe or branch, and graft it as you would doe any other cyon taken from a bearing tree,and looke what rare fruite,eyther Peare or Apple,the kernell was of that you fowed,or Peach or Plum &c.the ftone was fet, fuch fruite fhall you haue within two or three yeares at the moft after the grafting, if it take,and the ftocke be good. And thus may you fee fruit in farre leffe time then to ftay vntill the tree from a kernell or ftone beareth fruit of it felfe.

Zz Chap.

Chap. IIII.

The diuers manners of grafting all forts of fruits vfed in our Land.

THe moft vfuall manner of grafting in the ftocke is fo common and well known in this Land to euery one that hath any thing to doe with trees or an Orchard, that I think I fhall take vpon mee a needleffe worke to fet downe that is fo well knowne to moft; yet how common foeuer it is, fome directions may profit euery one, without which it is not eafily learned. And I doe not fo much fpend my time and paines herein for their fakes that haue knowledge, but for fuch as not knowing would faine be taught priuately, I meane, to reade the rules of the arte fet downe in priuate, when they would refufe to learne of a Gardiner, or other by fight: and yet I difcommend not that way vnto them to learne by fight; for one may fee more in an inftant by fight, then he fhall learn by his own practice in a great while, efpecially if he be a little practifed before he fee a cunning hand to doe it. There are many other kindes of grafting, which fhall be fpoken of hereafter, and peraduenture euen they that know it well, may learne fomething they knew not before.

1. The grafting in the ftocke, is, to fet the fprigge of a good fruit into the body or ftocke of another tree, bee it wilde or other, bee it young or old, to caufe that tree to bring forth fuch fruit as the tree bore from whence you took the fprigge, and not fuch as the ftocke or tree would haue borne, if it had not beene grafted, and is performed in this manner : Looke what tree or ftocke you will chufe to graft on, you muft with a fmall fine fawe and very fharpe, whip off, or cut off the head or toppe thereof at what height you eyther thinke beft for your purpofe, or conuenient for the tree : for if you graft a great tree, you cannot without endangering the whole, cut it downe fo low to the ground, as you may without danger doe a fmall tree, or one that is of a reafonable fize; and yet the lower or neerer the ground you graft a young tree, the fafer it is both for your ftocke and graft, becaufe the fappe fhall not afcend high, but foone giue vigour to the graft to take and fhoote quickly : After you haue cut off the toppe of your ftocke, cut or fmooth the head thereof with a fharpe knife, that it may be as plaine and fmooth as you can, and then cleaue it with a hammer or mallet, and with a ftrong knife, cleauer or cheffell, either in the middle of it if it be fmall, or of a reafonable fize, or on the fides an inch or more within the barke, if it be great : into both fides of the cleft put your grafts, or into one if the ftocke bee fmaller; which grafts muft bee made fit for the purpofe on this fafhion : Hauing made choife of your grafts from the toppe branches efpecially, or from the fides of that tree wherof you would haue the fruit, and that they be of a reafonable good fize, not too fmall or too great for your ftockes, and of one or the fame yeares fhoote; (and yet many doe cut an inch or more of the olde wood with the fprigge of the laft yeares growth, and fo graft the old and young together (but both are good, and the old wood no better then the young) cut your graft not too long, but with two, three or foure eyes or buds at the moft, which at the lower or bigger end for an inch long or more (for the greater ftockes, and an inch or leffe for the leffer fort) muft be fo cut, that it be very thin on the one fide from the fhoulders downward, and thicker on the other, and thin alfo at the end, that it may goe downe clofe into the cleft, and reft at the fhoulders on the head of the ftocke : but take heede that in cutting your grafts your knife bee very fharpe that you doe not rayfe any of the barke, eyther at the fides or the end, for feare of lofing both your paines and graft, and ftocke too peraduenture; and let not your grafts bee made long before you fet them, or elfe put the ends of them in water to keepe them frefh and cleane : when you fet them you muft open the cleft of your ftocke with a wedge or cheffell as moft doe, that the graft may goe eafily into it, and that the barke of both graft and ftocke may ioyne clofe the one to the other, which without ftirring or difplacing muft bee fo left in the cleft, and the wedge or cheffell gently pulled forth; but becaufe in the doing hereof confifteth in a manner the whole loffe or gaine of your paines, graft and ftocke, to preuent which inconuenience I doe vfe an iron Inftrument, the forme wherof is fhowne in the following

lowing page, marked with the letter A, crooked at both ends, and broade like vnto a cheffell, the one bigger, and the other leffer, to fit all forts of ftockes, and the iron handle fomewhat long betweene them both, that being thruft or knocked downe into the cleft, you may with your left hand open it as wide as is fit to let in your graft, without ftrayning, which being placed, this iron may bee pulled or knocked vp againe without any mouing of your graft : when you haue thus done, you muft lay a good handfull or more (according to the bigneffe of your ftocke) of foft and well moiftned clay or loame, well tempered together with fhort cut hey or horfe dung, vpon the head of your ftocke, as lowe or fomewhat lower then the cleft, to keepe out all winde, raine or ayre from your graft vntill Midfomer at the leaft, that the graft be fhot forth fomewhat ftrongly, which then if you pleafe may be remoued, and the cleft at the head only filled with a little clay to keepe out eatewigs, or other things that may hurt your graft.

A. The Iron Inftrument with cheffels at each end, the one bigger and the other leffer, to keepe the cleft of the Tree open vntill the graft bee placed in the ftocke, which with a knock vpwards will be eafily taken away.

B. The fmall Penne-knife with a broad and thinne ended hafte, to raife the fides both of the bud and the down-right flit in the body or arme of a Tree to be grafted in the bud.

C. A pen or quil cut halfe round to take off a bud from the branch.

D. An Iuory Inftrument made to the fame fafhion.

E. A fhielde of braffe made hollow before to be put into the flit, to keepe it open vntill the bud be put into its place.

F. The manner of grafting called incifing or fplicing.

G. A Ladder made with a ftoole at the toppe, to ferue both to graft higher or lower, and alfo to gather fruit without fpoyling or hurting any buddes or branches of Trees.

1. The firft flit in the body or arme of a Tree to be grafted in the bud with the croffe cut at the head.

2. The fame flit opened on both fides, ready to receiue the budde fhould be put therein : thefe fmall peeces ferue as well as trees to fhew the manner and order of the grafting.

3. The branch of a Tree with one budde cut ready to be taken off, and another not yet touched.

4. The bud cleane taken off from the branch, both the forefide and backfide.

5. The graft or bud now put into the ftocke or tree you intend to be grafted : but the binding thereof is omitted.

2. Inarching is another manner of grafting in the ftocke, and is more troublefome, and more cafuall alfo then the former, and is rather a curiofity then any way of good fpeede, certainety or profit, and therefore vfed but of a few. Yet to fhew you, the

man-

manner thereof, it is thus: Hauing a tree well growne, bee it high or low, yet the lower the better, with young branches well spread, they vse to set stockes round about it, or on the one side as you please ; into which stockes they ingratt the young branches of the well growne tree as they are growing, before they cut them from the tree) by bowing downe the branch they intend to graff, and putting it into the stocke, hauing first cut off the head thereof, and cut a notch in the middle of the head a little slope on both sides, wherein the branch must be fitted : let the branch be cut thinne on the vnderside, only of that length as may suffice to fit the notch in the stocke, leauing about halfe a yarde length of the branch, to rise aboue or beyond the stocke, which beeing bound on, and clayed ouer or couered with red or greene soft waxe, they let so abide, that if it take in the stocke they cut off the branch a little below the grafting place in Nouember following, and remouing the stocke, they haue thus gained a grafted and growne tree the first yeare : but it is vsually seene, that where one branch taketh, three doe misse : yet this manner of grafting was much in vse for May Cherries, when they were first known to vs, and the way thought to be a rare manner of grafting to encrease them, vntill a better way was found out, which now is so common and good also, that this is not now scarce thought vpon.

3. Another kinde of grafting in the stocke is called of some whipping, of some splicing, of others incising, and of others packing on (and as I heare, is much vsed in the West parts especially, and also in the North parts of this Land) and is performed in this manner : Take and slice the branch of a tree (so as the branch be not too bigge) or else a young tree of two, or three, or foure yeares growth at the most, quite off slope wise, about an inch and a halfe long or more, and cut a deep notch in the middle thereof, then fit into it a graft iust of that size or bignesse, cut on both sides with shoulders, and thin at the end, that it may ioyne close in the notch, and neyther bigger or lesser, but that the barke of the one may bee fitted iust to the barke of the other, the figure wherof is expressed at the letters E. F. which shew the one to be with a shoulder & the other without ; binde them gently together with bast, and put clay or waxe ouer the place, vntill it be taken : this is much vsed of late dayes for such young trees as are risen of stones or kernels after the second or third yeares growth, and thriue very well in that it not only saueth much time, but diuers checks by remouing and grafting.

4. Inoculating or grafting in the budde is another manner of grafting, which is the taking of a budde from one tree, and putting it into the barke of another tree, to the end, that thereby you may haue of the same kinde of fruit the tree bare from whence the budde was taken ; and although it bee sufficiently knowne in many places of this Land, yet as I vnderstand, good Gardiners in the North parts, and likewise in some other places, can scarce tell what it meaneth, or at the least how to doe it well. It is performed after a different fashion from the former, although they all tend vnto one end, which is the propagating of trees. You must for this purpose obserue, that for those trees you would graft, either with, or vpon , you choose a fit time in Summer, when the sappe is well risen, and your graft well shot, that the barke will rise easily and cleanly, both of stocke and graft, which time I cannot appoint, becauſe both the years doe differ in earlinesse , and the seuerall parts or countries of this Land likewise one from another, but most vsually in these Southerne parts, from the beginning of Iune vnto the end of it, or to the middle of Iuly, or either somewhat before or after. First (as I said) hauing taken the fittest time of the yeare, you must take especiall care, that your grafts be well growne, and of the same yeares shoote , and also that the buds or eyes haue but single leaues at them , as neare as you can : for I would vtterly refuse those buds that haue aboue two leaues as vnprofitable, either in Peaches or any other fruit ; and therefore see that your grafts or cyons bee taken from the chiefest place of the tree, that is, either from the toppe, or from a sunnie side thereof, and not from the contrarie side if you may otherwise, nor from any vnder-boughes ; for seeing your graft is so small a thing, you had neede take the more care that it be the best and faireſt. You must to take off this eye or budde from the sprigge, haue a small sharpe pen-knife, the end of the haft being made flat and thinne, like a cheſſell or wedge , the figure whereof is set forth at the letter B, and a pen or goose quill cut, to be lesse then halfe round, and to be broad at the end, but not sharpe pointed like a penne , or else such a peece of bone or Iuorie made in that fashion as the quill is , to bee thinne, hollow, or

halfe round, the figures of both which are marked with the letters C, D. with your knife cut the barke of the bud (hauing firſt cut off the leafe, leauing onely the ſhort foote ſtalke thereof at the bud) about a ſtrawes breadth aboue the eye thereof halfe round, and then from that round or ouerthwart cut, with your knife cut it downe on both ſides of the eye, cloſe to the bud ſlopewiſe about an inch long or thereabouts, that it bee broad at the head aboue the eye, and pointing at the end like a ſheild or ſcutcheon; and then cutting away the reſt of the barke from about it, with the thinne flat end of the haft of your knife raiſe vp both ſides of your bud a little, and with your quill or bone put vnder the barke, raiſe your budde, and thruſt it quite off, beginning at the toppe or head of your eye; but ſee that you thruſt it off cloſe to the wood of the branch or ſprigge, and that you doe not leaue the eye of the budde behinde ſticking vpon the branch; for if that eye be left or loſt, your bud is worth nothing; you muſt caſt it away, and cut another that may haue that eye abiding within the budde on the inſide: you may perceiue if that eye be wanting, if you ſee an emptie hole in the place where the eye ſhould be, to fill it vp on the inſide thereof; thus hauing taken off your bud well and cleanly, which is ſet forth vnto you at the figures 3 and 4. preſently ſet it on the tree you would graft (for your ſmall bud can abide no delay, leſt by taking the ayre too long it become dry, and nothing worth) in this manner: Cut the barke of your tree you would graft in a ſmooth place, at what height you pleaſe, firſt aboue or ouerthwart, and then downe right in the middle thereof, more then an inch long, the figure whereof you ſhall haue at the figure 1. and then raiſe vp both ſides of the barke, firſt one, and then another, with the flat and thinne haft end of your knife, a prettie way inwards (for if the barke will not riſe eaſily, the ſtocke is not then fit to graft vp-on) put in your budde into the cleft with the point downewards, holding the ſtalke of the leafe that is with the budde betweene your fingers of the one hand, and opening the cleft with the flat end of your knife with the other hand, that the head of your bud may be put cloſe vnder the ouerthwart cut in the ſtocke or tree (which muſt not be raiſed or ſtirred as the ſides are) & the eye of the bud ſtand iuſt in the middle of the ſlit that is downeright, and then cloſing the barke of the ſtocke or tree ſoftly vnto the bud thus put in with your fingers, let it be bound gently with a ſmall long peece of baſte, or other ſuch like ſoft thing, firſt aboue the eye, & then compaſſing it belowe as cloſe as you can, but not too hard in any caſe, vntil you haue bound it all ouer the ſlit you made, eſpecially the lower end, leſt any winde get into to dry and ſpoile it; and hauing tyed both ends thereof faſt, leaue it ſo for a fornight or ſomewhat more, in which ſpace it will take and hold, if it be well done, which you ſhall perceiue, if the bud abide green, and turne not blacke, when you haue vnlooſed the tying; for if it hold faſt to the tree, and be freſh and good, tye it vp gently againe, and ſo leaue it for a fortnight longer, or a moneth if you will, and then you may take away your binding cleane: this budde will (if no other miſchance happen vnto it) ſpring and ſhoote forth the next yeare, (and ſometimes the ſame yeare, but that is ſeldome) and therefore in the beginning of the yeare, cut off the head of the grafted tree about an handfull aboue the grafted place, vntill the graft be growne ſtrong, and then cut it off cloſe, that the head may be couered with the graft, and doe not ſuffer any buds to ſprout beſides the graft, either aboue or belowe it. If you graft diuers buds vpon one ſtocke (which is the beſt way) let that onely remaine and abide that ſhooteth beſt forth, and rubbe off, or take away the other: the ſeuerall parts of this grafting I haue cauſed to be expreſſed for your fur-ther information.

5. Grafting in the ſcutcheon is accounted another kinde of grafting, and differeth verie little from grafting in the budde: the difference chiefly conſiſteth in this, that in ſtead of the downe right ſlit, and that aboue ouerthwart, they take away iuſt ſo much barke of the great tree, as your bud is in bigneſſe, which vſually is a little larger then the former, and placing it therein, they binde it as formerly is ſaid: ſome vſe for this purpoſe a paire of compaſſes, to giue the true meaſure both of bud and ſtocke; this manner of grafting is moſt vſed vpon greater trees, whoſe young branches are too high to graft vpon in the former maner, and whoſe tops they cut off (for the moſt part) at the latter end of the next yeare after the bud is taken: both theſe waies were inuen-ted to ſaue the loſſe of trees, which are more endangered by grafting in the ſtocke,

then

then any of thefe waies; and befides, by thefe waies you may graft at a farre grea
height without loffe.

---

*Of the manner of grafting and propagating all
forts of Rofes.*

HAuing now fpoken of the grafting of trees, let mee adioyne the properties of
Rofes, which although they better fit a Garden then an Orchard, yet I could
not in a fitter place expreffe them then here, both for the name and affinity of
grafting,& becaufe I do not expreffe it in the firft part. All forts of Rofes may be grafted (although all forts are not, fome feruing rather for ftockes for others to be grafted
on) as eafily as any other tree,& is only performed, by inoculating in the fame maner I
haue fet downe in the former Chapter of grafting trees in the bud; for both ftocke and
budde muft bee dealt with after the fame fafhion. And although fome haue boafted of
grafting Rofes by flicing or whipping, as they call it, or in the ftocke, after the firft
manner, fet downe in the former Chapter, yet I thinke it rather a bragge, not hauing
feene or heard any true effect proceede from that relation. The fweete Briar or Eglantine, the white and the Damaske Rofes, are the chiefeft ftockes to graft vpon. And if
you graft lowe or neare the ground, you may by laying downe that graft within the
ground, after it hath bin fhot out well, and of a years growth, by pinning it faft downe
with fhort ftickes, a thwart or acroffe, caufe that grafted branch, by taking roote, to
become a naturall Rofe, fuch as the graft was, which being feparated and tranfplanted
after it hath taken root wel, will profper as well as any naturall fucker. And in this maner, by laying downe branchefe at length into the ground, if they be full of fpreading
fmall branches, you may increafe all forts of Rofes quickly and plentifully; for they
will fhoote forth rootes at the ioynt of euery branch: But as for the manner of grafting
white Rofes or Damaske vpon Broome ftalkes or Barbary bufhes, to caufe them to
bring forth double yellow Rofes, or vpon a Willowe, to beare greene Rofes, they
are all idle conceits, as impoffible to be effected, as other things, whereof I haue fpoken in the ninth Chapter of my firft part, concerning a Garden of flowers, vnto which
I referre you to be fatisfied with the reafons there alledged. And it is the more needleffe, becaufe we haue a naturall double yellow Rofe of it owne growing. The fowing
of the feedes of Rofes (which are fometimes found vpon moft forts of Rofes, although
not euery yeare, and in euerie place) hath bin formerly much vfed; but now the laying
downe of the young fhootes is a way for increafe fo much vfed, being fafe and verie
fpeedie to take, efpecially for thofe Rofes that are not fo apt to giue fuckers, that it
hath almoft taken quite away the vfe of fowing of the feedes of Rofes, which yet if
anie one bee difpofed to make the triall, they muft gather the feede out of the round
heads, from amongft the doune, wherein they lye verie like vnto the berries of the
Eglantine or fweete Briar bufh, and efpecially of thofe Rofes that bee of the more fingle kindes, which are more apt to giue berries for feed then the more double, although
fometimes the double Rofes yeeld the like heads or berries. Their time of fowing is
in the end of September (yet fome referue them vntill February) and their manner
of nourfing is to bee tranfplanted, after the firft or fecond yeares growth, and tended carefully, that while they are young they be not loft for want of moifture in the
dry time of Summer.

CHAP. VI.

Certaine rules and obseruations in and after grafting, not remembred in the former Chapter.

THe time of some manners of grafting being not mentioned before, must here be spoken of. For the grafting of all sorts of trees in the stocke, the most vsuall time is from the middle of February vntill the middle of March, as the yeare and the countrie is more forward or backward, with vs about London wee neuer passe midde March : but because the May Cherrie is first ripe, and therefore of a very forward nature, it doth require to be grafted somewhat sooner then others. The time of gathering likewise, or cutting your grafts for grafting in the stocke, is to be obserued, that they bee not long gathered before they bee grafted, for feare of being too dry, which I commend, howsoeuer diuers say, if they be long kept they are not the worse; and therefore if you be forced to haue your grafts from tarre, or by some other chance to keepe them long, be carefull to keepe them moist, by keeping their ends stucke in moist clay; but if neare hand, neglect no time I say after the cutting of them for their grafting, but either the same, or the next day, or verie speedily after, in the meane time being put into the ground to keepe them fresh. The grafts taken from old trees, because they are stronger, and shoote forth sooner, are to bee sooner grafted then those that are taken from younger trees : of a good branch may bee made two, and sometimes three grafts sufficient for anie reasonable stocke. For whipping, the time is somewhat later then grafting in the stocke, because it is performed on younger trees, which (as I said before) doe not so early bud or shoote forth as the elder. Inarching likewise is performed much about the later end of the grafting time in the stocke; for being both kindes thereof they require the same time of the yeare. The times of the other manners of graftings are before expressed, to bee when they haue shot forth young branches, from whence your buds must be taken; and therefore need not here againe to be repeated. If a graft in the stocke doth happen not to shoote forth when others do (so as it holdeth green) it may perchance shoot out a moneth or two after, & do well, or else after Midsummer, when a second time of shooting, or the after Spring appeareth: but haue an especiall care, that you take not such a graft that shal haue nothing but buds for flowers vpon it, and not an eye or bud for leaues (which you must be carefull to distinguish) for such a graft after it hath shot out the flowers must of necessitie dye, not hauing wherewith to maintaine it selfe. Also if your good graft doe misse, and not take, it doth hazzard your stocke at the first time, yet manie stockes doe recouer to be grafted the second time; but twice to faile is deadly, which is not so in the inoculating of buds in the greene tree : for if you faile therein three, or three times three, yet euerie wound being small, and the tree still growing greene, will quickly recouer it, and not be afterwards seen. Some vse to graft in the stocke the same yeare they remoue the stocke, to saue time, & a second checke by grafting; but I like better both in grafting in the stocke, and in the bud also, that your trees might be planted in the places where you would haue them growe, for a yeare or two at the least before you graft them, that after grafting there should be no remouall, I neede not be tedious, nor yet I hope verie sollicitous to remember many other triuiall, or at the least common knowne things in this matter. First, for the time to remoue trees, young or old, grafted or vngrafted, to be from a fortnight after Michaelmas vntill Candlemas, or if neede be, somewhat after, yet the sooner your remoue is, the better your trees will thriue, except it be in a very moist ground. For the manner or way to set them : *viz.* in the high and dry grounds set them deeper, both to haue the more moisture, and to be the better defended from windes; and in the lower and moister grounds shallower, and that the earth be mellow, well turned vp, and that the finer earth bee put among the small rootes, wherein they may spread, and afterwards gently troden downe, that no hollownesse remaine among the rootes: as also that after setting (if the time be not ouermoist) there may be some water powred to the rootes, to moisten and fasten them the better; and in the dry time of Summer, after the setting, let them not want moisture, if you will

haue

haue them thriue and profper ; for the want thereof at that time, hath often killed manie a likely tree. To ftake and fence them alfo if neede bee after they are new fet, and fo to continue for two or three yeares after, is verie expedient, left windes or other cafualties fpoile your paines, and ouerthrow your hopes. And likewife to defend your grafts from birds lighting on them, to breake or difplace them, to fticke fome prickes or fharpe pointed ftickes longer then your graft into your clay, that fo they may be a fure defence of it : As alfo to tye fome woollen cloathes about the lower end of your ftockes, or thruft in fome thornes into the ground about the rootes, to defend them from hauing their barkes eaten by Conies, or hurt by fome other noifome vermine.

Chap. VII.

Obferuations for the drefsing and well keeping of Trees and an Orchard in good order.

THere are two manner of waies to dreffe and keepe trees in good order, that they may bee both gracefull and fruitfull ; the one is for wall-trees, the other is for ftandards : for as their formes are different, fo is their keeping or ordering. Wall trees, becaufe they are grafted lowe, and that their branches muft be plafht or tackt vnto the wall to faften them, are to be fo kept, that all their branches may be fuffered to growe, that fhoote forth on either fide of the bodie, and led either along the wall, or vpright, and one to lappe ouer or vnder another as is conuenient, and ftill with peeces of lifts, parings of felt, peeces of foft leather, or other fuch like foft thing compaffing the armes or branches, faftened with fmall or great nailes, as neede requireth, to the wals, onely thofe buds or branches are to be nipped or cut off, that fhoot forward, and will not fo handfomely be brought into conformity, as is fitting ; yet if the branches growe too thicke, to hinder the good of the reft, or too high for the wall, they may, nay they muft be cut away or lopped off : and if anie dead branches alfo happen to be on the trees, they muft be cut away, that the reft may haue the more libertie to thriue. Diuers alfo by carefully nipping away the wafte and fuperfluous buds, doe keepe their trees in conformity, without much cutting. The time to pruine or plafh, or tye vp wall trees, is vfually from the fall of the leafe, to the beginning of the yeare, when they begin to bloffome, and moft efpecially a little before or after Chriftmas : but in any cafe not too late, for feare of rubbing off their buds. Some I know doe plafh and tye vp their wall trees after bearing time, while the leaues are greene, and their reafon is, the buds are not fo eafie or apt to bee rubbed from the branches at that time, as at Chriftmas, when they are more growne : but the leaues muft needes be very cumberfome, to hinder much both the orderly placing, and clofe faftening of them to the wall. This labour you muft performe euery yeare in its due time ; for if you fhall negle& and iouerflip it, you fhall haue much more trouble, to bring them into a fit order againe, then at the firft. The ftandard trees in an Orchard muft be kept in another order ; for whereas the former are fuffered to fpread at large, thefe muft be pruined both from fuperfluous branches that ouerload the trees, & make them leffe fruitfull, as well as leffe fightly, and the vnder or water boughes likewife, that drawe much nourifhment from the trees, and yet themfelues little the better for it, I meane to giue fruit. If therefore your Orchard confift of young trees, with a little care and paines it may bee kept in that comely order and proportion it was firft deftined vnto ; but if it confift of old growne trees, they will not without a great deale of care and paines be brought into fuch conformitie, as is befitting good and comely trees : for the marke of thofe boughes or branches that are cut off from young trees, will quickly be healed againe, the barke growing quickly ouer them, whereby they are not worfe for their cutting ; but an old tree if you cut off a bough, you muft cut it clofe and cleanly, and lay a fearcloth of tallow, waxe, and a little pitch melted together vpon the place, to keepe off both the winde, funne, and raine, vntill the barke haue couered it ouer againe : and in this manner you muft deale with all fuch fhort ftumps of branches, as are either broken fhort off with the winde, or by carelefneffe or want of skill, or elfe fuch armes or branches as are broken off clofe, or fliued from the body of the tree : for the raine beating and falling into fuch a place, will in fhort time rotte your tree, or put it in danger, befides the deformity. Some vfe to fill vp fuch an hole with well tempered clay, and tacke a cloth or a peece of leather ouer it vntill it be recouered, and this is alfo not amiffe. Your young trees, if they ftand in anie good ground, will bee plentifull enough in fhooting forth branches ; bee carefull therefore if they growe too thicke, that you pruine away fuch as growe too clofe (and will, if they be fuffered, fpoile one another) as they may be beft fpared, that fo the funne, ayre, and raine may haue free acceffe to all your branches, which will make them beare the more plentifully, and ripen them the fooner and the more kindly. If anie boughes growe at the toppe too high, cut them alfo away, that your trees may rather fpread then growe too high. And fo likewife for the vnder boughes, or anie other that by the weight of fruit fall or hang downe, cut them off at the halfe, and they will afterwards rife and fhoote vpwards. You fhall obferue, that at all thofe places where anie branches haue been cut away, the fappe will euer bee readie to put forth : if therefore you would haue no more branches rife from that place, rubbe off or nippe off fuch buddes as are not to your minde, when they are new fhot : and thus you may keep your trees in good order with a little paines, after you haue thus pruined and dreffed them. One other thing I would aduertife you of, and that is how to preferue a fainting or decaying tree which is readie to perifh, if it be not gone too farre or paft cure, take a good quantitie of oxe or horfe bloud, mixe therewith a reafonable quantitie of fheepe or pigeons dung, which being laid to the roote, will by the often raines and much watering recouer it felfe, if there bee anie poffibilitie ; but this muft bee done in Ianuarie or Februarie at the furtheft.

CHAP. VIII.

Diuers other obferuations to be remembred in the well keeping of an Orchard.

THere be diuers other things to be mentioned, whereof care muft be had, either to doe or auoide, which I thinke fit in this Chapter promifcuoufly to fet downe, that there may be nothing wanting to furnifh you with fufficient knowledge of the care, paines, and cafualties that befall an Orchard : for it hath many enemies, and euery one laboureth as much as in them lye, to fpoile you of your pleafure, or profit, or both, which muft bee both fpeedily and carefully preuented and helped ; and they are thefe : Moffe, Caterpillars, Ants, Earwigs, Snailes, Moales, and Birds. If Moffe begin to ouergrowe your trees, looke to it betimes, left it make your trees barren: Some vfe to hacke, and croffehacke, or cut the barke of the bodies of their trees, to caufe it fall away; but I feare it may endanger your trees. Others do either rubbe it off with a haire cloth, or with a long peece of wood formed like a knife, at the end of a long fticke or pole, which if it bee vfed cauteloufly without hurting the buds, I like better. Caterpillars, fome fmoake them with burning wet ftrawe or hay, or fuch like ftuffe vnder the trees; but I doe not greatly like of that way : others cut off the boughes whereon they breed, and tread them vnder their feete, but that will fpoile too manie branches; and fome kill them with their hands : but fome doe vfe a new deuifed way, that is, a pompe made of lattin or tin, fpout-fafhion, which being fet in a tubbe of water vnder or neare your trees, they will caufe the water to rife through it with fuch a force, and through the branches, that it will wafh them off quickly. To deftroy Ants, that eate your fruit before and when it is ripe, fome vfe to annoint the bodies of their trees with tarre, that they may not creepe vp on the branches; but if that doe not helpe, or you will not vfe it, you muft be carefull to finde out their hill, and turne it vp, pouring in fcalding water, either in Summer, but efpecially if you can in Winter, and that will furely deftroy them. I haue fpoken of Earwigs in the firft part of this worke, entreating of the annoyances of Gilloflowers, and therefore I referre you thereunto : yet one way more I

will

will here relate which some doe vse, and that is with hollow canes of halfe a yard long or more, open at both ends for them to creepe in, and stucke or laid among the branches of your trees, will soone drawe into them many Earwigs, which you may soone kill, by knocking the cane a little vpon the ground, and treading on them with your foote. Snailes must be taken with your hands, and that euerie day, especially in the morning when they will be creeping abroad. Moales by running vnder your trees make them lesse fruitfull, and also put them in danger to be blowne downe, by leauing the ground hollow, that thereby the rootes haue not that strength in the ground, both to shoote and to hold, that otherwise they might haue. Some haue vsed to put Garlicke, and other such like things into their holes, thinking thereby to driue them away, but to no purpose: others haue tryed manie other waies; but no way doth auaile anie thing, but killing them either with a Moale spade, or a trappe made for the purpose as manie doe know: and they must bee watched at their principall hill, and trenched round, and so to be caught. Birds are another enemie both to your trees and fruit; for the Bullfinch will destroy all your stone fruit in the budde, before they flower, if you suffer them, and Crowes, &c. when your Cherries are ripe: for the smaller birds, Lime twigs set either neare your trees, or at the next water where they drinke, will helpe to catch them and destroy them. And for the greater birds, a stone bowe, a birding or fowling peece will helpe to lessen their number, and make the rest more quiet: or a mill with a clacke to scarre them away, vntill your fruit be gathered. Some other annoyances there are, as suckers that rise from the rootes of your trees, which must be taken away euerie yeare, and not suffered to growe anie thing great, for feare of robbing your trees of their liuelihood. Barke bound, is when a tree doth not shoote and encrease, by reason the barke is as it were drie, and will not suffer the sappe to passe vnto the branches: take a knife therefore, and slit the barke downe almost all the length of the tree in two or three places, and it will remedy that euill, and the tree will thriue and come forward the better after. Barke pilled is another euill that happeneth to some trees, as well young as old, either by reason of casuall hurts, or by the gnawing of beasts, howsoeuer it bee, if it bee anie great hurt, lay a plaister thereon made of tallow, tarre, and a little pitch, and binde it thereto, letting it so abide vntill the wound bee healed: yet some doe only apply a little clay or loame bound on with ropes of hay. The Canker is a shrewd disease when it happeneth to a tree; for it will eate the barke round, and so kill the very heart in a little space. It must be looked vnto in time before it hath runne too farre; most men doe wholly cut away as much as is fretted with the Canker, and then dresse it, or wet it with vinegar or Cowes pisse, or Cowes dung and vrine, &c. vntill it be destroyed, and after healed againe with your salue before appointed. There are yet some other enemies to an Orchard: for if your fence be not of bricke or stone, but either a mudde wall, or a quicke set or dead hedge, then looke to it the more carefully, and preuent the comming in of either horse, or kine, sheepe, goates, or deere, hare, or conie; for some of them will breake through or ouer to barke your trees, and the least hole almost in the hedge will giue admittance to hares and conies to doe the like. To preuent all which, your care must be continuall to watch them or auoide them, and to stoppe vp their entrance. A dogge is a good seruant for many such purposes, and so is a stone bowe, and a peece to make vse of as occasion shall serue. But if you will take that medicine for a Canker spoken of before, which is Cowes dung and vrine mixed together, and with a brush wash your trees often to a reasonable height, will keepe hares and conies from eating or barking your trees. Great and cold windes doe often make a great spoile in an Orchard, but great trees planted without the compasse thereof, as Wall-nuts, Oakes, Elmes, Ashes, and the like, will stand it in great stead, to defend it both early and late. Thus haue I shewed you most of the euils that may happen to an Orchard, and the meanes to helpe them, and because the number is great and daily growing, the care and paines must be continuall, the more earnest and diligent, lest you lose that in a moment that hath been growing many yeares, or at the least the profit or beauty of some yeares fruit.

CHAP.

Chap. IX.

The manner and way how to plant, order, and keepe other trees that
beare greene leaues continually.

THe way to order thofe trees that beare their leaues greene continually, is differing from all others that doe not fo : for neyther are they to bee planted or
remoued at the time that all other trees are fet, nor doe they require that manner of dreffing, pruining and keeping, that others doe. And although many ignorant
perfons and Gardiners doe remoue Bay trees, and are fo likewife perfwaded that all
other trees of that nature, that is, that carry their greene leaues continually, may
bee remoued in Autumne or Winter, as well as all other trees may bee; yet it
is certaine it is a great chance if they doe thriue and profper that are fet at that
time, or rather it is found by experience, that fcarce one of ten profpereth well that
are fo ordered. Now in regard that there be diuers trees and fhrubs mentioned here in
this booke that beare euer greene leaues, wherein there is very great beauty, and many
take pleafure in them ; as the ordinary Bay, the Rofe Bay, and the Cherry Bay trees,
the Indian Figge, the Cypreffe, the Pinetree, the Mirtle and dwarfe Boxe, and many
others ; I will here fhew you how to plant and order them, as is fitteft for them. For
in that they doe not fhed their greene leaues in winter as other trees doe, you may in
reafon be perfwaded that they are of another nature; and fo they are indeede: for feeing they all grow naturally in warme Countries, and are from thence brought vnto vs,
we muft both plant them in a warmer place, and tranfplant them in a warmer time then
other trees be, or elfe it is a great hazzard if they doe not perifh and dye, the cold and
frofts in the winter being able to pierce them through, if they fhould bee tranfplanted
in winter, before they haue taken roote. You muft obferue and take this therefore for
a certaine rule, that you alwaies remoue fuch trees or fhrubbes as are euer greene in the
fpring of the yeare, and at no time elfe if you will doe well, that is, from the end of
March, or beginning of Aprill, vnto the middle or end of May, efpecially your more
dainty and tender plants, fhadowing them alfo for a while from the heate of the Sun,
and giuing them a little water vpon their planting or tranfplanting; but fuch water as
hath not prefently been drawn from a Well or Pumpe, for that will go neer to kill any
plant, but fuch water as hath ftood in the open ayre for a day at the leaft, if not two or
three. Yet for dwarfe Boxe I confeffe it may endure one moneth to be earlier planted
then the reft, becaufe it is both a more hardy and lowe plant, and thereby not fo much
fubieɛt to the extremitie of the colde : but if you fhould plant it before winter, the
frofts would raife it out of the ground, becaufe it cannot fo foone at that time of the
yeare take roote, and thereby put it in danger to be loft. Moreouer all of them will
not abide the extremitie of our winter frofts, and therefore you muft of neceffity houfe
fome of them, as the Rofe Bay, Mirtle, and fome others, but the other forts being fet
where they may bee fomewhat defended from the cold windes, froftes, and fnow in
winter, with fome couering or fhelter for the time, will reafonably well endure and
beare their fruit, or the moft of them. If any be defirous to be furnifhed with ftore of
thefe kinds of trees that will be nourfed vp in our Country, he may by fowing the feed
of them in fquare or long wooden boxes or chefts made for that purpofe, gaine plenty
of them : but hee muft be carefull to couer them in winter with fome ftraw or fearne,
or beane hame, or fuch like thing layd vpon croffe fticks to beare it vp from the plants,
and after two or three yeares that they are growne fomewhat great and ftrong, they
may bee tranfplanted into fuch places you meane they fhall abide : yet it is not amiffe
to defend them the firft yeare after they are tranfplanted, for their more fecuritie : the
feedes that are moft vfually fowen with vs, are, the Cypreffe tree, the Pine tree, the
Baye, the Pyracantha or prickly Corall tree, and the Mirtle : the Rofe Bay I haue had
alfo rifen from the feede that was frefh, and brought me from Spaine. But as for Orenge
trees, becaufe they are fo hardly preferued in this our cold climate (vnleffe it bee with
fome that doe beftow the houfing of them, befides a great deale more of care and re-
fpeɛt vnto them) from the bitterneffe of our cold long winter weather (although their
kernels

kernels being put into the ground in the Spring or Summer, and if care bee had of them and conuenient keeping, will abide, and by grafting the good fruite on the crab ftocke they may bee in time nurfed vp) I doe not make any other efpeciall account of them, nor giue you any further relation of their ordering. Now for the ordering of thefe trees after they are eyther planted of young fets, or tranfplanted from the fæde, it is thus : Firft for Bay trees, the moft vfuall way is to let them grow vp high to bee trees, and many plant them on the North or Eaft fide of their houfes that they may not bee fcorched with the Sunne; but the bitter winters which we often haue, doe pinch them fhrewdly, infomuch that it killeth euen well growne trees fometimes dowe to the roote : but fome doe make a hedge of them being planted in order, and keep them low by lopping of them continually, which will make them bufh and fpread. The Cy-preffe tree is neuer lopped, but fuffered to grow with all the branches from a foote a-boue the ground, if it may be, ftraight vpright; for that is his natiue grace and greateft beautie, and therefore the more branches doe dye that they muft bee cut away, the more you deforme his propertie. The Pine tree may be vfed in the fame manner, but yet it wil better endure to fuftaine pruining then the Cypreffe, without any fuch defor-mitie. The Laurocerafus or Cherry Bay may be diuerfly formed, that is, it may be ei-ther made to grow into a tall tree by fhredding ftill away the vnder branches, or elfe by fuffering all the branches to grow to be a low or hedge bufh, & both by the fuckers and by laying downe the lower branches into the earth, you may foone haue much increafe ; but this way will caufe it to bee the longer before it beare anie fruit. The Rofe Baye will verie hardlie bee encreafed either by fuckers or by layers, but muft bee fuffered to grow without lopping, topping or cutting. The Pyracantha or Prickly Corall tree may bee made to grow into a reafonable tall tree by fhredding away the lower branches, or it may be fuffered to grow lowe into an hedge bufh, by fuffering all the branches to grow continually, you may alfo propagate it by the fuckers, or by lay-ing downe the lower branches. The Myrtle of all forts abideth a low bufh fpreading his branches full of fweete leaues and flowers, without anie great encreafe of it felfe, yet fometimes it giueth fuckers or fhootes from the rootes : but for the more fpeedie propagating of them, fome doe put the cuttings of them into the earth, and thereby in-creafe them. There are fome other trees that are not of any great refpect, as the Yew tree, and the Savine bufh, both which may be encreafed by the cuttings, and therefore I need not make any further relation or amplification of them, and to fay thus much of them all, is (I thinke) fufficient for this Worke.

CHAP. X.

The ordering, curing, and propagating Vines of all forts.

IN moft places of this countrie there is fmall care or paines taken about the orde-ring of Vines : it fufficeth for the moft part with them that haue anie, to make a frame for it to fpread vpon aboue a mans height, or to tacke it to a wall or win-dow, &c. and fo to let it hang downe with the branches and fruit, vntill the weight thereof, and the force of windes doe teare it downe oftentimes, and fpoile the grapes : and this way doth fomewhat refemble that courfe that the Vineyard keepers obferue in the hot countries of Syria, Spaine, and Italy, and in the furtheft parts of France as I hear likewife : for in moft of thefe hot countries they vfe to plant an Oliue betweene two Vines, and let them runne thereupon. But manie of the other parts of France, &c. doe not fuffer anie trees to growe among their Vines ; and therefore they plant them thicke, and pruine them much and often, and keepe them lowe in comparifon of the other way, faftening them to pearches or poles to hold them vp. And according to that fafhion many haue aduentured to make Vineyards in England, not onely in thefe later daies, but in ancient times, as may wel witneffe the fundrie places in this Land, en-tituled by the name of Vineyards; and I haue read, that manie Monafteries in this King-dome hauing Vineyards, had as much wine made therefrom, as fufficed their couents yeare by yeare : but long fince they haue been deftroyed, and the knowledge how to order a Vineyard is alfo vtterly perifhed with them. For although diuers, both Nobles and Gentlemen, haue in thefe later times endeauoured to plant and make Vine-yards, and to that purpofe haue caufed French men, being skilfull in keeping and dref-fing of Vines, to be brought ouer to performe it, yet either their skill failed them, or their Vines were not good, or (the moft likely) the foile was not fitting; for they could neuer make anie wine that was worth the drinking, being fo fmall and heartleffe, that they foone gaue ouer their practice. And indeede the foile is a maine matter to bee chiefly confidered to feate a Vineyard vpon: for euen in Franke and other hot coun-tries, according to the nature of the foile, fo is the rellifh, ftrength, and durabilitie of the wine. Now although I think it a fruitleffe labour for any man to ftriue in thefe daies to make a good Vineyard in England, in regard not only of the want of knowledge, to make choife of the fitteft ground for fuch Vines as you would plant therupon, but alfo of the true maner of ordering them in our country; but moft chiefly & aboue all others, that our years in thefe times do not fal out to be fo kindly and hot, to ripen the grapes, to make anie good wine as formerly they haue done ; yet I thinke it not amiffe, to giue you inftructions how to order fuch Vines as you may nourfe vp for the pleafure of the fruit, to eate the grapes being ripe, or to preferue and keepe them to bee eaten almoft all the winter following : And this may be done without any great or extraordinarie paines. Some doe make a lowe wall, and plant their Vines againft it, and keepe them much about the height thereof, not fuffering them to rife much higher : but if the high bricke or ftone wals of your Garden or Orchard haue buttreffes thereat, or if you caufe fuch to bee made, that they bee fomewhat broade forwards, you may the more conueniently plant Vines of diuers forts at them, and by fticking down a couple of good ftakes at euery buttreffe, of eight or ten foot high aboue ground, tacking a few lathes acroffe vpon thofe ftakes, you may therunto tye your Vines, & carry them ther-on at your pleafure : but you muft be carefull to cut them euery year, but not too late, and fo keepe them downe, and from farre fpreading, that they neuer runne much be-yond the frame which you fet at the buttreffes : as alfo in your cutting you neuer leaue too many ioynts, nor yet too few, but at the third or fourth ioint at the moft cut them off. I doe aduife you to thefe frames made with ftakes and lathes, for the better ripe-ning of your grapes : for in the blooming time, if the branches of your vines bee too neare the wall, the reflection of the Sunne in the day time, and the colde in the night, doe oftentimes fpoile a great deale of fruit, by piercing and withering the tender foot-ftalkes of the grapes, before they are formed, whereas when the bloffomes are paft, and the fruit growing of fome bigneffe, then all the heate and reflection you can giue them is fit, and therefore cut away fome of the branches with the leaues, to admit the more Sunne to ripen the fruit. For the diuers forts of grapes I haue fet them downe in the Booke following, with briefe notes vpon euerie of them, whether white or blacke, fmall or great, early or late ripe; fo that I neede not here make the fame relation again. There doth happen fome difeafes to Vines fometimes, which that you may helpe, I thinke it conuenient to informe you what they are, and how to remedy them when you fhall be troubled with any fuch. The firft is a luxurious fpreading of branches and but little or no fruit : for remedie whereof, cut the branches fomewhat more neere then vfuall, and bare the roote, but take heed of wounding or hurting it, and in the hole put either fome good old rotten ftable dung of Horfes, or elfe fome Oxe blood new taken from the beafts, and that in the middle of Ianuarie or beginning of Februa-rie, which being well tempered and turned in with the earth, let it fo abide, which no doubt, when the comfort of the blood or dung is well foaked to the bottome by the raines that fall thereon, will caufe your Vine to fructifie againe. Another fault is, when a Vine doth not bring the fruit to ripeneffe, but either it withereth before it be growne of any bigneffe, or prefently after the blooming : the place or the earth where fuch a Vine ftandeth, affuredly is too cold, and therefore if the fault bee not in the place, which cannot bee helped without remouing to a better, digge out a good quantity of that earth, and put into the place thereof fome good frefh ground well heartned with dung, and fome fand mixed therewith (but not falt or falt water, as fome doe aduife, nor yet vrine, as others would haue) and this will hearten and ftrengthen your Vine to beare out the frut vnto maturitie. When the leaues of a Vine in the end of Summer or in Autumne, vntimely doe turne either yellow or red, it is a great figne the earth is

too hot and drie ; you muſt therefore in ſtead of dung and ſand, as in the former defect is ſaid, put in ſome freſh loame or ſhort clay, well mixed together with ſome of
the earth, and ſo let them abide, that the froſts may mellow them. And laſtly, a Vine
ſometimes beareth ſome ſtore of grapes, but they are too many for it to bring to ripeneſſe ; you ſhall therefore helpe ſuch a Vine (which no doubt is of ſome excellent
kinde, for they are moſt vſually ſubiect to this fault) by nipping away the bloſſomes
from the branches, and leauing but one or two bunches at the moſt vpon a branch, vntill the Vine be growne older, and thereby ſtronger, and by this meane inured to beare
out all the grapes to ripeneſſe. Theſe be all the diſeaſes I know doe happen to Vines :
for the bleeding of a Vine it ſeldome happeneth of it ſelfe, but commeth either by
cutting it vntimely, that is, too late in the yeare, (for after Ianuarie, if you will be well
aduiſed, cut not any Vine) or by ſome caſuall or wilfull breaking of an arme or a
branch. This bleeding in ſome is vnto death, in others it ſtayeth after a certaine ſpace
of it ſelfe : To helpe this inconuenience, ſome haue ſeared the place where it bleedeth
with an hot iron, which in many haue done but a little good ; others haue bound the
barke cloſe with packe-thred to ſtay it ; and ſome haue tied ouer the place, being firſt
dried as well as may bee, a plaiſter made with waxe roſſen and turpentine while it is
warme. Now for the propagating of them : You muſt take the faireſt and goaleſt ſhot
branches of one yeares growth, and cut them off with a peece of the old wood vnto it,
and theſe being put into the ground before the end of Ianuarie at the furtheſt, will
ſhoote forth, and take roote, and ſo become Vines of the ſame kinde from whence
you tooke them. This is the moſt ſpeedy way to haue increaſe : for the laying downe
of branches to take roote, doth not yeelde ſuch ſtore ſo plentifully, nor doe ſuckers
riſe from the rootes ſo aboundantly ; yet both theſe waies doe yeelde Vines, that being taken from the old ſtockes will become young plants, fit to bee diſpoſed of as any
ſhall thinke meete.

Chap. XI.

*The way to order and preſerue grapes, fit to be eaten almoſt all the Winter
long, and ſometimes vnto the Spring.*

ALthough it bee common and vſuall in the parts beyond the Sea to dry their
grapes in the Sunne, thereby to preſerue them all the year, as the Raiſins of the
Sunne are, which cannot bee done in our Countrie for the want of ſufficient
heate thereof at that time : or otherwiſe to ſcald them in hot water (as I heare) and
afterwards to dry them, and ſo keepe them all the yeare, as our Malaga Raiſins are prepared that are packed vp into Frayles : yet I doe intend to ſhew you ſome other waies
to preſerue the grapes of our Countrie freſh, that they may be eaten in the winter both
before and after Chriſtmas with as much delight and pleaſure almoſt, as when they
were new gathered. One way is, when you haue gathered your grapes you intend to
keepe, which muſt be in a dry time, and that all the ſhrunke, dried, or euill grapes in e-
uery bunch be picked away, and hauing prouided a veſſell to hold them, be it of wood
or ſtone which you will, and a ſufficient quantitie of faire and cleane drie ſand ; make
ſtratum ſuper ſtratum of your grapes and the ſand, that is, a lay of ſand in the bottome
firſt, and a lay of grapes vpon them, and a lay or ſtrowing againe of ſand vpon thoſe
grapes, ſo that the ſand may couer euery lay of grapes a fingers breadth in thickneſſe,
which being done one vpon another vntill the veſſell be full, and a lay of ſand vppermoſt, let the veſſell be ſtopped cloſe, and ſet by vntill you pleaſe to ſpend them, being
kept in ſome drie place and in no ſellar : let them bee waſhed cleane in faire water to
take away the ſand from ſo many you will ſpend at a time. Another way is (which Camerarius ſetteth downe he was informed the Turkes vſe to keepe grapes all the winter
vnto the next ſummer) to take ſo much meale of Muſtard ſeede, as will ſerue to ſtrow
vpon grapes, vntill they haue filled their veſſels, whereon afterwards they poure new
wine before it hath boiled, to fill vp their veſſels therwith, and being ſtopped vp cloſe,
they keepe them a certaine time, and ſelling them with their liquour to them that will

vfe them, they doe wafh the feedes or meale from them when they vfe them. Another
way is, that hauing gathered the faireft ripe grapes, they are to be caft vpon threds or
ftrings that are faftened at both ends to the fide walks of a chamber, neere vnto the feeling thereof, that no one bunch touch another, which will bee fo kept a great while,
yet the chamber muft be well defended from the frofts, and cold windes that pierce in
at the windowes, left they perifh the fooner : and fome will dippe the ends of the
branches they hang vp firft in molten pitch, thinking by fearing vp the ends to keepe
the bunches the better; but I doe not fee any great likelihood therein. Your chamber
or clofet you appoint out for this purpofe muft alfo bee kept fomewhat warme, but e-
fpecially in the more cold and froftie time of the yeare, left it fpoile all your coft and
paines, and fruftrate you of all your hopes : but although the frofts fhould pierce and
fpoile fome of the grapes on a bunch, yet if you be carefull to keepe the place warme,
the fewer will be fpoiled. And thus haue I fhewed you the beft directions to order this
Orchard rightly, and all the waies I know are vfed in our Countrie to keep grapes good
anie long time after the gathering, in regard wee haue not that comfort of a hotter Sun
to preferue them by its heate.

The fruits themfelues fhall follow euerie one in their order; the lower fhrubbes or
bufhes firft, and the greater afterwards.

THE THIRD PART
CALLED
THE ORCHARD,

Ontaining all forts of trees bearing fruit for mans vfe to eate, proper and fit for to plant an Orchard in our climate and countrie : I
bound it with this limitation, becaufe both Dates, Oliues, and o-
other fruits, are planted in the Orchards of Spaine, Italy, and o-
ther hot countries, which will not abide in ours. Yet herein I will
declare whatfoeuer Art, ftriuing with Nature, can caufe to profper
with vs, that whofoeuer will, may fee what can bee effected in our
countrie. And firft to begin with the lower fhrubbes or bufhes, and after afcend to
the higher trees.

CHAP. I.

Rubus Idæus. Rafpis.

THe Rafpis berrie is of two forts, white and red, not differing in the forme either of bufh, leafe, or berry, but onely in the colour and tafte of the fruit. The
Rafpis bufh hath tender whitifh ftemmes, with reddifh fmall prickes like haires
fet round about them, efpecially at the firft when they are young ; but when they grow
old they become more wooddy and firme, without any fhew of thornes or prickles
vpon them, and hath onely a little hairineffe that couereth them : the leaues are fomewhat rough or rugged, and wrinkled, ftanding three or fiue vpon a ftalke, fomewhat
like vnto Rofes, but greater, and of a grayer greene colour : the flowers are fmall,
made of fine whitifh round leaues, with a dafh as it were of blufh caft ouer them, many
ftanding together, yet euery one vpon his owne ftalke, at the tops of the branches; after which come vp fmall berries, fomewhat bigger then Strawberries, and longer, either red or white, made of many graines, more eminent then in the Strawberry, with
a kinde of dounineffe caft ouer them, of a pleafant tafte, yet fomewhat fowre, and nothing fo pleafant as the Strawberrie. The white Rafpis is a little more pleafant then
the red, wherein there is fmall feede inclofed : the rootes creepe vnder ground verie
farre, and fhoote vp againe in many places, much encreafing thereby.

There is another whofe ftemme and branches are wholly without prickles : the
fruit is red, and fomewhat longer, and a little more fharpe.

The Vfe of Rafpis.

The leaues of Rafpis may be vfed for want of Bramble leaues in gargles,
and other decoctions that are cooling and drying, although not fully to
that effect.

A a a 3

The

The Conserue or Syrupe made of the berries, is effectuall to coole an hot stomacke, helping to refresh and quicken vp those that are ouercome with faintnesse.

The berries are eaten in the Summer time, as an afternoones dish, to please the taste of the sicke as well as the sound.

The iuyce and the distilled water of the berries are verie comfortable and cordiall.

It is generally held of many, but how true I know not, that the red wine that is vsually so'd at the Vintners, is made of the berries of Raspis that grow in colder countries, which giueth it a kinde of harshnesse : And also that of the same berries growing in hotter climates, which giueth vnto the wine a more pleasant sweetnesse, is made that wine which the Vintners call Alligant : but we haue a Vine or Grape come to vs vnder the name of the Alligant Grape, as you shall finde it set downe hereafter among the Grapes; and therefore it is likely to be but an opinion, and no truth in this, as it may be also in the other.

Chap. II.

Ribes rubra, alba, nigra. Currans red, white, and blacke.

THe bushes that beare those berries, which are vsually called red Currans, are not those Currans either blew or red, that are sold at the Grocers, nor any kind thereof; for that they are the grapes of a certaine Vine, as shall be shewed by and by : but a farre differing kinde of berry, whereof there are three sorts, red, white, and blacke.

The red Curran bush is of two sorts, and groweth to the height of a man, hauing sometimes a stemme of two inches thicknesse, and diuers armes and branches, couered with a smooth, darke, **brownish barke**, without anie pricke or thorne at all vpon anie part thereof, whereon doe growe large cornered blackish greene leaues, cut in on the edges, seeming to be made of fiue parts, almost like a Vine leafe, the ends a little pointing out, and standing one aboue another on both sides of the branches : the flowers are little and hollow, comming forth at the ioynts of the leaues, growing many together on a long stalke, hanging downe aboue a fingers length, and of an herbie colour : after which come small round fruit or berries, greene at the first, and red as a Cherry when they are ripe, of a pleasant and tart taste : the other differeth not in anie other thing then in the berries, being twice as bigge as the former : the roote is wooddy, and spreadeth diuersly.

The white Curran bush riseth vsually both higher then the red, and straighter or more vpright, bigger also in the stemme, and couered with a whiter barke : the leaues are cornered, somewhat like the former, but not so large : the flowers are small and hollow like the other, hanging downe in the same manner on long stalkes, being of a whiter colour : the berries likewise growe on the long stalkes, somewhat thicker set together, and of a cleare white colour, with a little blacke head, so transparent that the seedes may be easily seene thorough them, and of a more pleasant winie taste then the red by much.

The blacke Curran bush riseth higher then the white, with more plentifull branches, and more pliant and twiggie : the stemme and the elder branches being couered with a brownish barke, and the younger with a paler : the flowers are also like vnto little bottles as the others be, of a greenish purple colour, which turne into blacke berries, of the bignesse of the smaller red Currans : the leaues are somewhat like vnto the leaues of the red Currans, but not so large : both branches, leaues, and fruit haue a kind of stinking sent with them, yet they are not vnwholsome, but the berries are eaten of many, without offending either taste or smell.

The Vse of Currans.

The red Currans are vsually eaten when they are ripe, as a refreshing to an

1 *Rubus Idæus*. The Raspis. 2 *Ribes fructu rubro vel albo*. White or red Currans. 3 *Grossularia vulgaris*. The ordinary Gooseberry 4 *Grossularia fructu rubro*. The great red Gooseberry. 5 *Grossularia aculeata*. The prickly Gooseberry. 6 *Oxyacantha seu Berberis*. The Barbary bush. 7 *Auellana Byzantina*. The Filberd of Constantinople. 8 *Auellana rubra nostras*. The best red Filberd.

hot ftomacke in the heate of the yeare, which by the tartneffe is much delighted. Some preferue them, and conferue them alfo as other fruits, and fpend them at neede.

The white Currans, by reafon of the more pleafant winie tafte, are more accepted and defired, as alfo becaufe they are more daintie, and leffe common.

Some vfe both the leaues and berries of the blacke Currans in fawces, and other meates, and are well pleafed both with the fauour and tafte thereof, although many miflike it.

CHAP. III.

Vua Crifpa fiue Groffularia. Goofeberries or Feaberries.

WEe haue diuers forts of Goofeberries, befides the common kinde, which is of three forts, fmall, great, and long. For wee haue three red Goofeberries, a blew and a greene.

The common Goofeberrie, or Feaberrie bufh, as it is called in diuers Countries of England, hath oftentimes a great ftemme, couered with a fmooth darke coloured bark, without anie thorne thereon, but the elder branches haue here and there fome on them; and the younger are whitifh, armed with verie fharpe and cruell crooked thorns, which no mans hand can well auoide that doth handle them, whereon are fet verie greene and fmall cornered leaues cut in, of the fafhion almoft of Smallage, or Hawthorne leaues, but broad at the ftalke : the flowers come forth fingle, at euerie ioynt of the leafe one or two, of a purplifh greene colour, hollow and turning vp the brims a little: the berries follow, bearing the flowers on the heads of them, which are of a pale greene at the firft, and of a greenifh yellow colour when they are ripe, ftriped in diuers places, and cleare, almoft tranfparent, in which the feede lyeth. In fome thefe berries are fmall and round ; in others much greater ; a third is great, but longer then the other : all of them haue a pleafant winie tafte, acceptable to the ftomacke of anie (but the long kinde hath both the thicker skin, and the worfer tafte of the other) and none haue been diftempered by the eating of them, that euer I could heare of.

The firft of the red Goofeberries is better knowne I thinke then the reft, and by reafon of the fmall bearing not much regarded ; the ftemme is fomewhat bigge, and couered with a fmooth darke coloured barke, the younger branches are whiter, and without anie thorne or pricke at all, fo long, weake, fmall, and flender, that they lye vpon the ground, and will there roote againe : the leaues are like vnto the former Goofeberries, but larger : the flowers and berries ftand fingle, and not manie to bee found anie yeare vpon them, but are fomewhat long, and are as great as the ordinarie Goofeberry, of a darke brownifh red colour, almoft blackifh when they are ripe, and of a fweetifh tafte, but without any great delight.

The fecond red Goofeberry rifeth vp with a more ftraight ftemme, couered with a brownifh barke ; the young branches are ftraight likewife, and whitifh, and grow not fo thicke vpon it as the former red kinde, and without any thorne alfo vpon them : the leaues are like vnto the former red, but fmaller : the berries ftand fingly at the leaues as Goofeberries doe, and are of a fine red colour when they are ripe, but change with ftanding to be of a darker red colour, of the bigneffe of the fmall ordinary Goofeberry, of a pretty tart tafte, and fomewhat fweete withall.

The third red Goofeberry which is the greateft, and knowne but vnto few, is fo like vnto the common great Goofeberry, that it is hardly diftinguifhed : the fruit or berries grow as plentifully on the branches as the ordinary, and are as great & round as the great ordinary kinde, but reddifh, and fome of them paler, with red ftripes.

The blew Goofeberry rifeth vp to bee a bufh like vnto the red Curran, and of the fame bigneffe and height, with broader and redder leaues at the firft fhooting out, then the fecond red Goofeberry : the berries are more fparingly fet on the branches, then on the fmall red, and much about the fame bigneffe, or rather leffer, of the colour of a Damfon, with an ouerfhadowing of a blewifh colour vpon them, as the Damfon hath, before it be handled or wiped away.

The

The greene prickly Gooseberry is very like vnto the ordinary Gooseberry in stemme and branches, but that they are not stored with so many sharpe prickles; but the young shootes are more plentifull in small prickles about, and the greene leafe is a little smaller: the flowers are alike, and so are the berries, being of a middle size, and not very great, greene when they are thorough ripe as well as before, but mellower, and hauing a few small short prickles, like small short haires vpon them, which are harmlesse, and without danger to anie the most dainty and tender palate that is, and of a verie good pleasant taste. The seede hereof hath produced bushes bearing berries, hauing few or no prickles vpon them.

The Vse of Gooseberries.

The berries of the ordinary Gooseberries, while they are small, greene, and hard, are much vsed to bee boyled or scalded to make sawce, both for fish and flesh of diuers sorts, for the sicke sometimes as well as the sound, as also before they be neere ripe, to bake into tarts, or otherwise, after manie fashions, as the cunning of the Cooke, or the pleasure of his commanders will appoint. They are a fit dish for women with childe to stay their longings, and to procure an appetite vnto meate.

The other sorts are not vsed in Cookery that I know, but serue to bee eaten at pleasure; but in regard they are not so tart before maturity as the former, they are not put to those vses they be.

Chap. IIII.

Oxyacantha, sed potius Berberis. Barberries.

THe Barberry bush groweth oftentimes with very high stemmes, almost two mens height, but vsually somewhat lower, with manie shootes from the roote, couered with a whitish rinde or barke, and yellow vnderneath, the wood being white and pithy in the middle: the leaues are small, long, and very greene, nicked or finely dented about the edges, with three small white sharpe thornes, for the most part set together at the setting on of the leaues: the flowers doe growe vpon long clustering stalkes, small, round, and yellow, sweete in smell while they are fresh, which turne into small, long, and round berries, white at the first, and very red when they are ,ripe, of a sharpe sowre taste, fit to set their teeth on edge that eate them: the roote is yellow, spreading far vnder the vpper part of the ground, but not very deepe.

There is (as it is thought) another kinde, whose berries are thrice as bigge as the former, which I confesse I haue not seene, and know not whether it be true or no: for it may peraduenture be but the same, the goodnesse of the ground and ayre where they growe, and the youngnesse of the bushes causing that largenesse, as I haue obserued in the same kinde, to yeeld greater berries.

There is said to be also another kinde, whose berries should be without stones or seede within them, not differing else in anie thing from the former: but becaufe I haue long heard of it, and cannot vnderstand by all the inquirie I haue made, that any hath seene such a fruit, I rest doubtfull of it.

The Vse of Barberries.

Some doe vse the leaues of Barberries in the stead of Sorrell, to make sawce for meate, and by reason of their sowrenesse are of the same quality.

The berries are vsed to be pickled, to serue to trimme or set out dishes of fish and flesh in broth, or otherwise, as also sometime to bee boyled in the broth, to giue it a sharpe rellish, and many other wayes, as a Master Cooke can better tell then my selfe.

The

The berries are preferued and conferued to giue to ficke bodies, to helpe to coole any heate in the ftomacke or mouth, and quicken the appetite.

The depurate iuyce is a fine menftrue to diffolue many things, and to verie good purpofe, if it be cunningly handled by an Artift.

The yellow inner barke of the branches, or of the rootes, are vfed to be boyled in Ale, or other drinkes, to be giuen to thofe that haue the yellow iaundife : As alfo for them that haue anie fluxes of choller, to helpe to ftay and binde.

Clufius fetteth downe a fecret that hee had of a friend, of a cleane differing propertie, which was, that if the yellow barke were laid in fteepe in white wine for the fpace of three houres, and afterwards drunke, it would purge one very wonderfully.

Chap. V.

Nux Auellana. The Filberd.

THe Filberd tree that is planted in Orchards, is very like vnto the Hafell nut tree that groweth wilde in the woods, growing vpright, parted into many boughes and tough plyable twigges, without knots, couered with a brownifh, fpeckled, fmooth, thinne rinde, and greene vnderneath : the leaues are broad, large, wrinkled, and full of veines, cut in on the edges into deepe dents, but not into any gafhes, of a darke greene colour on the vpperfide, and of a grayifh afh colour vnderneath : it hath fmall and long catkins in ftead of flowers, that come forth in the Winter, when as they are firme and clofe, and in the Spring open themfelues fomewhat more, growing longer, and of a brownifh yellow colour : the nuts come not vpon thofe ftalkes that bore thofe catkins, but by themfelues, and are wholly inclofed in long, thicke, rough huskes, bearded as it were at the vpper ends, or cut into diuers long iagges, much more then the wood nut : the nut hath a thinne and fomewhat hard fhell, but not fo thicke and hard as the wood nut, in fome longer then in other, and in the long kinde, one hath the skinne white that couereth the kernels, and another red.

There is another fort of the round kinde that came from Conftantinople, whofe huske is more cut, torne, or iagged, both aboue and belowe, then any of our country ; the barke alfo is whiter, and more rugged then ours, and the leaues fomewhat larger.

We haue had from Virginia Hafell nuts, that haue beene fmaller, rounder, browner, thinner fheld, and more pointed at the end then ours : I know not if any hath planted of them, or if they differ in leafe or any thing elfe.

The Vfe of Filberds.

Filberds are eaten as the beft kinde of Hafell nuts, at bankets among other dainty fruits, according to the feafon of the yeare, or otherwife, as euery one pleafe : But Macer hath a Verfe, expreffing prettily the nature of thefe nuts, which is,

Ex minimis nucibus nulli datur efca falubris.

that is, There is no wholfome food or nourifhment had from thefe fmall kinde of nuts.

Yet they are vfed fometime phyfically to be rofted, and made into a Lohoc or Electuary, that is vfed for the cough or cold. And it is thought of fome, that Mithridates meant the kernels of thefe nuts, to be vfed with Figs and Rue for his Antidote, and not of Walnuts.

CHAP. VI.

Vitis. The Vine.

THere is ſo great diuerſities of Grapes, and ſo conſequently of Vines that bear them, that I cannot giue you names to all that here grow with vs : for Iohn Tradeſcante my verie good friend, ſo often before remembred, hath aſſured me, that he hath twentie ſorts growing with him, that hee neuer knew how or by what name to call them. One deſcription therefore ſhall ſerue (as I vſe to doe in ſuch varieties) for all the reſt, with the names afterwards, of as many as we can giue, and the ſeuerall formes, colours and proportions of the grapes.

The manured Vine, in the places where it hath abiden long time, groweth to haue a great bodie, ſtemme or trunke, ſometimes of the bigneſſe of a mans arme, ſleeue and all, ſpreading branches if it bee ſuffered without end or meaſure, but vſually ſtored with many armes or branches, both old and new, but weake, and therefore muſt bee ſuſtained ; whereof the old are couered with a thin ſcaly rinde, which will often chap and peele off of it ſelfe ; the youngeſt being of a reddiſh colour, ſmooth and firme, with a hollowneſſe or pith in the middle : from the ioints of the young branches, and ſometimes from the bodie of the elder, breake out on euerie ſide broade greene leaues, cut on the edges into fiue diuiſions for the moſt part, and beſides notched or dented about : right againſt the leafe, and likewiſe at other ioynts of the branches, come forth long twining or claſping tendrels, winding themſelues about any thing ſtandeth next vnto them : at the bottome of theſe leaues come forth cluſters of ſmall greeniſh yellow bloomes or flowers, and after them the berries, growing in the ſame manner in cluſters, but of diuers formes, colours, taſtes and greatneſſe. For ſome grapes are great, others leſſe, ſome very ſmall (as the Currans that the Grocers ſell) ſome white, ſome red, blew, blacke, or partie-coloured, ſome are are as it were ſquare, others round : ſome the cluſters are cloſe, others open, ſome are ſweete, others ſower or harſh, or of ſome other mixed taſte ; euerie one differing from others, verie notably either in taſte, colour or forme ; within euerie one of which grapes, (and yet there is a grape without ſtones) are contained one, two, or more kernels or ſtones, ſome of them being ſmal, others greater : the rootes ſpread far and deepe. They that keepe their Vines in the beſt order, doe cut them low, not ſuffering them to grow high, or with too many branches, whereby they grow the better, take vp the leſſer roome, and bring their grapes fairer and ſweeter.

The kinds of Vines and Grapes.

Our ordinarie Grape both white and red, which excelleth Crabs for veriuice, and is not fit for wine with vs.

The white Muſcadine Grape is a verie great Grape, ſweete and firme, ſome of the bunches haue weighed ſixe pound, and ſome of the grapes halfe an ounce.

The redde Muſcadine is as great as the white, and chiefly differeth in colour.

The Burlet is a very great white Grape, but fitter for veriuice then wine for the moſt part ; yet when a hot yeare happeneth fit for it, the Grape is pleaſant.

The little blacke Grape that is ripe very early.

The Raiſin of the Sunne Grape is a very great Grape, and very great cluſters, of a reddiſh colour when it is ripe with vs, yet in an extraordinarie hot yeare, it hath got a little blewneſſe caſt ouer it by the heat: but naturally verie blew.

The Curran Grape (or the Grape of Corinth) is the leaſt Grape of all, and beareth both few, and verie ſeldome with vs, but in reaſonable great cluſters, and of a blackiſh blew colour, when they are ripe with vs, and very ſweet sweete. There is another sort of them that are red or browne, and of a sower taste, nothing so sweete.

The Greeke wine Grape is a blackish Grape, and very sweete.

The Frontignack is a white Grape, of a verie sweete and delicate taste, as the wine declareth, that smelleth as it were of Muske.

The square Grape is reported to beare a Grape not fully round, but sided, or as it were square, whereby it became so called.

The Damasco Grape is a great white grape, very sweete, and is the true *Vva Zibeba*, that the Apothecaries should vse in the *Trochisci Ciphi*: and such wee haue had in former times come ouer vnto vs in great, long and round white boxes, containing halfe an hundred weight a peece.

The Russet Grape is a reasonable faire grape, exceeding sweet and whitish, with a thicke skinne, crusted ouer with a shew of ash colour.

The white long Grape is like vnto a Pigeons egge, or as it were pointed pendent like a Pearle.

The partie-coloured Grape is a reasonable great Grape, and discoloured when it is ripe, sometimes the whole bunches, and sometimes but some of the grapes being parted whitish, and blacke halfe through, verie variably.

The Rhenish wine Grape is a white Grape, and endureth the cold of winter when it commeth earely, more then the Muscadine before set downe, and is nothing so sweete.

The White wine Grape is verie like vnto the Rhine Grape, the soile only and climate adding more sweetnesse vnto the one then to the other.

The Claret wine Grape is altogether like the white Grape, but that it is not white, but of a reddish colour, which lying bruised vpon the skins before they are pressed, giue that Claret tincture to the wine.

The Teint is a Grape of a deeper or darker colour, whose iuice is of so deepe a colour, that it serueth to colonr other wine.

The Bursarobe is a faire sweete white Grape of much esteeme about Paris.

The Alligant is a verie sweete Grape, giuing so deep and liuely a coloured red wine, that no other whatsoeuer is comparable to it, and therfore vsually called Spaniards blood.

The blew or blacke Grape of Orleans is another blacke Grape, giuing a darke coloured sweete wine much commended in those parts.

The Grape without stones is also a kinde by it selfe, and groweth naturally neere Ascalon, as Brochard affirmeth, the wine whereof is redde, and of a good taste.

The Virginia Vine, whereof I must needes make mention among other Vines, beareth small Grapes without any great store of iuice therein, and the stone within it bigger then in any other Grape: naturally it runneth on the ground, and beareth little.

The Vse of Vines, Grapes, and other parts that come of them.

The greene leaues of the Vine are cooling and binding, and therefore good to put among other herbes that make gargles and lotions for sore mouthes.

And also to put into the broths and drinke of those that haue hot burning feauers, or any other inflammation.

They stay (as it is held for true) womens longings, if they be either taken inwardly, or applyed outwardly.

Wine is vsually taken both for drinke and medicine, and is often put into sawces, broths, cawdles, and gellies that are giuen to the sicke. As also into diuers Physicall drinkes, to be as a *vehiculum* for the properties of the ingredients.

It is distilled likewise after diuers manners, with diuers things, for diuers & sundry waters to drinke, & for diuers purposes both inward and outward.

As

1 *Vuæ nigræ minores.* The small blacke Grape. 2 *Vuæ cæruleæ maiores.* The great blew Grape. 3 *Vuæ Muschetelliæ.* The Muscadine Grape. 4 *Vuæ suauciores.* The Burlet Grape. 5 *Vuæ insolatæ.* The Rayfins of the sunne Grape. 6 *Ficus.* The figge tree bush.

Also diftilled of it felfe, is called Spirit of wine, which ferueth to diffolue, and to draw out the tincture of diuers things, and for many other purpofes.

The iuice or veriuice that is made of greene hard grapes, before they be ripe, is vfed of the Apothecaries to be made into a Syrupe, that is very good to coole and refrefh a faint ftomacke.

And being made of the riper grapes is the beft veriuice, farre exceeding that which is made of crabs, to be kept all the yeare, to be put both into meates and medicines.

The grapes of the beft forts of Vines are preffed into wine by fome in thefe dayes with vs, and much more as I verily beleeue in times paft, as by the name of Vineyard giuen to many places in this Kingdome, efpecially where Abbies and Monafteries ftood, may bee coniectured: but the wine of late made hath beene but fmall, and not durable, like that which commeth from beyond Sea, whether our vnkindly yeares, or the want of skill, or a conuenient place for a Vineyard be the caufe, I cannot well tell you.

Grapes of all forts are familiarly eaten when they are ripe, of the ficke fometimes as well as the found.

The dryed grapes which we call great Rayfins, and the Currans which we call fmall Rayfins, are much vfed both for meates, broths, and fawces, in diuers manners, as this Countrey in generall aboue any other, wherein many thoufands of Frailes full, Pipes, Hogs-heads, and Buts full are fpent yearly, that it breedeth a wonder in them of thofe parts where they growe and prouide them, how we could fpend fo many.

The Rayfins of the Sunne are the beft dryed grapes, next vnto the Damafco, and are very wholfome to eate fafting, both to nourifh, and to helpe to loofen the belly.

The dryed Lees of wine called Argoll or Tartar, is put to the vfe of the Goldfmith, Dyer, and Apothecary, who doe all vfe it in feuerall manners, uery one in his art.

Of it the Apothecaries make *Cremor Tartari*, a fine medicine to bee vfed, as the Phyfitian can beft appoint, and doth helpe to purge humours by the ftoole.

Thereof likewife they make a kinde of water or oyle, fit to bee vfed, to take away freckles, fpots, or any fuch deformities of the face or skinne, and to make it fmooth. It caufeth likewife haire to growe more aboundantly in thofe places where it naturally fhould growe.

The liquor of the Vine that runneth forth when it is cut, is commended to be good againft the ftone wherefoeuer it be; but that liquor that is taken from the end of the branches when they are burnt, is moft effectuall to take away fpots and markes, ring-wormes and tetters in any place.

CHAP. VII.

Ficus. The Figge tree.

THe Figge trees that are nourfed vp in our country are of three forts, whereof two are high; the one bearing againft a wall goodly fweete and delicate Figs, called Figs of Algarua, and is blewifh when it is ripe: the other tall kinde is nothing fo good, neither doth beare ripe Figges fo kindly and well, and peraduenture may be the white ordinary kinde that commeth from Spaine. The third is a dwarfe kinde of Figge tree, not growing much higher then to a mans body or fhoulders, bearing excellent good Figges and blew, but not fo large as the firft kinde.

The Figge trees of all thefe three kindes are in leaues and growing one like vnto another, fauing for their height, colour, and fweetneffe of the fruit, hauing many armes or branches, hollow or pithy in the middle, bearing very large leaues, and fomewhat thicke, diuided fometimes into three, but vfually into fiue fections, of a darke greene colour on the vpperfide, and whitifh vnderneath, yeelding a milkie iuyce when it is
broken,

broken, as the branches also or the figges when they are greene : the fruit breaketh out from the branches without anie blossome, contrary to all other trees of our Orchard, being round and long, fashioned very like vnto a small Peare, full of small white graines or kernels within it, of a very sweete taste when it is ripe, and very mellow or soft, that it can hardly be carried farre without bruising.

The other two sorts you may easily know and vnderstand, by so much as hath been said of them. Take only this more of the Figge tree, That it you plant it not against a bricke wall, or the wall of an house, &c. it will not ripen so kindly. The dwarfe Figge tree is more tender, and is therefore planted in great square tubs, to be remoued into the sunne in the Summer time, and into the house in Winter.

The Vse of Figges.

Figges are serued to the table with Raysins of the Sunne, and blanched Almonds, for a Lenten dish.

The Figs that growe with vs when they are ripe, and fresh gathered, are eaten of diuers with a little salt and pepper, as a dainty banquet to entertaine a freind, which seldome passeth without a cup of wine to wash them downe.

In Italy (as I haue beene enformed by diuers Gentlemen that haue liued there to study physicke) they eate them in the same manner, but dare not eate many for feare of a feuer to follow, they doe account them to be such breeders of bloud, and heaters of it likewise.

The Figges that are brought vs from Spaine, are vsed to make Ptisan drinkes, and diuers other things, that are giuen them that haue coughes or colds.

It is one of the ingredients also with Nuts and Rice, into Mithridates counterpoison.

The small Figges that growe with vs, and will not ripen, are preserued by the Comfitmakers, and candid also, to serue as other moist or candid banquetting stuffe.

Chap. VIII.

Sorbus. The Seruice tree.

THere are two kindes of Seruice trees that are planted in Orchards with vs, and there is also a wilde kinde like vnto the later of them, with Ashen leaues, found in the woods growing of it selfe, whose fruit is not gathered, nor vsed to bee eaten of any but birds. And there is another kinde also growing wilde abroad in many places, taken by the Country people where it groweth, to be a Seruice tree, and is called in Latine, *Aria Theophrasti*, whose leaues are large, somewhat like Nut tree leaues, but greene aboue, and grayish vnderneath : some doe vse the fruit as Seruices, and for the same purposes to good effect, yet both of these wilde kindes wee leaue for another worke, and here declare vnto you onely those two sorts are nourfed vp in our Orchards.

The more common or ordinary Seruice tree with vs, is a reasonable great tree, couered with a smooth barke, spread into many great armes, whereon are set large leaues, very much cut in on the edges, almost like vnto a Vine leafe, or rather like vnto that kind of Maple, that is vsually called the Sycomore tree with vs : the flowers are white, and growe many clustering together, which after bring forth small browne berries when they are ripe, of the bignesse almost of Hasell nuts, with a small tuft, as if it were a crowne on the head, wherein are small blacke kernels.

The other kinde, which is more rare with vs, and brought into this Land by Iohn Tradescante, heretofore often remembred, hath diuers winged leaues, many set together like vnto an Ashen leafe, but smaller, and euery one endented about the edges : the flowers growe in long clusters, but nothing so many, or so close set as the wilde kinde : the fruit of this tree is in some round like an Apple, and in others a little longer

like

like a Peare, but of a more pleafant tafte then the ordinarie kinde, when they are ripe and mellowed, as they vfe to doe with both thefe kindes, and with Medlars.

The Vfe of Seruices.

They are gathered when they growe to be neare ripe (and that is neuer before they haue felt fome frofts) and being tyed together, are either hung vp in fome warme roome, to ripen them thoroughly, that they may bee eaten, or (as fome vfe to doe) lay them in ftrawe, chaffe, or branne, to ripen them.

They are binding, fit to be taken of them that haue any fcouring or laske, to helpe to ftay the fluxe; but take heed, left if you binde too much, more paine and danger may come thereof then of the fcouring.

Chap. IX.

Mefpilus. The Medlar tree.

THere are three forts of Medlers : The greater and the leffer Englifh, and the Neapolitan.

The great and the fmall Englifh Medlar differ not one from the other in any thing, but in the fize of the fruit, except that the fmall kinde hath fome prickes or thornes vpon it, which the great one hath not, bearing diuers boughes or armes, from whence breake forth diuers branches, whereon are fet long and fomewhat narrow leaues, many ftanding together; in the middle whereof, at the end of the branch, commeth the flower, which is great and white, made of fiue leaues, broad at the ends, with a nicke in the middle of euery one; after which commeth the fruit, being round, and of a pale brownifh colour, bearing a crowne of thofe fmall leaues at the toppe, which were the huske of the flower before, the middle thereof being fomewhat hollow, and is harfh, able to choake any that fhall eate it before it be made mellow, wherein there are certaine flat and hard kernels.

The Medlar of Naples groweth likewife to bee a reafonable great tree, fpreading forth armes and branches, whereon are fet many gafhed leaues, fomewhat like vnto Hawthorne leaues, but greater, and likewife diuers thornes in many places : the flowers are of an herbie greene colour, and fmall, which turne into fmaller fruit then the former, and rounder alfo, but with a fmall head or crowne at the toppe like vnto it, and is of a more fweete and pleafant tafte then the other, with three feeds only therein ordinarily.

The Vfe of Medlars.

Medlars are vfed in the fame manner that Seruices are, that is, to be eaten when they are mellowed, and are for the fame purpofes to binde the body when there is a caufe : yet they as well as the Seruices, are often eaten by them that haue no neede of binding, and but onely for the pleafant fweetneffe of them when they are made mellow, and fometimes come as a difh of ripe fruit at their fit feafon, to be ferued with other forts to the table.

Chap. X.

Lotus. The Lote or Nettle tree.

THe firft kinde of Lote tree, whereof Diofcorides maketh mention, is but of one kinde; but there are fome other trees fpoken of by Theophraftus, that may be referred thereunto, which may bee accounted as baftard kindes thereof, of which I meane to entreate in this Chapter, hauing giuen you before the defcription
of

1 *Sorbus legitima*. The true Seruice tree. 2 *Sorbus vulgaris fiue Torminalis*. The ordinary Seruice tree. 3 *Mespilus vulgaris*. The common Me lar tree. 4 *Mespilus Aronia*. The Medlar of Naples. 5 *Lotus arbor*. The Nettle tree. 6 *Lotus Virginiana*. The Pishania or Virginia Plumme. 7 *Cornus mas*. The C rnell Cherry tree.

of another kinde hereof (by the opinion of good Authors) vnder the name of *Lauro-cerasus.*

The first or true Lote tree groweth to be a tree of a great height, whose bodie and elder branches are couered with a smooth darke greene barke, the leaues are somewhat rough in handling, of a darke greene colour, long pointed, and somewhat deepe dented about the edges, somewhat like vnto a Nettle leafe, and oftentimes growe yellow toward Autumne : the flowers stand here and there scattered vpon the branches, after which come round berries like vnto Cherries, hanging downewards vpon long foot-stalkes, greene at the first, and whitish afterwards; but when they are ripe they become reddish, and if they be suffered to hang too long on the branches, they grow blackish, of a pleasant austere taste, not to be misliked, wherein is a hard round stone.

The second, which is a bastard kinde, and called *Guaiacum Patauinum*, groweth to bee a faire tree, with a smooth darke greene barke, shooting out many faire great boughes, and also slender greene branches, beset with faire broad greene leaues, almost like vnto the leaues of the Cornell tree, but larger : the flowers growe along the branches close vnto them, without any or with a very short foote-stalke consisting of foure greene leaues, which are as the huske, containing within it a purplish flower, made of foure leaues somewhat reddish : the fruit standeth in the middle of the green huske, greene at the first, and very harsh, but red and round when it is ripe, and somewhat like a Plumme, with a small point or pricke at the head thereof, and of a reasonable pleasant taste or rellish, wherein are contained flat and thicke browne seeds or kernels, like vnto the kernels of *Cassia Fistula*, somewhat hard, and not so stonie, but that it may somewhat easily be cut with a knife.

The third is called in Virginia *Pishamin*, The Virginia Plumme (if it be not all one with the former Guaiacana, whereof I am more then halfe perswaded) hath growne with vs of the kernels that were sent out of Virginia, into great trees, whose wood is very hard and brittle, and somewhat white withall : the branches are many, and grow slender to the end, couered with a very thinne greenish bark, whereon doe grow many faire broad greene leaues, without dent or notch on the edges, and so like vnto the former *Guaiacum*, that I verily thinke it (as I before said) to bee the same. It hath not yet borne flower or fruit in our Countrey that I can vnderstand : but the fruit, as it was sent to vs, is in forme and bignesse like vnto a Date, couered with a blackish skinne, set in a huske of foure hard leaues, very firme like vnto a Date, and almost as sweete, with great flat and thicke kernels within them, very like vnto the former, but larger.

The Vse of these Lote trees.

The first sort is eaten as an helper to coole and binde the body : the last, as Captaine Smith relateth in the discouery of Virginia, if the fruit be eaten while it is greene, and not ripe, is able by the harsh and binding taste and quality to draw ones mouth awry (euen as it is said of the former Guaiacana) but when it is thorough ripe it is pleasant, as I said before.

Chap. XI.

Cornus mas. The Cornell tree.

THe Cornell tree that is planted in Orchards, being the male (for the female is an hedge bush) is of two sorts, the one bearing red, the other whiter berries, which is very rare yet in our country, and not differing else.

It groweth to a reasonable bignesse and height, yet neuer to any great tree, the wood whereof is very hard, like vnto horne, and thereof it obtained the name : the body and branches are couered with a rugged barke, and spreadeth reasonable well, hauing somewhat smooth leaues, full of veines, plaine, and not dented on the edges : the flowers are many small yellow tufts, as it were of short haires or threads set together, which come forth before any leafe, and fall away likewise before any leafe bee much open : the fruit are long and round berries, of the bignesse of small Oliues, with an

hard

hard round ſtone within them, like vnto an Oliue ſtone, and are of a yellowiſh red when they are ripe, of a reaſonable pleaſant taſte, yet ſomewhat auſtere withall.

The white (as I ſaid) is like vnto the red, but onely that his fruit is more white when it is ripe.

The Vſe of the Cornelles.

They helpe to binde the body, and to ſtay laskes, and by reaſon of the pleaſantneſſe in them when they are ripe, they are much deſired.

They are alſo preſerued and eaten, both for rarity and delight, and for the purpoſe aforeſaid.

Chap. XII.

Ceraſus. The Cherry tree.

THere are ſo many varieties and differences of Cherries, that I know not well how to expreſſe them vnto you, without a large relation of their ſeuerall formes. I will therefore endeauour after one generall deſcription (as my cuſtome is in many other the like variable fruits) to giue as briefe and ſhort notes vpon all the reſt, as I can both for leafe and fruit, that ſo you may the better know what the fruit is, when you haue the name.

The Engliſh Cherrie tree groweth in time to be of a reaſonable bigneſſe and height, ſpreading great armes, and alſo ſmall twiggy branches plentifully ; the leaues whereof are not verie large or long, but nicked or dented about the edges : the flowers come forth two or three or foure at the moſt together, at a knot or ioynt, euerie one by it ſelfe, vpon his owne ſmall and long footeſtalke, conſiſting of fiue white leaues, with ſome thredes in the middle ; after which come round berries, greene at the firſt, and red when they are through ripe, of a meane bigneſſe, and of a pleaſant ſweete taſte, ſomewhat tart withall, with a hard white ſtone within it, whoſe kernell is ſomewhat bitter, but not vnpleaſant.

The Flanders Cherrie differeth not from the Engliſh, but that it is ſomewhat larger, and the Cherry ſomewhat greater and ſweeter, and not ſo ſower.

The early Flanders Cherry is more rathe or early ripe, almoſt as ſoone as the May Cherry, eſpecially planted againſt a wall, and of many falſe knaues or Gardiners are ſold for May Cherrie trees.

The May Cherrie in a ſtandard beareth ripe fruite later then planted againſt a wall, where the berries will be red in the verie beginning of May ſometimes.

The Arch-Dukes Cherrie is one of the faireſt and beſt cherries wee haue, being of a very red colour when it is ripe, and a little long more then round, and ſomewhat pointed at the end, of the beſt relliſh of any Cherrie whatſoeuer, and of a firme ſubſtance; ſcarce one of twentie of our Nurſerie men doe ſell the right, but giue one for another: for it is an inherent qualitie almoſt hereditarie with moſt of them, to ſell any man an ordinary fruit for whatſoeuer rare fruit he ſhall aske for : ſo little they are to be truſted.

The ounce Cherrie hath the greateſt and broadeſt leafe of any other cherrie, but beareth the ſmalleſt ſtore of cherries euerie yeare that any doth, and yet bloſſometh well : the fruit alſo is nothing anſwerable to the name being not verie great, of a pale yellowiſh red, neere the colour of Amber, and therefore ſome haue called it, the Amber Cherrie.

The great leafed Cherrie is thought of diuers to bee the Ounce Cherrie, becauſe it hath almoſt as great a leafe as the former : but the fruit of this alſo doth not anſwer the expectation of ſo great a leafe, being but of a meane bigneſſe, and a ſmall bearer, yet of a pale reddiſh colour.

The true Gaſcoign Cherry is known but to a few, for our Nurſery men do ſo charge the names of moſt fruits they ſell, that they deliuer but very few true names to any : In former times before our wilde blacke Cherrie was found to grow plentifully in our owne woods in many places of this Land, the French continually ſtored vs with wilde ſtockes to graft vpon, which then were called Gaſcoigne ſtocks, but ſince they haue ſo

refined

termed another red Cherrie, and obtruded it vpon their cuſtomers: but the true is one or our late ripe white Cherries, euen as Gerard ſaith, it is a great cherrie and ſpotted: and this is that Cherrie I ſo commend to be a fit ſtocke to graft May cherries vpon.

The Morello Cherrie is of a reaſonable bigneſſe, of a darke red colour when they are full ripe, and hang long on, of a ſweetiſh ſower taſte, the pulpe or ſubſtance is red, and ſomewhat firme : it they be dryed they will haue a fine ſharpe or ſower taſte very delectable.

The Hartlippe Cherrie is ſo called of the place where the beſt of this kinde is nourſed vp, being betweene Sittingbourne and Chattam in Kent, and is the biggeſt of our Engliſh kindes.

The ſmaller Lacure or Hart Cherrie is a reaſonable faire Cherrie, full aboue, and a little pointing downward, after the faſhion of an heart, as it is vſually painted, blackiſh when it is full ripe, and leſſer then the next.

The great Lacure or Hart Cherrie differeth not in forme, but in greatneſſe, being vſually twice as great as the former, and of a reddiſh blacke colour alſo : both of them are of a firme ſubſtance, and reaſonable ſweete. Some doe call the white cherrie, the White hart cherrie.

The Luke Wardes Cherrie hath a reaſonable large leafe, and a larger flower then many other : the cherries grow with long ſtalkes, and a ſtone of a meane ſize within them, of a darke reddiſh colour when they are full ripe, of a reaſonable good relliſh, and beareth well.

The Corone Cherrie hath a leafe little differing from the Luke Wardes cherrie, the fruit when it is ripe, is of a faire deepe red colour, of a good bigneſſe, and of a verie good taſte, neither verie ſweete or ſower: the pulpe or iuice will ſtaine the hands.

The Vrinall Cherrie in a moſt fruitfull yeare is a ſmall bearer, hauing many yeares none, and the beſt but a few ; yet doth bloſſome plentifully euery yeare for the moſt part : the cherrie is long and round, like vnto an Vrinall, from whence it tooke his name ; reddiſh when it is full ripe, and of an indifferent ſweete relliſh.

The Agriot Cherrie is but a ſmall Cherrie, of a deepe redde colour when it is ripe, which is late ; of a fine ſharpe taſte, moſt pleaſant and wholſome to the ſtomacke of all other cherries, as well while they are freſh as being dryed, which manner they much vſe in France, and keepe them for the vſe both of the ſicke and found at all times.

The Biguarre Cherrie is a fair cherrie, much ſpotted with white ſpots vpon the pale red berry, and ſometimes diſcoloured halfe white and halfe reddiſh, of a reaſonable good relliſh.

The Morocco Cherrie hath a large white bloſſome, and an indifferent big berrie, long and round, with a long ſtalke of a darke reddiſh purple colour, a little tending to a blew when it is full ripe, of a firme ſubſtance : the iuice is of a blackiſh red, diſcolouring the hands or lips, and of a pleaſant taſte : Some doe thinke that this and the Morello be both one.

The Naples Cherrie is alſo thought to bee all one with the Morello or Morocco.

The white Spaniſh Cherrie is an indifferent good bearer, the leafe and bloſſome ſomewhat large, and like the Luke Wardes cherrie : the cherries are reaſonable faire berries, with long ſtalkes and great ſtones, white on the outſide, with ſome redneſſe, on the one ſide of a firme ſubſtance, and reaſonable ſweet, but with a little aciditie, and is one of the late ripe ones : But there is another late ripe white Cherry, which ſome call the Gaſcoigne, before remembred.

The Flanders cluſter Cherrie is of two ſorts, one greater then another : the greater kinde hath an indifferent large leafe ; the bloſſomes haue many threds within them, ſhewing as it were many parts, which after turne into cluſters of berries, foure, fiue or ſixe together, and but with one ſtalke vnder them, as if they grew one out of another, and ſometimes they will beare but two or three, and moſt of them but one cherry on a ſtalke, which are red when they are ripe, very tender, and wateriſh ſweete in eating.

The leſſer is in all things like the greater, but ſmaller, which maketh the difference.

The wilde cluſter or birds cluſter Cherry beareth many bloſſomes ſet all along the ſtalkes, and cherries after them in the ſame maner, like a long thinne bunch of grapes, and therefore called of ſome the Grape cherry : there are of them both red and blacke.

The

1 *Cerasus praecox.* The May Cherry. 2 *Cerasus Batauica.* The Flanders Cherry. 3 *Cerasus Hispanica siue alba.* The white Cherry. 4 *Cerasus platophyllos.* The great leafed Cherry. 5 *Cerasus Luca Wards.* Luke Wards Cherry. 6 *Cerasus Neapolitana.* The Naples Cherry. 7 *Cerasus Cordata.* The Heart Cherry. 8 *Cerasus maculata.* The bignarre or spotted Cherry. 9 *Cerasus acuminaramosa.* The wilde cluster Cherry. 10 *Cerasus Corymbifera.* The Flanders cluster Cherry. 11 *Cerasus Archiducu.* The Archdukes Cherry. 12 *Chamaecerasus.* The dwarfe Cherry.

The soft shield Cherrie is a small red cherrie when it is ripe, hauing the stone within it so soft and tender, that it may easily be broken in the eating of the cherrie.

Iohn Tradescantes Cherrie is most vsually sold by our Nursery Gardiners, for the Archdukes cherrie, because they haue more plenty thereof, and will better be increased, and because it is so faire and good a cherrie that it may be obtruded without much discontent: it is a reasonable good bearer, a faire great berrie, deepe coloured, and a little pointed.

The Baccalaos or New-found-land Cherrie hath a shining long leafe, most like vnto a Peach leafe, the blossomes come very many together as it were in an vmbell, which is such a cluster as is neither like the Flanders cluster, nor the wilde cluster cherrie blossome: it bringeth forth berries standing in the same manner euerie one vpon his own footestalke, being no bigger then the largest berrie of the red Curran tree or bush, of a pale or waterish red colour when it is ripe.

The strange long cluster Cherrie, or *Padus Theophrasti Dalechampio* is reckoned by the Author of that great Herball that goeth vnder his name, among the sorts of cherries ; and so must I vntill a fitter place be found for it. It groweth in time to be a great tree, with a sad coloured barke both on the bodie and branches, whereon doe grow many leaues, somewhat broade, shorter, harder, and a little more crumpled then any cherrie leafe: the blossomes are very small, and of a pale or whitish colour, smelling very sweete and strong, or rather heady, like Orenge flowers, growing on small long branches, very like the toppe of flowers vpon the Laburnum or Beane trefoile trees : after which come small blacke berries, growing together all along the long stalke, like vnto the wilde cluster or birds cherrie mentioned before, but not much bigger then tares, with small stones within them, and little or no sustance vpon them : the French call the tree *Putier*, because the wood thereof stinketh, and make it to be wonderfull that the blossomes of the tree should be so sweete, and the wood so stinking.

The Cullen Cherrie is a darke red cherrie like the **Agriot**, which they of those parts neere Cullen and Vtrecht &c. vse to put into their drinke, to giue it the deeper colour.

The great Hungarian Cherrie of Zwerts is like both in leafe and fruite vnto the Morello cherrie, but much greater and fairer, and a far better bearer : for from a small branch hath beene gathered a pound of cherries, and this is vsuall continually, and not accidentally, most of them foure inches in compasse about, and very many of them more of a faire deepe red colour, and very sweete, excelling the Arch-Dukes cherry, or any other whatsoeuer.

The Cameleon or strange changeable Cherry deseruedly hath this name, although of mine owne imposition, not only because it beareth vsually both blossomes, greene and ripe fruit at one time thereupon, but that the fruit will be of many formes ; some round, some as it were square, and some bunched forth on one side or another, abiding constant in no fashion, but for the most part shewing forth all these diuersities euerie yeare growing vpon it : the fruit is of a very red colour, and good taste.

The great Rose Cherry, or double blossomd Cherry differeth not in any thing from the English Cherrie, but only in the blossomes, which are very thicke of white leaues, as great and double as the double white Crowfoote, before remembred, and somtimes out of the middle of them will spring another smaller flower, but double also ; this seldome beareth fruit, but when it doth I suppose it commeth from those blossomes are the least double, and is red, no bigger then our ordinary English cherrie.

The lesser Rose or double blossomd Cherrie beareth double flowers also, but not so thicke and double as the former ; but beareth fruit more plentifully, of the same colour and bignesse with the former.

The Dwarfe Cherrie is of two sorts ; one whose branches fall downe low, round about the body of it, with small greene leaues, and fruit as small, of a deep red colour.

The other, whose branches, although small, grow more vpright, hauing greener shining leaues: the fruit is little bigger then the former, red also when it is ripe, with a little point at the end : both of them of a sweetish rellish, but more sower.

The great bearing Cherry of Master Millen is a reasonable great red cherry, bearing very plentifully, although it bee planted against a North wall, yet it will bee late ripe, but of an indifferent sweet and good rellish.

The long finger Cherry is another small long red one, being long & round like a finger, wherof it took the name: this is not the Vrinall cherry before, but differing from it.

The

The Vſe of Cherries.

All theſe ſorts of Cherries ſerue wholly to pleaſe the palate, and are eaten at all times, both before and after meales.

All Cherries are cold, yet the ſower more then the ſweete ; and although the ſweete doe moſt pleaſe, yet the ſower are more wholſome, if there bee regard taken in the vſing.

The Agriot or ſower Cherries are in France much vſed to bee dryed (as is ſaid before) as Pruines are, and ſo ſerue to miniſtred to be the ſick in all hot diſeaſes, as feuers &c. being both boyled in their drinkes, and taken now and then of themſelues, which by reaſon of their tartneſſe, doe pleaſe the ſtomacke paſſing well.

The Gum of the Cherrie tree is commended to bee good for thoſe are troubled with the grauell or ſtone. It is alſo good for the cough being diſſolued in liquour, and ſtirreth vp an appetite. The diſtilled water of the blacke Cherries, the ſtones being broken among them, is vſed for the ſame purpoſe, for the grauell, ſtone, and winde.

Chap. XIII.

Prunus. The Plumme tree.

THere are many more varieties of Plummes then of Cherries, ſo that I muſt follow the ſame order with theſe that I did with them, euen giue you their names apart, with briefe notes vpon them, and one deſcription to ſerue for all the reſt. And in this recitall I ſhall leaue out the Apricockes which are certainly a kind of Plum, of an eſpeciall difference, and not of a Peach, as Galen and ſome others haue thought, and ſet them in a chapter by themſelues, and only in this ſet down thoſe fruits are vſually called Plums.

The Plum tree (eſpecially diuers of them) riſeth in time to bee a reaſonable tall and great tree, whoſe bodie and greater armes are couered with a more rugged barke, yet in ſome more or leſſe, the younger branches being ſmooth in all, the leaues are ſomewhat rounder then thoſe of the Cherrie tree, and much differing among themſelues, ſome being longer, or larger, or rounder then others, and many that are exerciſed herein, can tell by the leafe what Plum the tree beareth (I ſpeake this of many, not of all) as in many Cherries they can doe the like : the flowers are white, conſiſting of fiue leaues: the fruit is as variable in forme, as in taſte or colour, ſome being ovall, or Peare faſhion or Almond like, or ſphericall or round, ſome firme, ſome ſoft and wateriſh, ſome ſweete, ſome ſower or harſh, or differing from all theſe taſtes : and ſome white, others blacke, ſome red, others yellow, ſome purple, others blew, as they ſhall bee briefly ſet downe vnto you in the following lines, where I meane not to inſert any the wilde or hedge fruit, but thoſe only are fit for an Orchard, to be ſtored with good fruit: and of all which ſorts, the choyſeſt for goodneſſe, and rareſt for knowledge, are to be had of my very good friend Maſter Iohn Tradeſcante, who hath wonderfully laboured to obtaine all the rareſt fruits hee can heare off in any place of Chriſtendome, Turky, yea or the whole world ; as alſo with Maſter Iohn Millen, dwelling in Olde ſtreete, who from Iohn Tradeſcante and all others that haue had good fruit, hath ſtored himſelfe with the beſt only, and he can ſufficiently furniſh any.

The Amber Primordian Plumme is an indifferent faire Plumme, early ripe, of a pale yellowiſh colour, and of a wateriſh taſte, not pleaſing.

The red Primordian Plumme is of a reaſonable ſize, long and round, reddiſh on the outſide, of a more dry taſte, and ripe with the firſt ſorts in the beginning of Auguſt.

The blew Primordian is a ſmall plumme, almoſt like the Damaſcene, and is ſubiect to drop off from the tree before it be ripe.

The white Date Plum is no very good plum.

The

The red Date plumme is a great long red pointed plumme, and late ripe, little better then the white.

The blacke Muffell plumme is a good plumme, reafonable drye, and tafteth well.

The red Muffell Plumme is fomewhat flat as well as round, of a very good tafte, and is ripe about the middle of Auguft.

The white Muffell plumme is like the redde, but fomewhat fmaller, and of a whitifh greene colour, but not fo well tafted.

The Imperiall plum is a great long reddifh plum, very waterifh, and ripeneth fomewhat late.

The Gaunt plum is a great round reddifh plum, ripe fomewhat late, and eateth waterifh.

The red Pefcod plum is a reafonable good plum.

The white Pefcod plum is a reafonable good rellifhed plumme, but fomewhat waterifh.

The greene Pefcod plum is a reafonable big and long pointed plum, and ripe in the beginning of September.

The Orenge plum is a yellowifh plum, moift, and fomewhat fweetifh.

The Morocco plumme is blacke like a Damfon, well tafted, and fomewhat drye in eating.

The Dine plum is a late ripe plum, great and whitifh, fpeckled all ouer.

The Turkie plum is a large long blackifh plum, and fomewhat flat like the Muffell plum, a well rellifhed dry plum.

The Nutmeg plumme is no bigger then a Damfon, and is of a greenifh yellow colour when it is ripe, which is with vs about Bartholmew tide, and is a good plum.

The Perdigon plumme is a dainty good plumme, early blackifh, and well rellifhed.

The Verdoch plum is a great fine greene fhining plum fit to preferue.

The Ienua plum is the white Date plum, before remembred.

The Barberry plum is a great early blacke plum, and well tafted.

The Pruneola plum is a fmall white plum, of a fine tart tafte : it was wont to bee vfually brought ouer in fmall round boxes, and fold moft commonly at the Comfitmakers, (cut in twaine, the ftone caft away) at a very deere rate: the tree groweth and beareth well with vs.

The Shepway Bulleis is of a darke blewifh brown colour, of a larger fize then the ordinary, and of a fharpe tafte, but not fo good as the common.

The white and the blacke Bulleis are common in moft Countries, being fmall round plums, leffer then Damfons, fharper in tafte, and later ripe.

The Flufhing Bulleis groweth with his fruite thicke cluftring together like grapes.

The Winter Creke is the lateft ripe plum of all forts, it groweth plentifully about Bifhops Hatfield.

The white Peare plum early ripe, is of a pale yellowifh greene colour.

The late ripe white Peare plum is a greater and longer plum, greenifh white, and is not ripe vntill it be neere the end of September, both waterifh plums.

The blacke Peare plum is like vnto the white Peare plumme, but that the colour is blackifh when it is ripe, and is of a very good rellifh, more firme and drye then the other.

The red Peare plumme is of the fame fafhion and goodneffe, but is the worft of the three.

The white Wheate plum is a waterifh fulfome plum.

The red Wheate plum is like the other for tafte.

The Bowle plum is flat and round, yet flatter on the one fide then on the other, which caufed the name, and is a very good rellifhed blacke plum.

The Friars plumme is a very good plum, well tafted, and comming cleane from the ftone, being blacke when it is ripe, and fome whitifh fpots vpon it.

The Catalonia plum is a very good plum.

The don Alteza is alfo a very good plum.

The Mufcadine plum, fome call the Queene mother plumme, and fome the Cherry plum, is a faire red plum, of a reafonable bigneffe, and ripe about Bartholmew tide.

The Chriftian plum, called alfo the Nutmeg plum ; the tree groweth very fhrubby,

and

1 *Prunum imperiale*. The Imperiall Plum. 2 *Prunum Turcicum*, The Turkey Plum. 3 *Prunum præcox rubrum*. The red Primordian Plum. 4 *Prunum Myrobalanum*. The Musell Plum. 5 *Prunum Ambertum*. The Amber Plum. 6 *Prunum Reginum*. The Queene mother Plum. 7 *Prunum viride*. The green Oysterly Plum. 8 *Prunum Arantiacum*, The Orenge Plum. 9 *Prunum Myristicum*. The Nutmeg Plum. 10 *Prunum Siliquastrum* The Peskod Plum. 11 *Prunum Gaudanum*. The Giant Plum. 12 *Prunum Dactyliteum*. The Date Plum. 13 *Prunum Tyrium præcox*. The early pecre Plum.

and will abide good for six weekes at the least after it is gathered, and after all other plums are spent.

The Cherry plum remembred before, speaking of the Muscadine plum, is a very good plum, but small.

The Amber plum is a round plum, as yellow on the outside almost as yellow waxe, of a sowre vnpleasant taste that which I tasted, but I thinke it was not the right; for I haue seene and tasted another of the same bignesse, of a paler colour, farre better relli-shed, and a firmer substance, comming cleane from the stone like an Apricocke.

The Apricocke plum is a good plum when it is in its perfection, but that is seldome; for it doth most vsually cracke, thereby diminishing much of its goodnesse, and be-sides yeeldeth gumme at the crackes.

The Eason plum is a little red plum, but very good in taste.

The Violet plum is a small and long blackish blew plum, ripe about Bartholmew tide, a very good dry eating fruit.

The Grape plum is the Flushing Bulleis before remembred

The Dennie plum is called also the Cheston, or the Friars plum before remembred.

The Damaske Violet plum, or Queen mother plum spoken of before.

The blacke Damascene plum is a very good dry plum, and of a darke blew colour when it is ripe.

The white Damson is nothing so well rellished as the other.

The great Damson or Damaske plum is greater then the ordinary Damson, and sweeter in taste.

The blew Damson well knowne, a good fruit.

The Coferers plum is flat, like vnto a Peare plum, it is early ripe and blacke, of a very good rellish.

The Margate plum the worst of an hundred.

The green Oysterly plum is a reasonable great plum, of a whitish green colour when it is ripe, of a moist and sweete taste, reasonable good.

The red Mirobalane plum groweth to be a great tree quickly, spreading very thicke and farre, very like the blacke Thorne or Sloe bush: the fruit is red, earlier ripe, and of a better taste then the white.

The white Mirobalane plum is in most things like the former red, but the fruit is of a whitish yellow colour, and very pleasant, especially if it be not ouer ripe: both these had need to be plashed against a wall, or else they will hardly beare ripe fruit.

The Oliue plum is very like a greene Oliue, both for colour and bignesse, and grow-eth lowe on a small bushing tree, and ripeneth late, but is the best of all the sorts of greene plums.

The white diapred plum of Malta, scarce knowne to any in our Land but Iohn Tra-descante, is a very good plum, and striped all ouer like diaper, and thereby so called.

The blacke diapred plum is like the Damascene plum, being blacke with spots, as small as pins points vpon it, of a very good rellish.

The Peake plum is a long whitish plum, and very good.

The Pishamin or Virginia plum is called a plum, but vtterly differeth from all sorts of plums, the description whereof may truely enforme you, as it is set downe in the tenth Chapter going before, whereunto I referre you.

The Vse of Plums.

The great Damaske or Damson Plummes are dryed in France in great quantities, and brought ouer vnto vs in Hogs-heads, and other great ves-fels, and are those Prunes that are vsually sold at the Grocers, vnder the name of Damaske Prunes: the blacke Bulleis also are those (being dryed in the same manner) that they call French Prunes, and by their tartnesse are thought to binde, as the other, being sweet, to loosen the body.

The Bruneola Plumme, by reason of his pleasant tartnesse, is much ac-counted of, and being dryed, the stones taken from them, are brought ouer to vs in small boxes, and sold deere at the Comfitmakers, where they very often accompany all other sorts of banquetting stuffes.

Some

Some of these Plums, because of their firmnesse, are vndoubtedly more wholsome then others that are sweete and waterish, and cause lesse offence in their stomackes that eate them ; and therefore are preserued with Sugar, to be kept all the yeare. None of them all is vsed in medicines so much as the great Damson or Damaske Prune, although all of them for the most part doe coole, lenifie, and draw forth choller, and thereby are fittest to be vsed of such as haue chollericke Agues.

Chap. XIIII.

Mala Armeniaca siue Præcocia. Apricockes.

THe Apricocke (as I said) is without question a kinde of Plumme, rather then a Peach, both the flower being white, and the stone of the fruit smooth also, like a Plumme, and yet because of the excellencie of the fruit, and the difference therein from all other Plummes, I haue thought it meete to entreate thereof by it selfe, and shew you the varieties haue been obserued in these times.

The Apricocke tree riseth vp to a very great height, either standing by it selfe (where it beareth not so kindly, and very little in our country) or planted against a wall, as it is most vsuall, hauing a great stemme or body, and likewise many great armes or branches, couered with a smooth barke : the leaues are large, broad, and almost round, but pointed at the ends, and finely dented about the edges : the flowers are white, as the Plumme tree blossomes, but somewhat larger, and rounder set : the fruit is round, with a cleft on the one side, somewhat like vnto a Peach, being of a yellowish colour as well on the inside as outside, of a firme or fast substance, and dry, not ouer-moist in the eating, and very pleasant in taste, containing within it a broad and flat stone, somewhat round and smooth, not rugged as the Peach stone, with a pleasant sweete kernell (yet some haue reported, that there is such as haue their kernels bitter, which I did neuer see or know) and is ripe almost with our first or earliest Plummes, and thereof it tooke the name of *Præcox* ; and it may bee was the earliest of all others was then knowne, when that name was giuen.

The great Apricocke, which some call the long Apricocke, is the greatest and fairest of all the rest.

The smaller Apricocke, which some call the small round Apricocke, is thought to be small, because it first sprang from a stone : but that is not so ; for the kinde it selfe being inoculated, will bee alwaies small, and neuer halfe so faire and great as the former.

The white Apricocke hath his leaues more folded together, as if it were halfe double : it beareth but seldome, and very few, which differ not from the ordinary, but in being more white, without any red when it is ripe.

The Mascoline Apricocke hath a finer greene leafe, and thinner then the former, and beareth very seldome any store of fruit, which differeth in nothing from the first, but that it is a little more delicate.

The long Mascoline Apricocke hath his fruit growing a little longer then the former, and differeth in nothing else.

The Argier Apricocke is a smaller fruit then any of the other, and yellow, but as sweete and delicate as any of them, hauing a blackish stone within it, little bigger then a Lacure Cherry stone : this with many other sorts Iohn Tradescante brought with him returning from the Argier voyage, whither hee went voluntary with the Fleete, that went against the Pyrates in the yeare 1620.

The Vse of Apricockes.

Apricockes are eaten oftentimes in the same manner that other dainty Plummes are, betweene meales of themselues, or among other fruit at banquets.

They

They are also preferued and candid, as it pleaseth Gentlewomen to be-
stowe their time and charge, or the Comfitmaker to sort among other candid fruits.

Some likewise dry them, like vnto Peares, Apples, Damsons, and other
Plummes.

Matthiolus doth wonderfully commend the oyle drawne from the kernels of the stones, to annoint the inflamed *hæmorrhoides* or piles, the swel-
lings of vlcers, the roughnesse of the tongue and throate, and likewise the
paines of the eares.

Chap. XV.

Mala Persica. Peaches.

AS I ordered the Cherries and Plummes, so I intend to deale with Peaches, because their varieties are many, and more knowne in these dayes then in former
times: but because the Nectorin is a differing kinde of Peach, I must deale
with it as I did with the Apricocke among the Plummes, that is, place it in a Chapter
by it selfe.

The Peach tree of it selfe groweth not vsually altogether so great, or high as the A-
pricocke, because it is lesse durable, but yet spreadeth with faire great branches, from
whence spring smaller and slenderer reddish twigges, whereon are set long narrow
greene leaues, dented about the edges: the blossomes are greater then of any Plumme,
of a deepe blush or light purple colour: after which commeth the fruit, which is round,
and sometimes as great as a reasonable Apple or Pippin (I speake of some sorts; for
there be some kindes that are much smaller) with a furrow or cleft on the one side, and
couered with a freese or cotton on the outside, of colour either russet, or red, or yellow, or of a blackish red colour; of differing substances and tastes also, some being
firme, others waterish, some cleauing fast to the stone on the inside, others parting
from it more or lesse easily, one excelling another very farre, wherein is contained a
rugged stone, with many chinkes or clefts in it, the kernell whereof is bitter: the roots
growe neither deepe nor farre; and therefore are subiect to the winds, standing alone,
and not against a wall. It sooner waxeth old and decayeth, being sprung of a stone,
then being inoculated on a Plumme stocke, whereby it is more durable.

The great white Peach is white on the outside as the meate is also, and is a good well
rellished fruit.

The small white Peach is all one with the greater, but differeth in size.

The Carnation Peach is of three sorts, two are round, and the third long; they are
all of a whitish colour, shadowed ouer with red, and more red on the side is next the
sunne: the lesser round is the more common, and the later ripe.

The grand Carnation Peach is like the former round Peach, but greater, and is as
late ripe, that is, in the beginning of September.

The red Peach is an exceeding well rellished fruit.

The russet Peach is one of the most ordinary Peaches in the Kingdome, being of a
russet colour on the outside, and but of a reasonable rellish, farre meaner then many
other.

The Island Peach is a faire Peach, and of a very good rellish.

The Newington Peach is a very good Peach, and of an excellent good rellish, being
of a whitish greene colour on the outside, yet halfe reddish, and is ripe about Bartholmew tide.

The yellow Peach is of a deepe yellow colour; there be hereof diuers sorts, some
good and some bad.

The St. Iames Peach is the same with the Queenes Peach, here belowe set downe,
although some would make them differing.

The Melocotone Peach is a yellow faire Peach, but differing from the former yellow both in forme and taste, in that this hath a small crooked end or point for the most
part, it is ripe before them, and better rellished then any of them.

The

1 *Malus Armeniaca siue Praecocia.* The Apricocke. 2 *Malus Persica Melocotonea.* The Melocotune Peach. 3 *Persica Moschatellina.* The Nutmeg Peach.
4 *Persica nigra.* The blacke Peach. 5 *Persica Carnea longa.* The long Carnation Peach. 6 *Persica Reginea.* The Queenes Peach. 7 *Nuciuersica.* e Al-
biaa I. 8 *Persica du Troas.* The Peach du Troas, 9 *Nuciuersica rubra optima.* The best Romane red Nectorin. 10 *Nuciuersica rubra altera.* The vnited
red Nectorin with a painting blossome.

The Peach *du Troas* is a long and great whitish yellow Peach, red on the outside, early ripe, and is another kinde of Nutmeg Peach.

The Queenes Peach is a faire great yellowish browne Peach, shadowed as it were ouer with deepe red, and is ripe at Bartholmew tide, of a very pleasant good taste.

The Romane Peach is a very good Peach, and well rellished.

The Durasme or Spanish Peach is of a darke yellowish red colour on the outside, and white within.

The blacke Peach is a great large Peach, of a very darke browne colour on the outside, it is of a waterish taste, and late ripe.

The Alberza Peach is late ripe, and of a reasonable good taste.

The Almond Peach, so called, because the kernell of the stone is sweete, like the Almond, and the fruit also somewhat pointed like the Almond in the huske; it is early ripe, and like the Newington Peach, but lesser.

The Man Peach is of two sorts, the one longer then the other, both of them are good Peaches, but the shorter is the better rellished.

The Cherry Peach is a small Peach, but well tasted.

The Nutmeg Peach is of two sorts, one that will be hard when it is ripe, and eateth not so pleasantly as the other, which will bee soft and mellow; they are both small Peaches, hauing very little or no resemblance at all to a Nutmeg, except in being a little longer then round, and are early ripe.

Many other sorts of Peaches there are, whereunto wee can giue no especiall name; and therefore I passe them ouer in silence.

The Vse of Peaches.

Those Peaches that are very moist and waterish (as many of them are) and not firme, doe soone putrefie in the stomacke, causing surfeits oftentimes; and therefore euery one had neede bee carefull, what and in what manner they eate them : yet they are much and often well accepted with all the Gentry of the Kingdome.

The leaues, because of their bitternesse, serue well being boyled in Ale or Milke, to be giuen vnto children that haue wormes, to help to kill them, and doe gently open the belly, if there be a sufficient quantity vsed.

The flowers haue the like operation, that is, to purge the body somewhat more forceably then Damaske Roses; a Syrupe therefore made of the flowers is very good.

The kernels of the Peach stones are oftentimes vsed to be giuen to them that cannot well make water, or are troubled with the stone; for it openeth the stoppings of the vritory passages, whereby much ease ensueth.

Chap. XVI.

Nuciperfica. Nectorins.

I Presume that the name *Nuciperfica* doth most rightly belong vnto that kinde of Peach, which we call Nectorins, and although they haue beene with vs not many yeares, yet haue they beene knowne both in Italy to Matthiolus, and others before him, who it seemeth knew no other then the yellow Nectorin, as Dalechampius also: But we at this day doe know fiue seuerall sorts of Nectorins, as they shall be presently set downe; and as in the former fruits, so in this, I will giue you the description of one, and briefe notes of the rest.

The Nectorin is a tree of no great bignesse, most vsually lesser then the Peach tree, his body and elder boughes being whitish, the younger branches very red, whereon grow narrow long greene leaues, so like vnto Peach leaues, that none can well distinguish them, vnlesse it be in this, that they are somewhat lesser : the blossomes are all reddish, as the Peach, but one of a differing fashion from all the other, as I shall shew you by and by : the fruit that followeth is smaller, rounder, and smoother then Peaches, without any cleft on the side, and without any douny cotton or freeze at all ; and herein is like vnto the outer greene rinde of the Wallnut, whereof as I am perſwaded it tooke the name, of a faſt and firme meate, and very delicate in taſte, eſpecially the beſt kindes, with a rugged ſtone within it, and a bitter kernell.

The Muske Nectorin, ſo called, becauſe it being a kinde of the beſt red Nectorins, both ſmelleth and eateth as if the fruit were ſteeped in Muske : ſome thinke that this and the next Romane Nectorin are all one.

The Romane red Nectorin, or cluſter Nectorin, hath a large or great purpliſh bloſſome, like vnto a Peach, reddiſh at the bottome on the outſide, and greeniſh within : the fruit is of a fine red colour on the outſide, and groweth in cluſters, two or three at a ioynt together, of an excellent good taſte.

The baſtard red Nectorin hath a ſmaller or pincking bloſſome, more like threads then leaues, neither ſo large nor open as the former, and yellowiſh within at the bottome : the fruit is red on the outſide, and groweth neuer but one at a ioynt ; it is a good fruit, but eateth a little more rawiſh then the other, euen when it is full ripe.

The yellow Nectorin is of two ſorts, the one an excellent fruit, mellow, and of a very good relliſh ; the other hard, and no way comparable to it.

The greene Nectorin, great and ſmall ; for ſuch I haue ſeene abiding conſtant, although both planted in one ground : they are both of one goodneſſe, and accounted with moſt to be the beſt relliſhed Nectorin of all others.

The white Nectorin is ſaid to bee differing from the other, in that it will bee more white on the outſide when it is ripe, then either the yellow or greene : but I haue not yet ſeene it.

The Vſe of Nectorins.

The fruit is more firme then the Peach, and more delectable in taſte ; and is therefore of more eſteeme, and that worthily.

Chap. XVII.

Amygdala. Almonds.

THe Almond alſo may be reckoned vnto the ſtock or kindred of the Peaches, it is ſo like both in leafe and bloſſome, and ſomewhat alſo in the fruit, for the outward forme, although it hath onely a dry skinne, and no pulpe or meate to bee eaten : but the kernell of the ſtone or ſhell, which is called the Almond, maketh recompenſe of that defect, whereof ſome are ſweete, ſome bitter, ſome great, ſome ſmall, ſome long, and ſome ſhort.

The Almond tree groweth vpright, higher and greater then any Peach ; and is therefore vſually planted by it ſelfe, and not againſt a wall, whoſe body ſometime exceedeth any mans fadome, whereby it ſheweth to be of longer continuance, bearing large armes, and ſmaller branches alſo, but brittle, whereon are ſet long and narrow leaues, like vnto the Peach tree : the bloſſomes are purpliſh, like vnto Peach bloſſoms, but paler : the fruit is ſomewhat like a Peach for the forme of the skinne or outſide, which is rough, but not with any ſuch cleft therein, or with any pulpe or meate fit to bee eaten, but is a thicke dry skinne when it is ripe, couering the ſtone or ſhell, which is ſmooth and not rugged, and is either long and great, or ſmall, or thicke and ſhort, according as the nut or kernell within it is, which is ſweete both in the greater and ſmaller, and onely one ſmaller kinde which is bitter : yet this I haue obſerued, that all the Almond trees that I haue ſeene growe in England, both of the ſweete and bitter kindes, beare Almonds thicke and ſhort, and not long, as that ſort which is called the Iorden Almond.

The Vſe of Almonds.

They are vſed many wayes, and for many purpoſes, either eaten alone with Figges, or Rayſins of the Sunne, or made into paſte with Sugar and Roſewater for Marchpanes, or put among Floure, Egges, and Sugar, to
make

make Mackerons, or crufted ouer with Sugar, to make Comfits, or mixed with Rofewater and Sugar, to make Butter, or with Barley water, to make Milke, and many other waies, as euery one lift, that hath skill in fuch things.

The oyle alfo of Almonds is vfed many waies, both inwardly and outwardly, for many purpofes; as the oyle of fweete Almonds mixt with poudered white Sugar Candy, for coughes and hoarfeneffe, and to be drunk alone, or with fome other thing (as the Syrupe of Marfh Mallowes) for the ftone, to open and lenifie the paffages, and make them flipperie, that the ftone may paffe the eafier. And alfo for women in Child bed after their fore trauell. And outwardly either by it felfe, or with oyle of Tartar to make a creame, to lenifie the skin, parched with the winde or otherwife, or to annoint the ftomacke either alone, or with other things to helpe a cold.

The oyle of bitter Almonds is much vfed to be dropped into their eares that are hard of hearing, to helpe to open them. And as it is thought, doth more fcoure and cleanfe the skin then the fweet oyle doth, and is therefore more vfed of many for that purpofe, as the Almonds themfelues are.

CHAP. XVIII.

Mala Arantia. Orenges.

I Bring here to your confideration, as you fee, the Orenge tree alone, without mentioning the Citron or Lemmon trees, in regard of the experience we haue feen made of them in diuers places : For the Orenge tree hath abiden with fome extraordinary looking and tending of it, when as neither of the other would by any meanes be preferued any long time. If therefore any be defirous to keepe this tree, he muft fo prouide for it, that it be preferued from any cold, either in the winter or fpring, and expofed to the comfort of the funne in fummer. And for that purpofe fome keepe them in great fquare boxes, and lift them to and fro by iron hooks on the fides, or caufe them to be rowled by trundels, or fmall wheeles vnder them, to place them in an houfe, or clofe gallerie for the winter time : others plant them againft a bricke wall in the ground, and defend them by a fhed of boardes, couered ouer with feare-cloth in the winter, and by the warmth of a ftoue, or other fuch thing, giue them fome comfort in the colder times : but no tent or meane prouifion will preferue them.

The Orenge tree in the warme Countries groweth very high, but with vs (or elfe it is a dwarfe kinde thereof) rifeth not very high : the barke of the elder ftemmes being of a darke colour, and the young branches very greene, whereon grow here and there fome few thornes : the leaues are faire, large, and very greene, in forme almoft like a Bay leafe, but that it hath a fmall eare, or peece of a leafe, fafhioned like vnto an heart vnder euery one of them, with many fmall holes to be feene in them, if you hold them vp betweene you and the light, of a fweet but ftrong fmell, naturally not falling away, but alwaies abiding on, or vntill new be come vp, bearing greene leaues continually : the flowers are whitifh, of a very ftrong and heady fent ; after which come fmall round fruit, greene at the firft, while they are fmall, and not neere maturitie, but being grown and ripe, are (as all men know) red on the outfide, fome more pale then others, and fome kindes of a deeper yellowifh red, according to the climate, and as it receiueth the heate of the funne, wherein is contained fower or fweete iuice, and thicke white kernels among it : it beareth in the warme Countries both bloffomes and greene fruit continually vpon it, and ripe fruit alfo with them for the beft part of the yeare, but efpecially in Autumne and Winter.

The Vfe of Orenges.

Orenges are vfed as fawce for many forts of meates, in refpect of their fweete fowerneffe, giuing a rellifh of delight, whereinfoeuer they are vied.

The inner pulpe or iuice doth ferue in agues and hot difeafes, and in Summer to coole the heate of deiected ftomackes, or fainting fpirits.

The

The dryed rinde, by reaſon of the ſweete and ſtrong ſent, ſerueth to bee put among other things to make ſweet pouthers.

The outer rindes, when they are clenſed from all the inner pulpe and skins, are preſerued in Sugar, after the bitterneſſe by often ſteepings hath been taken away,& do ſerue either as Succots,and banquetting ſtuffes, or as ornaments to ſet out diſhes for the table, or to giue a relliſh vnto meats, whether baked or boyled : Phyſically they helpe to warme a cold ſtomack, and to digeſt or breake winde therein : or they are candid with Sugar, and ſerue with other dryed Iunquets.

The water of Orange flowers is oftentimes vſed as a great perfume for gloues, to waſhe them,or in ſtead of Roſe-water to mixe with other things.

It is vſed to bee drunke by ſome, to preuent or to helpe any peſtilentiall feuer.

The oyntment that is made of the flowers, is very comfortable both for the ſtomache, againſt the could or cough,or for the head,for paines and diſineſſe.

The kernels or ſeede beeing caſt into the ground in the ſpring time, will quickely grow vp, (but will not abide the winter with vs, to bee kept for growing trees) and when they are of a finger length high,being pluckt vp, and put among ſallats, will giue them a maruellous fine aromaticke or ſpicy taſte, very acceptable.

The ſeed or kernels are a little cordiall, although nothing ſo much as the kernels of the Pomecitron.

Chap. XIX.

Poma. Apples.

THe ſorts of Apples are ſo many, and infinite almoſt as I may ſay,that I cannot giue you the names of all, though I haue endeauoured to giue a great many, and I thinke it almoſt impoſſible for any one,to attaine to the full perfection of knowledge herein, not onely in regard of the multiplicitie of faſhions, colours and taſtes, but in that ſome are more familiar to one Countrey then to another, being of a better or worſe taſte in one place then in another, and therefore diuerſly called : I will therefore as I haue done before,giue you the deſcription of the Tree in generall,as alſo of the Paradiſe or dwarfe Apple, becauſe of ſome eſpeciall difference, and afterwards the names of as many, with their faſhions, as haue come to my knowledge, either by ſight or relation : for I doe confeſſe I haue not ſeene all that I here ſet downe, but vſe the helpe of ſome friends, and therefore if it happen that the ſeuerall names doe not anſwer vnto ſeuerall ſorts, but that the ſame fruit may bee called by one name in one Country, that is called by another elſewhere, excuſe it I pray you ; for in ſuch a number,ſuch a fault may eſcape vnknowne.

The Apple tree for the moſt part is neyther very high, great or ſtraight, but rather vſually boweth and ſpreadeth(although in ſome places it groweth fairer and ſtraighter then in others) hauing long and great armes or boughes, and from them ſmaller branches, whereon doe grow ſomewhat broade, and long greene leaues, nicked about the edges : the flowers are large and white, with bluſh coloured ſides, conſiſting of fiue leaues : the fruit (as I ſaid) is of diuers formes, colours and taſtes,and likewiſe of a very variable durabilitie; for ſome muſt be eaten preſently after they are gathered,and they are for the moſt part the earlieſt ripe; others will abide longer vpon the trees, before they bee fit to be gathered; ſome alſo will be ſo hard when others are gathered, that they will not be fit to be eaten, for one, two or three months after they bee gathered; and ſome will abide good but one,two or three moneths,and no more; and ſome will be beſt,after a quarter or halfe a yeares lying, vnto the end of that yeare or the next.

The Paradiſe or dwarfe Apple tree groweth nothing ſo high as the former,and many times not much higher then a man may reach, hauing leaues and flowers altogether like the other, the fruit is a faire yellow Apple,and reaſonable great,but very light and ſpongy or looſe, and of a bitteriſh ſweet taſte, nothing pleaſant. And theſe faults alſo are incident vnto this tree, that both bodie and branches are much ſubiect vnto cancker, which will quickely eate it round, and kill it ; beſides it will haue many bunches, or tuberous ſwellings in many places, which grow as it were ſcabby or rough, and will ſoone cauſe it to periſh : the roote ſendeth forth many ſhootes and ſuckers, whereby it may be much increaſed. But this benefit may be had of it, to recompence the former faults, That being a dwarfe Tree, whatſoeuer fruit ſhall bee grafted on it, will keepe the graft low like vnto it ſelfe, and yet beare fruit reaſonable well. And this is a pretty way to haue Pippins, Pomewaters, or any other ſort of Apples (as I haue had my ſelfe, and alſo ſeene with others) growing low, that if any will, they may make a hedge rowe of theſe low fruits, planted in an Orchard all along by a walke ſide : but take this Caueat, if you will auoide the danger of the cancker and knots, which ſpoile the tree, to graft it hard vnto the ground, that therby you may giue as little of the nature of the ſtock thereunto as poſſibly you can, which wil vndoubtedly help it very much.

The kindes or ſorts of Apples.

The Summer pippin is a very good apple firſt ripe, and therefore to bee firſt ſpent, becauſe it will not abide ſo long as the other.

The French pippin is alſo a good fruit and yellow.

The Golding pippin is the greateſt and beſt of all ſorts of pippins.

The Ruſſet pippin is as good an apple as moſt of the other ſorts of pippins.

The ſpotted pippin is the moſt durable pippin of all the other ſorts.

The ordinary yellow pippin is like the other, and as good; for indeed I know no ſort of pippins but are excellent good well relliſhed fruites.

The great pearemaine differeth little either in taſte or durabilitie from the pippin, and therefore next vnto it is accounted the beſt of all apples.

The ſummer pearemaine is of equall goodneſſe with the former, or rather a little more pleaſing, eſpecially for the time of its eating, which will not bee ſo long laſting, but is ſpent and gone when the other beginneth to be good to eate.

The Ruſſetting is alſo a firme and a very good apple, not ſo wateriſh as the pippin or pearemaine, and will laſt the beſt part of the year, but will be very mellow at the laſt, or rather halfe dryed.

The Broading is a very good apple.

The Pomewater is an excellent good and great whitiſh apple, full of ſap or moiſture, ſomewhat pleaſant ſharpe, but a little bitter withall : it will not laſt long, the winter froſts ſoone cauſing it to rot, and periſh.

The Flower of Kent is a faire yellowiſh greene apple both good and great.

The Gilloflower apple is a fine apple, and finely ſpotted.

The Marligo is the ſame, that is called the Marigold apple, it is a middle ſized apple, very yellow on the outſide, ſhadowed ouer as it were with red, and more red on one ſide, a reaſonable well relliſhed fruit.

The Blandrill is a good apple.

The Dauie Gentle is a very good apple

The Gruntlin is ſomewhat a long apple, ſmaller at the crowne then at the ſtalke, and is a reaſonable good apple.

The gray Coſterd is a good great apple, ſomewhat whitiſh on the outſide, and abideth the winter.

The greene Coſterd is like the other, but greener on the outſide continually.

The Haruy apple is a faire great goodly apple, and very well relliſhed.

The Dowſe apple is a ſweetiſh apple not much accounted of.

The Pome-paris is a very good apple.

The Belle boon of two ſorts winter and ſummer, both of them good apples, and fair fruit to look on, being yellow and of a meane bigneſſe.

The pound Royall is a very great apple, of a very good and ſharpe taſte.

The Doues Bill a ſmall apple.

The Deuſan or apple Iohn is a delicate fine fruit, well relliſhed when it beginneth to be fit to be eaten, and endureth good longer then any other apple.

The Maſter William is greater then a pippin, but of no very good relliſh.

The Maſter Iohn is a better taſted apple then the other by much.

The

The Spicing is a well tasted fruite.

Pome de Rambures
Pome de Capanda } all faire and good apples brought from France.
Pome de Calual

The Queene apple is of two sorts, both of them great faire red apples, and well rel-lished, but the greater is the best.

The Bastard Queene apple is like the other for forme and colour, but not so good in taste : some call this the bardfield Queening.

The Boughton or greening is a very good and well tasted apple.

The Leathercoate apple is a good winter apple, of no great bignesse, but of a very good and sharpe taste.

The Pot apple is a plaine Country apple.

The Cowsnout is no very good fruit.

The Gildiling apple is a yellow one, not much accounted.

The Cats head apple tooke the name of the likenesse, and is a reasonable good apple and great.

The Kentish Codlin is a faire great greenish apple, very good to eate when it is ripe; but the best to coddle of all other apples.

The Stoken apple is a reasonable good apple.

The Geneting apple is a very pleasant and good apple.

The Worcester apple is a very good apple, as bigge as a Pomewater.

Donime Couadis is a French apple, and of a good rellish.

The French Goodwin is a very good apple.

The old wife is a very good, and well rellished apple.

The towne Crab is an hard apple, not so good to be eaten rawe as roasted, but ex-cellent to make Cider.

The Virgilling apple is a reasonable good apple.

The Crowes egge is no good rellished fruit, but nourced vp in some places of the common people.

The Sugar apple is so called of the sweetnesse.

Sops in wine is so named both of the pleasantnesse of the fruit, and beautie of the apple.

The womans breast apple is a great apple.

The blacke apple or pippin is a very good eating apple, and very like a Pearemaine, both for forme and bignesse, but of a blacke sooty colour.

Tweenty sorts of Sweetings and none good.

The Peare apple is a small fruit, but well rellished being ripe, and is for shape very like vnto a small short Peare, and greene.

The Paradise apple is a faire goodly yellow apple, but light and spongy, and of a bit-terish sweet taite, not to be commended.

The apple without blossome, so called because although it haue a small shew of a blossome, yet they are but small threds rather than leaues, neuer shewing to bee like a flower, and therefore termed without blossome : the apple is neyther good eating nor baking fruit.

Wildings and Crabs are without number or vse in our Orchard, being to be had out of the woods, fields and hedges rather then any where else.

The Vse of Apples.

The best sorts of Apples serue at the last course for the table, in most mens houses of account, where, if there grow any rare or excellent fruit, it is then set forth to be seene and tasted.

Diuers other sorts serue to bake, either for the Masters Table, or the meynes sustenance, either in pyes or pans, or else stewed in dishes with Rose-water and Sugar, and Cinamon or Ginger cast vpon.

Some kinds are fittest to roast in the winter time, to warme a cup of wine, ale or beere ; or to be eaten alone, for the nature of some fruit is neuer so good, or worth the eating, as when they are roasted.

Some

Some forts are fitteft to fcald for Codlins, and are taken to coole the fto-macke, as well as to pleafe the tafte, hauing Rofewater and Sugar put to them.

Some forts are beft to make Cider of, as in the Weft Countrey of England great quantities, yea many Hogfheads and Tunnes full are made, efpe-cially to bee carried to the Sea in long voyages, and is found by experience to bee of excellent vfe, to mixe with water for beuerage. It is vfually feene that thofe fruits that are neither fit to eate raw, roafted, nor baked, are fitteft for Cider, and make the beft.

The iuice of Apples likewife, as of pippins, and pearemaines, is of very good vfe in Melancholicke difeafes, helping to procure mirth, and to expell heauineffe.

The diftilled water of the fame Apples is of the like effect.

There is a fine fweet oyntment made of Apples called *Pomatum*, which is much vfed to helpe chapt lips, or hands, or for the face, or any other part of the skinne that is rough with winde, or any other accident, to fupple them, and make them fmooth.

CHAP. XX.

Cydonia. Quinces.

WEe haue fome diuerfities of Quinces, although not many, yet more then our elder times were acquainted with, which fhall be here expreffed. The Quince tree groweth oftentimes to the height and bigneffe of a good Apple tree, but more vfually lower, with crooked and fpreading armes and branches farre abroad, the leaues are fomewhat round, and like the leaues of the Apple tree, but thicker, harder, fuller of veines, and white on the vnderfide: the bloffomes or flowers are white, now and then dafht ouer with blufh, being large and open, like vnto a fingle Rofe: the fruit followeth, which when it is ripe is yellow, and couered with a white cotton or freeze, which in the younger is thicker and more plentifull, but waxeth leffe and leffe, as the fruit ripeneth, being bunched out many times in feuerall places, and round, efpecially about the head, fome greater, others fmaller, fome round like an Apple, others long like a Peare, of a ftrong heady fent, accounted not wholfome or long to be endured, and of no durabilitie to keepe, in the middle whereof is a core, with many blackifh feedes or kernels therein, lying clofe together in cels, and compaffed with a kinde of cleare gelly, which is eafier feene in the fcalded fruit, then in the raw.

The Englifh Quince is the ordinarie Apple Quince, fet downe before, and is of fo harfh a tafte being greene, that no man can endure to eate it rawe, but eyther boyled, ftewed, roafted or baked; all which waies it is very good.

The Portingall Apple Quince is a great yellow Quince, feldome comming to bee whole and faire without chapping; this is fo pleafant being frefh gathered, that it may be eaten like vnto an Apple without offence.

The Portingall Peare Quince is not fit to be eaten rawe like the former, but muft be vfed after fome of the waies the Englifh Quince is appointed, and fo it will make more dainty difhes then the Englifh, becaufe it is leffe harfh, will bee more tender, and take leffe fugar for the ordering then the Englifh kinde.

The Barbary Quince is like in goodneffe vnto the Portingall Quince laft fpoken of, but leffer in bigneffe.

The Lyons Quince.

The Brunfwicke Quince.

The Vfe of Quinces.

There is no fruit growing in this Land that is of fo many excellent vfes as this, feruing as well to make many difhes of meate for the table, as for

D d d ban-

banquets, and much more for the Physicall vertues, whereof to write at large is neither conuenient for mee, nor for this worke: I will onely briefly recite some, as it were to giue you a taste of that plenty remaineth therein, to bee conuerted into sundry formes: as first for the table, while they are fresh (and all the yeare long after being pickled vp) to be baked, as a dainty dish, being well and orderly cookt. And being preserued whole in Sugar, either white or red, serue likewise, not onely as an after dish to close vp the stomacke, but is placed among other Preserues by Ladies and Gentlewomen, and bestowed on their friends to entertaine them, and among other sorts of Preserues at Banquets. Codiniacke also and Marmilade, Ielly and Paste, are all made of Quinces, chiefly for delight and pleasure, although they haue also with them some physicall properties.

We haue for the vse of physicke, both Iuyce and Syrupe, both Conserue and Condite, both binding and loosening medicines, both inward and outward, and all made of Quinces.

The Ielly or Muccilage of the seedes, is often vsed to be laid vpon womens breasts, to heale them being sore or rawe, by their childrens default giuing them sucke.

Athenæus reciteth in his third booke, that one Philarchus found, that the smell of Quinces tooke away the strength of a certaine poison, called *Phariacum.* And the Spaniards haue also found, that the strength of the iuyce of white Ellebor (which the Hunters vse as a poyson to dippe their arrow heads in, that they shoote at wilde beasts to kill them) is quite taken away, if it stand within the compasse of the smell of Quinces. And also that Grapes, being hung vp to bee kept, and spent in Winter, doe quickly rot with the smell of a Quince.

Chap. XXI.

Pyra. Peares.

THe variety of peares is as much or more then of apples, and I thinke it is as hard in this, as before in apples, for any to be so exquisite, as that hee could number vp all the sorts that are to be had: for wee haue in our country so manie, as I shall giue you the names of by and by, and are hitherto come to our knowledge: but I verily beleeue that there be many, both in our country, and in others, that we haue not yet knowne or heard of; for euery yeare almost wee attaine to the knowledge of some, we knew not of before. Take therefore, according to the manner before held, the description of one, with the seuerall names of the rest, vntill a more exact discourse be had of them, euery one apart.

The Peare tree groweth more slowly, but higher, and more vpright then the apple tree, and not lesse in the bulke of the body: his branches spread not so farre or wide, but growe vprighter and closer: the leaues are somewhat broader and rounder, greene aboue, and whiter vnderneath then those of the apple tree: the flowers are whiter and greater: the fruit is longer then round for the most part, smaller at the stalke, and greater at the head, of so many differing formes, colours, and tastes, that hardly can one distinguish rightly betweene them, the times also being as variable in the gathering and spending of them, as in apples: the roote groweth deeper then the apple tree, and therefore abideth longer, and giueth a faster, closer, & smoother gentle wood, easie to be wrought vpon.

The kindes of Peares.

The Summer bon Chretien is somewhat a long peare, with a greene and yellow russetish coate, and will haue sometimes red sides; it is ripe at Michaelmas: some vse to dry them as they doe Prunes, and keepe them all the yeare after. I haue not seene or heard any more Summer kindes hereof then this one, and needeth no wall to nourse it as the other,

The

1 *Malus Cotonea.* The Quince tree. 2 *Cydonium Lusitanicum.* The Portugall Quince 3 *Pyrus.* The Peare tree. 4 *Pyrum Pompeianum, sive Cucume sium byemale.* The Winter Bon Chretien. 5 *Pyrum pictum vel striatum.* The painted or striped Peare of Ierusalem. 6 *Pyrum Palatinale.* The Burgomot Peare. 7 *Pyrum Cucumerinum sive Pompeianum aestivum.* The Summer Bon Chretien. 8 *Pyrum Volemum.* The best Warden. 9 *Pyrum Librale.* The pound Peare. 10 *Pyrum Windsorianum.* The Windsor Peare. 11 *Pyrum Cnamomineum.* The Gratiola Peare. 12 *Pyrum Caryophyllatum.* The Gilloflower Peare.

Ddd 2

The Winter bon Chretien is of many forts, fome greater, others leſſer, and all good; but the greateſt and beſt is that kinde that groweth at Syon: All the kinds of this Winter fruit muſt be planted againſt a wall, or elſe they will both ſeldome beare, and bring fewer alſo to ripeneſſe, comparable to the wall fruit : the kindes alſo are according to their laſting ; for ſome will endure good much longer then others.

The Summer Bergomot is an excellent well relliſhed peare, flattiſh, & ſhort, not long like others, of a meane bigneſſe, and of a darke yellowiſh greene colour on the outſide.

The Winter Bergomot is of two or three forts, being all of them ſmall fruit, ſomewhat greener on the outſide then the Summer kindes ; all of them very delicate and good in their due time : for ſome will not be fit to bee eaten when others are will-nigh ſpent, euery of them outlaſting another by a moneth or more.

The Diego peare is but a ſmall peare, but an excellent well relliſhed fruit, taſting as if Muske had been put among it ; many of them growe together, as it were in cluſters.

The Duetete or double headed peare , ſo called of the forme , is a very good peare, not very great, of a ruſſettiſh browne colour on the outſide.

The Primating peare is a good moiſt peare, and early ripe.

The Geneting peare is a very good early ripe peare.

The greene Cheſill is a delicate mellow peare, euen melting as it were in the mouth of the eater, although greeniſh on the outſide.

The Catherine peare is knowne to all I thinke to be a yellow red ſided peare , of a full wateriſh ſweete taſte, and ripe with the foremoſt.

The King Catherine is greater then the other, and of the ſame goodneſſe, or rather better.

The Ruſſet Catherine is a very good middle ſized peare.

The Windſor peare is an excellent good peare, well knowne to moſt perſons, and of a reaſonable greatneſſe : it will beare fruit ſome times twice in a yeare (and as it is ſaid) three times in ſome places.

The Norwich peare is of two ſorts, Summer and Winter, both of them good fruit, each in their ſeaſon.

The Worſter peare is blackiſh, a farre better peare to bake (when as it will be like a Warden, and as good) then to eate rawe ; yet ſo it is not to be miſliked.

The Muske peare is like vnto a Catherine peare for bigneſſe, colour, and forme, but farre more excellent in taſte, as the very name importeth.

The Roſewater peare is a goodly faire peare, and of a delicate taſte.

The Sugar peare is an early peare, very ſweete, but wateriſh.

The Summer Popperin ⎰both of them are very good firme dry peares, ſomewhat
The Winter Popperin ⎱ſpotted, and browniſh on the outſide.

The greene Popperin is a winter fruit, of equall goodneſſe with the former.

The Soueraingne peare, that which I haue ſeene and taſted, and ſo termed vnto me, was a ſmall browniſh yellow peare, but of a moſt dainty taſte ; but ſome doe take a kind of Bon Chretien, called the Elizabeth peare, to be the Soueraigne peare ; how truely let others iudge.

The Kings peare is a very good and well taſted peare.

The peare Royall is a great peare, and of a good relliſh.

The Warwicke peare is a reaſonable faire and good peare.

The Greenfield peare is a very good peare, of a middle ſize.

The Lewes peare is a browniſh greene peare, ripe about the end of September, a reaſonable well relliſhed fruit, and very moiſt.

The Biſhop peare is a middle ſized peare, of a reaſonable good taſte, not very wateriſh ; but this property is oftentimes ſeene in it, that before the fruit is gathered, (but more vſually thoſe that fall of themſelues, and the reſt within a while after they are gathered)it will be rotten at the core, when there wil not be a ſpot or blemiſh to be ſeene on the outſide, or in all the peare, vntill you come neare the core.

The Wilford peare is a good and a faire peare.

The Bell peare a very good greene peare.

The Portingall peare is a great peare, but more goodly in ſhew then good indeed.

The Gratiola peare is a kinde of Bon Chretien, called the Cowcumber peare, or Spinola's peare.

The Rowling peare is a good peare, but hard, and not good before it bee a little rowled or bruiſed, to make it eate the more mellow.

The

The Pimpe peare is as great as the Windfor peare, but rounder, and of a very good rellifh.

The Turnep peare is a hard winter peare, not fo good to eate rawe, as it is to bake.

The Arundell peare is moſt plentifull in Suffolke, and there commended to be a verie good peare.

The Berry peare is a Summer peare, reafonable faire and great, and of fo good and wholfome a tafte, that few or none take harme by eating neuer fo many of them.

The Sand peare is a reafonable good peare, but fmall.

The Morley peare is a very good peare, like in forme and colour vnto the Windfor, but fome what grayer.

The peare pricke is very like vnto the Greenfield peare, being both faire, great, and good.

The good Rewell is a reafonable great peare, as good to bake as to eate rawe, and both wayes it is a good fruit.

The Hawkes bill peare is of a middle fize, fomewhat like vnto the Rowling peare.

The Petworth peare is a winter peare, and is great, fomewhat long, faire, and good.

The Slipper peare is a reafonable good peare.

The Robert peare is a very good peare, plentifull in Suffolke and Norfolke.

The pound peare is a reafonable good peare, both to eate rawe, and to bake.

The ten pound peare, or the hundred pound peare, the trueſt and beſt, is the beſt Bon Chretien of Syon, fo called, becaufe the grafts coſt the Maſter fo much the fetching by the meſſengers expences, when he brought nothing elfe.

The Gilloflower peare is a winter peare, faire in fhew, but hard, and not fit to bee eaten rawe, but very good to bake.

The peare Couteau is neither good one way nor other.

The Binfee peare is a reafonable good winter peare, of a ruffetifh colour, and a fmall fruit : but will abide good a long while.

The Pucell is a greene peare, of an indifferent good tafte.

The blacke Sorrell is a reafonable great long peare, of a darke red colour on the outfide.

The red Sorrell is of a redder colour, elfe like the other.

The Surrine is no very good peare.

The Summer Haſting is a little greene peare, of an indifferent good rellifh.

Peare Gergonell is an early peare, fomewhat long, and of a very pleafant tafte.

The white Genneting is a reafonable good peare, yet not equall to the other.

The Sweater is fomewhat like the Windfor for colour and bigneſſe, but nothing neare of fo good a tafte.

The bloud red peare is of a darke red colour on the outfide, but piercing very little into the inner pulpe.

The Hony peare is a long greene Summer peare.

The Winter peare is of many forts, but this is onely fo called, to bee diſtinguifhed from all other Winter peares, which haue feuerall names giuen them, and is a very good peare.

The Warden or Luke Wards peare of two forts, both white and red, both great and fmall.

The Spanifh Warden is greater then either of both the former, and better alfo.

The peare of Ierufalem, or the ſtript peare, whofe barke while it is young, is as plainly feene to be ſtript with greene, red, and yellow, as the fruit it felfe is alfo, and is of a very good tafte : being baked alfo, it is as red as the beſt Warden, whereof Maſter William Ward of Eſſex hath aſſured mee, who is the chiefe keeper of the Kings Granary at Whitehall.

Hereof likewife there is a wilde kinde no bigger then ones thumbe, and ſtriped in the like manner, but much more.

The Choke peares, and other wilde peares, both great and fmall, as they are not to furnifh our Orchard, but the Woods, Forreſts, Fields, and Hedges, fo wee leaue them to their naturall places, and to them that keep them, and make good vfe of them.

The Vfe of Peares.

The moſt excellent forts of Peares, ferue (as I faid before of Apples) to

make

make an after-course for their masters table, where the goodnesse of his Orchard is tryed. They are dryed also, and so are an excellent repaste, if they be of the best kindes, fit for the purpose.

They are eaten familiarly of all sorts of people, of some for delight, and of others for nourishment, being baked, stewed, or scalded.

The red Warden and the Spanish Warden are reckoned among the most excellent of Peares, either to bake or to roast, for the sicke or for the sound: And indeede, the Quince and the Warden are the two onely fruits are permitted to the sicke, to eate at any time.

Perry, which is the iuyce of Peares pressed out, is a drinke much esteemed as well as Cyder, to be both drunke at home, and carried to the Sea, and found to be of good vse in long voyages.

The Perry made of Choke Peares, notwithstanding the harshnesse, and euill taste, both of the fruit when it is greene, as also of the iuyce when it is new made, doth yet after a few moneths become as milde and pleasant as wine, and will hardly bee knowne by the sight or taste from it : this hath beene found true by often experience; and therefore wee may admire the goodnesse of God, that hath giuen such facility to so wilde fruits, altogether thought vselesse, to become vsefull, and apply the benefit thereof both to the comfort of our soules and bodies.

For the Physicall properties, if we doe as Galen teacheth vs, *in secundo Alimentorum*, referre the qualities of Peares to their seuerall tastes, as before he had done in Apples, we shall not neede to make a new worke; those that are harsh and sowre doe coole and binde, sweet do nourish and warme, and those betweene these, to haue middle vertues, answerable to their temperatures, &c.

Much more might be said, both of this and the other kinds of fruits; but let this suffice for this place and worke, vntill a more exact be accomplished.

CHAP. XXII.

Nux Iuglans. The Wallnut.

ALthough the Wallnut tree bee often planted in the middle of great Court-yards, where by reason of his great spreading armes it taketh vp a great deale of roome, his shadow reaching farre, so that scarce any thing can well grow neare it; yet because it is likewise planted in fit places or corners of Orchards, and that it beareth fruit or nuts, often brought to the table, especially while they are fresheft, sweeteft, and fittest to be eaten, let not my Orchard want his company, or you the knowledge of it. Some doe thinke that there are many sorts of them, because some are much greater then others, and some longer then others, and some haue a more frangible shell then others; but I am certainly perswaded, that the soyle and climate where they grow, are the whole and onely cause of the varieties and differences. Indeed Virginia hath sent vnto vs two sorts of Wallnuts, the one blacke, the other white, whereof as yet wee haue no further knowledge. And I know that Clusius reporteth, he tooke vp at a banquet a long Wallnut, differing in forme and tendernesse of shell from others, which being set, grew and bore farre tenderer leaues then the other, and a little snipt about the edges, which (as I said) might alter with the soyle and climate: and besides you may obserue, that many of Clusius differences are very nice, and so I leaue it.

The Wallnut tree groweth very high and great, with a large and thicke body or trunke, couered with a thicke clouen whitish greene barke, tending to an ash-colour; the armes are great, and spread farre, breaking out into smaller branches, whereon doe grow long & large leaues, fiue or seuen set together one against another, with an odde one at the end, somewhat like vnto Ashen leaues, but farre larger, and not so many on a stalke, smooth, and somewhat reddish at the first springing, and tender also, of a reasonable good sent, but more strong and headie when they growe old: the fruit or nut is great and round, growing close to the stalkes of the leaues, either by couples or by three

three set together, couered with a double shell, that is to say, with a greene thicke and soft outer rinde, and an inner hard shell, within which the white kernell is contained, couered with a thinne yellow rinde or peeling, which is more easily peeled away while it is greene then afterwards, and is as it were parted into foure quarters, with a thinne wooddy peece parting it at the head, very sweete and pleasant while it is fresh, and for a while after the gathering; but the elder they growe, the harder and more oily: the catkins or blowings are long and yellow, made of many scaly leaues set close together, which come forth early in the Spring, and when they open and fall away, vpon their stalkes arise certaine small flowers, which turne into so many nuts.

The Vse of Wallnuts.

They are often serued to the table with other fruits while they abide fresh and sweete; and therefore many to keepe them fresh a long time haue deuised many wayes, as to put them into great pots, and bury them in the ground, and so take them out as they spend them, which is a very good way, and will keepe them long.

The small young nuts while they are tender, being preserued or candid, are vsed among other sorts of candid fruits, that serue at banquets.

The iuyce of the outer greene huskes are held to be a soueraigne remedy against either poyson, or plague, or pestilentiall feuer.

The distilled water of the huskes drunke with a little vinegar, if the fits growe hot and tedious, is an approued remedy for the same.

The water distilled from the leaues, is effectuall to be applyed to fluent or running vlcers, to dry and binde the humours.

Some haue vsed the pouder of the catkins in white wine, for the suffocation or strangling of the mother.

The oyle of Wallnuts is vsed to varnish Ioyners workes. As also is accounted farre to excell Linseede oyle, to mixe a white colour withall, that the colour bee not dimmed. It is of excellent vse for the coldnesse, hardnesse and contracting of the sinewes and ioynts, to warme, supple, and to extend them.

Chap. XXIII.

Castanea Equina. The Horse Chesnut.

ALthough the ordinary Chesnut is not a tree planted in Orchards, but left to Woods, Parkes, and other such like places; yet wee haue another sort which wee haue nourced vp from the nuts sent vs from Turky, of a greater and more pleasant aspect for the faire leaues, and of as good vse for the fruit. It groweth in time to be a great tree, spreading with great armes and branches, whereon are set at seuerall distances goodly faire great greene leaues, diuided into six, seuen, or nine parts or leaues, euery one of them nicked about the edges, very like vnto the leaues of *Ricnus*, or *Palma Christi*, and almost as great: it beareth at the ends of the branches many flowers set together vpon a long stalke, consisting of foure white leaues a peece, with many threads in the middle, which afterwards turne into nuts, like vnto the ordinary Chesnuts, but set in rougher and more prickly huskes: the nuts themselues being rounder and blacker, with a white spot at the head of each, formed somewhat like an heart, and of a little sweeter taste.

The Vse of this Chesnut.

It serueth to binde and stop any maner of fluxe, be it of bloud or humours, either of the belly or stomacke; as also the much spitting of bloud. They are roasted and eaten as the ordinary sort, to make them taste the better.

They are vsually in Turkie giuen to horses in their prouender, to cure them of coughes, and helpe them being broken winded.

CHAP.

Chap. XXIIII.

Morus. The Mulberrie.

THere are two forts of Mulberries fufficiently known to moft, the blackifh and the white : but wee haue had brought vs from Virginia another fort, which is of greater refpect then eyther of the other two, not onely in regard of the raritie, but of the vfe, as you fhall prefently vnderftand.

1. *Morus nigra.* The blacke Mulberrie.

The blacke Mulberrie tree groweth oftentimes tall and great, and oftentimes alfo crooked, and fpreading abroade, rather then high; for it is fubiect to abide what forme you will conforme it vnto : if by fuffering it to grow, it will mount vp, and if you will binde it, or plafh the boughes, they will fo abide, and be carried ouer arbours, or other things as you will haue it. The bodie groweth in time to bee very great, couered with a rugged or thicke barke, the armes or branches being fmoother, whereon doe grow round thicke leaues pointed at the ends, and nicked about the edges, and in fome there are to be feene deep gafhes, making it feeme fomewhat like the Vine leafe: the flowers are certaine fhort dounie catkings, which turne into greene berries at the firft, afterwards red, and when they are full ripe blacke, made of many graines fet together, like vnto the blacke berrie, but longer and greater : before they are ripe, they haue an auftere and harfh tafte, but when they are full ripe, they are more fweete and pleafant ; the iuice whereof is fo red, that it will ftaine the hands of them that handle and eate them.

2. *Morus alba.* The white Mulberrie.

The white Mulberrie tree groweth not with vs to that greatneffe or bulke of bodie that the blacke doth, but runneth vp higher, flenderer, more knotty, hard and brittle, with thinner fpreade armes and branches : the leaues are like the former, but not fo thicke fet on the branches, nor fo hard in handling, a little paler alfo, hauing fomewhat longer ftalkes: the fruit is fmaller and clofer fet together, greene, and fomewhat harfh before they be ripe, but of a wonderfull fweetneffe, almoft ready to procure loathing when they are thorough ripe, and white, with fuch like feede in them as in the former, but fmaller.

3. *Morus Virginiana.* The Virginia Mulberrie.

The Virginia Mulberry tree groweth quickely with vs to be a very great tree, fpreading many armes and branches, whereon grow faire great leaues, very like vnto the leaues of the white Mulberrie tree : the berry or fruit is longer and redder then either of the other, and of a very pleafant tafte.

The Vfe of Mulberries.

The greateft and moft efpeciall vfe of the planting of white Mulberries, is for the feeding of Silke wormes, for which purpofe all the Eafterne Countries, as Perfia, Syria, Armenia, Arabia &c. and alfo the hither part of Turkie, Spaine alfo and Italie, and many other hot Countries doe nourifh them, becaufe it is beft for that purpofe, the wormes feeding thereon, giuing the fineft and beft filke ; yet fome are confident that the leaues of the blacke will doe as much good as the white : but that refpect muft be had to change your feede, becaufe therein lyeth the greateft myfterie. But there is a Booke or Tractate printed, declaring the whole vfe of whatfoeuer can belong vnto them : I will therefore referre them thereunto, that

                                                              would

1 *Nux Iuglans.* The Wallnut. 2 *Castanea equina.* The horse Chesnut. 3 *Morus nigra vel alba.* The Mulberry. 4 *Morus Virginiana.*
The Virginia Mulberry. 5 *Laurus vulgaris.* The ordinary Bay tree. 6 *Laurea Cerasus Virginiana.* The Virginia Cherry Bay.

would further vnderstand of that matter.

Mulberries are not much defired to be eaten, although they be somewhat pleasant, both for that they staine their fingers and lips that eate them, and doe quickly putrefie in the stomacke, if they bee not taken before meate.

They haue yet a Physicall vfe, which is by reafon of the aftringent quality while they are red, and before they bee ripe, for fore mouthes and throats, or the like, whereunto alfo the Syrup, called Diamoron, is effectuall.

Corollarium.
A COROLLARIE
To this Orchard.

Here are certaine other trees that beare no fruit fit to bee eaten, which yet are often feene planted in Orchards, and other fit and conuenient places bout an houfe, whereof fome are of efpeciall vfe, as the Bay tree &c. others for their beauty and fhadow are fit for walkes or arbours; fome being euer green are moft fit for hedge-rowes; and fome others more for their raritie then for any other great vfe, wherof I thought good to entreat apart by themfelues, and bring them after the fruit trees of this Orchard, as an ornament to accomplifh the fame.

1. *Laurus.* The Bay tree.

THere are to bee reckoned vp fiue kindes of Bay trees, three whereof haue been entreated of in the firft part, a fourth wee will only bring here to your confideration, which is that kinde that is vfually planted in euery mans yard or orchard, for their vfe throughout the whole land, the other we will leaue to bee confidered of in that place is fit for it.

The Bay tree rifeth vp oftentimes to carry the face of a tree of a meane bigneffe in our Countrey (although much greater in the hoter) and oftentimes fhooteth vp with many fuckers from the roote, fhewing it felfe more like to a tall fhrubbe or hedge-bufh, then a tree, hauing many branches, the young ones whereof are fometimes reddifh, but moft vfually of a light or frefh greene colour, when the ftemme and elder boughes are couered with a darke greene barke: the leaues are fomewhat broad, and long pointed as it were at both the ends, hard and fometimes crumpled on the edges, of a darke greene colour aboue, and of a yellowifh greene vnderneath, in fmell fweet, in tafte bitter, and abiding euer greene: the flowers are yellow and moffie, which turne into berries that are a little long as well as round, whofe fhell or outermoft peele is greene at the firft, and blacke when it is ripe; wherein is contained an hard bitter kernell, which cleaueth in two parts.

The Vfe of Bayes.

The Bay leaues are of as neceffary vfe as any other in Garden or Orchard; for they ferue both for pleafure and profit, both for ornament and for vfe, both for honeft Ciuill vfes, and for Phyficke, yea both for the ficke and for the found, both for the liuing and for the dead: And fo much might be faid of this one tree, that if it were all told, would as well weary the Reader, as the Relater: but to explaine my felfe; It ferueth to adorne the houfe of God as well as of man: to procure warmth, comfort and ftrength to the limmes of men and women, by bathings and annoyntings outward, and by drinkes &c. inward to the ftomacke, and other parts: to feafon veffels &c. wherein are preferued our meates, as well as our drinkes: to crowne or encircle

circle as with a garland, the heads of the liuing, and to ſticke and decke forth
the bodies of the dead : ſo that from the cradle to the graue we haue ſtill vſe
of it, we haue ſtill neede of it.

The berries likewiſe ſerue for ſtitches inward, and for paines outward,
that come of cold eyther in the ioynts, ſinewes, or other places.

2. *Laurea Ceraſus, ſiue Laurus Virginiana.* The Virginian Bay, or Cherry Baye.

THis Virginian (whether you will call it a Baye, or a Cherrie, or a Cherrie Bay,
I leaue it to euery ones free will and iudgement, but yet I thinke I may as well
call it a Bay as others a Cherrie, neither of them being anſwerable to the tree,
which neyther beareth ſuch berries as are like Cherries, neither beareth euer greene
leaues like the Bay : if it may therefore bee called the Virginia Cherry Bay, for a diſtinction from the former Bay Cherry that beareth faire blacke Cherries, it will more
fitly agree thereunto, vntill a more proper may be impoſed) riſeth vp to be a tree of a
reaſonable height, the ſtemme or bodie thereof being almoſt as great as a mans legge,
ſpreading forth into diuers armes or boughes, and they againe into diuers ſmall branches, whereon are ſet without order diuers faire broade greene leaues, ſomewhat like
vnto the former Bay leaues, but more limber and gentle, and not ſo hard in handling,
broader alſo, and for the moſt part ending in a point, but in many ſomewhat round
pointed, very finely notched or toothed about the edges, of a bitter taſte, very neere
reſembling the taſte of the Bay leafe, but of little or no ſent at all, either greene or dryed, which fall away euery autumne, and ſpring afreſh euery yeare : the bloſſomes are
ſmall and white, many growing together vpon a long ſtalke, ſomewhat like the Bird
Cherry bloſſomes, but ſmaller, and come forth at the ends of the young branches,
which after turne into ſmall berries, euery one ſet in a ſmall cup or huske, greene at the
firſt, and blacke when they are ripe, of the bigneſſe of a ſmall peaſe, of a ſtrong bitter
taſte, and ſomewhat aromaticall withall, but without any fleſhy ſubſtance like a Cherry
at all vpon it ; for it is altogether like a berry.

The Vſe of this Virginia Cherry Bay.

Being a ſtranger in our Land, and poſſeſſed but of a very few, I doe not
heare that there hath beene any triall made thereof what properties are in
it : let this therefore ſuffice for this preſent, to haue ſhewed you the deſcription and forme thereof, vntill we can learne further of his vſes.

3. *Pinus.* The Pine tree.

MY purpoſe in this place is not to ſhew you all the diuerſities of Pine trees, or
of the reſt that follow, but of that one kinde is planted in many places of our
Land for ornament and delight, and there doth reaſonably well abide : take
it therefore into this Orchard, for the raritie and beautie of it, though we haue little o-
ther vſe of it.

The Pine tree groweth with vs, though ſlowely, to a very great height in many
places, with a great ſtraight bodie, couered with a grayiſh greene barke, the younger
branches are ſet round about, with very narrow long whitiſh greene leaues, which fall
away from the elder, but abide on the younger, being both winter and ſummer alwaies greene. It hath growing in ſundry places on the branches, certaine great hard
wooddy clogs (called of ſome apples, of others nuts) compoſed of many hard wooddy ſcales, or tuberous knobs, which abide for the moſt part alwaies greene in our
Countrey, and hardly become browniſh, as in other Countries, where they haue more
heat and comfort of the Sun, and where the ſcales open themſelues ; wherein are contained white long and round kernels, very ſweete while they are freſh, but quickely
growing oylely and rancide.

The

The Vſe of the Pine apples and kernels.

The Cones or Apples are vſed of diuers Vintners in this City, being painted, to expreſſe a bunch of grapes, whereunto they are very like, and are hung vp in their buſhes, as alſo to faſten keyes vnto them, as is ſeene in many places.

The kernels within the hard ſhels, while they are freſh or newly taken out, are vſed many waies, both with Apothecaries, Comfit-makers, and Cookes : for of them are made medicines, good to leniſie the pipes and paſſages of the lungs and throate, when it is hoarſe. Of them are made Comfits, Paſtes, Marchpanes, and diuers other ſuch like : And with them a cunning Cooke can make diuers Keck ſhoſes for his Maſters table.

Matthiolus commendeth the water of the greene apples diſtilled, to take away the wrinkles in the face, to abate the ouer-ſwelling breaſts of Maidens, by fomenting them after with linnen clothes, wet in the water; and to reſtore ſuch as are rauiſht into better termes.

4. *Abies.* The Firre tree.

THe Firre tree groweth naturally higher then any other tree in theſe parts of Chriſtendome where no Cedars grow, and euen equalling or ouer-topping the Pine : the ſtemme or bodie is bare without branches for a great height, if they bee elder trees, and then branching forth at one place of the bodie foure wayes in manner of a croſſe, thoſe boughes againe hauing two branches at euery ioynt, on which are ſet on all ſides very thicke together many ſmall narrow long hard whitiſh greene leaues, and while they are young tending to yellowneſſe, but nothing ſo long or hard or ſharpe pointed as the Pine tree leaues, growing ſmaller and ſhorter to the end of the branches : the bloomings are certaine ſmall long ſcaly catkins, of a yellowiſh colour, comming forth at the ioynts of the branches, which fall away : the cones are ſmaller and longer then of the Pine tree, wherein are ſmall three ſquare ſeede contained, not halfe ſo big as the Pine kernels.

The Vſe of the Firre tree.

The vſe of this tree is growne with vs of late daies to bee more frequent for the building of houſes then euer before : for hereof (namely of Deale timber and Deale boords) are framed many houſes, and their floores, without the helpe of any other timber or boord of any other tree almoſt ; as alſo for many other workes and purpoſes. The yellow Roſſen that is vſed as well to make ſalues as for many other common vſes, is taken from this tree, as the Pitch is both from the Pitch and Pine trees, and is boyled to make it to bee hard, but was at the firſt a yellow thin cleere Turpentine, and is that beſt ſort of common Turpentine is altogether in vſe with vs, as alſo another more thicke, whitiſh, and troubled, both which are vſed in ſalues, both for man and beaſt (but not inwardly as the cleere white Venice Turpentine is) and ſerueth both to draw, cleanſe and heale. Dodonæus ſeemeth to ſay, that the cleere white Turpentine, called Venice Turpentine, is drawn from the Firre : but Matthiolus confuteth that opinion, which Fulſius alſo held before him.

5. *Ilex arbor.* The euer-greene Oake.

THe *Ilex* or euer-greene Oake riſeth in time to be a very great tree, but very long and ſlow in growing (as is to be ſeene in the Kings priuy Garden at Whitehall, growing iuſt againſt the backe gate that openeth into the way going to Weſtminſter, and in ſome other places) ſpreading many fair large great armes and branches, whereon are ſet ſmall and hard greene leaues, ſomewhat endented or cornered, and
prickly

1 *Pinus.* The Pine tree. 2 *Abies.* The Firre tree. 3 *Ilex.* The euer greene Oake. 4 *Cupressus.* The Cipresse tree. 5 *Arcutus.* The
Strawberry tree. 6 *Alaternus.* The euer greene Privet.

prickly on the edges, especially in the young trees, and sometimes on those branches that are young and newly sprung forth from the elder rootes, but else in a manner all smooth in the elder growne, abiding greene all the winter as well as summer, and are of a grayish greene on the vnderside. It beareth in the spring time certaine slender long branches (like as other Okes doe) with small yellowish mossie flowers on them, which fall away, and are vnprofitable, the acornes not growing from those places, but from others which are like vnto those of our ordinary Oake, but smaller and blacker, and set in a more rugged huske or cuppe. This and no other kinde of *Ilex* doe I know to grow in all our land in any Garden or Orchard: for that kind with long and narrower leaues, and not prickly, growing so plentifully as Matthiolus saith in Tuscane, I haue not seen: and it is very probable to bee the same that Plinie remembreth to haue the leafe of an Oliue ; but not as some would haue it, that *Smilax* Theophrastus maketh mention of in his third Booke and sixteenth Chapter of his Historie of Plants, which the Arcadians so called, and had the leafe of the *Ilex*, but not prickly : for Theophrastus saith, the timber of *Smilax* is smooth and soft, and this of the *Ilex* is harder, and stronger then an Oake.

The Vse of the *Ilex* or euer-greene Oake.

Seeing this is to be accounted among the kindes of Oake (and all Oakes by Dioscorides his opinion are binding) it is also of the same qualitie, but a little weaker, and may serue to strengthen weake members. The young tops and leaues are also vsed in gargles for the mouth and throate.

---

6 *Cupressus.* The Cypresse tree.

THe Cypresse tree that is nourced vp by vs, in our Country, doth grow in those places where it hath beene long planted, to a very great height, whose bodie and boughes are couered with a reddish ash-coloured bark ; the branches grow not spreading, but vpright close vnto the bodie, bushing thicke below, and small vpwards, spire fashion, those below reaching neere halfe the way to them aboue, whereon doe grow euer greene leaues, small, long and flat, of a resinous sweete smell, and strong taste, somewhat bitter : the fruit, which are called nuts, grow here and there among the boughes, sticking close vnto them, which are small, and clouen into diuers parts, but close while they are young, of a russetish browne colour ; wherein are contained small browne seede, but not so small as motes in the Sunne, as Matthiolus and others make them to be.

The Vse of the Cypresse tree.

For the goodly proportion this tree beareth, as also for his euer-greene head, it is and hath beene of great account with all Princes, both beyond, and on this side of the Sea, to plant them in rowes on both sides of some spatious walke, which by reason of their high growing, and little spreading, must be planted the thicker together, and so they giue a goodly, pleasant and sweet shadow : or else alone, if they haue not many, in the middle of some quarter, or as they thinke meete. The wood thereof is firme and durable, or neuer decaying, of a brown yellow colour, and of a strong sweete smell, whereof Chests or Boxes are made to keepe apparell, linnen, furres, and other things, to preserue them from moths, and to giue them a good smell.

Many Physicall properties, both wood, leaues and nuts haue, which here is not my purpose to vnfold, but only to tell you, that the leaues being boyled in wine, and drunke, helpe the difficultie of making vrine, and that the nuts are binding, fit to bee vsed to stay fluxes or laskes, and good also for ruptures.

7. *Arbutus*

7. *Arbutus.* The Strawberry tree.

THe Strawberry tree groweth but flowly, and rifeth not to the height of any great tree, no not in France, Italy, or Spaine: and with vs the coldneſſe of our country doth the more abate his vigour, ſo that it ſeldome rifeth to the height of a man: the barke of the body is rough, and ſmooth in the younger branches: the leaues are faire and greene, very like vnto Baye leaues, finely dented or ſnipped about the edges, abiding alwayes greene thereon both Winter and Summer: the flowers come forth at the end of the branches vpon long ſtalkes, not cluſtering thicke together, but in long bunches, and are ſmall, white, and hollow, like a little bottle, or the flower of Lilly Conually, which after turne into rough or rugged berries, moſt like vnto Strawberries (which hath giuen the name to the tree) ſomewhat reddiſh when they are ripe, of a harſh taſte, nothing pleaſant, wherein are contained many ſmall feedes: It hardly bringeth his fruit to ripeneſſe in our countrey; for in their naturall places they ripen not vntill Winter, which there is much milder then with vs.

The Vſe of the Strawberry tree.

Amatus Lufitanus I thinke is the firſt that euer recorded, that the water diſtilled from the leaues and flowers hereof, ſhould bee very powerfull againſt the plague and poyfons: for all the ancient Writers doe report, that the fruit hereof being eaten, is an enemy to the ſtomacke and head. And Cluſius likewiſe ſetteth downe, that at Liſhbone, and other places in Portingall where they are frequent, they are chiefly eaten, but of the poorer ſort, women and boyes. They are ſomewhat aſtringent or binding, and therefore may well ſerue for fluxes. It is chiefly nourſed with vs for the beauty and rareneſſe of the tree; for that it beareth his leaues alwayes green.

8. *Alaternus.* The euer greene Priuet.

THe tree which we haue growing in our country called *Alaternus*, groweth not to be a tree of any height; but abiding lowe, ſpreadeth forth many branches, whereon are ſet diuers ſmall and hard greene leaues, ſomewhat round for the forme, and endented a little by the edges: it beareth many ſmall whitiſh greene flowers at the ioynts of the ſtalkes, and ſetting on of the lower leaues cluſtering thicke together, which after turne into ſmall blacke berries, wherein are contained many ſmall graines or feedes: the beauty and verdure of theſe leaues abiding ſo freſh all the yeare, doth cauſe it to be of the greater reſpect; and therefore findeth place in their Gardens onely, that are curious conſeruers of all natures beauties.

The Vſe of the euer greene Priuet.

It is ſeldome vſed for any Phyficall property, neither with vs, nor in the places where it is naturall and plentifull: but as Cluſius reporteth, hee learned that the Portingall Fiſhermen do dye their nets red with the decoction of the barke hereof, and that the Dyers in thoſe parts doe vſe the ſmall peeces of the wood to ſtrike a blackiſh blew colour.

9. *Celaſtrus Theophraſti Cluſio.* Cluſius his Celaſtrus.

ALthough the Collectour (who is thought to be Ioannes Molineus of the great Herball or Hiſtory of plants, and generally bearing Daleſchampius name, becauſe the finding and relation of diuers herbes therein expreſſed, is appropiate to him, and printed at Lyons) of all our moderne Writers doth firſt of all others appoint the *Celaſtrus*, whereof Theophraſtus onely among all the ancient Writers of

plants

plants maketh mention, to be the firſt *Aliternus* that Cluſius hath ſet forth in his Hiſtory of rarer plants: yet I finde, that Cluſius himſelfe before his death doth appropiate that *Celaſtrus* of Theophraſtus to another plant, growing in the Garden at Leyden, which formerly of diuers had beene taken to be a kinde of *Laurus Tinus*, or the wilde Baye; but he impugning that opinion for diuers reſpects, decyphreth out that Leyden tree in the ſame manner that I doe: and becauſe it is not onely faire, in bearing his leaues alwayes greene, but rare alſo, being nourſed vp in our Land in very few places, but principally with a good old Lady, the widow of Sir Iohn Leuſon, dwelling neere Rocheſter in Kent; I thought it fit to commend it for an ornament, to adorne this our Garden and Orchard. It groweth vp to the height of a reaſonable tree, the body whereof is couered with a darke coloured barke, as the elder branches are in like manner; the younger branches being greene, whereon are ſet diuers leaues thicke together, two alwayes at a ioynt, one againſt another, of a ſad but faire greene colour on the vpperſide, and paler vnderneath, which are little or nothing at all ſnipped about the edges, as large as the leaues of the *Laurus Tinus*, or wilde Baye tree: at the end of the young branches breake forth between the leaues diuers ſmall ſtalkes, with foure or fiue flowers on each of them, of a yellowiſh greene colour, which turne into ſmall berries, of the bigneſſe of blacke Cherries, greene at the firſt, and red when they begin to be ripe, but growing blacke if they hang too long vpon the branches, wherein is contained a hard ſhell, and a white hard kernell within it, couered with a yellowiſh skin. This abideth (as I ſaid before) with greene leaues as well Winter as Summer; and therefore fitteſt to be planted among other of the ſame nature, to make an euer greene hedge.

The Vſe of Cluſius his Celaſtrus.

Being ſo great a ſtranger in this part of the Chriſtian world, I know none hath made tryall of what property it is, but that the taſte of the leaues is ſomewhat bitter.

10. *Pyracantha.* The euer greene Hawthorne, or prickly Corall tree.

THis euer greene ſhrubbe is ſo fine an ornament to a Garden or Orchard, either to be nourſed vp into a ſmall tree by it ſelfe, by pruining and taking away the ſuckers and vnder branches, or by ſuffering it to grow with ſuckers, thicke and plaſhing the branches into a hedge, for that it is plyable to be ordered either way; that I could not but giue you the knowledge thereof, with the deſcription in this manner. The younger branches are couered with a ſmooth darke blewiſh greene barke, and the elder with a more aſh coloured, thicke ſet with leaues without order, ſome greater and others ſmaller, ſomewhat like both in forme and bigneſſe vnto the leaues of the Barberry tree, but ſomewhat larger, and more ſnipt about the edges, of a deeper greene colour alſo, and with ſmall long thornes ſcattered here & there vpon the branches: the flowers come forth as well at the ends of the branches, as at diuers places at the ioynts of the leaues, ſtanding thicke together, of a pale whitiſh colour, a little daſht ouer with a ſhew of bluſh, conſiſting of fiue leaues a peece, with ſome ſmall threads in the middle, which turne into berries, very like vnto Hawthorne berries, but much redder and dryer, almoſt like poliſhed Corall, wherein are contained foure or fiue ſmall yellowiſh white three ſquare ſeede, ſomewhat ſhining. It is thought to be the *Oxyacanthæ* of Dioſcorides; but ſeeing Dioſcorides doth explaine the forme of the leafe in his Chapter of Medlars, which he concealed in the Chapter of *Oxyacantha*, it cannot be the ſame: for *Meſpilus Anthedon* of Theophraſtus, or *Aronia* of Dioſcorides, hath the leafe of *Oxyacantha*, as Dioſcorides ſaith, or of Smalladge, as Theophraſtus, which cannot agree to this Thorne, but doth moſt liuely delineate out our white Thorne or Hawthorne, that now there is no doubt, but that *Oxyacantha* of Dioſcorides is the Hawthorne tree or buſh.

The Vſe of this Corall tree.

Although Lobel maketh mention of this tree to grow both in Italy, and

1 *Celastrus Theophrasti Clusio,* Clusius his Celastrus. 2 *Pyracantha.* The euer green prickly Corall tree. 3 *Taxus.* The Yewe tree. 4 *Buxus arbor.* The Boxe tree. 5 *Buxus humilis.* The lowe or dwarfe Boxe. 6 *Sabina.* The sauine tree. 7 *Paliurus* Christs thorne. 8 *Larix.* The Larch tree.

Prouence in France, in some of their hedges, yet he saith it is neglected in the naturall places, and to be of no vse with them : neither doe I heare, that it is applyed to any Physicall vse with vs, but (as I before said) it is preserued with diuers as an ornament to a Garden or Orchard, by reason of his euer greene leaues, and red berries among them, being a pleasant spectacle, and fit to be brought into the forme of an hedge, as one please to lead it.

11. *Taxus.* The Yewe tree.

THe Yewe tree groweth with vs in many places to bee a reasonable great tree, but in hoter countries much bigger, couered with a reddish gray scaly barke ; the younger branches are reddish likewise, whereon grow many winged leaues, that is, many narrow long darke greene leaues, set on both sides of a long stalke or branch, neuer dying or falling away, but abiding on perpetually, except it be on the elder boughes : the flowers are small, growing by the leaues, which turne into round red berries, like vnto red Asparagus berries, in taste sweetish, with a little bitternesse, and causing no harme to them for any thing hath been knowne in our country,

The Vse of the Yewe tree.

It is found planted both in the corners of Orchards, and against the windowes of Houses, to be both a shadow and an ornament, in being alwayes greene, and to decke vp Houses in Winter : but ancient Writers haue euer reckoned it to be dangerous at the least, if not deadly.

12. *Buxus.* The Boxe tree.

THe Boxe tree in some places is a reasonable tall tree, yet growing slowly ; the trunke or body whereof is of the bignesse of a mans thigh, which is the biggest that euer I saw : but sometimes, and in other places it groweth much lower, vsually not aboue a yard, or a yard and a halfe high, on the backe sides of many Houses, and in the Orchards likewise : the leaues are small, thicke and hard, and still the greater or lesser the tree is, the greater or lesser are the leaues, round pointed, and of a fresh shining greene colour : the flowers are small and greenish, which turne into heads or berries, with foure hornes, whittish on the outside, and with reddish seede within them.

Buxus aureus. / Gilded Boxe. There is another kinde hereof but lately come to our knowledge, which differeth not in any thing from the former, but onely that all the leaues haue a yellow list or gard about the edge of them on the vpperside, and none on the lower, which maketh it seeme very beautifull, and is therefore called gilded Boxe.

Buxus humilis. / Dwarfe Boxe. We haue yet another kinde of Boxe, growing small and lowe, not aboue halfe a foote, or a foote high at the most, vnlesse it be neglected, which then doth grow a little more shrubby, bearing the like leaues, but smaller, according to the growth, and of a deeper greene colour : I could neuer know that this kinde euer bore flower or seede, but is propagated by slipping the roote, which encreaseth very much.

The Vse of Boxe.

The wood of the Boxe tree is vsed in many kindes of small works among Turners, because it is hard, close, and firme, and as some haue said, the roots much more, in regard of the diuers waues and crooked veines running through it. It hath no Physicall vse among the most and best Physitians, although some haue reported it to stay fluxes, and to be as good as the wood of *Guaiacum,* or *Lignum vitæ* for the French disease. The leaues and branches serue both Summer and Winter to decke vp houses ; and are many times giuen to horses for the bots.

The lowe or dwarfe Boxe is of excellent vse to border vp a knot, or the

lon

long beds in a Garden, being a maruailous fine ornament thereunto , in regard it both groweth lowe, is euer greene , and by cutting may bee kept in
what maner euery one pleafe, as I haue before fpoken more largely.

13. *Sabina.* The Sauine tree or bush.

THe Sauine tree or bufh that is moft vfuall in our country , is a fmall lowe bufh,
not fo high as a man in any place , nor fo bigge in the ftemme or trunke as a
mans arme, with many crooked bending boughes and branches , whereon are
fet many fmall, fhort, hard, and prickly leaues, of a darke green colour, frefh and green
both Winter and Summer : it is reported , that in the naturall places it beareth fmall
blacke berries, like vnto Iuniper, but with vs it was neuer knowne to beare any.

The Vfe of Sauine.

It is planted in out-yards, backfides, or voide places of Orchards, as well
to caft clothes thereon to diy , as for medicines both for men and horfes :
being made into an oyle , it is good to annoint childrens bellies for to kill
the Wormes : and the powder thereof mixed with Hogs greafe, to annoint
the running fores or fcabs in their heads ; but beware how you giue it inwardly to men, women, or children. It is often put into horfes drenches,
to helpe to cure them of the bots, and other difeafes.

14. *Paliurus.* Chrifts thorne.

THis thorny fhrubbe (wherewith as it is thought, our Sauiour Chrift was crow-
ned, becaufe as thofe that haue trauelled through Paleftina and Iudæa, doe report no other thorne doth grow therein fo frequent , or fo apt to be writhed)
rifeth in fome places to a reafonable height, but in our country feldome exceedeth the
height of a man , bearing many flender branches , full of leaues , fet on either fide
thereof one by one, which are fomewhat broad and round , yet pointed , and full of
veines, thicke fet alfo with fmall thornes, euen at the foote of euery branch, and at the
foote of euery leafe one or two, fome ftanding vpright, others a little bending downe:
the flowers are fmall and yellow , ftanding for the moft part at the end of the branches, many growing vpon a long ftalke, which after turne into round , flat, and hard
fhelly fruit , yet couered with a foft flefhy skinne , within which are included two or
three hard, fmall, and browne flat feeds, lying in feuerall partitions. The leaues hereof
fall away euery yeare, and fpring forth afrefh againe the next May following. The rarity and beauty of this fhrubbe , but chiefly (as I thinke) the name hath caufed this to
be much accounted of with all louers of plants.

The Vfe of Chrifts thorne.

Wee haue fo few of thefe fhrubbes growing in our country, and thofe
that are, doe, for any thing I can vnderftand, neuer beare fruit with vs ; that
there is no other vfe made hereof then to delight the owners: but this is certainly receiued for the *Paliurus* of Diofcorides and Theophraftus, and
thought alfo by Matthiolus to be the very true *Rhamnus tertius* of Diofcorides. Matthiolus alfo feemeth to contradict the opinion is held by the
Phyfitians of Mompelier, and others, that it cannot be the *Paliurus* of Theophraftus. It is held to be effectuall to belpe to breake the ftone, both in the
bladder, reines, and kidneyes: the leaues and young branches haue an aftringent quality, and good againft poyfons and the bitings of ferpents.

15. *Larix.*

15. *Larix.* The Larch tree.

THe Larch tree, where it naturally groweth, riseth vp to be as tall as the Pine or Firre tree, but in our Land being rare, and noursed vp but with a few, and those onely louers of rarities, it groweth both slowly, and becommeth not high: the barke hereof is very rugged and thicke, the boughes and branches grow one aboue another in a very comely order, hauing diuers small yellowish knobs or bunches set thereon at seuerall distances; from whence doe yearely shoote forth many small, long, and narrow smooth leaues together, both shorter and smaller, and not so hard or sharpe pointed as either the Pine or Firre tree leaues, which doe not abide the Winter as they doe, but fall away euery yeare, as other trees which shed their leaues, and gaine fresh euery Spring: the blossomes are very beautifull and delectable, being of an excellent fine crimson colour, which standing among the greene leaues, allure the eyes of the beholders to regard it with the more desire: it also beareth in the naturall places (but not in our Land that I could heare) small soft cones or fruit, somewhat like vnto Cypresse nuts, when they are greene and close.

The Vse of the Larch tree.

The coles of the wood hereof (because it is so hard and durable as none more) is held to be of most force being fired, to cause the Iron oare to melt, which none other would doe so well. Matthiolus contesteth against Fuchsius, for deeming the Venice Turpentine to be the liquid Rossen of the Firre tree, which he assureth vpon his owne experience and certaine knowledge, to be drawne from this Larch tree, and none other, which cleere Turpintine is altogether vsed inwardly, and no other, except that of the true Turpintine tree, and is very effectuall to cleanse the reines, kidneyes, and bladder, both of grauell and the stone, and to prouoke vrine: it is also of especiall property for the *gonorrhea*, or running of the reines, as it is called, with some powder of white Amber mixed therewith, taken for certaine dayes together. Taken also in an Electuary, it is singular good for to expectorate rotten flegme, and to helpe the consumption of the lungs. It is vsed in plaisters and salues, as the best sort of Turpintine. The Agaricke that is vsed in physicke, is taken from the bodies and armes of this tree. And Matthiolus doth much insist against Brasauolus, that thought other trees had produced Agaricke, affirming them to be hard *Fungi*, or Mushroms (such as wee call Touch-wood) wherwith many vse to take fire, strooke thereinto from steele.

16. *Tilia.* The Line or Linden tree.

THere are two sorts of Line trees, the male and the female; but because the male is rare to be seene, and the female is more familiar, I will onely giue you the description of the female, and leaue the other.

The female Line tree groweth exceeding high and great, like vnto an Elme, with many large spreading boughes, couered with a smooth barke, the innermost being very plyant and bending from whence come smaller branches, all of them so plyable, that they may bee led or carried into any forme you please: the leaues thereon are very faire, broad, and round, somewhat like vnto Elme leaues, but fairer, smoother, and of a fresher greene colour, dented finely about the edges, and ending in a sharpe point: the flowers are white, and of a good smell, many standing together at the top of a stalke, which runneth all along the middle ribbe of a small long whitish leafe; after which come small round berries, wherein is contained small blackish seede: this tree is wholly neglected by those that haue them, or dwell neere them, because they suppose it to be fruitlesse, in regard it beareth chaffie huskes, which in many places fall away, without giuing ripe seede.

The

1 *Tilia fœmina.* The Line or Linden tree. 2 *Tamarix.* The Tamariske tree. 3 *Acer maius latifolium.* The Sycomore tree. 4 *Staphylodendron.* The bladder nut. 5 *Rhus Myrtifolia.* The Mirtle leafed Sumach. 6 *Rhus Virginiana,* The Buckshorne tree. 7 *Vitis sen Hedera Virginensis.* The Virginia Vine or rather Iuie.

The Vſe of the Line tree.

It is planted both to make goodly Arbours, and Summer banquetting houſes, either belowe vpon the ground, the boughes ſeruing very hand-ſomely to plaſh round about it, or vp higher, for a ſecond aboue it, and a third alſo : for the more it is depreſſed, the better it will grow. And I haue ſeene at Cobham in Kent, a tall or great bodied Line tree, bare withont boughes for eight foote high, and then the branches were ſpread round a-bout ſo orderly, as if it were done by art, and brought to compaſſe that middle Arbour : And from thoſe boughes the body was bare againe for eight or nine foote (wherein might bee placed halfe an hundred men at the leaſt, as there might be likewiſe in that vnderneath this)& then another rowe of branches to encompaſſe a third Arbour, with ſtayres made for the pur-poſe to this and that vnderneath it : vpon the boughes were laid boards to tread vpon, which was the goodlieſt ſpectacle mine eyes euer beheld for one tree to carry.

The coles of the wood are the beſt to make Gunpowder. And being kindled, and quenched in vinegar, are good to diſſolue clotted bloud in thoſe that are bruiſed with a fall. The inner barke being ſteeped in water yeeldeth a ſlimie iuyce, which is found by experience, to be very profitable for them that haue been burnt with fire.

17. *Tamarix.* Tamariſke tree.

THe Tamariſke tree that is common in our country, although in ſome places it doth not grow great, yet I haue ſeene it in ſome other, to be as great as a great apple tree in the body, bearing great arms; from whoſe ſmaller branches ſpring forth young ſlender red ſhootes, ſet with many very fine, ſmall, and ſhort leaues, a lit-tle criſped, like vnto the leaues of Sauine, not hard or rough, but ſoft and greene : the flowers be white moſſie threads, which turne into dounie ſeede, that is carried away with the winde.

Tamariſcus fo-lijs albidis. White Tama-riſke. There is another kinde hereof very beautifull and rare, not to be ſeene in this Land I thinke, but with Mr. William Ward, the Kings ſeruant in his Granary, before remem-bred, who brought me a ſmall twigge to ſee from his houſe at Boram in Eſſex, whoſe branches are all red while they are young, and all the leaues white, abiding ſo all the Summer long, without changing into any ſhew of greene like the other, and ſo abideth conſtant yeare after yeare, yet ſhedding the leaues in Winter like the other.

The Vſe of Tamariſke.

The greateſt vſe of Tamariſke is for ſpleneticke diſeaſes, either the leaues or the barke made into drinkes; or the wood made into ſmall Cans or Cups to drinke in.

18. *Acer maius latifolium.* The great Maple or Sycomore tree.

THe Sycomore tree, as we vſually call it (and is the greateſt kind of Maple, che-riſhed in our Land onely in Orchards, or elſewhere for ſhade and walkes, both here in England, and in ſome other countries alſo) groweth quickly to bee a faire ſpreading great tree, with many boughes and branches, whoſe barke is ſomewhat ſmooth : the leaues are very great, large, and ſmooth, cut into foure or fiue diuiſions, and ending into ſo many corners, euery one ſtanding on a long reddiſh ſtalke: the bloo-mings are of a yellowiſh greene colour, growing many together on each ſide of a long ſtalke, which afrer turne into long and broad winged ſeede, two alwaies ſtanding toge-ther on a ſtalke, and bunched out in the middle, where the ſeed or kernell lyeth, very like vnto the common Maple growing wilde abroad, but many more together, and larger.

The

The Vſe of the Sycomore tree.

It is altogether planted for ſhady walkes, and hath no other vſe with vs that I know.

---

19. *Nux Veſicaria.* The bladder Nut.

THis tree groweth not very high, but is of a meane ſtature, when it is preſerued and pruined to grow vpright, or elſe it ſhooteth forth many twigges from the rootes, and ſo is fit to plant in a hedge rowe, as it is vſed in ſome places : the body and armes are couered with a whitiſh greene barke : the branches and leaues on them are like vnto the Elder, hauing three or fiue leaues ſet one againſt another, with one of them at the end, each whereof is nicked or dented about the edges: the flowers are ſweete and white, many growing together on a long ſtalke, hanging downeward, in forme reſembling a ſmall Daffodill, hauing a ſmall round cup in the middle, and leaues about it : after which come the fruit, incloſed in ruſſetiſh greene bladders, containing one or two browniſh nuts, leſſer then Haſell nuts, whoſe outer ſhell is not hard and woody, like the ſhell of a nut, but tough, and hard withall, not eaſie to breake, within which is a greene kernell, ſweetiſh at the firſt, but lothſome afterwards, ready to procure caſting, and yet liked of ſome people, who can well endure to eate them.

The Vſe of the Bladder Nut.

The greateſt vſe that I know the tree or his fruit is put vnto, is, that it is receiued into an Orchard, either for the rarity of the kinde, being ſuffered to grow into a tree, or (as I ſaid before) to make an hedge, being let grow into ſuckers.

Some Quackſaluers haue vſed theſe nuts as a medicine of rare vertue for the ſtone, but what good they haue done, I neuer yet could learne.

---

20. *Rhus Myrtifolia.* The Mirtle leafed Sumach.

THis lowe ſhrubbe groweth ſeldome to the height of a man, hauing many ſlender branches, and long winged leaues ſet thereon, euery one whereof is of the bigneſſe of the broad or large Mirtle leafe, and ſet by couples all the length of the ribbe, running through the middle of them. It beareth diuers flowers at the tops of the branches, made of many purple threads, which turne into ſmall blacke berries, wherein are contained ſmall, white, and rough ſeed, ſomewhat like vnto Grape kernels or ſtones. This vſeth to dye down to the ground in my Garden euery Winter, and riſe vp again euery Spring, whether the nature thereof were ſo, or the coldneſſe of our climate the cauſe therof, I am not well aſſured. It is alſo rare, and to be ſeen but with a few.

The Vſe of this Sumach.

It is vſed to thicken or tanne leather or hides, in the ſame manner that the ordinary Sumach doth ; as alſo to ſtay fluxes both in men and women.

---

21. *Rhus Virginiana.* The Virginia Sumach, or Buckes horne tree of Virginia.

THis ſtrange tree becommeth in ſome places to bee of a reaſonable height and bigneſſe, the wood whereof is white, ſoft, and pithy in the middle, like vnto an Elder, couered with a darke coloured barke, ſomewhat ſmooth : the young branches that are of the laſt yeares growth are ſomewhat reddiſh or browne, very ſoft

and

and smooth in handling, and so like vnto the Veluet head of a Deere, that if one were cut off from the tree, and shewed by it selfe, it might soone deceiue a right good Woodman, and as they grow seeme most like thereunto, yeelding a yellowish milke when it is broken, which in a small time becommeth thicke like a gumme : the leaues grow without order on the branches, but are themselues set in a seemly order on each side of a middle ribbe, seuen, nine, ten, or more on a side, and one at the end, each whereof are somewhat broad and long, of a darke greene colour on the vpperside, and paler greene vnderneath, finely snipped or toothed round about the edges : at the ends of the branches come forth long and thicke browne tufts, very soft, and as it were woolly in handling, made all of short threads or thrums ; from among which appeare many small flowers, much more red or crimson then the tufts, which turne into a very small seede : the roote shooteth forth young suckers farre away, and round about, whereby it is mightily encreased.

The Vse of this Sumach.

It is onely kept as a rarity and ornament to a Garden or Orchard, no bo-die, that I can heare of, hauing made any tryall of the Physicall properties.

---

22. *Vitis, seu potius Hedera Virginensis.* The Virginia Vine, or rather Iuie.

THis slender, but tall climing Virginia Vine (as it was first called ; but Iuie, as it doth better resemble) riseth out of the ground with diuers stems, none much bigger then a mans thumbe, many lesse ; from whence shoote forth many long weake branches, not able to stand vpright, vnlesse they be sustained : yet planted neere vnto a wall or pale, the branches at seuerall distances of the leaues will shoote forth small short tendrels, not twining themselues about any thing, but ending into foure, fiue, or six, or more small short and somewhat broad clawes, which will fasten like a hand with fingers so close thereunto, that it will bring part of the wall, morter, or board away with it, if it be pulled from it, and thereby stay it selfe, to climbe vp to the toppe of the highest chimney of a house, being planted thereat : the leaues are crum-pled, or rather folded together at the first comming forth, and very red, which after growing forth, are very faire, large, and greene, diuided into foure, fiue, six, or seuen leaues, standing together vpon a small foote-stalke, set without order on the branches, at the ends whereof, as also at other places sometime, come forth diuers short tufts of buds for flowers ; but we could neuer see them open themselues, to shew what manner of flower it would be, or what fruit would follow in our country : the roote spreadeth here and there, and not very deepe.

The Vse of this Virginian.

We know of no other vse, but to furnish a Garden, and to encrease the number of rarities.

And thus haue I finished this worke, and furnished it with whatsoeuer Art and Na-ture concurring, could effect to bring delight to those that liue in our Climate, and take pleasure in such things ; which how well or ill done, I must abide euery ones censure : the iudicious and courteous I onely respect, let Momus bite his lips, and eate his heart ; and so Farewell.

FINIS.

Index omnium ſtirpium

quæ in hoc opere continentur.

Fſf

Cneorum

A

A Table of the English names of such Plants as are contained in this Booke.

The

THE

A Table of the Vertues and Properties of the Hearbes contained in this Booke.

To

Faults escaped in some Copies.

FOlio 8. line 14. for own reade home. f.12.l.27. for trouble reade treble. f.42. l.5. reade, like vnto that of a Lilly f.66.l.41 for χόρτα read χρίνα, or λείρια. f.73. l.37. for top of the flower, read cup. f.134.l.16. for composed reade compassed. f.150.l.4.for hath, reade haue, and line 5.for is are. f.173.l.12. put out these wordes, the inside, in the beginning of the line. f.189.l.38. reade Bini florum ordinibus. f.218.l.19.reade goulons, and line 28 pratensis. f.272.line 36. read Pothos. f.276.l.12.Chelidonia. f.281.l.37.for hath, haue. and l.28.Mastuerzo. f.284.l.15.Vicenza. f.287.l.39. Citrina. f.290 l.39.reade prouoke, and, helps. f.319.l.37.for Melancholicke, reade Flegmaticke. f.330.331.333. reade Eryngium in all places f.336.l.8.reade, and not very flat. f.356.l.31.Americanum. f.357.416. Cervicaria. f.458.l.4. reade, before it c1 1 haue. f.372.l.9.blot out, except it. f.389.for spockes, reade smockes. f.393.l.3. in the margent for cæruleo, read plena. f.397.l.10.reade dwarfe. f.424.l.45. Hirculus. f.428.l.20.Tarentina. f.431.l.10.Cyprium. l.12. Amomum. f.439.l.17.for Dioscorides, reade Theophrastus. f.442.l.3. for castings reade purgings. f.509.l.35. reade χρυσόλη f.513.l.24 transferre all that clause of Onions vnto the other side, vnder the vse of Onions. f.516.l.37.transferre these words, [Bauhinus vpon Matthiolus calleth it Solanum tuberosum esculentum] vnto the former Potatoes of Virginia. f.510 l.13. for swelleth, read smelleth. f.541.l.51.reade, after your stockes rayfed from stones. f.566.l.20.for as, read and. and l.19. euery one. f.567.l.24.for Rice, read Rue. f.575 l.8.reade serue to be ministred to the sicke. f.588 l.3. Capsndz. f.594.l.18.for facility, read faculty. f.595. l.39. reade Ricinus. f.600.l.4, Fuchsius

LONDON,
Printed by HVMFREY LOWNES *and* ROBERT YOVNG
at the signe of the Starre *on* Bread-street hill.
1629.